# BIOPHYSICAL BONE BEHAVIOR

# BIOPHYSICAL BONE BEHAVIOR

## PRINCIPLES AND APPLICATIONS

**Jitendra Behari**

*Jawaharlal Nehru University, India*

John Wiley & Sons (Asia) Pte Ltd

Visit our Home Page on www.wiley.com



***Other Wiley Editorial Offices***

John Wiley & Sons, Ltd, The Atrium, Southern Gate, Chichester, West Sussex, PO19 8SQ, UK

John Wiley & Sons Inc., 111 River Street, Hoboken, NJ 07030, USA

Jossey-Bass, 989 Market Street, San Francisco, CA 94103-1741, USA

Wiley-VCH Verlag GmbH, Boschstrasse 12, D-69469 Weinheim, Germany

John Wiley & Sons Australia Ltd, 42 McDougall Street, Milton, Queensland 4064, Australia

John Wiley & Sons Canada Ltd, 5353 Dundas Street West, Suite 400, Toronto, ONT, M9B 6H8, Canada

Wiley also publishes its books in a variety of electronic formats. Some content that appears in print may not be available in electronic books.

***Library of Congress Cataloging-in-Publication Data***

Behari, Jitendra.
Biophysical bone behavior/Jitendra Behari.
p. ; cm.
Includes bibliographical references and index.
ISBN 978-0-470-82400-9 (HB: alk. paper) 1. Bones. 2. Biophysics. 3. Bones–Pathophysiology. I. Title.
[DNLM: 1. Bone and Bones–physiology. 2. Electric Stimulation Therapy–methods. 3. Fracture Healing–physiology. 4. Osteoporosis–physiopathology. WE 200 B419b 2009]
QP88.2.B437 2009
612.7′5–dc22

2008045584

ISBN 978-0-470-82400-9 (HB)

Typeset in 10/12pt Times by Thomson Digital, Noida, India.
Printed and bound in Singapore by Markono Print Media Pte Ltd, Singapore.
This book is printed on acid-free paper responsibly manufactured from sustainable forestry in which at least two trees are planted for each one used for paper production.

# Contents

# Preface

Biophysical parameters governing bone growth and development have been of continuing concern for various reasons. However, the study of the solid state nature of bone in its clinical environment needs emphasis. Many data are available on the subject and a coherent approach has been lacking.

This book is an effort to correlate the solid-state properties of bone with its clinical behavior, particularly concerning bone fracture healing and osteoporosis. Its characterization, treatment and diagnosis have been discussed in detail. In turn, it may be predicted that by altering the solid-state properties, the latter phenomena can be controlled. This then sets a relationship between the two. An approach of this type is far from being fully understood. Presently, an attempt has been made to sequentially put forth bone physical properties to explain its overall behavior. This first chapter essentially deals with the basics of bone biophysics. The following two chapters deal with piezoelectricity and its solid state behavior. The possibility of controlling these parameters by external stimulus is discussed in subsequent chapters. Bioelectricity in bone followed by its stimulation in fracture healing and its concept are elaborated.

Several workers have developed models to support experimental data to explain the physical, electrical and mechanical properties of bone. An understanding of physical phenomena involved in bone disease is of paramount importance. Though one of recent origin this issue is now being widely recognized. This book aims to fill the gap, establishing its link with clinical behavior. While the treatment of fracture healing by conventional methods has long been in practice, the use of electrostimulation has been successfully reported for only about the last two decades, and its use is likely to increase. The treatment of induced osteoporosis add importance to this issue. This has been discussed in the light of exogenous biochemical supplement, is also considered along with electrical stimulation. The combination may significantly decelerate the process of osteoporosis. An emerging aspect related to nanoparticles is also included. A greater depth in understanding the mechanism of such interactions may go a long way in improving the quality of human life. Last but not the least non-invasive techniques of measuring osteoporosis are discussed in some detail in the last chapter.

The references are available on the book's companion website at the following URL: www.wiley.com/go/behari

# Preface

# Acknowledgements

It is difficult to find the words to thank all those who have been instrumental in their help and support in bringing this work into the present form. The concept of putting this book into its present shape emerged during a series of conversations with several scientists in different areas related to the present subject. This work was initiated at my post-doctoral stage and subsequently progressed with the involvement of several doctoral students and research staff. The data have since been revised and updated.

A major part of the writing and updating the manuscript was completed during sabbatical leave granted by Jawaharlal Nehru University, New Delhi, which is thankfully acknowledged. The final editing and compilation was completed with the assistance of Dr Paul Raj and Mr KK Kesari. The patience of Mr Balwant Singh in typing this manuscript is thankfully appreciated. Last but not least the cooperation and constant inspiration from my family has been a source of continuing strength in fulfilling this task.

It was a pleasing experience to have constantly interacted with Mr James Murphy of John Wiley. His inputs were helpful in the preparation of the manuscript.

J. Behari

# About the Book

The present book is the culmination of an effort to bring together biophysical phenomena in bone, relating to growth and its relation to electrical behavior. The book is primarily intended as a bridge between physics and biology of bone, leading to its clinical applications: electro stimulations in fracture healing and osteoporosis. The book is divided into seven chapters. The first chapter deals with the basic biological aspects and the parameters relating to growth and repair. Chapter 2 covers bone behavior in its relationship with mechanical properties (piezoelectric phenomena). Chapter 3 deals with the bioelectric phenomena in bone and sets out goals for *in vivo* experimentation. Chapter 4 is concerned with semi-conduction properties, (mechanism of change transport), Hall effect, photoelectric effect, electric effects, PN junction properties and several others aspects.

Subsequent to the above an effort is made to relate these properties with bioelectric phenomena in bone and its relation to bone fracture healing. The bioelectric response of bone to field stimulation, from DC to radio frequency, is discussed.

The related issue of osteoporosis is discussed in Chapter 6 in some detail, dealing with the definition of osteoporosis and the parameters used to characterize this along with biochemical and physical methods for treatment. Finally, Chapter 7 deals with non-invasive techniques for the measurement of bone osteoporosis. This last chapter covers various techniques, for example, dual energy X-ray absorptiometry, acoustic and ultrasonography, dual photon absorptiometry and others.

The book is based upon the author's own research work in the area and subsequent three major review articles published in *Critical Review of Biomedical Engineering* (Behari, 1990, 1994) and *Progress in Biophysics and Molecular Biology* (Behari 1991). Work of other researchers is mainly from their published papers and also as discuss by other workers. The material has been updated to state of the art status.

This book is intended to be a reference book for students, researchers, those involved in teaching in universities, medical institutes (biomechanics) and hospitals (orthopedic surgeons). It is intended to provide a smooth journey from basics to clinical applications, adjusted for scientists from various backgrounds. The author will feel his efforts have been rewarded if it meets this goal.

# 1

# Elements of Bone Biophysics

## 1.1 Introduction

Bone is a dynamic tissue that, when properly organized and distributed in a whole bone, acts as both a mechanically competent skeletal structure and a physiological unit. Bone is a very rigid anisotropic material and a specialized connective tissue, which forms the basis of the skeleton, and, as such, its functions are manifold and at times too complex to understand by any simplistic model. One of its apparent roles is to provide support to the body, which is carried out by cortical bone throughout the skeleton and by peripheral cancellous bone. While it is well established that geometrical characteristics contribute to bone strength, their dependence upon specific behavioral, hormonal and metabolic factors is unknown. In combination with the associated musculature, the bones of the skeleton also provide a means of physical support, locomotion and related movement. Evolution has led to a complex, multiphase, heterogeneous and anisotropic microstructure. Another important role is that as a reservoir for a multitude of inorganic ions (calcium, phosphorous, etc.) that are subsequently recruited by various physiological systems. Subsidiary functions are participation in plasma calcium homeostasis and support of hematopoiesis, which is carried out mainly by central cancellous bone (Parfitt, 2001). It is well known that in the event of calcium deprivation it is the mass of the skeleton that is sacrificed at the expense of other functions.

Bone is never static, it is a living structure, responding and adapting itself to the applied load and has the capacity to remodel. The cells of the skeleton act continuously to maintain the remodeling (Figure 1.1). It is thus in a constant state of dynamic equilibrium both in terms of its composition and structure and responds to external mechanical forces (or their absence) by adopting changes in its normal architecture. The mechanical properties of bone depend on the load direction. The loading regimen should be dynamic (Figure 1.2). Bone possess various physical, solid-state and electro-mechanical properties. These properties are characteristics of bone and are modified under the action of external stimulus and changes of Ca and P metabolism. However, the mechanism at work and their biological significance are not completely understood. Also, the functional significance of bone renewal is still a matter of speculation.

*Biophysical Bone Behavior* Jitendra Behari

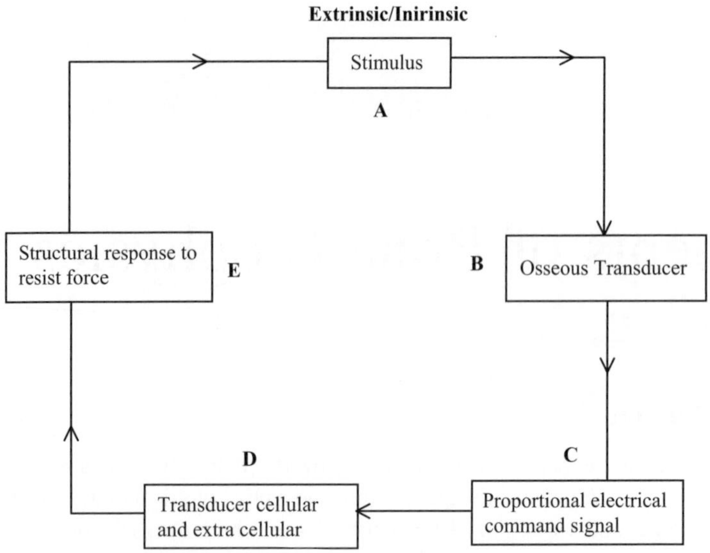


**Figure 1.1** Basic response pattern of bone cells to extrinsic/intrinsic stimulus

The mechanical competence of bone is determined by its density, the architecture and its intrinsic material properties. The conventional view holds that the intrinsic chemical, structural and mechanical properties of bone remain invariant and, therefore, that the old and young, or osteoporotic and normal bone, are not substantially different. It has been suggested that once the bone architecture is determined (possibly by genetic processes) its mechanical behavior can

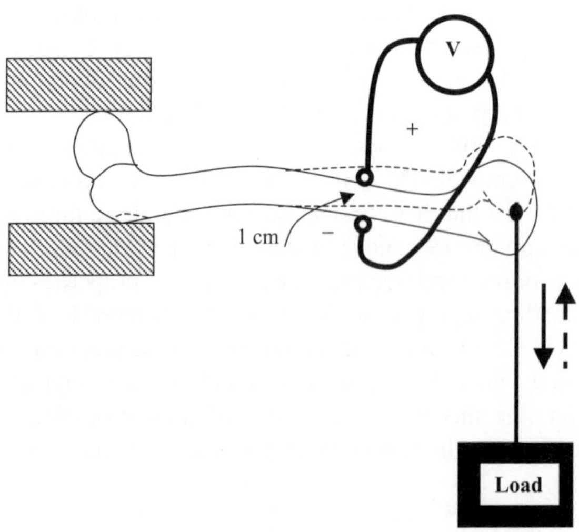


**Figure 1.2** Basic model experimental design for detecting electrical polarization of living/non-living bone subjected to dynamic loading. The magnitude and resulting wave shape depend upon the duration and type of loading

be extrapolated – a view that is still open to question. The general understanding is one of morphological adaptation. Namely, that peak mechanical strains in the tissue trigger a biological response. *In vitro* results support the notion that bone formation *in vivo* is stimulated by dynamic rather than static loads (Rubin and Lanyon, 1987) and that low magnitude, high frequency mechanical stimuli may act as stimulator for high amplitude, low frequency stimuli, provided that the cells are stimulated in a preconditioned state. Bioelectrically, it is now accepted that bone remodeling is dependent upon load-induced voltages and, hence, also that understanding the ability of bone to generate piezoelectricity is one of the keys to understanding bone physiology: an issue discussed in Chapter 2.

The physical hardness, which is a unique characteristic of bone, poses particular problems requiring special laboratory procedures to achieve a high standard of tissue section preparation. For instance, before attempting conventional histological methods, it is generally necessary to "soften" bone (and other calcified tissue) by removing the mineralized component. This procedure requires special treatment, and the time it adds to the overall tissue processing cycle will inevitably cause some delays in the assessment of diagnostic specimens. There are specific techniques that enable the preparation of sections from calcified tissue, without the need to remove the mineralized phase. Apart from bone, many other kinds of mineralized specimens may be encountered since virtually any normally soft tissue can become calcified, resulting in a disease process.

Indeed, human bone has a complex hierarchical microstructure (Yeni *et al.*, 1997) that can be considered at several dimensional scales (Rho, Kuhn Spearing and Zioupos, 1998). At the shortest length of scale it is composed of type-I mineralized collagen fibers (up to 15 μm long, 50–70 nm in diameter) bound and impregnated with carbonated apatite nanocrystals (tens of nm long and wide, 2–3 nm thick). Several skeletal tissues participate in this mechanical objective of transmission and protection: bone cartilage, tendons, ligaments and muscles. Bone mainly determines the global structural stiffness and strength, whereas other tissues transmit loads between bones. The mechanical properties of bone are a result of a compromise between the need for a certain stiffness (to reduce strain and achieve a more efficient kinematics) and the need for enough ductility to absorb impacts (to reduce the risk of fracture and minimize skeletal weight).

## 1.2 Structural Aspect of Bone

It is now commonly accepted that bone tissue changes its morphology in response to mechanical forces. An important role of the musculoskelatal system is to transmit forces from one part of the body to another under controlled conditions of stress and strain. This also offers protection to vital organs (e.g., lungs, brain, etc.) in addition to its other more specified functions. Several skeletal tissues participate in this mechanical objective of transmission and protection: bone, cartilage, tendons, ligaments and muscles. Bone exists in two main forms: cortical and trabecular bone. Trabecular bone has a greater surface area than cortical bone – mineral is drawn from the former during short-term calcium and phosphorous deficiencies to maintain equilibrium within the body. Trabecular bone is located on the end of long bones, in vertebral and envelopes the marrow (Figure 1.3a and b) (Jee, 1988; Wasserman, 1984; Hays and Swemson, 1984). Bone contains an outer casing of cortical bone, surrounding the medullary cavity. The latter contains a network of bone fibers, the trabecular bone. The spaces within the trabecular structure are filled with red hematopoietic or yellow (fatty) marrow. The relative

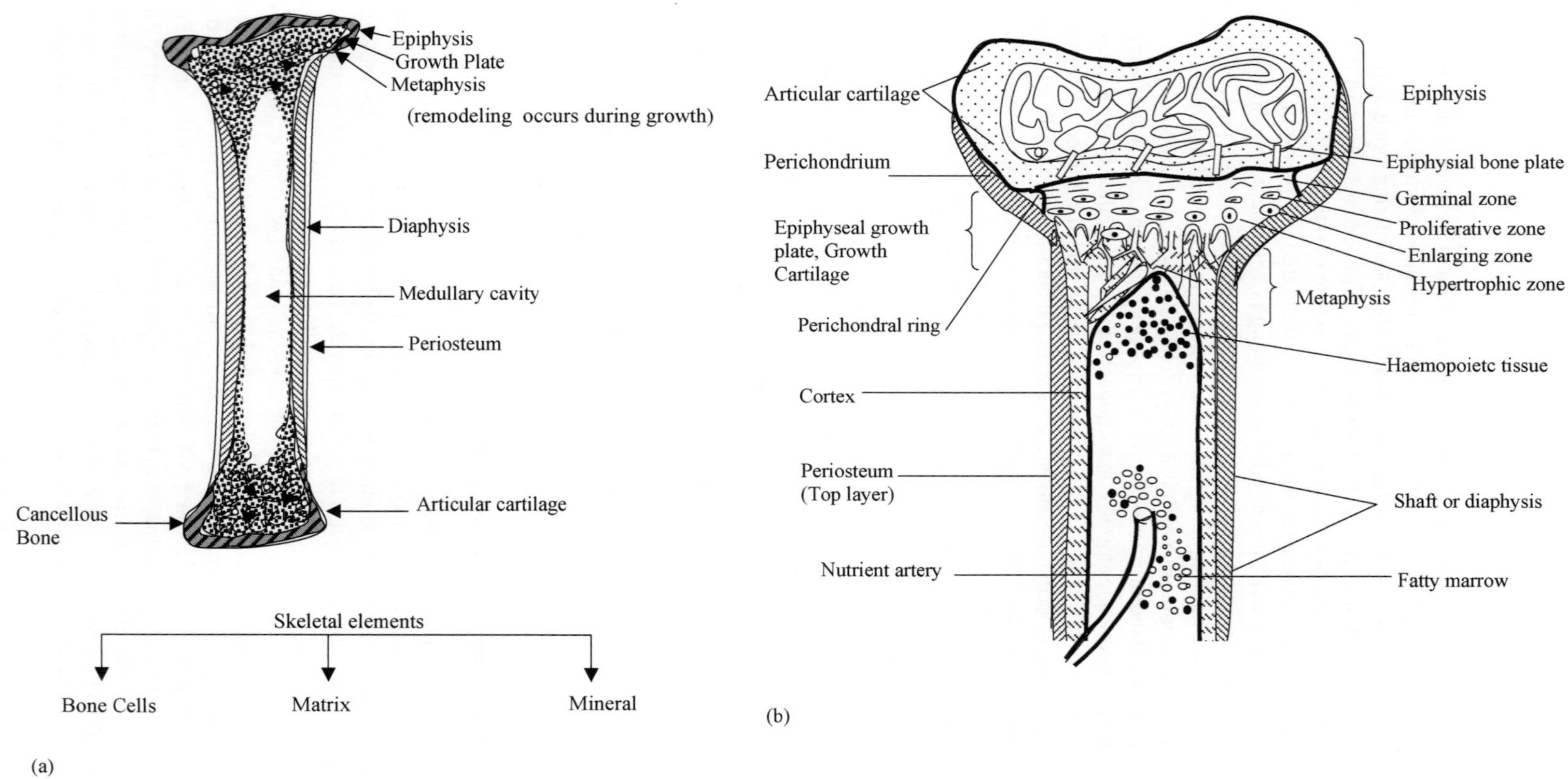


**Figure 1.3** (a) Diagrammatic representation of the parts of a long bone. (b) Features in developing long bone. Adopted from Little, 1973 (Reproduced with permission from K. Little, *Bone Behavior*, Academic Press, New York. ©1973, Elsevier B.V.)

composition of these four components is dependent on the bone location and also on the age of the skeleton.

Long bones are made up of the diaphysis, or shaft and the epiphysis, or end of the bone (Figure 1.3b). The diaphysis and epiphysis are connected by the metaphysis. In young animals, the epiphysis and metaphysis are separated by a thick cartilaginous tissue known as the growth plate. The long plate elongates the growth plate, as new bone is formed at the base of the plate where blood vessels infiltrate the cartilage matrix. In mature animals, the growth plate is replaced by trabecular bone and the bone ceases to grow in length. Both cortical and trabecular bone are made up of lamellae, which consist of circular collagen fibers and mineralized components.

In contrast to trabecular bone, which is more like a woven net, cortical bone has a solid mass, has a very low porosity and its anisotropy is mainly controlled by lamellar and osteonal orientation (Figure 1.4a and b). Bone anisotropy is defined by means of a second-order tensor (fabric tensor) that defines the principal values and directions of the bone mass distribution (Odgaard, Jensen and Gundersen, 1990; Whitehouse, 1974; Whitehouse and Dyson, 1974; Odgaard, 2001). In fact, structural anisotropy has a direct influence on stiffness properties as well as on strength. The average strength of a compact human bone in longitudinal compression tests is 105 MPa, and in transversal compression test it is 131 MPa (Reilly and Burnstein, 1975). There is evidence that resting cells in the cambium layer of the periosteum are coupled through junctional complexes to osteocytes within the underlying, osseous substance (Weinger and Holtrop, 1974). These junctions are very sensitive to minute changes in their electrical or calcium ion environment (Cooper and Keller, 1984; Flagg-Newton and Lowenstein, 1980; Lowenstein *et al.*, 1978; Ravel *et al.*, 1980; Sheridan *et al.*, 1978). Furthermore, they can act as a conduit both for macromolecular transfer and, probably, electrical signaling. When a local change in the strain pattern reaches a threshold level, which is strain rate, cycle and duration dependent, these cells are activated. A transition, actively from a sub-divided to an active state (or vice versa), will stimulate an appropriate adaptive remodeling response (Woo *et al.*, 1981), while localized adaptation can be seen within a single individual subjected to differential exercise (Jones *et al.*, 1997). The mechanism of this adaptive phenomenon remains unknown, though it has been suggested that microdamage within the extracellular matrix, which might accumulate following long periods of exercise, could stimulate a reparative process (Burr *et al.*, 1993). A cascade of events can be visualized from a single-threshold stimulus to produce a new bone (modeling form of osteogenesis) (Tornberg and Bassett, 1977).

Human long bone is composed of woven bone before an individual reaches the age of 4. During this time, as later, the bone diameter at the mid-diaphyses increases principally by new bone deposition at periosteal surfaces and old bone resorption at endosteal surfaces. Thus, microstructural information for growing woven bone may be studied discretely by examining the mid-diaphyses at different locations across the thickness of bone from periosteum to endosteum and representing different stages of both mineral and tissue age and maturation. Collagen provides the fundamental organic matrix framework for the deposition of apatite in all vertebrate mineralized tissue with the exception of dental enamel and eggshell (Lowenstam and Weiner, 1989). The overall organization of collagen fibrils in woven bone differs considerably from those in lamellar bone. In the latter, each lamella consist of highly oriented, densely packed collagen fibrils. The collagen fibrils adjacent lamellae differ by 90°. Woven bone, in contrast, has no lamellar structure. The collagen fibrils do not lie parallel to one

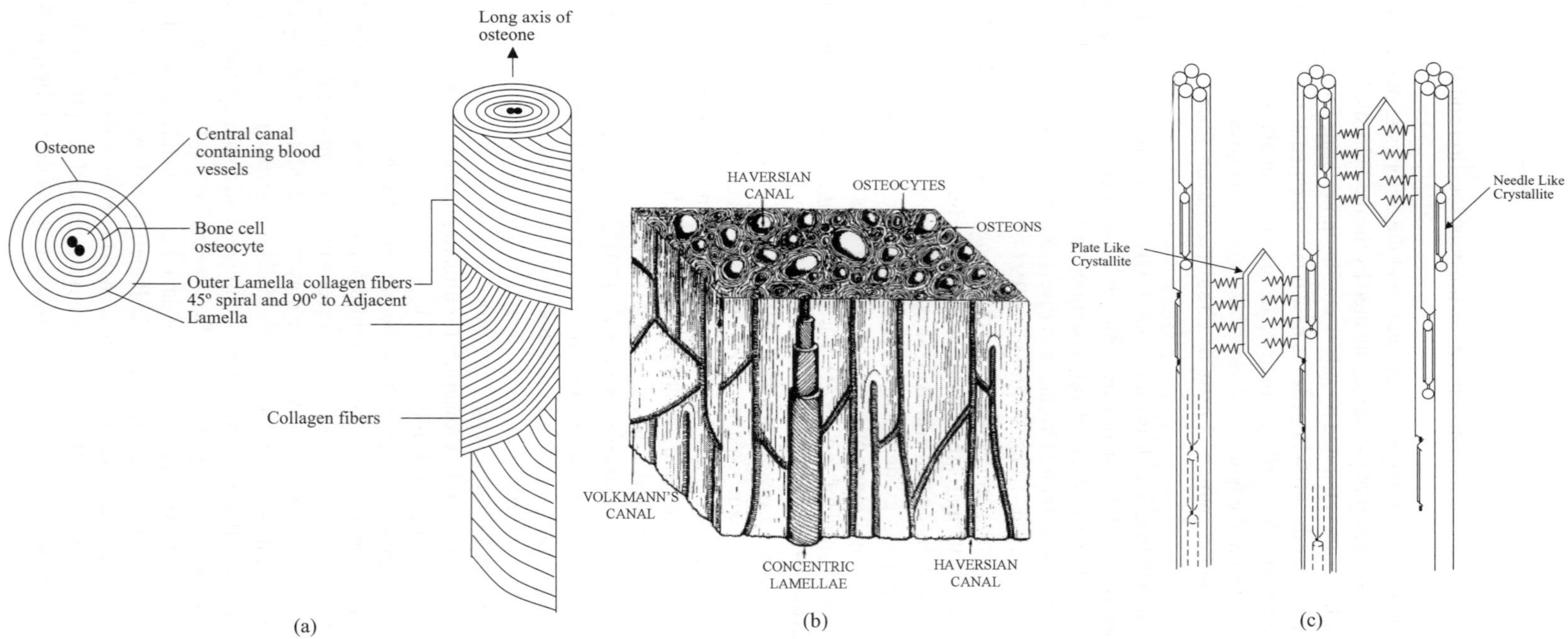


**Figure 1.4** (a) Microscopic structure of compact bone; (b) microstructure of adult human compact bone (vertical axis is parallel to the long axis of the bone) (Saha, 1977); (c) mineralizing bone collagen crystallites embedded in the crosslinks (Reproduced with permission from S. Lees and C. Davidson. "A theory relating sonic velocity to mineral content in bone." In: Linzer, M, ed. *Ultrasonic Tissue Characterization* Vol. II, Special Publication 525. Government Printing Office, Washington, DC ©1979)

another, though they usually show some degree of preferential orientation parallel to the long axis of the bone. In newly deposited woven bone the collagen fibrils are interwoven and dispersed (Su *et al.*, 1997). The study of longitudinal and cross-sectional woven bone sections by transmission electron microscopy has confirmed the existence of apatite crystals in the extrafibrillar spaces of collagen (Figure 1.4c). Within the layer of collagen, lacunae, canaliculi, the cavities and channels comprise the communication system for bone. One concentric unit of lamellae and its associated lacunae and canaliculi is known as an osteon or Haversian system (Jee, 1988; Lawrence and Fowler, 1997).

The properties of the cells in bone are primarily controlled by agents that affect cell membranes. These membranes are of two main types: (i) The rough membranes, which are known as the endoplasmic reticulum; these are so-called because of an array of ribosome particles on the surfaces of the membranes. (ii) The smooth membranes, which comprise the cell surface, nuclear envelope and their connecting membranes. The endoplasmic reticulum is responsible for the production of intercellular matrices. The surface membranes are for cell mobility and the smooth membranes for cell division. The stages of division are pre-division (inter phase), followed by prophase, metaphase, anaphase and telophase. The remainder of the division process is dependent upon the pre-division stage, which is the only one that requires a source of energy.

The structural integrity of bone, like any other form of matter, living or dead, is maintained by atomic forces, which are electrical in nature. The various physical properties of bone show that there may be a close relationship between biophysical properties and physiological processes involved in bone growth and remodeling (Cochran, 1966; Frost, 1980; Glimcher, 1976; Bassett, 1971). The present concept of bone as consisting of collagen fibers, hydroxyapatite crystal, a small amount of proteoglycans, non-collagenous proteins and water (Weiner and Wagner, 1998; Robinson and Elliot, 1957; Martin, 1984; Lucchinetti, 2001) is expected to explain mineralization, along with suggestions of new composite materials. The intervention of external agents to artificially control the growth pattern of bone by applying mechanical stress or electrostimulation or treatment of Ca and P minerals in isolation or in combination provides another reason for basic studies on its structure and properties.

### *1.2.1 Elementary Constituents of Bone*

Bone is a highly specialized form of connective tissue. The main elements of bone tissue may be considered to consist of, besides osteocytes, a crystalline mineral phase (hydroxyapatite), an amorphous mineral phase, a crystalline organic phase (collagen), an amorphous organic phase and liquids. The biphasic behavior of bone composite materials arises from inorganic crystals of calcium phosphate disposed within the collagen fibril of an organic matrix (Glimcher and Krane, 1968). Inorganic components (mainly hydroxyapatite) are mainly responsible for the compression strength and stiffness, while organic components provide the corresponding tension properties. This composition varies with species, age, sex, the specific bone and its pathogenic state. An important aspect that also characterizes this peculiar mechanical behavior of bone is its hierarchical organization. Weiner and Wagner (1998) have described this, stating from the nanometric level and ending at the macroscopic level, relating the latter to the mechanical properties.

## *1.2.2 The Fibers*

The organic matrix has two chief components: the collagenous fibers and ground substances. The ratio of organic to inorganic substance in bone is 35–65% (Glimcher, 1959). The organic component is often called osteoid. The organic part of bone is about 95% collagen by volume. Collagen is a generic term for a class of protein that make up much of the body tissue and are found in bone and cartilage. Osteoid consists of proteins and carbohydrates that are secreted outside cells. These proteins and carbohydrates include type I collagen fibers, proteoglycans and glycoproteins. These components are found throughout the connective tissues of the body and provide tissues with strength and also flexibility. The mineral occupies 35% of the volume. Since collagen is such an important body constituent its chemistry has been studied extensively. Because of the need to determine the presence of collagen in tissues and to identify its extent, the sonic and elastic properties of some types of collagen have remained an attractive subject of investigation.

Collagen provides connective tissue with strength and makes up a major portion of bone. A portion of cartilage supramolecular aggregate is associated with the extracellular matrix and is stabilized by the triple helix characteristic of collagen (Wasserman, 1984; Lawrence and Fowler, 1997; Marks and Popoff, 1988; Van der Rest and Garrone, 1991). Collagen molecules have a repeating pattern of glycine-proline-hydroxyproline-glycine-x-y amino acids, where x and y denote any other amino acid. Hydroxyproline and hydroxylysine are two amino acids found predominantly in connective tissue and are prevalent in collagen. For static reasons glycine is the only amino acid that can be in the center of the super helix of a collagen molecule.

All collagens are highly structured materials in a multilevel hierarchal order. The molecular weight of the basic collagen molecule, defined as tropocollagen, is approximately 300 000 daltons, which is a very large molecule even in organic chemistry. The molecule is much like a piece of spaghetti, being about 300 nm long and having a wet diameter of 1.5 mm. Bone collagen differs from other body tissue collagens by an extensive network of intermolecular crosslinks that render it insoluble in even the most potent of solvents. In contrast, tendon collagen is reported to be mostly lacking in intermolecular bonds but has many hydrogen bonded intramolecular links, and so that it can be dissolved and readily reconstituted. Figure 1.4b shows a schematic diagram. While other tissues like arteries and joints calcify, the special structural characteristics of bone collagen cause it to calcify in a unique manner (Lees and Davidson, 1977).

The collagen molecule has a triple-helix structure. The triple helix itself shows in a right-handed spiral around a common central axis (Ramachandran, 1963) that gives rise to an X-ray diffraction pattern indicating the structural similarity of collagen in bone, tendon and skin (Figure 1.5a–c). The collagen macromolecules are organized to form fibrils, which in turn are organized into fibers. Fibers are woven into lamellae, which in human compact bone tend to form cylindrical Haversian systems. The other components in the organic matrix include some other proteins, along with the acid mucopolysaccharides, and these constitute a part of ground substance. Some of the mucopolysaccharides are present in the form of mucoproteins (non-collagenous). The exact anatomical location of these components at a macromolecular level is not certain and their state of aggregation and polymerization is not well known (Glimcher, 1959). Collagen fibers, which are composed of 400–1200 Å diameter fibrils with a cross bonded structure, account for nearly one-third of the dry weight of the bone matrix

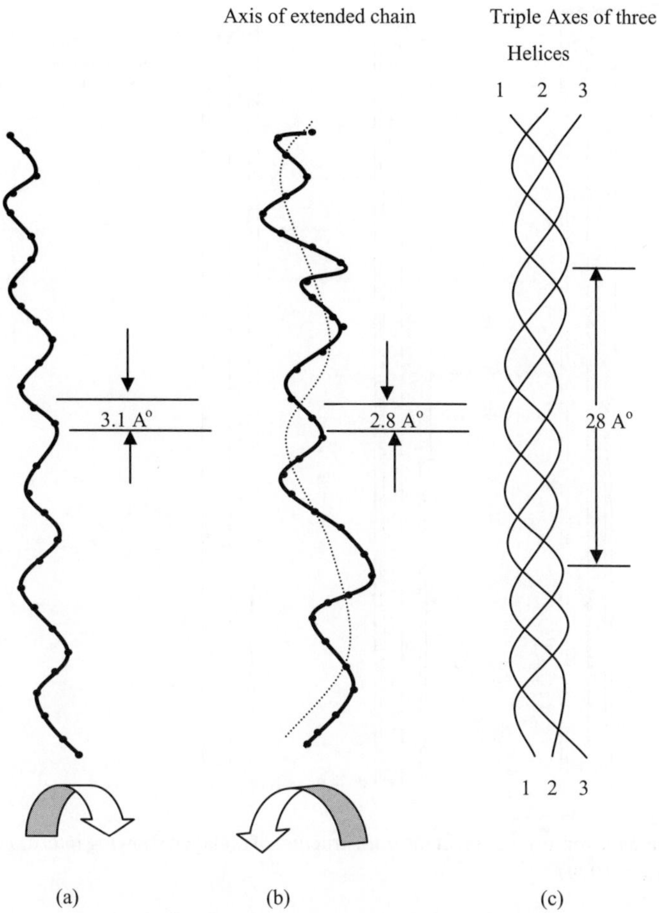


**Figure 1.5** (a) The $\alpha$-helix is left handed. Each spot is a residue and the repeat length is 3.1 Å. (b) The $\alpha$-helix is twisted to the right and the length between residues is reduced to 2.8 Å; the repeat of the supercoil is 10 times the residue length (28 Å). (c) Three $\alpha$-helices rethreaded to form a single tropocollagen (TC) unit. (Adapted from Lees and Davidson, 1977/79)

(Figure 1.6). Collagen molecules form the tropocollagen macromolecules, which consist of three polypeptide chains, each being twisted into a left-handed helix, and the three chains being wound round a common axis in the form of a right-handed superhelix. It forms a long, thin rod about 2800 Å long and 14 Å in diameter. Tropocollagen molecules are organized into fibrils extracellularly. The lamellae are built by collagen fibers and the inorganic salts are deposited into this organic polymerized protein crystals. An understanding of the nature of bone at the molecular level is important not only for a basic understanding but also for the treatment of degenerative bone diseases.

About 2% of the volume mature compact bone consists of cells, the remainder being extracellular matrix. Of this matrix 70% is occupied by mineral, about 2% is organic and

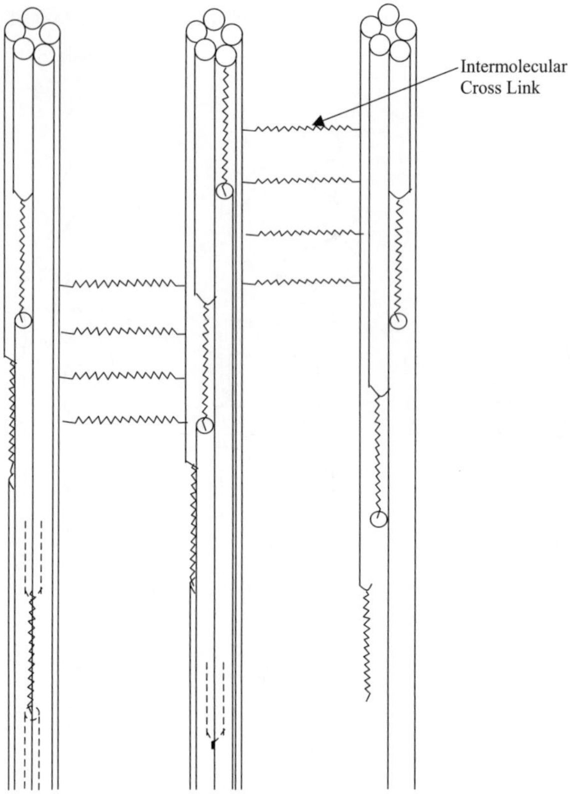


**Figure 1.6** Schematic representation of the ultrastructure of collagen showing intermolecular crosslinks (Lees and Davidson, 1979)

the remainder is water. Of the organic material 90% is collagen, about 1% is proteoglycan and the rest is a series of matrix proteins. The collagen of bone is almost all type I. The proteoglycans of bone are all of the small non-aggregating type (decorin and biglycan) and contain only chondroitin sulfate. The matrix proteins are characterized by their anionic nature, being rich in phosphate (phosphoprotein, e.g., osteonectin) sialic acid (sialophoproteins, e.g., osteonectin) or $\gamma$-carboxyglutamic acid (Gla proteins, e.g., osteocalcin). Growth factors such as TGF-$\beta$ and BMP (bone morphogenic protein) are also stored within the bone matrix.

As collagen is a polymer, bone may be considered to be a two-phase, mineral-filled polymer. Collagen is laid down first and then it becomes mineralized when the HAP crystallites are deposited (Figure 1.4c). When bone is demineralized the collagen matrix can be recovered as a rubbery solid, having the form and shape of the original bone. However, when collagen is removed the mineral structure left behind is a weak solid that easily crumbles into a powder. It may be concluded that the collagen forms a continuous medium but the mineral does not. The mineral fills voids in the well-laid collagen structures.

Collagen has long been studied and much is now understood about its chemistry and ultrastructure. Type I collagen is the most abundant form of the molecule; it is found in bone, tendon and skin and many other tissues. This molecule has an [a-1(I)2a-2(I)]] conformation. Type II collagen is largely found in cartilage and between vertebral and has an $[a\text{-}1(II)]_3$ structure. Type I and II collagens are examples of fibrial-forming collagens. Other collagen types, such as basement membrane microfibrillar, anchoring fibril, hexagonal network-forming fibril associated with interrupted triple helices transmembrane and multiplexin collagens, have different types of structures. These collagens have various subunits that help to differentiate each type of collagen (Lawrence and Fowler, 1997; Van der Rest and Garrone, 1991; Van der Mark, 1999). Type I collagen is the most prevalent type (90%) in bone. Additionally, collagen makes up about one-third of the dry weight of bone organic matrix (Pritchard, 1972; Wasserman, 1984; Lawrence and Fowler, 1997). Collagen molecules have many complex characteristics: such as three-dimensional order, the chemical character of the intermolecular crosslinks and the location (Ramachandran, 1967; Gallop and Paz, 1975; Veiss, 1974).

### *1.2.3 Collagen Synthesis*

Collagen is synthesized intracellularly as preprocollagen, followed by hydroxylation of proline and glycosylation of hydroxylysine to yield procollagen. The ends are cleaved and procollagen is expelled from the cell (Wasserman, 1984). Extracellularly, the procollagen molecules undergo hydrolyses to form tropocollagen, which is the basis for matrix collagen fibrils. The polypeptide chains of tropocollagen are wound in a left-handed fashion, formed when every third amino acid residue is turned into the center (Van der Rest and Garrone, 1991).

Fibril forming types of collagen molecules like those in bone are arranged parallel to each other, with spaces strategically placed between fibers once the molecules have been cleaved at the N- and C-telopeptides to form collagen fibrils extracellularly. The ends of the fibers can then become crosslinked. Crosslinkage between collagen molecules at hydroxylysine and lysine sites provides collagen with its stabilization and forms the striation in the collagen fibrils. Collagen, in conjunction with its crosslinking, yields a scaffold for minerals to bind in the matrix (Wasserman, 1984; Van der Rest and Garrone, 1991; Lawrence and Fowler, 1997; Marks and Popoff, 1988).

Pyridinoline (PYD) is a 3-hydroxypyridinum derivative consisting of three amino acids and carboxylysine residues; lysyl oxidase converts these amino acid residues into an aldehyde, (Last *et al.*, 1990). An analog crosslink, deoxypyridinoline (DPD), is formed in a similar manner as PYD, but the helix rather than hydroxylysine (Ogawa *et al.*, 1982). Eyre *et al.* (1988) have found that hydroxypyridinium residues persist in human bone through adulthood, thereby establishing them as final crosslinks.

Collagen exhibits many levels of order, terminating in a fibrous structure. The successive levels of organization have been explained by Lees and Davidson (1977). The smallest element is the $\alpha$-helix, which spontaneously combines in triplets to form a superhelix molecule, the tropocollagen (TC) unit. In turn, TC forms microfibril ropes, which in their turn form three-dimensional fibrils. Some characteristics of these elements are listed below:

- The basic element, the $\alpha$-helix chain, has a molecular weight of about 100 000 daltons. Several types of chain have been identified, depending on the amino acid composition.

The $\alpha_1$ chain has 1052 amino residues, of which 1011 are in triplet sets where the first term is glycine. The N-terminus has 16 nonhelical residues beginning with an amino ($NH_2$) group while the C end has 25 nonhelical residues terminating in a COOH group. Gallop and Paz (1975) as well as Hulmes *et al.* (1973) have prepared the residue map.

- The hydrogen-bonded state between the $\alpha$-helices, of which TC is the basic molecule, is stable.
- A TC unit is 4.4 D units long and packs into a structure exhibiting a repeated 0.6 D unit gap between TC ends. According to Hodge (1967) no two quarter stagger gaps can be adjacent, which puts a restriction on the three-dimensional arrangement (Segrest and Cunningham, 1973).
- The nonhelical chains at each end of the TC unit and some of the adjacent glycine coded triplets are probably the links between molecule ends but may also serve to link helices within ends – they may also serve to link helices within a single molecule, indicating that the, inter- and intramolecular links are similar. However, many of the intramolecular links are hydrogen bonds, while many of the intermolecular links between parallel molecules are chains covalently bonded as side chains to residues in the backbone of the molecule.
- Several three-dimensional structures have been proposed for the microfibrils, fibrils and fibers. Figure 1.5c shows a TC unit (Smith, 1968) as modified by Miller and Parry (1973). The model shows a 67 nm axial stagger, 72° azimuthal displacement between axes of nearest neighbors and a 43 nm hole between collinear TC units. It incorporates a repeat pattern after $5 \times 67$ nm and a helix with a fivefold screw axis (five subunits per screw turn characterized by the holes). The pentagonal unit is 4 nm in diameter while the diameters of the wet TC unit and the lumen are 1.5 and 1.1 nm, respectively.
- Miller and Parry (1973) have indicated that the five-strand rope must be twisted to produce a fourfold symmetrical supercoil, where the holes are 90° apart. However, there is still a pentagonal packing of the strands. The collinear TC units are inclined about 2.5° to the rope axis to accommodate the twist.
- The pattern of successive helical structures suggests that the microfibrils are coiled about each other while maintaining the four unit cell.

One way to characterize a molecule is to look for its resonance frequency. Enemeto and Krimm (1962) calculated a value for the macromolecular resonance of a very similar molecule, polyglycine II, which they found to be 41 GHz – a value far in excess of any reported for collagen or even for bone (26 GHz). Polyglycine II has a triple helix, hydrogen-bonded molecule and closely resembles collagen in its structure (Ramachandran, 1967).

### *1.2.4 Bone Matrix (Inorganic Component)*

Apatite crystals are one of the major constituents in bone and other mineralized vertebrate tissues (Lowenstam and Weiner, 1989). Their presence, which accounts for about 65 wt% of bone, provides most of the stiffness and strength of bone (Glimcher, 1992). With the normal increase of apatite deposition during tissue aging and maturation, bone mechanical properties increase greatly, an observation confirmed by measurement of the elastic modulus and microhardness of growing human long bone (Su *et al.*, 1997). Correlative to the increase in the diameter of the middiaphyses of long bone, occurring principally by periosteal bone deposition and the later resorption (Enlow, 1991), endosteal bone becomes more mineralized than the former. The elastic modulus and microhardness of endosteal bone can be up to three times higher than that of periosteal bone (Su *et al.*, 1997). Lees and Davidson (1977,1979) suggested that mineralization causes additional crosslinkages to be set up, some between TC units but more between the mineral, whether crystalline or amorphous, and the TC units. The collagen is stiffened by decreasing the crosslink length and the mineral fills. The mineral TC links may be reversible so that in the event of demineralization the collagen becomes less stiff.

The first role of the mineral component in calcified tissue in this hypothesis is to stiffen the collagen by decreasing the length of the crosslinks and increasing the crosslinking density proportionally to the amount of mineral present. At 34% volume, all the crosslinking is shortened to the minimum length and the maximum crosslinking density is attained. The additional strength that is obtained by additional mineral content is in accordance with the Reuss model (Chapter 2).

The general structure of the inorganic component of the bone matrix is mainly hydroxyapatite; a small amount of non-crystalline form of calcium phosphate also exists in the mineral phase. The plate-like hydroxyapatite crystals, roughly 400 Å long, with a comparable width (200–350 Å), and 25–50 Å thick (Robinson and Watson, 1955, 1952; Johansen and Stone, 1997), are embedded in the collagen fibers with orientations approximately parallel to the fibers. This molecule can generally be found around the collagen fibers in a crystalline structure. The organic matrix of the bone and the apatite crystals make a coherent unit that is vital to bone tissue function. The tropocollagen molecules create heterogeneous centers for crystallization of the bone minerals. These centers are usually placed at the points where the more polar heads of collagen fibrils are located.

The structure of hydroxyapatite can be derived from the spatial organization of a small number of the constituent ions (Posner, Perloff and Diorio, 1958). The unit cell for this is a right rhombic prism, when stacked in the manner of a simple hexagonal lattice. The length along an edge of the basal plane of the cells ($a$) is 9.432 Å and the cell height ($C$) is 6.881 Å. The spatial symmetry is symbolized as $P6_3/m$. There is considerable evidence indicating that biological factors are important in establishing the size, shape and orientation of bone crystals. The $C$-axis of the apatite crystallites is parallel to the collagen fibers (Engstrom and Zetterstrom, 1951).

The structural basis for the observed chemical stoichiometry of bone mineral involves the two major constituents, Ca and phosphate. The hydroxyapatite model would predict that the ideal composition of bone apatite should be $Ca_5(PO_4)_3OH$, with Ca and phosphate in Ca/P molar ratio of 1.67, 1.74 and 1.57 (Woodard, 1962; Eastoe, 1961). In studies on the dynamics of mineralization, the calcium/phosphorous ratio is adopted as best reflecting the formation of hydroxyapatite crystals. The calcium to phosphorous ratio in an ideal hydroxyapatite crystal is 1.67 : 1 and is slightly lower *in vivo* conditions (McLean and Budy, 1964). Diet has a significant effect on the mineralization of bone tissue. The carbon/calcium plus phosphorous ratio reflects the degree to which the bone matrix is saturated with hydroxyapatite crystals (Krawczyk *et al.*, 2007). A value of this ratio exceeding 10 is typical and is indicative of poorly or non-mineralized status. When fibrous bone tissue predominates the value is usually below 5 and for most advanced remodeling process the ratio is close to 9.

The physical character of bone, including the morphology (Fratzl *et al.*, 1993; Heywood *et al.*, 1990), dimensions and distribution of its composite apatite crystals, affects its mechanical properties. These parameters from different mineralized tissues have been investigated over many years. Isolated fracture apatite crystals studied with transmission electron microscopy have been shown to be platelet-like and, on average, ~50–100 nm or more long and 25–50 nm wide. Relatively newly deposition crystals are ~4–6 nm thick. The crystals are associated with collagen in bone and most other calcifying vertebrate tissues, but the precise spatial relationship between these two components is difficult to ascertain. A major problem in this regard is that electron microscopy, used to image the constituents, provides two-dimensional information. Three-dimensional aspects can be incorporated by applying low-voltage electron microscopic tomography and graphic insertions. These methods have demonstrated an apatite platelets are oriented approximately parallel to the another end to the collagen fibrils. The crystals are separated by a minimum of 4.2 nm.

While information concerning the morphology and dimensions of apatite crystals and their location and distribution, in relation to collagen, can be availed, such data have been mostly obtained from avian tendon and lamellar bone. Comparatively less data are available regarding these points in woven bone and particularly in human woven bone. Woven bone differs from lamellar bone in the organization of its collagen fibrils, its cell population and its mechanical properties. Woven bone is defined as having randomly distributed collagen fibrils and is deposited only during initial formation in fracture repair. The tissue physical, chemical and biological nature of woven bone, including the morphology and detailed organization of its composite crystals, needs to be considered to elucidate possible mineralization mechanisms. This has applications in biomechanics and biomedicine. It has been reported that bone modulus and microhardness increase from periosteal bone to endosteal woven bone in human fetal tissue.

The experimental results (TEM and X-rays) appear to be in good agreement in that the smallest dimension of the bone apatite crystal is about 50 Å. Considerable variation in the length appears to exist; the values reported from TEM are consistently higher than those obtained from X-ray bone broadening studies. A possible explanation for this observation is that the bone crystals appear subdivided in the direction of elongation. There is periodicity of 50–60 Å (Fernandez-Moran and Engstrom, 1957) along the crystal length as a rod exhibited subunit of about 50 Å (Molnar, 1960). These observations (Ascenzi and Bonucci, 1966; Molnar, 1960) suggest that bone crystals are composed of chains of microcrystals fused in an end to end relationship. Such a fusion process would account for the variability in reported lengths and may also account for the divergent view on shape. The X-ray diffraction study (Posner *et al.*, 1963) reported that the largest dimension is probably less than 100 Å, which is consistent with this view of bone apatite as a mosaic of microcrystals rather than a continuously uniform single crystal.

Changes in the size of bone crystals with age have been noted on X-ray powdered rat bones (Menezel, Posner and Harper, 1965). The mean size increases with age of the animal up to maturity. The initial growth of the apatite crystals is quite rapid, occurring in a matter of minutes or less (Eanes and Posner, 1965), with the average crystal size at the end of the precipitation being generally smaller than that of the crystals in mature bone. The synthetic crystals, however, gradually increase in size with subsequent aging, until they exceed the average size of bone crystals (Eanes, 1965). The crystal growth with age in synthetic systems continues indefinitely, though at progressively reduced rates. Water is necessary for the ripening to proceed in these synthetic systems. Reducing the water/mineral ratio retards secondary growth and, at a sufficiently low ratio, ripening can no longer be detected. Robinson (1966) has emphasized the fact that with maturation the mineral content per unit volume of whole bone increases at the expense of displaced tissue water. The water/mineral ratio, therefore, decreases with an increase in the degree of mineralization. This drop in water content with maturation could stop the ripening of bone crystals and cause the constancy of crystal size in mature bone.

Macroscopically, bone consists of tissue and anisotropic amorphous non-crystalline mineral phase (Hancox and Boothroyd, 1965; Molnar, 1959; Fitton-Jackson and Randall, 1956; Robinson and Watson, 1955). Porosity varies between 5 and 95%, with most bone tissues having either very low or very high porosity. Accordingly, there are two types of bone tissue. One is trabecular or cancellous bone with 50–95% porosity, which is usually found in cuboidal bones, flat bones and at the end of long bones. The pores are interconnected and filled with a tissue composed of blood vessels, nerves and various type of cells. While the main function is to produce the basic blood cells, the bone matrix has the form of plates and struts called trabecular with a thickness of about 200 μm and a variable arrangement (Martin, Burr and Sharkey, 1998). The second type is cortical bone, with 5–10% porosity and a different type of pores (Cowin, 1999). The first evidence to indicate that this amorphous phase is a major component of bone mineral came from X-ray diffraction (Termine and Posner, 1967; Harper and Posner, 1966) by analyzing the intensity pattern of the crystalline portions. It has also been supported by several other authors (Robinson, Doty and Copper, 1973; White *et al.*, 1977). It is estimated that 40% (Harper and Posner, 1966) of the mineral in the femur of adult human, cow and rat is non-crystalline. The amorphous content of bone, however, varies with age. This age dependency in the fraction of amorphous mineral has also been demonstrated by IR techniques (Termine and Posner, 1966). The amorphous phase of bone, like apatite, is supposed to be composed of Ca, phosphate and carbonate. Certain X-ray diffraction studies indicate that the crystallites are needle like, others that they are platelets. There is evidence indicating that both forms are present, depending on the site of formation. There are three levels of porosity, all containing bone fluid. These include the vascular porosity associated with Volkmann canals and the Haversian lumens (∼10 μm) and the lacunar canalicular porosity associated with the fluid space surrounding the osteocytes (∼0.1 μm) and the collagen–hydroxyapatite porosity associated with the spaces between the crystallites of the mineral hydroxyapatite (∼20–60 nm). Movement of the bone fluid in the collagen–hydroxyapatite porosity is negligible because most of the bone water in that porosity is bound by interaction with ionic crystals (Neuman and Neuman, 1958).

One cannot understand the structure of an amorphous particle from a knowledge of the spatial organization of a small group of its constituent atoms. This is because the atoms are not arranged in a regular, periodic array that would enable one to define the whole space occupied by the particle by simple translational repetitions of a basic structure of atoms and hence define the lattice type. This does not mean that local ordering of ions does not exist in amorphous calcium phosphate. However, there are sufficient random variations in this order, on going from one local co-ordination ensemble to the next, that long-range periodic order is absent. Nonetheless, the spatial organization of the ions comprising the amorphous material in bone mineral is unknown.

Bone mineral not only has a high surface area per unit weight, but its surface is chemically reactive with its environment. Low-temperature gas adsorption methods and small angle X-ray scattering have shown that the surface area of human and bovine deproteinated bone mineral ranges from 100 to 200 $m^2 g^{-1}$. The high magnitude of surface area of bone mineral provides a large interface for mineral reactions.

By measuring the heats of adsorption of small molecules on bone surface, it is shown that bone tends to bond strongly with certain molecules. In fact, poorly crystallized synthetic hydroxyapatite has long been used in chromatographic adsorption columns because of the high bonding capacity its surfaces have for specific proteins and polynucleotides. Studies on surface bonding (Posner, Betts and Blumenthal, 1979) suggest that a chemical linkage probably exists between the mineral in bone and certain free polar groups of collagen. Bone mineralization is a cell-mediated process in which complex events take place in sequentially (Wuthier, 1982). It has been reported (Lawrence *et al.*, 1994) that the mineralization process of equine bone is almost complete (76%) by twelve months of age and that the bone mineral content of the equine third metacarpal does not differ by sex.

The fluid shear stress amplitudes and frequencies in bone can be determined theoretically from known physiological loading parameters. By applying Biot's theory of poroelasticity to bone, the predicted range of *in vivo* fluid shear stress ranges from 0.8 to 3 Pa due to loading, with induced strains ranging between 1000 and 3000 $\mu\varepsilon$ (Cowin, 1999; Weinbaum *et al.*, 1994). Several *in vitro* studies (Bacabac *et al.*, 2004; Bakker *et al.*, 2001; Chen *et al.*, 2000; Frangos and Johnson, 1995; Jacobs *et al.*, 1998; Johnson, McAllister and Frangos, 1996; Klein-Nulend *et al.*, 1995a, 1995b, 1995c) have confirmed that this range of fluid shear stress magnitudes is able to stimulate bone cells.

It has been suggested that the rate (determined by the frequency and amplitude) rather than the magnitude alone of the applied loading stimulus correlates to bone formation (Bacabac *et al.*, 2004; Mosley and Lanyon, 1998; Turner, 1998). This implies that bone formation is enhanced by dynamic rather than purely static loading. Thus, both the amplitude and the frequency of loading seem to be important parameters for bone formation. Indeed, it has been shown that low magnitude ($<10\,\mu\varepsilon$), high frequency (10–1000 Hz) loading can stimulate bone growth and inhibit disuse osteoporosis (Rubin *et al.*, 2001). High amplitude, low frequency stimuli are rare in the activities of daily life, whereas high frequency, low amplitude stimuli are more common (Fritton *et al.*, 2000). Weinbaum *et al.* (1994) suggested that low amplitude postural strains due to muscular contraction could be more effective than high-amplitude low-frequency strains due to locomotion, in maintaining bone mass. This suggests that bone cells are likely to be excited by low-amplitude postural strains due to muscular contractions, and also by high amplitude, low frequency strains due to locomotion. Such behavior might explain why astronauts in a microgravity environment lose bone mass despite exercise. The rate of loading seems to be a decisive factor in bone formation and maintenance. However, the nature of bone cells response to the rate of loading remains to be fully understood.

High impact physical activity, including jumps in unusual directions, has a great osteogenic potential in humans (Nordstrom, Pettersson and Lorentzon, 1998) and in osteopenci ovariectomized rats (Tanaka, Alam and Turner, 2002). High impact drop jumps significantly increase bone formation rates compared to that of baseline walking (Judex and Zernicke, 2000). Furthermore, an initial high stress rate, as in step wise increases fluid shear stress, has been shown to stimulate neonatal rat calvarial bone cells. Therefore, the osteogenic response to high impact activity might be related to the response of bone cells to a sudden increase (i.e., stress-kick) in fluid shear stress. The osteogenic benefits of high impact activity imply that the bone cell response to fluid shear stress is nonlinear.

## 1.3 Classification of Bone Tissues

Two types of bone are observed in the normal, mature human skeleton: compact bone and cancellous bone. Although macroscopically and microscopically different, the two forms are identical in their chemical composition.

### *1.3.1 Compact Bone*

Compact (cortical) bone is a solid mass that is found along the shafts of the long bones (femur, tibia, radius, ulna) and is the principal component of the flat bones (skull and ribs). It has an extremely dense physical structure arranged around Haversian systems and Volkmann's canals, which are responsible for providing cellular nutrition. Because of its strength, cortical bone plays a significant role in the support of body weight and in the protection of internal organs. Approximately 80% of skeletal mass is cortical bone. In general, the compact bone is the ivory-like surface layers of the mature bone.

The compact or cortical bone can be divided into three main categories: primary osteon bones, secondary osteon bones and surface bones. Primary osteon bones are formed in the vascular tunnels of the cancellous bone as cylinders and they are structured due to the merging of the fine cancellous cartilage and membrane bones of young subjects. Secondary osteon bones are established during the remodeling process in the tunnels that are opened by osteoclastic activities. These gradually increase throughout adult life as a result of remodeling. The internal remodeling is very active in childhood but slows during the older ages, being lowest at about the age of 30 (Pauwels, 1948). Surface bones are the primary formation of solid bone on the medullary and periosteal surfaces of any existing bone shaft.

The cross section of the shaft of a long bone can be divided into the following areas. One tenth of the outermost of the thickness of the shaft consists of approximately circumferential lamellae, the inside, next to the bone marrow tend, to be cancellous. The thickness between the outer and the inner portion consists of the secondary osteons, Haversian canals, interstitial lamellae, blood vessels, and so on. The secondary osteons are also called the Haversian system – mostly they look like hollow cylinders in cross section. The inner tube of the osteon is called the Haversian canal and the concentric lamellae structure. Usually, osteons are made of 5–30 concentric lamellae. In the regeneration or remodeling processes, the existing bone is first removed by osteoclasts in a circular fashion and then lined with partly mineralized concentric lamellae. The mineral content of these lamellae increases with increasing time and complete mineralization may take a very long time. Therefore, the cross section of a compact bone contains osteons of various ages. During the formation of new osteon, some of the old osteons with irregular shapes lie in the interstitial lamellae sections. Some of the older osteons may have been partly excavated and are replaced during the new osteon formation.

### *1.3.2 Fine Cancellous Bone*

Compared to the cortical bone, cancellous (trabecular/spongy) bone is mesh like, considerably finer and more delicate in appearance. Its physical arrangement of broad plates connected by thin struts provides for maximum support but with a minimum of raw material. Cancellous bone is found principally in the vertebral of the spinal column and at the epiphyses of the long bones.

Fine cancellous bones are found in the second ossification centers of the fetal skeleton and induce new bones of pathological nature. Generally, they consist of bone bundles with a honeycomb structure spanning the intertrabecular spaces containing both blood vessels and cells. The small spicules of solid matrix in spongy bone are called trabeculae; therefore, the spongy bone is often called trabecular bone.

In general, there are two kinds of the cancellous bones. The difference in their structure depends on the place where the bone is formed. If it is formed in the cartilage, it is called fine cancellous cartilage bone, while if it is situated near the membrane it is called fine cancellous membrane bone. The internal cancellous structure of many bones, especially the long bones of the leg, contributes to their strength with a little amount of material.

### *1.3.3 Coarse Cancellous Bone*

Coarse cancellous bone has a similar structure to that of compact bone, with some exceptions. It contains less bone cells and less osteons. The cells are nourished from the vessels in the intertrabecular marrow spaces, because the Haversian canal system is not capable of supplying blood to the cells.

Figure 1.3a shows the arrangement of cortical and trabecular bone in the proximal femur, where the outer dense cortical bone of the femoral shaft encases the finer trabecular structure of the cancellous bone. The trabeculae have adopted a preferential alignment along the direction of osteogenic response to mechanical loading. This is because electronegativity is an indication of the region of growth. It has been documented in compact bone in diaphyseal cross sections (Pauwels, 1976; Radin *et al.*, 1982; Currey, 1964; Lanyon and Rubin, 1985; Martin and Burr, 1989; Hou *et al.*, 1990; Gross *et al.*, 1997; Lieberman and Crompton, 1998; Martin, Burr and Sharkey, 1998).

## 1.4 Lamellation

Within each lamellae of the osteons the collagen fibers have one dominant direction; this direction varies from one lamellae to the next, giving the effect of a family of co-axial helices of different helix angles. In general it is assumed that there are three types of arrangement of fibers in the osteons. The first has the fibers in all lamellae roughly parallel to the axis of the osteon (90° helix angle). The second has the fibers in alternate lamellae longitudinal and circumferential (helix angle alternates between 90 and 0°). The third has the inclined fibers in each lamellae but in opposite directions in successive lamellae (the helix angle alternates between plus and minus approximately 45°).

Bones are heterogeneous in their organization and may be divided into two distinct structural regions. At the outer surface is the periosteum, which is the source of the cells responsible for growth in bone width. Adjacent to this is cortical bone and on the inner surface of the bone, adjacent to the marrow cavity, is trabecular bone. The inner surfaces of the bone are covered by cells that form the endosteum. Both trabecular and cortical bone are made of calcified lamellae in which osteocytes are embedded. The osteocytes are linked by a network of unclassified channels termed canaliculi (Figure 1.7a and b).

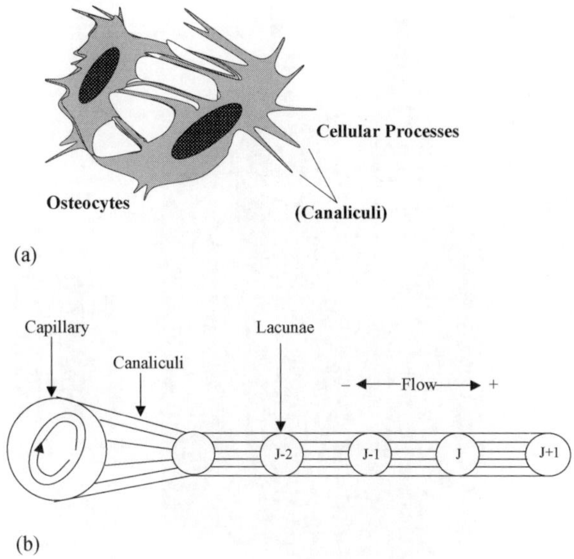


**Figure 1.7** (a) Schematic of osteocytes; (b) model of the Haversian canal-lacunae-canaliculi system

In cortical bone, the canaliculi are linked to the Haversian canals in which the bone vasculature resides (Figure 1.8a). Surrounding the blood vessels are concentric rings of lamellar bone, which form a unit termed an osteon or Haversian system. The Haversian canals run parallel to the axis of the bone and arise through bone remodeling. The vessels penetrate the bone from the periosteum to the marrow cavity through transverse Volkmann's canals. The bone surfaces are covered with cells, which include dormant bone lining cells, osteoblasts and osteoclasts. Like bone lining cells, osteocytes are inactive osteoblasts that are not buried in new bone. They remain on the surface during bone formation steps and can be reactivated in response to chemical and or mechanical stimulus (Miller and Jee, 1992). They are located in lacunae (Cowin, 1999) and communicate with the rest of cells via canaliculi (Figures 1.7b and 1.8b). Many authors (Cowin, Moss-Salentijn and Moss, 1991; Lanyon, 1993; Burger, 2000; Skerry *et al.*, 1989) suggest that osteocytes are mechanoreceptor cells that control bone remodeling, but this needs further confirmation. Furthermore, it is assumed that osteocytes, the only cells embedded in the bone matrix, are affected by processes that damage the bone matrix, interrupting their communication through canaliculi and thereby affecting their metabolic exchange. Fatigue microdamage may therefore create a situation resembling disuse at the level of the osteocyte cell body and lead to bone remodeling starting with osteoclast recruitment.

A cartilage model of the bone known as the analage cartilage is formed, which is approximately the same shape as the final bone. The analage cartilage is made up of cartilage cells with a surrounding immature intercellular matrix. In the long bones, during the first stages of conversion into bones, cartilage cells in the central part of the shaft increase their rate of proliferation, enlarge and become hypertrophic. Blood vessels penetrate into this altered cartilage and promote osteogenic activity. The osteogenic cells have the capacity to change

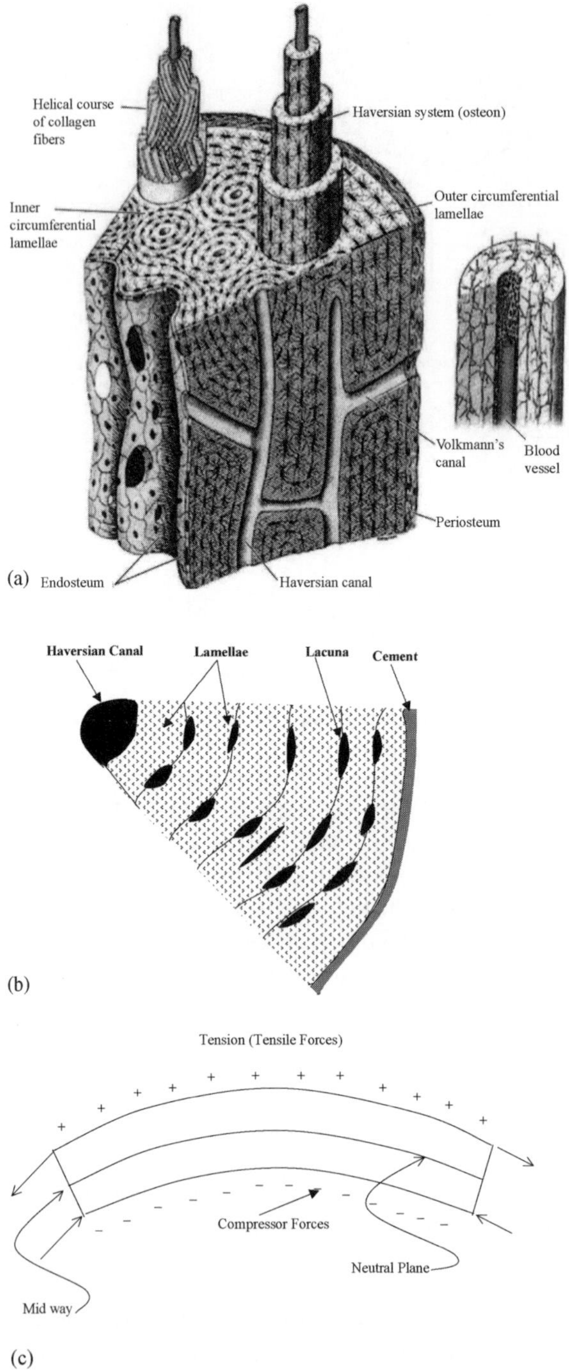


**Figure 1.8** (a) and (b) Schematic representation of bone anatomy at the microscopic level; (c) bone under bending, showing a negative charge on the concave surface and a positive charge on the convex surface – the polarities will be reversed if the direction of bending is reversed

their function, and the name given to a cell depends on its activity at the time. Cells in the process of laying down an osteoid matrix, which rapidly calcifies, are the osteoblasts. When the cells coalesce to form multinucleated cells, which remove bone, they are known as osteoclasts (Figure 1.9a and b). Their precursor cells have been given various names, one is macrophage.

The process of replacing cartilage by bone after vascular penetration is known as enchondral ossification. This proceeds from the central regions of the shaft towards the ends, the epiphyses. When the advancing ossification front reaches the epiphyseal region, then ossification commences separately within the epiphyses, to form the secondary bone nucleus. During childhood a region of cartilage remains at either end of the long bones, between the epiphysis and the metaphysis. It is in a highly organized state, contains vigorously proliferating cells and is known as the epiphyseal growth cartilage, or sometimes the epiphyseal growth plate. Regulating this process is the application of physical forces, which may either act directly or indirectly.

The zone of growth cartilage nearest the epiphysis is the germinal or resting zone. It is similar to the analage cartilage. Then comes the proliferative zone with columns of active rapidly dividing cells. These enlarge and form a calcifiable matrix that is laid down over the main structural part of the intercellular matrix. In the stages of most rapid growth they enlarge further to form the hypertrophic zone. As enchondral ossification takes place each hypertrophic space is invaded by an advancing sinusoid vessel. When the rate of growth allows, cartilage may be removed without enlargement of the cells to the hypertrophic state. The mode of removal takes place by multinucleated cells known as chondroclasts. These coalesce from similar cells to those that formed the osteoclasts. The different names denote the different function. In other cases enchondral ossification takes place with the formation of bone trabeculae. They may be very regularly arranged during the period of rapid growth which is accompanied by chondroclast activity. These trabeculae occupy the metaphyseal region of the bone.

During growth, while the epiphyseal growth cartilage provides for the growth in length, there is also an increase in diameter of the shaft of the long bones. New bone is laid down on the outer part of the cortex, the periosteal surface, while at the same time bone is removed from the inner surface of the cortex, the endosteal surface. In the case of periosteal bone formation cartilage is not formed as an intermediate stage in the process. Changes in the shape and distribution of the bone tissue occur both before and after growth has ceased, by the process of remodeling. One of the main factors regulating this process is the application of physical forces, which may act either directly on the walls of sinusoidal vessels in the bone or indirectly by altering blood flow and pressure. In the former case the remodeling consists of the removal of trabeculae followed by the formation of new ones in rather different positions. The latter frequently leads to the formation of osteons in the cortical bone in the shaft. More generally, this type of bone is known as compact bone. During osteon formation there is a proliferation of vessels, with accompanying osteoclasts. This channel – a Haversian canal – is then partly filled with new bone laid down concentrically around the vessels.

When cortical or trabecular bone is laid down the osteoblasts arise from the vessels walls, and are arranged in rows. Cortical bone that has been so formed in a regular array is known as lamellar bone. It is present in children, but when they take sufficient exercise it is all converted into osteonal bone by the time the adolescent changes are complete. Small animals, such as rabbits, tend to retain a lamellar bone structure throughout their life.

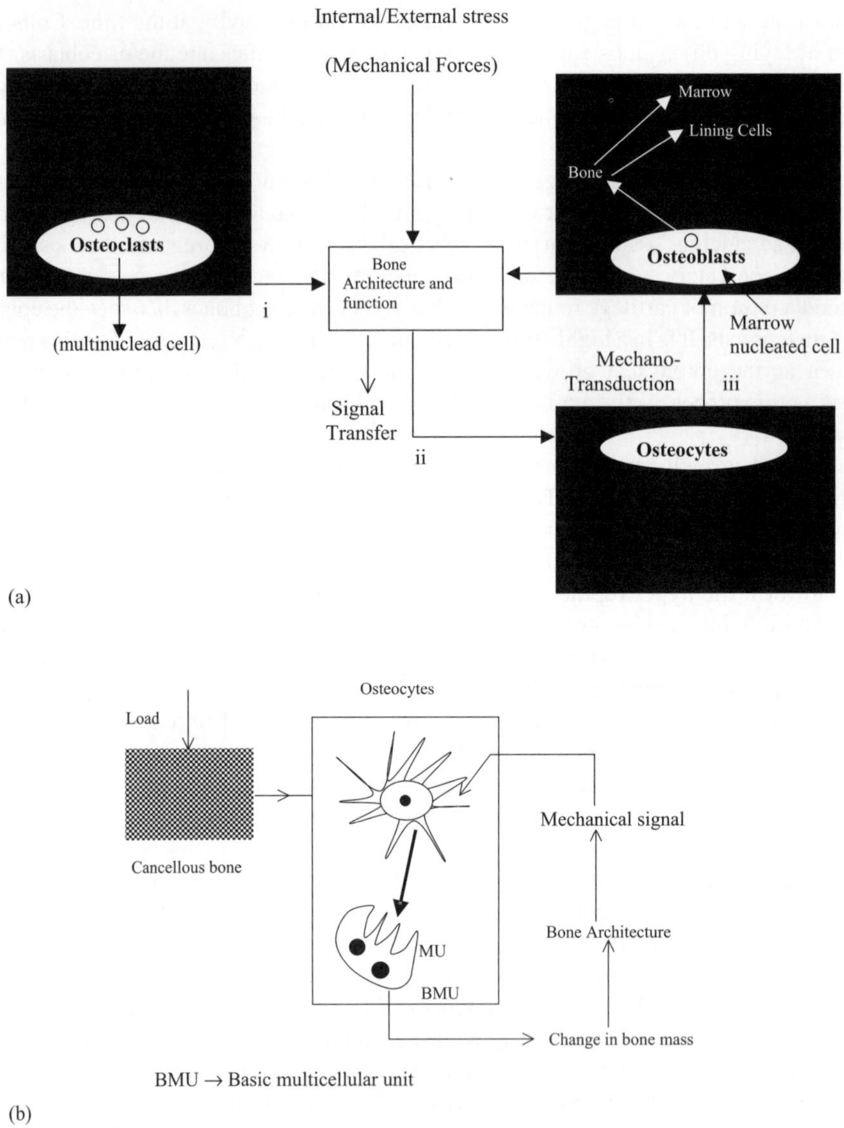


**Figure 1.9** (a) Probable signal transduction pathways: (i) osteoclasts resorb bone on trabecular surface locations; (ii) osteocytes sense a mechanical signal due to external load transfer through the architecture; the signal is transferred to the trabecular surface, where (iii) osteoblasts are initiated from bone. (b) Cellular response to bone stimulus

Bone, both the marrow, among the trabeculae and Haversian canals in compact bone are sinusoid vessels, which are of varying diameter. In normal marrow only about 1 in 10 of these is open at any given time. These blood vessels have a sheath of basement membrane or reticulum and flattened enthotheliad cells are seen at intervals on their walls. Wherever these sinusoid vessels are formed in the body, whether in bone or in the spleen and liver, their hemopoietic activity is possible. Their cells only act as bone precursors in the presence of a substance known as the osteogenic factor. Other blood vessels, whether arteries or veins, have move substantial walls. Even small capillaries tend to have a continuous layer of cells over their basement membrane. During proliferation, sinusoid vessels and capillaries display similar characteristics. Nerves are found accompanying may arteries.

The end of the long bones are covered by articular cartilage, while between the two articular surfaces, in a joint, is the synovial fluid, the remaining surfaces of the fluid-containing cavity being occupied by the synovial membrane. Cells on the surface of the synovial membrane secrete the synovial fluid. The shaft of the bone is covered by a sheath of connective tissue called the periosteum and outside this again there are usually muscles. Points of attachment of muscle to bone are known as muscle insertions. The tissue surrounding the growth cartilage is known as the perichondrial ring and fibrous tissue surrounding those parts of the articular cartilage that are not in contact with the synovial fluid is called the perichondrium.

Arrangements around other bones are of a related type. Between the vertebral bodies in the spine are the intervertebral dishes. Each side of a cartilage end plate, in the growing child, acts as a growth cartilage, while in the adult it has many of the characteristics of articular or hyaline cartilage. The central part of the disc is a type of fibrocartilage known as the nucleus pulposus and surrounding it is an oriented fibrous tissue known as the annulus fibrosus of the bone in the vertebral bodies. A part of this is formed by enchondral ossification, while the remainder is formed by periosteal ossification. The whole area of a normal vertebral body is occupied by trabecular bone surrounded by hemopoietic tissue.

### *1.4.1 The Cement*

The cement can be considered as the amorphous continuum phase in which the discrete fibers and inorganic crystals are embedded. In essence, what remains after removing the fibers and the inorganic phase constitutes the cement. This cement is composed mostly of mucopolysaccharides, glycoproteins, lipids, carbonate and citrate. The other constituents include sodium, magnesium and fluoride. The ground substance and the cement lines are made of cement and they show viscous and plastic behavior.

## 1.5 Role of Bone Water

Next to osteoid and mineral, water occupies the largest volume fraction in bone. It is probably because of this that the nature and role of water in calcified tissues has been a subject of serious investigations (Neuman and Neuman, 1958; Robinson and Elliot, 1957; Timmins and Wall, 1977).

Water not only resides within the vascular canals, lacunae and canaliculi but also exists within the collagen matrix and the mineral apatite (i.e., extracellular matrix). The estimated maximum water content in the vascular-lacunar-canalicular space is 12% of the total volume of bone, a value higher than the 8% calculated by others (Zhang, Weinbaum and Cowin, 1998). Robinson (1960) reported that water could exist in two fractions, one driven off at 50 °C, associated with marrow-vascular-osteoid, and the other driven off at 100 °C, associated with the calcified matrix. Water distribution in bone not only exists as mobile water in pores but also has other forms, which interact with bone tissue at different energy levels.

Water interacts with the collagen and mineral phases of bone in several ways. Firstly, the polarity of water facilitates its bonding with the hydrophilic groups of the collagen protein (glycine, hydroxyproline, carboxyl and hydroxylysine) and the charged groups, $PO_4^-$ or $Ca^{2+}$, of bone mineral. Secondly, studies on the hydration of collagenous tissue (human dura mater and rat-tail tendons) with dynamic mechanical spectroscopy indicate that water does bond with collagen at two levels (Nomura *et al.*, 1977; Pineri, Escoubes and Roche, 1978). Thus, collagen has structural water and loosely bound water. The former results from hydrogen bonding within the triple helix of collagen molecules (due to the hydroxyl group of hydroxyproline) and requires more energy to remove than the latter, which arises from hydrogen bonding with the polar side chains of collagen fibrils. Thirdly, there are two types of water interaction in the mineral phase, lattice water and surface bound water. Both X-ray diffraction and infrared (IR) spectroscopy of heat-treated synthetic, precipitated apatites found that water bound to surface crystals is lost at a lower temperature (<200 °C) than water inserted into the lattice structure (between 200 and 400 °C) (Le Geros, Bonel and Legros, 1978). Nonetheless, the spectra intensities from nuclear magnetic resonance of both surface water and lattice water decrease when bone dries up to 120 °C, with more lattice water remaining at higher temperatures (Casciani, 1971). Based on the energy characteristics of water with collagen and mineral, it is suggested that water removal from bone is related to energy level as follows: (i) mobile water molecules require less energy to evaporate than water on bone surfaces, (ii) the removal of the loosely bound water (via hydrogen bonding) requires less energy than the water molecules trapped inside collagen molecules, which in turn requires similar or less energy than water molecules bound to the surface charges of mineral apatite (more ionic in nature) and (iii) water that is imbedded in the lattice of hydroxyapatite (about 35 mg of water per g mineral more covalent in nature) requires the highest energy to dislodge (Neuman and Neuman, 1958).

This water cannot be displaced by simple drying at 100 °C. The work of Timmins and Wall (1977) indicates that removal of water by thermal dehydration is rather gradual. Neuman and Neuman (1958) have further pointed out that there is a 10 nm thick hydration layer (bound water) retained at centrifuges of 10 000*g* around each hydroxyapatite crystal.

In the unmineralized state, as the bone matrix is laid down by osteoblasts, the collagen fibers produced contain a large volume fraction of water (up to 60%). During calcification, while apatite crystals are deposited in the organic matrix, the osteoid water is gradually displaced and reduced down to a 20% volume fraction. The spin–lattice and spin–spin relaxation times of proton and deuteron (water adsorbed on collagen fibers) decrease with the decrease in water content. By comparing these changes with the change in the quadrupole splitting of heavy water, it is concluded that shorter relaxation times correspond to a higher degree of ordering for water in collagen. This fraction drops to 10% in senile bone (Robinson, 1952), corresponding to reduced impact strength in old age. It is thus plausible that the volume occupied by mineral cannot vary in a manner independent of that of collagen bound water since the total volume has to be preserved. Hence, any change in the mineral fraction will result in a corresponding change in the water fraction.

It has been suggested (Fung and Tautmann, 1971; Chapman *et al.*, 1971) that a small fraction of water adsorbed on collagen is highly ordered and the rest moves isotropically. It may be speculated that the ordered water molecules would have a longer correlation time and relax must faster than the isotropic water molecules and a change in their ratio would cause a change in the observed relaxation times. The spin–lattice relaxation time ($T_1$) decreases with decreasing amount of water in collagen. The spin–spin relaxation time ($T_2$, for $^2H_2O$ on collagen) also decreases with decreasing water content. The relaxation times are strongly frequency dependent.

The loosely associated water is what is referred to as "bulk water" (free water) that fills the pores of the calcified matrix making up the Haversian and lacuno-canalicular system. It is this fraction of water that has been shown to confer the unique viscoelastic properties of bone, which are largely lost after drying. In its natural fully hydrated state, stress-induced deformation upon application of a load is damped by the resistive forces experienced by the fluid in the lacuno-canalicular system (Garner *et al.*, 2000). Notably, the binding state of pore water depends on its proximity to the surface and thus is expected to be greatest in the canaliculi measuring less than 1 μm in diameter. This water has an additional critical role in that during mineralization and demineralization ions need to be transported to and from the osteoid sites. Water thus provides a medium for flow and diffusion driven ion transport (Neuman and Neuman, 1958). The study of pore water dynamics can provide detailed insight into transport phenomena occurring within the matrix. It is also possible that pressure-induced fluid flow through the lacuno-canalicular system is one possible mechanism of mechanotransduction regulating bone formation (Figure 1.10).

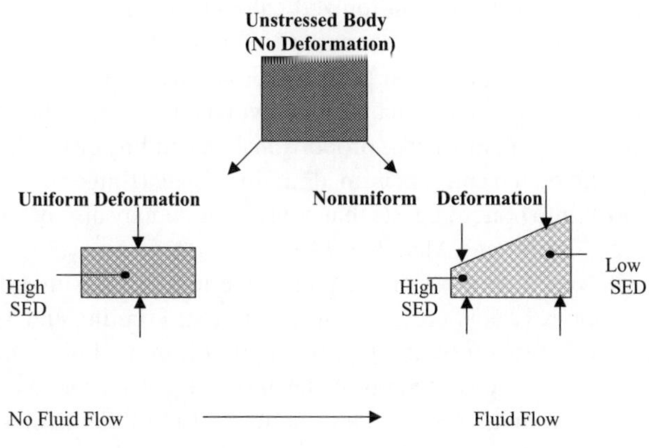


**Figure 1.10** Effects of different mechanical parameters and their gradients, for strain energy density (SED). A high volumetric strain gradient favors liquid flow (Ruimerman *et al.*, 2005)

## 1.6 Bone Metabolism

Bone metabolism is regulated by a wide variety of hormones, cytokines and other factors. Metabolism is also regulated by the activity of five different type of bone cells: osteoprogenitor, bone lining cells, osteoblasts, osteocytes and osteoclasts. Osteoprogenitor cells consist of preosteoblasts, which undergo mitosis and differentiation to become osteoblasts.

### *1.6.1 Ca and P Metabolism*

Calcium and phosphorus are essential elements required for various physiological and biochemical functions. Calcium is needed for processes such as bone formation, blood clotting, nerve conduction, muscle contraction, intercellular communication secretion and membrane permeability. Phosphorus is required for bone formation and the phosphoryl moiety is part of biologically significant molecules, such as ATP, DNA and RNA. The source of Ca and P for human beings and animals is food.

The level of Ca in the plasma is maintained by the diet and varies from 9.1 to 10.7 mg per 100 mL (Smith *et al.*, 1960; Peacock and Nordin, 1973). There are considerable diurnal variations in any individual. The value goes up in the day when calcium is ingested in the diet and it falls during the night when there is no dietary intake. The variation may be as much as 10% of 1 mg per 100 mL. The normal calcium level is maintained by control of absorption in the gut and excretion in the urine. Bone calcium acts only as an emergency reservoir in the state of calcium deficiency and possibly during the early morning fast calcium is mobilized from the skeleton to maintain the plasma calcium concentration.

Calcium is present in plasma in three forms: ionized, protein bound and combined with citrate and other organic acids. The ionized calcium is diffusible, the citrate calcium is not ionized but is also diffusible, while the calcium proteinate is neither ionized nor diffusible. The distribution of calcium (Lingarde, 1972; Walser, 1961) and its fall is affected by age. The $P_H$ (i.e. $[H^+]$ ion concetration) level is a controlling parameter which affects $Ca^{2+}$. Increased protein calcium binding counteracts $Ca^{2+}$ homeostatic control mechanism as the $Ca^{2+}$ decreases. The amount of diffusible non-ionized calcium is small and is generally believed to be of little clinical interest except in patients (Lingarde, 1973).

Interpretation of variations in the total plasma calcium depends on how much calcium is bound by the plasma proteins. The plasma protein involved is largely albumin; 1 g of albumin binds approximately 0.7 mg calcium and is proportional, depending on the $P_H$ and other factors. Some 80–90% of protein bound calcium is an albumin chelate (Lingarde, 1973). The fact that albumin may also move into bone suggests that consideration may also need to be given to the bound calcium (Owen, Triffitt and Melick, 1973).

In the contrast with calcium, most of the phosphate in plasma is in a diffusible form as $HPO_4$ or $H_2PO_4$ and about 12% is protein bound (Bijvoet, Froeling and Sluys Veer, 1972). The normal plasma level varies form 2.4 to 4.04 mg in different clinical states. The relative proportions of these two ions types depends on the plasma $P_H$. There are many biological uses of phosphate, extending from its role as an inorganic to its multiple involvements as organic phosphate in both enzymatic and structural proteins as well as in nucleic acids. The maintenance of calcium and phosphorus homeostasis involves delicate interrelationships of absorption by intestine, accretion and reabsorption by bone tissue and, finally, urinary excretion by the kidneys.

The calcium and phosphate are made available in the plasma – again a delicate balancing operation occurs, between accretion and mobilization in the bone and excretion by the kidney. The dietary intake or availability of calcium and phosphorus can diminish or be increased. It is possible to tip the balance in favor of increased bone mobilization or increased bone accretions to meet the stringent prerequisite of a constant serum calcium level. Thus the serum calcium may become elevated through stimulation of bone calcium mobilization or through PTH stimulation of the tabular reabsorption of Ca in the kidney. The PTH stimulates phosphate excretion, so that as serum calcium concentration increases there is usually an associated fall in the plasma phosphate level. In contrast, if serum calcium levels become too elevated, the action of calcitonin may come into play.

The dynamics of phosphate metabolism are not particularly different from those of calcium. Some 70% of the phosphate in the diet is absorbed. Absorption of phosphate is interrelated in a complex fashion with the presence of calcium and can be stimulated by a low calcium diet as well as vitamin D and its metabolites. The intestinal absorption of phosphate is inhibited by high dietary calcium levels. Phosphate in the body is also partitioned among three major pools: kidney ultrafiltrate, the readily exchangeable fraction of bone and the intercellular compartment in various soft tissues.

The central consideration of Ca and phosphorus homeostatic is the fact that the endoskeleton of higher organisms serves two important functions (Cummings *et al.*, 1995): A metabolic role concerned with the maintenance of dynamic steady state not only of Ca and P but also of many other ionic constituents for the body's extracellular fluids (Kanis *et al.*, 1994). This provides a mechanical or supportive function to locomotion and protection of the organism. A characteristic feature of bone that enables it to fulfill both these functions is that the skeleton normally undergoes a process of continual remodeling throughout life. This remodeling process is governed by changes in the mechanical stresses on the skeleton as well as by effects of metabolic regulators.

## 1.7 Osteoporosis

Osteoporosis is defined mainly according to pathological and anatomical principles and is considered a clinical entity caused by aging (Pommer, 1885;Albright, Smith and Richardson, 1941). The progressive diminution in bone volume starts at the beginning of adult life (Atkinson, 1964), is accelerated after middle age and appears to involve, sooner or later, both trabecular and compact bone. This is identified relating to metabolic bone disease. Therefore, this type of aging osteoporosis constitutes a major problem and has an immense social dimension.

Bone is a body of uniform strength and its construction seem to be optimally designed. The adaptational processes are similar to a feedback mechanism, which is controlled by mechanical stress. They concern the remodeling of the tissue as well as the exchange of mineral salts. After initial formation of bone, the tissue remains very dynamic. During long bone formation certain areas need to be shaped as the bone continues to grow. Therefore, in modeling, formation and resorption are separate and may occur at different sites. If the two processes are not coupled, a drift of the bone can occur, resulting in an eccentric shaft. During growth, modeling occurs continuously and there is an ultimate net gain of bone via this process (Jee, 1988; Lawrence and Fowler, 1997; Wasserman, 1984). Modeling and remodeling increases bone strength and stiffness by adding bone to areas where deformation (stress gradient) will be the greatest, the periosteum (Kimmel *et al.*, 1993).

The balance between bone architecture and mechanical loading is a dynamic one, where bone is constantly resorbed and new bone is formed. In osteoporosis, however, bone is inadequately maintained, leading to inferior quality of bone architecture and osteopenia (Kleerekoper *et al.*, 1985; Parfitt, 1984). This is characterized by decreased bone density, which correlates to a decrease in estrogen levels. In the postmenopausal stage, the production of follicle steps are the primary source of estrogen. In the absence of estrogen, bone resorption increases due to an increased number of active osteoclasts. In males the corresponding role is that of testosterone. In osteoporosis there is a loss of both hydroxyapatite and osteoid components and bone becomes much thinner and more fragile. It may be that many serious fractures occur in the vertebrae of osteoporotic patients, resulting at times in a bent-over posture.

Most workers now agree that aging osteoporosis results from an increased rate of bone resorption. This can be inferred from the normal rates of bone formation or accretion that most workers have observed in osteoporosis (Dymling, 1964; Heaney and Whedon, 1958). Jowsey (1960) demonstrated an increase in the resorbing surfaces of bone in the elderly. Moreover, Bordier (1964) has shown increased osteoclasis in the iliac crest biopsies of patients with osteoporosis. The cause of this increased bone resorption is uncertain, but there are two main hypothesis. The first is that there is a primary change in the bone matrix, which leads to an inevitable bone breakdown analogous to that occurs in other tissues with age. It has been pointed out that with pathologic osteoporosis the skeleton is, as it were, aging prematurely. This hypothesis describes certain biochemical (Casuccio, 1962), histologic and biophysical (Little and Kelley, 1962) changes in osteoporotic bone that might be interpreted in this sense, as well as certain associated skin changes (McConkey *et al.*, 1963).

Many types of genes are involved in bone metabolism. Such genes include those coding for structural proteins, such as type I collagen and the vitamin D receptor and genes involved in sex steroid metabolism. These categories of genes and others are referred to as osteoporosis candidate genes (Gennari and Brandi, 2001). Polymorphism has been discovered in these genes and investigations into their associations with a male osteoporotic phenotype is sought (Melton, 2002).

A loss of bone mass is accompanied by a negative balance of bone remodeling. However, different parts of the skeleton are not affected to the same degree. The clinical syndrome of spinal osteoporosis is characterized by the occurrence of nontraumatic vertebrae fractures. The observation that bone mass in the general population decreases with age has led to the suggestion that osteoporosis represents an extreme form of the normal aging process (Newton-John and Morgan, 1968). Therefore, we speak of a deficiency in bone mass in comparison with a healthy control group of corresponding age and sex, while the quality of bone substance is normal. According to this definition, the term osteoporosis clearly marks a pathological condition, which has to be sharply distinguished from physiological bone loss with increasing age. The age-dependent atrophy is sometimes called physiological osteoporosis, as opposed to a pathological osteoporosis (McLean and Wrist, 1968). The breaking strength of bone is linearly related to its mineral content and porosity (Saha, 1977), hence the measurement of bone mineral content at the actual site of fracture appears to be the most accurate method of determining fracture risk (Arnold, 1973; Chalmers and Weaver,

1966). Several methods of quantifying BMC (bone mineral content) of the appendicular skeleton have been developed (Mazess, 1979). These measurements, however, have not provided good discrimination between subjects with spinal osteoporosis and age and sex matched controls (Johnson *et al.*, 1968). Because appendicular bone is predominantly trabecular, several authors (Nordin, 1971; Mazess, 1979; Wahner, Riggs and Beabout, 1977) have hypothesized that subjects with spinal osteoporosis have lost disproportionately large amounts of trabecular bone.

## 1.8 Bone Cells

Cellular dynamics play a major role in natural and stress induced bone remodeling and consequently are also important in normal and in bone diseases. There are five different type of cells controlling bone metabolism, namely, osteoprogenitor, bone lining cells, osteoblasts, osteocytes and osteoclasts. Hormones growth factors, ion concentrations and nutrition regulate all five bone cells. Mesenchymal stem cells can also differentiate into bone marrow cells, which can then differentiate into blood forming cells (Marks and Popoff, 1988; Lawrence and Fowler, 1997). The role of cell dynamics in bone remodeling and repair of bone fracture is of major concern (Binderman, Somjen and Shimshoni, 1986; Frost, 1964a). Bone adapts to changes in its mechanical environment based on the principle of cellular accommodation. For bone to adapt to a new mechanical loading state, bone cells must have some memory of their previous state in order to determine that the new state is different and to respond accordingly (Schriefer *et al.*, 2005). Bone cells process loading information locally because of poor innervations of bone tissue and, unlike many mechano receptor cells, they do not depend on the central nervous system to integrate and distribute information about mechanical signals. Frost (1983) has proposed that adaptive changes in bone shape occur when minimum effective strain (MES) thresholds are surpassed. It is assumed that when a strain threshold is surpassed the sensor cells will gradually accommodate to the new state, either by cytoskeletal reorganization or by changing the extracellular microenvironment. Using this definition of cellular accommodation, a set point can be determined simply by summation of the past history of daily strain stimuli (Turner, 1999). Bone formation (or resorption) would then be dependent upon the difference between the new strain stimulus and the ever changing set point. The temporal skeletal adaptive response will depend, to some extent, upon how the strain history is integrated into cellular memory. It is further assumed that the memory of the cells previous loading decays exponentially. Lanyon (1992) assumes a value of the MES that sets the threshold for a response. However, for the cellular response to be bone specific the MES must vary from bone to bone. Furthermore, the MES is location dependent within bones. Most long bones are loaded in bending (Bertham and Biewener, 1988), creating a strain gradient across bone section that has a value of zero along the neutral axis (Figure 1.8c). To avoid excessive bone resorption at the neutral axis, the MES varies across the bone section.

Each mechanosensitive cell must have some strain threshold above which a mechanical signal causes a cellular response. It is well known that many cell types, including osteoblasts, recognize their cytoskeletons in response to a mechanical stimulus. Cytoskeletal reorganization in turn changes a cell's mechanosensitivity, allowing the cells to accommodate to the strain environment. In addition, cellular mechanosensitivity may be altered through reworking the local environment around the cell (Rubin, Judex and Hodijiargyrou, 2002). Osteocytes are considered to be part of the mechanosensory apparatus in bone and hence any change in their extracellular microenvironment could affect their mechanosensitivity. The extracellular environment of osteocytes does not remain constant as these cells actively form and remove layers of matrix on the surface of their lacunae (McKee and Nanci, 1996).

In principle, there are four types of functionally distinct cells that are known to play major roles in bone physiology: mesenchymal cells, osteoclast, osteoblast and osteocyte. Figure 1.9a and b shows their role in controlling bone dynamics. The mesenchymal cells have an outstanding capacity for proliferation and are capable of further differentiation into osteoclasts and osteoblasts (Hall, 1965).

Bone cells respond to mechanical stimuli with various biological signals, and a number of these signals have been utilized as measures of their responsiveness to oscillatory fluid flow (OFF) (You *et al.* 2000, 2001; Kurokouchi *et al.*, 2001). Batra *et al.* (2005) have reported that insertion of 10 and 155 rest periods into an OFF profile results in greater $[Ca^{2+}]_I$ response magnitudes then those found in cells exposed to continuous OFF. The intracellular calcium magnitude has been shown to control cell processes in other cell types (Thomas *et al.*, 1996; Dolmetsch *et al.*, 1997; Berridge *et al.*, 1998; Dolmetsch, 2003).

### *1.8.1 Osteoblasts*

At the bone surface, osteoblasts are recruited from the environment or by reactivation of lining cells (Chow *et al.*, 1998; Dobnig and Turner, 1995) to form bone in response to the total stimulus they receive. Bone lining cells cover the surface of bone. They also serve as ion barriers between the canaliculi and interstitial fluids. Distraction osteogenesis is predominantly achieved by intramembraneous bone formation that is characterized by direct differentiation of mesenchymal cells into osteoblasts without the occurrence of cartilage tissues (Aronson and Harp, 1990; Bizarov *et al.*, 1984). The osteoblast can then secrete matrix, which will become calcified. This increases the amount of bone in the area. This differs principally from secondary bone healing where perivascular mesenchymal cells are induced to differentiate into chondroblasts (Brighton, 1984), and with further development of the callus undergo endochondral ossification (Schenk, 1992). Mesenchymal stem cells concentrate in the area of new bone formation and differentiate into osteoblasts. This then increases the amount of bone in the area.

Osteoblasts, the bone-forming cells, become entrapped in the lacunae and canaliculi and secrete intracellular bone matrix, known as osteoid. Osteoblasts originate from osteogenic lineage and secrete minerals, the amorphous ground substance osteocalcin, osteonectin, water and type I collagen into the matrix; osteoblasts have rough endoplasmic reticulum. Osteoblasts surround themselves with osteoid, mature into osteocytes and reside in the lacunae till death (Jee, 1988; Lawrence and Fowler, 1997). Osteoblasts make all the proteins and carbohydrates in osteoid and secrete them to form a gel in the extracellular environment around the cell.

Osteoblasts are derived from stromal stem cells (fibroblast colon forming units, F-CFU) that differentiate into osteoprogenitor cells. The same stem cells can also give rise to fibroblasts, endothelial cells or adipocytes. In more mature bone, osteoprogenitor cells are present at the periosteum and stem cells are present in bone marrow. The number of stromal stem cells in the bone marrow decreases with age, while the number of adipocytes increases. The osteoprogenitor cells may differentiate into chondrocytes or osteoblasts depending to a large degree on the environmental conditions. In regions with a good blood supply (via the presence of capillaries) the development of osteoblasts is promoted, whereas in regions of inadequate blood supply differentiation into chondrocytes takes place.

### *1.8.2 Osteoblast Differentiation*

Studies of the pathophysiology and genetics of skeletal cell differentiation have provided insight into the mechanism by which osteoblasts arise from mesenchymal precursors and osteoclasts arise from hematopoietic precursors (Ducy *et al.*, 1997; Karsenty and Wagner, 2002; Karsenty, 2003). The differentiation of osteoblasts is controlled by the transcription factor Runx2 and also the transcription factor osterix, which functions downstream of Runx2 (Karsenty and Wagner, 2002; Nakashima *et al.*, 2002). In the absence of either Runx2 or osterix, no osteoblasts are formed. Mice and humans that carry only a single copy of Runx2 have abnormalities that are specific to certain bones rather than the entire skeleton. Overexpression of Runx2 leads to a decrease in bone mass (Geoffroy et al., 2002). A role for polymorphism of these transcription factors in osteoporosis has not yet been identified.

LRP5 (low density lipoprotein receptor-related protein 5) deficiencies lead to the development of osteoporosis in both mice and human (Gong *et al.*, 2001; Kato *et al.*, 2002) and also to chronic inflammatory disorders (Raisz, 2003). There are also suggestions that some of the anabolic effects of parathyroid hormone (PTH) are exerted through suppression of Dickkopf1 expression (Guo, Bringhurst and Kronenberg, 2004). Mutations in human LRP5 that interfere with the Dickkopf1 inhibition are suggested to cause a high bone mass phenotype (Boyden *et al.*, 2002).

The function of mature osteoblasts, including the ability to synthesize extracellular matrix proteins, also requires LRP5 as well as the signaling protein ATF4 (Yang *et al.*, 2004) and bone morphogenetic proteins (BMPs). Members of a family of secreted growth factors provide important tissue specific signals to preosteoblasts that are essential for full osteogenic differentiation (Chen, Zhao and Mundy, 2004; Mundy, 2002). Thus, it has been possible to use a reporter system driven by the BMP-2 promoter to screen a set of chemical compounds and natural products for potential bone anabolic agents (Mundy, 2002). There is evidence for interactions among the possible pathways (Mbalaviele *et al.*, 2005; Yang *et al.*, 2004).

A changing pattern of cell adhesion molecules (CAMs) expression may also characterize the aging process at the tissue level. One can speculate that age-related changes in the type and/or amount of CAM expressed may impair cell–cell communication, either directly or through abnormal expression of connexins. Cell–cell communication is necessary to maintain the work of osteoblasts synchronous within basic multicellular units. An improper signal exchange may cause the remodeling units to become asynchronous. This may disrupt or slow down bone turnover. Ultimately, age-related abnormalities in the expression of CAMs and/or connexins may impair the ability of osteoblasts precursors to differentiate, with a consequent decrease in osteoblast number. This is a commonly accepted cause and phenomena in osteoporosis. This phenomena of aging on the skeletal tissue may lead to decreased bone formation. Additionally, because cell–cell contact and communication may be required for osteoclast development

(Suda, Takahashi and Martin, 1992), abnormal expression of CAMs may also lead to decreased osteoclast activity and translate into reduced activation frequency of the remodeling cycle at the tissue level.

The identification of the critical role for the Wnt signaling pathway in regulating osteoblast function is of particular interest. It plays an important role in determining bone mass and strength (Little *et al.*, 2002; Boyden *et al.*, 2002; Van Wesenbreeck *et al.*, 2003; Gong *et al.*, 2001). LDL receptor related protein 5 (LRP5) interacts with the frizzled receptor to transduce signaling by Wint ligands. A mutation of LRP5 that leads to constitutive inactivation can result in an increase in bone density (Little *et al.*, 2002; Boyden *et al.*, 2002). Another key predictor is the rate of bone remodeling, increased rates of osteoclastic bone resorption, measured by the level of collagen breakdown products, and increased bone formation, measured by bone-specific alkaline phosphatase and osteocalcin. Procollagen peptide levels are also associated with an increase in risk of bone loss and fragility fractures.

### *1.8.3 Osteoclast*

Osteoclasts are identified as multinucleated cells that can move about in tissue, are physiologically specialized, do not undergo cell division and are considered the daughter cells of a mesenchymal cell (Frost, 1963). They are monocyte-macrophase derivatives that are responsible for degrading bone, and are derived from preosteoclasts in the bone marrow. Active osteoclasts cells are usually found in resorption pits or Howship's lacunae. Their specialized role is central to the process that continuously removes and replaces segments of the skeleton in the higher vertebrates. Osteoclasts allow skeletal mineral to be used to manage extracellular calcium activity. This allows solid skeletal structure to be replaced by hollow architecture that has a superior strength-to-weight ratio. Osteoclasts tend to dissolve bone mineral by acid secretion and secrete specialized proteinases that in turn degrade the organic matrix, mainly type I collagen. Osteoclastic differentiation is normally balanced with bone formation, which is a function of stromal cell-derived osteoblasts. Interactions between osteoclast precursors and bone-forming cells are believed to control osteoclast differentiation under most circumstances, preserving bone architecture over many cycles of bone replacement. The junction of osteoclast and bone forms a ruffled border, which forms an impermeable microenvironment and enhances resorption.

Osteoclasts are assumed to be recruited by osteocyte apoptosis due to microdamage or possibly cracks (Bronckers *et al.*, 1996; Noble *et al.*, 1997; Parfitt *et al.*, 1983; Verborgt, Gibson and Schaffler, 2000). Ruimerman *et al.* (2005) assumed that, due to daily loading conditions, microcracks and damage occur at spatially random locations, occurring anywhere at any time. This describes the coupling between formation and resorption as an effect of mechanical stress transfer. Regulatory biomechanical factors related to this are still to be elucidated.

Osteoclasts are apparently giant cells with the ability to resorb mineralized tissue (Katagiri and Takahashi, 2002; Suda *et al.*, 1999). They evolve out of the fusion of mononuclear cells of exterosseal origin (Suda *et al.*, 1999). The osteoclast initials bone resorption by demineralizing the bone (Remedios, 1999). Osteoclasts appear only rarely in intact adults bone tissue (Miyamoto and Suda, 2003). However, areas with high bone turnover are characterized by an increasing number of osteoclasts. This regularly occurs in growth plates of juvenile bones but is especially pronounced in bone fractures.

The bone resorbing osteoclasts participate in bone healing. The osteoclasts are drawn to a resorption area by signals sent by osteoblasts (Bonewald, 2002). Osteoclasts are always found at or near the bone surface (Amling and Delling, 1996). In contrast to osteoblasts, which are densely populated and normally appear in layers on the bone surface, osteoclasts resorb misaligned or redundant bone tissue during bone healing (Cruess and Dumont, 1975). A single osteoclast can resorb the amount of mineralized tissue synthesized by 100 osteoblasts (Remedios, 1999).

### *1.8.4 Osteoclast Differentiation*

Osteoclast differentiation requires the binding of macrophage colony-stimulating factor to its receptor as well as the binding of the soluble differentiation factor receptor activator of NFB ligand (RANKL) to its receptor (RANK) on osteoclast precursor cells (Teitelbaum and Ross, 2003). This process is also regulated by a secreted decoy receptor of RANKL, osteoprotegerin (OPG), which functions as a paracrine inhibitor of osteoclast formation (Udagawa *et al.*, 2000). Osteoprotegerin, originally identified as a novel secreted member of the TNFR super family, was later found to inhibit spontaneous or induced bone resorption and cause osteopetrosis (the converse of osteoporosis). Osteoprotegerin acts as a decoy receptor that binds to RANKL and prevents it from interacting with its receptor RANKL and osteoprotegerin, which are both produced by osteoblasts at different stages of maturity (Gori *et al.*, 2000), thereby accounting for some of the signals in osteoblast–osteoclast communication. In conjunction with macrophage colony-stimulating factor, the RANK-RANKL-osteoprotegerin system regulates osteoclast differentiation. Thus, mice and humans deficient in osteoprotegerin have a high rate of bone loss (Whyte *et al.*, 2002). As in other high turnover states, anti-resorptive agents can still reduce both bone formation and resorption and compensate for osteoprotegerin deficiency. The mechanism of how the signal(s) couples resorption and formation remains elusive, although several anabolic ligands, such as BMPs and TGF, are stored in bone matrix as one is formed and are released at the sites of bone resorption and can thus act on osteoblasts and precursors in the vicinity. Signals from osteoblasts to osteoclasts can be provided by RANKL and osteoprotegerin. In this case preosteoblasts express a high level of RANKL relative to osteoprotegerin, which stimulates osteoclast differentiation and function. More mature osteoblasts, by contrast, express high levels of osteoprotegerin relative to RANKL, which inhibits osteoclast differentiation and function (Boyden *et al.*, 2002). The production of RANKL and OPG is regulated by agents that signal through CAMP, the vitamin D receptor (VDR) and gp130 (Suda *et al.*, 1999; Boyle *et al.*, 2003). RANKL produced by cells of the osteoblast lineage binds to its receptor.

The precise nature of the mechanism by which osteocytes translate mechanical signals into bone formation stimuli is still unknown. Ruimerman *et al.* (2005) have used the strain energy density (SED) rate as the relevant osteocyte signal (Figure 1.10). That alternative choices of the osteocyte stimulus signal may affect the results considerably is obvious as the precise distribution of the mechanical signals throughout the structure can differ significantly. The osteoclast can transform into an osteoblast and vice versa (Frost, 1963) (Figure 1.11). The osteocyte has been referred to as the inhabiting cell of formed bones. Coupled with the bone formation process is osteoclastic activity on the innermost surface of the cortical bone. This results in an increased diameter of the diaphyses, without altering the width of the cortical bone (Wasserman, 1984).

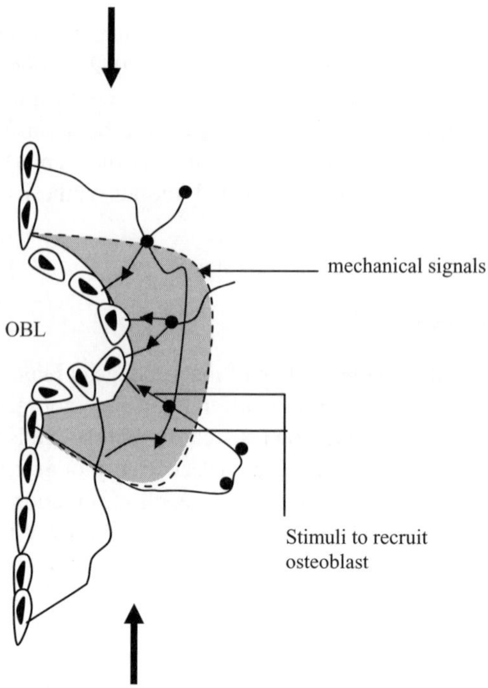


**Figure 1.11** A schematic diagram of Osteocytes recruiting osteoblasts cells. (Adapted from Ruimerman *et al.*, 2005)

It can thus be said that stresses applied from time to time play a definite role in the development of bone. Basically, these induce an electrical activity that is sensed as a command signal by cells or their environment that then respond accordingly. However, the nature of the command signal remains to be understood. Elevated external loads increase trabecular thickness (Jee *et al.*, 1990; Li, Luidier and Schaffler, 2003) and reduced loads cause trabecular thinning and loss of connectivity (Mosekilde, 1990). Evidently, bone is a part of system controlled by feedback from the applied stresses. The intrinsic and extrinsic control mechanism in bones, based on the generation of piezoelectric changes emanating from changes in pressures on or movements of crystals, remains, therefore, an attractive point of investigation. As an extrapolation of this, experimental and theoretical studies have shown that bone cells residing in intact matrix may be activated to initiate bone adaptation processes by interstitial fluid flow through the lacuno-canalicular network (Weinbaum *et al.*, 1994; Cowin *et al.*, 1995; Knothe Tate *et al.*, 2000).

During life, bones adapt their mass and structure to the average prevailing mechanical loads, to resist mechanical failure with a minimum of material expense. It is presently believed that this process of adaptation is governed by osteocytes, which respond to the loading induced flow of interstitial fluid through the lacuno-canalicular network (Burger and Klein-Nulend, 1999). When bones are loaded, the resulting deformation will drive the thin layer of interstitial fluid surrounding the network of osteocytes within the calcified bone matrix to flow from regions under high pressure to regions under low pressure (Piekarski and Munro, 1977; Weinbaum *et al.*, 1994). The loading-induced movement of labeled molecules demonstrates fluid flow in the mineralized matrix of bone both *in vivo* (Knothe Tate, Niederer and Knothe, 1998) and in *ex vivo* (Knothe Tate and Knothe, 2000; Klein-Nulend *et al.*, 1995a, 1995b). Osteocytes, and to a lesser extent also osteoblasts, respond to fluid flow stimulation *in vitro* (biochemical messengers) that is transduced through the canalicular network to the trabecular surface (Bacabac *et al.*, 2004; Bakker *et al.*, 2001; Chen *et al.*, 2000; Frangos and Johnson, 1995; Jacobs *et al.*, 1998; Johnson, McAllister and Frangos, 1996; Klein-Nulend *et al.*, 1995a, 1995b, 1995c; Burger and Klein-Nulend, 1999; Cowin, Moss-Salentijn and Moss, 1991; Knothe Tate *et al.*, 2002; Martin, 2000; Mullender and Huiskes, 1995; Nicolella and Lankford, 2002; Skerry *et al.*, 1989; Ruimerman *et al.*, 2005).

Osteoblasts are sensitive to fluid shear stress (Bakker *et al.*, 2001; Jacobs *et al.*, 1998) rather than to a streaming potential mediated by the transport of ions with the flow (Bakker *et al.*, 2001; Hung *et al.*, 1996). Bone cells are also more responsive to shear stress by fluid flow stimulation than to direct mechanical strain by substrate stretching (Mullender *et al.*, 2003; Owan *et al.*, 1997; Smalt *et al.*, 1997). Bone cells respond to fluid flow with increased nitric oxide (NO) and prostaglandin $E_2$ production, which are essential for the induction of new bone formation in relation to mechanical loading *in vivo* (Forwood, 1996; Turner *et al.*, 1996). In particular, MC3T3-EI osteoblastic cells produce NO in response to fluid shear stress in a rate-dependent manner (Bakker *et al.*, 2001). However, how bone cell sensitivity is modulated by the various flow parameters (amplitude, frequency, duration, etc.) is yet to be fully understood. It is suggestive that, as an alternative pathway, mechanical loading induces fluid flow due to the cyclic nature of the applied loads *in vivo*. This indicates that flow through the canalicular network is the actual trigger for metabolism (Burger and Klein-Nulend, 1999). Osteocytes are stimulated by a relatively small fluid shear stress acting on the membranes of their osteocytic processes (Weinbaum *et al.*, 1994). These authors have shown that in the physiological frequency range (1–20 Hz), associated with either locomotion (1–2 Hz) or the maintenance of posture (15–20 Hz), the fluid shear stress is nearly proportional to the product of frequency and strain. The impact of loosing osteocytes in bone may be enormous (Parfitt, 1993). In human bone, osteocyte cell death can occur in association with age and both osteoporosis and osteoarthritis, leading to increased bone fragility (Dunstan *et al.*, 1990; Dunstan, Somers and Evans, 1993; Frost, 1960).

Several *in vivo* studies have shown that bone can recover its responsiveness and can respond to mechanical stimuli with the same magnitude as earlier exposures to loading with the insertion of an appropriate length of rest periods (Forwood *et al.*, 1996; Robling *et al.*, 2000). Modulating loading parameters such as duration and magnitude may be a key factor in allowing bone to recover its mechanosensitivity. Understanding the mechanism required to restore mechanosensitivity in bone will be important for optimizing the osteogenic potential of mechanically induced bone formation. Robling *et al.* (2000) found, in a four-point bending model of rat tibia, that the insertion of a recovery period between loading cycles led to significantly higher relative bone formation rates than with continuously loaded limbs. Srinivasan *et al.* (2002) found that although continuous low magnitude loading is a non-osteogenic stimulus, the insertion of a 10s rest period between each loading cycle results in a potent anabolic response. These studies show that the introduction of short-term recovery periods, introduced into mechanical loading regiments *in vivo*, can have an osteogenetic effect at the tissue level. This corresponds to the fact that static loading (or approximating to that) has an effect on bone behavior. However, the cell–cell response to this loading pattern remains to be fully understood (Batra *et al.*, 2005). The quantum difference in the effect between the dynamic and near static loading on osteogenesis may be important to examine: if it resets the modeling pattern.

Ramnaraine, Pan and Clonisy (2006) found that treatment of NCD (cytosine deaminase fusion gene) transduced osteoclasts can promote killing of cancer cells. This introduces the possibility of developing osteoclast based treatment of primary bone cancers and breast cancers.

### 1.8.5 The Osteocytes

Osteocytes are mature bone cells that have many long cytoplasmic branches projecting from the main body (Figure 1.7a). Osteocytes are cells of osteoblast lineage that reside within the lacunae in the bone matrix. They communicate with each other and with the bone surface via cellular processes extending through canaliculi. While the genetic influence may dictate the general structural plan of bone but the environmental influence and growth modify it significantly. The osteocytes have a round cell body and many long fine extensions for cytoplasmic processes. Many types of osteocytes exist in the different parts of the bone matrix. They may vary in geometry and cytoplasmic detail as well as in the regularity and density with which they are placed in the matrix. Their main functional role is to facilitate the exchange of materials between tissue fluids and the bone matrix. In the normal adult skeleton, a balance exists between bone resorption and bone formation. The osteoclastic and osteoblastic activity cause a continuous turnover in the bone tissue but not an appreciable change in bone mass or a significant change in bone tissue structure. In a pathological case (e.g., osteoporosis), these activities become unbalanced and one of the processes dominates (resorption) and in turn causes a change in the bone structure.

### 1.8.6 Mathematical Formulation

The interrelationship of bone cell types and a summary of the theoretical formulation can be illustrated with a regulation scheme. Mathematical equations are introduced to quantify the relationship (Huiskes *et al.*, 2000; Ruimerman *et al.*, 2001). The change in bone mass at a particular trabecular surface location $x$ at time $t$ is determined by:

$$\frac{\mathrm{d}m_{\mathrm{tot}}(x,t)}{\mathrm{d}t} = \frac{\mathrm{d}m_{\mathrm{bl}}(x,t)}{\mathrm{d}t} - \frac{\mathrm{d}m_{\mathrm{cl}}(x,t)}{\mathrm{d}t} \tag{1.1}$$

with osteoblast bone formation, $\mathrm{d}m_{\mathrm{bl}}(x,t)/\mathrm{d}t$, and osteoclast bone resorption, $\mathrm{d}m_{\mathrm{cl}}(x,t)/\mathrm{d}t$.

Osteoblast activity at the trabecular surface is controlled by osteocyte bone formation stimuli. For a total stimulus $P$ (mol mm$^{-2}$ day$^{-1}$) that exceeds a certain threshold value, $k_{\mathrm{tr}}$(mol mm$^{-2}$ day$^{-1}$), the osteoblast tissue formation rate becomes:

$$\frac{\mathrm{d}m_{bl}(x,t)}{\mathrm{d}t} = \tau[P(x,t) - K_{\mathrm{tr}}] \tag{1.2}$$

where $\tau$ (mm$^5$ mol$^{-1}$) is a proportionality factor that regulates the formation rate relative to the formation stimulus. All osteocytes $N$ within the influence region contribute to the bone formation stimulus $P$ on trabecular surface to cation X, depending on their mechanosensitivity $\mu_i$, their distance $d$ to surface location $x$ and the signal $R$ (SED rate) the osteocytes sense in their location $x_i$; hence:

$$\begin{aligned} P(x,t) &= \sum_{i=1}^{N} f(x,x_i)\mu_i R(x,t), \quad \text{with} \\ f(x,x_i) &= \mathrm{e}^{-d(x,x_i)/D} \end{aligned} \tag{1.3}$$

where $f(x,x_\mathrm{i})$ is an exponential function that describes the signal intensity relative to distance $d$ and a decay parameter $D$.

The overall osteoclast bone resorption is described by:

$$\frac{\mathrm{d}m_{\mathrm{cl}}(x,t)}{\mathrm{d}t} - r_{\mathrm{cl}} \tag{1.4}$$

where $r_{\mathrm{cl}}$ ($\mathrm{mm}^3\,\mathrm{day}^{-1}$) represents a stochastic function that describes that portions of tissue are removed randomly from the surface.

Bone essentially receives all its nutrition via its extensive vascular system and when blood flow ceases the osteocytes entrapped in their calcified matrix rapidly die. Nutrients move to the osteocytes through the canaliculi by diffusion from the Haversian canals. Fluid flow through the canaliculi is not very efficient but osteocytes may be within 0.1–0.2 mm of blood vessel for adequate nutrition.

Examined in this way bone behaves like a dynamic structure in which the structural elements and the overall architectural plans are being continuously modified by proper cells, proper stimulations and proper nutrition. Bone undergoes substantial changes in structure, shape and composition according to the mechanical and physiological environment. Bone adaptability allows for efficient repair, which in turn helps to prevent fractures. Bone adaptation to mechanical loading depends on the duration and magnitude of the applied loads (Rubin and Lanyon, 1984; Forwood and Turner, 1994; Forwood *et al.*, 1996). However, application of continuous dynamic mechanical loads *in vivo* in animal models suggests that the osteogenic response of bone saturates with continued long-term mechanical loading (Turner *et al.*, 1994; Umemura *et al.*, 1997; Turner, 1998). However, the mechanisms behind bone's osteogenic saturation in bone to loading are not well understood. Bone as living tissue consists of cells (e.g., osteoblasts, osteoclasts and osteocytes) and their products, the extracellular substance (Hancox, 1972), and the cells that produce it belong to the connective series. The activity observed during the formation of bone is carried out by osteoblasts and is known as "osteogenic activity." The osteogenic cells have the capacity to change their function – the name given to a cell depends on its activity at that time (Little, 1973). Osteoblasts are the differentiated mesenchymal cells that produce bone, and are created at the periosteum layer or stromal tissue of bone marrow. Bone lining cells are inactive osteoblasts, remain on the surface when bone formation stops and can be reactivated in response to chemical and/or mechanical (electrical) stimuli (Miller and Jee, 1992). Like bone lining cells, osteocytes are former osteoblasts and are buried in the bone matrix and are located in lacunae (Figure 1.11) and communicate with other cells via canaliculi (Figure 1.7a and b). Look at in this way, matrix disruption may be expected to directly injure osteocytes, disrupting their attachments to bone matrix. This intervention is expected to interfere with their communication through canaliculi or alter their metabolic exchange. Fatigue micro damage may therefore create a situation resembling disuse at the level of the osteocyte cell body and lead to bone remodeling. These are responsible for the maintenance of bone, as a living tissue.

The presence of nerve terminals in bone is well established (Chenu, 2002), and the interactions between neurons and bone cells have come under intense study (Spencer, Hitchcoek and Genever, 2004; Chenu, 2002; Madsen *et al.*, 1996; Kingery *et al.*, 2003; Suyama *et al.*, 2003). Bone cells express a wide range of neurotransmitter receptors and transporters. This includes glutamate, $\gamma$-aminobutyric acid, purines, pyrimidines, (*S*)-hydroxytryptamine, catecholamines and neuropeptides (Chenu, 2002).

Intramembranous growth occurs on bone surfaces or "in membrane" such as under the periosteum of long and flat bones. There are two forms of bone growth – endochondral and intramembranous ossification. The compound (bone or cartilage) that provides the starting material for bone formation characterizes each of these forms of osteogenesis. Endochondral ossification indicates growth occurring in the cartilage, which is made up of chondrocytes and occurs in long bones at the growth plate. The growth plate is composed of zones. The first zone, known as the resting zone, is near the epiphysis of the bone and is made up of hyaline cartilage formed by chondrocytes. Although the exact function of this zone is unclear, chondrocytes are the likely recruit into the maturation process during growth. The hypertrophic zone begins when chondrocytes differentiate. Finally, in the zone of calcification, terminally differentiated chondrocytes, near the metaphysis, begin to lay down mineral into the matrix (Jee, 1988). The chondrocytes in this region undergo apoptosis. Endothelial and phagocytic cells initiate vascularization in the empty lacunae (Jacobson, Weil and Raff, 1997). In the chondro-osseous function, where calcification and chondrocyte cell death have begun, the marrow begins to form the primary spongiosa. In this region, osteoblast and surrounding matrix of the woven bone continue to release osteoid. Additionally, the marrow carries osteoclasts and resorbs the bone. This synergistic effect of formation and resorption helps to shape the long bone.

## 1.9 Bone Remodeling

The growth and development of bone to a stage of maturity is controlled by the combination of intrinsic and extrinsic forces. The mass and architecture in bones are governed, to some extent, by adaptive mechanisms sensitive to their mechanical environment.

The idea that mechanical forces shaped the architecture of the skeleton emerged in the work of Roux (1905) and Meyer (1967). However, Wolff (1892) stated that when bone is bent under a mechanical load it modifies its structure so as to resist external pressure by bony apposition in the concavity and by resorption in the convexity. Supported by much experimental data, the idea that mechanical stresses affect the form of bone is now generally accepted (Thompson, 1917; Frost,1964b, 1964c). A reduction in bone formation (Wronski and Morey, 1983; Shaw *et al.*, 1988; Morey and Baylink, 1978), mineral content (Vico *et al.*, 1987; Rambout and Goode, 1985; Russell and Simmons, 1985; Turner *et al.*, 1985; Cann and Adachi, 1983), and bone matrix protein production (Patterson-Buckendahl *et al.*, 1985) result from the skeletal unweighting associated with spaceflight. Conversely, increased skeletal loading through exercise has been shown to increase bone mass (Bassey and Ramsdale, 1994; Smith and Gilligan, 1990; Eisman *et al.*, 1990) and retard bone loss caused by postmenopausal osteoporosis (Krolner *et al.*, 1983;

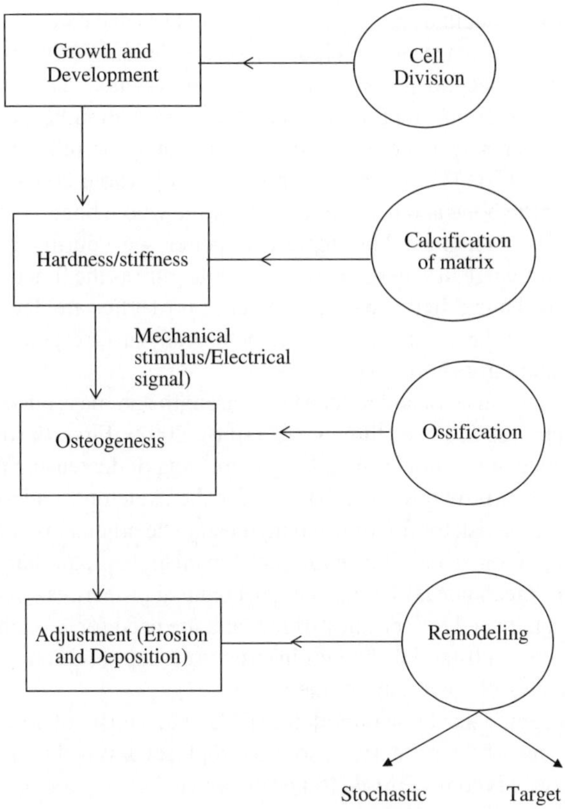


**Figure 1.12** A schematic presentation of broad steps in Broad steps in bone growth, development and in remodeling

Simkin, Ayalon and Leichter, 1987; Prince *et al.*, 1991; Chesnut, 1993). Figure 1.12 shows a sequence of bone mechanical stimuli leading to consequent bone remodeling.

The general hypothesis is that animals that experience higher levels of load-bearing activity are predicted to have larger articular surface areas (ASAs) relative to the body mass. This may be summarized in the modified Wolff's law as:

> *The form of a bone being given, the bone elements place or displace themselves in the direction of functional forces and increase or decrease their mass to reflect the amount of functional forces (Figure 1.1).*

The results indicate that the effects of mechanical loading from moderate exercise are not only complex but also differ substantially in diaphyses and the subchondral articular surfaces of epiphyses (Lieberman, Devlin and Pearson, 2001). Bone is not homogenous, both morphologically and mechanically, but interestingly inhomogeneities do not disturb the functional adaptation (Bennok, 1972). On the contrary, they contribute towards it. Most experiments and theory on bone adaptation are concerned with the size and shape of the bone. This is termed modeling and is probably produced by uncoordinated activity of bone cells. In contrast, "remodeling" occurs in cancellous bone, in which osteoclasts and osteoblasts work together in coordinated sequence to replace bone (Figure 1.9b). This leaves the total amount of bone unaltered, in the from of secondary osteons (Haversian systems). There has been much less concern about the adaptations that bone may have, in particular there has been little consideration of whether hard bone material properties may be related to the loads falling on the bone (Currey, 2003). The mechanical properties of whole bones depend both on the material properties of the bone and the architecture of the whole bone. There may be adaptive links between these two features. This feature is better seen during growth: the material properties of bone can change in synergy with the architecture as the function of the bone either changes or is about to change. In this case the material properties are determined over a long, evolutionary, time, while the architecture of the bones may also respond to the strains in the bone over a correspondingly short period.

Bone remodeling can conserve and reduce bone strength and mass but does not increase the same. Remodeling process achieves three goals (Burn, 2002). First it provides a way for the body to alter the balance of essential minerals by increasing or decreasing the concentration of these in serum. Second it provides a mechanism for the skeleton to adopt to its mechanical environment, reducing the risk for fracture and increasing the organisms chances for passing its genes to the next generation. Third, it provides a mechanism to repair damage created in bone by repetitive cycles of mechanical loading. The mechanical properties of bone are a result of a compromise between the need for certain stiffness and the need for enough ductility to absorb impacts. Bone undergoes substantial changes in structure, shape and composition according to the mechanical and physiological environment.

It is now widely accepted that bone remodeling is the mechanism of bone replacement in the vertebrate skeleton. One of the primary reasons for replacement is the functional capacity of bone, which in some way is compromised if it is allowed to become too old (Parfitt, 2001). The remodeling process arranges a given amount of calcified material to support the naturally occurring loads with the largest possible safety factor. The remodeling apparatus can accommodate circadian fluctuations in calcium balance (Parfitt, 1993) and can also cater to a temporary need for additional calcium lasting for a few months (Parfitt, 1981) and accomplishes slow thickening of trabeculae during growth (Parfitt, 2000). This also entails slow elimination of surplus bone in response to the age related decline in physical activity (Parfitt, 1990). Frost's mechanostat theory is unique in its distinction between modeling and remodeling processes, the threshold for activating lamellar or woven bone formation and its application to the etiology of osteopenia and osteoporosis. Frost (1987) suggested that certain hormones and biochemical agents might overcome this mechanical setting of bone to alter the boundaries of the pathological window. This allows normal mechanical usage to increase bone mass and bone strength in a significant manner.

Modeling can increase bone "mass" and hence its strength. It increases these when bone strains equal or exceed a threshold range. Formation and resorption drifts use osteoblasts and osteoclasts, respectively, to determine the cross sectional shape and size of bones and trabeculae and their longitudinal shapes and to increase their strength and mass. The modeling is best in evidence during growth and slows down on attainment of adulthood. If strain is greater than a particular threshold the modeling takes place otherwise it does not:

$$\text{Strain} < \text{Threshold} < \text{off}$$
$$\text{Strain} > \text{Threshold} > \text{on}$$

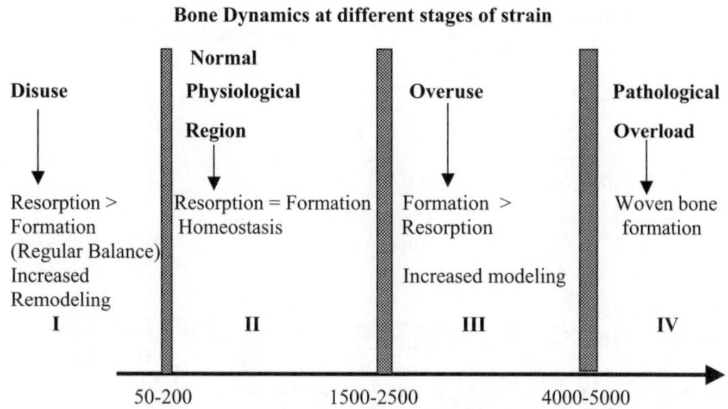


**Figure 1.13** Mechanostat theory according to Burr and Martin (1992). In the disuse window (I), strains are low and bone is lost due to increased remodeling. In the normal physiological window, bone is at a state of normal turnover and bone equilibrium is maintained (II). In overuse (III), bone formation exceeds resorption. The site depends upon strain. This causes laying down of woven bone (IV)

Bone that is strong enough to keep strains from reaching or exceeding the threshold would satisfy the above criterion (Frost, 1997). Peak bone strains from voluntary activities should be larger in growing subjects than in adults. These range between 2000 and 4000 microstrain in most growing subjects, but between 800 and 1300 microstrain in most adults (Frost, 1997) (Figure 1.13).

The mechanisms by which the mechanostat works are not fully understood; however, they require some form of cellular mechanotransduction. Mechanotransduction, or the conversion of a biophysical force into a cellular response, is an essential mechanism for a wide variety of physiological functions that allow living organisms to respond to the mechanical environment. Mechanotransduction in bone necessarily includes four distinct phases: (i) mechanocoupling, the transduction of mechanical force applied to the bone into a local mechanical signal perceived by a sensor cell; (ii) biochemical coupling, the transduction of a local mechanical signal into a biochemical signal and, ultimately, gene expression; (iii) transmission of signal from the sensor cell to the effecter cell, that is, the cell that will actually form or remove bone; and (iv) the effecter cell response, the final tissue level response. on the basis of mehano-transduction mechanics, forms the model for bone adaptation.

Bone remodeling is essentially a surface based event – the metabolic activity of cancellous bone is tenfold greater than at the cortical sites. At any given instant, approximately 10–15% of the bone surface is undergoing remodeling, the remaining surface being relatively quiescent. In the stage of dynamic equilibrium the process of calcium resorption (5 mmol) and new matrix formation are equal, as in the case of healthy adults (normal bone). Mineralization of bone occurs after matrix production and hence the skeletal demands for calcium are governed by the rate of matrix synthesis. This is consistent with clinical findings that severe calcium deficiency is associated with severe osteomalacia, but not with the formation of new bone. There is evidence that low calcium diets decrease formation in growing rats, but in adults most evidence suggests that calcium deficiency increases bone formation by increasing the turnover of bone. This is suggestive that, under normal

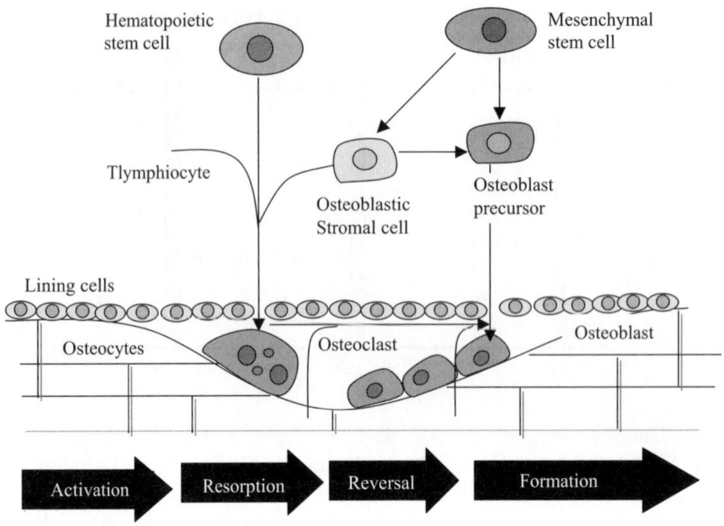


**Figure 1.14** BMU with possible estrogen action. Bone remodeling on the surface of trabecular bones is illustrated (Raisz, 2005). (Reproduced with permission from Lawrence G. Raisz, Pathogenesis of osteoporosis: concepts, conflicts, and prospects, *Journal of Clinical Investigation*, **115**(12), 3318–3325 ©2005, The American Society for Clinical Investigation)

conditions, the skeletal requirement of calcium is governed by the rate of matrix synthesis rather than by the availability of calcium.

From Wolff's law it can be easily extrapolated that living cells are constantly subjected to mechanical stimulations arising from the external environment, mechanical properties of bone and internal physiological conditions (Figure 1.14). Depending on the magnitude, direction and distribution of these mechanical stimuli, cells can respond in various ways. The mechanism by which the mechanical signal is translated into biological and chemical responses in a cell is a matter of continuing investigation (Wang, Liu and Shiau, 1993; Ingber, 2003) Apart from mechanical loads induced within or outside the body many chemicals also affect the mechanical properties of living cells. In turn, the mechanical properties of individual cells can control the structural integrity of the whole tissue. This may arise from the mechanical interactions between cells and the surrounding extracellular matrix (Wakatsuki *et al.*, 2000). On the other hand, mechanical loads exerted at the tissue level are possibly transmitted to individual cells and can influence their physiological functions (Guilak, Ratcliffe and Mow, 1995; Guilak and Mow, 2000).

The remodeling process is not performed by each cell but by group of cells functioning as organized unit, which are named "basic multicultural units" (BMU) (Frost, 1963). They operate on bone periosteum, endosteum, trabecular surfaces and cortical bone, replacing old bone with new bone in discrete packets. The cycle of remodeling stages may be described as:

$$\text{Activation} \rightarrow \text{Resorption} \rightarrow \text{Reversal} \rightarrow \text{Formation} \rightarrow \text{Quiescence}$$

Upon activation of bone surface, osteoclasts are recruited from the bone marrow and migrate to the site of activation, where they remove mineral and matrix in the resorption period

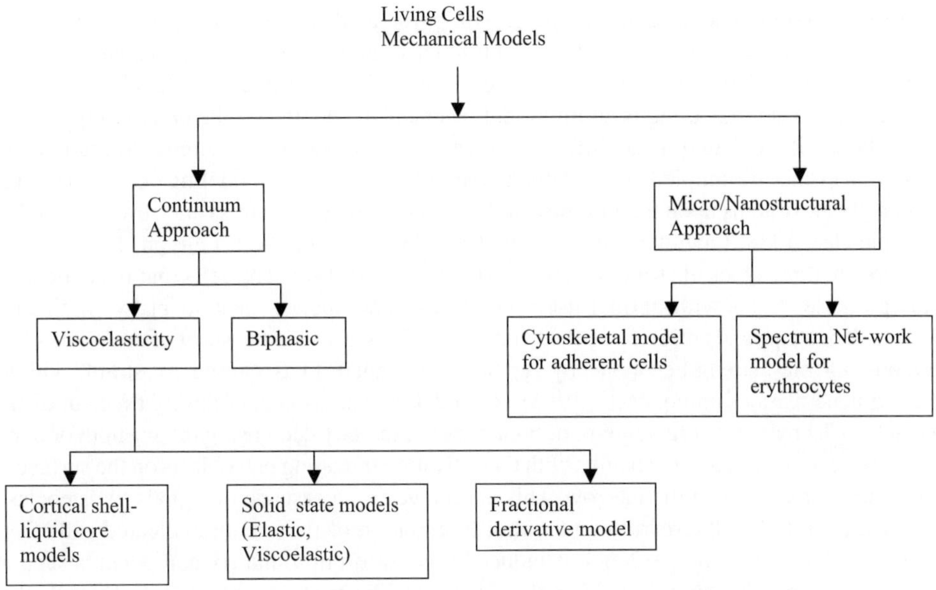


**Figure 1.15** Living cells subjected to mechanical stimulations and their broad response. (Adopted from Lin *et al.*, 2006)

(Figure 1.15). The reversal phase is characterized by the repression of osteoclastic activity and deposition of cement into the resorption cavity. Finally, new bone formation begins with matrix formation and sequential mineralization. Once this cycle is completed the surface of the bone will return to the quiescence level (Parfitt, 1984). The site of resorption and formation, where the remodeling occurs, is referred to as the bone remodeling unit. Furthermore, this turnover cycle is usually completed in about four months. Trabecular bone remodeling occurs on the surface, while cortical bone remodeling occurs by osteoblastic apposition occurring during Haversian canal formation (Jee, 1988). Initiation of the remodeling cycle begins with activation of osteoprogenitor cells in a localized area in bone. The stimulus produces mitotic division in the precursor cells, producing osteoclasts that gradually resorb bone. After resorption subsides, the reversal phase begins. New osteoblasts appear and lay down similar amounts of bone as previously resorbed. On the completion of the cycle, cellular activity returns to its resting state, leaving the remaining osteoblasts as inactive lining cells or surface osteocytes.

The process from activation to quiescence involves the disappearance of bone lining cells and their replacement by osteoclasts that generate resorption lacunae on the endosteal surface of bone over a 2–4-week interval. The resorption phase is then terminated, probably by osteoclast apoptosis and, after a brief reversal phase, osteoblasts are recruited that fill in the resorption cavity with new bone (Riggs, Khosla and Melton, 2002). The net result, the replacement of a packet of old bone with new bone, may be due to electrostimulation, mechanical stimulus, hormonal changes (estrogen deprivation or in response to endogenous parathyroid hormone uses), cytokine stimulation, growth hormone surges, glucocorticoid excess or changes in serum calcium level, and may be a combination of some of these.

Bone formed by modeling during the growth phase tends to be structurally inferior to the bone that is formed during remodeling. Additionally, adult bone tends to change over time, becoming damaged by microfractures as a result of loading or reduction in quality as it ages. For these reasons remodeling is an important mechanism adopted by the body to replace and repair bone. Remodeling is a different process than modeling, whereby formation and resorption events are coupled and isolated to the same location. Remodeling occurs to replace the Haversian systems upon their demise and to respond to altered stress and strains placed on the bone (Jee, 1988; Lawrence and Fowler, 1997; Wasserman, 1984; Lanyon, 1994).

There are three types of strain: tensile, compressive and shear. The cells that make bone up undergo these strains with normal use. Therefore, loaded bones must adapt by continually altering their shape and architecture (Lanyon, 1994). This is also in the spirit of the Wolff's law. Remodeling may strengthen bone by replacing damaged areas caused by strain, without altering bone mass (Kimmel *et al.*, 1993). Remodeling can also occur during times of disuse and acts to decrease the total amount of bone present, thereby decreasing the strength of bone.

The bone remodeling cycle begins with the activation of resting osteoblasts on the surface of bone and stromal cells in the marrow. This is followed by a cascade of signals to osteoclasts designed to stimulate the recruitment and differentiation of these multinucleated cells from hemoietic stem cells. After osteoclast-induced bone resorption, matrix components such as transforming growth factor-beta (TGF-$\beta$) and insulin-like growth factor-1 (IGF-I), as well as collagen and osteocalcin, are released into the micro-environment. The growth factors released by resorption contribute to the recruitment of new osteoblasts to the bone surface, which begin the process of collagen synthesis and biomineralization. In healthy adults, as many as two million remodeling sites may be active at any given time, and it is estimated that nearly one quarter of all cancellous bone is remodeled each year. In general, resorption takes only 10–13 days, while formation is a comparatively longer process, taking upwards of 3 months. In an ideal equilibrium situation the amount of bone resorbed by the end of the cycle equals the amount reformed. It is, however, also apparent that some individuals have impaired peak bone and acquisition. This scenario may be more common than previously appreciated and almost certainly represents inherited or acquired alterations in the rate of either bone formation or bone resorption during a critical period when several hormones acting synchronously leads to a marked increase in bone mass.

Some investigators (Mori and Burr, 1993; Parfitt, 1996a, 1996b) have suggested that there may be two kinds of bone remodeling: one that is stochastic (site independent though not totally random) and a second that is target specific. However, related aspects and their control mechanisms remain to be explored. Experimental studies in which damage is created in canine bone by cycling loading have demonstrated that remodeling sites associated with microcracks are more probable (four to six times) than expected by chance alone (site independent) (Burr *et al.*,1985, 1993; Mori and Burr, 1993). It is estimated that remodeling events occur subsequent to microdamage and the local initiation of repair. One can thus infer that damage leads to remodeling. Bentolila *et al.* (1998) have noted an increase in the number of a typical appearing osteocytes following fatigue loading in rat bone.

Tanaka, Alam and Turner (2002) have shown that stochastic resonance enhances bone formation, by which nonlinear systems are able to amplify a response to a small periodic stimuli with the aid of noise (Gammaitoni, 1998). Stochastic resonance has been used to explain various phenomena in biological systems (Gammaitoni, 1998; Pierson and Moss, 1995), including the activity of cells under microgravity (Kondepudi, 1991). During exercise, as in active sports, bone is subjected to quick variations in loading, which is equivalent to a noisy loading environment. Therefore, the possibility arises that bone cells are more responsive to fluid shear stress mediated by high impact activity compared to low impact activity. If so, the bone cell response to fluid shear stress might be nonlinear, that is, requiring an initial stress-kick. The initial stress-kick occurs during the quick transition from zero to a nonzero stress initiating a fluid shear stress stimulation of cells. NO production has been used as a parameter for bone cell activation since it is an early mediator of mechanical loading induced bone formation (Fox, Chambers and Chow, 1996) and it has been shown to be essential to adapt bone formation *in vivo* (Turner *et al.*, 1996).

One study (Bacabac *et al.*, 2005) suggests that the bone cell response to fluid shear stress is rate dependent, but that an initial stress-kick is required for the cells to respond. It has been reported that increasing either the shear stress amplitude or frequency, without increasing the average stress, enhanced NO production by bone cells *in vitro*. Bone cells respond similarly to similar rates of fluid shear stress, despite different frequencies (5 Hz, 21.99 Pa Hz or 9 Hz, 17.53 Pa Hz). Hence, the rate of fluid shear stress (i.e., $2\pi \times f \times \tau$) is proportional to the product of the fluid shear stress amplitude and frequency (Bacabac *et al.*, 2004).

Osteoblasts and osteocytes are mainly involved in the bone remodeling process (Frost, 1990). At the cellular level, bone remodeling can be traced to the combined action of the osteoblast and the osteoclast. Bone remodeling consists of a strict coupling of resorption and formation that continues throughout life and is necessary not only for skeletal growth but also in the maintenance of normal bone structure and function (Frost, 1964a; Harris and Heaney, 1969; Parfitt,1982, 2002; Martin and Rodan, 2001; Eriksen, 1986). Throughout life, bone is continuously being remodeled with resorption of old bone (catabolic process) performed by osteoclasts and deposition of new bone (anabolic process) performed by osteoblasts. The activities of osteoblasts and osteoclasts are combined into defined anatomical spaces (BMUs) (Frost, 1983). It can be stated that bone remodeling takes place in focal remodeling units that consist of osteoblasts, osteoclasts and their precursors, in which resorption and formation are coupled. Bone resorption is likely the initial event that occurs in response to local mechanical stress signals. The sequence of a cycle of resorption → formation on all surfaces appears to be valid for cortical bone remodeling as well as for trabecular bone. Hence, in densely compact bone the remodeling process must begin with bone resorption to produce surfaces for the apposition of new bone.

Several key components of the remodeling cycle are susceptible to systemic and local alterations. These, when perturbed, lead to a deleterious change in bone mass. The osteoblast–osteoclast interaction mechanism may have relevance for therapeutics as well as for understanding the dynamics of bone turnover. In particular, the activation of remodeling and the recruitment of osteoclasts represent the two most vulnerable sites in the cycle. The remodeling cycle clearly favors resorption over formation as the process of laying down new bone requires the combination of several processes. The osteoblast also functions to lay down collagen and regulate the mineralization of previously resorbed lacunae in the skeletal matrix.

The differentiation of mesenchymal stem cells become osteoblasts and rest on the surface of the remodeling space (Lian and Stein, 1999). Recruitment of the stem cells to osteoblasts, rather than adipocytes, is a critical step in bone formation and requires a series of factors that enhance differentiation. One of the most important components of this process is cbfal, a transcription factor that is essential in the early differentiative pathway of stem cells to bone and away from adipocytes (Ducy *et al.*, 1997). With the activation of resting osteoblasts, osteoblastic cells begin to synthesize several types of collagen as well as elaborating a series of growth factors in the skeletal matrix, where they are stored in latent forms. These are then released during subsequent remodeling cycles.

In addition to endogenous (cellular) signals, external signals (e.g., PTH, GH, interleukin-1 and oestrogen deprivation) to resting osteoblasts and stromal cells cause these cells to release cytokines [interleukins such as IL-1, -6 and -11, as well as m-CSF, tumor necrosis factor (TNF) and TGF-6]. These enhance the recruitment and differentiate the function of multinucleated giant cells destined to become bone-desorbing cells (Lian and Stein, 1999; Lorenzo and Raisz, 1999; Udagawa *et al.*, 1999). It is now accepted that osteoprotogluin (OPG) is a member of the TNF receptor super family, its role in bone remodeling being to act as a dummy receptor for the ligand now known as osteoprotegrin ligand (OPGL, TRANCE a RAKL). OPGL is a surface peptide that, when expressed on the osteoblast, can bind to the true OPG receptor on osteoclasts and initiate the cell–cell contact necessary for osteoclast activation and subsequent bone resorption (Suda, Takahashi and Martin, 1992).

The fundamental issue behind the choice of the stimulus signal is whether (re)modeling signals originate in the osteocytes themselves or whether they are an effect of increased flow through the osteocyte network. In the former case, they could be triggered by the osteocyte lacunar stress and strain concentrations, which can be much higher 15× the tissue (continuum) level (Kufahl and Saha, 1990; Burger and Klein-Nulend, 1999; Nicolella and Lankford, 2002). In this case the time derivatives of SED, maximal principal strain (MPS) or volumetric strain (VS) could be relevant controlling parameters. The osteocytes may signal the osteoblasts and osteoclasts to change their bone remodeling activities (Burger and Klein-Nulend, 1999; Smith, Burger and Huyghe, 2002). Osteocyte signals are important in the mechanostat, the gravity sensing device that modulates bone formation.

The osteocytes within bone, together with the osteocytes lining external and internal surfaces and resting osteoblasts, form a continuous syncytium of cells lining all surfaces, including the larger trabecular spaces as well as lacunae and tiny canaliculi. The cellular continuum is organized into distinct functional subunits that do not always communicate directly with each other. The cells in each subunit may respond independently from neighboring units. Each of these pockets of cells may be termed as BMU, which can be considered to represent an organizational unit of metabolic activity that remains after a remodeling unit has completed its task. Rasmussen and Bordier (1974) have suggested that the term BMU be used during the remodeling phase.

BMU is most readily demonstrated in cortical bone and possibly extends to cancellous bone (Parfitt, 1994). The BMU exists and moves in three dimensions, excavating and refilling a tunnel through cortical bone or a trench across the surface of cancellous bone. A cortical BMU travels for about 400 μm at about 20 μm per day (Parfitt, 1994). A cancellous BMU travels about half this distance at about half the speed (Parfitt, 1996). It is suggested that one important local stimulus may be obtained from extracellular nucleotides. This can exist locally within the bone microenvironment to initiate intracellular signaling via P2Y receptors, which sensitizes cells to the action of PTH. This is one mechanism for integrating local and systemic responses in bone.

There is evidence that osteocyte apoptosis is an important factor in initiating new remodeling sites. The prevailing view is that an intact osteocyte-canalicular system inhibits the recruitment or activation of osteoclasts, and that disruption of the network by microcracks, for instance, releases the normal inhibition to resorption (Burger and Klein-Nulend, 1999). A positive correlation has been shown between osteocyte apoptosis and bone resorption by osteoclasts in growing bone (Bronckers *et al.*, 1996) and between osteocyte apoptosis and increased activation frequency in estrogen-deficient osteoporosis (Noble *et al.*, 1997; Tomkinson *et al.*, 1997).

The quantum concept of remodeling in which discrete microscopic packets of bone are removed and replaced throughout the skeleton is now recognized as the predominant form of turnover in the adult skeleton. The primary event in remodeling is activation, during which lining cells retract from the bone surface to allow access of osteoclasts or their precursors to the underlying mineralized matrix. The focal nature of remodeling indicates that activation of this process is sensitive to local stimuli, including mechanical strain. In addition, systemic factors, in particular parathyroid hormone (PTH), are known to increase the rate of activation (Mundy, 1998). The visible bone damage is initiated at the molecular level and is then carried over to the submicroscopic level, proceeding and ending at the microscopic level (Parfitt, 2002). However, the signal pathways of the process and its detection threshold remain within the realm of investigation. Plasma calcium homeostasis requires rapid ion exchange (Parfitt, 1993), which is impeded by progression of secondary mineralization and loss of water from the surface mineral. It would be advantageous for hypermineralized surface bone to be preferentially replaced, but the existence of such targeted remodeling has not been established. A turnover rate greatly exceeding that required to maintain mechanical equilibrium could obviate the need for targeted remodeling. Such a need may, however, exist if turnover is at the low end of the normal range (Han and Mould, 1990). Thus, remodeling at foci is the combined action of systemic factors and local phenomena.

Both osteoblasts and osteoclasts are targets for nucleotide activation through multiple P2 receptor types (Arnett and King, 1997; Bowler *et al.*, 1995; Gallinaro, Reimer and Dixon, 1995; Kumagi, Sacktor and Filburn, 1991; Naemsch *et al.*, 1999; Reimer and Dixon, 1992; Weibe, Sims and Dixon, 1999). These receptors couple to multiple signal transduction cascades, including inositol triphosphate ($IP_3$)-mediated intracellular calcium release (P2Y receptors) or nonselective outward cation currents (P2X receptors). These signaling cascades, dependently and independently of mitogen-activated protein kinase activation, induce expression of genes such as c-fos (Bowler *et al.*, 1999), a transcription factor. The importance of this, in the early regulation of remodeling processes is reflected in the skeletal pathologies associated with its deletion (Wang *et al.*, 1992) or overexpression (Johnson, Spiegelman and Papaioannou, 1992). Nucleotides have been reported to modulate osteoblast induced bone formation (Jones *et al.*, 1997) and proliferation (Shimegi, 1996) and also modulate osteoclastogenesis and bone resorption (Arnett and King, 1997; Buckley *et al.*, 2000; Morrison *et al.*, 1998).

The ability of ATP (adenosine triphosphate) to effect neurotransmitter responses from certain nerve terminals has been proposed by Burnstock (1972). Since that time, a large family of receptors has been defined, both pharmacologically and molecularly, which transduce signals arising from nucleotide stimulation (Boarder *et al.*, 1994; Fredholm *et al.*, 1994). These P2 receptors have been subdivided into two classes: P2Y, which consists of five distinct receptors coupled to heteromeric G proteins, and P2X, which contains seven distinct receptors that function as cationic gated channels. Physiologically, P2X receptors are activated exclusively by adenine nucleotides, whereas P2Y receptors are responsive to adenine or uridine nucleotides or, in some cases, to both. Receptors are coupled to multiple specific cellular functions in processes as diverse as neurotransmission, wound healing, morphogenesis and apoptosis (Burnstock, 1993).

Some of the notable observations can be summarized thus:

- Cells of the osteoblast and osteoclast lineages express functional P2 receptors.
- Activation of P2 receptors results in the induction of *c-fos*, a transcription factor that plays a pivotal role in the regulation of bone remodeling.
- Extracellular nucleotides exhibit a potent synergy with PTH, enhancing intracellular calcium release ($[Ca^{2+}]_i$), downstream signaling and gene activation (Bowler *et al.*, 1999; Dixon and Sims, 2000; Jones *et al.*, 1997; Yu and Ferrier, 1993).
- Bone cells release ATP into the extracellular environment and this release is enhanced by mechanical strain and other combinations of regulatory parameters, leading to healing processes in bone.

For P2 receptors to modulate the activity of bone cells, ATP must exist at a sufficient concentration in the surrounding microenvironment to initiate receptor activation. All cells contain high intracellular ATP concentrations (1–5 mmol $L^{-1}$), as well as the capacity to release ATP following trauma. In addition, increased blood flow associated with the general inflammatory response could greatly increase the nucleotide concentration available upon platelet aggregation. Hence, at sites of tissue injury, wounding or fracture, high local ATP concentrations are present to activate P2 receptors. Nevertheless, it is obvious that nucleotides must exist transiently in the bone microenvironment without cell damage to be physiologically relevant regulators of bone remodeling. A real-time detection system has been used that relies on the high yield chemiluminescent reaction generated by luciferin and luciferase in the presence of ATP (Cobbold and Lee, 1991; Koop and Cobbold, 1993). This demonstrated that human osteoblasts constitutively release ATP into their extracellular environment in the nanomolar range. Released ATP probably reaches much higher concentrations near the membrane, because the presence of nucleotidases at the cell surface, membrane trapping and unstirred layer effects will all influence bulk phase measurement. This concept has been confirmed in a study in which membrane anchored luciferase revealed micromolar ATP concentrations at the surface of platelets (Beigi *et al.*, 1999).

In addition to receptors responsive to ATP, osteoblasts express P2Y (Albert *et al.*, 1997; Arnett and King, 1997) and P2Y receptors (Bowler, Bilbe and Gallagher, 1998; Maier *et al.*, 1997) are preferentially or selectively activated by uridine nucleotides. Some studies have demonstrated conclusively that cells release uridine triphosphate (UTP) into their extracellular environment, providing locally released agonist capable of activating these P2-receptor family members (Lazarowski *et al.*, 1997). Opinions differ as to why cells are so responsive to nucleotides. One hypothesis suggests that this may be a means of regulating ATP concentration at the plasma membrane, where it is required for the activation of ATP-dependent channels and enzymes (Schwiebert, 1999). Others propose that nucleotide release, through autocrine signaling, represents a universal key determinant in establishing the "set point" for activation of signal transduction pathways (Ostrom, Gregorian and Insel, 2000). It appears that nucleotide release is fundamental in defining basal cellular activity. Nucleotide-activated cellular activity is visualized as a wave of mobilized $[Ca^{2+}]_i$ spreading outward through the cell population from a single point of ATP/receptor contact (Osipchuk and Cahalan, 1992). This mechanism would seem to be important in cells and tissues that lack communication via gap junctions, such as chondrocytes and cartilage. In agreement with this model, chondrocytes abundantly express P2Y receptors (Leong, Russell and Caswell, 1994) and can release ATP (Figure 1.16) (Lloyd *et al.*, 1999).

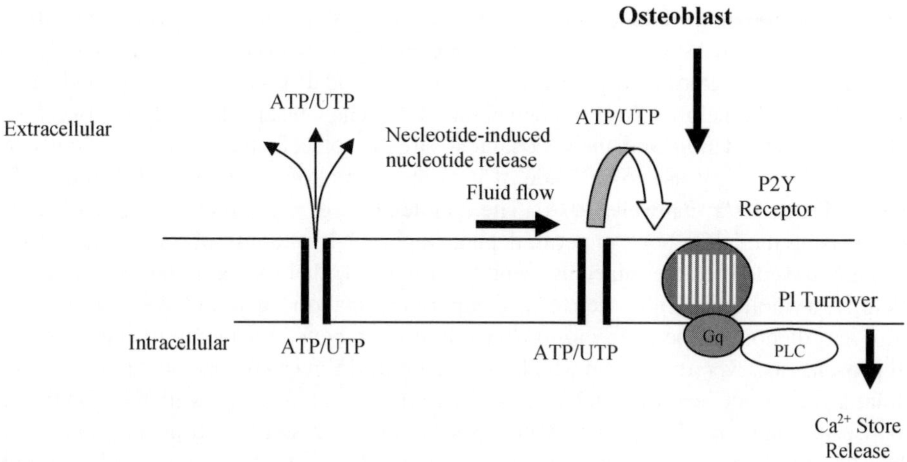


**Figure 1.16** Mechanism of P2 receptors modulating bone cell activity. (Reproduced with permission from W.B. Bowler *et al.*, Extracellular nucleotide signaling: a mechanism for integrating local and systemic responses in the activation of bone remodeling, *Bone*, **28**(5), 507–512 ©2001, Elsevier B.V.)

Further evidence for a distinct, nonlytic mechanism of nucleotide release lies in the observation that it can be regulated positively (and presumably negatively) (Figure 1.16) (Bodin, Bialy and Burnstock, 1991; Bowler *et al.*,1998a, 1998b). In a positive feedback loop, UTP acts through P2Y (Albert *et al.*, 1997) receptors to upregulate ATP release from primary human osteoblasts (Bowler *et al.*, 1998a, 1998b). In addition, ATP and UTP release is modulated positively by fluid flow (Bowler, Bilbe and Gallagher, 1998; Lazarowski *et al.*, 1997). Indeed, recent evidence suggests that nucleotide release occurs as a consequence of fluid forces generated following cellular manipulation in almost all *in vitro* experimental systems (Lazarowski *et al.*, 1997; Ostrom, Gregorian and Insel, 2000). The importance of this finding becomes particularly apparent in the context of the bone microenvironment, where shear forces resulting from fluid flow within trabeculae and canicular spaces are thought to transmit mechanical forces (Hung *et al.*, 1995; Korenga *et al.*, 1994; Reich, Gay and Frangos, 1990; Turner *et al.*, 1994). These observations indicate that a cellular mechanism exists to release ATP into the extracellular environment at concentrations sufficient to activate cell surface P2 receptors and that this release can be modulated by mechanical and agonist stimulation.

Once released, ATP is rapidly broken down by ectonucleotidases that is bound on the extracellular side of osteoblast membranes (Caswell and Russell, 1988). The half-life for ATP released from human SaOS-2 osteosarcoma cells is estimated to be circa 50 s. This short half-life restricts the action of ATP to that of a localized signal. However, the ability of nucleotides to induce their own release provides a mechanism for sustaining and possibly propagating a more widespread response (Bodin and Burnstock, 1996; Bowler, Bilbe and Gallagher, 1998). ATP release and subsequent paracrine signaling in osteoblasts might provide a mechanism for cell-to-cell communication independently of gap junction (Jorgensen *et al.*, 1997). Elevated calcium levels can also activate various intracellular signaling systems in different cell types. One that is activated in osteoblasts upon P2Y receptor stimulation is the extracellular signal regulated (ERK) cascade.

The self-remodeling of long bone in child with midshaft fracture during angulation in response to mechanical stress is an example of the effects of environmental (epigenetic) factors on the genetically determined processes, development and tissue remodeling and may be explained with the help of Wolff's law (Figure 1.1). The causal relationship between bone modeling or remodeling and the mechanical stresses placed upon it is sufficiently well understood for clinical use to be made of it in the correction of congenital disorder of the skeleton (Frost, 1973; Pauwels, 1976). The maintenance of a continuous supply of reactive bone, housing the labile bone mineral, depends upon the continuity of remodeling of bone, which is believed to be determined by weight bearing and other forces acting externally on the skeleton as a whole, and is now seen to have important metabolic functions (McLean and Wrist, 1961). In orthodontics the ability to move teeth through bone in response to applied stress is well proven. However, the way in which the mechanical forces are mediated to the tissues at cellular level has not been well understood. The remodeling that occurs with the growth and in response to changing skeletal stress consists largely of resorption from one site to, and redeposition of bone on, another. The bone formation and resorption rates are closely linked and synchronized (Sissons, 1971; Lee, 1964; Manson and Walters, 1963; Harris, 1960) and have been found to be 10–15 μm weekly. Information on the rate of remodeling or turnover of the tissue has also been well documented (Lee, Marshall and Sisson, 1965; Marotti, 1963; Vander Hoeft, Kelly and Peterson, 1962; Frost, 1960). This may be of interest in understanding diseases of bone as well as its behavior under normal conditions.

## 1.10 Biochemical Markers of Bone and Collagen

An intermediate event in bone cell mechanotransduction is prostaglandin $E_2$ ($PGE_2$) synthesis and release. $PGE_2$ is released by osteoblastic cells in response to mechanical loading and has been implicated in regulating bone turnover *in vivo*. (Nolan *et al.*, 1983; Feyen *et al.*, 1984; Rodan *et al.*, 1986; Imamura *et al.*, 1990). Prostaglandins can regulate bone turnover by both stimulating bone formation and regulating bone resorption (Dietrich, Goodson and Raisz, 1975; Raisz and Fall, 1990; Jee *et al.*, 1991). Other studies have shown that bone cells respond to OFF with increase in $PGE_2$ release (Donahue *et al.*, 2003a; Saunders *et al.*, 2003).

Osteocalcin (OC) is a non-collagenous protein synthesized largely by the osteoblast; it contains three amino acids of $\gamma$-carboxyglutamic acid and is therefore known as bone Gla protein (Fraher, 1993). Osteocalcin is vitamin K dependent (Price, 1982) binds Ca and may be involved in the control of mineralization (Fraher, 1993; Lepage *et al.*, 1991). The Gla residues specific to OC are formed after translation of the protein occurs by a carboxylase enzyme system. In the presence of Ca, these Gla residues undergo conformational changes that allow OC to bind to hydroxyapatite and then accumulate in bone matrix (Lian and Gundberg, 1988). Furthermore, OC has a high affinity for hydroxyapatite (Price, 1982). The protein exists in a 1 : 1 ratio with collagen in bone and is proportional to hydroxyapatite (Lian and Gundberg, 1988). OC may regulate the size of the hydroxyapatite crystal. Osteocalcin is cleared through proteolysis and is cleared from circulation by the kidneys (Gomez *et al.*, 1995). Serum concentrations of OC are probably a product of diffusion into the circulatory system prior to binding to the hydroxyapatite (Price, 1982). Osteocalcin is mostly formed by osteoblast activity and thus is highly correlated with bone formation (Lian and Gundberg, 1988; Kannus *et al.*, 1996). Therefore, serum OC is invariably used as marker of bone synthesis.

It is reported that when animals are exposed to natural light OC undergoes a circadian rhythm (Lepage *et al.*, 1991). Osteocalcin concentrations in normal female standard bred horses were found to be consistent during the day from 0700 to 1900. The concentration then fell in the early evening hours (1900–2000) and increased at night until a peak at 0500 the next morning (Lepage *et al.*, 1991). These authors concluded that blood samples should be taken during daylight hours, since this period experienced the least fluctuation in OC concentrations. However, in another study, where horses were exposed to only fluorescent lighting, a circadian rhythm was not observed (Hope *et al.*, 1993).

Carboxy-terminal pyridinoline crosslinked telopeptides of type I collagen (ICTP) are a kDa portion of type I collagen that is released during resorption of bone (Hassager *et al.*, 1994) and can be quantitated in serum and urine (Risteli *et al.*, 1993). Once type I collagen is broken down into ICTP it can not be used again for synthesis of new collagen in bone and is cleared from the body in circulation through the kidneys (Tahtela and Tholix, 1996; Risteli *et al.*, 1993). Concentrations of ICTP decrease with the age of a horse, up until about four years of age, after which there are no significant differences (Price *et al.*, 1995). A diurnal cycle has also been reported with ICTP concentrations (Risteli *et al.*, 1993).

Finally, PYD and DPD are crosslinking amino acids residues of collagen released from bone matrix during turnover. An enzyme-linked immunoassay (ELISA) exists for PYD and DPD crosslinks in serum and urine. Since PYD is formed in cartilage and bone, PYD and DYD can be used as biochemical markers to analyze metabolic bone or cartilage diseases. Increased urinary PYD is seen in patients with osteoarthritis and rheumatoid arthritis (Robins *et al.*, 1986).

## 1.11 Summary

The concept of bone behavior consists of three parts. The first considers the solid state behavior of "*in vitro*" bone while the second is concerned with studying inherent bioelectricity under external stimuli with different possible combinations. This finally culminater (third part) onto its clinical behavior. Initially, however, there is need to understand basic bone properties such as piezoelectricity and its intermixing with stress generated potential. These two together define the physical aspects of total bone behavior.

The solid state behavior of bone appears to be one of the most baffling observations in solid state biology. It has been reported that bone has wide variety of properties ranging from semiconduction to piezoelectricity. Many semiconductor properties have been studied in detail in our laboratory (Behari and Andrabi, 1981; Andrabi *et al.*, 1980). In the present context the optical behavior is seen over a fairly wide band of frequencies. The choice of properties to be examined is dictated by the fact that it is important to characterize the solid state of the materials. It provides a clue as to the mechanism of charge transport, band gap and nature of the charge carrier besides unfolding possibilities for its use as an optical sensor. The surface morphological changes arising after UV exposure are studied by scanning electron microscopy to examine the changes at the microscopic level.

The above discussions show that bone growth and remodeling can be controlled by external stimulations. Though several successful attempts, mainly reported from Professor Bassett's laboratory, are in clinical use, the search for optimizing the clinical usefulness of these is still on. It may be felt that this task will be considerably facilitated by knowledge of the basic mechanism at work. This task has been undertaken (Behari, 1991) in two ways. First, the response of bone to electrical stimulations in wide range extending from DC to MHz frequencies regions, is scanned. So far only the DC and low frequencies stimulations have been examined.

The physiological state of bone under Ca and P deficiency is expected to provide data of considerable clinical importance. The model adopted is to stimulate the pathological conditions in a colony of rats extending over both sexes. The physical state of bone is characterized by measurement of various parameters, using, for example, X-ray, infrared spectra, transmission electron microscopy, fluorescence, spectrochemical analysis, and with various biochemical messengers. Accompanying biological changes are measured by an estimation of blood parameters, for example, Hb cell counts, $P_H$, $P_{CO2}$, $P_{O2}$, Ca and P. Various biophysical parameters used to characterize bone osteoporosis are elaborated in Chapter 4. In addition, various established techniques of measuring bone osteoporosis are discussed.

# 2

# Piezoelectricity in Bone

## 2.1 Introduction

It has long been known that tension or compression of bone can effect its formation and resorption. The mechanism is not fully understood, although piezoelectric effects on bone crystal or matrix have been identified that might affect bone cells and alter their activity (Bassett, 1968). Piezoelectricity is a bilinear interaction between electrical and mechanical variables, while pyroelectricity is the interaction between electrical and thermal variables. A linear system is one in which the variables for two different phenomena are linearly coupled. Because of the orientation of their molecular elements, pyroelectric crystals posses a permanent electric dipole moment, in which the direction coincides with that of a polar crystal axis. Any change in temperature modifies the length of the pyroelectric body and alters its elementary dipole moments (i.e., the pyroelectric effect). In the same way, any mechanical change in length or any deformation, of a piezoelectric body produces a modification of its dipole moment (i.e., piezoelectric effect). Thus, every pyroelectric system is at the same time piezoelectric.

Bone has electrical properties that depend upon mechanical loading. It has been suggested that the electrical potentials might mediate biological processes (Bassett and Becker, 1962). Studies of the mechano-electrical response of bone under a wide range of conditions have shown it to be a feature of living, freshly excised and dry bone. It is a bone material property and is possibly unrelated (or at least not shown otherwise) to the cellular or other *in vivo* factors.

The direct piezoelectric effect is that electric polarization is produced by mechanical stress. Closely related to it is the converse effect. In this process a crystal becomes strained. Both effects are manifestations of the same fundamental property of the crystal. Investigation of the converse piezoelectric effect is a way to quantify the existence of this phenomena. The spatial condition required for a crystal to be piezoelectric is that it lacks a center of symmetry. The permanent electric dipoles already present in the unit cell will not therefore cancel each other out when the crystal is deformed but will produce a net resultant dipole moment. Not all crystal classes satisfy this relationship. The electrical charges bound within the lattice of the material

*Biophysical Bone Behavior* Jitendra Behari

contract with the frequency of the applied field to produce a mechanical stress. As a consequence of the above, dimensional changes (converse effect) are predictably associated with the application of an applied electric field (Foresbergh, 1956). Piezoelectricity is thus also a cross effect between elasticity and dielectricity (Nye, 1960). Dielectricity is the relation between electric field and polarization.

It may also be asked whether the electrical phenomenon is of primary importance in determining cell regulation and hence growth or healing characteristics. The charges measured may be only an indication of structural deformation (strain), that is, of secondary importance in being descriptive of conformational changes that have occurred in response to other factors. The healing of bone, *in vivo* and other biological processes *in vitro*, can though be influenced by externally applied electrical fields.

## 2.2 Piezoelectric Effect

Wolff's law, that bone macro- and microstructures adapt to their mechanical environments, is a long-held principle of bone biology (reviews in Currey, 1984; Lanyon and Rubin, 1985; Martin and Burr, 1989; Bertram and Swartz, 1991; Lieberman and Crompton, 1998; Martin, Burr and Sharkey, 1998). Osteogenic responses to mechanical loading have mostly been documented in trabecular bone in diaphyseal cross sections and in epiphyses (e.g., Pauwels, 1976; Radin *et al.*, 1982; Currey, 1984; Lanyon and Rubin, 1985; Martin and Burr, 1989; Hou *et al.*, 1990; Gross *et al.*, 1997; Lieberman and Crompton, 1998; Martin, Burr and Sharkey, 1998). In addition, problems arising from electrostriction and thermal dilatation need to be resolved and understood.

Bone adapts to its mechanical environment (Wolff and Das Gesetzder, 1986), and responds to cyclic rather than static strain. In an isolated turkey ulna preparation a few cycles of load per day are enough to prevent disuse osteoporosis (Rubin and Lanyon, 1984). The precise mechanism for the response is not fully understood, but it has been proposed that strain produces transient pressure changes in bone, leading to interstitial fluid flow, and this in turn produces shear forces acting on the osteocyte cell membrane (Reich, Gay and Frangos, 1990). This theory has been modified to provide a mechanism for amplification of the signal detected by the osteocytes (You *et al.*, 2001). Both shear and mechanical load lead to a release of nitric oxide (NO) from bone. NO is a small uncharged free radical with diverse multifunctional signaling properties in many tissue, including regulation of blood pressure. NO has been shown to have many effects on bone, namely, a reduction in osteoclastic motility and much cytokine action on bone is mediated by NO production by osteoblasts (Hukkanen, Polak and Hughes, 1998).

In bone, two dissimilar materials (collagen and calcium hydroxyapatite) are cemented together by polysaccharides, whose role is very significant in explaining the mechanical properties of bone. This has led to the suggestion that bone may be considered as a two-phase composite (Currey, 1964). Piezoelectric properties are central to mechanical, electrical, bone growth and the remodeling, and for any other tissue, and for understanding life processes in general.

The mechanical properties of bone are a function of position and orientation (Pope and Outwater, 1974). In general, bone obeys Hooke's law with certain limitations. Beyond the boundaries of linear behavior, Hooke's law is not true. Dry bone (Evans, 1969) behaves completely in accordance with Hooke's law, whereas wet bone, because of its water content, ceases to follow Hooke's law beyond a certain limit of stress. The long bone changes from being highly anisotropic at the mid-diaphysis to essentially isotropic at the epiphyses. The above authors have concluded that the strength and structure of compact bone indicate that these properties are adapted to the bone's physiological role. X-ray analysis confirms that there is an increase in density at sites of high stress and a corresponding decrease at lower stress sites. In this respect, bone behaves like a system controlled by feedback. An applied force causes stresses and strains in bones; the latter are the stimulus for remodeling processes. In the equilibrium state both the osteoblastic and osteoclastic activities are equal (Bennok, 1972).

Several authors have concluded, from observed relationships between polarization and stress, that the effect is truly piezoelectric, resulting from the slipping of collagen-fibers past one another (Fukada and Yasuda, 1957; Shamos, Lavine and Shamos, 1963). However, calcified bone (Bassett, 1962) was found to lose its capacity to generate DC potentials due to applied stresses. Shamos and Lavine (1964) attributed the piezoelectric effects in bone to the organic constituent, that is, collagen. The effect is highly directional, with a maximum for shear stresses and minimum for pure compressive or tensile stresses. Bone in cantilever form and under stress acquires a positive charge on the convex surface (maximum tension) and a negative charge at the concave side (maximum compression) (Fukada and Yasuda, 1957). Eriksson (1974), however, postulated that for wet bones this stress induced electrical polarization arises from streaming potentials caused by a unidirectional flow of extracellular bone fluid in transversely running pores of ultramicroscopic dimensions. It is proposed that channels that run in transverse directions will be increased in diameter on the convex side (if extracellular fluid in bone carries a net positive charge) and simultaneously decreased in diameter on the concave side when the bone is flexed, leading to a flow of fluid towards the convex surface.

Viewed in this way, bone behaves like a biotransducer with unique characteristic. A related parameter, the coupling coefficient ($k$), is a kind of characteristic index for the performance of a transducer. The value of $k^2$ is a measure of the magnitude of the transducer bandwidth. For composite materials, the macroscopic piezoelectric constant is a function of the piezoelectric moduli, elastic moduli, dielectric constants and geometry of the different phases (Date, 1972). Electrostriction is the most common effect of higher order electro-mechanical coupling. In non-piezoelectric substances, a strain induced by an electric field is proportional to a square of the field strength. This is interpreted from the strain dependence of the dielectric constant in electromagnetic theory. The moisture content is important for the piezoelectric and dielectric properties (Lakes, Harper and Katz, 1977).

The existence of piezoelectricity in bone has aroused great interest because it seems to provide an important key in understanding bone physiology and the mechanism controlling osteogenesis. Several authors have also predicted and showed this effect in biological tissues (Anderson and Eriksson, 1968; Shamos and Lavine, 1967). Stresses of about $10^7$ N m$^{-2}$ applied to bone produce some 10 mV in suitably attached electrodes (Damask, 1978; Steinberg *et al.*, 1973). This generation of an electrical potential difference of bone under mechanical stress has been variously ascribed to piezoelectricity (Becker, Bassett and Bachman, 1964; Fukada and Yasuda, 1957; Williams and Breger, 1975) and to streaming potentials (Gross and Williams, 1982; Guzelsu and Saha, 1984; Pollack *et al.*, 1984). Marino and Becker (1964) suggested a mechanism by which bone's piezoelectric signal could regulate bone growth. The piezoelectric-like effect in bone is now generally accepted as the mechanism by which bone cells sense the stress-state in their vicinity and add or remove bone to meet the demand of the situation (Lakes, Harper and Katz, 1977; Reinish and Nowick, 1975; Marino and Becker, 1975). This raises the question of a possible relationship in the well-understood behavior of dry bone and that of wet bone. The bioelectric potential is now closely linked with possible bone formation and resorption (Bassett, 1989). Piezoelectric inhomogeneity in bone has been indicated by the work of McElhaney (1967) and is also suggested by the microstructural variations in density and orientation of the piezoelectrically active constituents, namely, the fibers of collagen (Fukada and Yasuda, 1964; Williams, 1974).

Dielectric effects in bone (including conductivity) have a strong influence on the piezoelectric effect, especially in composites, and the presence of a finite conductivity introduces a frequency (or time) dependence into the measured voltages. A simple system is represented by a dielectric (a capacitor) in parallel with a resistance, where the ratio $\varepsilon'/\sigma$ is simply the $RC$ time constant. In this case $\varepsilon(\omega)$ can be a complex quantity; the real part ($\varepsilon'$) represents the dielectric constant and the imaginary part ($\varepsilon'' = \sigma/\omega$) represents the conductivity divided by the frequency ($\omega$). Then:

$$\varepsilon = \varepsilon' - i\frac{\sigma}{\omega}$$

As the resistance and the capacitance depend on the frequency, so $\varepsilon'$ is frequency dependent (Figure 2.1) (Reinish and Nowick, 1979; Lakes, Harper and Katz, 1977). Several authors have measured the electrical and dielectric properties of cortical bone (Chakkalakal *et al.*, 1980; Chakkalakal and Johnson, 1981; Kosterich, Foster and Pollack, 1983; Lakes, Harper and Katz, 1977; Reddy and Saha, 1984; Singh and Behari, 1984a). The dielectric measurement of bone (Freeman, 1967) shows that the bound water exists in the structure of bone.

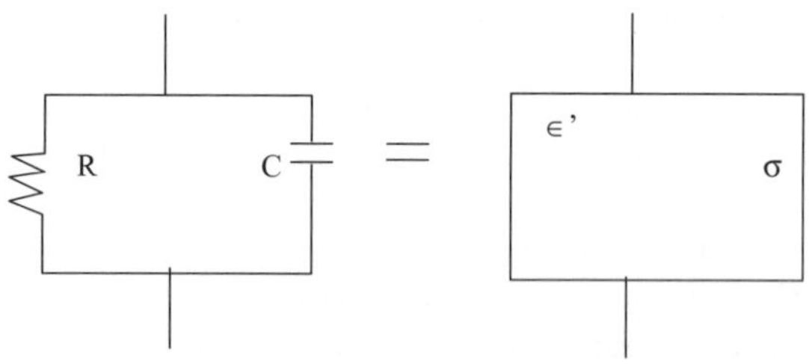


**Figure 2.1** Dependence of a resistance and capacitance combination on frequency; $\in'$ is thus frequency dependent

The mechanical behavior of a composite at the level of extracellular matrix is a function of the individual phases (matrix and filler), their interaction and adhesion. Other key parameters are crystal size, filler aspect ratio and volume fraction of mineralization. Such information may be useful in describing the mechanism of fracture and fatigue phenomena. The complex nature of bone microstructure may necessitate that, ultimately, some combination of mechanical models will be needed. An understanding at the molecular level may also provide useful information for subsequent histologic-based models.

The stiffness, toughness and compressive strength of a polymeric solid such as collagen can be enhanced by incorporation of a second hard inorganic phase. Kerner (1956) has given predictions of stiffness for spherically symmetric inclusions. Formulations have been applicable over a wide range of volume fraction of the inorganic phase. In this theory, the properties of a matrix phase are changed by the presence of a particle with different properties. Direct application of Kerner's equations to bone would be limited because the hydroxyapatite inclusions are not spherical. For calculation purposes, the bulk modulus ($K$) and shear modulus (μ) of collagen may be estimated from Young's modulus ($E$) and the Poisson ratio ($\nu$) using known identities for isotropic solids. A relationship between these parameters is:

$$K = \frac{E}{3(1-\nu)} \tag{2.1}$$

$$\mu = \frac{E}{2+2\nu} \tag{2.2}$$

The filled composite model approximates cortical bone most closely where there is a maximum concentration of hydroxyapatite and collagen matrix and a lack of fat and hematopoietic elements. In cortical bone, the volume fraction of filler or percent mineralization is approximately 0.65 (Currey, 1975). Paracrystalline aggregates of hydroxyapatite from bone have an aspect ratio of approximately 10 and are 4.0–4.5 nm wide and 40 nm long (Ascenzi, 1988).

### 2.2.1 Properties Relating to Piezoelectricity

The thesis that bone components behave as a pressure sensitive semiconductor junction can be examined by applying stress to vacuum stored rectangular bone samples ($7.0 \times 1.1 \times 0.2$ cm) by means of a tensometer. Strain can be measured with strain gauges bonded to opposite faces of the sample and bone resistance by electrodes embedded into the sample and secured with silver conductive paint. A constant direct voltage is applied and the current measured with a picoammeter. The electrical resistance is measured in both $x$ and $y$ directions.

The effect of ionic fluids can be studied using cylindrical samples (0.5–1.5 cm long, 0.3–0.4 cm diameter). They are immersed in deionized water for two weeks to ensure the absence of ions and then, after drying in vacuo, silver electrodes are painted on the end faces. The electrodes are sealed from water ingress by coating with epoxy resin followed by encapsulation with silicone rubber. Samples are conditioned by immersion in solutions of known conductivity buffered to known pH with standard buffers, in a strictly controlled temperature environment. A strain rate of $5 \times 10^{-4}\,\mathrm{mm\,s^{-1}}$ can be used, neglecting the viscoelastic behavior.

A converse effect can be examined by applying an electrical stimulus over a wide frequency range (up to MHz) and examining the response, using a strain gauge. A good correlation is realized by the adopted relationship. A cut of around (13–14 MHz) is obtained (Figure 2.2a and b) (Jha and Behari, 1993).

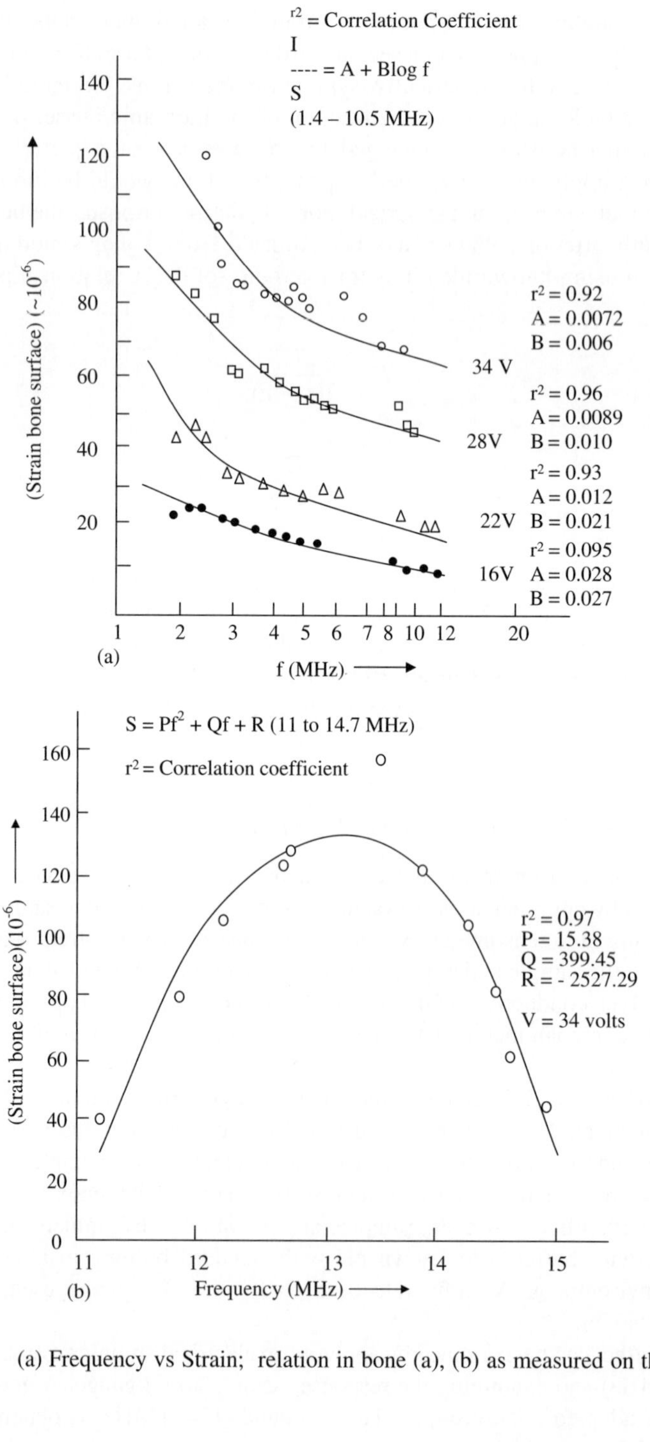


**Figure 2.2** (a) Frequency vs Strain; relation in bone (a), (b) as measured on the surface

## 2.3 Physical Concept of Piezoelectricity

Piezoelectric media are intrinsically anisotropic. Piezoelectricity provides a coupling between elastic and dielectric phenomena. In a quasi-static case, the importance of coupling between the mechanical and electrical fields has been emphasized by Guzelsu (1978). Since bone is an anisotropic material containing oriented crystallites, the piezoelectric data are expressed in terms of the piezoelectric matrix:

$$P_i = \sum_{j=1}^{6} d_{ij} T_j \tag{2.3}$$

where $d$ is the constant of proportionality between the applied mechanical stress $T$ and the generated electric polarization $P$ (Figure 2.3). The $d$ matrix contains 18 coefficients relating to the piezoelectric polarization in different directions with the applied tensile and shear stress in different directions; $i = 1, 2, 3$ corresponds to $x$, $z$, coordinates and where the macroscopic field $E$ across the sample is zero. The six $j$ indices correspond to three tensile stress and three shear stresses. In general the $d$ coefficients are complex quantities, $d^* = d' - jd''$, where the real part is proportional to the charge generated in phase with the applied stress and the imaginary part is proportional to the charge generated out of phase with the applied stress. When $d''$ is nonzero, a relaxation dispersion is present and measurements as a function of frequency become a necessity. The equations for the piezoelectric relations in a crystal producing a single mode of motion can be written in the form:

$$\begin{aligned} S_2 &= S_{22}^{E}.T_2 + d_{21}E_1 \\ D_1 &= d_{21}T_2 + \epsilon_1^{T} E_1 \end{aligned} \tag{2.4}$$

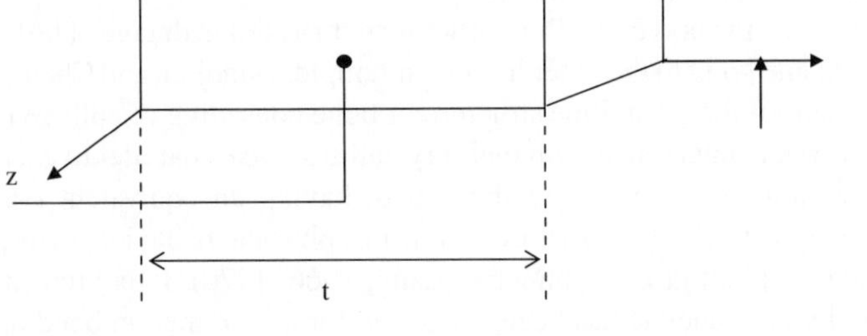


**Figure 2.3** Relationship between applied mechanical stress T and the generated electric polarization in an anisotropic crystal

where $S_2$ and $T_2$ are the extensional strain and stress, respectively; $S_{22}^{\mathrm{E}} = S_{11}^{\mathrm{E}}$, the elastic compliance (i.e., inverse of Young's modulus along the length measured at constant electric field), $d_{21}$ is the piezoelectric constant relating the strain with applied field $E_1$; $D_1$ is the electric displacement; and $\in_1^T$ is the dielectric constant measured at constant stress. These equations suffice to determine the static and low frequency behavior of piezoelectric crystals. The constant strain dielectric constant ($\in_1^{\mathrm{S}}$) is related to the constant stress dielectric constant ($\in_1^{\mathrm{T}}$) by the relation:

$$\in_1^{\mathrm{S}} = \in_1^{\mathrm{T}} (1 - k^2) \tag{2.5}$$

where $k$ is the electromechanical coupling factor.

The square of the coupling factor is written as:

$$k^2 = \frac{d_{21}^2}{S_{22}^{\mathrm{E}} \in_1^{\mathrm{T}}} \tag{2.6}$$

By solving Equations 2.6, it is readily shown that the clamped dielectric constant ($\in_1^{\mathrm{S}}$), obtained by setting $S_2 = 0$, and elastic compliance of constant displacement ($S^{\mathrm{D}}$), obtained by setting $D_1 = 0$, are related to the constant stress dielectric constant ($\in_1^{\mathrm{T}}$) and the elastic compliance at constant field ($S_{22}^{\mathrm{E}}$) by the equation:

$$\frac{\in_1^{\mathrm{S}}}{\in_1^{\mathrm{T}}} = \frac{S_{22}^{\mathrm{D}}}{S_{22}^{\mathrm{E}}} = 1 - k^2 \tag{2.7}$$

### *2.3.1 Piezoelectric Theory*

The phenomenon of stress generated potential (SGP) in bone has been interpreted as piezoelectric and originates in its organic component (Marino and Becker, 1971). However, the correlation between the observed sign of potential in bending and the sign fixed by specimen orientation has not been established. Obviously, there is some aspect of bone that decouples the sign of the strain gradient potential (SGP) in the PZE matrix from the macroscopic specimen coordinates. The problem was solved by Korostoff (1979). To solve these problems there has been growing interest in the electromechanical properties of bone, the stiffness constant and the piezoelectricity.

Bone has a microscopic polycrystalline structure with a certain degree of ordering that varies substantially from one point to the other in a given sample. Gundjian and Chen (1974) assumed the microscopic unit of the crystalline structure of bone consisting of collagen molecules and hydroxyapatite crystals intervened with their crystallites *c*-axes parallel to each other. A bone crystallite unit is thus a composite of these two, having an equivalent *c*-axes of infinite rotational symmetry [$T(\infty)$]. This structure results in a physical behavior corresponding to that of a single crystal of point group symmetry (Lang, 1966, 1970). In conformity with this, an idealized parallel plate lamellae has been suggested for a specimen in bending, in which the prevailing collagen orientations in a local region are alternatively at approximate right angle to each other. Each lamellar region is composed to be described by the same PZE matrix. Acting upon these crossed (90°) orientations of collagen are the tensile stresses parallel to the plane but at some angle to the collagen axes.

The relationship between piezoelectric polarization $P$ and the mechanical stress $\sigma$ is expressed in terms of a series of piezoelectric moduli, given by:

$$\begin{aligned} P_i &= d^{ijk}\sigma_{jk} \\ \text{and}\, \boldsymbol{d}_{ij} &= \begin{matrix} 0 & 0 & 0 & d_{14} & 0 & 0 \\ 0 & 0 & 0 & 0 & -d_{14} & 0 \\ 0 & 0 & 0 & 0 & 0 & 0 \end{matrix} \end{aligned} \tag{2.8}$$

The resulting polarizations are perpendicular to the lamellar plane and they are antiparallel for adjacent planes. An antiparallel arrangement would result in cancellation of polarization to give zero net polarization perpendicular to the lamellar plane, while experimentally one observes a net unidirectional polarization. The apparent disagreement between this theoretical postulate and experimental finding is averted by suggesting a possible mechanism for reorientation on the antiparallel polarization configuration to a parallel configuration. These energies are estimated to be $+36$ and $-36\,\mathrm{J\,m^{-2}}$, respectively, for a polarization strength of $2\times10^{-6}\,\mathrm{C\,m^{-2}}$ (Korostoff, 1979). The higher energy for the antiparallel configuration derives from the interactions of the electric field due to the polarization in a given lamella with the oppositely directed polarization of the adjacent lamella. The concurrent energy increase for strain and demagnetizing field that accompany the change from antiparallel to parallel alignment is very small. Thus, the antiparallel state predicted by the PZE sensor is unstable, and yields to the equilibrium configuration in which the polarizations are parallel. In this theory the process of redirection of polarization from the unstable antiparallel configuration thus decouples the macroscopic bone specimen from the rigid polarity imposed by the PZE matrix. It is further postulated that the trigger is based upon the net molecular surface charge of bone, which is normally negative (Nieders, Weiss and Cudney, 1970; Eriksson, 1976). The bone has been idealized as composed of surfaces with net negative charge. In a bent configuration, the polarization vector can be thought of as adding vectorially to the antiparallel piezoelectric polarization to enhance that direction. It is seen that the trigger polarization is about 0.1% of the PZE polarization.

## 2.4 Sound Propagated in a Piezoelectric Medium

Sound propagation in a non-piezoelectric medium has been fully elucidated. When a wave vector direction, or a wave normal direction, is specified, three sound modes are found, one (quasi-)longitudinal and two (quasi-)transverse polarized that are perpendicularly to one another. The sound velocity is given by:

$$v = (C_{\mathrm{eff}}/\rho)^{1/2} \tag{2.9}$$

with an effective elastic constant ($C_{\mathrm{eff}}$) for each mode. This constant is given an eigenvalue for a Christoffel equation or a velocity equation.

When the medium is piezoelectric, it is necessary to take electrical conditions into account because an electric field is associated with the sound. Accordingly, Maxwell's equations of electromagnetic theory are considered together with the elastic equations of motion. However, the velocity of sound is very small compared with that of an electromagnetic wave. Therefore, a polarization wave cannot be propagated in the form of an electromagnetic wave. The electric field is only associated with the sound as an electrostatic field. Accordingly, $B$ is usually disregarded in favor of rot $E$. This assumption is called a quasi-electrostatic approximation. If the medium is insulating, the electrical conditions are given by:

$$\operatorname{div} D = 0 \text{ and } \operatorname{rot} E = 0 \tag{2.10}$$

A coupling coefficient is usually defined for the $\omega \rightarrow 0$ limit on the basis of an appropriate piezoelectric relation. Nevertheless, the transducer has electrical and mechanical terminals for energy supply and delivery. The fundamental or first resonance occurs at a position corresponding to the half-wavelength of the sound and harmonic resonances occur successively at odd multiples of the fundamental frequency. Accordingly, it could be said that the resonance itself represents a real mechanical aspect of the sound.

## 2.5 Equivalent Single-Crystal Structure of Bone

Gundjian and Chen (1974) further suggested that the equivalent $c$-axis of the composite collagen–hydroxyapatite crystallite is not oriented perfectly throughout a macroscopic bone sample but is distributed around a distribution symmetry axis with a finite angular dispersion around this axis. The "standard equivalent single crystal structure" (SESCS) of bone is thus considered a hypothetical structure in which the bone crystallites are all perfectly oriented in the direction of the distribution symmetry axis. Since the measured values of the physical properties of natural bone correspond to an average over a large aggregate of differently oriented bone crystallites, they are expected to be sensitively dependent on the actual crystallite orientation distribution function of the sample. Consequently, natural variations for different samples will produce corresponding variations in the measured values of the physical properties, thus making impossible the identification and interpretation of the more fundamental intrinsic changes in the values of these properties.

It is thus apparent that the standardized value of the physical properties will correspond directly to those of the individual crystallites. This will, therefore, exhibit only changes resulting from actual physical changes in these crystallites. The angular position of the crystallite distribution asymmetry axis, as well as the distribution function, vary slightly with position in a given bovine femur sample. The width is found to be of the order of 30°.

## 2.6 Piezoelectric Properties of Dry Compact Bones

Most experiments have been carried out under zero electric field or zero stress. Here we select the former case – a well-accepted assumption by researchers in this field. Thus, Equation 2.7 is to be applied for the description of the piezoelectric body under investigation. The elastic properties are uncoupled from the electrical properties under the assumption of zero electric field. In general, for a anisotropic crystal the piezoelectric constants, in matrix form, can be represented by (Cady, 1964; Nye, 1960):

$$\begin{bmatrix} d_{11} & d_{12} & d_{13} & d_{14} & d_{15} & d_{16} \\ d_{21} & d_{22} & d_{23} & d_{24} & d_{25} & d_{26} \\ d_{31} & d_{32} & d_{33} & d_{34} & d_{35} & d_{36} \end{bmatrix} \tag{2.11}$$

While working on the dynamical properties of bones, Yasuda, Noguchi and Sato (1955) observed the piezoelectric behavior of bone tissue. However, the systematic investigation of direct and converse piezoelectric effects in bone tissue starts with the work of Fukada and Yasuda (1957). In this work the authors showed that the polarization is linearly related with the pressure (direct effect), while a similar relationship is also valid between electric field and the strain (converse effect). Using dry specimens cut from the femur of a man and ox, they observed that the piezoelectric surface charge appears only when the shearing force is applied parallel to the collagen fibers, that is, when the collagens slip with respect to each other, and the magnitude of the piezoelectric constants depend on the angle between the applied pressure and the axis of the femur, as well as the angle between the gravitational field and the femur axis in the standing position of the subject. As a result of their experimental observations, Fukada and Yasuda (1957) showed that, for bone, the matrix in Equation 2.11 reduces to:

$$\begin{bmatrix} 0 & 0 & 0 & d_{14} & 0 & 0 \\ 0 & 0 & 0 & 0 & -d_{14} & 0 \\ 0 & 0 & 0 & 0 & 0 & 0 \end{bmatrix} \tag{2.12}$$

which fits to the hexagonal $(D_6)+$ crystal class (Cady, 1964). Figure 2.4b shows that the piezoelectric constants of the femur of a man depends on the angle between the pressure direction and the bone axis. By examining this, one can see that the material symmetry axis of the femur is inclined about 10° to the right-hand side of the femur axis. In later work on collagen, Fukada and Yasuda (1964) showed that the material symmetry of a horse femur has the hexagonal polar $(C_6)$ structure, which revises their former observation on human femur (Fukada and Yasuda, 1957) and in which case the non-shear components $d_{11}$, $d_{12}$, $d_{13}$, $d_{21}$, $d_{22}$, $d_{23}$ of the matrix $\boldsymbol{d}_{\text{im}}$ are taken to be zero. Anderson and Eriksson (1968, 1970), testing wet and dry beef bone samples (the latter being dried over silicon gel), showed that the non-shear portion of the matrix $\boldsymbol{d}_{\text{im}}$ is not necessarily zero (they have, rather, finite values). This inconsistency, however, results from the differences in selecting the reference coordinate system in measuring the material coefficients. As was pointed out by Fukada and Yasuda (1957), taking the values given in by Eriksson (1974) for dried Achilles tendon collagen as the reference values and then rotating the coordinate system by 10°, they observed that the calculated coefficients relative to the coordinate system gives coefficients similar to those of Anderson and Eriksson (1970). This provides a measure of the agreement between the experimental and calculated results. Their measurement further showed that, taking the specimen texture into account, the coefficients measured for pure collagen of dry states and dry bone tissue are quite similar. From this observation it is concluded that the electrical signals, obtained by stressing dry bone samples, are derived from the collagen portion of the bone tissue, which is the same as observed by Bassett and Becker (1962).

Liboff *et al.*, (1971) and Liboff and Furst (1974) studied the piezoelectric properties of dry bovine femur in vacuo at 38 °C and obtained the matrix ($\boldsymbol{d}_{\mathrm{im}}$) for their sample crystals. Similar results have been obtained by Bur (1976), who measured the piezoelectric constants of bone samples with 75% relative humidity. The results of the latter measurements are not consistent with these of Fukada and Yasuda (1957, 1964) and Anderson and Eriksson (1970). Although it has been emphasized (Liboff, Shamos and De Virgilio, 1971; Liboff and Furst, 1974; Bur, 1976) that bone has a very complex anisotropic and heterogeneous structure, it is generally assumed that it has hexagonal polar ($C_6$) symmetry.

McElhaney (1967), he measured the charge distribution on the surface of a femur under an axial load. Observing the charge density resulting from the piezoelectric effect and varying from one point to another on the femur, it was concluded that this is the result of a non-uniform distribution of collagens in the osteons.

## *2.6.1 Piezoelectric Properties of Dry and Wet Collagens*

### 2.6.1.1 Structure of Collagen

It is now possible to study ultrasonic wave propagation in bone with two concepts in mind. First, that for a specific two-phase mineral-filled polymer the longitudinal modulus of the composite can be found from the Reuss formalism and, second, the longitudinal modulus of the matrix can be affected by the presence of the filler.

Several attempts have been made to describe the biomechanical properties of bone (Ascenzi, 1988), but with marginal success. Often this description does not go beyond the histological level. Others have studied the forces in bone macroscopically, as exemplified in models using finite element analysis (Manhanian and Piziali, 1988).

Fukada and Yasuda (1964) have shown that the crystal symmetry of collagen fits to hexagonal polar ($C_6$) symmetry, which is given as follows:

$$\begin{bmatrix} 0 & 0 & 0 & d_{14} & d_{15} & 0 \\ 0 & 0 & 0 & d_{15} & -d_{14} & 0 \\ d_{31} & d_{31} & d_{33} & 0 & 0 & 0 \end{bmatrix} \tag{2.13}$$

They measured the direct and converse piezoelectric effects in dry Achilles tendon of bovine and horse, which are composed of well-oriented collagen fibers. They also showed that, for dry collagen, the polarization results from the displacement of hydrogen bonds formed in the polypeptide chains of collagen crystals. Measurement of collagen (Lang, 1966), separated from bone by use of the nitric acid, shows that the piezoelectric (as well as pyroelectric) behavior of bone arises from the structure of collagen molecules. Similar tests by Shamos and Lavine (1967) indicated that the fibrous structure of collagen in bones and the fibrous molecular proteins in the soft tissue are the main causes of the piezoelectric behavior of biological tissues. The streaming potential and wet bone experiments of Anderson and Eriksson (1968, 1970) indicated that wet collagen has a different structure to dry collagen, because the bound water in the material changes its crystal symmetry to the point where no piezoelectric properties are observed. Since the solid matrix of bone tissue puts constraints on the collagen, the water molecules absorbed by collagen in tendon cannot be the same as in the bone tissue. Studies by Liboff, Shamos and De Virgilio, 1971 and Liboff and Furst, 1974 show that a certain minimum amount of water concentration, which increases the crystal symmetry, is required to maintain overall structural integrity. Owing to the variability in behavior of collagen in the wet and dry states, wet bone has a different piezoelectric symmetry relation (Anderson and Eriksson, 1970).

### *2.6.2 Measurement of Piezoelectricity in Bone*

The measurement of piezoelectricity in bone and collagen and its quantification is beset with several problems:

1. Bone impedance is large, meaning that low voltage measurements show fluctuations and hence are unreliable (Gizdulich and Calsolaro, 1993);
2. The capacitive effects of the bone–electrode interface (Gizdulich and Aschero, 1993);
3. Electrode polarization (Schwan, 1963).

The first quantitative measurement of the piezoelectric constant of bone by means of both the direct and converse piezoelectric effect were performed by Fukada and Yasuda (1957). In their method (Figure 2.4a) the direct effect was studied by the static method. Pressure was applied on the side plane of the specimen by a lever mechanism and the resultant electric charge that appeared on the square plane was detected by galvanometer deflection. The piezoelectric constants were found to depend on the pressure direction and bone geometrical axis (Figure 2.4b). These authors measured the converse piezoelectric effect by a method of comparison. When the electric field is applied in the direction of the electric polarization the resultant mechanical strain is produced in the direction of applied force. Due to the converse piezoelectric effect the oscillational force is transmitted to the piezoelectric crystal. In the direct and converse effect both the relationship between stress and applied electric field turns out to be linear (Figures 2.5 and 2.6). Table 2.1 shows the corresponding piezoelectric constant of bone for human samples. Fukada and Yasuda (1957) also studied the variation of the complex dielectric constant ($\varepsilon'$ and $\varepsilon''$), at 1 kHz, with temperature and humidity. As reported, these parameters increase sharply with increasing hydration.

Marino, Becker and Bachman (1967) measured the dielectric constant ($\varepsilon'$, $\varepsilon''$) of human cortical bone at 1–100 kHz and found that these parameters alter sharply at a critical hydration level, which ranges from 0.037 to 0.048 g g$^{-1}$. Tomaselli and Shamos (1973) working in the frequency range 0.1–100 kHz at room temperature found the critical hydration of bovine Achilles tendon to be 0.054–0.115 g g$^{-1}$. Marino *et al.* (1971) have measured the piezoelectric constant in bone and its two constituents and reported that the mineral exhibits no piezoelectric behavior (Figure 2.7).

Reinish and Nowick (1976) have investigated collagen hydration with X-ray diffraction and reported that absorption of water by the helical structure of the collagen molecule becomes saturated at a moisture content of 26 wt% (RH = 60%). The piezoelectric properties of bone are generally attributed to the collagen, because tendon (almost pure

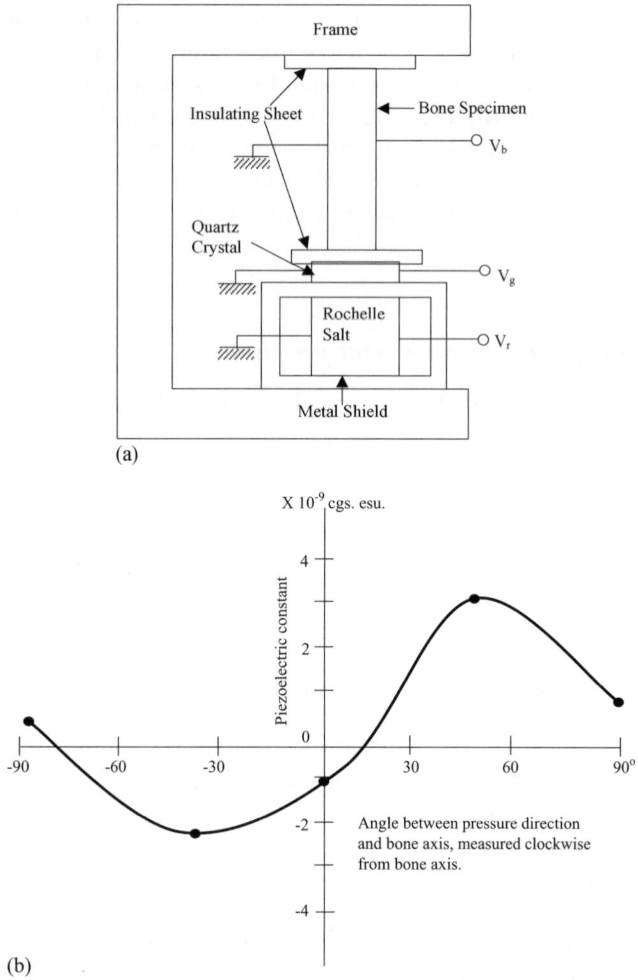


**Figure 2.4** (a) Schematic diagram of a device for measuring the piezoelectricity in bone. (Reproduced with permission from E. Fukada and I. Yasuda, "On the piezoelectric effect of bone," *Journal of the Physical Society of Japan*, **120**, 1158–1162 ©1957, The Physical Society of Japan.) (Fukada and Yasuda, 1957.) (b) Dependence of the piezoelectric constant of the femur of man on the angle between the pressure direction and the bone axis (Fukada and Yasuda, 1957)

collagen) is piezoelectric (Fukada and Yasuda, 1964), while the crystal structure of hydroxyapatite is 6/m, which is centrosymmetric. For collagen with 10% weight moisture content, there exist two kinds of relaxations, which depend upon moisture content and the frequency. It is concluded that the relaxations at 0 °C result from reoverietal motion of the collagen molecules' side chains, whereas relaxation around −80 °C is caused by low frequency oscillation of the collagen's main chain. This seems to be correlated with the amount of adsorbed water.

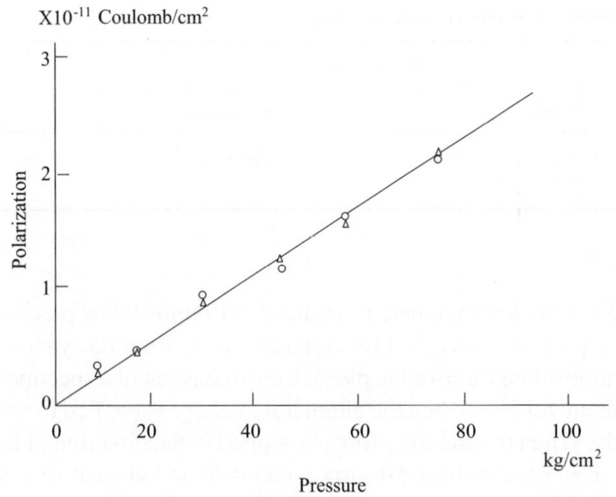


**Figure 2.5** Direct piezoelectric effect of bone

For collagen, $\varepsilon'$ and $\varepsilon''$ increased with increasing hydration, while $d'$ at $-150\,°C$ increased below critical hydration (hc) but decreased above hc with increasing hydration. For bone the dependence of piezoelectric constants on hydration and temperature reflects a two-phase structure consisting of collagen and mineral hydroxyapatite. The critical hydration for bone is $0.04\,g\,g^{-1}$. The piezoelectric constant ($d = d' - jd''$) gives the ratio of polarization to stress at 45 °C to the direction of fibrillar orientation in the collagen ($=d_{14/2}$).

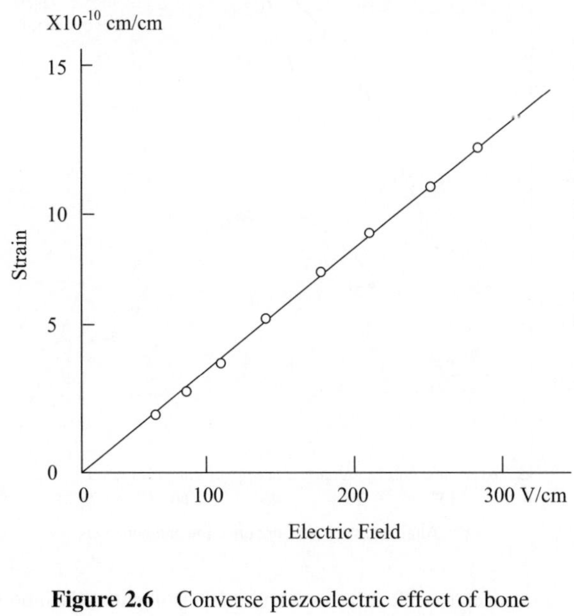


**Figure 2.6** Converse piezoelectric effect of bone

**Table 2.1** Piezoelectric constant of bone of man

| | Direct effect | | Converse effect |
|---|---|---|---|
| | Static | Dynamic | Dynamic |
| 1 | $2.0 \times 10^{-9}$ | $2.9 \times 10^{-9}$ | $2.9 \times 10^{-9}$ |
| 2 | $2.2 \times 10^{-9}$ | $3.6 \times 10^{-9}$ | $3.5 \times 10^{-9}$ |

From the above it is evident that there is a sonic contribution to the piezoelectric response in bone. The converse piezoelectric effect for collagen has been similarly measured (Fukada and Yasuda, 1964). The absolute value of the piezoelectric constant of a specimen is determined by the ratio of two output voltages when the alternating voltage is applied to specimen and quartz crystal respectively. When the shearing force is applied in the direction of fiber axes the upper end becomes polarized negatively and the lower end is polarized positively. When the pressure is applied in a perpendicular direction to the fiber axis, similar polarization takes place. If the direction of force is reversed, the sign of electric polarization is also reversed. Converse piezoelectric effects are also observed when the electric field is applied in the direction of electric polarization.

The dielectric relaxation and mechanical properties of collagen and bone have been examined by Meda and Fukada (1982). These authors determined the piezoelectric elastic and dielectric constants of collagen at a selected hydration and temperature range (−150–50 °C) at a frequency of 10 Hz.

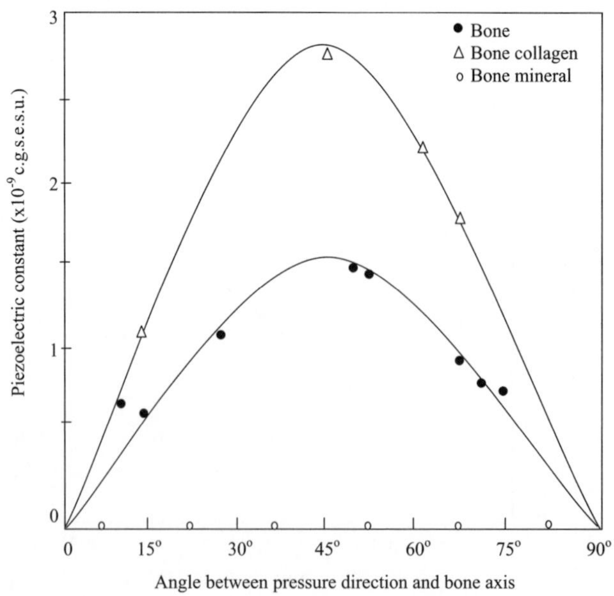


**Figure 2.7** Piezoelectric effect in bone and its constituents (Marino *et al.*, 1971)

Pfeiffer (1977) has described a method of measuring the frequency response of piezoelectric coefficients of the mineralized collagen fibril, including the ground substance of adult human cortical bone. Reinish and Nowick (1974) have found that the piezoelectric polarization at 100% humidity decreases to 30% of the value obtained for dried bone. This model makes use of the fact that:

1. Bone exhibits mechanical isotropy but piezoelectric behavior is anisotropic.
2. The symmetry for the mineralized fibril is $C_\infty$.

Sedlin and Hirsch (1966) and McElhaney (1967) have pointed out that cortical bone is a viscoelastic material and hence material data become time dependent. This implies that the piezoelectric coefficients for direct and inverse effects are not identical. If the stress is applied periodically, the material parameters become frequency dependent.

The magnitude of piezoelectric constants increases with increasing molecular orientation for bone tendon and extracted collagen film (Fukada, 1974). The single crystal equivalent parameters, as reported by Gundjian and Chen (1974), show a greater increase than the averaged data for bone reported by Fukada (1968), although the results of Pfeiffer (1977) are greater still. The difference can be explained best by the increase at low frequencies and, on the other hand, by different organization of the lamellae in bovine and human femur samples.

Measurements of the piezoelectric properties of bone (Holt, Martin and Advani, 1977; Liboff and Furst, 1974; Bur, 1976; Fukada, 1974) indicate that:

1. Although the piezoelectric element of bone (the collagen molecule) has a hexagonally symmetric crystal structure, bone itself exhibits no apparent symmetry in its matrix of piezoelectric coefficients.
2. Inhomogeneities in bone composition and/or structure result in considerable variations in the piezoelectric matrices, measured in different specimens, even though the specimens may come from neighboring sites in the same bone.

## 2.7 Bone Structure and Piezoelectric Properties

The bone samples used in the work reported were all taken from the shafts of femurs and tibias of mammals and primarily from mature bovine femurs and tibias. Such bone is called "compact" or "cortical" bone. Compact bone is a composite consisting of calcium phosphates (about 70–75 wt% in dry bone) and collagen (about 20–25 wt% in dry bone) (Miller and Martin, 1968 and Herring, 1968).

Collagen, a fibrous protein, is a major component of most connective tissue, for example, tendon, skin and ligament, and has remained an attractive subject of investigation. As mentioned in Chapter 1, collagen fibers are made of tropocollagen (TC) molecules. The TC molecule is about 2900 Å long and 15 Å in diameter and is itself a triple helix of three polypeptide chains (Figure 1.5) (Kuhn, 1969). In solution the TC molecule exhibits the large dipole moment characteristic of polypeptides (Bernengo, Rous and Herbage, 1974) and the existence of a pyroelectric effect in dry collagen indicates that the TC molecule may possess a permanent dipole moment when dry. Although collagen will denature at 64 °C in solution, the collagen in dry bone does not denature until about 200 °C (Liboff and Shamos, 1973). Fukada and Hara (1969) have measured the temperature dependence of the piezoelectric moduli in bone and found that no irreversible changes take place, other than effects associated with drying, when the samples are heated to 150 °C.

The collagen fibers form layers or lamella that are typically 7 μm thick (Robinson, Doty and Copper, 1973). The fibers within a lamella appear to be approximately parallel, but different lamella often have differing fiber directions. This difference in fiber orientation can be detected with a polarizing microscope. The lamella make several different types of structural units. They may form concentric cylinders about a Haversian canal, a channel which in living bone contains blood vessels. The canals range from 25 to 80 μm in diameter and the canal plus concentric lamella, a unit called an "osteon," typically has a diameter of 120–230 μm (Pfeiffer, 1977). In other instances the lamella will have a flat or wavy appearance known as "primary" bone. Compact bone contains osteons and primary bone in the interstitial regions between the osteons.

Mature, remodeled bone is characterized by the presence of a large numbers of osteons, while younger bone has fewer osteons and more primary bone. In bovine specimens the earlier lamellar bone is usually replaced by osteons from the inside out. The osteons form first near the endosteal, or inner, surface of the bone and gradually spread towards the periosteal, or outer, surface (Carter, Hayes and Schurman, 1976 and Smith, 1960).

X-Ray diffraction studies of bone have shown that the hydroxyapatite crystals have a preferred orientation down the length of the bone axis (Eanes, 1973). Since the hydroxyapatite crystals are parallel to the collagen fiber axis, this observation implies that there is a preferred or a net orientation of the collagen fibers down the length of the bone. This conclusion is confirmed by microscopic examination. However, the orientation is only approximate and deviations from parallelism may have important consequences for the electrical properties of bone.

Fukada and Yasuda (1957, 1964) showed that bone was piezoelectric when they measured both the direct effect, using uniaxial compression, and the converse effect in samples cut from dried femurs of oxen, humans and horses. The piezoelectric tensor they found had hexagonal polar symmetry with the polar axis nearly parallel to the long axis of the bone. It has the following form in matrix notation:

$$\begin{bmatrix} 0 & 0 & 0 & d_{14} & d_{15} & 0 \\ 0 & 0 & 0 & d_{15} & -d_{14} & 0 \\ d_{31} & d_{31} & d_{33} & 0 & 0 & 0 \end{bmatrix} \tag{2.14}$$

For horse femur they measured $d_{14} = -65$, $d_{15} = 0.13$, $d_{31} = 0.01$ and $d_{33} = 0.01$ in units of $10^{-8}$ cgs-esu. (Fukada and Yasuda, 1964). However, more recent piezoelectric measurements using uniaxial compression indicate that the symmetry of bovine bone cannot be described so simply. Liboff, Shamos and De Virgilio (1971) measured the piezoelectric matrix for specimens cut from bovine femurs and found the following matrix:

$$\begin{bmatrix} 0.06 & 0.27 & -0.18 & 6.09 & 0.33 & \\ -0.48 & 0.00 & 0.54 & 0.33 & -3.90 & \\ -0.57 & 0.51 & 0.15 & & & \end{bmatrix}$$

where the units are in $10^{-9}$ cgs-esu, the $z$-axis is along the length of the femur and the $y$-axis is in the radial direction.

Two experimental studies have demonstrated the existence of a unique electrical axis in bone. Bone is found to be pyroelectric (Lang, 1966, 1969), and the electrical signals generated in bone are measured by the application of hydrostatic pressure (Athenstaedt, 1968, 1969, 1970). Athenstaedt performed his measurements on a large number of samples and found that the direction of the piezoelectric axis varied both with position in the bone and with the age of the animal. He also found a component of the piezoelectric axis perpendicular to the collagen fiber direction in some collagenous tissues (Athenstaedt, 1970).

## 2.8 Piezoelectric Transducers

A transducer has electrical and mechanical terminals for energy supply and delivery (Ikeda, 1990). In the range of ultrasonic frequencies, piezoelectric transducers are used at all levels of intensity. These transducers make use of piezoelectric components, such as plates or other suitable configurations that generate a charge on preferred surfaces under the influence of stresses that change dimensions when they are subjected to an electric field. The electrical characteristics of transducers are influenced by their mechanical properties and by the types of mechanical loading to which they are subjected. Since these influences have the characteristics of electrical impedances, they may be represented in a circuit diagram that also includes all the electrical parameters of the transducer. A circuit diagram of this type is called an "equivalent" circuit (Figure 2.8a). This, like any other electrical circuit, is useful for estimating the efficiency and power handling capabilities and for analyzing its performance. An equivalent circuit is based on the assumption that the transducer operates in a linear manner and, therefore, all circuit values are constants. However, when nonlinearities occur in the stress–strain curve of materials, in the electrical–mechanical interaction at various driving potentials, there occurs variations in the acoustical load impedances at various levels of intensity.

The suitability of a material for use in transducers for producing ultrasound is dependent upon several factors, the relative importance of which is determined by the application. These factors include the magnitude of the piezoelectric coefficients, stability, dielectric properties, electrical breakdown strength and tensile properties of the material. A simple piezoelectric transducer, consisting of a plate vibrating in the thickness mode, may radiate from both faces (symmetrical load) or from only one face (air backed). An air-backed transducer has only one half of the radiating surface of a symmetrically loaded transducer. Thus, if a transformation factor $\alpha$ is used to convert mechanical impedances into electrical impedances for a symmetrically loaded transducer, a factor of $2\alpha$ must be used for the air-backed transducer.

In a piezoelectric material, the imposed stress is in phase with the impressed voltage. The piezoelectric element is an electrical condenser of capacitance $C_0$ by virtue of the dielectric nature of the transducer material. The electrical resistance of the condenser is negligible. The capacitance $C_0$ appears in parallel with a series branch that includes the converted mechanical impedances. These mechanical impedances consist of (i) a resistances $R_e$ corresponding to the losses in the transducer (negligible in high-$Q$ piezoelectric transducers); (ii) a load resistance or radiation impedance $Z_R = P_0 C_0 S$, where $P_0 C_0$ is the characteristic acoustic impedance of the load and $S$ is the area of the radiating surface; (iii) an inductance $M$ due to the mass of the transducer; and (iv) a capacitance $1/K$ due to the compliance of the transducer. The equivalent circuit assumes that $R_e$ is negligible (Figure 2.8b).

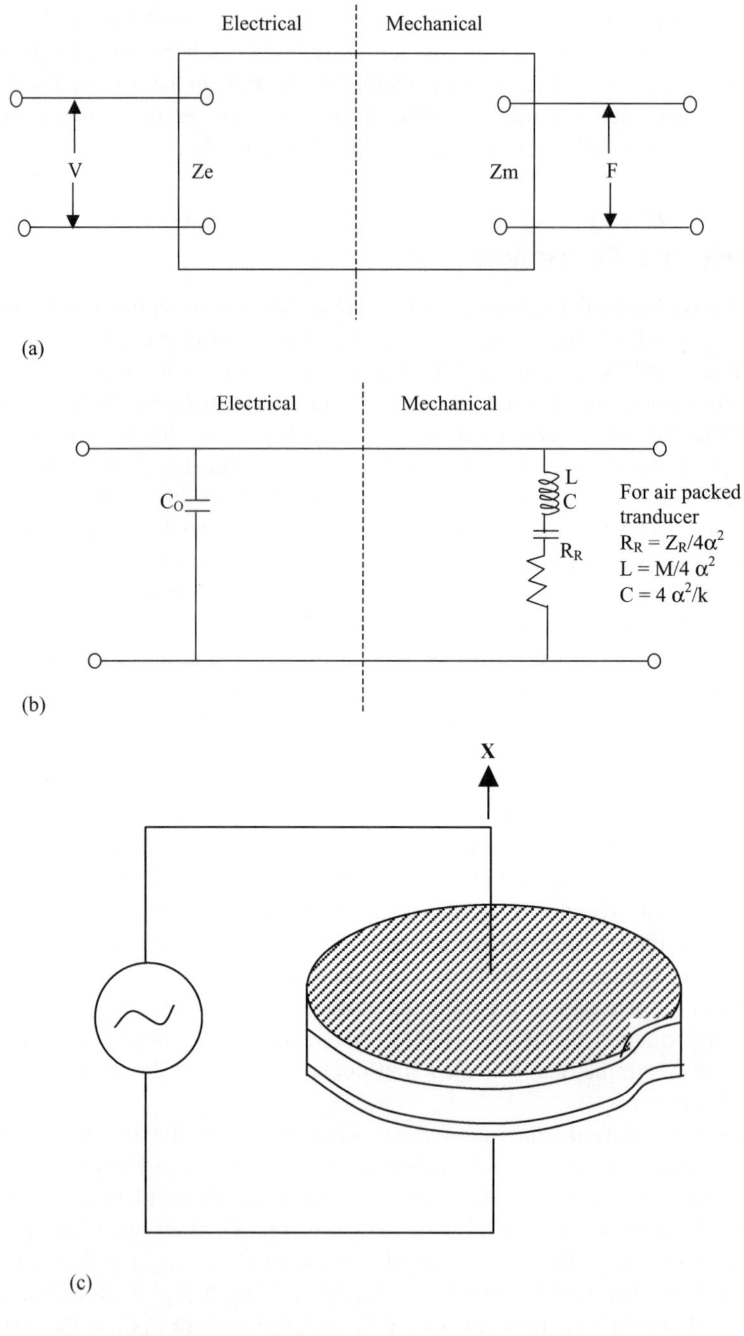


**Figure 2.8** (a) Four terminal network representing the equivalent circuit of a transducer; $Z_e$ and $Z_M$ are the electrical and mechanical impedance, respectively, and *F* is the force. (b) The equivalent electrical circuit for a piezoelectric plate transducer operating near resonance where mass, elastic compliance and mechanical damping are transformed into equivalent electrical elements *L*, *C*, *R* by piezoelectric effect; $C_O$ is the capacitance of the transducer at the resonant frequency under static conditions; $R_{rad}$ is the radiation resistance. (c) Thickness longitudinal vibration of the *L*-effect

The piezoelectric element is electrically excited to vibrate in a plate thickness mode to radiate acoustic energy into a medium of characteristic impedance *PC* that is in contact with the element over one of the larger faces. It is also assumed that the medium in contact with the opposite face has an extremely low impedance, so that practically no radiation is emitted from this face, as compared with the other face. This condition is nearly fulfilled if the former medium is liquid and the other face is air backed. In these formulations it is assumed that no losses occur in the holder and that the element is essentially unrestrained as far as the vibration is concerned, except by the medium into which it radiates.

### 2.8.1 Transducer Vibration

In the vibration of a transducer various electrical and mechanical situations can be simulated. The vibration behavior depends upon the nature of coupling, that is, the longitudinal (L) or transverse (T) effect.

### 2.8.2 Transverse-Effect Transducer

In this case, the fundamental or first resonance occurs at a position corresponding to the half-wavelength of the sound and harmonic resonances occur successively at odd multiples of the fundamental frequency. Accordingly, it can be said that resonance itself represent a real mechanical aspect of the sound.

When interpreted with the equivalent circuit, a series resonance of $L$ and $C_n$ in the piezoelectric branch give rise to this resonance. In contrast, the anti-resonance corresponds to a parallel resonance of $LC_n$ and a capacitance close to $C_0$. Thus, the resonant angular frequency of the $n$th harmonic is given by:

$$W^2\text{R}_n = 1/LC_n$$

The anti-resonant angular frequency by (Ikeda, 1990):

$$W^2A_n = \frac{1}{L}\left(\frac{1}{C_n} + \frac{1}{\overline{C_0}}\right)$$

where:

$$\overline{C_0} \approx C_0 + \tfrac{1}{4}C_n$$

The value of $\overline{C_0}$ is estimated by examining the admittance of other piezoelectric branches. Around the $n$th overtone frequency, the higher branches look capacitive and the lower inductive.

Figure 2.8c shows the thickness longitudinal vibration of a plate with L-effect coupling for an X-cut plate under constant voltage excitation. The transverse vibration may be excited though, but this resonant frequency is assumed to be far below the resonance of the thickness vibration.

In the L-effect mode of vibration, the first anti-resonance corresponds to the sounds half-wavelength and each harmonic resonant frequency is equal to an odd multiple of this frequency.

Accordingly, in this transducer, the anti-resonance reflects, as it were, the real mechanical situation for the sound. This is in contrast to the situation in the transducer with T-effect coupling.

## 2.9 Ferroelectricity in Bone

Considering the composite material of bone, interest has arisen in investigating the origin of the mechano-electrical phenomena. However, its physiological consequences remain to be fully understood. The strain-dependent potentials are a consequence of its ferroelectric nature and have an important role in normal stimulated bone growth and fracture treatment.

Pyroelectric materials produce electric charges on the surface during uniform heating, and an additional structural requirement is that the crystal should have a single polar axis. It should also be present in the absence of external electric fields. The discovery of pyroelectric properties in bone (Lang, 1966; Athenstaedt, 1969) has further focused our understanding of the role of piezoelectricity in a wider context. It is therefore important to investigate the ferroelectric properties of bone. The ferroelectric materials are spontaneously polarized dielectric materials. The crystal contains several domains and in each of these the polarization has a specific direction. The direction varies from one domain to another, which can be changed by the application of small magnitude of external electric fields. The structural implications are that bone possess a domain structure with respect to charge distributions. The charged dipoles are arranged in domains. Their mutual orientation within the domains can be changed by the application of an external electric field. Applied mechanical forces can also produce a perturbation and can lead to a non-equilibrium situation in which the distortion of the dipoles leads to the generation of a net electric field. It is the deformation of the domain structure that is likely to be of greater importance than the magnitude of the force. This further emphasizes the significance of strain rather than stress. Support for an internal arrangement of dipoles into a domain structure also comes from observations that bone can behave as an electret and that these can be effected by physical methods (Andrabi and Behari, 1981). These authors have emphasized the importance of persistent charge storage effects. Both electret and ferroelectric behavior both require a grouping of dipoles in domains that can be affected by applied electric fields. A micro-domain can be energized separately within an organized layer, for example, in a membrane. However, their role in controlling biological processes in bone and in biological systems, in general, remains unclear.

Ferroelectric studies show (Hastings, El Messiery and Rokowski, 1981) that the *D–E* relationship of dry cortical bone is characterized by a hysteresis loop comparable to that of a weakly ferroelectric material (Figure 2.9a). This is repeatable with different samples irrespective of age or site and orientation in the tibia. Therefore, ferroelectricity is an intrinsic property of dry bone and confirms its piezoelectric character. Such results from moist bone are often unreliable because of the presence of leakage currents.

The presence and hence role of ferroelectricity in moist bone and *in vivo* is, and probably will always remain, within the realm of speculation. The dimensions of the collagen matrix will change with extent of hydration and it may possibly alter intrinsic properties related to basic structural elements. It has been so-reported for mechanical properties. A dry matrix may, however, differ from its hydrated counterpart and the difference remains to be explained from a purely physical point of view.

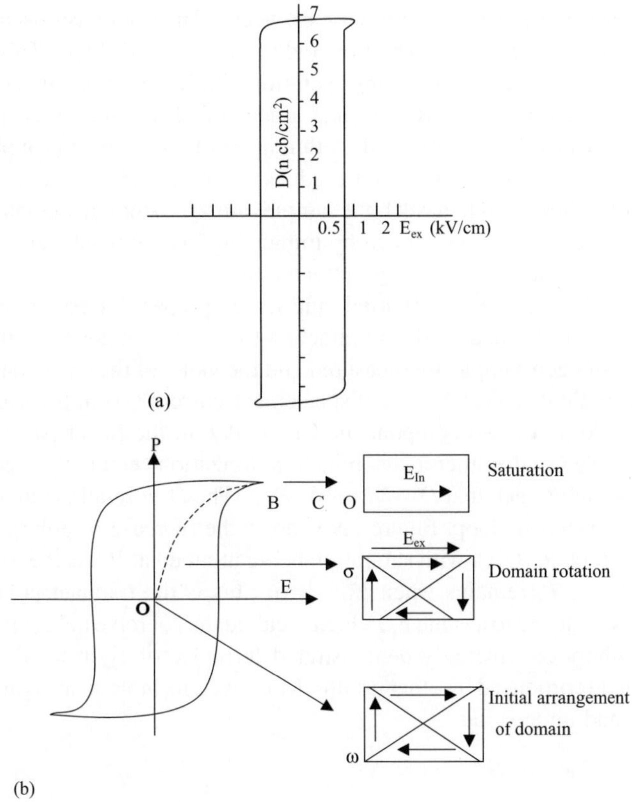


**Figure 2.9** (a) Voltage displacement response of a known ferroelectric crystal; (b) type of hysteresis loop for ferroelectric materials, showing the relationship between externally applied field $E$ and internal polarization $P$ (Hastings and Mahmud, 1991). (Reproduced with kind permission from G.W. Hastings and F.A. Mahmud, The electromechanical properties of fluid-filled bone: a new dimension, *Journal of Materials Science: Materials in Medicine*, **2**(2), 118–124 (p.121, Figure 2) © 1991, Springer Science + Business Media)

Hence, local domains may have a very important role in monitoring the constant mechanical strains to which bones are subjected and in controlling or triggering the transport of calcium and other ions at the cellular level. Application of small electric field to effect physical behavior (PN junction, photoelectric effect, Hall effect, etc.) offers support to these findings. Once a transport mechanism has set in, there are various built-in amplification systems that may increase the effectiveness of the process. The role of such processes in bone fracture healing and osteoporosis appears to be important.

## 2.9.1 Experimental Details

Hastings, El Messiery and Rokowski (1981) and Andrabi and Behari (1981) used normal human bone obtained from tibiae following amputation and sacrificing animals respectively.

The samples are dried at room temperature for six weeks. The test pieces used in the different experiments are cut at the desired orientation and to the required shape. This permits a very light loading of the sample during the cutting operation. The bone surface at the site of electrode attachment is polished with progressively finer grades of glass paper, washed with distilled water, acetone, water again and then dried. A thin layer of silver paint is applied and copper connecting leads are attached using the paint as adhesive. A standard procedure is used to attach gauges to bone with adhesive. When wet bone samples are to be studied, the gauges are attached first to dry bone, which is then rehydrated by immersing in deionized water. Test pieces are stored in vacuo for two days prior to experimentation.

To measure ferroelectricity, vacuum dried samples are prepared in rectangular form (0.5 to 8 mm thick) and electrodes and leads are attached to the two opposite, parallel faces. The thickness and area of each sample are measured and the value of the capacitance determined. The applied electric field ($E$) produces a displacement current ($D$) in the sample and this is supposed to be made up of two components. One is due to the linear part of the dielectric constant of the sample ($\varepsilon_r$) the other to a nonlinear polarization vector ($P_{ferr}$) corresponding to the ferroelectric domains such that $D = \varepsilon_r E + 4\pi P_{ferr}$. The $D$–$E$ relationship for ferroelectric materials shows a hysteresis loop. Figure 2.9b shows the increase of polarization occurring when domains are rotated (OA), with saturation being attained at B. As the external field $E$ is reduced a polarization $P_r$ remains when $E$ is zero. This is the remnant polarization and to remove it $E$ must be reduced to a value $E_c$, which is called the coercive field (Figure 2.9a and b).

While these authors convincingly demonstrated ferroelectricity in bone, it has not been explicitly examined/reproduced by other groups. Moreover, its biological significance remains to be established and understood.

## 2.10 Two-Phase Mineral-Filled Plastic Composites

### *2.10.1 Material Properties*

The composition of bone tissue is extremely complex compared to most engineering composites. The complex nature of bone microstructure suggests the need for mechanical models. An understanding at the molecular level may also provide useful information for subsequent histologic based models. The organization of bone within a Haversian system consists of a central canal surrounded by concentric lamella (Figure 1.8a). Lamellae can be observed with the light microscope. Bone is elastically isotropic and its modules in the longitudinal axis are greater than in the directions normal to it (Reilly and Burnstein, 1975). This has effect of increasing somewhat the stress concentrating effect of voids if the load on the bone is directed primarily along its length (Green and Taylor, 1945; Kaltakci, 1995).

In nature, the strongest and most efficient materials, bone, wood and fiber glass, have two phases (Katz, 1971). They are made up of two different substances with contrasting properties of strength and elasticity. Materially, the strength of any tissue is derivable from two sources, its elastic and plastic components. For a calcified tissue, elasticity is related to the mineral phase of the tissue and plasticity to the matrix (Burstein *et al.*, 1977; Currey, 1989; Jepsen *et al.*, 1992). Knowledge of the orientation of the hydroxyapatite (HAP) crystallites with respect to the collagen fibers is important in predicting the orientation of the principal axis of stress in bone tissue. These are (i) compound bar, (ii) pre-stressed material and (iii) a two-phase material.

To explain the mechanical properties of bone, Currey (1964) suggested a two-phase composite. This initial work applied the simple rule of mixtures wherein the modulus of the composite is calculated by averaging the proportional contributions of each phase. Reported data of Lees and Davidson (1977), Lees (1979) suggest that bony tissues considered as a two-phase composite systems of HAP and COL (collagen) are described as a family of Reuss solids, where the soft matrix (COL) become stiffer as the hard filler (mineral, HAP) is added until a maximum elastic modulus is reached (Figure 2.10). It is postulated that an increased density of cross linking is associated with increased HAP content and that HAP crystallites provide rigid bases for shortened links to stiffen the composite by reversible enzyme-directed processes (Figure 2.10). These investigations open up the possibility of obtaining a desired set of bone material properties by mixing these constituents in different proportions.

Currey (1964), Bonfield and Li (1967) and Katz (1971) have tried to model bone as a simple composite material using the linear rule of mixtures from a strictly volumetric viewpoint. If one assumes that the stresses are the same in the materials, one has the Reuss model and the longitudinal modulus of the composite as:

$$E_{\mathrm{R}} = \frac{E_{\mathrm{f}} E_{\mathrm{m}}}{E_{\mathrm{f}} V_{\mathrm{m}} + E_{\mathrm{m}} V_{\mathrm{f}}} \tag{2.15}$$

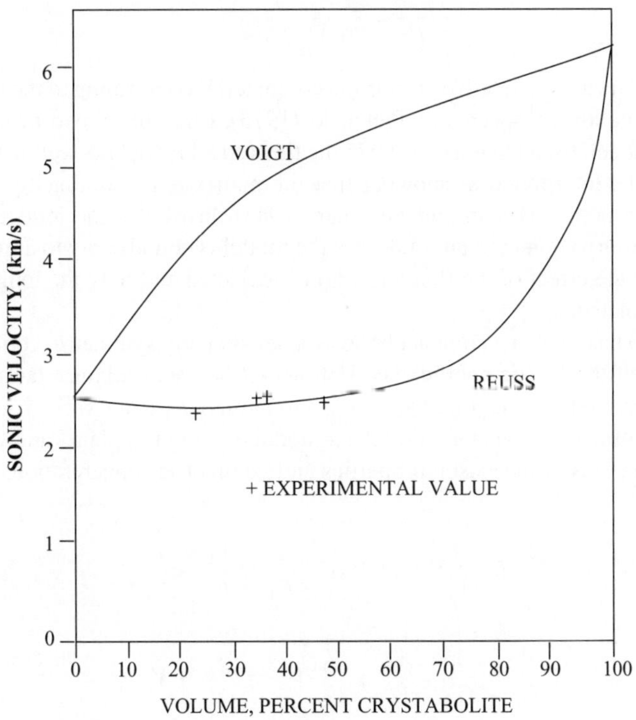


**Figure 2.10** Epoxy-crystabolite composite system (Lees and Davidson, 1977). (Reproduced with permission from S. Lees and C. Davidson. A theory relating sonic velocity to mineral content in bone. In: Linzer, M, ed. *Ultrasonic Tissue Characterization* Vol. II, Special Publication 525. Government Printing Office, Washington, DC © 1979)

where $V_m$ and $V_f$ are, respectively, the volumetric concentrations of matrix (collagen) and fibers (hydroxyapatite). If, on the other hand, one makes the strains equal in the two materials, one has the Voight model, for which:

$$E_V = E_f V_f + E_m V_m \tag{2.16}$$

The density of the composite is given by:

$$P = p_m \nu_m + p_f \nu_f$$

The Reuss model gives a lower bound to the modulus, the Voight model an upper bound. For instance, assuming $V_f = 0.41$, $V_m = 0.59$, $E_f = 11.5 \times 10^{11}$ dyne cm$^{-2}$ (1 dyne $= 10^{-5}$ N) and $E_m = 0.124 \times 10^{11}$ dyne cm$^{-2}$, one has $E_R = 0.209 \times 10^{11}$ dyne cm$^{-2}$ and $E_V = 4.788 \times 10^{11}$ dyne cm$^{-2}$. The range of experimentally determined values of elastic modulus of bone lies between these limits.

Hirsch (1962) has proposed a model that combines the Reuss and Voight models and expresses the elastic modulus as:

$$\frac{1}{E_R} = \frac{\chi}{E_V} + \frac{(1-\chi)}{E_R} \tag{2.17}$$

where $\chi$ and $(1 - \chi)$ are the relative proportions of material conforming to the upper and lower boundaries of solutions, respectively. Piekarski (1973), using his experimental data and the Hirsch (1962) model, found that if $\chi = 0.925$ the theoretical model fits well with experimental data. This may be interpreted as showing that the matrix is predominantly an equal strain system. Another model (Hashin and Strikman, 1963) distributes the mineral and collagen phases in an arbitrary geometry, but improves the modulus bounds only to a limited extent. In all these models the effect of the Poisson ratio is neglected and only the longitudinal elastic modulus is calculated.

Turhan (1977) has treated a compact bone as a hexagonally symmetric composite material and calculated all the elastic coefficients. Following the usual mixture laws for composite materials he expressed $E_L$ (longitudinal Young's modulus), $E_T$ (transverse Young's modulus), $G_L$ (shear modulus in the 2–3 plane), $G_T$ (shear modulus in the 1–2 plane) and $V_L$ (longitudinal Poisson ratio) in terms of the elastic properties and volumetric concentrations of constituents as:

$$E_L = V_f E_f + V_m E_m, \frac{1}{E_T} = \frac{V_f}{E_f} + \frac{V_m}{E_m} \tag{2.18}$$

$$\frac{1}{G_L} = \frac{1}{G_T} = \frac{V_f}{G_f} + \frac{V_m}{G_m} V_L = V_f \nu_f + V_m \nu_m \tag{2.19}$$

However, notably, the longitudinal and transverse moduli correspond to elastic moduli in Voigt and Reuss models, respectively.

To explain bone structure, Currey (1969) has suggested that the microcrystallites are fused together to make the hard material more fibrous, which is contrary to the findings of other workers (Ascenzi, Bonucci and Bocciarelli, 1965; Glimcher and Krane, 1968). Bone contains various other components in addition to the two major constituents, though in small proportions. These include ground substances, namely mucopolysaccharides, and various trace components, for example, citrate, Na, Mg, K, Cl, Zn, Sr, Pb, Cu and Ba. Some of these function through involvement in metabolism and so there is a need to understand the mechanisms of the physical and biological changes they produce. For example, vanadium and strontium cause a marked stimulation of mineralization in bone while calcium changes the electrical activity in membrane by forming insoluble salts (Schroeder, 1960). Absorbed lead exchanges for calcium in bone mineral (Neuman and Newman, 1953). Furthermore, the observations that trace elements in bone may be associated with either the organic or the inorganic components and influence the electrical or physical properties of bone demand further investigations of the mechanical properties of bone doped with varying proportions of additives. Some studies of this type have been undertaken by Singh and Behari (1984a).

The longitudinal sonic velocity has two values: one for the Voigt and other for the Reuss model. The Reuss calculations shown by Hill (1952) provide a rigorous lower bound to the elastic modulus of a mixture. When the modules of the components are almost equal there is very little difference between the models in predicting the correct behavior and the average of the two may be a good approximation.

In a two-component bone composite material mineral phase (mainly hydroxyapatite) confers the strength (Zioupos 2001) and stiffness (Currey 1988) and the organic matrix (mostly type I collagen) primarily influences the toughness of bone (Wang *et al.* 2001, Zioupos 2001, Zioupos *et al.* 1999). Mineral and collagen each contribute to the bones competency and to microarchitecture (e.g. porosity and trabecular connectivity) (Figure 2.11). They also contribte to the microstructure (e.g. the curvature of diaphysis) and thickness of the cortical shell besides in vivo microdamage (e.g. microcracks and diffuse cracks) and their interaction with water is responsible to effect the mechanical behavior of bone. Thus bone may also be considered a fluid structure in which the distribution of water affects the mechanical properties of bone (Nyman *et al.* 2006). Currey (1969) suggested that the crystallites are much longer than observed in order to attribute a fibrous mineral structure to bone to account for the observed elastic and strength properties. Figure (1.4C) indicates a representation of mineralized bone collagen showing crystallites embedding the cross links.

This two-component rule is useful in estimating the modulus of unidirectional continuous fiber composites but is not considered a proper model for bone. Katz (1977, 1980, 1981) has developed a model of Haversian bone such that individual osteons are treated as circular coherent cylindrical entities packed in a hexagonal manner. The osteons are built up of concentric lamellae containing ground substance and hydroxyapatite. One of the objectives in this model is to account for the anisotropy in the modulus of bone. The smallest elements that are treated are those that are observed microscopically. There are, however, interactions at the molecular level that are important in determining the mechanical properties of bone. These interactions would take place within each lamella. One approach is to apply composite theory to the smallest constituent microstructural elements that make up bone and treat them as a filled polymeric composite. The structure–function relationships in bone are addressed at essentially the level of the extracellular matrix. The mechanical behavior of a composite at this level is a function of the properties of the individual phases (matrix and filler), their interaction and adhesion. Other key parameters are crystal size, filler aspect ratio and volume fraction of

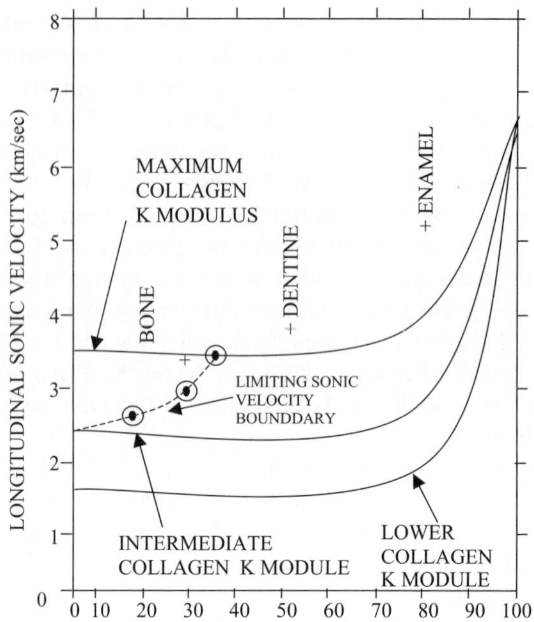


**Figure 2.11** Collagen–HAP composite system (Lees and Davidson, 1977). (Reproduced with permission from S. Lees and C. Davidson. A theory relating sonic velocity to mineral content in bone. In: Linzer, M, ed. *Ultrasonic Tissue Characterization* Vol. II, Special Publication 525. Government Printing Office, Washington, DC © 1979)

mineralization. Such information may play an important role in describing the mechanism of bone fracture and fatigue.

Williams and Breger (1974, 1975) have extended the linear, phenomenological theory of piezoelectricity, in which the polarization is assumed to be proportional not only to the stress but also to the gradient of the stress. With fourth rank tensor formalism it is possible to derive an expression that predicts the observed symmetry properties of the transverse voltage and its dependence on electrode location for a bent cantilever beam.

Williams and Breger (1975) demonstrated that piezoelectrically homogeneous specimens of quartz and tourmaline do not model bone or tendon. It is shown that specially fabricated poled ceramics can do so successfully and thus places the electromechanical response of dry bone and tendon on a solid theoretical foundation. When subjected to fluid shear stress, endothelial cells align parallel to the direction of flow (Dartsch and Betz, 1989). This response to mechanical stimulation is accompanied by an increase in filamentous actin (F-actin) stress fibers, which also align in the direction of flow (Dartsch and Betz, 1989). Interestingly, when endothelial cells, fibroblasts or osteoblasts are grown on flexible, silicone-bottomed culture plates and subjected to cyclic biaxial deformation the cells align perpendicular to the major vector of strain (Buckley *et al.*, 1988; Dartsch and Betz, 1989). This realignment is also accompanied by an increase in cytoskeletal stress fibers aligned in the same direction as the cell. These observations suggest that different types of mechanical strain produce different cellular responses.

## *2.10.2 Bone Mechanical Properties*

### 2.10.2.1 Theoretical Considerations

Figure 2.12 depicts an idealized composite stress–strain relationship under tension. At low levels of tensile strain, stress is proportional to the strain and the constant of proportionality is known as Young's modulus. Lower levels of strain represent the elastic region before any irreversible damage is induced. At higher levels of strain the yield point is reached, which represents the onset of plastic deformation. In bone, it occurs at a strain level of approximately 1% (Crowninshield and Pope, 1974). For a filled polymeric composite, interfacial debonding represents the primary mode of failure (Trachte and Dibenedetto, 1971). This is most likely an important mechanism in bone, although failure can take place also between larger elements, such as those at the lamella level. The critical strain release rate or energy required to extend a crack exhibits a positive correlation with mineralization (Wright and Hayes, 1977). This suggest that increased levels of particle–matrix bonding results in increased fracture energy. Fracture toughness is increased by increasing the filler concentration or by improved adhesion (Trachte and Dibenedetto, 1971; Jancar and Kucera, 1990).

For the purpose of studying the mechanical properties of bone, the crystalline and amorphous mineral phases may be assumed to be lumped together and the amorphous organic phase, which is found in small amounts in the bone, may also be lumped together with collagen. With these assumptions and classifications presented in the work by Welch (1970) at the ultrastructural level, compact bone may be considered to consist of hydroxyapatites (minerals), collagen (organic) and liquid (pores). With the help of electron microscopy, it is revealed that the crystals are actually within the fibers. Differences in making further assumptions on the structure of bone has led investigators to suggest different analytical models for the mechanical behavior of compact bone. These are (i) the elastic composite model and (ii) the porous elastic solid. In the composite model, the organic phase and liquids (pores) are lumped together to

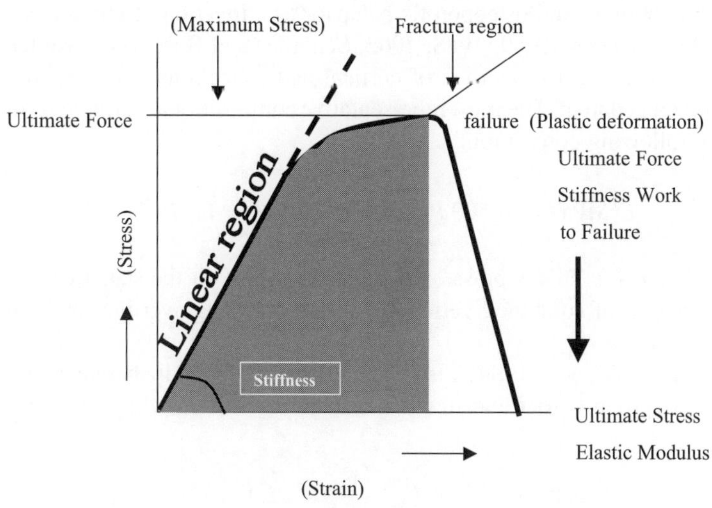


**Figure 2.12** Stress–strain relationship in a solid medium

form a matrix material of the composite reinforced by hydroxyapatite fibers. However, in the porous elastic solid model, the organic and mineral phases are assumed to be lumped together to form the basic material of elastic solids in which the pores are randomly distributed.

For ellipsoidal fillers, Eshelby (1957) calculated the elastic field for a dilute suspension. Chow (1977, 1978, 1980) subsequently generalized Eshelby's theory to estimate the five elastic constants of a filled composite at a finite volume concentration of filler particles. Chow's general theory reduces to Kerner's equations for spherical fillers. Particle shape is characterized by the aspect ratio, $\rho$, which is the ratio of the major to minor axes. For a composite model of bone the theory is very useful, as it addresses the principal relevant parameters such as filler dimension, interfacial debonding, elastic constants of the constituent phases and volume fraction of filler or mineralization.

The mechanical properties of bone depend upon composition and structure. However, composition is variable in living tissues. It changes in terms of the mechanical environment (stimulation), ageing, disease, nutrition and many other unaccounted factors. Efforts have been made to correlate mechanical properties with composition (Carter and Hayes, 1977; Carter and Spengler, 1978; Gibson, 1985; Goldstein, 1987; Rice, Cowin and Bowman, 1988; Martin, 1991). Vose and Kubala (1959) tried to establish a relation between composition and mechanical properties, thereby obtaining a correlation between ultimate binding strength and mineral content. Carter and Hayes (1977) found that elastic modules and the strength of trabecular and cortical bone are closely related to the cube and the sequence of the apparent wet bone density, respectively. Bone that is exposed to high mechanical strain will add bone tissue to reduce the strains to an acceptable level. Bone tissue subjected to high mechanical strains will increase the lamellar bone formation rate (Turner *et al.*, 1994; Chow, Jagger and Chambers, 1993). Rubin, Bain and Mcleod (1990) have observed that high frequency (15–30 Hz) strain peaks are functionally important because bone adapts more easily to strains applied at these frequencies than at low frequencies (1–2 Hz) (Turner *et al.*, 1995). However, it has been shown that in humans the higher frequency harmonies of walking are greater than 15 Hz (Antonson and Mann, 1985). The mechanical properties of bone have significant clinical importance, especially in understanding fracture and osteoporotic behavior as a function of mineralization.

Several authors (Currey, 1969, 1988, 2002; Schaffler and Burr, 1988; Keller, 1994) have shown that the mechanical properties of cortical and cancellous bone depend on apparent density and mineral content. The most representative compositional variable is the ash density, which has the following correlation:

$$E(\text{MPa}) = 10500\, P_a^{2.57\pm0.04} \sigma_C(\text{MPa}) = 117 P_{\text{a}}^{1.93\pm0.04} \tag{2.20}$$

where $P_{\text{a}}$ is ash density. This expression explains over 96% of the statistical variation in the mechanical behavior of combined vertebral and femoral data over the range of ash density (0.03–1.22 g cc$^{-1}$).

Keyak, Lee and Skinner (1994) also studied the relationship between the mechanical properties and ash property for trabecular bone, obtaining the following expression ($r^2 = 0.92$):

$$\begin{aligned} E(\text{MPa}) &= 33900\, P_a^{2.2} \quad \text{if} \quad P_{\text{a}} \leq 0.27\ \text{g cm}^{-3} \\ \sigma_C(\text{MPa}) &= 137\, P_a^{1.88} \quad \text{if} \quad P_{\text{a}} \leq 0.317\ \text{g cm}^{-3} \end{aligned} \tag{2.21}$$

One limitation of these models is that they do not separate the influence of bone volume fraction from the ash fraction. Hernandez *et al.* (2001) express the apparent density as a function of the bone volume fraction [bone volume ($B_V$)/total volume ($T_V$)] from the ash fraction ($\alpha$):

$$P = \frac{B_V}{T_V} P_t = \frac{B_V}{T_V}(1.41 + 1.29\alpha) \tag{2.22}$$

where $P_t$ is the true density of the bone, which is linearly related to the ash fraction ($\alpha$). They determined the elastic modules and compression strength, independently of bone volume fraction and ash fraction, ($r^2 = 0.97$):

$$E(\text{MPa}) = 84370\left(\frac{B_V}{T_V}\right)^{2.58\pm0.02} \alpha^{2.74\pm0.13}$$

$$\sigma_C(\text{MPa}) = 749.33\left(\frac{B_V}{T_V}\right)^{1.92\pm0.02} \alpha^{2.79\pm0.09} \tag{2.23}$$

In these expressions, the exponent related to ash fraction is larger than that associated with bone volume fraction, suggesting that a change in mineral content will produce a larger change in mechanical properties.

Pietruszczak, Inglis and Pande (1999) have included the directional dependence of strength using the expression:

$$\sigma_C(\vec{e}) = \sigma_{Co}\left(\frac{1 - \vec{n}(e)}{1 - \vec{n}_0}\right)Y \tag{2.24}$$

where $n_0$, $\sigma_{Co}$ are reference properties (the first is the average porosity, and the second the corresponding strength for the level of porosity), $Y$ is a constant in the interval (1,2) and $n(e)$ is the directional porosity that can be approximated as:

$$\vec{n}(e) \approx n(1 + \boldsymbol{or}_{ij} l_i l_j) \tag{2.25}$$

where $n$ is the average porosity, $l_i$ the orientation in relation to a fixed Cartesian coordinate system and $\boldsymbol{or}_{ij}$ is defined as a symmetric traceless tensor, which describes the distribution of voids.

Interrelationships of trabecular bone electrical and dielectric properties with mechanical properties and density need to be explicitly examined and established. Since the trabecular bone is an inhomogeneous composite material, containing porous bony matrix filled with fluid and being structurally and mechanically anisotropic, its electrical characteristics are complex (Chakkalakal and Johnson, 1981). Several *in vitro* studies have been carried out to clarify the electrical and dielectric properties of either compact, trabecular or whole bone (Freeman, 1967;Cochran, Pawluk and Bassett, 1968; Behari, Guha and Agarwal, 1974; Chakkalakal and Johnson, 1981; Singh and Saha, 1987; Saha and Williams, 1989). To separate the effect of the fluid phase from that of the solid matrix, studies on dry bone and fluid saturated tissue (Kosterich, Foster and Pollack, 1983) have been made. These studies have shown that the high frequency limit of relative permittivity depends on the water content: the main source of dielectric dispersion in long bone and in biological tissues in general. Conflicting results exist on the interrelationships between bone density and electrical or dielectric properties. In contrast to the findings of De Mercato and Garcia-Sanchez (1988, 1992), Williams and Saha (1996) found that there is a statistically significant and positive, linear correlation between the bone density and specific capacitance, which is strongly and positively related to relative permittivity. Investigation of the properties of bone in three orthogonal directions revealed a strong electrical and dielectric anisotropy of compact bone (Reddy and Saha, 1984) and a transversely isotropic nature of cancellous bone (Saha and Williams, 1989). In addition, *in vitro* studies and a few *in vivo* studies exist (Tzukert *et al.*, 1983; Rubinnacci and Brigatti, 1984; Skinner *et al.*, 2001). In an *in vitro* study, the bone healing and corticotomy have been reported to be successfully followed by measuring electrical impedance (Skinner *et al.*, 2001).

Since most bones are composed of compact and trabecular tissue, knowledge of the electrical and dielectric properties of both components is desirable. Several investigators (Sierpowska *et al.*, 2003; Williams and Saha, 1996) have reported that bone relative permittivity and phase angle show a strong linear correlation with volumetric bone mineral density. De Mercato and Garcia-Sanchez (1988) found that the relative permittivity decreased with increasing compactness of bovine bone, that is, they proposed a negative correlation between the bone mineral density and relative permittivity.

The absolute values of electric and dielectric parameters reported by Sierpowska *et al.* (2003) agree with those of Saha and Williams (1989) and Gabriel, Gabriel and Corthout (1996). Both absolute values of electrical and dielectric parameters, as well as the correlations between the mechanical properties or BMD with relative permittivity, depend significantly on the frequency. At frequencies over 6 kHz, the relative permittivity predicted well both the mechanical properties and the BMD value. This variation is attributed to site-dependent variation of electrical and dielectric parameters. The variation may be related to difference in tissue structure (e.g., size and orientation of trabeculae) as well as to differences in spatial distribution and density of the liquid phase (Kosterich, Foster and Pollack, 1983; De Mercato and Garcia-Sanchez, 1988, 1992).

Mineral and collagen each contribute to bone strength, while for other parameters, namely, microarchitecture (e.g., porosity and trabecular connectively), macrostructure (e.g., curvature of diaphysis and thickness of cortical shell), and *in vivo* microdamage (e.g., microcracks and diffuse cracks), their interaction with water is equally important to the mechanical behavior of bone. Thus, bone is also a fluid embedded material in which the distribution of water affects its mechanical properties. Some studies have demonstrated that the stiffness, tensile strength and hardness increases, whereas the strain at fracture and energy to fracture decreases, following the dehydration of bone tissues (Dempster and Liddicoat, 1952; Evans, 1973; Evans and Lebow, 1951; Sedlin and Hirsch, 1966; Smith and Walmsley, 1959; Yamada and Evans, 1970).

Reduced energy to fracture has also been observed for dehydrated dentine (Jameson, Hood and Tidmarsh, 1993). In addition, as trabecular bone loses water, its buckling behavior changes from ductile to brittle. The effects of drying on bone properties are well documented; however, the effect of hydration is less well known. Water is present in the microscopic pores, which increase in number and size with age, but also exists within the extracellular matrix of bone tissues. The distribution of water in bone appears to change throughout life. It has been reported that the distribution of water in bone tissues decreases with skeletal growth (Jonsson, Ranta and Stromberg, 1985) and with progressive mineralization (Robinson, 1975, 1979). The observation that mineral content increases with age, tapering at 60 years (Mueller, Trias and Ray, 1966; Timmins and Wall, 1977), implies that the amount of water in the tissue is likely to be reduced in the elderly skeleton. Furthermore, non-enzymatic, glycation induced collagen cross links increase with age (Wang *et al.*, 2002) and may decrease water interaction with collagen as seen in connective tissues (Kopp, Bonnet and Renou, 1989). Understanding the role of water distribution in the mechanical behavior of bone may provide insight into the susceptibility of bone to fracture in the elderly population.

The distribution of water within the tissue of bone – the amount of water bound to collagen, to mineral and the mobile water in the vascular-lacunar-canalicular cavities – may control the bone's mechanical behavior:

$$\text{Water loss}(\%\,\text{volume}) = 100 \times \frac{W_{\text{et}} - W_{\text{dry}}}{W_{et} - W_{\text{sub}}} \tag{2.26}$$

where $W_{\text{et}}$ = mass of wet specimen before drying, $W_{\text{dry}}$ = mass of the specimen after drying and $W_{\text{sub}}$ is the mass of weight bone when submerged in water.

The mechanical properties of bone show variation, depending upon the physical conditions: for example, Young's modulus for wet bone (load in the direction of bone axis) varies in the range $0.71–2.65 \times 10^{11}$ dyne cm$^{-2}$ (Guzelsu and Demiray, 1979). If we can obtain an insight into the determinants of bone strength, then better methods to image, assess and treat patients with abnormal bone can be identified. Vose and Kubala (1959) have reported a rapid increase in breaking strength of bone from 6.5 kg mm$^{-2}$ for an 8% increase in mineral content. Currey (1969) has also reported a gain in Young's modulus by a factor of three as the ash content increases from 63 to 68%.

There have been difficulties in attaining a proper understanding of the mechanical properties of bone as the elastic modulus of collagen is not well known and there has been no distinction among the various types. Firstly, some investigators have used estimates for tendon fibers rather than bone collagen. Except for Mason (1966), sonic velocity measurement on kangaroo tail tendon fibers, the estimates of elastic modulus are probably based on low strain rate tests. Secondly, the relationship between the chemical structure of collagen and the properties of bone has not been fully explored. Collagen is considered to be a structureless homogeneous isotropic continuum and the short-range order of body tissue collagen is often ignored. Thirdly, there is much to be understood as to how the HAP crystallites are laid down *vis-à-vis* the collagen molecule. Fourthly, estimates of the elastic modulus of bone have been mostly obtained by standard stress–strain tensors. However, bone is viscoelastic in nature. Strain rates and the magnitude of the strain in conventional stress–strain testing causes hysteresis and often permanent distortion. In ultrasonic velocity measurements the amplitude of displacement is of the order of an ångström and the corresponding period is much less than that of any relaxation time of the medium (μs).

Studies on the dynamic characteristics of bone have been aimed at elucidating some of the basic viscoelastic properties of bone and also at developing non-invasive techniques for diagnosing various diseases of bone (e.g., osteoporosis, fractures). These investigations can be divided into three categories: resonance, ultrasound and impact studies. Investigators studying the resonance properties of bone (Saha, 1982; Selle and Jurist, 1966; Jurist, 1970a; Jurist, 1972; Campbell and Jurist, 1971; Doherty, Bovill and Wilson, 1974; Jurist and Kianinan, 1973; Garner and Blackketter, 1975) have treated the long bones as a vibrating body that will resonate at certain frequencies that are related to the length, geometry and mass distribution of the bone. They have attempted to use the variation of mechanical properties of bone in certain disease states and thus variations in the resonance frequency and the impedance characteristics to obtain a measurement of such parameters for osteoporosis or the state of healing of fractures (Jurist, 1970b). However, in these methods, the vibration monitoring transducer is generally placed on thc skin. The vibration response thus measured is seriously affected by the properties of the soft tissue and by the pressure that is applied to the transducer itself (Lakes and Saha, 1977). Pelker and Saha (1983) have performed measurements on fresh and embalmed human long bones. These authors generated stress waves by the longitudinal impact of a steel ball on one end of the bone. The wave characteristics (velocity, attenuation and dispersion) were measured by using two bonded strain gauges (Figure 2.13). The authors claims to have obtained data of clinical significance.

Following Chow (1982), in a longitudinally oriented composite where the filling particles are firmly bonded to the matrix, Young's modulus is given by:

$$\frac{E_0}{E_{\mathrm{m}}} = 1 + \frac{(K_{\mathrm{r}}/K_{\mathrm{m}} - 1)G_1 + 2(\mu_{\mathrm{r}}/\mu_{\mathrm{m}} - 1)K_1}{2K_1G_3 + G_1K_3}\phi \tag{2.27}$$

where $E_0$ is the longitudinal modulus of the composite, $E_{\mathrm{m}}$ is the matrix modulus, $K$ is the bulk modulus, $\mu$ is the shear modulus and $\phi$ is the volume fraction of filler. The subscripts

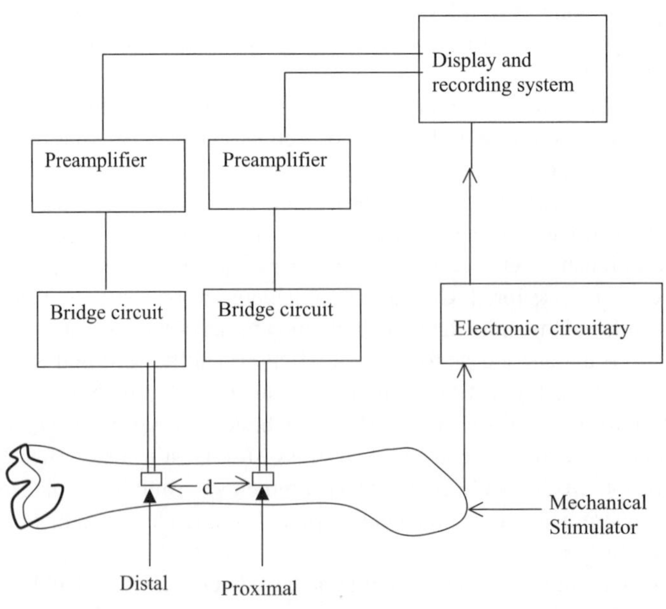


**Figure 2.13** Experimental set up for demonstration of stress wave propagation in bone (*in vitro*)

"f" and "m" refer to the filler and matrix, respectively. $K_i$ and $G_i$ are functions of $k$, $\rho$, $\phi$ and the Poisson ratio of the matrix. The filling particles are aligned along the long axis but are randomly distributed and thus are transversely isotropic. The experimental data of Reilly and Burnstein (1975) also support a transversely isotropic bone model. Because the composite strain is low with respect to the matrix phase, it does not carry much tensile load and transfers primarily shear stress to the high modulus filler particles. This can also be viewed as a function of the difference of modulus between the organic and inorganic phases.

The model can be further extended to predict the ultimate tensile strength (Chow, 1982). The ratio of the maximum stress within the composite to the applied stress is denoted by the stress concentration factor. If this local stress is sufficient, it may lead to evolution of a defect and subsequent failure of the composite. For aligned reinforcing particles, the stress concentration factor is highest at the ends of the particles, which is where an adhesion failure may take place. For a crack to form, the strain energy must exceed the energy of debonding (that required to create new surfaces of matrix and filler). Thus:

$$W \geq \gamma A$$

where $\gamma$ is the work of adhesion and $A$ is the surface created. From calculations of the strain energy required for crack formation, Chow (1982) derived a critical particle size, $d_c$. Inclusions larger than $d_c$ will provide flaws, or stress concentrations greater than the inherent flaw size, and reduce the ultimate strain. The critical size is given as:

$$d_c = 12\gamma / E_m e_m^2 [E_0] \tag{2.28}$$

where $e_m$ is the ultimate strain of the matrix, and $[E_0]$ is the effective Young's modulus of the composite:

$$[E_0] = \frac{E_0 - E_m}{E_m \phi} \tag{2.29}$$

The ultimate tensile strength at the point can then be calculated and is given by Chow (1982) as:

$$\sigma_u = [(12\gamma E_m / d)(1/[E_0] + \phi)]^{1/2} \tag{2.30}$$

The inverse square-root dependence of strength on particle size was reported earlier both theoretically (Nicholson, 1979) and experimentally (Leidner and Woodhams, 1974).

No adequate theory has yet been developed to model compression in a filled composite. While incorporating from Equations 2.26 and 2.28, data on the constituent phases of bone need to be defined. The properties of hydroxyapatite crystals have been reported by Katz and Ukraincik (1971). The bulk modulus is given as 89 GPa, the shear modulus as 44.5 GPa and the Poisson ratio as 0.22. Currey (1969) has estimated the true modulus of the collagenous to be approximately 1.47 GPa. This is a reasonable value, considering that the moduli of highly cross linked aromatic epoxy resin systems range between 2.5 and 3.5 GPa. The Poisson ratio for the organic matrix may be assumed to be 0.35, which is close to the value for polymeric materials.

To calculate the tensile strength, an estimate of the work of adhesion is needed. In the case of complete particle wetting, this should be close to the surface energy of hydroxyapatite. Kelly (1973) estimated the surface free energy of a solid ($\gamma$) as:

$$\gamma = Ea_0/10$$

where $E$ is Young's modulus and $a_0$ is the equilibrium separation between atomic planes. For hydroxyapatite, given that $a_0 = 0.94$ nm and $E = 114$ GPa (Katz and Ukraincik, 1971), $\gamma$ is estimated to be about $10^4$ erg cm$^{-2}$ (1 erg = $10^{-7}$ J).

The critical filler dimension in bone is estimated from Equation 2.28, which demands an estimate for the ultimate strain of the collagen matrix. While this is unknown, the calculation may be performed using a realistic range of 0.1–0.5 strain. Assuming a value of $10^4$ erg cm$^{-2}$ for the work of adhesion, the critical diameter is approximately 2–5 μm for above matrix strain. Thus, the effect of particle size is felt only when there are inclusions larger than this critical value. The measured sizes of hydroxyapatite particles are well below this value and they should, therefore, not degrade bone tensile strength.

In modeling the compact bone it is generally assumed that both matrix (collagen) and fibers (hydroxyapatite) are locally isotropic in their elastic properties. Thus, in constructing the composite model one has to know the Young's moduli and Poisson ratio of hydroxyapatite and collagen fibrils. For specificity one uses the subscripts "f" and "m" for the quantities associated with hydroxyapatite fibers and collagen matrix material, respectively. The elastic properties of hydroxyapatite have been determined by Gilmore and Katz (1968) and Katz and Ukraincik (1971). Table 2.2 lists the estimate of Currey (1964) for the Young's modulus of collagen and the computed values of Katz (1971) for the bulk and shear moduli ($K$ and $G$) of collagen, assuming that its Poisson ratio is 0.35.

### 2.10.2.2 Effect of Strain Rate

Figure 2.12 shows a relationship between stress and strain, yielding the Young's modulus. One of the important factors that contributes to the elastic properties of compact bone is the rate of strain (or the speed of loading). McElhaney and Byras (1965) and McElhaney (1966) were among the first to investigate the effects of strain rate on the elastic properties of bovine and human femora. It is generally observed that with increasing strain rate the Young's modulus

**Table 2.2** Elastic properties[a] of hydroxyapatite and collagen ($10^{11}$ dyne cm$^{-1}$)

| Material | $K$ | $G$ | $E$ | $\nu$ |
|---|---|---|---|---|
| Hydroxyapatite (Lang, 1969) (f) | 8.9 | 4.45 | 11.5 | 0.27 |
| Collagen (Wright and Hayes, 1976) (m) | 0.14 | 0.046 | 0.124 | 0.35 |

[a] $K$, $G$, $E$: bulk, shear and Young's moduli, respectively; $\nu$ is the Poisson ratio.

**Table 2.3** Effect of strain rate on elastic constants (tension) (Burstein, Reilly and Martens, 1976)

| | | Young's moduli ($10^{11}$ dyne cm$^{-1}$) | |
|---|---|---|---|
| Material | Strain rate ($s^{-1}$) | Wright and Hayes (1976) | Currey (1975) |
| Fresh bovine femur | 0.00 013 | — | 2.1 |
| | 0.00 053 | 1.77 | — |
| | 0.0053 | 1.99 | — |
| | 0.0055 | — | 2.13 |
| | 0.0183 | — | 2.47 |
| | 0.053 | 2.30 | — |
| | 0.174 | 2.23 | — |
| | 14.3 | 2.75 | — |
| | 59.2 | 2.38 | — |
| | 237 | 4.04 | — |

increases, while the Poisson ratio decreased (Table 2.3). Similar results have been found by other authors (Bird *et al.*, 1968; Bonfield and Clark, 1973; Panjabi, White and Southwick, 1973). Black and Korostoff (1973) tested cylindrical specimens cut from tibia at various strain rates. They observed that Young's modulus increases with increasing loading frequency (almost linearly), drops sharply by nearly 50% at about 200 Hz and then starts to increase at a more rapid rate. The presence of an anomalous reduction in modulus near 200 Hz suggests that there is a mechanical component in bone that has never been in non-physiological conditions. Jurist (1970) has made similar observations.

Crowninshield and Pope (1974) have shown that the elastic modulus of fresh bovine compact bone is comparatively less sensitive to strain rate when the specimen is tested under tension. However, performed similar tests on standardized longitudinal specimens of bovine compact bone at strain rates from $5.3 \times 10^{-4}$ to 237 s$^{-1}$ and obtained contrary results. They found, instead, a linear relationship (approximately) between Young's modulus and the logarithm of strain rate. This is somewhat similar to the results of McElhaney (1966). The results of Currey (1975) on fresh bovine femur also support the above findings. However, Currey introduced a linear regression line between Young's modulus, strain rate ($S$ in units of s$^{-1}$) and ash content ($C$%), for example:

$$E = (13.5 + 1.37C + 0.75S) \times 10^{10}\ \text{dyne cm}^{-2} \qquad (2.31)$$

The organic percentage of rat long bones increases during the growth period until the age of about 180 days and then tends to level off (Dickerson, 1962; Jankovich, 1970). In conformation with this the electrical conductivity of bone can be expected to increase with age during growth. Owing to increased bone density and mineralization, dry bone mineral is more conductive than dry collagen at room temperature (Becker, Bassett and Bachman, 1964).

Jankovich (1970) has studied the effects of age and whole body vibration on electrical conductivity in rat tibiae. This author exposed rats to 25 Hz vibration (12 h day$^{-1}$) at the normal

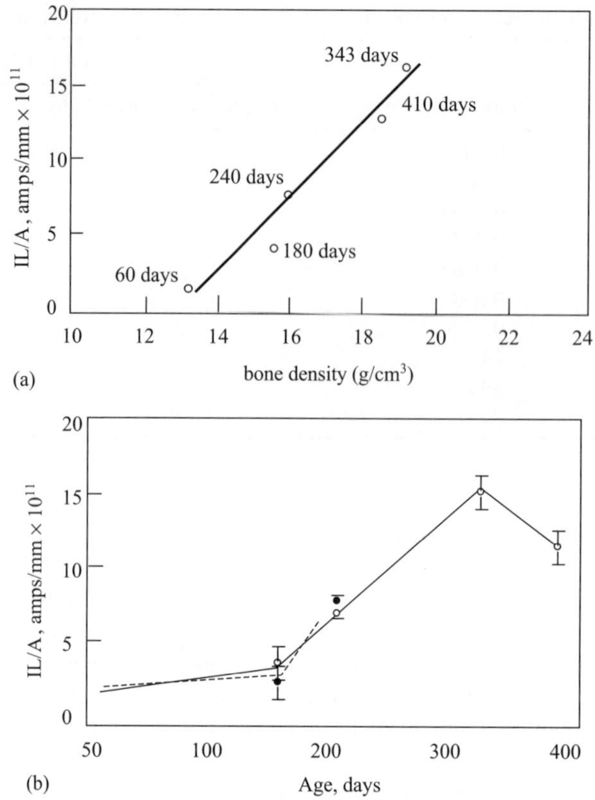


**Figure 2.14** (a) Average conductivity in rat tibiae as a function of calculated density change with age; *I*, *L* and *A* represent the current, length and area, respectively, of the cortical bone. (b) Conductivity as a function of age for vibrated (•) and control (○) rat tibiae; the labels on the *x*-axis are the same as in (a). The decrease in conductivity after 343 days is due to moving beyond the maturity age

level of acceleration due to gravity ($g$). Sixty-day-old rats were vibrated until the age of 180 and 240 days. The author concluded that there is an increase in conductivity until the age of 343 days (corresponding to maturity) (Figure 2.14a). This is largely because the inorganic portion of bone and bone density increases during the growth period. Conductivity, as well as bone growth, is unaffected in rat limbs following prolonged exposure to vibration (Swanson and Lafferty, 1972). There is a decrease in conductivity after the animal attains maturity (Figure 2.14b). These authors also reported an increase in average conductivity as a function of density change with age. Also, there is a linear relationship between average density and conductivity among the same group of animals.

Swanson and Lafferty (1972) further reported that conductivity is decreased following bone demineralization by immobilization. However, after release from the cast, normal values are recovered.

**Table 2.4** Effect of aging on elastic constants (Burstein *et al.*, 1976)

| Young's moduli ($10^{11}$ dyne cm$^{-1}$) | | | |
|---|---|---|---|
| Age (yr) | Femur (tension) | Femur (compression) | Tibia (tension) |
| 20–29 | 1.70 | 1.81 | 1.89 |
| 30–39 | 1.76 | 1.86 | 2.70 |
| 40–49 | 1.77 | 1.87 | 2.88 |
| 50–59 | 1.66 | 1.82 | 2.31 |

#### 2.10.2.3 Effect of Age

The elastic modulus of bone is also affected by the age of the subjects considered. Studies by Yamada (1973), Mather (1967) and Sedlin and Hirsch (1966) show that the Young's modulus of compact bone increases with the age of subject up to (approximately) the age of 30 but it decreases thereafter. Examining the data, after 30 years of age, one can approximately obtain a linear relationship between Young's modulus in the axial direction and age ($A$):

$$E = (1.78 - 0.6A) \times 10^{11} \text{dyne cm}^{-2} \tag{2.32}$$

Burstein, Reilly and Martens (1976) have studied the effects of aging on the elastic properties of human femur and tibia by grouping the subjects according to their ages as 20–29, 30–39, 40–49 and 50–59 and found results whose average value is given in Table 2.4. Interestingly, in all the cases examined the elastic moduli increase with age approximately up to 45 year and then starts to decline.

Expressing the elastic modulus of human femur as a function of bone density and measuring the latter as a function of age, Martin and Atkinson (1977) have shown that Young's modulus of bone increases up to the age of 50 and thereafter starts to decrease. In these studies only adult human femur is considered. The mechanical properties of children's wet bone have been studied by Vinz (1970, 1972) and Currey and Butler (1975). Vinz (1970, 1972) divided ages of his subjects into seven categories, more or less on a logarithmic scale: group I, 0–2 weeks, group II, 3–5 months; group III, 7–11 months; group IV, 1.5–2.2 years; group V, 4–13 years; group VI, 8–40 years and group VII, 70–85 years. He found that the elastic modulus increases from group I to group VI and fell in group VII. Using three-point bending test method Currey and Butler (1975) studied the Young's modulus of subjects between 2 and 48 years old and observed that it again increases with age.

#### 2.10.2.4 Effect of Mineral Content

An important factor that contributes to the values of elastic modulus is the mineral content of bone, which is also related to aging. Several authors (e.g., Ascenzi and Bonucci, 1968; Bonfield and Clark, 1973; and Currey, 1975) have found an increase in modulus of elasticity with increasing mineralization.

An inverse problem has been studied by Burstein *et al.* (1975), who showed that progressive surface decalcification of wet bovine Haversian cortical bone with dilute hydrochloric acid resulted in a progressive decrease in elastic modulus.

#### 2.10.2.5 Effect of Drying

Another important parameter that affects the elastic modulus of a material (bone) is the state of wetness of the specimen. Wertheim (1847) was the first to work out the effects of drying on the modulus of elasticity of human femora, finding that the elastic modulus of dry bone is larger than that of fresh bone. Rauber (1987) also studied a similar problem and observed that the elastic moduli of a longitudinal specimen cut from human femur are $2.1 \times 10^{11}$ (fresh) and $2.5 \times 10^{11}$ dyne cm$^{-2}$ (dry). The wetness of a specimen is particularly important in viscoelasticity and plasticity.

In addition to above-mentioned factors, other parameters also contribute to the elastic properties of material. Some of these are sex, race, species, density, porosity and histology (Evans, 1973). Moreover, contrary to the assumption of isotropy by (Petrov, 1975), the bone tissue is strongly anisotropic in its mechanical properties (Lang, 1970).

The physical structure of the wet bone tissue can be considered as the collection of two different phases, namely, the solid matrix phase and liquid phase (blood). During remodeling of wet bone tissue, the bone salts in the blood are transferred to the bone matrix and vice versa. Moreover, because of the piezoelectric properties of the matrix bone, the electrical potentials generated by external effects decay very rapidly through the attraction of monovalent ions (Na, Cl, $CO_3$) to charged surfaces (Bassett, 1971). These two distinct physical phenomena, taking place in the bone, suggest that the blood may be considered to consist of two species, which are the bone salts [$Ca_3(PO_4)_2$] and monovalent ions ($Na^+$, $Cl^-$, $HO_3{}^+$). In the light of the above, one can state that a suitable mechanical model for wet bone is a mixture of bone matrix, bone salts and conducting ions.

#### 2.10.2.6 Effect of Collagen Content and Collagen Crosslinking

The precise reasons for the age-related declines in strength and failure energy of whole bone is a matter of concern and hence of continuing interest. The role of collagen may be related either to the amount of collagen or to its molecular stability and crosslinking. Bailey and Knott (1999) have found that the age-related decline in collagen content in bone from the human iliac crest is nonlinearly correlated to maximum stress at failure (a strength criterion, $r^2 = 0.46$–$0.50$ in men and women, respectively) and to the modulus of elasticity ($r^2 = 0.36$–$0.48$ in men and women, respectively). This does not mean, of course, that changes in collagen content are necessarily responsible for the changes. Other changes with age, for example, the rate of bone turnover or the degree of mineralization, could affect both the collagen content and the mechanical properties of bone independently. In other locations, such as the femoral head and neck, no changes in collagen content are found (Bailey *et al.*, 1993).

It has been proposed that the interfibrillar pyrrole crosslinks have a greater influence on the bending strength of bone than the intrafibrillar pyridinoline crosslinks (Bailey and Knott, 1999). This is consistent with the decline in pyrrole crosslinks and the constant level of pyridinoline crosslinks in osteoporosis. It has also been speculated, however, that the intramolecular crosslinks are important for enhancement of bone toughness, whereas the intermolecular bonds may be less important for this purpose (Zioupos, Currey and Hamer, 1999). The maturity of the crosslinks may not be very important to bone's mechanical properties (Bailey and Knott, 1999; Zioupos, Currey and Hamer, 1999), although there is a decrease in the number of immature collagen crosslinks in newly formed collagen from osteoporotic subjects (Bailey and Knott, 1999). Wang *et al.* (2001) have suggested that the age-related degradation of collagen's mechanical properties is due to the increased concentration of pentosidine, a marker of non-enzymatic glycation of the collagen. Their work demonstrates that normal changes in collagen with age can have a significant effect on the properties of the bone.

The mineralization of bone and dentine is a process of filling the spaces in the microfibrils. About 60% of the mineral is crystalline, located within, on and between the collagen fibrils. The rest of the mineral (40%) is amorphous (Eanes, 1973). The crystalline matter is supposed to be hydroxyapatite. Glimcher and Krane (1968) have noted that the holes between TC units fill in with mineralized matter first and that the microcrystals in such tissue are about the same dimensions as holes. Gallop and Paz (1975) have noted that enzyme-directed covalent crosslinking is most often unidirectional and irreversible. These authors also reported that protein phosphorylations directed by specific kinases are reversed in many cases by specific phosphatases. In addition, certain proteins that contain senyl phosphate residues (as do bone and phosphorous) homeostate in reactions involving mineralization and the insertion and removal of phosphoryl groups. Veiss (1967) pointed out that there are several noncollagenous components that apparently play a part in crosslinkage.

## 2.11 Mechanical Properties of Cancellous Bone: Microscopic View

By taking a homogenous and isotropic approach within a small region, cancellous bone can be assumed to be a homogenous material that is not related to the structure of individual trabeculae. With this approach, only bone density is evaluated and the isotropy of the cancellous bone is ignored. The bone density can be defined in several ways. An apparent density has been used for many investigations, which is defined as the mass of bone tissue divided by the bulk volume of the test specimen including mineralized bone and bone marrow spaces (Goldstein, 1987; Kabel *et al.*, 1999). In histologic studies, bone density can be defined as the ratio of the bone area to the entire field area. It is sometimes called an area fraction. Radiographic density is also often used to define the bone density by calibration with the known hydroxyapatite (HA) density or normalization by density of the control area. The bone density can be defined as the volume of the bony tissue to the bulk volume of the test specimen including the bone marrow space.

Although the distinction between low-density cortical bone and high-density cancellous bone is somewhat arbitrary the stress–strain properties of the cancellous bone are markedly different from those of cortical bone. A linear relationship for bone density versus strength and elastic modulus has been reported for cortical bone. However, a nonlinear dependence of strength and elastic modulus of cancellous bone on bone density is observed (Goldstein, 1987). This nonlinearity has been well described as a power law by Carter and Hayes (1977). Their study reveals that the compressive strength of cancellous bone is related to the square of the apparent density and that the elastic modulus is also related to the apparent density by a power law function with an exponent range between 2 and 3. It is now accepted that the directionality and density of cancellous bone are related to the direction and magnitude of the load applied to the bone (Wolff, 1986). Anisotropy of the microstructure and mechanical properties of the cancellous bone are also well recognized (Goldstein, 1987).

The results of Nyman *et al.* (2006) indicate a relationship between the mechanical properties and water distribution in cortical bone. The results suggest that the removal of water from extracellular matrix, in addition to that removed from the void spaces within bone, affects the mechanical properties of bone.

## 2.12 Ultrasound and Bone Behavior

Piezoelectricity and converse piezoelectricity in bone having been established, it is now suggestive to go one step further: to the lack of a center of symmetry in bone and hence its ultrasonic behavior and frequency response. This suggests the use of bone as an ultrasonic transducer to establish it as an ultrasonic material. Over the last decade, ferroelectric ceramics have been accepted as the active material because of their high piezoelectric and electromechanical coupling coefficients. Unfortunately, some of their benefits are lost due to the mismatch between the ceramic and a biological medium. Greater incentive is thus given to exploit materials with lower acoustical impedance. Singh and Behari (1984a) have prepared various materials by mixing major bone components with various inorganic materials, yielding a group of biomaterials whose properties can be adjusted according to the demands of the situation. It is suggestive of the possibility that the bone matrix may behave as an ultrasonic transducer and respond to dynamic loading. Although bone composites have lower voltage constants and charge constants, the piezoelectric coupling coefficient is comparable and they have the added advantage of a lower dielectric constant and acoustical impedance (Singh and Behari, 1984a). It is, therefore, anticipated that bone composites will serve as an active material, as a good piezoelectric transducer, specifically at high frequency for miniature and wideband ultrasonic transmitters. This is discussed in more detail in Chapter 4.

### *2.12.1 Biochemical Coupling*

Immobilization produces a bone loss (Uhtoff and Jaworski, 1978) and is resorbed at the rate of $100\,\mu g\,day^{-1}$ (Johnson, 1964). This is sufficient to prove an important and interdependent relationship between the biomechanical activity of bone and the process of osteogenesis (bone formation) and, consequently, to its ultrasonic behavior.

Though the mechanism for the initial detection and conversion of mechanical force into a biochemical signal remains to be established several likely candidates have emerged. One possible transduction pathway is that the cells attach to the extracellular matrix through binding to membrane-spanning glycoproteins called integrins. Integrins in turn attach to the actin cytoskeleton through several actin-associated proteins such as vinculin, talin, tensin and $\alpha$-actinin (Pavalko *et al.*, 1991). The cytoskeleton has been shown to form a network, connecting the extracellular matrix with the nucleus and the cytoplasmic constituents of the cell (Bockholt and Burridge, 1993). Modeling and experimental evidence indicate that the cell generates an internal force through the cytoskeleton, which exerts a tension on the extracellular matrix (Sims, Karp and Ingber, 1992; Ingber *et al.*, 1993; Ingber, 1993). This internal tension, similar in concept to the architectural system of tensegrity (Fuller, 1975), produces forces on the adhesion sites of the cell in excess of those produced by exogenous mechanical stimuli (Ingber, 1993). Without attachment these internal forces would produce a spherical cell. The binding of integrins to the matrix proteins must, therefore, have to overcome the tensional forces of the cell, evoking changes in the cytoskeletal structure. Owing to the tension of the cytoskeleton, physical stimulus is transmitted to the nucleus and may alter the gene expression. Indeed, cellular attachment to the extracellular matrix plays an important role in the regulation of cellular proliferation, differentiation, morphogenesis and gene expression (Ingber, 1993; McClay and Ettensohn, 1987; Ingber and Folkman, 1989; Mooney *et al.*, 1992; Ben-Ze'ev *et al.*, 1988). Recruitment and/or differentiation of osteoblasts and osteoclasts similarly are modulated by cellular adhesion to the extracellular matrix (Suda, Takahashi and Martin, 1992; Tenenbaum, 1992) and attachment of the osteoblasts to specific extracellular matrix proteins appears to depend on the differentiated state of the cell (Manduca *et al.*, 1993; Vukicevic *et al.*, 1990). Experimental evidence suggests that modulation of cellular function by attachment to the extracellular matrix occurs through changes in the cytoskeleton, which, in turn, alters phenotypic expression. Chondrocytes assume a flattened morphology and do not express differentiation markers when grown in culture. However, when treated with cytochalasin B, a mold metabolite that induces repolymerization of filamentous actin, the cells assume a spherical shape and produce type IV collagen (Brown and Benya, 1988). Furthermore, type IV collagen production is stimulated even when cytochalasin B is given at concentrations that induce actin repolymerization but do not alter cell shape (Benya, Brown and Padilla, 1988). These observations suggest that it is the modulation of the cytoskeleton that mediates alterations in gene expression during cell adhesion and mechanical stimulation.

Wang *et al.* (2001) has attempted to address the difficulty of assessing the direct effect of collagen, independent of mineral, on the biomechanical properties of bone using a model of heat-induced collagen denaturation in human cadaver femurs. Collagen denaturation without a change in bone mineral significantly decreases bone' toughness and overall strength, while having a minimal effects on elastic modulus. This underscores the positive contribution that collagen makes to increasing the energy required for bone failure.

There have been attempts to focus on the importance of bone's strength and stiffness, both as a structure and as a tissue. These are the properties that are defined mechanically. Collagen does not contribute much to the strength and stiffness of the whole bone, or bone matrix. This is evident from the respective elastic moduli of collagen ($\sim$1.5 GPa) and hydroxyapatite (114 GPa) (Zioupos, Currey and Hamer, 1999). Collagen is, however, crucial in determining the amount of energy required to cause matrix failure (Wang *et al.*, 2001). This mechanical property –the failure energy of the material independent of its size or geometry – is defined by the area under the stress–strain curve and is called the modulus of toughness. This fracture mechanics property is called the stress intensity factor [$K_c$]. It is related to the modulus of toughness, but is a different property of bone. The modulus of toughness is reported as N m$^{-2}$ or J m$^{-3}$, whereas $K_c$ is presented in units of N m$^{-1}$ or J m$^{-2}$ (Martin, Burr and Sharkey, 1998; Norman, Vashisth and Burr, 1995; Turner and Burr, 2001). Collagen may be the primary toughening mechanism in bone, having greater effects on bone toughness (Wang *et al.*, 1998; Wang *et al.*, 2002) than on strength or stiffness (Zioupos, 2001). Indeed, when bone is subjected to ionizing radiation that specifically damages the collagen (producing crosslinking) bone toughness shows a decline (Currey *et al.*, 1997).

With advancing age, an inherent fragility in bone tissue allows it to be damaged more easily from cyclic loading (Courtney, Hayes and Gibson, 1996). Changes in the bone collagen with age and in osteoporosis are partly responsible for this increased fragility. In cyclic load tests performed *ex vivo*, cracks are initiated in specimens from older women (mean age $72 \pm 6$ years), but not in bone from younger women (mean age $26 \pm 5$ years), even with an equivalent 34% decline in elastic modulus in both groups (Courtney, Hayes and Gibson, 1996). This suggests that microdamage accumulation in bone stems from some inherent fragility in the tissue. In studies using a baboon model, the percentage of denatured collagen as compared with total collagen content is significantly related to failure energy and to the fracture toughness of the tissue (Wang, Athanasiou and Agrawal, 1998a; Wang *et al.*, 1998b). This means that collagen in bone is a primary arrestor of cracks, inhibiting their growth to critical level. This may be one explanation for the observation that aging has a more profound effect on the plastic deformation of bone than it has on elastic deformation (McCalden *et al.*, 1993).

It is generally thought that one component of the increased tissue fragility in older people is as a result of the slowing of bone turnover as one ages (Birkenhager-Frenkel and Nigg, 1993;Burstein, Reilly and Martens, 1976; Currey, 1979; Currey, Brear and Zioupos, 1996; Dickenson, Huton and Stott, 1981). It could also be related to molecular changes in either the inorganic (Grynpas, 1993; Paschalis *et al.*, 1997a, 1997b; Simmons, Pritzker and Grynpas, 1991) or organic (Bailey *et al.*, 1993; Danielsen *et al.*, 1994; Mehta, Oz and Antich, 1998; Oxlund, Mosekilde and Ortoft, 1996) fractions of the matrix (Batge *et al.*, 1992). The age-related reduction in post-yield deformation (i.e., less energy in the post-yield region required for fracture) occurs in the collagen moiety. As demonstrated by Wang *et al.* (2002), this may contribute equally to age-related fragility by decreasing overall fracture toughness (Currey, 1988; Norman, Nivargikar and Burr, 1996; Zioupos, 2001; Zioupos and Currey, 1998) and increasing risk of fracture, in men and women. The age-related reduction in the ability of bone to absorb energy prior to failure has clinical significance. This makes osteoporotic bone more prone to failure from any impact load, such as one resulting from a fall. Collagen changes that reduce bone's energy absorption capability, therefore, are possibly a primary factor increasing the risk of fracture in older women with low bone mass.

## 2.13 Traveling Wave Characteristics

To understood many facts concerning the impact loading of bones, Pelker and Saha (1983) measured some of the traveling wave characteristics (velocity, attenuation coefficient and dispersion) for a single compressive pulse. These have been correlated with the mineral density, porosity and cross-sectional area of the bone for normal bone and wet and dry embalmed bone. The wave velocity for fresh bone is 3377 m s$^{-1}$ and for embalmed bone 3228 m s$^{-1}$. Owing to the viscoelastic, porosity and non-uniform geometry of a long bone, any stress wave will attenuate as it propagates along the length of the bone (Guzelsu and Saha, 1981; Saha, 1982). The attenuation coefficient can be calculated from the peak strains measured at each location using the relationship (Kolsky, 1963):

$$\alpha = \frac{1}{X} \ln(E/E_0) \tag{2.33}$$

where $\alpha$ = attenuation coefficient, $E_0$ = the strain measured at the proximal strain gauge end, $E$ = the strain measured at the distal strain gauge site and $X$ = the path length between the strain gauges. These authors have further reported that for fresh bone the attenuation coefficient was $0.023 \pm 0.030\,\mathrm{m}^{-1}$ and for embalmed bone $0.058 \pm 0.035\,\mathrm{cm}^{-1}$.

The dispersion is determined by measuring the broadening of the wave pulse. The "broadening factor" is calculated from the ratio of the pulse width at the distal strain gauge station to the width at the proximal end:

$$\text{Dispersion} = \frac{(T_{1/2})_2}{(T_{1/2})_1} \tag{2.34}$$

where $(T^1/_2)_1$ = the pulse width at the half-maximum at the proximal strain gauge station and $(T^1/_2)_2$ = pulse width at half-maximum at the distal strain gauge station. For measured porosities ranging from 3.4 to 25.7% and equivalent mineral densities in the range 0.97–1.57 g $\mathrm{cm}^{-3}$. The dispersion for fresh bone is $1.07 \pm 0.12$ and for embalmed bone $0.97 \pm 0.07$. A correlation between the cross sectional area ($A$) and the attenuation coefficient ($\alpha$) is given by the linear equation:

$$A = 0.746 + 94.3\alpha$$

and the parabolic relationship:

$$A = 1.57 + 135\alpha - 908\alpha^2 \tag{2.35}$$

These imply that by measuring the attenuation of a traveling wave one could extrapolate the cross-sectional area of a bone. It is thus possible to estimate the cortical thinning or thickening that occur in a diseased bone. A theoretical analysis reveals that the wave velocity depends on the frequency and the diameter of the bone (Pelker and Saha, 1975; Pelker, 1976; Guzelsu and Saha, 1981). Another correlation is the parabolic relationship between the porosity of the bone ($F$) and the attenuation coefficient ($\alpha$):

$$\alpha = 0.012 - 0.010\,F + 0.000\,28\,F^2 \tag{2.36}$$

This suggests a manner of calculating the porosity of bone by measuring the attenuation of a stress wave. Measuring the velocity or pulse dispersion might also prove to be feasible techniques for assessing this, as suggested by their relationship with porosity and mineral density, respectively.

The technique, used to study the wave propagation characteristics, is a variance of the bar method. It is applicable to the study of wave propagation in a cylindrical bar when the pulse length is much longer than the radius of the bar and if the bar is not stressed beyond the proportional limit. If these criteria are met, then the wave will propagate down the bar without distortion. The maximum strain created here is well within the elastic limit for bone. For a wave velocity of $3 \times 10^3\,\mathrm{m\,s}^{-1}$, this corresponds to a length of 15 cm (50 μs pulse). This is significantly greater than the average radius of the specimens, which is approximately 1.5 cm.

A wave velocity of approximately $3200\,\mathrm{m\,s}^{-1}$ is not significantly different in fresh and embalmed bone. Also, bone drying does not significantly affect the velocity. Several authors (Ko, 1953; Yahoo, 1952) have reported an increase in the elastic modules by 20–30%. From the elementary theory of elasticity ($\nu = \sqrt{E/\sigma}$) the velocity should be increased 9–14% with drying. It can be imagined that such a change is masked by the biologic variation between specimens.

## 2.14 Viscoelasticity in Bone

Many observations show that bone is anisotropic and shows viscoelastic behavior (Katz, 1971, 1980; Currey, 1969; Abendschein and Hyatt, 1970). In general, bone behaves according to Hook's law for low values of strain. For higher strains, both dry and wet bone cease to follow Hook's law and behave as viscous materials.

Bone, which has a large amount of fibrous proteins, is probably held together to a significant extend by hydrogen bonding. The pressure generated by loading is time dependent because water moves. Therefore, bone exhibits viscoelastic behavior and so water moving in the pores certainly influences viscoelasticity, as has been observed when collagen is denatured (Yamashita *et al.*, 2002). Viscoelasticity is now recognized as an important property of bone (Currey, 1965; Lakes and Katz, 1979; Sasaki and Yoshikawa, 1993).

Bonfield and Li (1967) pointed out that the initial inelastic flow occurs in the collagen of bone. Since the bone mineral (hydroxyapatite) is a crystalline substance and not likely to be affected by room-temperature drying, it is assumed that the collagen content of bone is the substance that is primarily targeted by changes in moisture content. Therefore, any change in properties due to drying are attributable to the changes in collagen. Laird and Kingsbury (1973) found that bone specimens become more rigid as they lose their moisture content. Thus one would expect that the water content of bone would add significantly to its viscoelastic properties. This appears to be borne out by the measurements of attenuation or damping coefficients. However, the differences are not statistically significant. A higher damping coefficient for wet bone has been reported by Saha and Haines (1979).

A successful theoretical treatment of the viscoelasticity of bone requires a quantitative treatment. The relaxation modulus, $M(t)$, of a material is described as:

$$M(t) = M_E + \psi(t) \tag{2.37}$$

where $M_E$ is the modulus in an equilibrium state at $t \sim \infty$, and $\psi(t)$ is the relaxation function. The conventional model used to describe viscoelasticity is the Debye model, which gives the exponential relaxation function $\exp[-t(\tau)]$, where $\tau$ is the characteristic time of the relaxation. However, very few solids and liquids show Debye type relaxation. It is generally believed that all of the relaxation phenomena of the Debye type in condensed matter can be described by a few simple relaxation functions. Several models for the relaxation mechanism have been proposed, which lead to the universal description of a relaxation function, $\psi(t)$.

Sasaki and Yoshikawa, 1993 have carried out investigations to (i) classify the stress relaxation of bone as Debye or non-Debye type and (ii) to check the adaptability of the proposed universal description for the relaxation function, that is, whether it is non-Debye type. For a viscoelastic material, a relaxation function of the shape:

$$\psi(t) = \exp(-t/\tau) \tag{2.38}$$

has been proposed, where $\tau$ is the relaxation time. As Equation 2.38 is consistent with Debye's dielectric decay function, such a relaxation process is called a Debye-type relaxation. Such a viscoelastic response in an actual material is defined as non-Debye relaxation. The conventional approach to non-Debye relaxation is to consider a distribution of Debye-type relaxing elements, each with its own relaxation time. Then the relaxation function is expressed by the relaxation spectrum, $H(\ln\tau)\,\mathrm{d}(\ln\tau)$ as given below (Ferry, 1980):

$$\psi(t) = \int_{-\infty}^{+\infty} H(\ln\tau)\exp(-t/\tau)\mathrm{d}(\ln\tau) \tag{2.39}$$

Several empirical relaxation functions for non-Debye relaxation process have been proposed and applied to specific samples of materials. Non-Debye relaxations are obtained when relaxation is governed by a continuum or hierarchy of sequential cooperative process. (Ngai, 1979).

Generally, the relaxation function of stress relaxation in bone and bone collagen is described as a combination of a general Debye-type relaxation function (process II) and the KWW function (process I). The ultimate strength of bone is greater under compression than under torsion and tension (Nielson, 1975; Reilly and Burnstein, 1975). In KWW-type stress relaxation of bone and bone collagen, the viscoelastic properties of collagen seem to play an important role.

On the basis of the empirical formula for the relaxation modulus (Sasaki and Yashikawa 1993) deduced the relaxation mechanism of bone. These authors have reported that each relaxation process has two stages, which are referred as process I and process II in order of time. Process II is described by a simple exponential decay while process I is Kohlraush-Williams and Watts (KWW) function. This is shown in Figure 2.15. (Klafter and Shlesinger 1986, Palmer *et al.* 1984). For the Debye type of the relaxation, a slippage of osteons along a cement line has been considered as an important factor (Lakes and Saha 1979).

On the basis of this equation the relaxation mechanism in bone and bone collagen is identified. Figure 2.15 shows a schematic description of the relaxation observed in bone. For the Debye-type the relaxation, a slippage of osteons along a cement line has been considered as an important factor (Lakes and Saha, 1979).

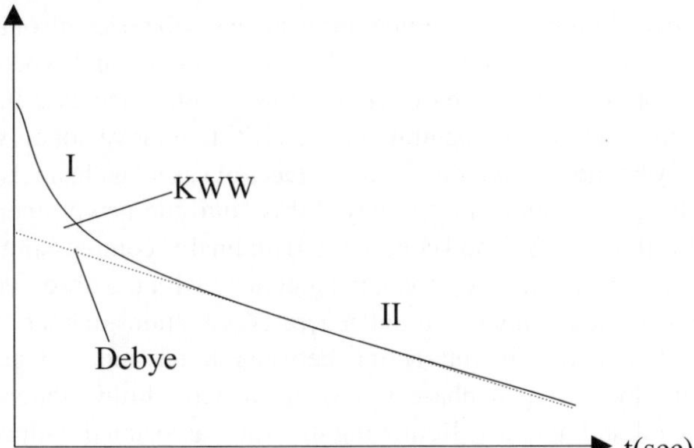


**Figure 2.15** Schematic diagram of the stress relaxation of bone and bone collagen. During the initial stress relaxation, the KWW- and Debye-type relaxation processes are mixed. Later, the KWW-type relaxation diminishes and merges into the Debye-type relaxation process

The measured physical properties (cross sectional area, mineral density and velocity) are statistically intercorrelated with the traveling wave characteristics (velocity, dispersion and attenuation or damping coefficient). Linear and parabolic correlations are found to be statistically significant.

Several techniques have been considered for applying these concepts to clinical applications. However, none have yet received widespread support. This is owing to the shortcomings inherent in most of the techniques used to date, such as the soft tissue affecting the measured response (Lakes and Saha, 1977). To overcome these problems some investigators have tried "semi-invasive" designs such as using a percutaneous pin to directly stimulate and monitor the bone (Sonstegard and Matthews, 1976). Non-invasive systems in use have good sensitivity based on the earlier developed techniques (Lakes and Saha, 1978; Behari and Singh, 1980).

Dehydration affects the viscoelasticity of bone: compared with wet bone, dry bone has less anelastic deformation (i.e., less recoverable strain from creep) (Currey, 1965), a lower loss factor $\tan\delta$ (Yamashita *et al.*, 2001; Yamashita *et al.*, 2002) and a much higher relaxation rate (Sasaki and Enyo, 1995). Despite the documented effects of drying on bone properties, more needs to be known about the underlying mechanisms of such changes at the microscopic level.

## 2.15 Discussion

It is suggested that drying at room temperature in a vacuum oven may affect the mechanical properties of bone. Assuming that water in pores does not affect the mechanical properties (which is likely when measured with monotonic loading at a constant strain rate), these results suggest that water may be lost from some phases of extracellular matrix even at room temperature.

Nyman *et al.* (2006) observed that (i) for each drying temperature the rate of water loss between the third and fourth hour in the vacuum oven was less than 2%; (ii) an increase in the drying temperature always causes an increase in the loss of water; (iii) there is a nonlinear relationship between strength and water loss; and (iv) water loss affected toughness, even when it did not exceed the amount that could exist in the vascular-lacunar-canalicular space.

Water loss caused by drying at room temperature increases the strength of bone. The stiffness of collagen increases (Nomura *et al.*, 1977; Pineri, Escoubes and Roche, 1978) and the molecular diameter of collagen decreases (Lees, 1981) with a decrease in hydration. There appears to be enough energy in the vacuum oven at 21 °C to remove not only mobile water but also the water loosely bound to collagen. As the surfaces of vascular channels dry, water loosely bound in the nearby extracellular matrix may diffuse into the newly opened space. Subsequently, collagen fibrils can stiffen and contract longitudinally, compressing the mineral phase and thereby increasing the strength of bone, in agreement with the observed data. It has been suggested that the increase in mineralization generates pre-strains in bone (Yeni *et al.*, 2002). This may involve a reduction in interaction between water and collagen. Under normal hydrated conditions, the collagen phase is thought to have little influence on strength in comparison to mineral and porosity. Removing the water associated with collagen, however, would impart a contribution to its strength. Strength decreases when bone is dehydrated, when sufficient energy is applied to remove both the water–collagen interaction and the water–mineral interaction. Loss of water from the mineral phase may affect its strength, though the loss of lattice water changes the size of the bone mineral crystal (i.e., the distance between neighboring lattice sites decreases), as has been observed in dehydrated enamel and precipitated apatites (LeGeros, Bonel and Legros, 1978).

A loss of plasticity (or post-yield toughness) occurs upon room temperature drying in which only the water–collagen interaction is believed to be mainly affected. Collagen is known to influence toughness (Wang *et al.*, 2002), and its interaction with water appears to be important in giving bone post-yield deformation behavior. At higher drying temperatures, there is also a decrease in work to fracture, suggesting that water bound to the mineral phase also affects the toughness of bone. Of course, this decrease is related to the decrease in strength and also to the decrease in strain at fracture. The decrease in the energy absorbing capacity of bone with an increase in mineral content, as observed by Currey, Brear and Zioupos (1996), may be due to a decrease in the interaction between water and mineral. Demineralization experiments have shown that there is less water in more mineralized bone (Broz, Simske and Greenberg, 1995; Lees, 1981).

The modulus of elasticity has a linear association with water loss. There is a general trend of increasing stiffness with water loss. Thus, the strength and stiffness of bone has an inverse relationship when water is removed from the extracellular matrix by drying at 70 °C. As found in earlier mechanical tests (Dempster and Liddicoat, 1952; Evans and Lebow, 1951; Smith and Walmsley, 1959; Yamada and Evans, 1970), dehydration increases the stiffness and decreases the toughness of bone. Also in agreement with earlier tests, it has been observed that dehydration decreases strain at fracture, correlating with the decrease in toughness.

Strength, however, does not necessarily increase after dehydration but does so in uniaxial tension and compression tests reported in earlier studies (Dempster and Liddicoat, 1952; Evans, 1973; Evans and Lebow, 1951). In contrast, the three-point bending tests of Sedlin and Hirsch (1966) found no significant change in maximum stress for mid-femoral cortical bone tested after 1 week of incubation when compared to wet bone of similar type. As an explanation for these discrepancies, those studies observing an increase in strength and stiffness did not heat the specimens; therefore, the applied energy was not sufficient to remove the water associated with the lattice. It seems that water removal by low-temperature drying increases strength (as collagen stiffens) while greater water removal at higher temperatures decreases strength (as the mineral lattice changes). In any event, the failure mechanism of bone is complex because of its hierarchical architecture and interacting constituents, but water certainly has a role in affecting the mechanical properties of bone.

The use of heat to determine the water distribution in bone does not necessarily reveal the exact distribution of water in pores and bone constituents. Emerging techniques based on NMR (Fernandez-Seara *et al.*, 2004; Fernandez-Seara, Wehrli and Wehrli, 2002; Wang and Ni, 2003) may provide the means to distinguish between water in pores and water in the bone matrix. That the pores in control specimens are actually filled with water can not be established. Thus, the maximum volume of mobile water (12%) may have been overestimated. Regardless, room temperature drying likely removes water at least from the collagen matrix, given its effect on strength (which increased). Room temperature drying only causes a loss of mobile water and strength would decrease, as there would not be any fluid pressure to resist deformation.

It may be concluded that the loss of water from the extracellular matrix of bone tissue affects the mechanical properties of bone. Based on the relationships between these properties and water loss, a working model for the effects of water distribution on mechanical behavior of bone runs as follows. Water bound to the collagen fibrils provides post-yield toughness to bone. Removing this water though increases both strength and stiffness but decreases bone toughness. Whether water bound to the surface of mineral increases or decreases strength is unclear. Nonetheless, loss of lattice water within the apatite likely decreases strength and toughness. Age-related changes in bone strength and toughness (indicators of bone fragility) may also reflect age-related changes in water distribution. It may be inferred that, with aging, the amount of water in the tissue decreases while the amount in the pores increases. Therefore, a loss of toughness may reflect a decrease in water bound to collagen due to mineralization, or an increase in collagen crosslinks that displaces water in the collagen matrix.

# 3

# Bioelectric Phenomena in Bone

## 3.1 Macroscopic Stress-Generated Potentials of Moist Bone

A clearer understanding of electrokinetic phenomena in bone is necessary to an understanding of life processes. Fundamental to this is a knowledge of stress-generated potentials, under simulated physiological conditions, that though small in magnitude are yet important. These potentials may play an important role in controlling bone growth and remodeling process. The electrokinetic phenomenon of streaming potential has generally been accepted as the dominant mechanism for the electrical potentials observed in wet bone upon deformation (Johnson *et al.*, 1980; Gross and Williams, 1982; Otter, Goheen and Williams, 1988; Walsh and Guzelsu, 1991; Guzelsu and Walsh, 1990). A piezoelectric contribution in stress-generated potentials produced upon mechanical deformation may also exist (Hastings and Mahmud, 1991). Both mechanisms are present at full hydration, with the piezoelectric effect leading the streaming potential in the time domain (Hastings and Mahmud, 1991). However, streaming potentials arise due to the movement of ions in the fluid flow. They may also arise due to fluid flow under a pressure gradient (Figure 3.1a and b).

## 3.2 Mechanism of Biopotential Generation

In an aqueous medium, the surface charges on a protein-containing surface are opposed by a layer of absorbed counterions of opposite polarity electrostatically bound to the solid surface (Figure 3.2). To a first approximation, this electrostatic double layer may be regarded as a parallel plate capacitance. In addition, a diffuse, weakly held layer of ions extends away from the interface into the bulk of the liquid. This diffuse layer contains both positive and negative ions in unequal numbers. As a consequence, this diffuse layer is electrostatically charged (Figure 3.3). The motion of a liquid through a porous membrane under the influence of an applied electric field is called as electro-osmosis. The converse of this is called a streaming potential (Figure 3.4).

When the liquid is forced to move against such a surface, the region of the diffuse layer outside the hydrodynamic slip plane is carried along with the rest of the liquid. A movement of charged liquid generates an electric current and therefore causes a difference in potential

*Biophysical Bone Behavior* Jitendra Behari

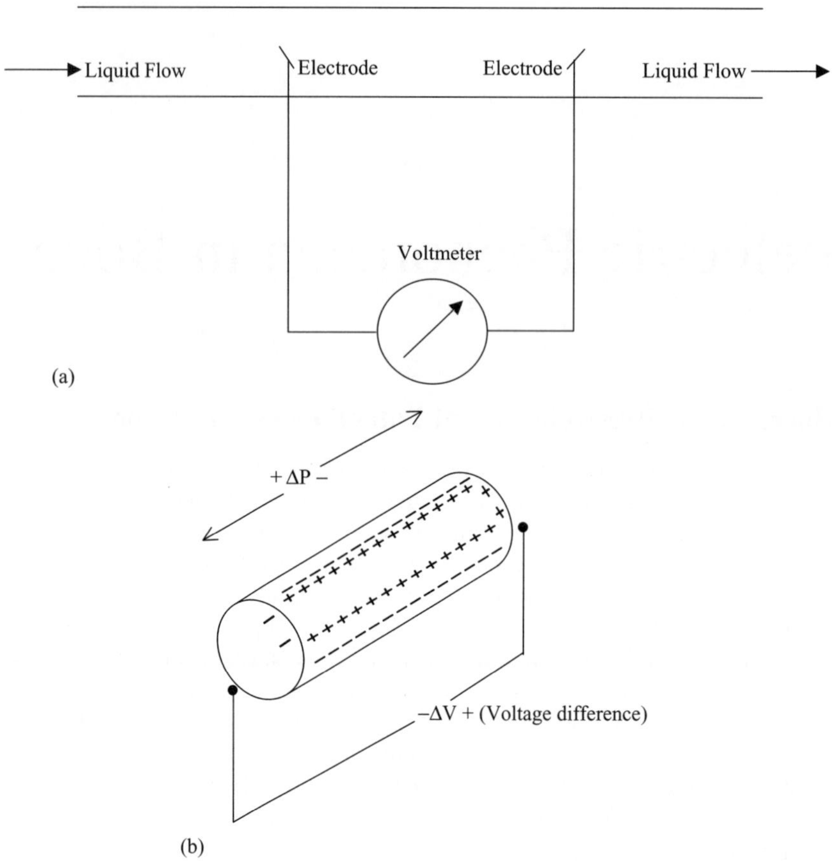


**Figure 3.1** (a) Schematic illustration of the measurement of a streaming potential between two electrodes placed in a liquid flowing through a capillary; (b) model for the generation of a streaming potential (SP) flow through cylindrical channels under a pressure gradient

between two electrodes placed in the streaming liquid. Correspondingly, these voltages are therefore referred to as "streaming potentials." The magnitude of this current depends on the rate of motion. Wet collagen, which is not piezoelectric (Anderson and Eriksson, 1968; Fukada and Yasuda, 1964), can not generate bioelectric signals in the wet bone. However, it has been pointed out (Anderson and Eriksson, 1970) that collagen in wet bone cannot behave like free wet collagen in tendon. Wet bone tissue, therefore, shows piezoelectric behavior as well as the streaming point effect. The potential difference between the bulk of the liquid and that which exists at the hydrodynamic slip plane is termed the zeta potential. All electrokinetic phenomena may be related to the zeta potential for a particular solid–liquid interface. For a laminar flow of a liquid under pressure ($P$) through a capillary, the streaming potential ($E_s$) across the ends of the capillary is given by:

$$E_s = \frac{\in P\xi}{4\pi\eta\, k_0} \tag{3.1}$$

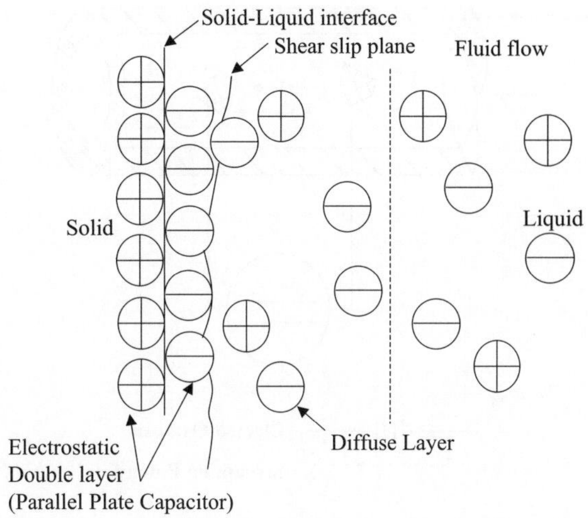


**Figure 3.2** Schematic representation of an electric double layer at a solid–liquid interface

where $\xi$ is the zeta potential, $\in$ represents the dielectric constant of the liquid, $P$ is the pressure across the capillary, $\eta$ designates the viscosity of the liquid and $k_0$ is the specific conductance of the liquid. Streaming potentials are, therefore, directly proportional to the pressure that causes the liquid to move, the zeta potential and the dielectric constant. They are inversely proportional to the viscosity and the specific conductance of the liquid. It is evident from the above relationship that streaming potentials are independent of the size, shape and length of the capillaries. When the radius of the capillary becomes so small that the thickness of the

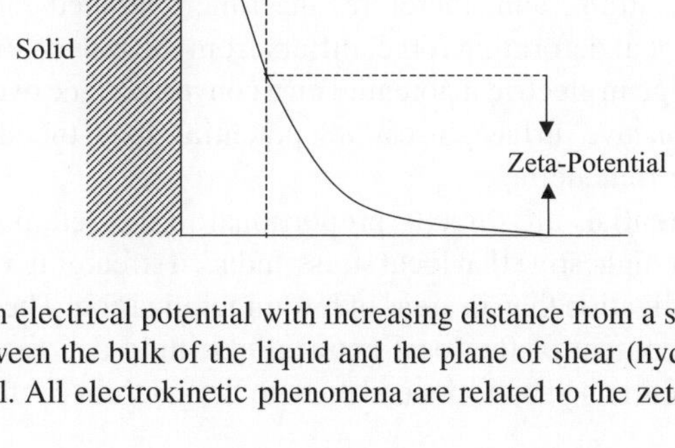


**Figure 3.3** Variation in electrical potential with increasing distance from a solid–liquid junction. The potential difference between the bulk of the liquid and the plane of shear (hydrodynamic slip plane) is termed the zeta potential. All electrokinetic phenomena are related to the zeta potential

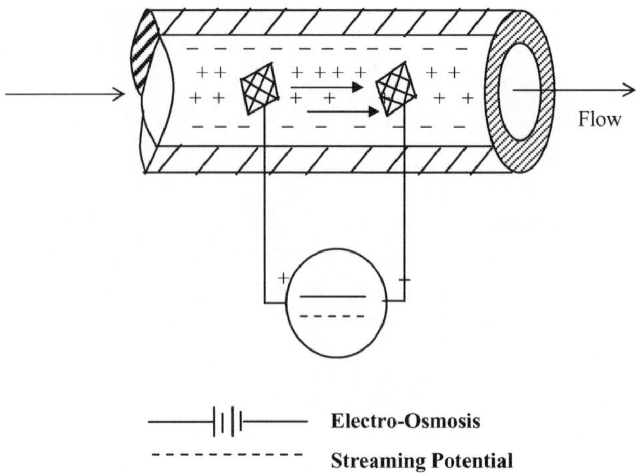


**Figure 3.4** Generation of electrical signal as a result of fluid flow

double layer is not negligible, deviations result from the above formula. For most capillaries in bone, this limitation is acceptable.

The electrical current owing to streaming is directly proportional to the cross sectional area of the capillaries, so that the total current that flows in a thick piece of bone is greater than the total current that flows in a thin piece. However, because the electrical resistance is inversely proportional to the cross sectional area and because the streaming potential and streaming current are related by Ohm's law, streaming potentials are independent of the diameter and length of the capillaries. The movement of the fluid in bone causes an electrical current and, therefore, is directly responsible for measured streaming potentials. As a result, within the limits of laminar flow, the faster the movement of the liquid the greater the magnitude of the streaming potential. It is of no consequence to the phenomenon whether the fluid is caused to flow on application or release of the stress. The slope of the streaming potential versus pressure curve has been analyzed for pressure levels less than 69 kPa where the flow conditions are satisfied (circa 10 psi) (Walsh and Guzelsu, 1991). The fluid flow assumptions necessary for the use of this equation to calculate zeta potentials from intact streaming potential experiments have been examined previously and verified in bone permeability experiments (Walsh and Guzelsu, 1991). To produce bending, any small transverse canaliculi are differentially strained. The stress distribution in a living bone is time dependent. If the time averaged rate of strain, and, therefore, the time averaged rate of fluid flow upon application of a functional deforming force, differs from the rate of strain when the stress is released, the total change in electrical potential on a convex surface over a period of time will differ from that on a concave surface. Streaming potentials may, therefore, have a direct role in stress induced bone remodeling.

Since streaming potentials are directly proportional to the zeta potential, with all other variables constant, it is suggestive that local stress-induced streaming voltages will be greater in areas of high mineralization than in areas of low mineralization. These will also increase in value as the calcium content of the local extracellular fluid decreases. As a result, stress induced streaming potentials in bone depend on both the calcium content of the extracellular fluid and on the effective calcium content of the surface against which the fluid moves. Consequently, when the degree of mineralization is locally increased at the expense of ionic calcium in the local extracellular fluid, the magnitude of localized stress induced streaming potentials rises considerably due to these two effects. Streaming potentials and the electrical double layer are dependent on the properties of the solid as well as the liquid phases. In bone, the solid phase is composed of the mineralized organic matrix of type I collagenous matrix and non-collagenous proteins embedded with bone mineral (hydroxyapatite analogue) $[Ca_{10}(PO_4)_6(OH)_2]$ (Posner and Betts, 1975). In this context, wet bone obviously has a component of electrokinetic potentials, though its exact nature and contribution remains to be understood.

## 3.3 Stress-Generated Potentials (SGPs) in Bone

In principle, the potentials measured as a result of bending are termed as stress generated potentials (SGPs) and were first measured in bone by Yasuda (1954). Because of its crystal structure, these were attributed to piezoelectricity in bone. In dry bone, piezoelectric voltages are of the order of few hundred millivolts (Williams, 1982). Fukada and Yasuda (1957) also demonstrated that the electric field was linearly related to strain (at 200 Hz), in line with the prediction of classical piezoelectric theory.

The much smaller potentials generated by bending moist cortical bone ($\sim$100 $\mu$V) are now generally accepted to be electrokinetic (streaming) potentials (rather than electrostatic, piezoelectric). Examined in this way, it is suggestive that the potentials produced either by bending hydrated bone or by forcing fluid through bone can be termed electrokinetic (Carquiglini *et al.*, 1967). Conversely, SGPs for bone soaked in solutions of pH near the isoelectric point for tendon (Anderson and Eriksson, 1970) undergo several order of magnitude change in potential as the bone is dried. Streaming potentials (SPs) and streaming currents (SCs) in bone generated as a result of mechanical loading may send signals to cells to undergo remodeling (Becker, Bassett and Bachman, 1964; Cowin, Moss-Salentijn and Moss, 1991; McLeod and Rubin, 1992; Pollack, 1984). The electrokinetic phenomena (streaming potentials) can be used to investigate pore structure, fluid composition and possibly the molecular interactions in a given system (Berkenbilt *et al.*, 1995; MacGinitie *et al.*,1993a, 1993b; Otter *et al.*, 1994; Pollack *et al.*, 1977). Since the deformation of the high-modulus stiff mineral phase of bone will be very small, the transient streaming potential behavior is a function of the electrical double layer and possible deformation of adsorbed ions in the slip plane caused by fluid flow.

Experimental evidence indicates that SGP is due to flow in very small pores, with a radius of less than several hundred ångström (MacGinitie, Gluzband and Grodzinsky, 1994; Salzstein and Pollack, 1987; Scott and Korostoff, 1990). Some studies support the hypothesis (Johnson *et al.*, 1982; Petrov, Pollack and Blagoeva, 1989) that fluid relaxes through small pores to large vascular canals, or other large pores, so that the SP magnitude and kinetics depend on large as well as small pore structures (Cochran *et al.*, 1993; MacGinitie *et al.*, 1993a, 1993b; Otter *et al.*, 1994; Starkebaum, Pollack and Korostoff, 1979; Steinberg, Singh and Mitchell, 1981). Since the SP is taken as a basic measurement parameter as a result of stress, in a clinical sense it is a clear indication of healing and remodeling as compared to normal bone (MacGinitie *et al.*, 1993a, 1993b). As the healing progresses the gap decreases and hence the impedance also decreases.

The smaller SP magnitude in laminar tissue for various flow types as assessed by comparison of magnitudes at each frequency is consistent with the hypothesis that vascular canals modulate SP magnitude. Further, the mean SP magnitude is smaller for greater vascular canal connectivity (for laminar tissue with tangential flow), as indicated by canal orientation across the electrode site, which is consistent with smaller SPs for healing and remodeling versus normal cortical bone (Cochran *et al.*, 1993; MacGinitie *et al.*, 1993a, 1993b; Steinberg, Singh and Mitchell, 1981) and for trabecular versus cortical bone (Wu *et al.*, 1993). For healing versus normal bone, a large increase in porosity produces a difference in SP kinetics (MacGinitie *et al.*, 1993a, 1993b). The lack of measurable difference in SP or SC kinetics indicates the small difference in connectivity between groups. Even for trabecular bone, the intertrabecular spaces produce a faster decay time than cortical bone (Wu *et al.*, 1993).

The results of MacGinitie *et al.* (1997) suggest the importance of SPs as remodeling signal. No difference has been observed for radial flow, a predominant flow for bone bending *in vivo*. For the same loads, laminar and Haversian bone would produce similar, uniform SPs. However, under some conditions tangential flow might be important, perhaps near muscle attachment sites. Haversian bone would produce larger SPs than laminar bone. Hence it can be said that SPs in normal bone depend on a large pore structure and fluid flow direction, but are independent of macroscopic bone impedance.

Gross and Williams (1982) found that potentials increased with pH between 6.2 and 8.2, in agreement with the findings of Anderson and Eriksson (1970) and with the streaming hypothesis. The relaxation time of the potential in response to step loading increases with increasing viscosity.

The measured potential magnitude normalized to strain decreases with NaCl concentration (and measured conductivity) for near physiological conductivities and with viscosity. The potential reaches a plateau at low NaCl concentrations ($<0.009\,\mathrm{mol\,L^{-1}}$), as expected theoretically for concentrations low enough for the double layer thickness to approach the pore radius. The relaxation time increases with viscosity and does not depend on NaCl concentration.

However, the $Z$ potential calculated by Gross and Williams (1982) from the measured SP is independent of concentration, while that of Pienkowski and Pollack (1983) shows a concentration dependence. In addition, Pienkowski and Pollack (1983) observed a sign change in $Z$ potential for high NaCl concentration and viscosities, which has not been reported by Gross and Williams (1982). These studies indicate that the dependence of stress generated potentials on ionic concentration, on viscosity and on pH and hydration is consistent with a streaming rather than a piezoelectric hypothesis. Piezoelectric fields may, however, be generated locally, resulting in local displacement currents with relaxation times on the order of microseconds, which are measured macroscopically (Williams, 1982).

## 3.4 Streaming Potentials and Currents of Normal Cortical Bone: Macroscopic Approach

Bending of wet cortical bone tissue results in deformation of the bone matrix and the subsequent fluid flow has been used to characterize the electromechanical properties of bone (Johnson *et al.*, 1980; Gross and Williams, 1982; Pienkowski and Pollack, 1983; Otter, Goheen and Williams, 1988; Hastings and Mahmud, 1991). The change in spatial charge density followed by ionic movement can generate piezoelectric and streaming potential responses (Hastings and Mahmud, 1991). Bone streaming potentials and currents are important – they may provide a signal to cells to remodel bone in response to mechanical loading (Becker, Bassett and Bachman, 1964; Cowin, Moss-Salentijn and Moss, 1991; McLeod and Rubin, 1992; Pollack *et al.*, 1984). Electrokinetic phenomena may also be used to probe pore structure, matrix and fluid composition and molecular interactions (MacGinitie *et al.*, 1993a, 1993b; Otter *et al.*, 1994; Pollack *et al.*, 1977). This is also suggestive of a relationship between electrokinetic phenomena and bone structure. Both SP and SC result from fluid flows driven by mechanical loading, past the negatively charged bone matrix in the charge double layer, past the negatively charged bone matrix. This flow produces a convective current in the same direction. Thus, SPs represent the balance of conductive and convective currents, while SCs represent the trabecular and resorption spaces in healing and remodeling. Osteoporotic bone appears to play a role in SP generation with more porous (lower impedance) bone having generally smaller SPs (MacGinitie *et al.*, 1993a, 1993b; Otter *et al.*, 1994; Otter *et al.*, 1995) and faster SP kinetics (MacGinitie *et al.*, 1993a, 1993b).

Streaming potentials have been considered to be the dominant mechanism and a piezoelectric contribution in fully hydrated bone has often been questioned, largely because of the absence of a piezoelectric response from wet collagen. However, Hastings and Mahmud (1991) have shown a piezoelectric response in fully hydrated bone saturated with deionized water. Therefore, a piezoelectric response may modify the streaming potentials generated and influence bone formation, repair and remodeling, though to a lesser extent. Stress-generated potentials do not allow the streaming potentials to arise due to direct mechanical pressure on the fluid phase alone. The SPs (Walsh and Guzelsu, 1991; Guzelsu and Walsh, 1990) may be considered to arise only from ion flow (without deformation of the surrounding matrix).

Alterations in the characteristics of *in vitro* fluid phase used in electrokinetic characterization of bone (pH, ionic strength, ion composition, viscosity) result in attenuation in the measured voltage (Gross and Williams, 1982; Pienkowski and Pollack, 1983; Otter, Goheen and Williams, 1988; Walsh and Guzelsu, 1991; Guzelsu and Walsh, 1990; Hastings and Mahmud, 1991). Therefore, changes in the composition of bone fluid *in vivo* may also modify the bioelectric potentials. The electrokinetic properties of cortical bone tissue reflect the compartmental organization of fluid spaces (Walsh and Guzelsu, 1991). Unique fluid spaces have been demonstrated in bone, associated with the vascular channel system (Haversian canals) and the mineralized matrix (Hughes *et al.*, 1978). Different electrokinetic potentials have been found to be associated with the vascular channel system (examined in direct streaming potential experiments) (Walsh and Guzelsu, 1991; Guzelsu and Walsh, 1990) and the mineralized matrix (examined in particle electrophoresis) (Eriksson, 1976; Beretta and Pollack, 1986; Guzelsu and Regimbal, 1990). These two fluid spaces have different organic constituents exposed to the fluid phase, as revealed by their different isoelectric points (Walsh and Guzelsu, 1991; Guzelsu and Walsh, 1990). Zeta potentials obtained from streaming potential experiments in increasing ionic strength NaCl solutions are reported to be much greater than those for bone particle electrophoresis (Beretta and Pollack, 1986). This supports the compartmental model for the electrokinetic properties of bone (Walsh and Guzelsu, 1991).

Positive streaming potentials, which stabilizes after a few seconds, are found at each pressure level in 0.145–0.6 M NaCl at pH 7.3 for intact samples. The positive streaming potentials indicate a negative zeta potential and an exposed organic interface. The slope of the streaming potential (mV) versus applied pressure gradient to the fluid phase (kPa) decreases in magnitude between 0145 and 0.6 M NaCl. This reflects an increase in conductivity and compression of the electrical double layer with an increase in ionic strength (Hunter, 1981).

Deproteinized samples provide a simplified system with only an inorganic bone mineral surface in a univalent ionic sodium chloride solution. The deproteinized samples (NaOCl treated, mineral alone) generated negative streaming potentials in 0.145 and 0.5 M NaCl at pH 7.3. The negative streaming potential at pH 7.3 indicates that the mineral phase has a positive zeta potential. This supports the possibility that the mineral phase is not directly exposed to the fluid phase but is shielded by the organic layer (Streeter and Brandi, 1990; Walsh and Guzelsu, 1991; Guzelsu and Walsh, 1990; Guzelsu and Regimbal, 1990). Streaming potentials apparently depend upon the saline concentration. The SP is reduced in magnitude for deproteinized samples from 0.145 to 0.5 M NaCl. Pienkowski and Pollack (1983) have reported bone SGPs in solutions of different concentrations and viscosities. These authors also demonstrated a reduction in SGP amplitude as the NaCl concentration is increased. The SGP polarity changes at NaCl concentrations above 0.7 M (Pienkowski and Pollack, 1983). The streaming potentials versus applied pressure for intact samples in 0.75 and 2.0 M NaCl differed from samples of relatively lower concentration (0.145–0.6 M NaCl). The intact cortical bone streaming potentials in 0.75 and 2.0 M NaCl under a continuous flow condition suggest that the SP sign inversion observed in a four-point bending experiment during a step loading profile in high concentration NaCl solutions is a transient phenomena. Samples equilibrated in 0.75 and 2.5 M NaCl indicate that at the onset of a new pressure level the slope of the streaming potential versus pressure curve changes (the zeta potential changes during this time period from positive to negative). Under constant fluid flow (due to direct pressure to the fluid phase) the SP stabilizes and remains constant at each pressure level. From this it is apparent that electrokinetic phenomena depend upon several parameters. Extrapolation of such data to *in vivo* is largely speculative. Figure 3.5 shows a typical variation of microelectrode potential with distance in both compression and tension modes.

The streaming potential polarity is determined by the polarity of the zeta potential and the direction of fluid flow. The differences between the SGP (Pienkowski and Pollack, 1983; Pollack *et al.*, 1984) and the SP results may reflect, in part, that the two techniques do not characterize the same surface (Walsh and Guzelsu, 1991). The overlap region might be an indication of the mechanical deformation of the bone matrix and a piezoelectric contribution to the SGP (Pienkowski and Pollack, 1983; Hastings and Mahmud, 1991; Pollack *et al.*, 1984). Streaming potentials are generated in response to direct application of mechanical pressure on the fluid phase. Deformation of the bone matrix at such pressure levels is assumed to be extremely small. At this stage, a piezoelectric contribution is minimal. This may also account for the differences observed in SGP work (Pienkowski and Pollack, 1983; Hastings and Mahmud, 1991; Pollack *et al.*, 1984) and SP results. The transient SP voltage recorded in high NaCl solutions may be characteristic of the electrical double layer itself and the constituents that comprise it in intact samples [ions and organic material (proteins) adsorbed to the mineral phase].

SPs in response to a step load typically look like an exponential decay in time. However, the magnitudes in response to sinusoidal loading increase with frequency monotonically towards an asymptote (on a linear scale) near 10 Hz (Salzstein and Pollack, 1987; Scott and Korostoff, 1990; MacGinitie *et al.*, 1994). SP phase relative to bending, for sinusoidal loading, decreases monotonically from $\pi/2$ radians at low frequencies towards zero near 10 Hz. Typical SPs measured across normal intact cortical bone at or near 10 Hz are on the order of a few $\mu V\ \mu\epsilon^{-1}$

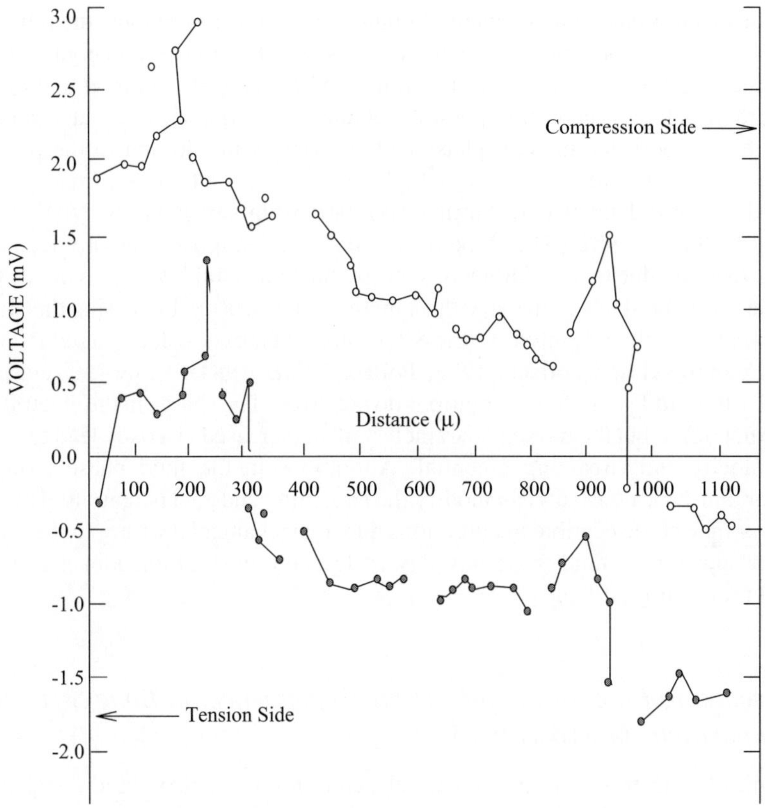


**Figure 3.5** Microelectrode potential versus distance. Open circles (○) are measured potentials relative to electrode on the compression side of specimen and filled circles (•) are relative to the tension side (Starkebaum, Pollack and Korostoff, 1979)

(where strain is assessed on the sample surface) *in vitro* and slightly less *in vivo* (MacGinitie *et al.*, 1993a, 1993b; Otter *et al.*, 1992). Several SP measurements show that the difference between *in vitro* and *in vivo* measurements is small (Cochran, Pawluk and Bassett, 1968; Steinberg *et al.*, 1968).

From the above it can be concluded that the $Z$ potential can be altered by changing the pH or otherwise changing the fluid ionic composition to modulate the matrix change. This is because the conductivity ($\sigma$) depends on ionic concentration. As the SP depends on stress driven fluid flow and fluid relaxation in response to stress depends on viscosity and not on ionic concentration, SGP relaxation times are expected to depend accordingly on these parameters. (Gross and Williams, 1982; Pienkowski and Pollack, 1983). Cignitti *et al.* (1970) measured the potentials produced by step loading of bone and an imposed pressure difference across bone samples kept in fluids with different conductivities, composition and pH. They measured potentials across cortical (rabbit tibias) and trabecular bone (lamb and ox ribs) and showed that a component of the signal generated by bending varied with conductivity and increased pH.

While the electrokinetic phenomenon, in intact bone appears obvious, the mineral phase seems to be shielded by organic layers at high ionic strengths. However, in high ionic strength solutions the zeta potentials of control samples (where both the mineral and organic are present) are negative. Therefore, possible changes in organic secondary and tertiary structures may expose the mineral phase and contribute, in addition to the organic constituents, to the electrokinetic behavior in high ionic strength solutions. Streaming potential per applied pressure demonstrates an inverse relationship versus viscosity (Equation 3.1) (Gross and Williams, 1982). The slope of the SPs versus applied pressure decreases with increasing viscosity due to the addition of sucrose. Sustained fluid flow does not result in any measurable transient SP phenomena with an increase in viscosity. Increasing the viscosity of the solution reduces the magnitude of the SP per applied pressure due to a reduction in flow velocity (Pienkowski and Pollack, 1983; Pollack *et al.*, 1984). However, zeta potentials calculated in 0.5 and 1.5 M sucrose solutions do not differ. The charge in the electrical double layer does not differ, but the physical parameters of an increased viscosity lead to a reduction in flow velocity and streaming potential. Alterations in the fluid phase (conductivity, viscosity or dielectric constant) can modify the streaming and zeta potentials. The electrical double layer may be susceptible to alterations due to mechanical pressure to the fluid phase. Localized changes in the fluid phase may play a role in altering the biofeedback signal and the repair and remodeling of bone.

### *3.4.1 Streaming Potential and Current Dependence on Bone Structure and Composition: Macroscopic View*

Macroscopically, electrokinetic phenomena depend on bone matrix, fluid composition and matrix structure. Spirt and Pollack (1993) reported that the *Z* potential of rat femur, as assessed by particle electrophoresis, decreased from 1.5 to 6 months and then increased between 12 and 24 months. The SP magnitude varies as bone matures (Pollack *et al.*, 1977) and may also vary with bone age after maturation (Spirt and Pollack, 1993).

Cochran *et al.* (1993) have measured SPs generated by sinusoidal bending at the callus of healing gap osteotomies in canine tibia *in vivo*, 6 or 12 weeks following surgical creation of the osteotomy. The SP magnitude is greater on the callus relative to an adjacent remodeling cortical bone site when the callus is immature and flexible. For more mature rigid callus, the callus is like the SP measured at the adjacent cortical bone site. SP normalized to surface strain is somewhat smaller for the callus ($0.11 \pm 0.11\ \mu\mathrm{V}\ \mu\epsilon^{-1}$), than for the adjacent cortical bone ($0.16 \pm 0.11\ \mu\mathrm{V}\ \mu\epsilon^{-1}$). Linear regression showed that SP normalized to strain for both callus and adjacent cortical bone sites tends to increase with decreasing porosity, as can be anticipated.

Using electrokinetic methods the difference in bone structure and composition between osteoporotic and normal cortical bone has been studied by several workers. Otter, Goheen and Williams (1988) studied the effects of disuse osteoporosis on SP using a canine tibia model. Disuse osteoporosis is produced by surgically removing a segment of the Achilles tendon, dividing the patella tendon and then maintaining the foot in dorsiflexon by banding for 12 weeks following surgery. SPs for the more porotic disuse bone are significantly smaller than for normal intact bone. The relaxation time for SPs generated by bending osteoporotic bone was also significantly shorter than for normal control bone. This implies that SP magnitude increases more slowly with bending frequency.

Cochran *et al.* (1991) have compared the Z potential of osteoporotic human bone (as assessed by clinical examination) to normal bone by using particle electrophoresis and found that the potentials are larger for osteoporotic bones. This increase suggests that measured SP magnitudes should also be larger for osteoporotic bone because the two are proportional. However, SPs also tend to decrease with bone porosity, as observed in healing bone (MacGinitie *et al.*, 1993a, 1993b; Cochran *et al.*, 1993) and osteoporotic bone (Otter *et al.*, 1994). The effect of a change in bone porosity may prevent measurement of an increase in SP magnitude due to an increase in *Z* potential. The increase in *Z* potential may be a response to compensate for the increase in porosity.

## 3.5 Microscopic Potentials and Models of SP Generation in Bone

Although tissue level fields have been measured under various conditions, cellular level fields are difficult to measure and interpret. Two approaches appears suggestive for determining fields at the cellular level: microscopic measurements and models to relate these fields to bone anatomic structure.

Microscopic measurements are the most direct approach for determining fields near cells. Several workers (Iannacone, Korostoff and Pollack, 1979; Starkebaum, Pollack and Korostoff, 1979) have measured microscopic SPs on the polished bone surface of fully hydrated samples tested in four-point bending and in compression. These SPs exhibit a nearly symmetric cusp-like dependence on distance from the Haversian canal, with local fields 1–2 orders of magnitude greater than the macrofields. The sign and magnitude of the cusp depends only on the macro stress for four-point bending (Starkebaum, Pollack and Korostoff, 1979) (Figure 3.5). An abrupt change in magnitude is observed at cement lines. Pienkowski and Pollack (1983) found that microscopic fields measured on the bone surface depended on fluid concentration in the same manner as macroscopic potentials. Models that have been developed relate SP generation to bone fluid spaces. These spaces include Haversian canals (radius 10–100 Å), canaliculi ($R \approx 0.05$–$0.1\ \mu$m) and interstitial matrix spaces ($R \approx 100$– 600 Å) (Chakkalakal, 1989; Salzstein and Pollack, 1987; Scott and Korostoff, 1990).

The homogeneous models of Salzstein *et al.* (1987) and Spirt and Pollack (1993) require, in addition, a no-flow boundary condition at the bone surface to obtain nonzero SPs, as postulated in another model (Johnson *et al.*, 1982) for measurements in a humid environment. For a no-flow boundary condition, there is a net pressure difference across a bone sample in uniform bending and therefore nonzero SP. This no-flow condition may not be valid, especially *in vivo*, where soft tissues surround the bone. The alternative, a free-flow boundary condition, has been assumed to be valid (Salzstein *et al.*, 1987) for fluid covered bone. This alternative boundary condition results in no net pressure difference across a sample in uniform bending, Suggesting an obvious: that for a homogeneous model there may be no macroscopic SP. The validity of the no-flow boundary condition has been suggested for SP measurements in a humidity chamber on the basis that the surface is not wetted and capillary pressure will keep the fluid within the bone (Salzstein *et al.*, 1987; Johnson *et al.*, 1982). This hypothesis has some experimental basis. Iannacone, Korostoff and Pollack (1979) have observed menisci in lacunae and Haversian canals at the bone sample surface during microscopic SP measurements. However, the SP magnitude and frequency response are the same whether measured with agar bridge contacting the bone, with a pool of fluid contacting the bone or with an AgCl electrode directly contacting a dry bone surface (MacGinitie *et al.*, 1993a, 1993b).

Petrov, Pollack and Blagoeva (1989) have developed a microscopic model for a homogeneous matrix surrounding a Haversian canal. The SP is calculated as a function of distance from the canal, based on the assumptions of zero pressure (instantaneous fluid relaxation) and zero potential in the Haversian canal, and no flow at the cement line. This model predicts a SP magnitude as a function of distance from the canal that is similar to microscopic measurements. Weinbaum, Cowin and Wu (1991) have developed a microscopic model in which flow canaliculi around the cellular processes gives rise to SPs. This model takes into account the cellular processes in the canaliculi.

Kowalchuk, Pollack and Corcoran (1993) have developed a model for the bone interface, based on zeta potential measurements of bone and hydroxyapatite particles, in various fluids and following various treatments. The proposed bone–fluid interface consists of a bone substrate (including mineral and proteins), a stationary layer (including proteins and ions) and the bone fluid. The zeta potential is assumed depending on all three components.

## 3.6 Stress-Generated Fields of Trabecular Bone

Weinbaum *et al.* (1994) have shown that in the physiological frequency range 1–20 Hz, associated with locomotion or the maintenance of posture (15–20 Hz), the fluid shear stress is nearly proportional to the product of frequency and range. These authors have emphasized that the lacunar-canalicular porosity determines the fluid pressure relaxation time and the occurrence of the SP.

Wu *et al.* (1993) have measured SPs across moist cubes of trabecular bone in sinusoidly non-uniform axial loading and compared these with SPs of cortical bone under the same conditions and from the same specimens. Trabecular bone SPs are more than one order of magnitude smaller than the cortical bone SPs ($\sim$0.001 versus $\sim$0.3 $\mu$V $\mu\varepsilon^{-1}$ at 20 Hz). Trabecular bone SPs also increased less rapidly with bending frequency than cortical bone SPs.

Trabecular bone anatomy suggests that the trabecular bone SP frequency response should be similar to that of cortical bone but that the SP magnitude should be smaller, because of the more porous structure of trabecular bone. Distances to the intertrabecular fluid space or Haversian canals within trabeculae are similar to the inter-Haversian distances in cortical bone (Lozupone and Favia, 1990; Sata *et al.*, 1986). Trabecular bone, however, has a much smaller resistivity than cortical bone [$\sim$0.2–1.0 k$\Omega$ cm (Saha and Williams, 1989) versus $\sim$15–55 k$\Omega$ cm (Singh and Saha, 1984)]. If the intertrabecular spaces in trabecular bone contribute to fluid relaxation as the Haversian canals do in cortical bone, then the SPs of trabecular bone (neglecting any contribution by the marrow) would have a similar frequency response but smaller magnitude than those of cortical bone. Measured differences in the frequency response between trabecular and cortical bone SPs cannot be explained in terms of fluid relaxation to the Haversian canals or intertrabecular spaces alone.

Gross and Williams (1982) and Pollack *et al.* (1984) have developed microscopic models for SP generation by flow in canaliculi to the Haversian system. The model due to Pollack *et al.* (1984) incorporated the idea of flow to the Haversian system and as a result explains the cusp-like dependence of microscopic potentials. Neither of these approaches, however, can be used as a predictive model for macroscopic potentials.

Salzstein *et al.* (1987) have developed a macroscopic biphasic continuum model for flow in small pores for rectangular bone samples in four-point bending. Spirt and Pollack (1993) extended the Salzstein model to a cylindrical geometry to model SPs produced by loading the diaphysis of whole long bones. The model of Salzstein *et al.* (1987) provides fits to macroscopic SP data as a function of frequency. These have been used to identify radii on the order of 105–350 Å (Salzstein and Pollack, 1987) and 200–600 Å (Scott and Korostoff, 1990) for the pores involved in SP generation. The size range suggests that these pores are interstitial pores. This estimate of the effective pore radius, $R$, is close to estimates ($R < 200$ Å) obtained by another method (MacGinitie *et al.*, 1994) and is consistent with pore radius estimates for macroscopic SPs measured across cortical bones ($R$ near 45 Å) (Pollack *et al.*, 1977) and $R \ll 0.1\ \mu m$ (Johnson *et al.*, 1982). However, the homogeneous model predicts that SP frequency response should depend on sample thickness. Contrary to experimental results (Otter *et al.*, 1994), frequency response is found to be independent of sample thickness for SPs measured across bovine bone strips in sinusoidal cantilever bending for thicknesses between 1 and 4 mm. These authors predict that the SP should be independent of sample thickness for greater than the inter-Haversian spacing. This hypothesis that fluid relaxes via Haversian canals could also explain the faster relaxation time for healing, remodeling and osteoporotic bone. The greater porosity ($R < 5\ \mu m$) visualized in these cases suggests, if this hypothesis is correct, that shorter relaxation times are possibly due to the shorter distance that the fluid travels between resorption lacunae.

Some models have assumed (Gross and Williams, 1982) that macroscopically measurable SPS arise from fluid flow within the canaliculi. Others have proposed (Williams, 1993; Spirt and Pollack, 1993) that SPS arise from flow to take bone surface through homogeneity. Weinbaum *et al.* (1994) have proposed that osteocytes can be stimulated by relatively small fluid shear stress acting on the membrane of their osteocytic processes. These authors have further pointed out that these stresses lead to the release of intracellular $Ca^{2+}$ that regulates the opening and closing of membrane ion channels in the communicating functions linking the osteocytic processes at their apical ends. In this manner, the gap functions modulate the intracellular potential and current that passes through the network of interconnected osteocytes and surface osteoblasts. This view has also been supported by Xia and Ferrier (1992). Table 3.1 shows a set of SGP generation data as measured by various workers.

**Table 3.1** Stress-generated potentials in whole bone

| Reference | Source of specimen | Technique | Results/comments |
|---|---|---|---|
| Bassett, 1962 | Animal tibia | Cantilever bending | 30 mV for 30 g stress |
| Shamos, Lavine and Shamos, 1963 | Human rib | Compression | 0.3 mV kg$^{-1}$ |
| Cochran, Pawluk and Bassett, 1968 | Animal tibia | Cantilever bending | 0.5–2.0 mV for 1–2 kg |
| Steinberg *et al.*, 1968 | Rabbit femur and tibia | Two-point compression | 1.3 mV |
| Steinberg *et al.*, 1974 | Rabbit tibia | Cantilever bending | 4.5 mV |
| Steinberg *et al.*, 1976 | Rat femur | Four-point bending | Temperature dependent |

Several authors (Walsh and Guzelsu, 1991; Guzelsu and Walsh, 1990), using a pressure controlled nitrogen gas supply to force fluid through intact bone samples, have determined the flow in their system through the vascular channel system. Streaming potentials measured using this technique are due to ion flow in response to direct application of mechanical pressure to the fluid phase (Hastings and Mahmud, 1991). This experimental technique differs from the stress generated potential binding where fluid motion is the result of mechanical deformation of the solid bone matrix, which generates a piezoelectric effect and streaming potentials (Hastings and Mahmud, 1991). The relationship between the applied pressure gradient and developed streaming potential allows the calculation of the zeta potential in order to characterize the electrokinetic properties of bone tissue. The surface of shear is considered to lie close to the surface of the solid phase where ions in the electrical double layer are stationary (Hunter, 1981).

## 3.7 Biopotential and Electrostimulation in Bone

As considered above for electrokinetic phenomena, the measurement of biopotentials in bone seems to provide directions to the bone growth and remodeling process. This arouses interest in their measurement under various conditions.

### *3.7.1 Electrode Implantation*

The experiments are preformed on adult rabbits weighing 1.5 to 2.0 kg. The method consists of measuring the biopotential and applying stimulation by implanting high-speed stainless electrodes inside the tibia and femur bone under general anesthesia. Six electrodes (of 1 mm diameter) insulted along their length, except the tips, are drilled through the bone with their ends touching its inner portion and clamped at their positions with dental cement. The electrodes (Figure 3.6a) on both tibia and femur are kept 1 cm apart.

### *3.7.2 Control Data*

Biopotentials can be measured by using a high input impedance digital electrometer. The reference electrode is taken as the lowest electrode "1," while the bone potentials are recorded between any two electrodes. Data on biopotential can also be measured after stimulation of electrical and electromagnetic fields, which are applied under various combinations.

Bioelectrical potentials (DC) have been found to vary in the range 60–120 mV using stainless steel electrodes. The potential decreases as one moves toward the femur from the tibia (electrodes 1–6) with respect to reference electrode 1 (Figure 3.6b). These findings confirm the established view that areas of active growth have higher electrical activity (Assima, 1968; Friendenberg and Brighton, 1966; Klapper and Stellard, 1974; Bassett, Pawluk and Becker, 1964). In studies carried out by Behari and Singh, 1981b, the full length of the electrode is insulated, except at the tip, suggesting that potentials are picked up from the bone's inner surface. The electrode material was chosen so as to reduce the polarization effect. In addition, since both electrodes are made of identical material, the polarization due to the galvanic cell is identical near both electrodes. This helps to reduce the polarization effect. Biopotential variation is insignificant at each electrode site ($p > 0.1$).

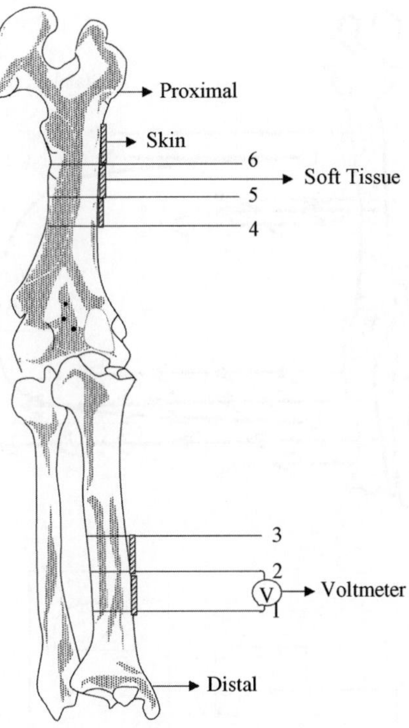


(a)

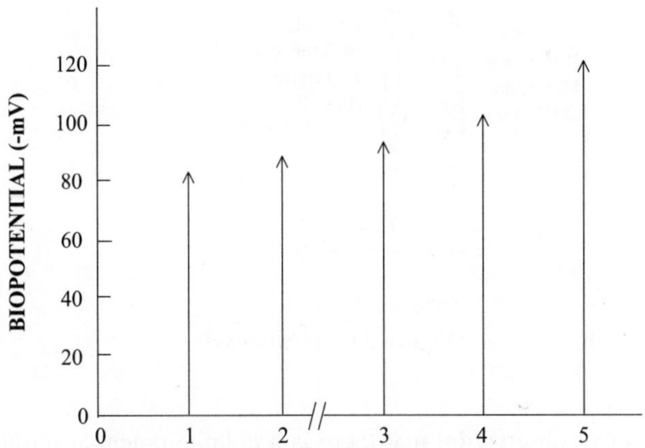


(b) **Distance (in cm) between electrodes**

**Figure 3.6** (a) Positions of electrodes in the femur and tibia; (b) biopotential between any two electrodes versus the corresponding distance curve for various electrodes; (c) electrode implantation (1–6) in bone and related measurement; (d) method of determining potentials at different levels on the rabbit tibia/femur; the reference electrode remains fixed while the recording electrode is placed at different levels along the tibia

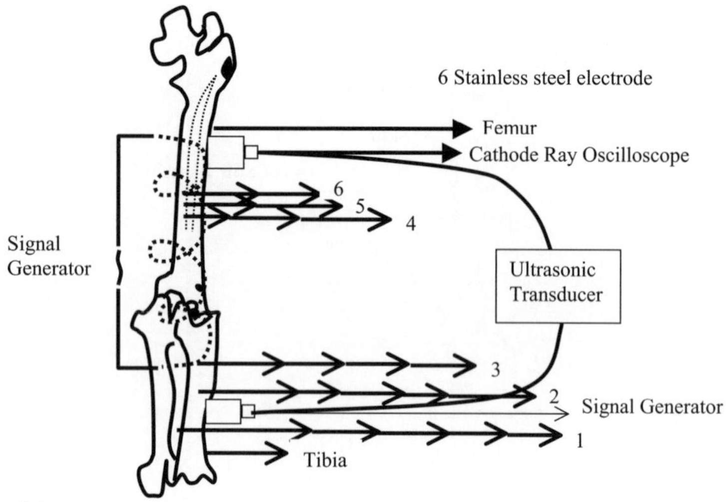


(c)

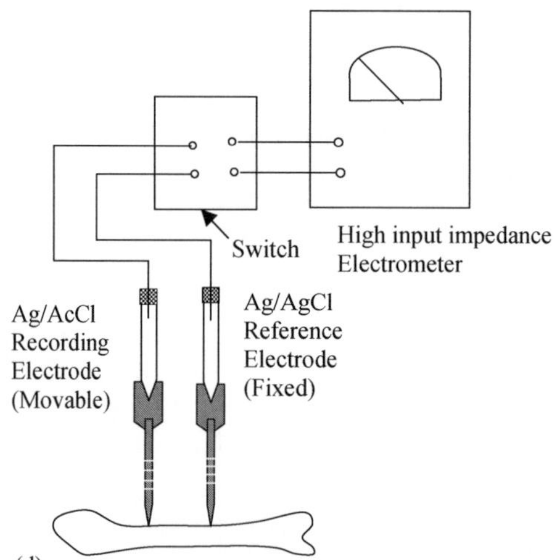


(d)

**Figure 3.6** *(Continued)*

Guided by the above observation that there is a definite potential profile as we go from electrodes 1 (reference electrode) to 6, the external stimulus can be applied so as to control the ongoing bioelectrical processes. This follows from the fact that higher electronegativity will result in enhanced bone growth processes. Reversing the direction of the external stimulus tends to reverse the potential profile along the segment of bone in question. Because a uniform change in potential variation is obtained through electrodes 1–6, it may be interpreted that bioelectrical bone behavior is spatially distributed along this region. No abnormal variations in electrical behavior under question have been reported when the experiments are repeated with electrode positions 1,3 and 1,6.

### 3.7.3 Pulsating Fields

The pulsed electromagnetic field is generated by applying the pulse with a pulse generator in the range of 1–10 V through a copper coil (22 gauge, 15 turns) implanted along the femur and tibia of an adult rabbit. Signals at different frequencies (0.1–10 MHz), pulse delay (0.1–100 μs) and pulse width (0.1–100 μs) were applied.

The biopotentials are recorded after the exposure of fields, where the instantaneous rise is recorded in all cases. The procedure is repeated for seven days and the enhanced values of biopotentials are recorded. Before each stimulation, the potential between 1,2 electrodes (Figure 3.6a and c) is measured. Figure 3.6d shows a method of measuring biopotentials on bone surface.

### 3.7.4 DC Stimulation

A constant current source of 100 $\mu$A is used for DC stimulation. Excitation is given through the implanted electrodes in tibia and femur. On applying DC stimulation (3.5 $\mu$A) there is a significant ($P < 0.001$) rise in voltage at each electrode position (with respect to the reference electrode).

### 3.7.5 Electromagnetic Field (50 Hz) Stimulation Along with Radio Frequency Field Coupling

The control value of the biopotential is measured between electrodes (Figure 3.6a and b), using a high input impedance digital electrometer. To examine the field exposure effect on bone biopotential the current is passed through the solenoid and the biopotential is measured. The solenoid coil used has 200 turns and is 8 cm in diameter (field = 1.8 gauss at the center). The tibia of the rabbit with implanted electrodes is placed at the center of the solenoid to record biopotentials under stimulated conditions. The biopotential is recorded with the field in "on" and "off" positions. Figure 3.7 indicates the effect of a low frequency electromagnetic field alone. Application of this field brings about an increase of bone potential that tends to return to original value once the field is turned off. This clearly indicates the effect of an induced field in "*in vivo*" bone and distinguishes it from DC stimulation (Rai, 1983).

DC and the AC (1.27 MHz) have been applied between electrode positions (1,5) and (1,4), respectively (Figure 3.8). These electrodes are implanted along the longitude in the cortical region of the bone as mentioned earlier. A DC stimulation of 25 and 50 $\mu$A (constant current) is used for excitations that fall in the stimulation range of physiological interest (Friendenberg *et al.*, 1970). A DC stimulation of 50 $\mu$A applied for 15 min raises the potential from +60 to about −500 mV, and an application of 1.27 MHz produces a diminution in the potential, which recovers almost immediately to the original value once the AC stimulation is turned off. This can be attributed to the charge fluctuation with the application of AC stimulation. There is usually a sharp decline in biopotential over a period of about 20 min (after turning off the DC stimulation), with the potential falling close to the basal value. As anticipated, a smaller DC source stimulation leads to a smaller rise in bone basal potential. In this case the 1.27 MHz excitation affords an almost instant rise, which tends to reach to a saturation value, and once the stimulation is turned off the bone potential returns to the original value. The change in potential due to AC excitation is independent of the polarity of the basal value voltages. The effect produced by AC stimulation remains the same, even when it is preceded by DC stimulation.

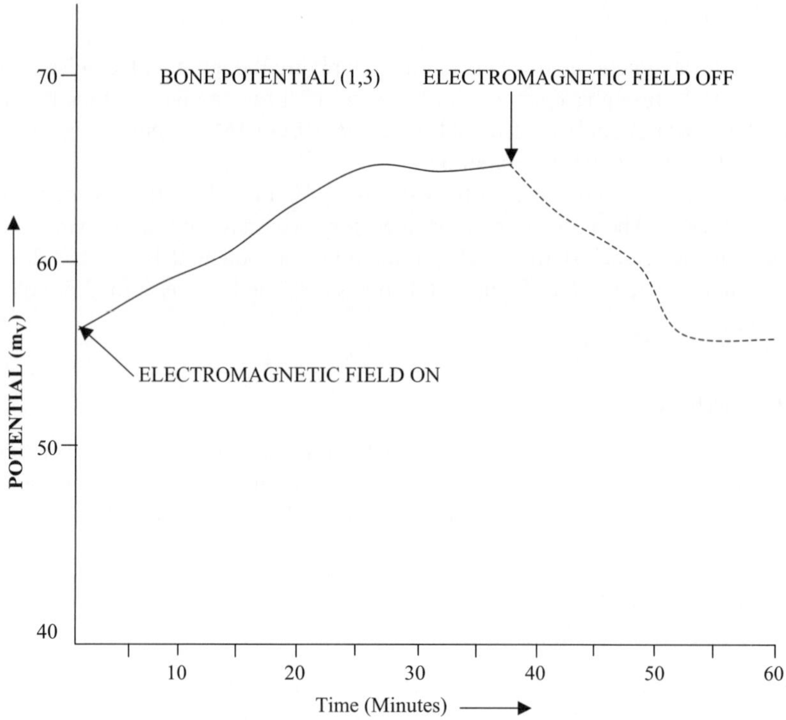


**Figure 3.7** Biopotential under electromagnetic field (50 Hz) stimulation

Figure 3.9 reflects the effect of coupling the electromagnetic field with the possible ultrasonic wave propagation generated by excitation of 1.27 MHz. Owing to 1.27 MHz excitation there is a rapid rise in bone potential followed by a slower rise, which tends to reach a saturation value. At this point an electromagnetic field (50 Hz) (applied through solenoid) tends to increase the biopotential further. Switching off the 1.27 MHz excitation brings about a fall in bone potential, which tends to fall further when the low frequency (50 Hz) field is switched off.

## 3.7.6 Continuous Fields

Another experiment employed a frequency of 1.27 MHz (45 V) to introduce an oscillating field through the implanted electrodes by the signal generator. The surface biopotentials are measured by using needle electrodes on tibia and femur, by removing all skin, muscles and fats from the surface of bone. Needle electrodes made of identical material (stainless steel) and having the same tip sizes are used. The insulation is used for insulation on the needle electrodes except at the tip. The immediate rise in biopotential in longitudinal and transverse directions is measured along the bone by an electrometer. A change in potential due to stimulation is considered to arise because of ultrasound propagation (Figure 3.9) (Rai, 1983).

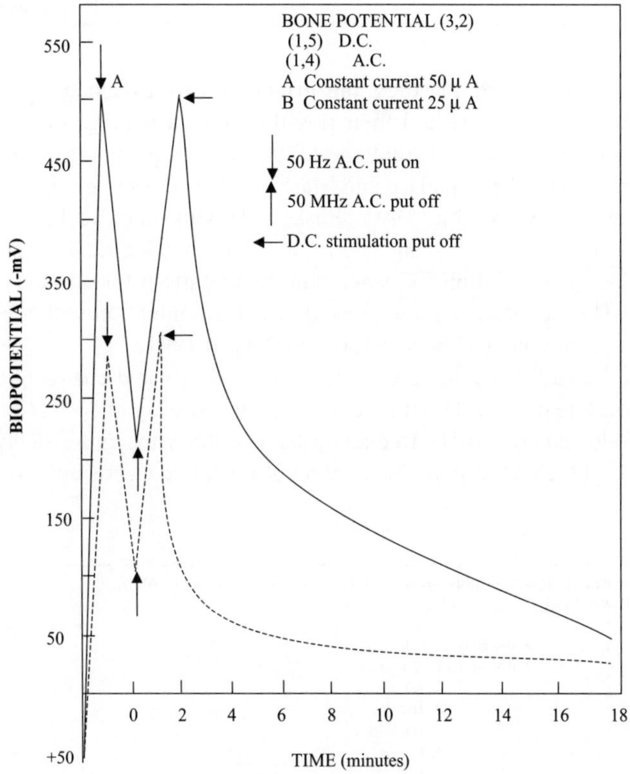


**Figure 3.8** Extremely low frequency electric field (50Hz) stimulation

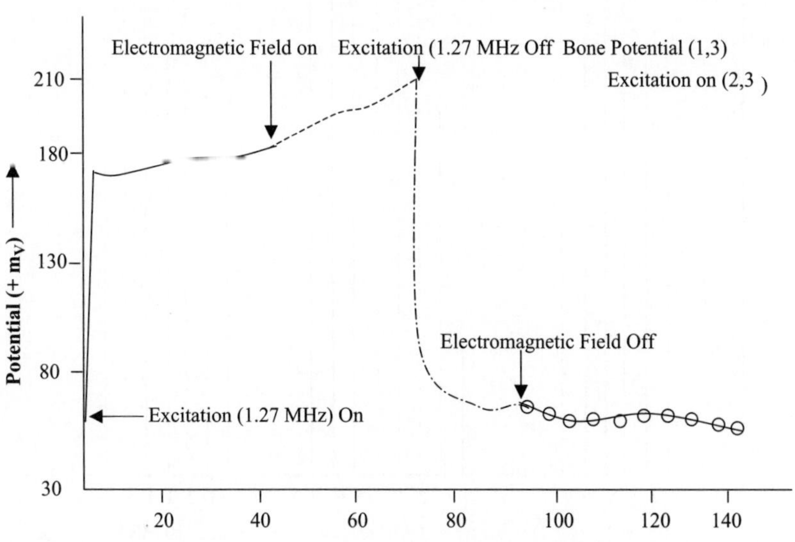


**Figure 3.9** Rise in bone potentials due to low and high frequency electromagnetic field coupling in the dielectric configuration as shown in figure 3.6a

These investigations aimed to study the effect of various electrical stimulations on the bioelectric potential of bone and their possible role in bone growth. Several authors have studied the effects of direct current and low frequency pulsating field stimulations on bone behavior *in vivo* (Becker, Spadaro and Marino, 1977; Bassett *et al.*, 1982; Ilfeld *et al.*, 1974; Levy, Lees and Stevenson, 1992; Bassett, Pawluk and Becker, 1964); the high frequency effect has received scant attention. The bioelectrical state measured by (0.1 MHz–10.0 MHz) stimulation follows a consistent pattern for all pulse delays applied (0.1 μs–100 ms). This suggests that the pulse delay in combination with a given frequency would optimize the biopotential bone response (Behari, 1991).

Figure 3.10 is drawn by keeping the pulse delay at 0.1 μs and changing the pulse width (0.1 μs to 100 ms). Here the 10 MHz stimulation and a pulse delay of 100 ms has the maximum effect, closely followed by 1 MHz frequency having the same pulse delay. Figures 3.10 and 3.11 show that the effect of pulse delay and pulse width depends upon various repetition

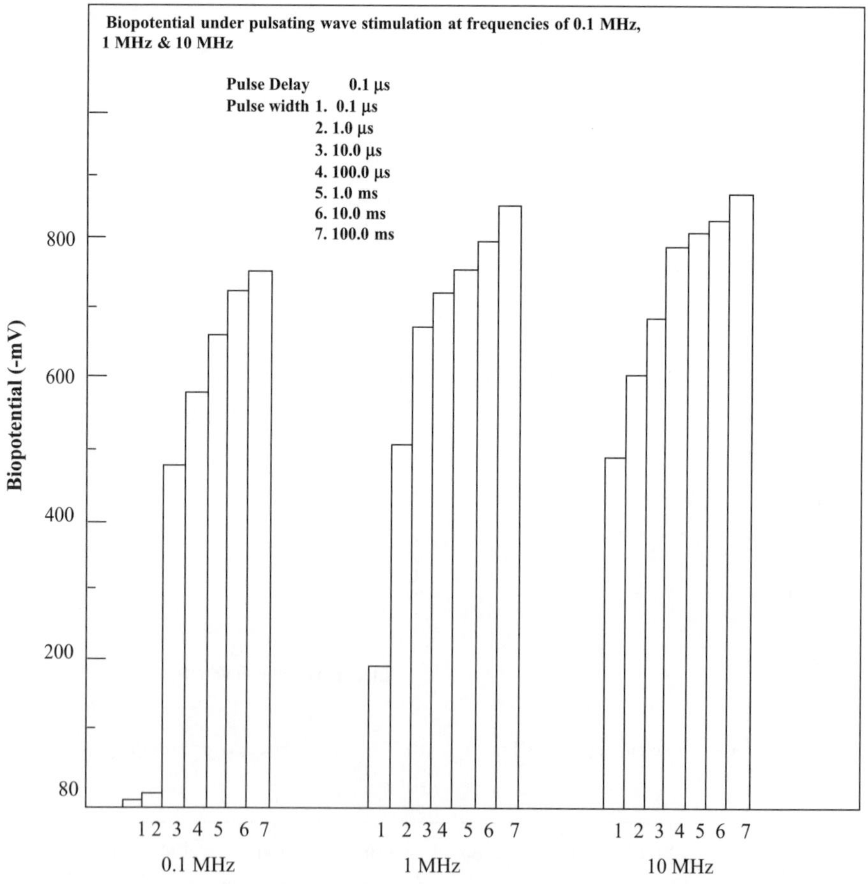


**Figure 3.10** Biopotential under pulsating wave stimulation at frequencies of 0.1 MHz, 1 MHz & 10 MHz

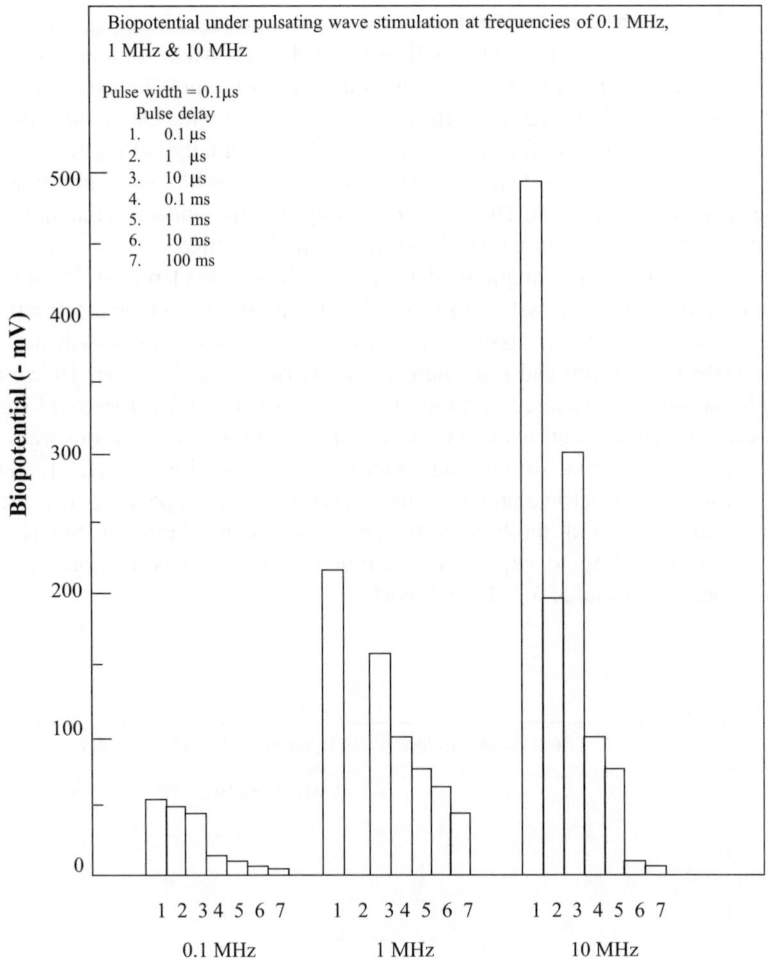


**Figure 3.11** Biopotential under pulsating wave stimulation at frequencies of 0.1 MHz, 1 MHz & 10 MHz

frequencies. The results support the view (Bassett *et al.*, 1982) that pulse induced currents depend on the characteristics of the pulse, for example, pulse design, shape, duration and repetition frequency. The pulse excitation is made more effective by choosing the pulse design to exert selective influences on ionic species near the cell and its membranes and produces frequency matching and response effects (Bassett, 1978). Selective action is not possible with DC. A similar trend has been observed by Friendenberg *et al.* (1974). Guided by these observations, it is concluded that the rise in biopotential depends upon the pulse width (pulse delay). Also, increasing the pulse width produces higher responses (Behari, 1991). These results also incorporates the fact that pulsed signals contain multiple messages and that the biological system response to this is more effective. While a DC stimulus is effectively a polarization, the pulsed signal is a signal transmission.

Figure 3.12 shows the measurement of surface potential with changing distance between two electrodes in tibia and femur. On application of the 1.27 MHz stimulation the potential uniformly decreases in the region extending from tibia to femur. The excitations are given on electrode positions 1,2 and 4,5. Near the tibia and femur joint the electrical activity is more than in other positions. The surface potential varies longitudinally but not transversely. This suggests that endochondrial ossification (Ogden, 1980) responsible for longitudinal growth is greater than appositional growth. The above observation further indicates that bone activity is very sensitive to the manner, in which the stimulations is applied.

An electric current causes augmented healing of bone (Spadaro, 1977; Norton, 1974; Klapper and Stellard, 1974; Hassler *et al.*, 1977). The ability of such an electrical current to mediate cell functions such as controlled healing, regeneration and growth has also been investigated (Becker, Bassett and Bachman, 1964; Harrington and Becker, 1973). Biopotentials from bone can be considered as a measure of electrical activity. Electrical signals have been measured at various points in the bone (tibia and femur). These properties were compared with those obtained after bone stimulation. Since bone possesses inherent electrical activity, it is treated as a source of electrical power. Hence measurement of impedance, loss factor and $Q$ parameter (quality factor) for the bone in the frequency range of interest (500 Hz–100 kHz) provides an understanding to explain the nature of bioelectricity in bone and assist in optimizing electrical stimulus for clinical work.

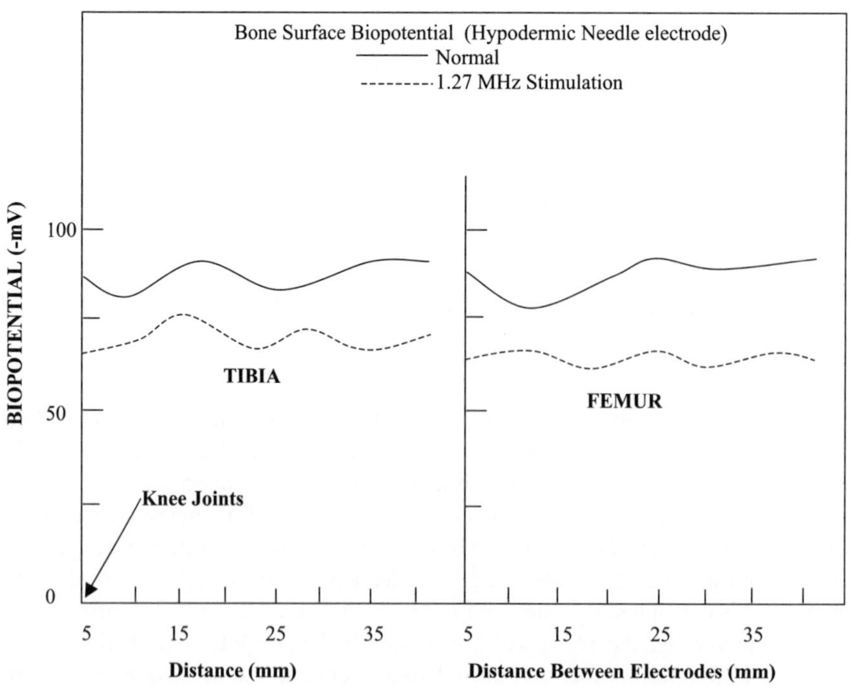


**Figure 3.12** Bone surface potential and 1.27 MHz stimulation

### 3.7.7 Impedance Measurements

The impedance between consecutive electrodes implanted in cortical bone is of the order of ~kΩ and, as anticipated, it decreases with increasing frequency (500 Hz–100 kHz) (Behari and Singh, 1981b). The impedance between different electrodes positions turns out to be of the same order of magnitude (Figure 3.13). The nature of the variation of impedance (and also of resistance and capacitance) with frequency is in agreement with the reported results of Geddes, 1972.

The voltage available across the electrodes is indicative of the fact that bone can be treated as a biogenerator. This concept can be used as a current source in controlling a biofeedback system.

The loss factor and $Q$ have been found in the frequency range 500 Hz–100 kHz; the loss factor in this frequency range falls in the range $10^{-8}$–$10^{-5}$. The corresponding $Q$ values are low (<3). This seems to be in agreement with the work of Kawai (1969) for other high resistivity materials.

Figure 3.14 shows the fall of charge between electrode positions (1,6) with time (for 45 min) after the stimulation was turned off. The process was repeated for seven days. Clearly, the charge values are higher for the seventh day, which may be attributed to the retention of charge by intact bone. At the end of the seven-day period the stimulation (10 $\mu$A) was stopped and the charge between respective electrodes was measured until it returned to the original value of $8.45 \times 10^{-9}$ coulomb (the charge available between electrodes 1 and 6 before stimulation). It took about seven days to fall to the original value from the saturated one ($2.40 \times 10^{-7}$ C). This has also been found to be true for the intermediate electrode (1 and 3) positions.

This is interpreted by suggesting that passing a higher constant current results in a higher degree of polarization and results in an increase in charge retention. It can be inferred

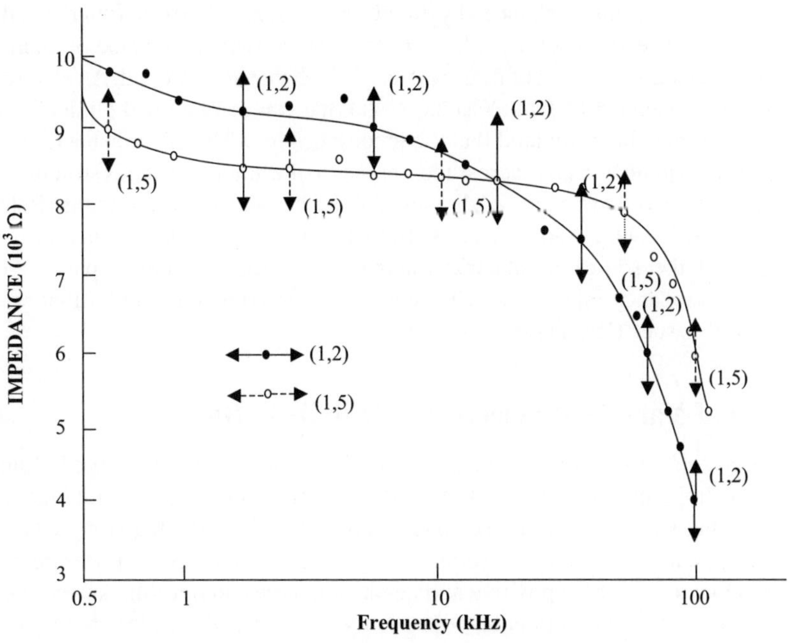


**Figure 3.13** Frequency vs impedance curve between electrodes (1,2) & (1,5)

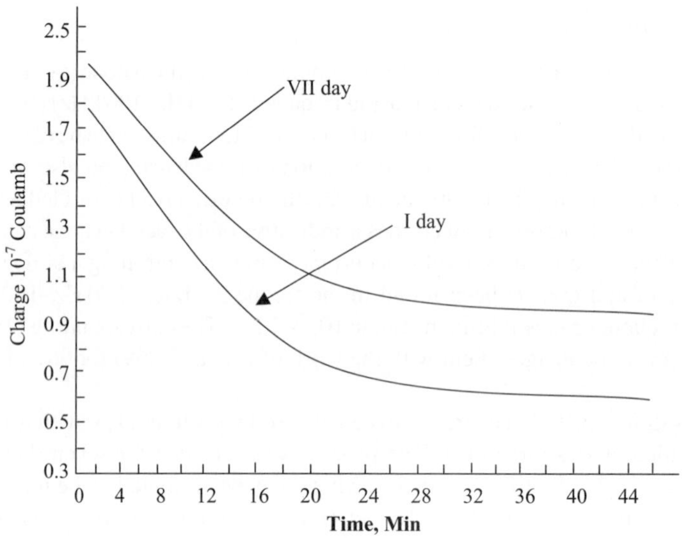


**Figure 3.14** Charge decay versus time between for electrode position (1,6) each day after stimulation. The experiment was repeated for seven days in succession. The saturation value on the seventh day was $2.40 \times 10^{-7}$ coulomb

that retaining a capacity of charge depends on the magnitude of the current (in the region of physiological interest) and the duration of stimulation. From this, the electrical activity of bone *in vivo* is clearly significantly enhanced by direct current. It has also been found that the area of active growth (lower extremity) has a higher electrical activity, and hence by increasing the activity of these electrical signals bone growth can be enhanced/controlled. An electret effect in bone *in vivo* is an established fact (Yasuda, 1977) that can be exploited clinically. From the above analysis, it may be postulated that the precise origin of DC electrical activities may be ionic gradients and ionic transport across membranes to the inhomogeneous structure of tissues that give rise to polarized molecules of dipoles. This agrees with reported results (Behari, Guha and Agarwal, 1974, 1975; Andrabi, 1978; Behari, Rai and Jha, 1979) wherein it has been pointed out that the source of electrical activities are dipoles, ionic spaces and proton conduction ($H^+$ vacancy migration). The behavior observed here is similar that reported for other biological tissues (Repachole *et al.*, 1971).

## 3.8 Origin of Various Bioelectric Potentials in Bone

Friendenberg *et al.* (1973) have recorded potentials from the anterior, medial and lateral surfaces of the tibia that indicated that the surface voltage is constant around the transverse axis of a bone but varies in relation to its long axis. This suggests that stress can not be the origin of such potentials, which is corroborated by the fact that tensile and compressive forces have been shown to distribute positive and negative charges on opposite surfaces of a bone. Further, the steady state potentials on bone are the sum total of the electromotive forces of all the cells in the localized area (Friedenberg *et al.*, 1973; Harlow, Friendenberg and Brighton, 1971).

The sources of potential in unstressed bone are cells. It is thus suggestive that electrical activity is an inherent property of bone and that it varies along its length. These results offer support to the findings of Athenstaedt (1979) and Sawyer and Pete (1953). Athenstaedt (1979) found the existence of a permanent electrostatic field (PESF) in collagenous structure. He pointed out that such potentials play a significant role in controlling the direction of biological growth processes. The generation and reorientation of the permanent dipoles as a result of external stimulus seem to be in a direction that controls the potential variation for enhancement of bone electrical activity. However, the mechanism underlying this phenomenon remains obscure. Sawyer and Pete (1953) have also suggested that the direct current potential may be related to the trabular shape of the bone.

Three types of potentials can in general be identified in living organisms: (i) bioelectrical, (ii) action and (iii) strain related. Bioelectric potentials are generally attributed to the difference in potentials and are a general property of all living tissue (Burr, 1950). At the time of occurrence of fractures there is a sharp rise in electronegativity, which tapers with the healing, stress and muscle contraction and decreases in magnitude following bone death (Friendenberg and Brighton, 1966). These are in the region of 10–20 mV. Action potentials are of much higher magnitude and arise from muscle contraction or nerve stimulation. These are of very short duration (milliseconds) and depend upon cellular viability. On the other hand, strain related potentials have their origin in the strained collagen fiber or hydroxyapatite crystal. These signals last for a duration for a period of the application or release or stress (Bassett and Becker, 1961; Becker and Murray, 1970), indicating a time dependance.

To establish possible clinical significance of the above-mentioned, more measurements on similar lines are required. This is supported by the fact that bone growth can be modulated by various parameters (Yasuda, 1977; Hassler *et al.*, 1977; Becker, Spadaro and Marino, 1977; Connolly, Hahn and Jardon, 1977; Engler *et al.*, 1978; Bassett, Pawluk and Pilla, 1974). Before formalizing such a relationship on a quantitative basis, precise measurement of electrical parameters must be carried under *in vivo* conditions. These parameters may be:

- bone biopotentials;
- surface potentials;
- cavity potentials;
- after electrical stimulation, the above potentials and electrical properties such as impedance values.

Such data are expected to lead to a better understanding of biological events (including pathological) that are important for living organisms and particularly in bone. Electrical signal generation and propagation is, in general, a key parameter in understanding the growth and development of life processes.

# 4

# Solid State Bone Behavior

## 4.1 Introduction

The solid state of bone more than any other biological tissue has been a subject of continuing investigation because of its varied behavior. The prime reason seems to be that it retains and reveals several of its physical properties, even after extraction from the body. Intrinsically, it is subjected to stresses and consequent generation of charges and currents that contribute to bone growth and development. Bone possesses various solid state properties that are governed by low fields and currents.

The concept of charge transport in biology is of utmost importance, for it holds key to understanding several biological phenomena. While mainstream thinking about life processes is oriented towards molecular and ionic reactions, key processes such as locomotion, photosynthesis, respiration and vision are electrically mediated. The idea that electron transport mechanisms and their control should be universal in biology is not new and, indeed, it preceded recent developments in electronics. Szent-Gyorgyi (1941a, 1941b) pointed out that electrons play an important role in cellular processes, especially in cell division and even in cancer. The author suggested that the proteins, long-chain active polymeric molecules, provide pathways for electron flow that may be regulated in the cell by semiconductor-like interaction of acceptor or donor molecules with the protein. The concept that proteins and collagen might be semiconductors has been pursued experimentally over many years (Cardew and Eley, 1959; Eley and Spirey, 1960; Eley, Lockhart and Richardson 1977, 1979; Rosenberg and Postow, 1969; Behari *et al.*, 1975). Such studies have shown that samples in the form of compressed discs have temperature sensitive conductivities with activation energies of around 2–3 eV.

In contrast, cell membrane systems are topologically closed and have a conductance many millions times less than the aqueous medium on either side of them. They consist of phospholipid molecular bilayers less than 50 Å thick, together with proteins that either completely span the membranes or are partially integrated with them. The proportion of phospholipid to protein varies with the type of membrane. Each phospholipid molecule has a hydrophilic phosphatidic head group and a hydrophobic hydrocarbon tail, the latter being responsible for giving the membrane its insulating properties.

*Biophysical Bone Behavior* Jitendra Behari

Though the early ideas of Szent-Gyorgyi that biological systems might act as semiconductors are still valid, a biosystem cannot be treated as an elemental crystalline semiconductor as this often leads to incorrect ideas about the character and transport process of electrons. In less periodic materials (such as biological matter) where electron states are localized, the passage of electrons and holes can be extremely slow, sequential and deliberate. In semiconductor crystals, by contrast, their passage is naturally haphazard and, until their free wanderings are brought under control by the erection of barriers and filters built at the limits of modern technology, they do not act intelligently or take on even the simplest attributes of a living system.

This field was further developed by earlier experiments conducted by several group of workers (e.g., Bassett, Pawluk and Becker, 1964; Bassett, Pawluk and Pilla, 1974; Becker, Bassett and Bachman, 1964; Behari, Guha and Agarwal, 1974, 1975; Friendenberg *et al.*, 1970) who observed that when electric current of different magnitudes is made to pass through bone (*in vivo*) osteogenic and osteoclastic like activity took place at the negative and positive electrodes, respectively. In addition, this conductivity can be varied by physical methods. The results of some of these experiments generally indicate that DC currents in the range 1–100 μA, or an applied voltage of 1.0–1.5 V, are sufficient to induce osteogenesis and, consequently, the efficiency of wound healing can be increased under induced electrical conditions. It is further suggested that the electrical environment of mesenchymal cells may control, to a large extent, their mitotic and functional activity. Because of the above, bone solid state (electrical) behavior remains a favorite subject of investigation.

Equipped with present knowledge of electrical stimulus induced osteogenesis and with the knowledge of piezoelectricity, it can be concluded that such behavior significantly affects the function of bone in the body.

## 4.2 Electrical Conduction in Bone

Interest in bioelectric bone behavior began with Bassett (1991) first reporting electrical polarization in cantilever beam samples of fresh bone and wood during dynamic deformation.

There is ample evidence that not only bone but also its two major constituents exhibit solid-state behavior. This finding is of primary importance for it may lead to a better understanding of basic phenomena in biology. The need is, therefore, to correlate the observed phenomena to behavior *in vivo* and to other known properties.

### *4.2.1 Bone as a Semiconductor*

Piezoelectric like effects in bone and the suggested role of semiconduction in biology (Szent-Gyorgyi, 1941) has significantly extended the scope of investigation in the field. The relevant solid state properties in relation to bone are subsequently taken into account (Shamos and Lavine, 1964; Becker and Brown, 1965; Bachman and Ellis, 1965; Becker, Bassett and Bachman, 1964; Becker and Marino, 1966; Lakes, Harper and Katz, 1977; Behari and Andrabi, 1978; Behari, Rai and Jha, 1979). The mechanism and nature of charge transport is the first such phenomena that needs to be investigated. The difficulty here lies in the complex composition of bone and the consequent difficulties in drawing energy level diagrams. To explain the electrical conductivity in bone Behari, Guha and Agarwal (1974, 1975) pursued an indirect approach.

The free electron resonance, analyzed by EPR spectroscopy, the photoconductivity (Becker and Brown, 1965) and fluorescence in bone (Bachman and Ellis, 1965) show that bone possesses solid state properties and behaves as a semiconductor.

Various solid state materials have been grouped on the basis of their electrical resistivity (reciprocal of conductivity). Conductors are loosely defined as materials having resistivities less than about $10^{-3}\,\Omega\,\mathrm{cm}$ (metals) and insulators are considered to include substances with resistivities of $10^{10}\,\Omega\,\mathrm{cm}$ (ebonite, glass, mica, etc.).

It was reported earlier that dry bone is a semi-insulator with a resistivity in the range of $10^{10}$–$10^{12}\,\Omega\,\mathrm{cm}$ and activation energy of greater than 6.7 eV (Shamos and Lavine, 1964). Wet bones, in contrast, have corresponding parameters in the range of $10^{8}$ ohm cm and <1 eV (Becker, Bassett and Bachman, 1964). The same authors also reported that bone has many characteristics of a semiconductor. These characteristics of a semiconductor are related to the semiconduction properties of each of the two major components of the bone matrix and structured water.

Another parameter that governs bone conductivity is temperature. In pursuance of these ideas, the temperature dependence of the resistivity of bone under varying field strengths has been measured. These studies offer some data relating to the transport phenomenon in bone. However, limited information is available. In an earlier attempt (Behari, Guha and Agarwal, 1974) the variation of resistance with temperature was measured between ambient temperature up to 60 °C. The slope of the straight line plot obtained gives the energy gap in the expression:

$$P = A\exp(E_g/2kT) \tag{4.1}$$

where $A$ = constant; $E_g$ = energy gap (in joules), $P$ = resistivity, $T$ = temperature (kelvin).

Values of $42.5 \pm 1.2$, $37.3 \pm 0.5$ and $41.8 \pm 0.6 \times 10^{-20}$ J were obtained for the three samples and are consistent with results reported elsewhere.

Figure 4.1 shows the voltage–current characteristic curve. The values of current as voltage increase up to 240 V are the same as those obtained for a decreasing voltage, but as reported earlier (Becker and Brown, 1965) the slope changes when the polarity is reversed. These results indicate the effect on bone conductivity of temperature. In another study tibiae from a middle aged human exhibited similar results to those of adult rat bone, with activation energy that is independent of temperature up to 322 K. Becker, Bassett and Bachman (1964) obtained activation energies of 1 eV for hydrated bone and 5 eV for collagen.

Dehydrated bone exhibits (semiconductor properties) conductivity that increases nearly exponentially with temperature. The activation energy of dehydrated rat tibiae decreases more rapidly with age below body core temperature than at body temperature while above body temperature the activation energy shows little variation with age. As the bone approaches maturity, the activation energy becomes independent of temperature, reflecting the effect of complete mineralization. The independence of activation energy on temperature at maturity indicates that the interaction between the bone apatite and collagen has changed. Brown and Aftergut (1962) have postulated that the substance with lower activation energy is preferentially excited to the triplet state and, as the temperature is increased, the higher activation energy substance reaches the triplet state for the case where no interaction occurs between the two components. Below the age of 240 days, the rat tibiae activation energy decreases with increasing temperature, indicating that the interaction between the apatite and collagen is of

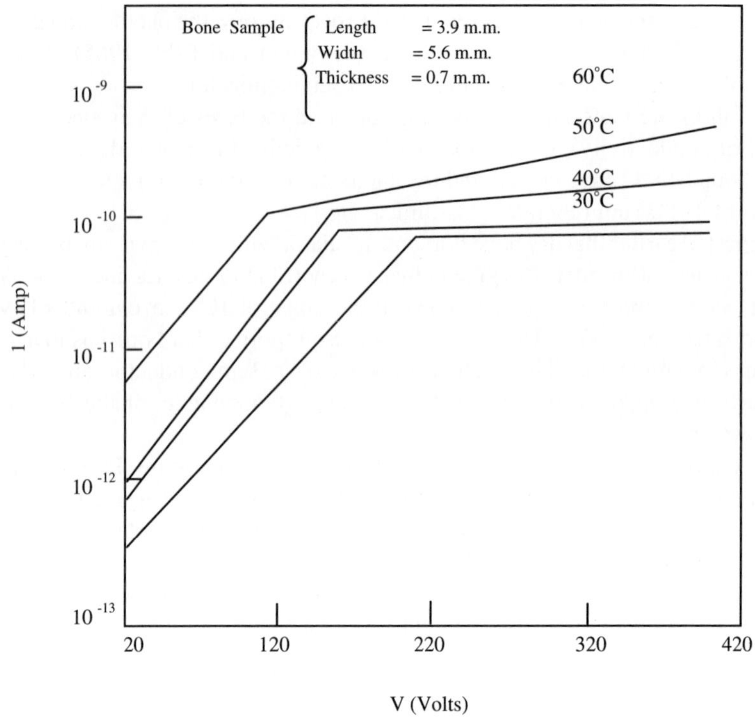


**Figure 4.1** *V–I* Characteristics of a bone sample at temperatures 30 °, 40 °, 50 ° and 60 °C (x axis represents Volts)

primary importance, while above this age the highly mineralized bone with a low activation energy behaves as a single component material with activation energy independent of temperature. On the other hand, hydrated bone behaves like a heterogeneous dielectric with a broad spectrum of relaxation times.

It can be assumed that charge carriers (during conduction process) are sensitive to radiation and this can affect the electrical properties of bone. Further, bone and its two major constituents (collagen and apatite along with water) are hydrogen-bonded structures (Ramachandran, 1967; Berendsen, 1962; Hamilton, 1968; Urback, 1969 and Behari and Guha, 1976). The findings of these authors suggest that the conductivity in all three cases (i.e., bone, collagen and apatite) decreases by one to two orders of magnitude upon exposure to UV radiation (Figure 4.2). They attributed this to the breaking of hydrogen bonds, which then offer impediments to charge transport. This further led them to suggest that protonic conduction may be predominant in such types of solids.

The foregoing discussion shows that a definite link exists between the bioelectrical activity, semiconduction and other solid state properties of bone. One event flips the other and at certain stages a series of cyclic phenomena takes place. Whereas, the bioelectrical properties are thought to be responsible for the development of bone, the solid state nature can evoke, sustain, store, convert and direct them as command signals. But the gap between the findings *in vivo* and *in vitro* needs to be more clearly understood and firmly cemented. Apart from this the nature of

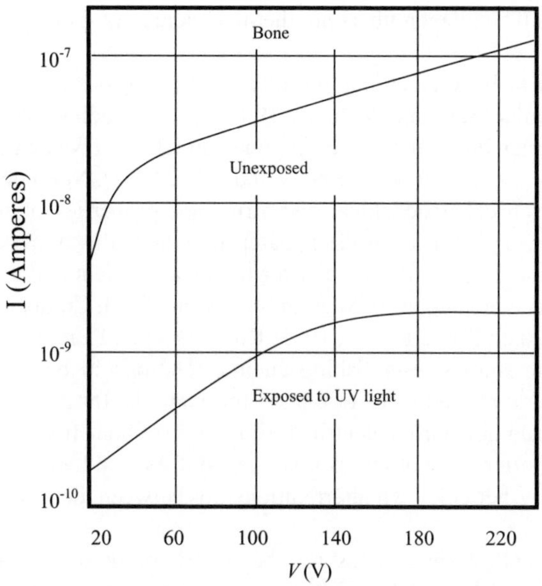


**Figure 4.2** *V–I* characteristics of bone; effect of exposure to ultraviolet light

the charge carriers and conduction mode is not yet well understood. Consequently, various experiments have been undertaken, both *in vivo* and *in vitro*, and efforts have been made to lessen the lacunae in the available data, and to work out other solid state and semiconductor properties in bone.

### *4.2.2 Bone Dielectric Properties*

Reinish (1975) found that the permittivity increased with increasing humidity and decreasing frequency. Dielectric values at low frequency are of interest because a pulsed field at these frequencies stimulates bone growth *in vivo* (Levy, 1974). Also, bone in the body is subjected to these frequencies in every day activity. Biologically important energy transfer processes may occur in bone at a frequency of a few hertz, corresponding to the rate at which the bone is stressed in walking. Marino, Becker and Bachman (1967) have performed dielectric measurements on bone and concluded that the effects of free and bound water can be distinguished and that a value for the critical hydration corresponding to filling of the bound water compartment in bone can be assigned.

Several *in vitro* studies have been carried out to clarify the electrical and dielectric properties of either compact, trabecular or whole bone (Freeman, 1967; Cochran, Pawluk and Bassett, 1968; Chakkalakal and Johnson, 1981; Singh and Saha, 1984; De Mercato and Garcia-Sanchez, 1988; Saha and Williams, 1989). To separate the effect of the fluid phase from that of the solid matrix, studies on dry bone (DeMercato and Garcia-Sanchez, 1922) and fluid saturated tissue (Kosterich, Foster and Pollack, 1983) have been undertaken. These studies have shown that the high frequency limit of relative permittivity depends on the water content of the tissue and that the collagen fibers are the main source of dielectric dispersion in a long bone.

The measurement of bone electrical and dielectric properties may provide a tool for the diagnosis of bone quality (Saha and Williams, 1989). Earlier studies investigating properties of compact and trabecular bone in three orthogonal directions revealed strong electrical and dielectric anisotropy of compact bone (Reddy and Saha, 1984). Not much is known about the relation between electrical, mechanical and structural properties of trabecular bone. The electrical and dielectric properties of tissue determine the pathway of current flow; a deeper understanding of these properties is crucial for the measurement and analysis of parameters obtained by impedance techniques (Foster and Schwan, 1989). Compared to *in vitro* studies, few *in vivo* studies exist (Tzukert *et al.*, 1983; Rubinnacci and Brigatti, 1984; Skinner *et al.*, 2001). In therapeutic studies, establishing an interrelationship between bone density and electrical and dielectric properties is necessary, for example, for the calculation of electrical field and current distribution during electrical stimulation of bone fracture (Williams and Saha, 1996). Bone healing after corticotomy is successfully followed by correlation with electrical impedance. Obviously, better known interrelationships between these properties may be of use in clinical practice.

Saha and Williams (1989) have measured the electric properties of wet human cancellous bone as a function of frequency. These authors reported that the resistivity decreased by 15% from 1 kHz to 1 MHz and then remained fairly constant for the frequency range up to 10 MHz. The log of the specific capacitance decreased as the log of the frequency increased, showing an inverse relationship, which changed less at higher frequencies (Figure 4.3). The specific impedance is almost equal to the resistivity at lower frequencies. Table 4.1 summarizes the electric properties of whole bone.

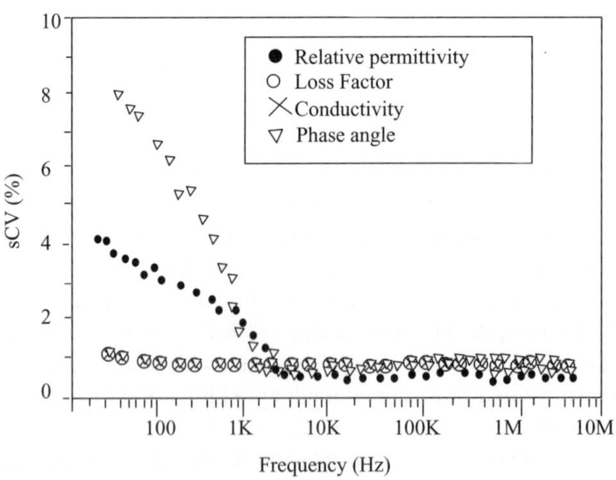


**Figure 4.3** Standardized CV [sCV (%)] of measured electrical and dielectric parameters as a function of frequency. At frequencies over 1 kHz, the sCV was less than 1% for all electrical and dielectric parameters (Reproduced with permission from J. Sierpowska *et al.*, Electrical and dielectric properties of bovine trabecular bone: relationships with mechanical properties and mineral density, *Physics in Medicine and Biology*, **48**, 775–786 © 2003, Institute of Physics and IOP Publishing Limited)

**Table 4.1** Electrical properties of whole bone (Behari, 1991)

| Sr. no. | Specimen and source | Range of measurement | Results | Techniques | Reference |
|---|---|---|---|---|---|
| 1 | Human femur | $V$ vs $I$ curve (DC) | Sample resistance increases with exposure to ultraviolet light, but decreases with increasing temperature | Two-point method | Behari, Guha and Agarwal (1974, 1975) |
| 2 | Rabbit femur human tibia | $V$ vs $I$ curve (DC) | $R = 2–5 \times 10^5\ \Omega\,cm^{-1}$ | Two-point method | Liboff *et al.* (1975) |
| 3 | Rabbit tibia (*in vivo*) | $V$ vs $I$ curve (DC) | $R = 3.5\ M\Omega$ for cortical bone | Two-point method | Stan, Sensen and Mulier (1976) |
| | | AC 0.01–1 kHz | $C = 20 \pm 8\ \mu F$ | | |
| 4 | Rabbit femur (*in vivo*) | $V$ vs $I$ curve (DC) | $R = 20\ k\Omega\,cm^{-1}$ | Two-point method | Sansen *et al.* (1978) |
| | | $V$ vs $I$ curve AC (00.01–1 kHz) | $C = 5\ \mu F\,mm^{-2}$ | Two-point method | |
| | | | $R_b = 2–3\ k\Omega\,cm^{-1}$ | Four-point method | |
| 5 | Sheep bone (fresh without periosteum) | $V$ vs $I$ curve | $R = 2–3\ k\Omega\,cm^{-1}$ | Four-point method | Durand *et al.* (1978) |
| | | AC (100 Hz–5 kHz) | | | |
| 6 | Rabbit femur (*in vivo*) | AC (500 Hz) | $Z = 8\ k\Omega\,cm^{-1}$ | Two-point method | Behari and Singh (1981a) |
| 7 | Bovine femur (fluid saturated) | DC step function | $P_1 = 4.5–4.8\ \Omega\,cm^{-1}$ | Direction-dependent | Chakkalakal and Johnson (1981) |
| 8 | Rat femur | AC(1 kHz) impedance analyzer, vector impedance meter | $P_1 = 12.9 \pm 2.7\ \Omega\,cm^{-1}$ fresh | Cell suitable for radial direction only | Kosterich, Foster and Pollack (1983) |
| | | | $P_1 = 4.8 \pm 0.1\ \Omega\,cm^{-1}$ fixed | | |
| 9 | Bovine femur (wet) | AC 10 kHz | $P_r = 54\ k\Omega\,cm$, $C_{sP,r} = 21.4\ f\,cm^{-1}$<br>$P_c = 36\ k\Omega\,cm$, $C_{sP,c} = 24.74\ f\,cm^{-1}$<br>$P_1 = 17\ k\Omega\,cm$, $C_{sP,1} = 60.87\ f\,cm^{-1}$ | Differential technique | Reddy and Saha (1984) |
| 10 | Goat femur (wet) | 400–1300 MHz | 150–240 $ms\,m^{-1}$ | Network analyzer | Ray and Behari (1986) |
| 11 | Wet cancellous bone (human) (RH = 100%) | 120 Hz–10 MHz | Mean resistivity ($\Omega\,cm^{-1}$) at 100 kHz<br>500 (longitudinal direction)<br>613 (anterior posterior direction)<br>609 (lateral-medial direction) | LCR meter | Saha and Williams (1989) |

**Table 4.2** Electrical properties (mean ± s.d.) of 30 cancellous bone specimens measured at 100 kHz

| | Longitudinal | Anterior-posterior | Lateral-medial |
|---|---|---|---|
| Resistivity ($\Omega\,cm^{-1}$) | 580 ± 170 | 613 ± 221 | 699 ± 242 |
| Specific capacitance ($pF\,cm^{-1}$) | 8.64 ± 2.53 | 15.25 ± 3.71 | 14.64 ± 4.56 |

Table 4.2 shows the means and standard deviations for the resistivity and specific capacitance of cancellous bone specimens measured in the three directions at 100 kHz. Both the specific capacitance and resistivity are somewhat lower in the longitudinal direction than in the anterior-posterior (AP) or lateral medial (LM) directions.

The transversely isotropic nature of cancellous bone in the distal tibia is interesting. The difference in capacitance values between the longitudinal and transverse directions is significant ($P < 0.01$) at 100 kHz and 1 MHz but not at the lower frequency of 10 Hz. The same authors have reported a highly significant ($P < 0.001$) and positive correlation between the resistivities in the longitudinal ($R_L$) and AP direction ($R_{AP}$). This relationship is expressed as:

$$R_{AP} = -1.70 + 12.3\,R_L \tag{4.2}$$

where $R_{AP}$ and $R_L$ are in $\Omega\,cm^{-1}$. These authors further stated that the variation in electrical properties for longitudinal and transverse directions can be explained in terms of the orientation of trabeculae in cancellous bone samples.

Since most studies have focused on compact bone, knowledge of the electrical and dielectric properties of both components is desirable. Sierpowska *et al.* (2003) have studied the relationships between bovine trabecular bone's electrical and dielectric properties, mechanical characteristics and mineral density (Figure 4.3). Measurements of trabecular bone electrical and dielectric properties are highly reproducible over a wide range of frequencies. At frequencies above 1 kHz, standardized coefficient of variations (SCVs) are less than 1% for all electrical and dielectric properties. Below 1 kHz the reproducibility is less. The relative permittivity showed the strongest linear correlation with mechanical properties ($r > 0.547$, $P < 0.01$, $n = 40$ at 50 kHz), and with $BMD_{vol}$ ($r = 0.866$ $P < 0.01$, $n = 40$ at 50 kHz).

In general, at frequencies over 6 kHz, linear correlations between relative permittivity and $BMD_{vol}$ or mechanical properties are found to be maximum, even though they are still dependent on frequency. $BMD_{vol}$, mechanical properties and dielectric and electrical properties show significant variation ($P < 0.05$). At 50 kHz, the relative permittivity reaches the highest value (381 ± 61) in the femoral head and the lowest (85 ± 35) in the greater trochanter. The highest $BMD_{vol}$ ($0.586 \pm 0075\ gm\,cm^{-3}$) is also found in the femoral head and the lowest ($0.198 \pm 0.032\ g\,cm^{-3}$) is in the greater trochanter. Similarly, all mechanical properties reached their highest and the lowest values in the femoral head and greater trochanter, respectively.

Following Williams and Saha (1996), one concludes that, relative permittivity and phase angle show a linear and phase angle showing linear correlation with $BMD_{vol}$. However, Sierpowska *et al.* (2003) suggests a stronger relation between the relative permittivity and density ($r = 0.866$) at 50 kHz than that at 100 kHz ($r = 0.617$). This difference may be related to the use of bovine femoral bone, instead of human tibial bone, studied by Williams and Saha (1996). Bovine trabecular bone is denser and stiffer than human bone (Jiang *et al.*, 1998; Toyras *et al.*, 2002). Initially, De Mercato and Garcia-Sanchez (1988) found that the relative permittivity decreased with increasing compactness of bovine bone, that is, they proposed a negative correlation between the bone mineral density and relative permittivity. This contradicts the findings of Sierpowska *et al.* (2003) and Williams and Saha (1996) of a positive density–permittivity relation.

In two samples from different anatomical sites and with similar $BMD_{vol}$, the permittivity is also highly similar (Sierpowska *et al.*, 2003). The conductivities, however, are different. Obviously, the conductivity is affected by different factors than the permittivity, and different properties of bone are reflected by these two electrical parameters.

While it is suggestive that both phases (mineralized tissue and marrow) contribute to the results, it may not be possible to determine the quantitative contribution of each phase, especially over a wide frequency range. The simple rule of mixtures for trabecular bone matrix with physiological solution or bone marrow as an inclusion cannot be applied successfully for either conductivity or permittivity. The effect of the wet phase in the composite structure of bone is smaller than the estimate based on the amount and the values of separated wet phase (Kosterich, Foster and Pollack, 1983). For example, the conductivity of bone is only 1% of the value of the immersion solution, although the content of the wet phase is 15–20% in volume. In the frequency range kHz–MHz other factors, such as microstructure or polarization by diffusional processes in double layers surrounding charged interfaces, may play a major role. Measurements of the electric properties of wet bone with and without marrow could provide more evidence about the role of the wet phase in the electrical properties of trabecular bone.

Phase angle, loss factor and especially conductivity are poor estimates of bone density or mechanical properties, as compared to the relative permittivity. The conductivity is mainly a consequence of ion movements through the fluid phase and may change without any change in mineral density. The absolute values of electric and dielectric parameters reported here are in agreement with other studies (Saha and Williams, 1989; Gabriel, Gabriel and Corthout, 1996). Both absolute values of electrical and dielectric parameters as well as the correlations between the mechanical properties or $BMD_{vol}$ with relative permittivity depend significantly on the frequency. The site-dependent variation of electrical and dielectric parameters may be related to differences in tissue structure (e.g., size and orientation of trabeculae) as well to differences in spatial distribution and density of the liquid phase (Kosterich, Foster and Pollack, 1983; De Mercato and Garcia-Sanchez, 1988, 1992).

The electrical parameters account for about 30% of the variation in mechanical properties (ultimate strength, Young's modulus). The correlations might possibly be even higher with some viscoelastic parameters of bone. Although the electrical parameters do not characterize mechanical properties of bone perfectly, they could still be used in some clinical cases. In this connection it may be mentioned that the electric parameter predicted mechanical characteristics of bovine are better than the broadband ultrasound attenuation (BUA) measurements (Hukulinen *et al.*, 2002), which are known to be feasible in osteoporosis diagnostics and predict risk of fracture for elderly people as effectively as BMD measurements (Kroger *et al.*, 1995; Schott *et al.*, 1995; Pfeifer, Pollaehne and Minne, 1997).

Tomaselli and Shamos (1973) have worked on the dielectric properties of hydrated collagen and the effect of water on the complex dielectric constant of bovine Achilles tendon in the frequency range 100–10 000 Hz. The experiments at room temperature show that the dry collagen exhibits almost no frequency or temperature dependence. When the water content of the collagen increases the dielectric constant and the loss factor increase simultaneously. They also showed a sharp rise at a hydration level, which may be related to occupation of the primary adsorption sites. The authors found that, at about 40 °C, the dielectric constant of hydrated collagen takes its maximum value for various frequencies. This result is attributed to (i) the state of bound water molecules changes to the free state and (ii) the water molecules diffuse out of the system. Their measurements for the dielectric constant of dry collagen at room temperature give the value 4.5. However, for elevated temperatures, the dehydration reduces this value to about 2.3. The above experiments verified that (Lang, 1966; Friendenberg *et al.*, 1970; Shamos and Lavine, 1967) the main cause of the piezoelectric behavior of bone tissue is the collagen molecules and that the difference in piezoelectric properties between dry and wet bones is also caused by the structure of the collagen (Bassett and Pawluk, 1972; Anderson and Eriksson, 1968; Liboff, Shamos and De Virgilio, 1971; Liboff and Furst, 1974; Tomaselli and Shamos, 1973).

Several other authors (Toyras, Kroger and Jurvelin, 1999; Toyras *et al.*, 2002) have shown linear correlations between the mechanical properties and BMD. The values of electrical properties for wet trabecular bone as measured lie between the values reported for compact bone and those for bone marrow (Smith and Foster, 1985; Gabriel, Gabriel and Corthout, 1996). The conductivity in bone with high BMD approaches that of compact bone. These authors have concluded that relative permittivity is, comparatively, a better estimate of bone density or mechanical properties. Sierpowska *et al.* (2003) further concluded that there is a linear correlation of trabecular density and mechanical bone properties with the relative permittivity. This suggests that measurement of electrical and dielectric properties may provide the means for quantitative diagnosis of bone status and that 50 kHz is an optimal choice in diagnostic measurement.

The significant linear correlations of trabecular bone density and mechanical properties with relative permittivity suggest that measurement of electrical and dielectric properties may provide the means for quantitative diagnostics of bone status. The correlations between the mechanical parameters and $BMD_{vol}$ are higher than those for electrical parameters. Electrical parameters may not provide more information on the bone strength or stiffness than that obtained with DEXA. At most they may offer some supporting data on bone quality and structure to supplement DEXA measurements. In special cases, such as during bone grafting or total hip replacement, there is no method for direct measurement of bone quality. Electrical measurements utilizing open ended probe technology (Whit Athey, Stuchly and Stuchly, 1982) are potentially feasible in this kind of application. In addition, better understanding of the relationships between bone density, mechanical, electrical and dielectric properties may help us in pathological situations to improve techniques related to electrical stimulation and follow up of healing processes in fractured bone (Williams and Saha, 1996).

## 4.3 Microwave Conductivity in Bone

Microwave dielectric constants provide clues as to the mechanism of interaction of an electromagnetic field with biological systems. The complex impedances of dry and fluid saturated bones can be measured using a Network analyzer system. The technique provides good phase resolution in the frequency range 400–1300 MHz. This provides a comprehensive view of the conductivity and permittivity of bone, and to further assess the nature of bonding in bone and its relation to the dielectric constant UV-irradiated bone has also been examined (Ray and Behari, 1985, 1986).

Specimens of bone were obtained from femurs of adult goat. The cylindrical samples of compact bone specimen used were machined from femur portions using a low speed saw. Continuous water irrigation was applied to minimize any thermal damage during the machining process. The final specimens were approximately 7 mm in diameter and 1 mm thick. A 3-mm hole was glued through the center of each sample and filled with conducting silver cement in each case. Proper care was taken during the drilling so that the temperature of the sample did not increase; any increase may affect the microscopic structure. For wet bone, the specimens were kept in 0.9% saline solution for 24 hours before experiment to achieve complete equilibrium at 17 °C and 72% humidity level. Additional liquid was then blotted out from each sample before the start of an experiment. After an experiment, each sample was exposed to UV light source (300 watt, 350 μm). The light was then turned off and the samples allowed to equilibrate with room temperature environment for 24 hours before making the measurements. To obtain collagen specimens, fresh bone samples were taken and kept in 15% formic acid at 52 °C for 18 hours with occasional stirring. This dissolves apatite and the chemical process leaves only collagen components along with some macromolecules and minerals. Thereafter the samples were treated similarly as those of whole bone.

The key instrument used is the H.P. Network Analyzer (8754A) and reflection transmission test set-up (8402A) (Figure 4.4). The sample is placed at the test port of a reflection

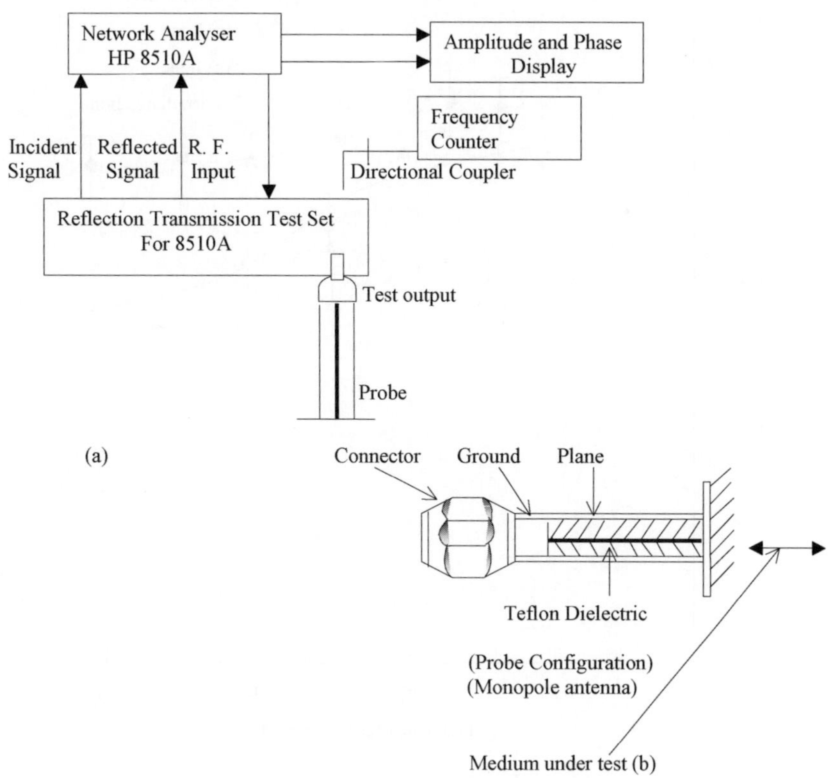


**Figure 4.4** Schematic diagram of instrumentation with a network-analyzer

transmission test set-up so that the silver-filled hole makes a continuation with the inner conductor of the test port. The reflection coefficient is calculated from the measured reflection loss and change in phase angle. The instrument is standardized at the reference plane of the test port using short circuit, open circuit and matched load at three different positions (da Silva and Mephun, 1978). From the reflection coefficient data the impedance is calculated. The permittivity ($\varepsilon'$) and conductivity ($\sigma$) are calculated from the following relationship in which bone is modeled as a right circular cylinder (Reddy and Saha, 1984):

$$\sigma = \frac{G}{2\pi l}\ln\frac{a}{b}$$

and:

$$\varepsilon' = \frac{C}{2\pi\varepsilon_p b}\ln\frac{a}{b} \tag{4.3}$$

where $\sigma = \omega\varepsilon_p\varepsilon''$, with $\varepsilon''$ being the relative loss factor permittivity of free space. In these equations $G$ and $C$ are the parallel equivalent conductance and capacitance of the sample, respectively, while $l$ is the length of sample; $a$ and $b$ are the outer and inner diameter, respectively, of the bone sample.

Figure 4.5 shows dielectric permittivities for dry, fluid saturated, UV-irradiated bone and for collagen in the frequency range 400–13 MHz. In each case a small distinct dispersion is

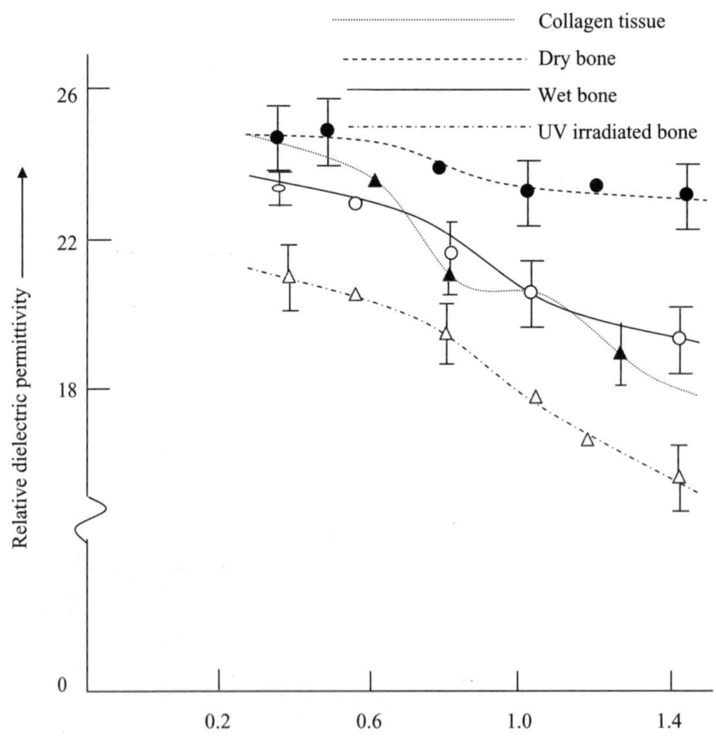


**Figure 4.5** Dielectric permittivity versus frequency of dry bone, fluid saturated bone, UV-irradiated bone and of collagen tissue at 17 °C and 72% humidity level

observed. The high frequency limit of the permittivity can be explained on the basis of the mixture theorem proposed by Maxwell (1873):

$$\varepsilon' = \varepsilon_b \frac{2\varepsilon_b + \varepsilon_w - 2\phi(\varepsilon_b - \varepsilon_w)}{2\varepsilon_b + \varepsilon_w + 2\phi(\varepsilon_b - \varepsilon_w)} \tag{4.4}$$

where $\varepsilon_b$ and $\varepsilon_w$ are the permittivities of the solid matrix and the electrolyte, respectively, and $\phi$ is the volume fraction of the electrolyte in the tissue. Assuming the permittivity of the bone matrix is 6 and that of fluid 79, it is readily observed that, as $\varepsilon_w \gg \varepsilon_b$, the permittivity will increase with increasing $\phi$. If we assume that, at this humidity level (72% at 17 °C), the water fraction of fully saturated bone is 15%, the same value adopted by other authors (Kosterich, Foster and Pollack, 1983), then $\varepsilon'$ will reach the limiting value of 10. In our case, the $\varepsilon'$ of 21.5 suggests that the dielectric relaxation of most of the fluid (free water) occurs above 1.3 GHz. Accepting the data of 400 MHz, a slight dispersion is observed that can be assigned to the presence of bound water, as has been done for other protein solutions (Grant, Keefe and Takashima, 1968). If a protein of volume $\nu_p$ and permittivity $\varepsilon_p$ is surrounded by a shell of hydration of permittivity $\varepsilon_\lambda$ the relationship is expressed as (Schwan, 1965):

$$\frac{\varepsilon_p - \varepsilon_\lambda}{\varepsilon_p + 2\varepsilon_\lambda} + \frac{\nu_p}{\nu}\frac{\varepsilon_p - \varepsilon_w}{\varepsilon_p + 2\varepsilon_\lambda} \tag{4.5}$$

Taking $\varepsilon_\lambda = 100$ gives that 2.6% water compared to protein volume will act as bound water. To obtain all the dielectric parameters, a range of values of relaxation wavelengths ($\lambda_s$) and distribution parameters ($\alpha$) are initially calculated for the dispersion range ($\delta$) by using the corresponding values of $\varepsilon'$ and $\varepsilon''$ in the equation:

$$\frac{\lambda\varepsilon''}{\varepsilon' - \varepsilon_\infty} = \lambda\left(\frac{\lambda_s}{\lambda}\right)^{1-\alpha}\left[1 + \left(\frac{\lambda_s}{\lambda}\right)^{1-\alpha}\frac{\alpha\pi}{2}\right] \tag{4.6}$$

Table 4.3 gives the dispersion parameters of bone in different physiological conditions and collagen at 17 °C and 72% humidity level and hence affords the most probable value of $\alpha$ and $\lambda_s$. The equation was derived by Grant, Buchanan and Cook (1957) assuming the Cole–Cole dielectric theory. Table 4.3 shows that the relaxation frequencies for dry and wet bone are close although for wet bone is larger than that of dry bone. This is due to the presence of an extra amount of water. For UV radiated bone and collagen the relaxation frequency differs greatly. This is expected because the amount of bound water attached to dry and wet bone under the two conditions will be different and hence the extent of dispersion will vary. It is anticipated that, due to exposure of UV light, the collagen molecules aggregate, leading to a decrease in dipole moment and consequently a decrease in relaxation frequency. For collagen, one of the

**Table 4.3** Dispersion parameters of bone in different physiological conditions and collagen at 17 °C and 72% humidity level

| Parameters | Dry bone | Wet bone | UV irradiated bone | Collagen |
|---|---|---|---|---|
| $\alpha$ | 0.04 | 0.04 | 0.03 | 0.01 |
| $\lambda_s$ (cm) | $29.33 \pm 3.8$ | $28.92 \pm 4.2$ | $37.44 \pm 6.7$ | $57.48 \pm 7.5$ |
| $f_s$ (MHz) | $1022 \pm 117$ | $1037 \pm 131$ | $801 \pm 160$ | $518 \pm 56.7$ |

components of bone, the structure and shape is unique. Naturally the relaxation frequency is entirely different from intact bone under different physiological conditions. However, the extent of dielectric dispersion is much higher compared to intact bone. This suggests that, owing to the removal of apatite by chemical process, the nature of binding of water with collagen may be altered, which leads to a larger dispersion. The depression of the permittivity curve at the high frequency end of this dispersion compared to that of pure water is a measure of the amount of water that is unable to turn in the electric field at a frequency near 1 GHz (Grant, Sheppard and South, 1978). Considering the essential role of water in the structure and functioning of biological systems, the comparative estimates of the quantity of bound water present may be of interest. The stability of collagen molecule is also affected by a chain of water molecules running parallel to the collagen helix. This suggests that relaxation of bound water alone can explain the observed dispersion in all four cases.

The dielectric decrement $\delta\varepsilon'$ is defined by:

$$\delta\varepsilon' = \frac{\varepsilon'_w - \varepsilon'}{c} \tag{4.7}$$

where $\varepsilon'_w$ is the dielectric constant of pure water at the appropriate wavelength and $c$ is the concentration in grams of protein per 100 mL of solution. The term bound water is taken to mean water bound to protein molecules by bonds of greater strength than the water bond existing in pure water. An estimation of hydration from dielectric results has the disadvantage that a dielectric mixture formula must be chosen. According to Fricke (1924):

$$\varepsilon' - \varepsilon_1 = KP(\varepsilon_2 - \varepsilon_1)$$

where $\varepsilon_1$ and $\varepsilon_2$ refer to solvent and solute, respectively, and $K$ is the shape factor of the solute, which lies range 1.0–1.8 in most cases. The term $P$ can be modified to the form (Grant, Keefe and Takashima, 1968):

$$100\,\delta\varepsilon' = K[\bar{\nu}(\varepsilon'_W - \varepsilon'_p) + W(\varepsilon'_W - \varepsilon'_B)] \tag{4.8}$$

where $\nu$ is the partial specific volume of the protein, which is chosen to be $0.73\,\mathrm{cm}^3\,\mathrm{g}^{-1}$ (Hunter, 1966), $W$ is the weight of the bound water per unit weight of protein and $\varepsilon'_p$, $\varepsilon'_w$ and $\varepsilon'_B$ are the permittivities of the protein, free water and bound water, respectively. Using above formula, estimations have been made for dry bone, wet bone, UV-irradiated bone and collagen. The most consistent values are obtained with $K = 1.6$. The term $\varepsilon_{SB}$ is the dielectric permittivity of bound water at a frequency much lower than the relaxation frequency and $\varepsilon_{11}$ is the dielectric permittivity of the sample at 1 GHz (Table 4.4).

**Table 4.4** Hydration and static dielectric permittivity of bound water attached with bone in different physiological conditions and collagen at 17 °C and 72% humidity level

| Sample | $W$ in gms/gm of protein | $\varepsilon_{SB}$ | $\varepsilon_{11}$ |
|---|---|---|---|
| Wet bone | 0.05 | 116.5 | 22.1 |
| Dry bone | 0.02 | 172.8 | 24.0 |
| UV irradiated bone | 0.07 | 63.2 | 19.8 |
| Collagen | 0.05 | 113.47 | 22.0 |

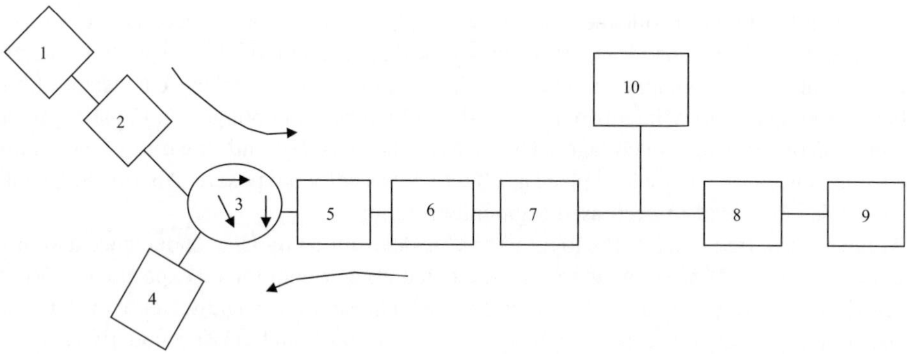


1. Klystron Power Supply
2. Klystron with Mount
3. Circulator
4. Matched load
5. Variable Attenuator
6. Frequency Meter
7. Slotted Section
8. E-H Tuner
9. Sample Holder (Cavity)
10. VSWR meter

**Figure 4.6** Schematic of the set-up for dielectric measurement using the cavity perturbation technique

For $K = 1.4$, the bound water fraction $W$ turns out to be 0.10, 0.15, 0.18 and 0.16 for dry bone, wet bone, UV-irradiated bone and collagen, respectively.

Behari *et al.* (1982) measured the dielectric constant of bone samples using cavity perturbation method (Figure 4.6) developed by Bethe and Schwinger (1943). They used the following expressions to calculation of real and imaginary part of $\varepsilon$:

$$\varepsilon' = 1 + \frac{1}{2}\frac{V_C}{V_S}\frac{(f_1 - f_2)}{f_2} \tag{4.9}$$

$$\varepsilon'' = \frac{1}{4}\frac{V_C}{V_S}\left\{\frac{1}{Q_2} - \frac{1}{Q_1}\right\}$$

where $V_S$ and $V_C$ are the volumes of the sample and the cavity, respectively; ($f_1$,$f_2$) and ($Q_1$, $Q_2$) are the resonance frequencies and quality factor without and with samples loaded in the cavity. These authors reported a decrease in $\varepsilon'$ and $\varepsilon''$ ranging from 19 to 27% and 44 to 61%, respectively. However the value of $\tan\delta$ ($=\varepsilon''/\varepsilon'$) remains the same before and after the exposure of ultraviolet light (300 W, 350 μm). Of the many possible mechanisms of interaction of biopolymers with electromagnetic radiation (i) monowater rotation and (ii) segmental rotation of biopolymers have been characterized as responsible for microwave interaction (Illinger, 1970). Monowater acts as a broadband microwave attenuator and thus partially accounts for the high dielectric loss factor.

Exposure of proteins to UV light (UVL) causes the rupture of some disulfide (–SS–) and H-bonds, which changes the tertiary and possibly the secondary structure of the proteins (McLaren and Shugar, 1964). The dielectric constant of solids depends simply on the displacement of charges inside the molecules. The presence of H-bonds in collagen and apatite (Hamilton, 1968; Ramachandran, 1967) will cause displacement of charges when the samples are placed in an electromagnetic field. The observed decrease in dielectric constant of bone, collagen and apatite may be attributed to H-bond breakage in collagen and apatite and also aggregation of molecules in collagen. Owing to breakage of H-bonds the transport of charges inside the molecules can not take place as effectively as before exposure. Also, an aggregation of molecules may lead to decreased segmental rotation.

At microwave frequencies, the loss factor $\varepsilon''$ is determined by DC conductance and dipole orientation (Smyth, 1955). Owing to a decrease in dipole orientation on exposure to UVL, the dielectric loss factor would decrease – as indicated by these results. The decrease in $\varepsilon''$ may also be partly due to the decrease in $\varepsilon''_{dc}$, which is directly proportional to DC conductivity. The DC conductivity of bone, collagen and apatite decreases upon exposure to UVL (Behari, Guha and Agarwal, 1975).

Collagen constitutes almost 35% by weight of bone and it has no disulfide bonding. Consequently, the only vulnerable points are the breakage of H-bonds and H-bonded water bridges in the secondary structure and even photolysis of peptide bonds between polypeptide chains. Some free radicals may also be produced, which may lead to aggregation of collagen molecules and increased crosslinking. In hydroxyapatite crystals, too, H-bonds will break down. After UVL exposure the dielectric permittivity decreases throughout the frequency range with respect to that of wet bone. The percentage reduction is 15%, as compared to 21% observed earlier (Behari, Kumar and Aruna, 1982). It is anticipated that, due to aggregation of collagen molecules, the dielectric permittivity of bone matrix has been reduced.

From the conductivity data it is seen that the wet bone has a larger conductivity than dry bone. The increase in conductivity from 150 to 240 ms $m^{-1}$ shows the relaxation phenomenon in this frequency region. Wet bone probably shows greater conductivity than dry bone because of its higher DC conductivity, which is primarily determined by the movement of ions through the fluid filled portion of the bone, including the Haversian canals, canaliculi and porous fractions. However, the contribution of bound water relaxation cannot be assessed from the present data. The percentage decrement of conductivity due to UV exposure is 33% compared to 50% observed by others (Behari, Kumar and Aruna, 1982) at 9.4 GHz. Kosterich, Foster and Pollack (1983) have found a sizable increase in conductivity at 100 MHz. Not much data are available for bone conductivity in the higher frequency range. In the frequency range of studies by Ray and Behari (1987) a relaxation phenomena due to "bound water" molecules attached to the protein molecules of bone was observed. This is identified as $\delta$ dispersion, which is usually smaller than $\beta$ dispersion (Schwan 1977). The relaxation frequency lies in the range 800–1100 MHz for dry bone, wet bone and UV-irradiated bone. The contribution from these two adjacent dispersions ($\beta$ and $\delta$) actually determines the conductivity in the frequency region 400–1300 MHz. The nature of the conductivity curve for dry bone, wet bone and UV saturated bone can be explained physically as due to $\beta$ and $\delta$ dispersion (rotation of polar protein molecules and bound water due to applied field). Separate contributions due to these two dispersions cannot be assessed due to the complexity of the system. For collagen, a major component of bone, the relaxation frequency turns out to be 500 MHz. A small discontinuity at 1 GHz observed with collagen may be attributed to the polarizability due to its side polar chain (Schwan, 1957). In the absence of hydroxyapatite and mucopolysaccharides (present in full bone) the dielectric behavior is expected to be markedly different.

## 4.4 Electret Phenomena

Among many others physical properties, bone possesses electret phenomena, which contribute toward callus formation. A body behaving like the electric analog of a permanent magnet is termed an "electret." Figure 4.7 depicts electret formation in the presence of an electric field. Role of electret phenomena in living system, though may have a significant role is yet to be fully understood, as the former would require a pre-poling.

The charge storage capability in bone has been studied by Mascarenhas (1974). The dried bone was shown to be able to store charge of the order of $10^{-8}$ C cm$^{-2}$. However, it is difficult, in practice, to measure such effects in physiologically moist bone with a high water content and the effect as yet has not been demonstrated to occur. Such a method of charge storage could have important consequences for the understanding of mineralization as well as the induction of new bone by external electric fields (Eriksson, 1974). However, Yasuda (1977) and Shiro *et al.* (1977) showed that a Teflon electret either wrapped around or laid near the rabbit femur gives rise to a callus formation.

Sources of electric fields in bone are dipoles, ionic space charge and the mechanism of charge transport includes protonic conduction ($H^+$ vacancy migration). Upon switching off the electric field a change in alignment takes place that accounts for the decay of charge with time. The effect in bone is due to the collagen component and is linked to a space charge type of

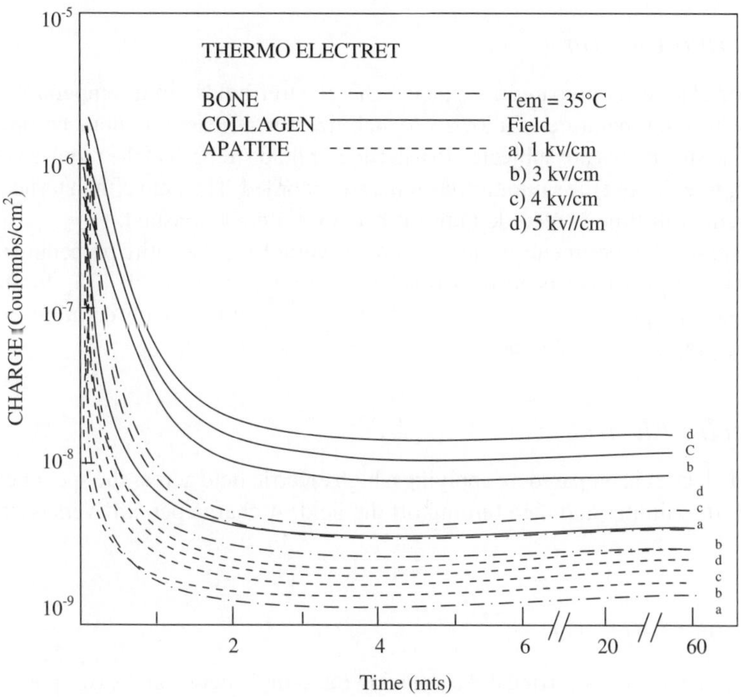


**Figure 4.7** Charge of area vs time plot for bone, collagen and apatite at 35 °C and at the indicated field (thermo electret) (Andrabi and Behari, 1980)

storage mechanism. Bound water in bone may also be responsible for electrical energy storage via that electret effect (Mascarenhas, 1974). A similar mechanism may be assumed to be present in collagen and apatite. The behavior observed here is similar to that reported for other dielectrics.

The formation of callus may be due to the transfer of dynamic energy into electrical energy (Hamburg *et al.*, 1971). Yasuda (1977) showed that the cathode side had more callus than the anode side when a current of 1 μA was passed through femur for three consecutive weeks. They further observed that for a current of between 1 and 100 μA bony callus was formed. When the current was higher than 100 μA contilagineous callus also appeared and when the current was further raised bone destruction took place. From this experience, Yasuda, Noguchi and Sato (1955) concluded that dynamic energy upon being transformed into electrical energy plays an important role in callus formation (Becker, 1972).

Several experiments (Andrabi, 1980) extended to bone and its two major components (collagen, and apatite) offer an interesting comparison. Apatite, which is considered to be a non-piezoelectric material, also exhibits the electret phenomenon and can retain the charge for a long time (five months). This may be attributed to the void and defects in the inorganic structure introduced by the method of extraction. Different types (viz. thermo, electro and magneto) of electrets of all three materials (bone, collagen and appetite) were made and consistent results were obtained. Figure 4.7 shows a typical data set for thermo electrets.

### 4.4.1 Thermo Electret

This type of electret is prepared by placing the electret holder in a temperature controlled chamber, whose temperature is allowed to stabilize for at least one hour before taking the readings. The specimens are subjected to a particular field strength at the required temperature for thirty minutes before the saturation currents are recorded. The field is then turned off and the fall of current with time is recorded until it becomes almost constant.

In a second set of experiments the temperature is varied and the entire procedure repeated. In this way the samples have been polarized at three temperatures, viz. 35, 45 and 65 °C at different field strengths, namely 1, 3, 4 and 5 kV $cm^{-1}$. From these observations a charge per $cm^2$ vs time graph can be obtained.

### 4.4.2 Electro Electret

This type of electret is prepared by applying a high electric field across the specimen sandwich for about thirty minutes and then turning off the field. A charge per $cm^2$ versus time graph is obtained.

### 4.4.3 Magneto Electret

In this case polarization is carried out by placing the sample between the two pole pieces of an electromagnet. A weak electric field and a high magnetic field of $\sim$16 K Gauss are applied across the specimen sandwich. After ten minutes, both the electric and magnetic fields are turned off and a charge per $cm^2$ versus time graph is obtained.

## 4.5 Hall Effect in Bone

Since the epoch making discovery of Fukada and Yasuda (1957) of the piezoelectricity of bone, many workers (Becker, Bassett and Bachman, 1964; Bassett, 1971; Williams and Breger, 1974) have pointed to the existence of several other electrical effects in bone. Further to this is the demands for knowledge of other solid state parameters such as the Hall effect and its optical counterpart (Figure 4.8a and b). This study further helps in understanding the mechanism of charge transport processes in bone. Since it is directly connected to the mobility, a comparison of Hall mobility and drift mobility is expected to offer insight into the mechanism of charge transport. This provides information on the number of charge carriers and their mobility. In organic semiconduction and also in the present case the mobility values may differ markedly. While the Hall mobility ($\mu_H$) is related to the motion of current carriers within the limits of a single crystallite, the drift mobility ($\mu_D$) is determined by the concentration of energy levels of the carriers' traps. Behari and Andrabi (1978) have measured these parameters, viz. the Hall

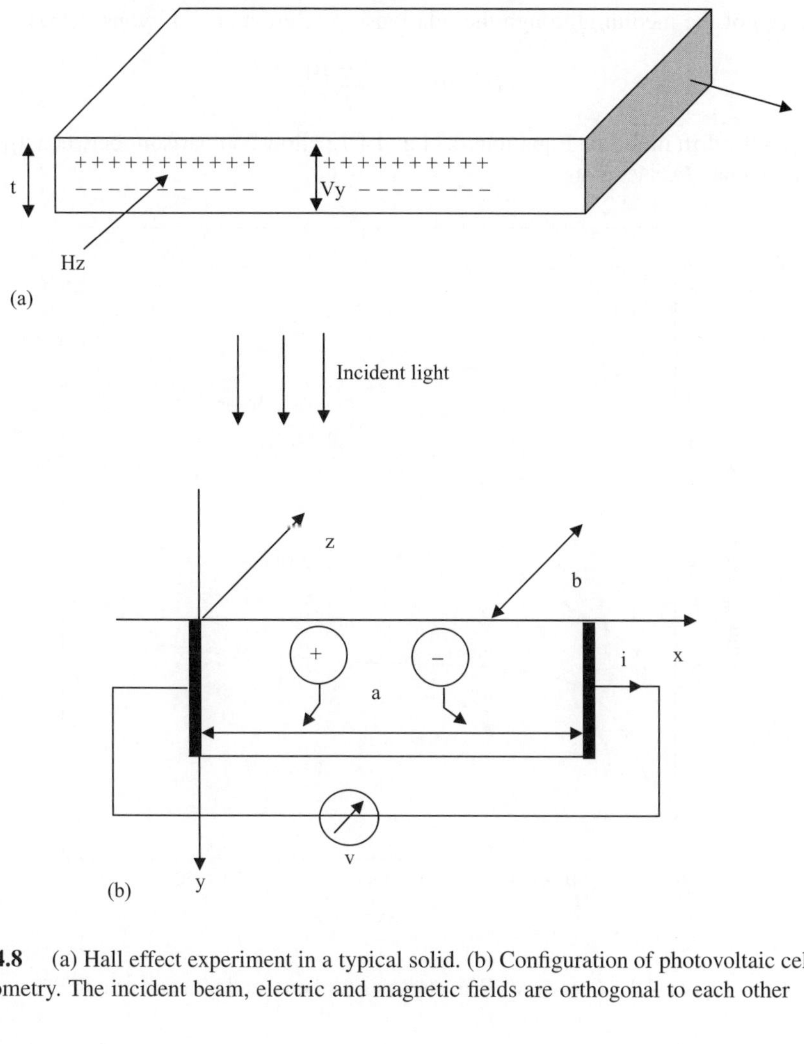


**Figure 4.8** (a) Hall effect experiment in a typical solid. (b) Configuration of photovoltaic cells in the Hall geometry. The incident beam, electric and magnetic fields are orthogonal to each other

coefficient, Hall mobility and drift mobility. A similar set of measurements were also carried out in collagen and apatite, at room temperature and humidity. This ensures that specimens have a normal component of full and bound water. In view of the structural similarly it is expected that the data will be similarly oriented.

The Hall coefficient ($R_H$) is obtained using the relationship (Kittel, 1974):]

$$R_H = \frac{V_y \times t}{I_x \times H_z} \tag{4.10}$$

where $V_y$ is the Hall voltage generated in the $y$ direction, when magnetic field ($H_z$) is applied in the $z$ direction, and $I_x$ refers to the direction of the current $I$, and $t$ is the thickness of the sample. The conductivity ($\sigma$) is related to the Hall mobility ($\mu_H$) through the relationship:

$$\mu_H = R_H \sigma \tag{4.11}$$

To obtain an indication of the order of magnitude of the drift mobility, the electrode contacts are taken so that the samples are in sandwich form. $I$ and $V^2$ are related through the dielectric constant ($\in$) of the medium through the relationship (Gutmann and Lyons, 1967):

$$I = 10^{-13} \frac{V^2 \in \mu_D}{t^3} \tag{4.12}$$

where $\mu_D$ is the drift mobility. Equations 4.11 and 4.12 allow comparisons between $\mu_H$ and $\mu_D$. Figure 4.9 shows $I$ vs $V^2$ plots.

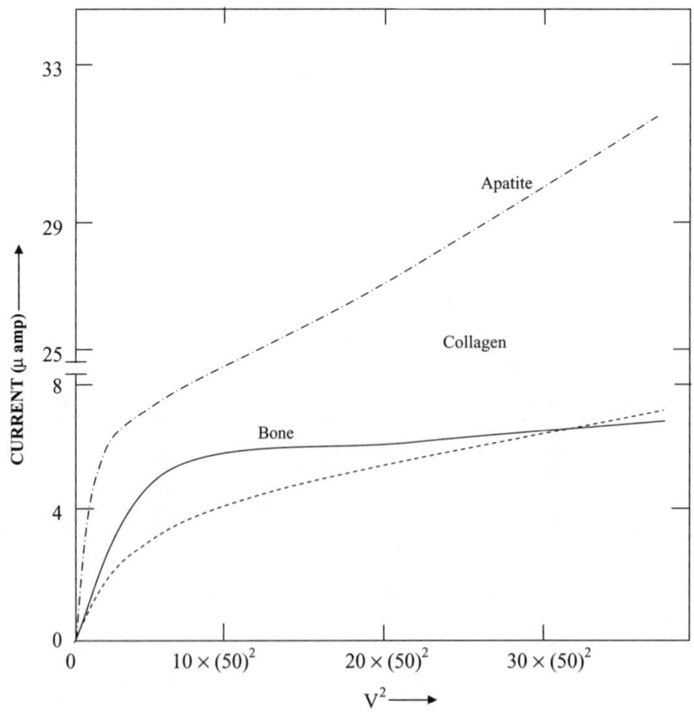


**Figure 4.9** $I$ versus $V^2$ characteristics for bone, collagen and apatite

**Table 4.5** Dielectric constant of bone, collagen and apatite at 9.2 GHz

| Sample | $\varepsilon'$ | $\varepsilon''$ | tan(δ) |
|---|---|---|---|
| Bone | $4.96 \pm 0.37$ | $1.72 \pm 0.20$ | 0.347 |
| Collagen | $3.85 \pm 0.34$ | $1.78 \pm 0.11$ | 0.459 |
| Apatite | $3.61 \pm 0.24$ | $1.75 \pm 0.17$ | 0.484 |

### *4.5.1 Hall Effect, Hall Mobility and Drift Mobility*

The resistivity of the sample is of the order of $10^6$–$10^8\ \Omega$ cm at room temperature, and decreases with increasing temperature. Also, as anticipated, the activation energy falls with increasing temperature, and hence the conductivity increases with rising temperature.

The Hall voltage, of the order of a few tens of a millivolt, has been recorded for bone, collagen and apatite. Table 4.6 presents the resistivity of these materials. For accuracy, the measurements were restricted to specimens of moderately high resistivity (an indication of hydration levels) for the application of the applied field. The sources of polarization or charge storage in the macromolecules are dipoles and the ionic space charge. Bound water and free water can both contribute (Mascarenhas, 1974).

Table 4.6 also lists the Hall coefficients ($R_H$). $R_H$ is strongly dependent on the resistivity and the thermodynamic history of the sample. The Hall mobility is evaluated by Equation 4.11. The small value of $\mu_H$ ($\sim 1\ cm^2\ V^{-1}\ s^{-1}$) suggests the inapplicability of the simple band model to explain the phenomenon in such materials. Eley and Spirey (1960) have suggested a band model for the propagation of electrons having low energy contributing to the conductivity. Also, as reported earlier (Behari, Guha and Agarwal, 1975) the conductivity decreases on exposure to UV light. This is attributed to the breaking of hydrogen bonds and a subsequent reduction in probability of $H^+$ transport. A diminution of the Hall voltage in the same specimen after exposure to UV light has also been reported, possibly because of the decrease in conductivity of the sample. These data suggest an important role of protonic transport in these cases. The low mobilities suggest that the carriers can move only within the limits of the regions of continuous conjugation. On application of the electric field the medium is polarized with the formation of dipoles. The dipoles are aligned along the direction of the electric field. On application of the magnetic field the dipoles undergo realignment, causing generation of a Hall voltage. A reversal in the direction of the magnetic field does not change the sign of the Hall voltage. A low value of $\mu_H$ further suggests the possible operation of a hopping type mechanism in such cases (Gutmann and Lyons, 1967). The distance between the atoms (or molecules) is less than their own dimension and ions move with the holes as a group in the regions of continuous conjugation separated by thin barriers. Protons and their holes may move together

**Table 4.6** Resistivity and corresponding drift mobility in bone and its two major constituents

| Sample | Resistivity ($\rho$, $10^8\ \Omega\ cm^{-1}$) | Mobility[a] ($\mu_D$, $10^{-4}\ cm^2\ V^{-1}$) |
|---|---|---|
| Bone | 10.8 | 0.52 |
| Collagen | 2.26 | 22.8 |
| Apatite | 11.24 | 94.3 |

[a]These mobilities refer to the highest points in the $I$ vs $V^2$ plots.

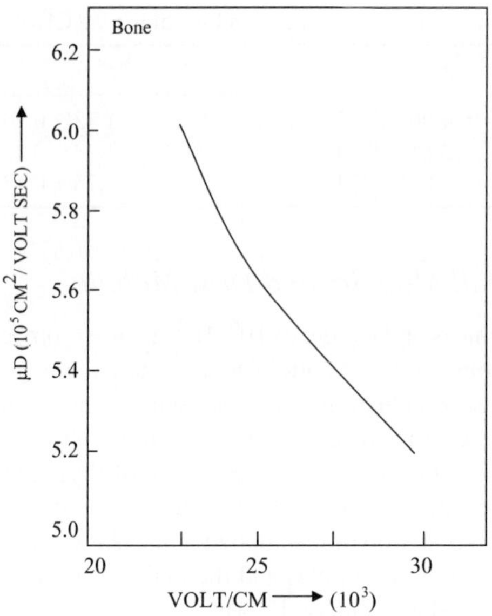


**Figure 4.10** Variation of drift mobility ($\mu_D$) with applied field for bone

through hydrogen bridges, as a pair. The order of the Hall voltage observed here is of the same order of magnitude as that for a P- or N-type semiconductor, though with a much higher applied electric and magnetic field. This may be due to the high resistivity of the samples. Figures 4.10–4.12 show the variation of drift velocity ($\mu_O$) ($<\mu_H$) with applied field.

## *4.5.2 Magnetic Field Dependence of the Hall Coefficient in Apatite*

Figure 4.13 shows an increase of $R_H$ with magnetic field; $R_H$ does not increase appreciably at lower fields but shows a sharp increase beyond 10 kG. The Hall voltage was not reversed by a

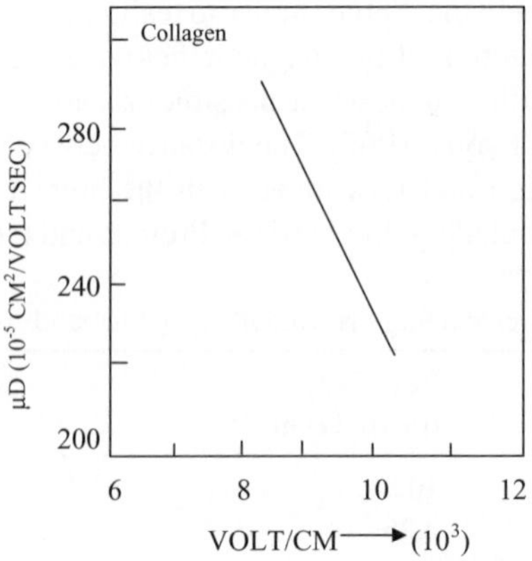


**Figure 4.11** Variation of drift mobility with applied field for collagen

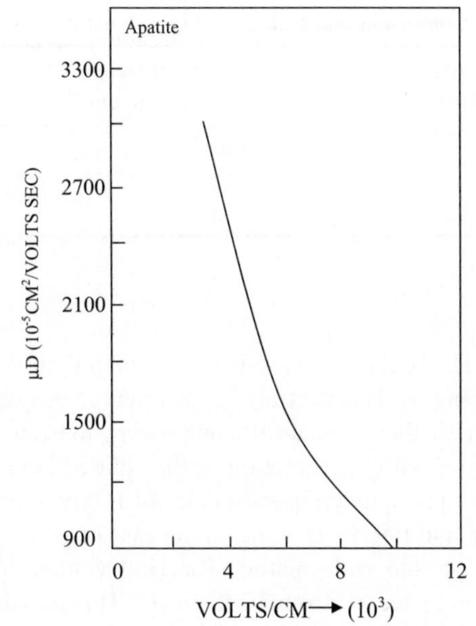


**Figure 4.12** Variation of drift mobility with applied field for apatite

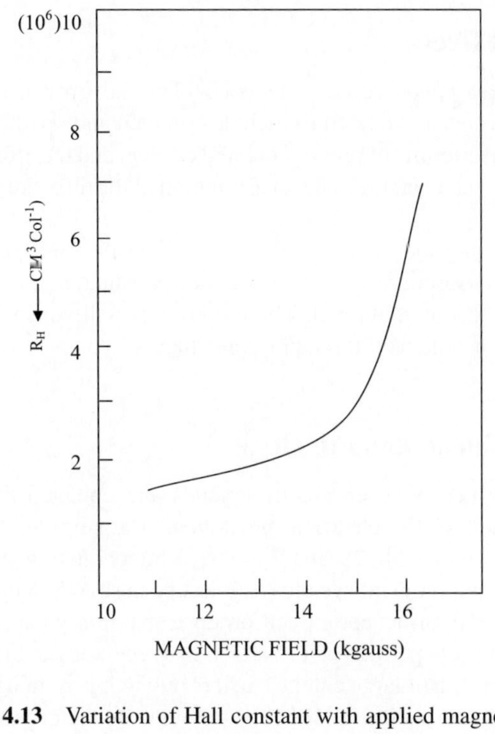


**Figure 4.13** Variation of Hall constant with applied magnetic field

**Table 4.7** Resistivity, Hall Coefficient and Hall Mobility in bone and its two major constituents

| Sample | Resistivity ($\rho\,\Omega$ cm) $10^6$ | Hall coefficient ($R_H$, $10^6$ cm$^3$ col$^{-1}$) | Mobility ($\mu_H$, cm$^2$ volt/sec) |
|---|---|---|---|
| Bone | 23.0 | 3.08 | 0.13 |
| Collagen | 1.61 | 1.26 | 0.78 |
| Apatite | 4.40 | 7.31 | 1.66 |

reversal in magnetic field. The Hall mobility remains less than 1 cm$^2$ V$^{-1}$ s$^{-1}$ at lower fields but increases at higher field strengths. The electret effects observed in apatite (Andrabi and Behari, 1981a) confirm the hypothesis that given a sufficient energy to excite the charge carriers they can be aligned in accordance with the direction of the applied field.

These results differ from previously reported data for P-type inorganic material (Dunlap, 1950; Willardson, Harman and Beer, 1954). $R_H$ in the case of germanium decreases with the field while the effect is opposite with apatite. The Hall voltage increases by an order of magnitude when the field is increased from 10 to 16 kG. This may be because in apatite the bonding is comparatively weak (Hamilton, 1968) so that the atoms (or group of atoms) are more mobile and hence they are more prone to the magnetic field variations. This is supported by the fact that mobility (both drift and Hall, Tables 4.6 and 4.7) is consistently higher for apatite than for bone and collagen.

## 4.6 Photovoltaic Effect

In pursuance of the above, photovoltaic effects (Becker and Brown, 1965) in bone received early attention. This has been extended to include collagen and apatite (Andrabi and Behari, 1981b). On this basis the lifetime of photogenerated carriers and the effect of magnetic field on photoconductivity have been obtained and an estimation of the diffusion length is subsequently obtained.

This suggests examining the bone photovoltaic cell in the Hall geometry (Figure 4.8b). Subsequently, the light absorption constant, extinction constant, mass absorption constant and efficiency of such cells can be obtained. These constants will consolidate the prospects for generating basic data for bone and its major constituents.

## 4.7 PN Junction Phenomena in Bone

The functional unit formed between apatite crystals and collagen fibril is a PN junction. Osteogenesis is attributed to the electrical phenomenon arising due to stress sensitivity of multiple PN junctions (Figure 4.14a and b). The semiconduction mechanism may be of functional significance in biological systems as it has been shown (Huggin and Young, 1962) that strong electron donors or acceptors can produce mammary cancer provided the steric configuration of the molecule permitted it to attach to a receptor site in the tissue, resulting in charge transfer. The injection of large charge carriers into living systems influences the growth pattern (Becker, Bassett and Bachman, 1964). If the semiconductor properties of bone can be

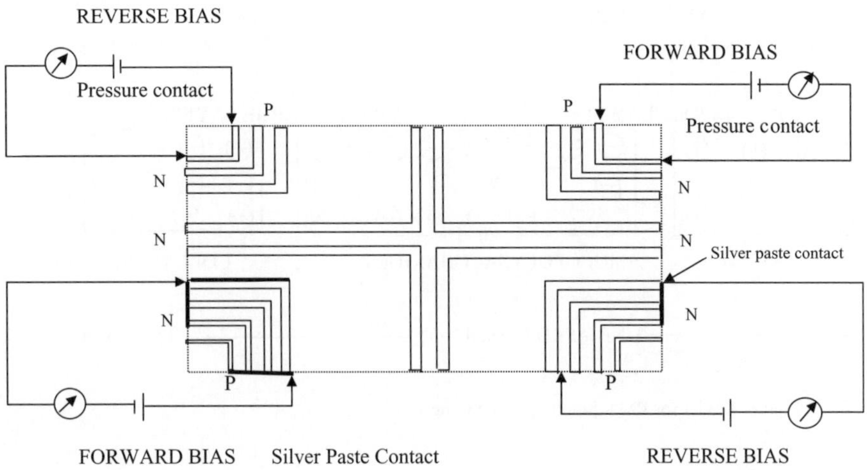


**Figure 4.14** (a) Electrode placement for measurements of PN junction characteristics and photoelectromagnetic studies; (b) a bone in the PN junction configuration

changed by stimulations, a structural change in bone is expected. It seems, therefore, that manipulation of a control system may eventually be directed to result in a clinically useful control of bone growth and architecture.

PN junction diode characteristics for bone and the behavior under reverse and forward bias can be obtained. It is suggested that mucopolysaccharides play an important role in controlling junction characteristics. On subjecting the junction to UVL or IR (infrared) light a photocurrent is generated. UV exposure produces a permanent change in behavior. Photoelectromagnetic effects in such a junction have been observed and it has been found that the magnetic field enhances the optical response of the junction (Figures 4.15 and 4.16).

The PN junction components are confirmed by the observation that a probable contact to apatite and collagen material surface produces a diode characteristic curve. In contrast, a probable collagen–collagen and apatite–apatite contact produced a purely linear *V–I* characteristic. Furthermore, a reversal of polarity in PN junction contact confirms the well-known reverse bias characteristic. The fact that diode characteristics are obtained only when the contact is across collagen and apatite complex confirms that their formation in full bone has a similar link to that in well-known inorganic PN junctions. Failure to obtain such a characteristic with collagen–collagen or apatite–apatite contact substantiates that there is no homogeneity at the molecular level. On the other hand, in PN junction contact, a breakdown region occurs that makes the *V–I* characteristic irreversible. Also, below a particular value (the breakdown voltage) the *V–I* diode characteristics are reversible. The samples are mounted on a holder placed between the pole pieces of an electromagnet such that the light from an IR radiation source (250 W) can be focused on the junction and the magnetic field applied parallel to the junction surface. To study junction diode characteristics, contacts (26 gauge needle) are established (Figure 4.15a and b).

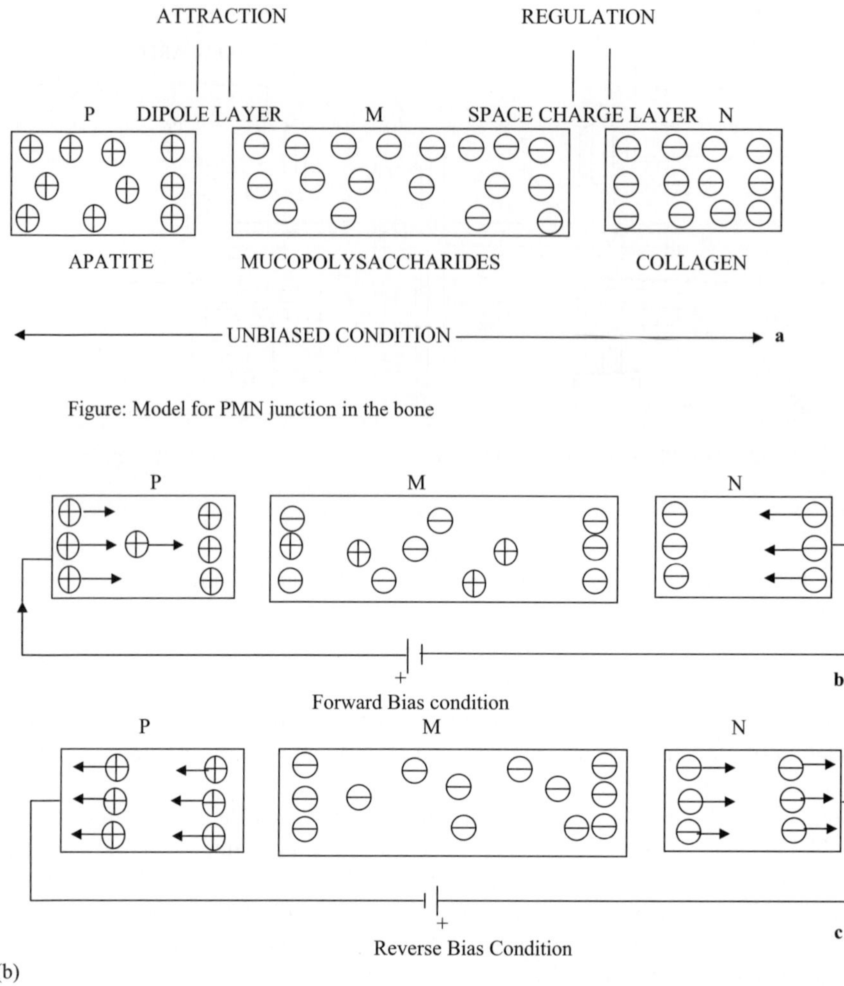


**Figure 4.14** *(Continued)*

For the photoelectromagnetic effect, a given bias voltage establishes the current and IR light is allowed to fall on the junction. A rise in current with time is observed and when it reaches to a saturation limit the light is then turned off and the fall of current with time is recorded. Such observations are repeated after equilibrium is reached. The experiments are repeated with various bias voltages.

In another set of experiments, the junction is exposed to IR/VL (VL = visible light) that results in generation of a photocurrent and its saturation value. At this stage a magnetic field ($\sim$16 kG) is turned on and off instantaneously. The resulting increase in current is monitored. For comparison, the magnetic field and light source (IR/VL) are turned off simultaneously and the fall of current with time is noted.

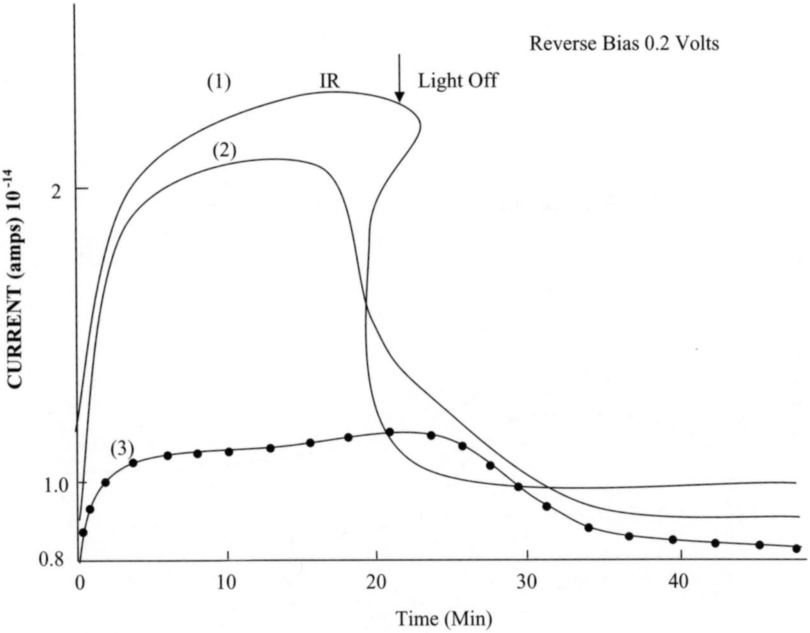


**Figure 4.15** Behavior of a PN junction at 0.2 V reverse bias under (1) IR; (2) indicates repeat characteristics after IR exposures; (3) indicates the PN junction behavior during IR exposure after a dose of UV light for 30 min

### *4.7.1 Breakdown Phenomenon of PN Junction*

When the junction is forward biased, that is, the collagen end is connected to the negative terminal of the battery and apatite to the positive terminal (Figure 4.14b), the current rapidly increases, thus showing a path of lesser resistance. In the reverse bias configuration (Figure 4.14b), the current first increases and later becomes constant in the region of 2–3 V. The current showed a sharp increase by about one to two orders of magnitude with a small further increase of voltage. The junction is said to be in the "breakdown" region beyond this point and its diode behavior is lost. The breakdown may be of the Zener effect or avalanche type.

The current–voltage relationship for a PN junction diode is written as:

$$I = I_0\left[\exp \phi \frac{Ve}{k_B T} - 1\right] \tag{4.13}$$

where $I_0$ = saturation current; $V$ = applied voltage; $k_B$ = the Boltzmann constant; $T$ = temperature (K).

Equation 4.13 is derived by assuming a band structure model. Though no such theoretical data are available or can be easily derived for structures like collagen and apatite and it may be assumed that they have a very complex geometry. The value of $\phi$ may be the ratio of drift mobility to Hall mobility ($10^{-3}$), which affects the carrier transport (Behari, Rai and Jha, 1979;

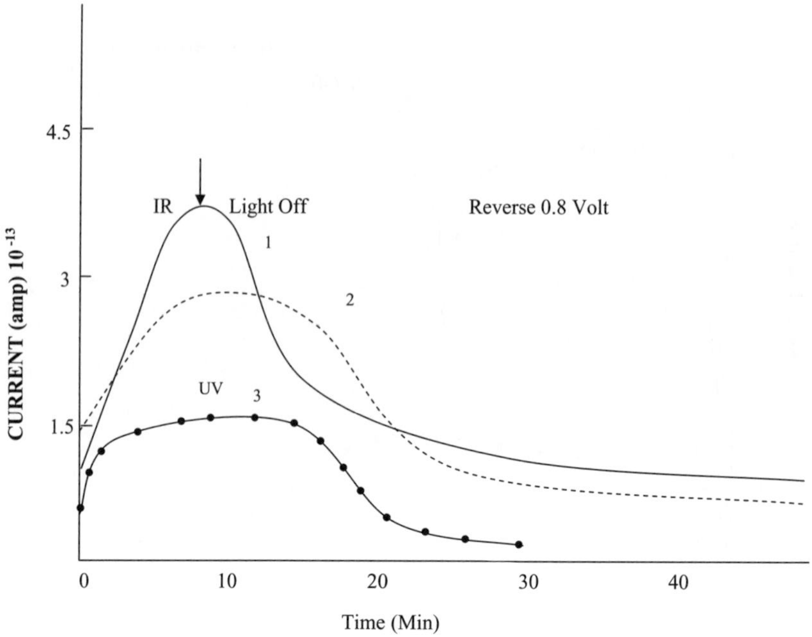


**Figure 4.16** Behavior of a PN junction at 0.8 V reverse bias under (1) IR light. The dotted line (2) indicates PN junction behavior during IR exposure. The line labeled (3) shows the PN junction behavior after a dose of UV light for 30 min

Behari and Andrabi, 1978). An exact conduction mechanism of a PN junction is difficult to offer owing to the complex structure of bone at the molecular level, though a qualitative picture can be presented on the following lines. The collagen and apatite are cemented together by mucopolysaccharides, which are electrically negative (Bassett, 1971). The crystalline collagen tends to have an abundance of electrons, while crystals of hydroxyapatite, a lack of them (Bassett, 1965) and as such the interface between collagen and apatite is a semiconductor junction of PN type. Further, tropocollagen, which makes up the native collagen fibril, is polar and there is a permanent electric moment in the direction of the longitudinal tropocollagen axis (Athenstaedt, 1970). Collagen fibril behaves like an electrical analogue of a permanent rod magnet (Ramachandran, 1967, 1968; Hodge, 1967). Mucopolysaccharides accelerate the process of calcification by regulating $Ca^{2+}$ supplies (Samachsam, 1969), which takes place inside the fibrils of collagen (Glimcher, 1959); therefore, it may be assumed that calcification generally takes place near the negative end of the native tropocollagen because the intermolecular forces between collagen and inorganic ions must be just strong enough to affect their interaction energies without firmly binding them to the collagen structure. Otherwise, if either calcium or phosphate ions were very strongly bound by collagen fibrils, collagen would act as a demineralizer similar to chelating agents, unless a second mechanism were involved that released the ions after they were bound. Hence, a layer of mucopolysaccharides along with proteins separates and cements the two main constituents of bone and as such acts as both inhibitor and promoter during calcification.

Negative charges on tropocollagen molecule are sites for calcification and the negatively charged mucopolysaccharides molecules are between collagen and apatite. Hence, it may be suggested that the negative charge on the "M" faces the negative charge on N during bone formation. The mucopolysaccharides form two different electrically charged layers, namely a dipole layer (attraction) on the apatite side and an unipolar space charge layer (repulsion) on the collagen side. No charge transfer or conduction takes place in the junction, say PM or PN, because a space charge layer may form near the PM junction as the negative charges from "M" will be attracted towards "P," which will then produce a negative player on the "P" side facing "M" and there is no possibility of conduction between M and N. The extent of this space charge layer between the two materials is known as the "depletion layer." The characteristics of junctions can be summarized as follows:

1. In an unbiased PMN junction there is no applied voltage and the junction is said to in dynamic equilibrium.
2. In a junction is forward biased (Figure 4.14b) the positive charge carriers, under the action of applied field, cross the potential barrier in P and M side and move towards M and N regions and in the process are accelerated, thus causing a net current to flow in the PN junction. The path is of least resistance and therefore the magnitude of current flow is greater.
3. In a junction that is reverse biased (Figure 4.14c), the potential barrier between P and M and M and N increases and it also widens and, therefore, the current flow is less. At a particular critical voltage the molecules are suddenly polarized, forming dipoles that are aligned in the direction of the applied field and an increase in current of about one to two orders of magnitude is observed; this is called the "breakdown" of the junction. Beyond this voltage the current remains practically constant. If the voltages applied cross this critical voltage then "thermal breakdown" occurs, which destroys the collagen apatite fiber (junction) properties and as such as the PN characteristics are not reproduced.

### *4.7.2 Behavior of the PN Junction Under IR and UV Conditions*

The behavior of a PN junction under reverse and forward conditions with different applied bias when subjected to IR radiations has been examined. The current first increases and then decreases (Figures 4.15–4.17). A faster decrease is observed when the light is turned off at constant current value. The magnitude of current with IR light was greater than after UVL exposure, indicating a change in structure at the molecular level introduced by the later. The PN junction characteristics of such samples are lost. These observations show that the energy of UVL is sufficient to bring about structural changes (probably by disruption of hydrogen bonds) whereas IR light changes the structure to a significantly lesser extent (Behari and Andrabi, 1978; Behari, Rai and Jha, 1979).

### *4.7.3 Photoelectromagnetic (PEM) Effect*

The PN junction of bone subjected to IR light and visible light (VL) shows an increase in photocurrent (Figure 4.17) that is further increased by the application of magnetic field. The phenomenon can be explained as follows.

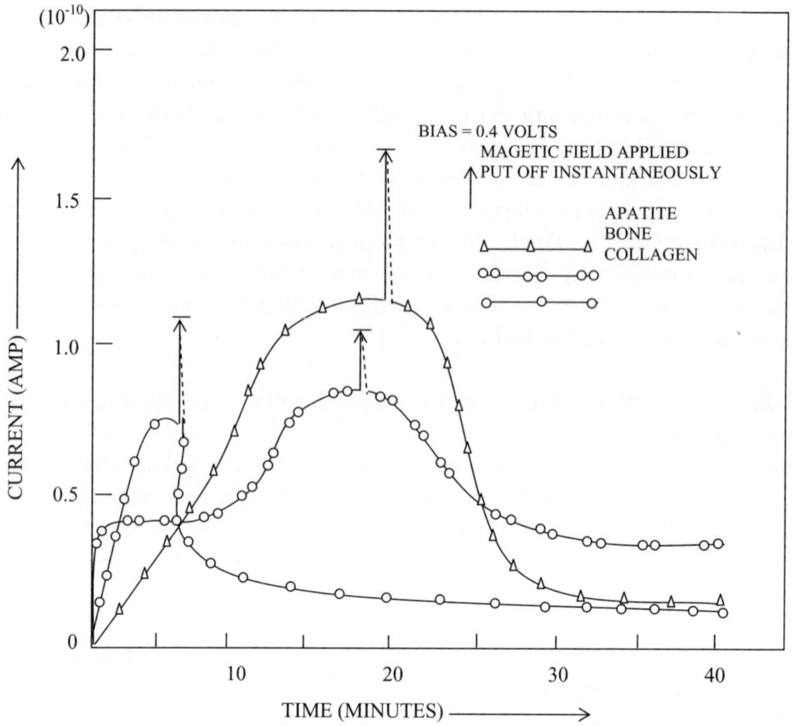


**Figure 4.17** Growth and decay of photocurrents before (due to exposure of IR light) and after the application of a magnetic field for bone, collagen and apatite; bias, 0.4 V

On exposure of junctions (PM, MN) to light they absorb photons of sufficient energy to effect excitation and migration of charge carriers and some impurities. Consequently, there are more charge carriers than in the absence of radiation and hence an increase of current is observed, which shows a steady state after some time. At this steady state value, turning on and off instantaneously a magnetic field of the order of 16 kG produces an enhanced photocurrent. This phenomenon can be explained by suggesting that the application of a magnetic field provides extra energy for the movement of dipoles and hence the enhancement of photocurrent. It is also observed that reversal in the direction of the magnetic field does not change the direction of "enhanced current." It may be suggested that the application of a field merely reverses the alignment of charge carriers.

On further exposure of light after the steady state value the current starts falling. This may be due to the breaking of hydrogen bonds. However, as pointed out earlier (Behari, Rai and Jha, 1979), the energy of these radiations is not sufficient to bring this process to completion. On exposing the sample to UVL and repeating the process the photoresponse of a PN junction becomes insignificantly less and the effect of the magnetic field is also not clearly perceptible under similar conditions. Thus, it may be suggested that UVL ruptures the arrangement of diodes (hydrogen bonds) in bone by causing a structural change at the molecular level.

The sensitivity of bone to various wavelengths of the electromagnetic spectrum has also been examined (Becker and Brown, 1965). The existence of absorption bands in the visible and near-infrared region (Behari and Guha, 1977) confirms the sensitivity of bone to these radiations. This is further augmented by the observation that a photovoltage is developed when bone is exposed to visible light and IR light. These results give an overall estimate of the phenomena occurring in bone at the molecular level. A related aspect seems to be the lifetime of photogenerated carriers in bone. This will clearly depend upon the structure of the solid and the type of charge carrier and is expected to help in elucidating the mechanism of photo-dynamics of charge carriers. The photoconductivity of bone and its two major constituents is sensitive to IR light (Behari, Rai and Jha, 1979).

### 4.7.4 Life Time of Charge Carriers

To obtain basic data on the solid state of bone (viz., lifetime of charge carrier, drift mobility) experiments have been performed *in vitro*. A pulse of light was taken from a strobometer and the flash duration adjusted to $\sim 10\,\mu s$. This short duration pulse was absorbed and photocarriers were generated. The flashing rate was 1800 rev $min^{-1}$ and the pulse duration and flashing rate were kept constant for all three materials. The average thickness of the sample was 0.5 mm. The decay pattern of the pulse was recorded and the time at which the pulse height fell to 36.8% of its peak value (Lindmayer and Wigley, 1971). This gives the lifetime of the photogenerated carriers (Figure 4.18). The curves in the figure are drawn on the same scale and their shapes are identical for all three materials. Calculation of the lifetime yields values of 15, 25 and 5 μs for bone, collagen and apatite, respectively.

$I$ versus $V^2$ plots are similar to those obtained for such relationships in high resistivity organic semiconductors. The mobility ($\mu_D$) has been evaluated for the highest field values, and its variation in that region is plotted in Figures 4.10–4.12; $\mu_D$ shows a strong field dependence. The

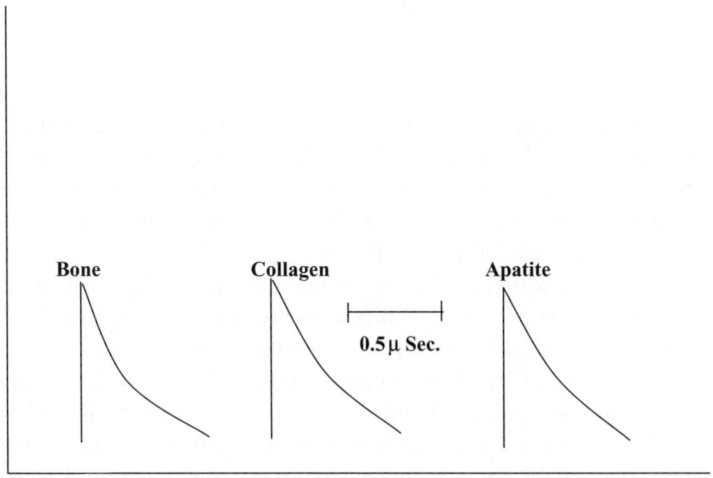


**Figure 4.18** Decay characteristics of injected charge carriers in bone, collagen and apatite

transit time ($\tau$) can be calculated using the relationship:

$$\tau = \frac{t^2}{\mu_D V} \tag{4.14}$$

which turns out to be 2.0, 3.5 and 9.5 m s$^{-1}$ in the case of bone, collagen and apatite, respectively, and is nearly independent of the magnitude of the electric field under study. The similar order of magnitude of these values for all three materials suggests that a similar mechanism is operative in all three cases, and this establishes the semiconduction behavior of bone and its components.

Tables 4.6 and 4.7 show comparative values of $\mu_H$ and $\mu_D$. As expected $\mu_H$ is higher than $\mu_D$ by several orders of magnitude. This provides qualitative support to these findings. An exact quantitative comparison of the various parameters in the three samples is not possible because different amounts of free and bound water may be present. However, their order of magnitude compares favorably and the magnitude of the evaluated parameters agrees with those of other organic semiconductors (Gutmann and Lyons, 1967) with resistivities in the range reported in these investigations:

$$\mu_D = \mu_0^\theta \tag{4.15}$$

where $\theta$ = fraction of the total space charge that remains free and $\mu_0$ = free carrier mobility. This yields $\mu_D \sim 10^{-4}\,cm^2\,V^{-1}\,s^{-1}$. The value of $\theta$ is found to be $\sim 10^{-3}$, suggesting that the band model is not applicable in these cases.

It is possible that, on exposure to light, a set of $H^+$-vacancy pairs (or ion-vacancy pairs) are generated and this group moves as a whole in the region of continuous conjugation, separated by thin barriers. Since the path length of the carriers is very small (smaller than the dimensions of the molecules themselves) they are quickly trapped at the sites of impurity and are then excited again.

The diffusion coefficient and diffusion length can be calculated from the Einstein diffusivity mobility relation (Lindmayer and Wigley, 1971):

$$D = \frac{\mu_D k_B T}{e} \tag{4.16}$$

and $L^2 = D\tau$

where $D$ = diffusion length, $\mu_D$ = drift mobility, $k_B$ = the Boltzmann constant, $T$ = absolute temperature, $e$ = electronic charge, $L$ = diffusion length and $\iota$ = the lifetime of the photo-generated carriers. Using the value of $\mu_D$ above, $D$ is calculated to be $\sim 10^{-4}\,cm^2\,s^{-1}$ and, knowing $\tau$, $L$ comes out as $\sim 10^{-5}$ cm. These data agree well with those reported for other biological materials (Gutmann and Lyons, 1967).

The initial rise in current with time may be attributed to the generation of charge carriers on exposure to light. The phenomenon of photoconductivity can be explained on the basis of electronic and protonic transport: electrons are excited from the traps due to the energy provided by the incoming photons, thereby contributing to the conduction phenomena. Bone and its components are also hydrogen-bonded solids. There is a probability of protons ($H^+$) moving by a hopping mechanism and contributing to the conduction. This is supported by the decline in photoconduction seen after prolonged exposure to IR radiation. This is attributed to the probability of hydrogen-bond breaking. This seems possible since the energy of the hydrogen bond is of the same order of magnitude as that of the incident photon. A heating effect would have produced an effect in the opposite direction. When the breaking of hydrogen bonds becomes dominant, the current starts falling with time.

The effect of the magnetic field on photoconductivity in bone and its constituents can be explained thus. Clearly, there is a rise in current as a result of applying the magnetic field and this may be attributed to the energy supplied to the charge carriers (by the magnetic field) over and above that gained due to the incident photon beam. On reversing the direction of the magnetic field the overall effect remains the same in magnitude and polarity, as also observed in Hall effect measurements. However, this phenomena is clearly distinguishable from a pure magnetoresistance effect. Application of a magnetic field of the same order as above with its direction parallel to the electric field and of the same value produces no change in the current. In addition, as one increases the electric field the corresponding magnetic field effect becomes smaller. It is thus concluded that the effects of magnetic fields (for this order of magnitude) can be detected only at low electric fields. In the studies described here the effect of the magnetic field was examined at one value only. This is intended to separate the contribution of the magnetic field from the photo effect. As pointed out, bone and both of its constituents are sensitive to IR light and hence the magnetic field effect is examined at the point where the light effect reaches its saturation value. The magnetic field is then switched off immediately (Figure 4.17). A similar phenomenon is observed in collagen and apatite.

The role of water is significant in biomaterials as it has been reported that for protein an absorbed monolayer may raise its conductivity by a factor of $10^8$ or so (Cardew and Eley, 1959). A significant amount of water is contained in bone *in vivo* and some is bound with collagen fibers. From experimental results Becker, Bassett and Bachman (1964) concluded that bone samples exposed to room humidity seem to have a thin film of free water on their surface that can not be removed by heating up to 100 °C since it produces irreversible damage to collagen fibers. The hydration experiments reveals that bound water is an important factor in collagen conductivity, which increases with increasing content of absorbed water, in agreement with results on other proteins (Rosenberg, 1962). However, bound water seems to be of much less functional importance for apatite and for the diode zone itself. The high conductivities for all bone components in partially rehydrated specimens are interpreted as representing re-acquisition of the complete bound water components by apatite–collagen complex. The decline in magnitude in conductivity when a specimen is left to re-equilibrate to normal humidity is, perhaps, the result of hygroscopic action producing a free-water compartment, providing a shunt resistance to the diodes. These results show the importance of bound water and free water in controlling the conductivity of bone.

## 4.8 Bone Electrical Parameters in Microstrip Line Configuration

Ray and Behari (1988) have designed a microstrip line with bone as a substrate material and measured the dielectric behavior in two orthogonal directions. They used a time domain reflectometer (TDR) to observe a pulse that is reflected from the impedance discontinuities of the microstrip line (Figure 4.19). The aim of a TDR system is to observe the shape and size of the input and output pulses, which can be used to calculate the dielectric constant at each frequency point of interest. The authors have measured the permeability of the microstrip substrate for a pulse of step rise time of 35 ps, which corresponds to a fundamental frequency of

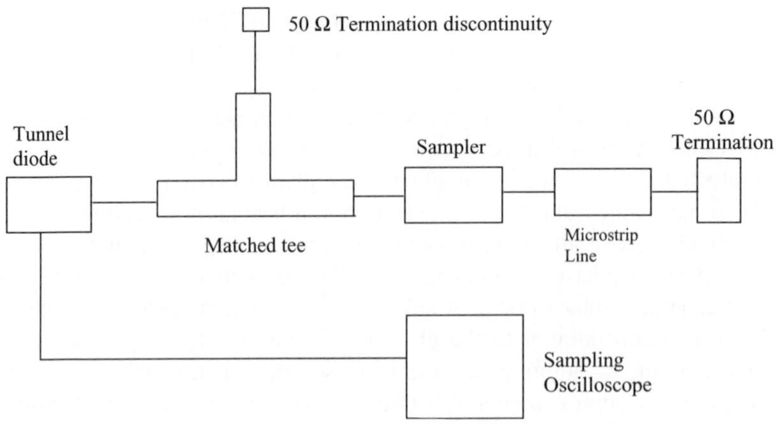


**Figure 4.19** Experimental arrangement for reflection measurement using a time domain reflectometer

4.5 GHz. The system can, however, respond to other frequencies. The advantage of the system is that despite the presence of multiple reflections, a TDR system is capable of the frequency domain where only aggregates of reflections can be traced.

## *4.8.1 Theoretical Formulation*

As we apply a step function to the microstrip line through a coaxial line of characteristic impedance 50 Ω, the pulse will be partially transmitted and partially reflected from the two ends of the microstrip line due to mismatch of impedance with the connectors at the two ends. The time difference between the two pulses coming from these surfaces of discontinuity is:

$$t = \frac{2x\sqrt{\varepsilon'_{\mathrm{eff}}}}{c} \tag{4.17}$$

where $\varepsilon'_{\mathrm{eff}}$ is the effective permittivity of the microstrip substrate and $x$ is the length of the microstrip line; $c$ is the velocity of electromagnetic radiation in free space (Grant, Sheppard and South, 1978). The size of the pulses gives information about the reflection coefficient, which is related by:

$$\rho_{1,2} = \frac{Z_1 - Z_2}{Z_1 + Z_2} \tag{4.18}$$

where $Z_1$ is the impedance of the coaxial line and $Z_2$ is that of the microstrip line: $Z_2$ is a function of frequency. By analyzing the pulses, it is possible to evaluate the frequency dependence of the dielectric constant of the substrate material. In microstrips, the dielectric substrate has to be considered as a capacitor with mixed dielectric. Thus, in this case it is partially filled. The effective filling fraction $q$ is related to the effective dielectric constant ($\varepsilon'_{\mathrm{eff}}$) by the following relation:

$$\varepsilon'_{\mathrm{eff}} = (1 - q) + \varepsilon' \tag{4.19}$$

Based on conformal transformation, the expression for dielectric filling $q$ for the wide strip line (where the width of the strip $w$ is $\gg d$, the thickness of the dielectric substrate) is expressed by the relationship:

$$q\nu = \nu - \ln\left(\frac{\nu+\mu}{\nu-\mu}\right) + \frac{0.732}{\varepsilon'}\left\{\ln\left(\frac{\nu+\mu}{\nu-\mu}\right) - \cosh^{-1}(0.358\nu + 0.953)\right\}$$
$$+ \frac{\varepsilon'-1}{\varepsilon'}\left(0.386 - \frac{0.5}{\nu-1}\right) \tag{4.20}$$

where $\nu = 1 + (1 + \mu^2)$ and $\mu$ is given by the expression:

$$\frac{W\pi}{2d} = \mu - \sinh^{-1}\mu \tag{4.21}$$

Given the known value of $\varepsilon'_{\text{eff}}$, $\varepsilon'$ (the relative dielectric constant of the microstrip substrate) can be determined from Equation 4.19.

The microstrip line is connected to the Network analyzer and the transmission loss is determined in the frequency range 4–11 GHz in each case. To determine $\varepsilon'_{\text{eff}}$ of the microstrip line the "time domain reflectometer" is used. The theoretical and experimental values of $Z_0$ agree fairly well and confirm the accuracy of measurements. The relative dielectric constant, relative loss factor, loss tangent, characteristic impedance and microstrip dispersions have calculated for several samples when the signal passes along the longitudinal and the transverse direction. In the longitudinal direction the relative dielectric constant for dry bone substrate is 11.5, which is in fair agreement ($\sim$10) with the findings of Kosterich, Foster and Pollack (1983). In the transverse direction, the permittivity value increases about threefold with respect to the above value. The transmission loss at the same frequency in the longitudinal direction is 4.19 dB cm$^{-1}$, whereas in the transverse direction it is 3.14 dB cm$^{-1}$. The change in transmission loss, dielectric parameter, and loss tangent may be attributed to the structural difference in the two orthogonal directions leading to anisotropy (Katz and Yoon, 1984). Microstrip dispersion in the longitudinal direction (0.725) is lower than that in the transverse direction (1.42). In the transverse direction, the dispersion is similar to that of Rexolite substrate (Yamashita, Atsulci and Hirahata, 1981). However, the loss is very much higher than for substrates conventionally used for microstrip fabrication.

## 4.9 Bone Physical Properties and Ultrasonic Transducer

The electromechanical properties of bone have been investigated by various workers (Cochran, Pawluk and Bassett, 1968; Abendschein and Hyatt, 1970, 1971; Gjelsvik, 1973). Knowledge of such properties is of potential importance in understanding bone remodeling dynamics and mechanisms involved therein, upon application of external fields. It is now established that bone is a heterogeneous, viscoelastic and anisotropic material and its physical properties depend on such factors as physical dimensions, density, rigidity, microhardness, mineral content, anisotropy, temperature, position and orientation of collagen (COL) fibers and hydration. Becker, Bassett and Bachman (1964) have reported that the conductivity can be expected to increase with age during growth owing to an increase in bone density and mineralization, probably because bone material is more conductive than dry COL at room temperature. Inasmuch as (i) the material properties of bone have been shown to be related to its growth and development (Lanyon and Bourn, 1979) and (ii) those parameters that affect the electrical properties of bone change with age (Marino and Becker, 1970), the study of physical properties at various stages of bone development has much importance. The electrical characteristics of bone materials in the high frequency region (1–100 MHz) have been examined and an equivalent electrical circuit for bone materials as ultrasonic transducers has been worked out (Singh and Behari, 1984b).

To highlight this aspect, various physical, dielectric, piezoelectric and electromechanical properties have been measured together with the ultrasonic propagation velocity in bone materials obtained by mixing COL and HAP in different proportions by weight. This study has been extended to examine the role of various foreign additives in bringing about changes in the above properties by adding 5–10% powdered additives, namely, $AlBr_3$, $Na_2CO_3$, $Li_2CO_3$, $SrCO_3$, $Sb_2O$, ZnO, $Nb_2O_5$, PZT, $Ba(OH)_2$ and $Pb(NO_3)_2$. The choice of a particular doping material is governed by various factors, for example, it should be compatible with bone matrix and have effects on various physical, dielectric, piezoelectric and electromechanical properties (Singh and Behari, 1984b).

Table 4.8 gives data on various physical (density, resistivity, Curie temperature), dielectric (dielectric constant, dielectric loss), piezoelectric (charge constant and voltage constant) and electromechanical (frequency constant and coupling coefficient) properties. In addition, the temperature dependence of the dielectric constant ($K^T$) and loss factor ($\tan\delta = \omega CR$) has been examined at 25–225 °C (Figure 4.20).

The variation of capacitance with temperature from 25 to 225 °C is also not the same for all bone compositions. In some cases (85% COL + 15% HAP, 95% COL + 5% HAP, 90% COL + 10% HAP, 75% COL + 25% HAP and 100% HAP) the rise in dielectric constant with temperature is very sharp. The maximum capacitance at the Curie temperature was observed at about 1500 pF for the 85% COL + 15% HAP composition, while for others it fell to around 400 pF. An identical trend was also observed in the variation of loss factor with temperature, which was in the range 0.03–0.3 (Singh and Behari, 1984b).

The density, resistivity and loss factor show definite enhancement as mineral content increases. However, there is no definite trend in the variation of dielectric constant with bone compositions. While some bone compositions (50% COL + 50% HAP, 40% COL + 60% HAP) have dielectric constant of about 22 others (100% COL, 95% COL + 5% HAP 35% COL + 65% HAP, 25% COL + 75% HAP and 15% COL + 85% HAP) have lower values (≈6). Furthermore, the resistivity is enhanced by a factor of $10^2$ after the test pieces are poled, although there is no such rise in dielectric constant. The resistivities and dielectric constants are higher for intact bone than for the samples made of powdered bone materials. Furthermore, heating denatures COL. The fall in density, resistivity, dielectric constant and loss factor may be attributed to the pores present in test pieces, as also observed by Katz (1971), and also to the presence of poly(vinyl alcohol). A similar observation has been reported with multiphase ceramic materials, wherein the rupture strength decreases as the porosity increases (Ryshkewitch, 1953; Coble and Kingery, 1956; McAdam, 1951; Balshin and Fedotor, 1964).

The piezoelectric behavior of an unpoled specimen may be attributed to the inherent property of the components. After crushing the specimen the microcrystallites still retain their identity, which is reflected in the bulk properties. This may result from a particular ordering of COL molecules. Piezoelectric effects have also been observed in electro-deposited and evaporated films of COL, suggesting that they originate at the level of the tropocollagen molecule or with

**Table 4.8** Properties of Doped Bone Specimens

| | Composition Sample[a] | | | | | | | | | | | |
|---|---|---|---|---|---|---|---|---|---|---|---|---|
| | 95% B + 5% $Na_2CO_3$ | | 95% B + 5% $Nb_2O_5$ | | 90%B + 10% ZnO | | 90% B + 10% PZT | | 90% B + Ba $(OH)_2$ | | 90% B + 10% $Pb(NO_3)_2$ | |
| | Unpoled | Poled | Unpoled | Poled | Unpoled | Poled | Unpoled | Poled | Unpoled | Poled | Unpoled | Poled |
| Density ($kg/m^3$) | 1685 | 16[illegible]6 | 1688 | 1686 | 1678 | 1786 | 1799 | 1802 | 1809 | 1820 | 1882 | 1885 |
| Resistivity ($\Omega$ cm) | $10^9$ | $10^{10}$ | $10^9$ | $10^{10}$ | 109 | $10^{11}$ | $10^9$ | $10^{11}$ | $10^9$ | $10^{12}$ | $10^{10}$ | $10^{11}$ |
| Dissipation factor (tan $\delta$) | 0.084 | 0.038 | 0.035 | 0.036 | 0.08 | 0.09 | 0.037 | 0.048 | 0.06 | 0.08 | 0.035 | 0.045 |
| Change constant d (C/N)$10^{-12}$ | 0.57 | 0.8[illegible] | 0.06 | 0.69 | 0.30 | 0.78 | 1.12 | 1.28 | 0.21 | 0.30 | 0.26 | 0.29 |
| Voltage constant g (Vm/N)$10^{-3}$ | 7.12 | 9.1[illegible] | 13.63 | 15.65 | 1.20 | 2.92 | 25.35 | 28.82 | 1.63 | 2.42 | 14.67 | 16.28 |
| Dielectric constant $K^T$ | 9 | 10 | 5 | 5 | 29 | 30 | 5 | 5 | 14 | 14 | 2 | 2 |
| Frequency constant $N_t$ (cps-meter) $2 \times 10^4$ | 4.0 | 4.0 | 4.9 | 4.9 | 4.8 | 4.8 | 5.1 | 5.1 | 5.1 | 5.2 | 4.8 | 4.8 |
| Coupling coefficient ($k$) | 0.48 | 0.49 | 0.40 | 0.42 | 0.52 | 0.55 | 0.32 | 0.34 | 0.43 | 0.45 | 0.32 | 0.34 |
| Curie temperatire (°C) | 150 | 150 | 135 | 135 | 130 | 130 | 180 | 180 | 165 | 165 | 175 | 180 |

[a]B = Bone (experiments refer to 30 °C).

Table shows the variation in bone properties due to additive materials.

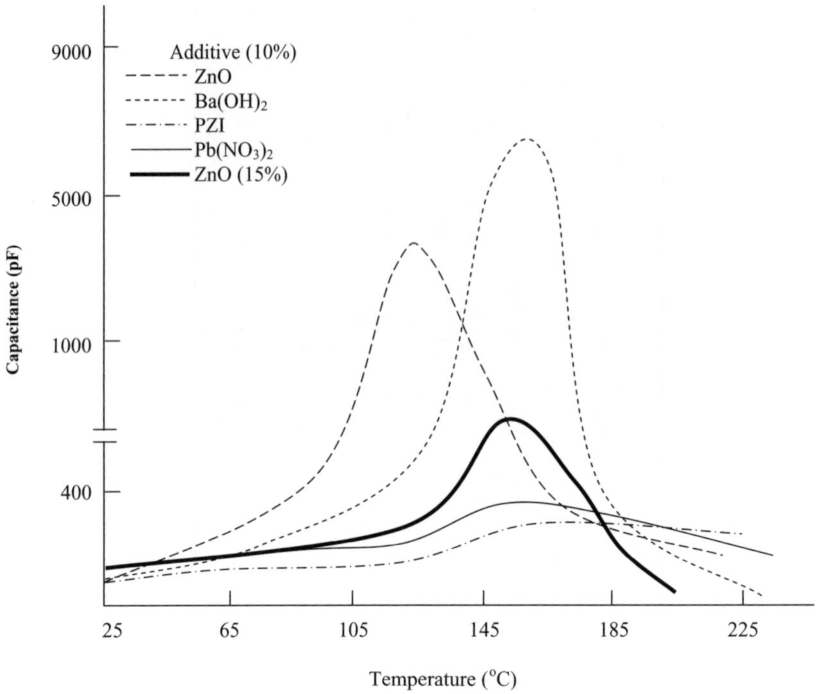


**Figure 4.20** Variation of capacitance with temperature for disc-shaped samples (1.2 cm in diameter and 2.00 mm thick) containing different additives

aggregated structures no more than 50 Å in diameter (Marino *et al.*, 1980). However, in such mixing as here, it is not possible to control the relative orientations in composites.

The rise in resistivity is due to polarization of the sample. It is known that bone and its two major constituents show electret phenomena comparable to those of any other dielectric (Mascarenhas, 1974; Andrabi and Behari, 1981). However, the actual rise in resistivity (extent of polarization) will depend upon the composition of the material in question. Though some role of electrets in controlling bone behavior is conceded (Yasuda, 1977), its effect and the mechanism at work is still a matter of investigation.

Figures 4.21 and 4.22 present the variation of phase angle and relative output voltage with frequency at 0.5–1.08 MHz. The impedance value may also be related to the nature of the material. Furthermore, the graph of phase angle (tan $\lambda$) vs frequency shows that the material is highly capacitive (sample diameter 10.0 mm and thickness 1.0 mm). On the other hand, in PCB-6 the phase angle is always positive and hence it behaves as an inductor in the region of concern. No significant variation in peak position was observed for bone compositions containing varying proportions of the minerals. The frequency constant is higher for samples made with pure powdered COL than with pure HAP; for other compositions, the frequency constant fell close to 7.2. The maximum value of the coupling coefficient ($k$) was about 0.53 and the minimum 0.36 for 90% COL + 10% HAP and 100% HAP, respectively. These parameters changed, though marginally, for poled samples.

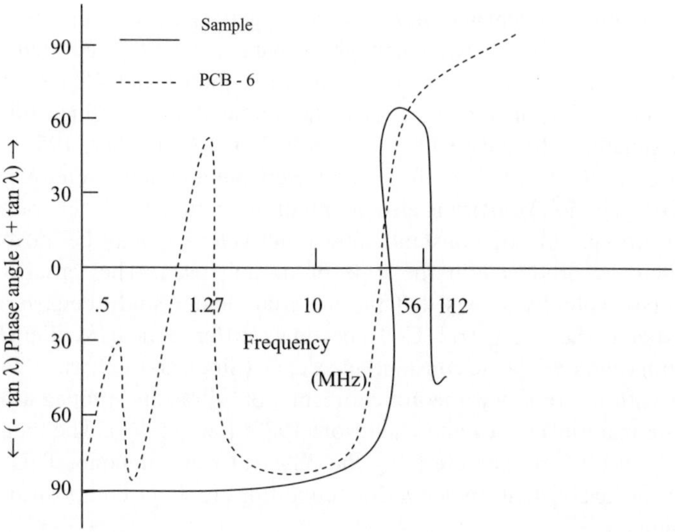


**Figure 4.21** Variation of phase angle (tan λ) with frequency for discs (10 mm in diameter and 1.00 mm thick) of bone, apatite, collagen and PCB-6. Curves for the first three materials are alike

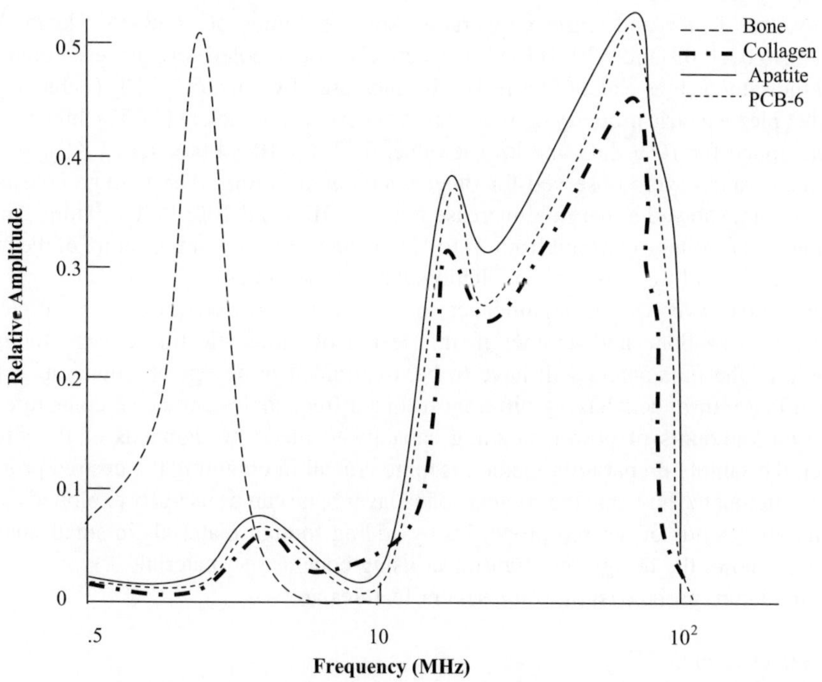


**Figure 4.22** Relative amplitude versus frequency for natural bone, apatite, collagen and PCB-6 (disc-shaped samples 10.0 mm in diameter, 1.0 mm thick)

Regarding electromechanical and piezoelectric properties, bone compositions corresponding to 50% COL + 50% HAP showed the highest charge and voltage constants, $3.64 \times 10^{-12}$ $\mathrm{C\,N^{-12}}$ and $24.75 \times 10^{-3}\,\mathrm{V\,m\,N^{-1}}$, respectively, for unpoled and $4.84 \times 10^{-12}\,\mathrm{C\,N^{-1}}$ and $32.56 \times 10^{-3}\,\mathrm{V\,m\,N^{-1}}$ for poled samples. Of all the combinations examined, this comes closest to the *in vivo* situation (30–35% COL + 65–70% HAP) (Glimcher, 1959). Changing the proportions of COL and HAP alters these parameters but a definite trend is not visible.

Ultrasound propagation velocity is also an important parameter for characterization. The increase in density and elastic constant (ultrasound velocity) may be attributed to cross-linking that may be produced in the type of mixing used. The variation for mineral proportions versus velocity shows that the materials under study behave as a family of Reuss solids wherein the soft matrix (COL) becomes stiffer as the hard filler (HAP) is added until a maximum velocity (elastic constant) is reached (Singh and Behari, 1984b). These data are consistent with the reported theoretical results on ultrasonic propagation velocity for other composite materials (Lees and Davidson, 1977; Lees, 1979). The data on ultrasonic velocity may be useful in characterizing the physical state of bone. This work offers a comparison of elastic constants under various conditions for bone compositions using a non-invasive technique.

There is quite a significant variation among different physical, dielectric piezoelectric and electromechanical properties in the presence of various additives. Addition of 10% ZnO or $Ba(OH)_2$ increased the dielectric constant to about 29 and 14, respectively. The dielectric constant was not much changed by other additives: $AlBr_3$, $Na_2CO_3$, $Li_2CO_3$, $SrCO_3$, $Sb_2O_3$, $Nb_2O_5$, PZT, $Pb(NO_3)_2$. For doped specimens the dielectric constant varied from 2 to 10. The density, resistivity, loss factor and Curie temperature showed values of $1668–1885\,\mathrm{kg\,m^{-3}}$, $10^9$–$10^{10}\,\Omega\,\mathrm{cm}$, 0.03–0.09 and 130–180 °C, respectively, for unpoled test pieces. Poled pieces showed the same values except that resistivity increased by a factor of $10^2$ (Table 4.8).

Another piezoelectric property of interest is the coupling coefficient ($k$). The highest values, 0.55, was found for 10% ZnO and lowest value, 0.32, for 10% $Pb(NO_3)_2$, PZT and $SrCO_3$; intermediate values were observed for other bone compositions. The voltage constant and charge constant showed a marginal increase for 5% $AlBr_3$ and 10% PZT mixture. The 10% ZnO reduced the voltage constant from 24 to 1.2. In view of the complex nature of the mixture product, it is difficult to make any prediction about these trends.

These results have twofold significance, one for the characterization of bone electrical properties *in vivo* bone and another in the design of ultrasonic transducers, though the magnitude of the parameters will have to be ascertained in any given situation. It is also suggested that transducers having different resonant frequencies can be so constituted as to have varying amounts of power handling capacity to meet the demands of the situation. However, the sample preparation method is quite crucial in obtaining the desired properties. This is an attempt to show that the contents of cadaver bone can be usefully exploited as a base to obtain other solids of desired properties by adding foreign materials in small quantities. Figure 4.23 shows the design of a transducer using bone as the material.

For a miniature probe design, some salient features are:

- small effective area
- broad bandwidth
- high acoustical transparency
- high immunity from radio frequency interference.

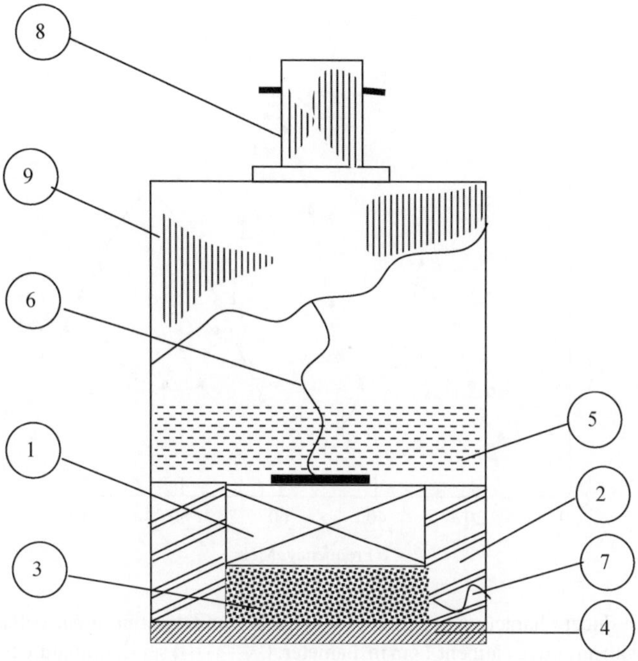


**Figure 4.23** Schematic diagram of the mounting of a large ultrasonic bone transducer: (1) piezoelectric bone element, (2) insulating epoxy, (3) quarter wave matcher, (4) thin window, (5) backing material, (6) wire connecting upper electrode, (7) earth plane, (8) electrical connector and (9) metallurgical housing sleeve

Some miniature transducers have been designed with bone composites as the active material for this purpose. A unique characteristic of such probes is their ability to work over a broad range of frequencies up to 100 MHz or even higher. Some piezoelectric polymer probes have been designed for ultrasonic absorption measurements and tested at correspondingly low frequencies (De Reggi *et al.*, 1981). These probes are constructed by rendering a small region of a thin sheet of a polymer [poly(vinylidene fluoride)]. By employing thinner sheets we can achieve a higher resonance frequency (up to 160 MHz for a 6 μm film). However, thinner films will have reduced pressure sensitivity as the active volume of the piezoelectric medium will be reduced.

An obvious extension of the above is to examine the field pattern of this bone transducer. This has been studied in deionized water, in 0.9% saline solution and in a phantom material mimicking soft tissues, near the resonance frequency of the transducer (36 MHz). Figure 4.24 shows a plot of normalized output of the bone matrix transducer and probe (sets I and II) over a frequency range of approximately 15–100 MHz. This plot is constructed by varying the frequency continuously and by measuring the acoustic pressure pulses using an identical transducer as receiver. In (set I) two peaks of maximum output, corresponding to thickness resonance frequencies of 36 MHz and 72 MHz, are observed. These correspond to the frequency and its higher harmonic. Similarly, for the miniature transducers, the maximum normalized output corresponds to 36 and 100 MHz. Although we take the same biomaterial as the active material in both cases, the resonant frequencies show a difference because of the

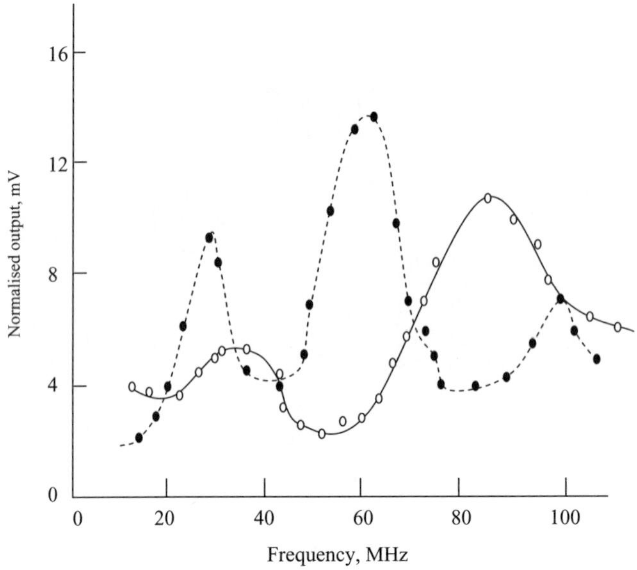


**Figure 4.24** Transducer characteristics (output voltage per unit external input voltage). (——) Set I: large transducers with an active element 1 cm in diameter; (-----------) set II: miniature transducers with an active element 3 mm in diameter

differing geometry of the transducer material. Output voltage amplitudes also differ in the two cases. It may be suggested that the phenomenon under observation is related to the microscopic nature of the material used therein (Singh and Behari, 1984a, 1984b). Owing to its biphasic and anisotropic structure, the stress is expected to be distributed non-uniformly over the entire medium of the transducer and, as such, a non-uniform field pattern might be expected from transducers employing bone matrix as an active material.

For calibration purposes a particular frequency (36 MHz) was chosen, close to the resonance frequency of the transducer. Figure 4.25 shows plots of the field patterns in deionized degassed water, 0.9% saline and the phantom material mimicking soft tissue structures. Here the plot is of output voltage as observed by the receiving miniature transducer, at different angular positions and on the circle of a constant radial distance from the center point. The output voltage is expressed in decibels with respect to that at the central point. It is found that the radiation field is not isotropic in any of these media. At 36 MHz the attenuation in 0.9% NaCl solution is higher than that of water. From the observed output voltages along the center axis of the transmitting transducer we have estimated the attenuation coefficient $\alpha$ for three media (at 22 °C) (Table 4.9). At 36 MHz, a detailed quantitative comparison is not possible. The attenuation coefficient is related to the frequency by the relationship $\alpha = af^n$. Taking $\alpha$ at 1 MHz for deionized degassed water at 22 °C (Hall, 1948), the calculated value of $\alpha$ at 36 MHz is $0.32 \pm 0.02\,\mathrm{dB\,cm^{-1}}$ and $n = 1.4$.

Attenuation coefficients of freshly excised biological tissues such as liver and fat muscle have been measured at the near resonance frequency (36 MHz) using these transducers. The observed attenuation coefficients are presented in Table 4.9 along with the corresponding data at 1 MHz reported by Goss, Johnston and Dunn (1978).

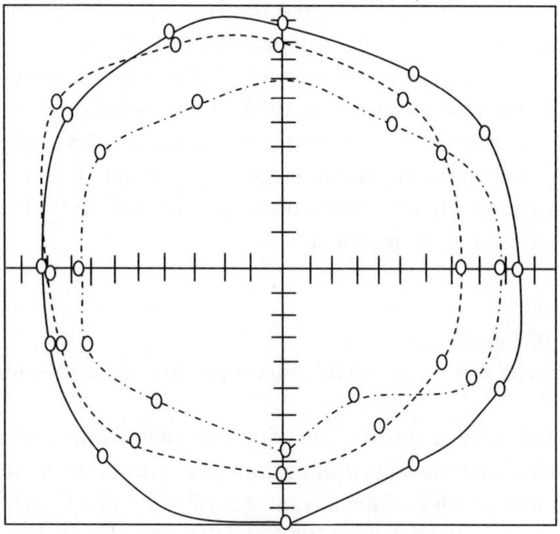


**Figure 4.25** Field pattern of an ultrasonic transducer in water (—), saline (0.9%, -------) and phantom (-·-·-·-·-·-) at a depth of 1 cm. Calibration is carried out in a circle of radius 3 cm with the origin at the center of the co-ordinates. Along the radius, one division = 1dB

Again, considering that the attenuation coefficient $\alpha$ for biological tissues depends on the frequency and is governed by the empirical relationship $\alpha = af^n$, where $a$ and $n$ are constants and are characteristic of the tissue, the value of $n$ turns out to be 1.32 for liver and 1.08 for fat and muscle. There is an almost linear relationship between frequency and attenuation coefficient for these two soft tissues. The viscous forces acting between a suitably chosen distribution of suspended particles and a suspending liquid can account for the experimentally observed linear relationship between the ultrasound attenuation coefficient and frequency (Fry and Dunn, 1962). The frequency band over which linearity is obtained is determined by the limits of the distribution of values for the parameters chosen to describe the structural element. This frequency band may extend beyond 36 MHz. The excess attenuation coefficient (in excess of

**Table 4.9** Attenuation constant $\alpha$ (dB cm$^{-1}$)

| Sample | Frequency (MHz) | | |
|---|---|---|---|
| | 36 | 10 | 1 |
| Degassed water | $0.32 \pm 0.02$ | $0.02^a$ | |
| 0.9% saline | $0.43 \pm 0.05$ | $4.00 \pm 0.24^b$ | $0.16 \pm 0.03$ |
| Fat muscle | $7.10 \pm 1.74$ | — | $0.26^c$c |
| Liver | $11.03 \pm 1.30$ | $1.20^a$ | $0.099^c$ |

[a]Hill (1986).
[b]Parry and Chivers (1979).
[c]Goss, Johnston and Dunn (1978).

the limiting high frequency value due to viscosity) may be attributed to the reaction of hydrogen phosphate ions ($HPO_4^{2-}$) with the protonated imidazole ring of the histidyl residue and the imidazole itself at the normal pH of the solution (Slutsky, 1980).

The attenuation constant on tissue specimens has also been measured at 100 MHz using a scanning laser acoustic microscope (Kessler, 1973). The method provides an estimate of phase velocity compared with group velocity obtained by the pulse echo method. By using multiple specimens of varying thickness the errors in making average attenuation measurements at 100 MHz with this instrument have been reduced significantly (Tervola *et al.*, 1985a, 1985b).

Some advantages of bone transducers are:

- wide frequency band
- low cost and ease of fabrication
- peak frequencies can be shifted by suitable mixing of composites (Singh and Behari, 1984b).

The significance of such transducer-like behavior may enable one to study the phenomena of high frequency ($> 10$ MHz) ultrasound propagation in a wide range of media and to investigate their biological significance. One of the advantages of using this higher frequency is that the resulting wavelength is expected to be less (0.05 and 0.1 mm). This is smaller than most normal tissue structures and much smaller than the lateral dimension of the ultrasonic beam. The probable use of ultrasonic waves at high frequencies for medical diagnostic instrumentation has necessitated our understanding of its interaction with human tissues. This knowledge is also vital in evaluating and establishing exposure safety standards, determining the hazard levels and the various effects in this frequency region. Also, their role in biological systems remains illusive.

# 5

# Bioelectric Phenomena: Electrostimulation and Fracture Healing

## 5.1 Introduction

Since Fukada and Yasuda (1957) demonstrated that dry bone behaved piezoelectrically, both with direct and converse effects, it has become apparent that potentials in the range 1–5 mV $cm^{-1}$ in cartilage, tendon and other structural tissues reflect at least two electrogenic events. Both piezoelectric and electrokinetic properties can contribute to the voltage waveforms that follow mechanical deformation of these tissues. When rates of deformation are rapid, the initial, high amplitude of the waveform appears to be shaped by piezoelectric response (Bassett and Pawluk, 1972; Reinish and Nowick, 1974, 1975, 1976). This suggest that bone electrical activity and hence its remodeling pattern can be influenced by external stimuli.

When bone is subjected to mechanical strain, a part of the energy is converted into other forms and ultimately dissipated as heat. There appear to be a correlation between healing quality and power dissipation (Hassler *et al.*, 1977). There is a rapid increase in bone growth with increasing power dissipation, reaching an optimum value around 35 μW This pattern of power dissipation also depends on the bone physiological state and its mode of function (Behari and Singh, 1981). This is a manifestation of the fact that bone can be treated as a biogenerator and, possibly, dissipated power is involved in bone growth and repair.

Presently, two main possibilities have been suggested to account for the response of bone to external stimuli. The first hypothesis assumes that the pressure transmitted through the tissue to the cell surface or its extension produces a deformation that is received by the cell (Bassett, 1968). The second widely accepted hypothesis states that mechanical stress generates an electrical signal, which sets in motion the subsequent events. This hypothesis is based on the fact that (i) in bone, exposed to mechanical forces, electrical signals are produced by piezoelectric effect/or streaming potential or electret (Bassett and Pawluk, 1972) and (ii) electrical current affects bone growth and remodeling (Becker and Murray, 1967; Friendenberg

*Biophysical Bone Behavior* Jitendra Behari

*et al.*, 1970; Lavine *et al.*, 1974; Marino and Becker, 1971; Mascarenhas, 1974). If one considers that the primary function of bone is to bear load, which is a physical function, then it is suggestive that the signal that directs bone growth, repair and remodeling, through action on cells, may also be of physical origin.

Viewed in the above context, initiatives have been made towards understanding the mechanism of bone growth by external stimulation. It implies that in response to a stimulus applied internally or externally the bone adjusts itself to meet the demands of the situation. Since then, several animal experiments and human trials have been carried out with varying degrees of success. In accordance with Wolff's law, Figure 1.1 presents two suggested possibilities.

All changes in environmental conditions (internal or external) are considered as stimuli that are converted into electrical stimulus. This in turn acts on the cells, causing them to proliferate and form the callus. By extrapolation, the final cause of callus formation and consequent fracture healing is the electrical stimulus. It seems possible that alterations in the physicochemical properties of bone influence the occurrence of piezoelectric and streaming potentials, which in turn influence the transformation of bone. The biphasic nature of the piezoelectric potential in intact bone may be the result of a series of low frequency relaxation phenomena that depend upon its structure and elastic properties.

A significant amount of literature has accumulated relating to cyclic mechanical deformation with dynamical electrical polarization (Fukada and Yasuda, 1957) and the control of bone cell activity (Bassett, 1984, 1989). It appears that electrical or mechanical stimulation is equally effective in maintaining or improving bone mass (McLeod and Rubin, 1992).

## 5.2 Biophysics of Fracture

### *5.2.1 Mechanisms of Bone Fracture*

Bone strength depends on its material properties and architecture (Martin and Burr, 1989). Genetic information in osteoblasts mainly determine bone material properties, including stiffness, ultimate strength, true density, proportional limit and yield point. These properties do not change with age, disease, sex and species. The architectural features that determine a whole bone's strength also include the amount of bone in a cross section.

Key to fractures are two main mechanisms: when the damage rate exceeds the remodeling/repair rate (excess damage) or when a normal damage rate is not repaired properly due to a defective remodeling/repair mechanism (deficient repair). At the macroscopic level damage is hardly visible before there is a large crack even though the mechanical functionality may have altered substantially during earlier stages. Since mechanically triggered osteogenesis is related to strain rate, level, distribution and duration, bone formation can also be site specific (Rubin and Lanyon, 1987). The whole bone, or only a given part, can undergo hypertrophy. For example, stress fractures occur frequently in the regions of teenage bone where residual, unremodeled woven or interstitial lamellar bone serves as a stress resin in the surrounding, (Johnson, 1964). Remodeling produces a structure that distributes stress evenly.

Bone fracture appears when an accidental load exceeds the physiological range, inducing stresses over the strength that bone tissue has achieved after adaptation during growth and developmental phase (traumatic fracture). There are two main causes of this type of fracture (Sloan and Holloway, 1981): an external impact produced, for example, by a fall, or fractures that occur "spontaneously" by a muscular contraction without trauma (Figure 5.1). The latter are quite common in elderly people with osteoporosis. Several authors suggest that the main cause of hip fracture is contraction of the iliopsoas muscle and gluteus medius (Yang *et al.*, 1996; Smith, 1953). Figure 5.2 shows the same effect in an osteoporotic bone.

When bone fracture occurs, the blood supply to the severed vessels ceases and the osteocytes in the immediate vicinity of these vessels die. This accounts for the fact that a cortical bone graft is essentially dead tissue. Dead bone grafts are useful not only for the support that they provide, but also for their ability of their matrix growth factors to promote resorption and subsequent new bone formation. Impairment of blood supply to a region of bone due to vessel trauma or occlusion results in osteonecrosis.

The kind of fracture that is often produced by normal loads acting on a bone that has been weakened by disease or age (Zioupos and Currey, 1998) is normally called pathologic. Most of these fractures are provoked by osteoporosis in the elderly and are more frequent in women than in men. In fact, the higher risk of bone fracture in the elderly is not only due to the progressive reduction of bone consistency (osteoporosis) and therefore strength but also to additional factors such as the inability of soft tissues to absorb the energy generated in a

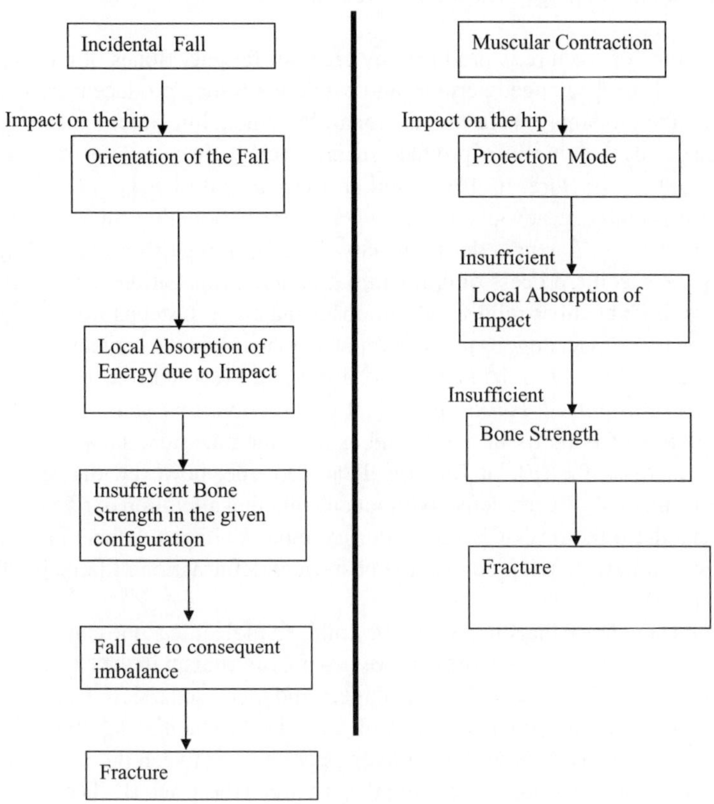


**Figure 5.1** Two, usually, probable modes of bone fracture

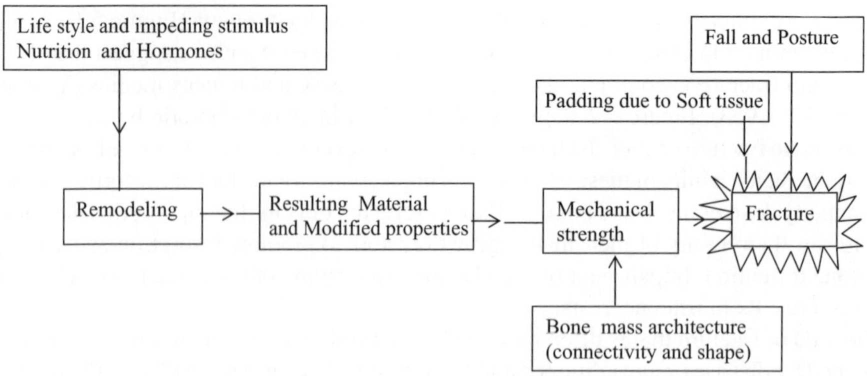


**Figure 5.2** Sequential arrangement of factors leading to osteoporotic fracture. It effects bone material properties and bone strength

fall and the change of the kinematic variables of the gait. Lotz and Hayes (1990) report that only a small amount of energy is needed to break a bone (i.e., 5% of the energy available in a fall).

The second type of fracture is produced by creep or fatigue. Bones often support more or less constant loads for prolonged periods, and cyclic loads may produce microdamage. If the accumulation of microdamage is faster than repair by remodeling, microcracks (or other kinds of microdamage) can multiply to produce macrocracks and complete fracture. Clinically, this is called a stress fracture. It also occurs at lower activity levels in bones weakened by osteoporosis, especially at advanced ages when bone remodeling almost stops. However, the prevention of stress fractures does not only depend on repair by remodeling but is also controlled by the specific process of crack initiation and propagation. The microstructure of cortical bone is similar to fiber-reinforced composite materials. Osteons are analogous to fibers, interstitial bone tissue is analogous to the composite matrix and the cement line acts as a weak interface where cracks may initiate (Burr, Schaffler and Frederickson, 1988).

It is generally recognized, based on clinical observation, that pathologies that affect the material properties of bone tissue by changing its mineralization, such as osteomalacia or osteopetrosis, increase the risk of fracture. It is also true, however, that diseases that are primarily the result of a collagen defect without any inherent alteration of tissue mineralization can likewise increase the risk of fracture. For example, collagen defects in a mouse model of osteogenesis imperfecta (OI) reduced the post-yield deformation of bone by 60%, (Jepsen *et al.*, 1996, 1997).

There is ample evidence that changes in the collagen molecule alone can alter fracture risk. Several reports have shown a direct link between a polymorphism in the COLIAI gene, which encodes the $\alpha 1(\mathrm{I})$ protein chain of type I collagen, and increased risk of fracture independent of low bone mineral density (BMD) (Langdahl *et al.*, 1998; Mann *et al.*, 2001). Using a meta-analysis, Mann *et al.* (2001) showed a stronger association between the COLIAI Spl binding site polymorphism and osteoporotic fracture than between these and BMD or body mass index (BMI) at both the lumbar spine and the femoral neck. In a separate study, Bernad *et al.* (2002) also showed the COLIAI (Spl) TT genotype to be associated with a 5.9-fold increase in fracture risk in postmenopausal women and a 4.8-fold increase when prevalence was adjusted for age, BMI and BMD. This polymorphism increases the collagen content of the tissue, and alters the ratio of the collagen $\alpha 1$(I) to $\alpha 2$(I) chain. Although the effect on fracture risk could be partly explained by the lower mineral content in heterozygotes (Keen *et al.*, 1999a, 1999b, Uitterlinden *et al.*, 1998), the fact that the effects are at least partly independent of bone mass (Bernad *et al.*, 2002) suggests a potential direct effect of the alteration in collagen on bone quality.

### *5.2.2 Mechanical Stimulation to Enhance Fracture Repair*

The concept that bone formation and remodeling is controlled by both physical and biochemical factors has long been known. Mechanical loads such as high-impact exercise, which produce dynamic strong bones, may also play an important role in controlling bone mass and strength. Several studies have shown the beneficial effects of physical training on site-specific bone mineral density in postmenopausal women (Danz *et al.*, 1998; Kerr *et al.*, 1996; Nelson *et al.*, 1994). The effects of exercise are more pronounced at the spine than at the femoral neck (Wallace and Cumming, 2000). Rats and canines have been used as experimental animal models for manually creating long bone fracture in surgically producing transverse oblique osteotomies (Aro *et al.*, 1989, 1990; Aro, Wahner and Chao, 1991; Behari *et al.*, 1991b). Gap and contact with and without static compression were used to study the effects of several fracture immobilization methods on healing patterns.

When loss or rigid internal fixation (intramedullary nail without interlock) and external fixation (smaller and fewer pins in unilateral frame) methods are used, fracture repair follows the secondary healing mechanism with an abundant amount of periosteal and endosteal callus but without osteonal migration (Rand *et al.*, 1981). Bone formation occurs secondary to endochondral and intramembranous ossification. The distribution of these two depends on the biologic environment and loading condition. The transformation from callus into mineralization woven bone occurs early in the healing period and increases its volume quickly to replace cartilage and undifferentiated tissue (Markel, Wikenheiser and Chao, 1990). This is responsible for the mechanical strength. Osteocytes, which are distributed throughout the bone, produce a signal through electrical potentials, activation of ion channels, flow of interstitial fluid and activation of signaling cascades, in direct proportion to the amount of mechanical loading (Martin, 2000).

It has been suggested that bone contains sensor cells, possibly osteocytes that monitor mechanical stress and activate adaptive biological processes to remodel the bone. It can be assumed that osteocytes are distributed in a network throughout the bone tissue. The external mechanical load is transmitted through the tissue whereby each osteocyte senses a particular mechanical signal (Mullender and Huiskes, 1995). After evaluation of the signal the osteocyte stimulates groups of osteoblasts and osteoclasts, the basic multicellular units (BMUs), to regulate net bone turnover. It is further assumed that the effect of each osteocyte on a BMU depends on the distance from the osteocyte. At each location of the bone tissue, the total stimulus is evaluated and bone mass is adapted according to its magnitude. This process is continued until a balance between the external loads and bone architecture is established, *vis-à-vis* the bone mass. An absence of this leads to decreased bone formation (Figure 5.2).

The amount of stimulus received by a BMU depends on the distance between the osteocyte and the location of the cells above (Figure 1.9b). Hence the stimulus, $F(x, t)$, at location $x$ at time $t$ is the sum of the stimuli received from all osteocytes:

$$F(x,t) = \sum_{i=1}^{N} f_i(x)[s_i(t) - k] \tag{5.1}$$

where $k$ is an arbitrary constant, $N$ = number of osteocytes and the spatial influence function is:

$$f_i(x) = \mathrm{e}^{-d_i(x)/d} \tag{5.2}$$

which describes the redirection of stimulus with increasing distance, $d_i(x)$ (mm), between osteocytes, $i$, and location, $x$. The parameter $d$ determines the gradient of the decline.

The change in relative density, $m(x, t)$, at location $x$ is governed by the local stimulus, $F(x,t)$. Hence:

$$\frac{\mathrm{d}m(x,t)}{\mathrm{d}t} = \tau F(x,t) \text{ with } 0.01 < \mu < (\xi, \tau) < 1 \tag{5.3}$$

where $\tau$ is a constant of proportionality concerning the rate of the process. The local elastic properties of the tissue [$E(x,t)$], are calculated from the local relative density using a cubic power law:

$$E(x,t) = Cm^3(x,t)$$

where $C$ is a constant.

Mechanical stimulation can induce fracture healing (alter its biologic pathway) (Aro and Chao, 1991; Brighton, 1984; Claes *et al.*, 1997; Wu *et al.*, 1984). Repetitive loading under small strain and high frequency or overloading through an elevated exercise regime causes bone hypertrophy (Goodship and Kenwright, 1985; Rubin *et al.*, 2001). The added bone formation is also related to the direction and magnitude of overloading, which will affect the internal state of stress of the repairing tissue. However, the regulating cellular mediations responsible for such a phenomenon at the cell membrane or cytoplasmic level can be linked directly to the mechanical stimulant; the most effective modality to maintain or enhance bone regeneration may be established for the treatment of different fractures especially in patients with deficient osteogenic potential attributable to either local or systematic factors. When the mechanisms for tissue formation at the cellular level are understood and well defined, physiologic conditions or pharmacologic agents may be developed to accomplish the same callus formation and bone regeneration effects without the mechanical interventions that are difficult to administer under adverse conditions. However, the cellular mechanisms and the potential mechanical receptors on the cell membrane that are sensitive to stress, inhibit electromechanical or stream potential signals remains to be identified. Chao *et al.* (2004) have concluded that under rigid internal or external fixation, the logic features of the fracture union matched those of the primary healing mechanism with direct osteonal migration across the fracture gap. However, periosteal and endosteal callus and new bone formation are common, depending on the loading condition and the consequent movement at the fracture site inherent to the specific immobilization method. To date, it can be said that the mechanical signal is primary to the fracture healing (callus formation) (Figure 5.3).

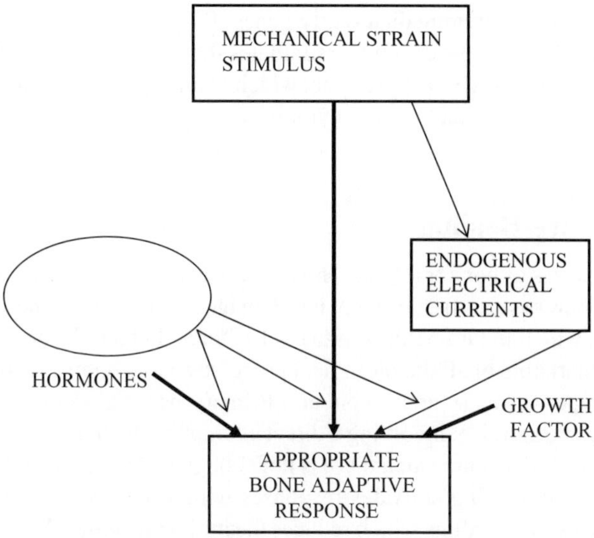


**Figure 5.3** Bone formation/Resorption (Remodeling); mechanical VS electrical stimulus (mechanical stimulus is primary)

The initial mechanotransduction event in cells contained in bone is that transactivation of target genes involves stimulation of mitogen activated protein kinase signaling cascades, as well as protein phosphorylation and dephosphorylation and ultimately results in the direct binding of transcription factors to gene promoters. The mechanical shear stress that is the stimulus in both systems causes transcriptional activation via distinct transcription factor binding sites that function as shear stress response elements (Braddock *et al.*, 1988). These elements are found in many promoters such as PDGF-B, TGF-β and FGF (Khachigian *et al.*, 1996) that are activated by injury. Transcription factors themselves, such as early growth response factor-1 (Egr-1) and stimulating protein-1 (Sp-1), are also responsive to mechanical stimuli such as shear stress (Schwachtgen *et al.*, 1998). At a fracture site, it appears that an optimum level of mechanical stress is a route to fast and effective healing. For example, distraction osteogenesis is a surgical procedure that may be used to correct limb length discrepancies or deformities or to treat non-union fracture following acute loss of functional bone, for instance after a crush injury. In this procedure, an osteotomy is performed followed by the application of an external fixator device. After ~1 week, distraction is performed at a controlled rate and the bone begins to grow and will eventually fill the gap. Studies in a rabbit model of distraction osteogenesis have shown that the expression of BMP-2, -4 and -7 is increased after distraction is initiated and is maintained throughout the distraction period (Rauch *et al.*, 2000). In addition, *in vitro* studies in rat osteoblasts exposed to pulsed electromagnetic fields have shown that expression of BMP-2 and -4 is increased, which is accompanied by an increase in mineralized bone nodule formation (Bodamyali *et al.*, 1998). These observations are consistent with physical forces inducing BMP expression and subsequent stimulation of new bone formation. Finally, it has been known for some time that cartilage regeneration may be enhanced by drill hole injury into subchondral bone. One possible explanation for this regenerative effect may be mechanotransduction, in which injury-induced stimulation of immediate-early genes (IEGs) drives the production of growth-stimulating proteins. BMP-4, for example, stimulates the production of the IEG Egr-1, which may feed back to stimulate BMP-4 expression, which in turn serves to promote tissue repair by activation of multiple growth factors and cytokines.

## 5.3 Bone Fracture Healing

There are several causes of bone fractures. However, in contrast with inert materials, bone can regenerate to form new osseous tissue where it is damaged or missing. Bone adaptability allows for efficient repair, which in turn helps to prevent fractures. In fact, the healing of a fracture is one of the most remarkable of all the biological processes in the body. Fracture healing can be divided into two types: primary and secondary. Primary healing occurs in cases of extreme stability and negligible gap size, involving a direct attempt by the bone to form itself (Einhorn, 1998). Secondary healing occurs when there is not enough stabilization and the gap size is not negligible. In this case, healing activates responses within the periosteum and external soft tissues that form an external callus, which reduces the initial movement by increasing stiffness. Secondary fracture healing has a series of sequential stages that can overlap to a certain extent, including inflammation, callus differentiation, ossification and remodeling. Most fractures are repaired by secondary healing, which does a more thorough job of replacing old and damaged bone.

The first stage begins after bone fracture. Once a fracture occurs, the basic healing process is auto-activated naturally to repair the site. Blood emanates from the ruptured vessels and a hemorrhage quickly fills the fracture gap space. Macrophages remove the dead tissue and generate initial granulation tissue for the migration of undifferentiated mesenchymal cells, originating an initial stabilizing callus. These cells proliferate and migrate from the surrounding soft tissue (Einhorn, 1998; McKibbin, 1978; Iwaki *et al.*, 1997; Sandberg, Aro and Vuorio, 1993).

The repair of a fractured bone takes time, depending upon the nature of the injury. For example, a long bone shaft fracture takes about 3–4 months to heal completely. As a result of immobilization and unloading of the affected bone, considerable restrictions in the usual daily activities are connected to the relatively long rehabilitation period needed after a fracture. Furthermore, the healing processes are disturbed in 5–10% of all fractures. This leads to, depending upon the geometric, mechanical and biologic factors, a delayed union or a non-union. (Einhorn, 1995). In these cases, the necessary additional treatments and prolonged reconvalescence time give rise to much morbidity. Consequently, treatment methods aimed at enhancing fracture healing are being developed. The ultimate goals of these methods are twofold: to lesser the healing time and to reduce the rate of impaired healing. Healing involves the differentiation of several tissues.

During normal fracture repair, there is an intermediate stage in bony bridging in which the newly formed tissues opposite the break are composed of fibrocartilage and the avascular tissue must be calcified before the final stages of healing (bone union) can occur. Cartilage is essentially composed of an avascular ancural and alymphatic hyaline matrix interspersed with a single cell type, the chondrocyte. The primary constituent of this matrix is aggrecan, an aggregating proteoglycan that interacts with hyaluronic acid and link protein and type II collagen, representing most of the cartilage fiber. This matrix is responsible for the ability of cartilage to facilitate articulation and to resist compressive forces, whereas the chondrocytes are responsible for matrix maintenance (Buckwalter and Mankin, 1997; Mow, Rateliffe and Poole, 1992). Following calcification, the fibrocartilage is invaded by new blood vessels from the bony areas flanking the gap. As the blood vessels operate the calcified tissue, cells begin its removal (chondroclasia) and its replacement by bone. The process is called endochondral ossification and is similar in its cellular dynamics to those that occur in the normal growth plate (epiphysis). Once bridging with fiber bone is accomplished, the ends of the fractured bone are connected and controlled mechanical stresses can result in a gradual maturation and strengthening of the break area.

Fracture healing is a natural process that can reconstitute injured tissue and recover its original function and form. The fracture evolution depends on many factors, including, mechanical loads, type of fracture, gap size, blood supply, hormones, growth factors, and so on. It is a complex process that demands the coordinated participation of immigration, differentiation and proliferation of inflammatory cells, angioblasts, fibroblasts, chondroblasts and osteoblasts, which synthesize and release bioactive substances of extracellular matrix components (e.g., different types of collagen and growth factors). Deviatoric stresses are the specific stimulus for the formation of fibrous connective tissue or bone, whereas hydrostatic stress controls the formation of cartilaginous tissue damaged or missing. Pauwels (1960) has suggested that bone ossification in the embryo and the growing child could occur in different forms: endochondral, intramembranous or appositional ossification. In the first, cartilage is formed, calcified and replaced by bone. In the second, bone is formed directly by osteoblasts (flat skull bones). In the third, ossification is adjacent to membrane layers of mesenchymal cells that differentiate into osteoblasts. When osteoblasts are not part of a membrane (i.e., endosteal or Haversian canal surface) ossification is called appositional. This ossification (appositional) is normally the only one found in healthy adult but the two types can be activated during the fracture healing process. Therefore, this process is important in understanding tissue repair and generation.

## *5.3.1 Histologic Fractures*

As osteogenesis increases the cross sectional and/or material property (strength), the load per unit area decreases, and strain may fall below threshold requirements. It is entirely likely that this response is triggered, and controlled, electromechanically, through transduction of mechanical energy into electrical energy and vice versa. Fracture healing of long bone involves the formation and resorption of the extracellular matrix. Cartilaginous callus is formed at the fracture site and is subsequently replaced by new bone during the endochondral ossification process. In this process, replacement of type II collagen, the major component of cartilage, with type I collagen, the predominant structural protein of bone, occurs rapidly. For fracture repair process to proceed, matrix resorption should coincide with matrix synthesis. Little is known about the replacement of matrix proteins. Yamagiwa *et al.* (1999) have concluded that MMP-13 plays an important role in the healing process of fractured bone in mice.

Mesenchymal cells may differentiate into chondrocytes, osteoblasts or fibroblasts, depending on the biological and mechanical conditions. These differentiated cells begin to synthesize the extracellular matrix of their corresponding tissue. Intramembranous woven bone is produced by direct differentiation of the stem cells into osteoblasts and appears adjacent to

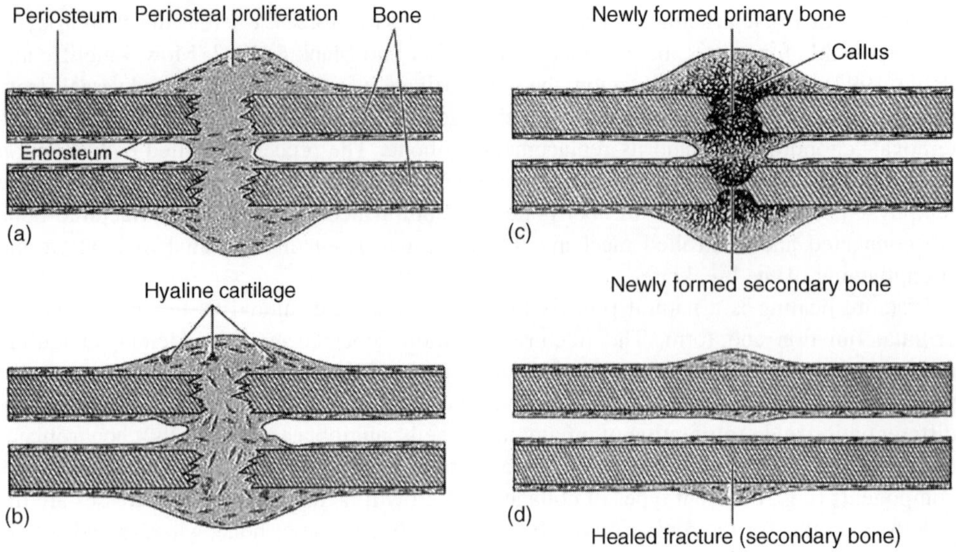


**Figure 5.4** Repair of a fractured bone by formation of new bone tissue through periosteal and endosteal cell proliferation

each side of the gap site, advancing to the center of the callus. At the same time, at the center of the callus, cartilage is formed by chondrogenesis, except immediately beside the gap where the stability is still very low and high relative displacement prevents the differentiation of mesenchymal cells (Figure 5.4). The ossification continues until all the cartilage has been replaced by bone and a bone bridge surrounds the fracture gap, achieving a good stabilization and sufficient stiffness. When the fracture is completely stabilized, mesenchymal cells begin to invade the gap (Figure 5.4). Once the gap has ossified, remodeling of the fracture site begins gradually in order to restore the original internal structure and shape (internal and external bone remodeling). The last stage is much longer than the previous one (1 year down to several weeks, depending on the animal species).

Three developmental features of the fracture healing process (Sandberg, Aro and Vuorio, 1993) inflammation, resorption and remodeling are observed in sequence. During the inflammatory phase (day 1 and 3 after the onset of fracture), the periosteum thickens at a distance of about 0.5 mm from the fracture site and proliferation of periosteal cells occurs. In the reparative phase (days 5–10), intramembranous bone formation is observed in the periosteal region. Chondrogenesis is also observed in the thickened periosteum. At day 5, hypertrophic chondrocytes appear beside cortical bone. At day 10, fracture callus grows and vascular cavities began to invade the callus. In the remodeling phase (days 14–21), new trabecular formation by endochondral ossification is observed. The amount of fracture callus peaks at day 14. In the newly formed marrow cavity, TRAP-positive cells and osteoblasts lining the newly formed trabeculae are observed. At day 21, the callus undergoes further remodeling. The fragment ends unite and the bone marrow space is regenerated. Figure 5.5 shows a schematic representation.

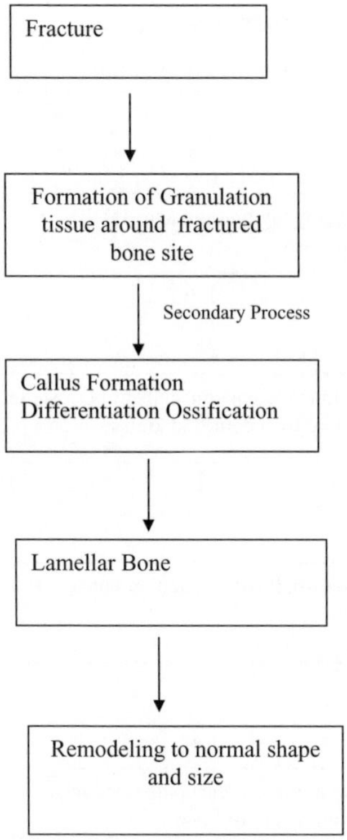


**Figure 5.5** Histological stages in bone fracture healing

The histological features and function of the fibrocartilage callus in non-union bone fractures are similar to those of the secondary cartilage in the mandibular condyle (Gerling, Sinclair and Roa, 1985). Kopman *et al.* (1987) has noted that molecular changes in proteoglycans in the fracture callus are similar to those seen in normal endochondral ossification.

The successful outcome of a healing fracture, wherein bone tissue differentiates both functionally and spatially, is a illustration of cell–cell communication (Doty, 1981; Lowenstein, 1981). Secondary fracture healing has a requirement that can overlap to a certain extent, including inflammation, callus differentiation, ossification and remodeling. Some authors (Perren, 1979; Perren and Cordey, 1980) have proposed that tissue differentiation is controlled by the resistance of various tissues to strain. Their main idea is that a tissue that ruptures or fails at a certain strain level cannot be formed in a region experiencing strains greater than this (Lacroix *et al.*, 2000).

Carter *et al.* (1998) have developed a tissue differentiation theory, which correlates new tissue formation with the local stress/strain history. They described qualitatively the relationship between the ossification pattern and the loading history, using finite elements to quantify the local stress/strain level, assuming that the tissue in the callus is formed by a single solid phase. They proposed several differentiation routes (Figure 5.6). When the tissue is subjected to high tensile strains (above the tension line) fibrous matrix is produced with cartilage or bone depending on the hydrostatic pressure level. Ament and Hofer (2000) have proposed a

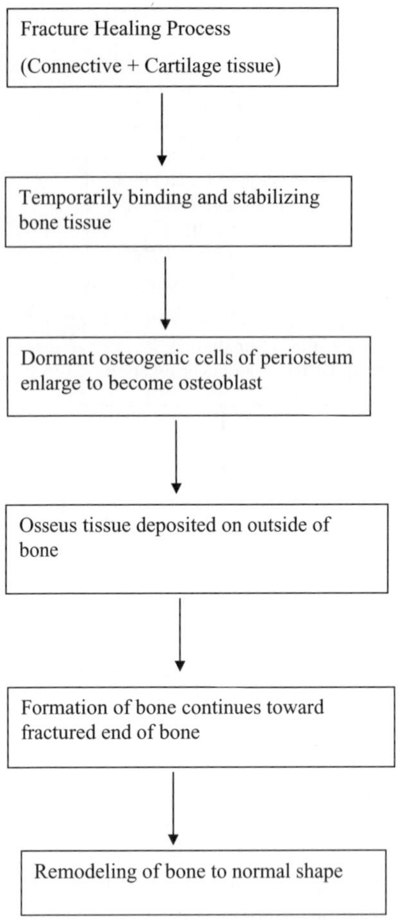


**Figure 5.6** Tissue differentiation in the fracture healing process

tissue regulation model based on a set of fuzzy logic rules derived from medical experiments, using the strain energy density as the mechanical stimulus that controls the process of cell differentiation. Many authors have also used computational models (mainly based on finite elements) to estimate local strains and stresses during the different stages of fracture healing (Carter *et al.*, 1998; Carter, Blenman and Beaupre 1988; Blenman, Carter and Beaupre, 1989; Cheal *et al.*, 1991; Claes and Heigele, 1999; Gardner *et al.*, 2000), since there is experimental evidence (Sandberg, Aro and Vuorio, 1993; Ashhurst, 1986) that tissue differentiation is dependent upon mechanical signal.

Kuiper *et al.* (1996) and Kuiper, Richardson and Ashto (2000) have developed a differentiation tissue theory using the tissue shear strain and fluid shear stress as the mechanical stimuli regulating tissue differentiation and the strain energy as the mechanical stimulus regulating the bone resorption. In an attempt to predict typical healing patterns, including callus formation, these authors used an axisymmetric biphasic model of finite elements of a fracture and applied movements on the cortical bone. Their results show that larger movements tend to increase callus size and caused a delayed bone healing. Lacroix *et al.* (Lacroix *et al.*, 2000; Lacroix and Prendergast, 2002) used the differentiation rules proposed by Prendergast, Huiskes and Soballe (1997) to predict different fracture healing patterns depending on the origin of the stem cells. The model can predict the callus resorption produced in the last stage of the fracture healing process, but can not predict callus growth during the initial reparative phase.

When bone is first formed after injury and local cell death it is irregularly arranged and has different properties. This type of bone is known as woven bone. Available evidence shows that cells of the sinusoidal vessel walls are the source of osteogenic cells, for both ordinary bone and woven bone. When bone tissue is formed it is accompanied by bone marrow. Marrow is of two main types. One is red marrow, containing hemopoietic tissue, which is found in those parts of the bone when there is cell activity, either because of growth or due to turnover. This tissue contains the precursors of the circulating blood cells, red cells or erythrocytes, leucocytes of various types and platelets. These platelets are formed from large cells known as megakaryocyte. The proportion and properties of the components of hemopoietic tissue are under hormonal control. However, in the more quiescent regions of bone, fatty marrow is found.

## *5.3.2 Growth Hormone (GH) Effect on Fracture Healing*

Koskinen *et al.* (1975) have demonstrated increased callus formation and faster stabilization in rat tibial fractures when growth hormone (GH) is taken everyday, and their beneficial effects have been demonstrated (Ashton and Dekel, 1983; Muhlbach *et al.*, 1972).

Some later reports have also demonstrated that intermittent treatment with PTH promotes osteogenesis in experimental fracture healing (Alkhiary *et al.*, 2005; Andreassen *et al.*, 2001; Holzer *et al.*, 1999; Nakajima *et al.*, 2002). Intermittent low-dose treatment with rhPTH (1–34) can increase hard callus formation (callus formed by intramembranous bone formation in periosteal osteoblasts and osteoprogenitor cells) and its mechanical strength without apparent systematic side effects (Nakajima *et al.*, 2002). These authors have reported that treatment with PTH stimulates proliferation and differentiation of osteoprogenitor cells and production of bone matrix proteins early in the fracture healing process. Thus, based on its osteogenic effects, treatment of fractures with intermittent low-dose rhPTH (1–34) may represent an effective strategy for the enhancement of fracture healing and could be the first systemic intervention for the repair of skeletal injuries.

Chondrogenesis, an essential component of endochondral ossification in long bones, is a key component of fracture healing. Growth factors and cytokines that are produced by the inflammatory response to skeletal injury support the chemotaxis of immature mesenchymal cells to the site of the fracture and promote their differentiation into chondrocytes. A well-organized development and maturation process progresses, leading to the formation of calcified cartilage that is subsequently replaced by bone (Nakajima *et al.*, 2001, 2003).

Intermittent low-dose treatment with rhPTH (1–34) increases the recruitment of mesenchymal cells into the chondrocyte lineage, proliferation of mesenchymal (chondroprogenitor) cells and synthesis of type II collagen. This results in the formation of a larger cartilaginous callous. The PTH-induced cartilaginous callus is rapidly replaced with bone (Nakazawa *et al.*, 2005).

### 5.3.3 Biological Principles

Bioelectrical potentials (DC) vary in the region (80–120 mV) for stainless steel electrodes. There is definite trend in this variation, as we move towards the femur region (electrodes 1–6) with respect to the reference electrode "1" (Figure 3.6a).

### 5.3.4 Cell Array Model for Repairing or Remodeling Bone

An essential requirement for an external signal $E(t)$ to achieve a physiologically meaningful bioeffect is that it must be detectable above the baseline electrical activity and the prevailing thermal noise, at the target. This suggests that the signal to noise ratio (SNR) must be in the range 0.5–1. Given the characteristic of the EMF target, $E(t)$ must have a requisite amplitude ($S$) in the desired frequency range to affect the kinetics of processes such as ion or ligand binding and ion transport in the presence of thermal noise forces. This effect can be understood using theoretical stimulations that evaluate SNR using either a single spherical cell or a cell array as the target. The cell array is the more general of the models. Cell–cell communication in the Haversian system of contact bone, and in each stage of repairing bone, provides an adequate template for a cell array of tissue model. In this case, the target is modeled with an electrical equivalent circuit, representing the electrical properties of the membrane capacitance and the rate constant of the ionic process of the cells involved in bone repair and remodeling (Figure 5.7). The cell array utilizes a distributed parameter electrical equivalent circuit to represent cells in gap junction contact.

The induced transmembrane voltage is $V(x,w)$ at any point $x$ along the array from an applied electric field, $E(w)$, where $w$ is the angular frequency and is the frequency domain representation of $E(t)$. It has been shown (Pilla, Nasser and Kaufman, 1994) that $V(x,w)$ is given by:

$$V(x, w) = -\text{abs}\left(\frac{E(w)\sin\text{h}(y\,x)}{Y\cos\text{h}(y\,\text{L})}\right) \tag{5.4}$$

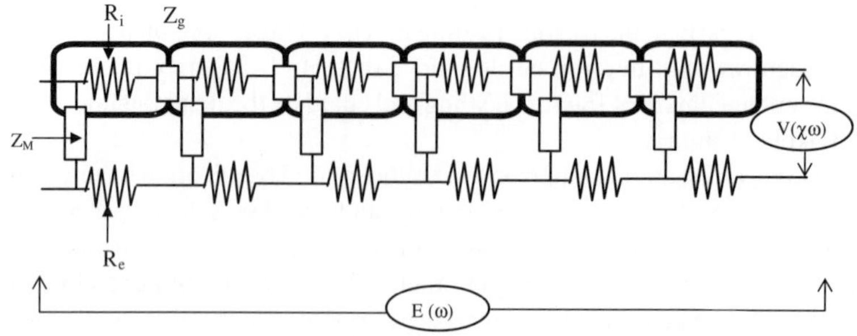


**Figure 5.7** Distributed-parameter cell array model. Cells are in gap junction contact via electrical impedance $Z_g$. This model is identical to a transmission line. The applied electric field $E(\omega)$ propagates throughout the array causing progressively higher changes in induced transmembrane voltage $V(\chi,\omega)$. $Z_M$, $R_i$, $R_e$ and $R_g$ are the membrane impedance and intracellular, extracellular and gap junction resistances per unit length, respectively

where $2L$ is the array length and $Y$ contains the membrane impedance:

$$Y = \left(\frac{(R_e + R_i + R_g)}{Z_M(w)}\right)^{1/2} \tag{5.5}$$

where $Z_M$, $R_i$, $R_e$ and $R_g$ are the membrane impedance and intracellular, extracellular, and gap junction resistances per unit length, respectively.

The voltage spectrum $E(w)$ depends upon the particular waveform characteristics of the induced electric field and determines the spectrum of the array response. Comparison of $V(x,w)$ with the transmembrane thermal noise voltage determines whether the cell tissue can detect the EMF signal. The SNR is given by:

$$\text{SNR} = \frac{|V(x,w)|}{\text{RMS}[S_n(w)]} \tag{5.6}$$

Where RMS is the root mean square value (in volts) of the thermal noise spectrum ($S_n$).

Equation 5.4 shows that transmembrane voltage is a function of both frequency ($w$) and the array length, $L$. As the length of array increases, for example, the dendritic formation of cells within the canaliculi of remodeling bone, SNR increases as $\tan h(yl)$, and the array acts as an increasingly low pass filter. This change can produce transmembrane voltages above the thermal threshold in a frequency range at amplitudes significantly lower than that expected for a single 10 μm-radius spherical cell. Equations 5.4 and 5.6 may be employed to evaluate the SNR for a target pathway utilizing $E(t)$ from any input waveform, even if the electrical equivalents are nonlinear, as in the Hodgkin–Huxley membrane model (Oin, Rubin and McLeod, 1998). Due to the nonlinear effects of the propagation of the pressure wave in the tissue, the major frequency components are below 1 kHz. Induced $E$ in this case is a single repetitive pseudorectangular pulse with slow (millisecond) rise and fall times. Induced SGP is most often a slowly decaying exponential-type function having peak amplitudes in the $\text{mV cm}^{-1}$ range and major frequency components centered in the extremely low frequency limit ($<100$ Hz).

## 5.4 Electromagnetic Field and Fracture Healing

Several authors (Bassett and Becker, 1962; McElhaney, 1967) suggest that *in vivo* bone development is related to physiologically generated electric fields. Knowledge of variations in the electrical characteristics of bone during growth (age and physical exercise) and atrophy (loss of muscular activity and immobilization) is therefore of potential importance in understanding bone remodeling dynamics and in controlling bone growth by the application of electric fields.

The ability to heal a fracture is a latent potential of the mesenchymal tissue in bone. Several authors have suggested (Pawluk and Becker, 1964; Yasuda, Noguchi and Sato, 1955) the ability to initiate and control this transient osteogenic response. This provides a means of accelerating the rate and capacity of fracture healing. Potential hazards associated with electric and magnetic fields, at extremely low frequency, especially those associated with overhead electric power lines, have been a matter of much debate (McCann *et al.*, 1993). There is some evidence that very high intensity electric fields are capable of genotoxic effects after *in vivo* exposure (Nordenson *et al.*, 1988). Similar results for *in vitro* exposure are lacking. A large volume of data on electromagnetic fields suggest a lack of genotoxicity for this type of energy.

Martini *et al.* (2003) have examined the effects of an electric field applied to osteoblast like human cells in culture. These authors found that after 24 hours of treatment there is a significant increase in osteocalcin and nitric oxide (NO) production: Biochemical pathways are activated when electrical stimulation is applied to bone cells. Signal transduction of the capacitated coupling has been shown to translocate $Ca^{2+}$ ions through cell-membrane, voltage-gated calcium channels, inducing an increase in cytosolic $Ca^{2+}$ (Brighton *et al.*, 2001). It has also been reported that capacitively coupled electric fields accelerate cell proliferation and differentiation in osteoblast-like cells *in vitro* and enhance cell synthesis. This leads to matrix formation and maturation (Harting, Joos and Wiesmann, 2000). It is further suggested that osteoblasts in culture are sensitive to electrical stimulation, resulting in enhancement of biomineralization (Wiesmann *et al.*, 2001).

Various PEMFs (pulsed electromagnetic fields) promote bone formation *in vivo* (endochondral ossification) and skeletal homeostasis and in clinical studies of fracture non-unions. An important effect is to decrease osteoclastic bone resorption *in vitro*. Investigations of the effect of EMFs on bone formation *in vitro* require the growth of osteoblastic cells and, particularly, the production of mineralizing osteoid and the formation of bone-like tissue (Bellows *et al.*, 1986; Sudo *et al.*, 1983; Whitson *et al.*, 1984; Williams *et al.*, 1980).

In essence, fracture healing proceeds along two different pathways: intramembranous ossification and endochondral ossification. During the healing process, various cellular and biochemical events occur: migration and proliferation of immature mesenchymal cells, differentiation to chondrogenic or osteogenic cells, production of proteins and preparation of vascular invasion. The control of such morphogenetic events requires regulatory interactions between different types of cells and between the cells and extracellular matrix. These interactions are mediated by many local and systemic regulatory factors, including growth and differentiation factors, hormones, cytokines and components of the extracellular matrix. Some molecular studies have revealed that certain peptide growth factors and hormones, such as TGF-$\beta$1 and $\beta$2 (Joyce *et al.*, 1990; Steinbrech *et al.*, 1999), bone morphogenetic protein (BMP)-2 (Onishi *et al.*, 1998; Sanyal *et al.*, 1999), metalloproteinase (MMP)-13 (Yamagiwa *et al.*, 1999) and 1$\alpha$,25-dihydroxyvitamin $D_3$ (Jingushi *et al.*, 1998) play important roles in fracture healing. Furthermore, Cyr61 (Hadjiargyrou, Ahrens and Rubin, 2000), a member of the CCN gene family, is reported to be one of the regulators of fracture healing.

### *5.4.1 Methods in Bone Fracture Healing*

Present day applications of electromagnetic energy for the purpose of bone fracture healing are in a way an extension of the concept of converse piezoelectricity. To date, various animal experiments, adopting different models and under various clinically stimulated conditions, have been found to produce positive results (Chao *et al.*, 2004). This probably arises because of the fact that cellular responses to a given form of energy (electromagnetic energy) is tissue specific and leads to bone healing. The fact that biopotentials affect the growth of bone suggests that the effect of externally introduced voltages or currents should be studied in greater detail.

Experimental studies have been made on the effects of direct and indirect electrical and magnetic stimulation on bone metabolism in normal and abnormal bones (Bassett, 1984;

Brighton, 1984; McLeod, Rubin and Lanyon, 1987). Low-intensity (non-thermal) non-invasive electromagnetic fields (EMFs), ultrasound (US) and mechanical loading (strain-generated potentials, SGPs) have been demonstrated to accelerate bone repair. Capacitive coupling using electrodes placed on the skin (non-invasive), DC stimulation using implanted electrodes (invasive) and electromagnetic stimulation by inductive coupling using time varying magnetic fields (non-invasive) are the three techniques frequently employed for accelerating bone fracture healings. Electrical fields and currents have been shown to stimulate the process. Figure 5.8a and b shows a general method for the generation and application of an electromagnetic field.

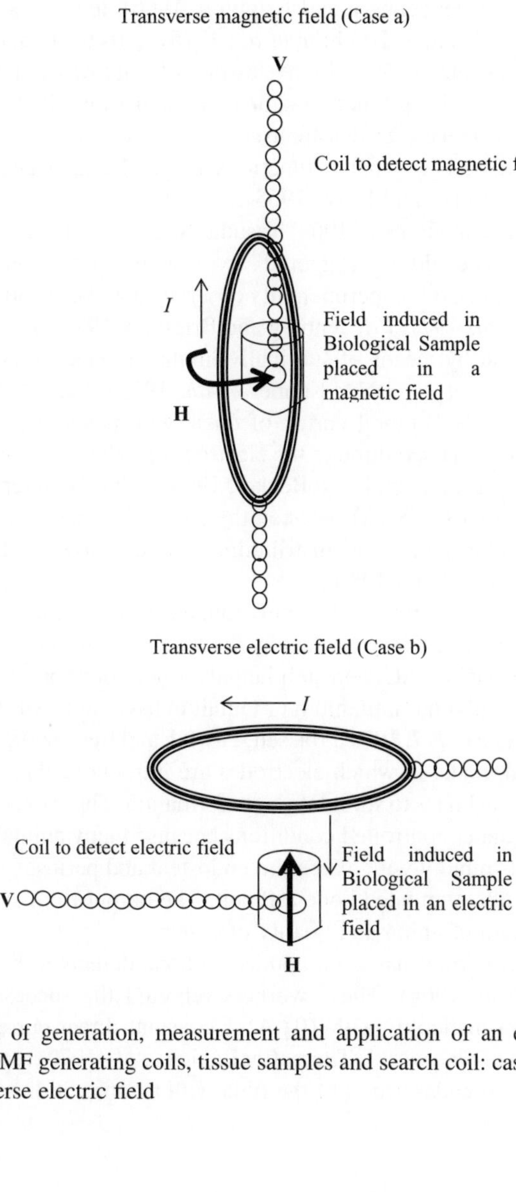


**Figure 5.8** Method of generation, measurement and application of an electromagnetic stimulus. Disposition of the PEMF generating coils, tissue samples and search coil: case a – transverse magnetic field; case b – transverse electric field

The hypothesis that stimulating bone and cartilage through the application of electric and magnetic fields will affect growth and repair has been investigated (Bassett, Pawluk and Pilla, 1974; Gupta *et al.*, 1991). It appears from these observations that electricity, whether it may be constant direct current, pulsed direct current or magnetically induced current can stimulate osteogenesis and consequently bone fracture healing. However, it is still under investigation as to which form of electrical energy is the most efficient stimulator for these purposes and why.

### *5.4.2 Stimulation by Constant Direct Current Sources*

The observation that the electric potential on the surface of long bones affects fracture, the diaphysis being positively charged with respect to the bone ends whereas the fracture site became electro-negative (Friendenberg and Brighton, 1966), led to a study of the effect of direct currents on fracture healing (Brighton *et al.*, 1975). This technique can produce union with the use of implantable electrodes, which is essentially an invasive method (Figure 5.9a). Electrodes are commonly used to promote fracture healing because thick soft tissues surround bone structures. Stainless steel electrodes are used. In pursuance of the above, bone is known to adapt its architecture to best carry out its functions, in which it also tends to provide skeletal support (Treharne, 1981; Epker and Frost, 1965).

Several authors (Pawluk and Becker, 1964; Yasuda, Noguchi and Sato, 1955) have reported that increased osteogenesis could be "triggered" by weak electric currents in the absence of significant mechanical inputs. It is experimentally observed that the negative pole causes faster ossification (Hambury *et al.*, 1971; Friedenberg and Brighton, 1974). In conformity with this, the introduction of current by means of surgically implanted electrodes has received widespread attention (Connolly *et al.*, 1974; Kenner *et al.*, 1975). Bassett, Pawluk and Becker implanted electrodes into the femoral cortex of dogs, with penetration into the medullary cavity. Weak direct currents passed through the electrodes produce bone growth and enhancement of biopotentials (Bassett, Pawluk and Becker, 1964). It has been reported that in rabbits the passage of direct current ($\sim 15\,\mu A$) between the electrodes, near the fractures, alters the biopotentials and accordingly the field distribution, pointing toward the possibility of an accelerated bone growth (Andrabi, 1980).

In the technique of exogenous physical stimulation, electrodes are used to promote fracture healing, because bony structure is surrounded by thick soft tissues. Originally, stainless steel electrodes to administer DC or AC were implanted to promote bone repair; however, less reactive materials such as platinum, titanium or tantalum have replaced steel as electrodes of choice. The work of Yasuda (1977, 1974), Bassett, Pawluk and Becker (1964) and Levy (1974) utilized the transcortical model in which electrodes are introduced through small holes into the medullary canal of long bones to stimulate bone formation. These experiments tend to give confusing results, even under controlled conditions because many animals tend to spontaneously repair such defects rapidly with exuberant endosteal and periosteal callus. This makes discernment of electrically stimulated bone difficult. The output from such experiments has focused on speed, strength of union and quality of callus.

Subsequently, a direct current has been applied to bone defects in human (Friendenberg *et al.*, 1971; Cieszynsktio, 1964). These workers reported the successful treatment of a non-union of the medial malleolus with $10\,\mu A$ of constant direct current delivered to the non-union site for a total of nine weeks. Thereafter Lavine *et al.* (1972) reported the successful treatment of a congenial pseudarthrosis of the tibia with a constant direct current.

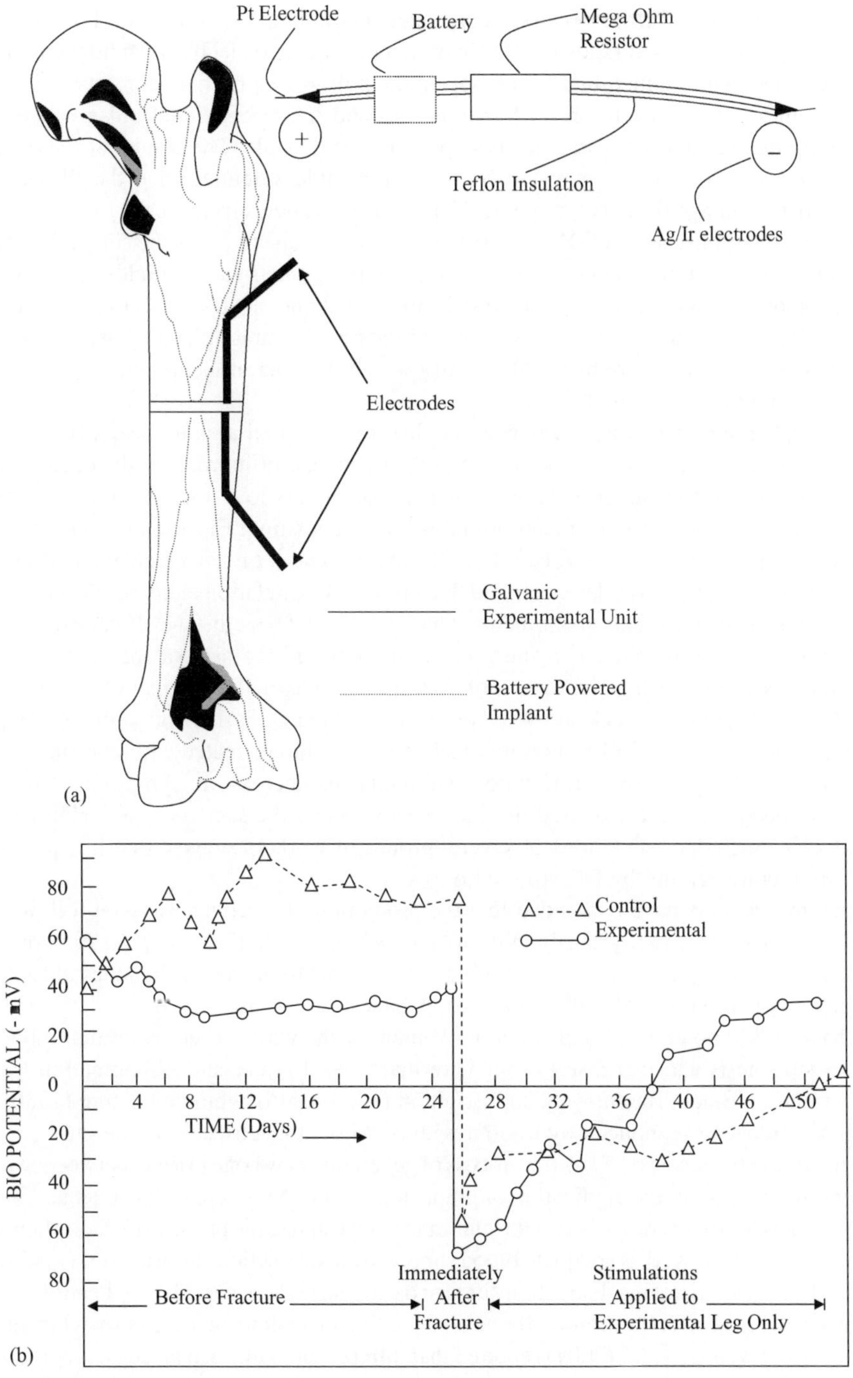


**Figure 5.9** (a) Stimulation by electrode placement in a fractured bone using DC stimulation; (b) variation of biopotentials with the introduction of injury and stimulations applied (Andrabi, 1980)

Jacobson, Weil and Raff (1997) have developed an implantable stimulator. For each assay, the cells exposed to pulsed fields via sterile electrodes traverse electric dish lids with the tips immersed in the culture medium. In this set-up the pulsed electric field consists of a burst of pulses, with a repetition rate of 1.5 bursts per second. Each individual pulse consists of a positive portion 65 μs wide, and a negative portion 195 μs wide. The results of these authors indicate that the field types generated by the implantable stimulator and the PEMF signal devices are not mutagenic and clastogenic. Neither do these fields increase the frequency of cell transformation or unscheduled DNA synthesis. This technique requires the surgical implantation of current by transcutaneous routes, which enables the electrodes to be close to the affected site to promote new bone growth. In fracture healing, the bones are then stabilized with internal plates and screws or intramedullary nails, and/or the limb is immobilized by using an external cast. Its other advantages are that it allows full weight bearing, uses portable equipment and does not require patient compliance.

Point application techniques have been applied over a small area to hasten bone fracture healing, to change existing bone structure and to enhance primordial limb regeneration in Rana Pipens. It has been suggested that the generation of very low amperage on the surface of bone in the presence of tropocollagen initiates bone deposition. Electric current stimulation provides a significant increase in new bone formation and a higher mechanical strength of healing. Animal studies have demonstrated that electrical stimulation increases the healing rate of a mandibular fracture. This phenomenon could be of great benefit. The efficacy of electrical stimulation is inversely related to the size of the gap and the size of the fractured bone. The success rate (84%) is high for the use of multiple cathodes in the treatment of non-unions in larger bones (Brighton, Pollack and Windsor, 1981). Schaden, Fischer and Sailler (2001) have reported a success rate of 76% in non-unions or delayed unions treated with one-time therapy that resulted in bony consolidation with a simultaneous decrease in symptoms. Non-union cases have been extensively treated in human patients and the success rate is claimed to be around 71%. Such demonstrations of several groups of workers suggest that it is possible to speed up fracture healing by DC stimulation.

Some investigators have contended that the promotion of osteogenesis seen following the insertion of a wire is partly attributable to the inflammatory (Ijiri *et al.*, 1995) or injury (Aronson *et al.*, 1988; Einhorn *et al.*, 1990) reaction of the tissue rather than the sole action of the electricity (Spadaro *et al.*, 1986; Spadaro, Albanese and Chase, 1990). Spadaro, Albanese and Chase (1990) noted that mechanical movement of the wire caused by animal movement led to osteogenesis and that insertion of wire that caused injury and inflammation leads to bone formation. Black, Baranowski and Brighton (1984) and Brighton and Hunt (1986) have described osteogenesis induced by trauma with or without electrical stimulation.

Distraction osteogenesis (DO) is the process of generating new bone in a gap between two bone segments in response to the application of graduated tensile stress across the bone gap. Various animal models have been described in the literature to investigate the process of cranio-facial DO. Osteocyte viability as well as adequate blood supply to the distraction site are an essential source of osteoblastic activity (at the distraction site) for osteogenesis to occur. The application of direct current to bone is an effort to stimulate bone regeneration in order to increase normal healing by a factor of two. Swennen *et al.* (2001) reported that different animal, such as dogs, rabbits, sheep, minipig, monkeys, rats and cats, have been used as an animal model to study the effect of an electric current on bone. The results of this study proved that goats are a suitable experimental animal model to study the effect of electricity on the healing of distraction osteogenesis.

The use of a computer program to assess the newly formed bone during different distraction phases after direct current stimulation indicated that electrical stimulation displays synergism on healing of distracted bone when it is simultaneously applied during activation and consolidation periods. This is in agreement with Hagiwara and Bell (2000) who reported that electrical stimulation during gradual distraction promotes new bone formation in the early retention period. The lack of an electric current effect during latency found in this study may be attributed to the lack of cells present between the distraction gap during latency, as the presence of mature osteoblasts in the gap tissue is a prerequisite for effective electric current stimulation. The results of El-Hakim *et al.* (2004) agree with those of Lavine *et al.* (1971), who reported that electric stimulation affects the functional activity rather than the total number of osteoblasts or the differentiation of an additional number of osteoblasts from the undifferentiated mesenchymal cells.

The results of the study by El-Hakim *et al.*, 2004 also agree with those reported by Tis *et al.* (2002), who found that the electrical current increases the size of the distraction callus and might alter the composition of regenerated bone. One possible mechanism for this to occur could involve stimulation of macrophage migration to the site of the wound. Electric fields cause macrophage migration on laminin or fibronectin coated substrates without inducing changes in cellular morphology. The migration velocity is not significantly altered but the random movement is suppressed. The electric current so-generated might stimulate the undifferentiated mesenchymal cells in bone marrow to proliferate and transform into osteoblasts. Block and Brister (1994) and Blank (1987) have reported that calcium gates enlarged during electric stimulation and that this resulted in more incorporation of calcium into the cells. Cane *et al.* (1996) suggested that an electric field might accelerate the healing process through enhancement of the enzymatic activity of the repaired tissue. Electric stimulation might affect the external callus (periosteal), intermedullary callus and the cortical callus as a result of osteoblastic stimulation, and bone cells respond to electromagnetic stimuli by modulating the activity of more primary activators such as hormones, growth factors, cytokines and mechanical forces. DC stimulation may be useful in mandibular distraction during activation and consolidation periods and may shorten the overall duration of lengthening, and the distractor can be removed earlier as early bone healing is expected.

Some mechanism other than the piezoelectric effect must also be considered in accelerating bone healing. Fluid transport may play a significant role not only in various aspects of bone metabolism (such as mineralization) but also in the electromechanical properties of bone. Electric potentials might control cell nutrition and proliferation, the proliferative ability of cells and the orientation of intra- and extracellular macromolecules. It may be that the an electric potential activates the cyclic adenosine monophosphate (AMP) system within the cell and directs it to proliferate and produce osteoid. The cathodal growth may modify the association–dissociation equilibrium to the extent that both phosphate and calcium ions migrate towards the cathode. In addition, collagen fibers become oriented in the electrical field, which might lead to a spatial arrangement that enhances calcification.

Mathematical analysis of electric fields inside the bone together with Fourier analysis of induced voltage waveforms produced by commonly used electrical stimulation waveforms has been performed. A hypothesis based on assigning different weightings to different frequencies for osteogenic response has been proposed. Using this hypothesis similar effective values of electric fields have been found in different systems. It is shown that the effective electric field rather than peak electric field is the main parameter responsible for osteogenesis. The results are in agreement with experimental findings made on human beings by some investigators (Gupta *et al.*, 1991). A delay in the Haversian remodeling within the first 2 weeks was found in the animals. This delay could not be detected in all electrically stimulated groups (Schubert, Kleditzsch and Wolf, 1986).

Exposure to certain low-frequency electrical fields may improve the clinical outcome of osteonecrosis of the femoral head (ONFH) by modifying the repair process. Modification of bone is generally governed by the characteristics of the electrical signal. A low frequency electrical signal decreases bone loss and maintains bone mass by decreasing osteoclastic bone removal. Studies of the effect of these fields upon osteoclast *in vitro* have shown a decrease in osteoclastic differentiation and lysosomal enzyme content. Finally, these fields have been shown to interfere with parathyroid hormone receptor signaling, associated with a decrease in osteoclastic differentiation. It is suggested that a low frequency electric field may lead to an increase in local vasculature and bone production so that the balance of repair may be altered in favor of bone conservation, maintenance of structural integrity and, ultimately, preservation of the femoral head (Aaron and Ciombor, 1997).

An experimental model has been devised for the consistent production of delayed bone healing of the tibia in adult dogs. A double blind trial was undertaken to evaluate the use of a commercially available direct current bone growth stimulator with this model. The stimulator produced a statistically significant acceleration of bone healing at four weeks in the experimental model. Osteogenesis was normal and no dysplastic, inflammatory or neoplastic changes were found. This research has shown that electrical stimulation of bone is safe and augments bone formation. The bone growth stimulator unit remains on trial, but in future it may alter the management of difficult orthopedic problems (Paterson *et al.* (1977).

The obvious advantage of using implanted electrodes in bony tissue is that they can be inserted directly into or close to the area to promote new bone growth. In fracture healing, the bones are then stabilized with internal plates and screws or intramedullary nails, and/or the limb is immobilized by using an external cast.

This clinically attractive idea is beset, though, with several difficulties:

1. The osteogenic effect is dose dependent within 5–20 μA and demands constancy of the applied current. The authors found that smaller or larger currents than the above produce no effect and cause cell necrosis, respectively. Also, the modalities required for optimum results, for example, type and size of electrical stimulation, density of the current applied, and the positioning of anode and cathodes, needs investigation.
2. In the case of possible application to large bones of men, multiple electrodes would be required to produce a clinically significant bulk of callus in a reasonable period of time.
3. The method is surgically invasive, and is therefore clinically unacceptable.
4. Electrode materials (platinum, iridium, stainless steel) are in intimate contact with cells, extracellular matrices and fluids. Under these conditions some electrolysis effect may develop, where toxic ions will build up in the extracellular environment. At high DC currents both anodic and cathodic effects would be undesirable.

Because of these shortcoming of this procedure other methods that are essentially non-invasive are being increasingly adopted. This involves two methods of field coupling to the body tissue, by using the specific nature of field characteristics.

### *5.4.3 Pulsed Electromagnetic Fields (PEMFs)*

Some stimulation of bone growth occurs with selected wave patterns, including sine wave and the use of ultrasound (Li, 1992; Bachner, Klion and Caruana, 1987; Pilla *et al.*, 1990; Heckman *et al.*, 1994). The induction of electric current in bone is not only speculated to prevent the bone loss of functional disuse but also induces new bone formation. The promise of successful non-invasive alternatives to accelerate bone healing is thus of continuing concern.

The healing rate of 24 acute tibial fractures with a pulsating asymmetric direct current was reported to be 30% faster than in a control group (Jorgensen, 1972). Krause and Lechner (1972) used an electromagnetically induced current to heal non-unions and pathological fractures. They reported the successful treatment of a congenital pseudarthrosis of the tibia by means of a pulsed electromagnetic field. These clinical protocols, using pulsating electromagnetic fields, were subsequently followed in the treatment of Paget's disease, osteoporosis, avascular necrosis and metastatic bone disease (Bassett, 1978). Low frequency AC currents (Klapper and Stellard, 1974; Iida *et al.*, 1956), pulsating DC (Richez *et al.*, 1972) and electric fields (Bassett, Pawluk and Pilla, 1974) have also been successfully employed for bone growth.

#### 5.4.3.1 Physical Principle of PEMF Generation and Interaction

When a conductor, such as a wire, is moved through a static magnetic field, due to a relative displacement, a current is produced. The magnitude of the current produced depends upon the strength of the field (*H*) and the rate of relative displacement. Since cells and tissues also have finite conductivity, and if similarly treated, induced currents will be produced, which are mainly ionic in nature.

Currents are induced inside in a stationary biologic specimen subjected to a time varying magnetic field (i.e., inductive coupling). With these currents are an associated electric field (*E*), which exists in air, tangential to the magnetic field (*B*) flux lines. Biological materials that display predominantly ionic conduction are affected mainly by the electric field component of a low energy time varying magnetic field. This being the case the positioning of biological materials in the magnetic field becomes significant, since the electric field arises as function of distance. By adopting an appropriate design of coils, magnetic fields of greater or lesser homogeneity can be created. For maximum uniformity, a double coil system consisting of a pair of circular, flat coils placed half their diameter apart can be used (Figure 5.10a and b). Such coils are termed "Helmholtz-aiding" and are characterized by a uniform distribution of magnetic flux lines within the volume.

Pulsing currents delivered by electrodes differ from DC and are rarely similar to those developed inductively. When the electrical microenvironment of a cell is modified by non-electrochemical capacitive coupling, one type of effect exists if the field is static and another if the field is dynamic (Bassett, 1968). Further, if the field is dynamic, different results may be expected depending on pulse characteristics such as amplitude, shape, duration and repetition rate (Bassett, 1978, 1971). The biological implications of electric currents induced by pulsing electromagnetic fields clearly far exceed those of constant currents (delivered by electrodes) because the two systems are fundamentally different. Inductively coupled pulsed electromagnetic fields (PEMFs) produce electric currents in tissue in the absence of electrodes (Bassett *et al.*, 1982; Bassett, 1978; Bassett, Pawluk and Pilla, 1974). They do not, therefore, give rise to the polarization, electrolytic by-products and foreign chemicals species normally associated

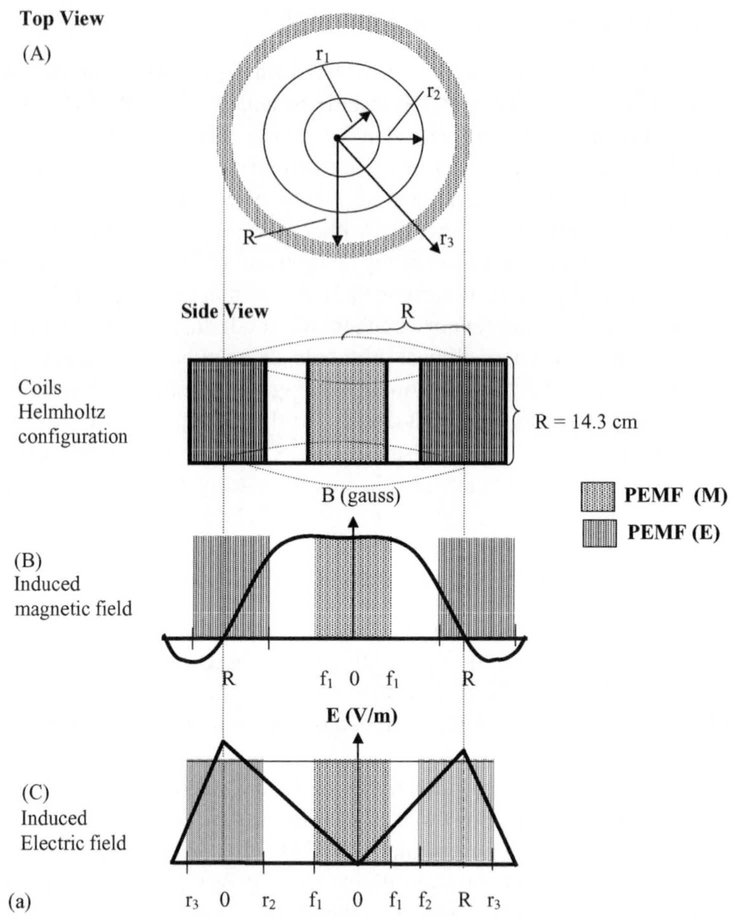


**Figure 5.10** (a) (A). Top and side views of PEMF apparatus constructed of concentric Lucite cylinders. Animals are placed in the inner chamber, PEMF-M and outer chamber, PEMF-E. The coils are arranged in a Helmholtz configuration. (B) Schematic representation of the induced magnetic field across the horizontal axis of the PEMF apparatus. (C) Schematic representation of the induced electrical field across the horizontal axis of the PEMF apparatus (Adopted from Bassett, 1991). (b) Electromagnetic coils connected to a pulse generator (inductive coupling)

with the Faradaic (redox) action of electrodes (Zengo *et al.*, 1976). Several commercially available intermedullary cathodes are now available to surgeons and less invasive techniques have also been developed.

PEMFs carry a broad band of frequencies that occupy a discrete portion of the lower end of the electromagnetic spectrum. Although their repetition rate falls in the ELF range (e.g., 1–100 Hz), their frequency content by discreet Fourier transform ranges from DC to $> 10$ MHz. These characteristic derive mainly from their waveform pattern and sequencing. Typically, these waveforms are asymmetric, biphasic and quasi-rectangular or quasi-triangular in shape, having various amplitudes. The patterns differ considerably from the usual action

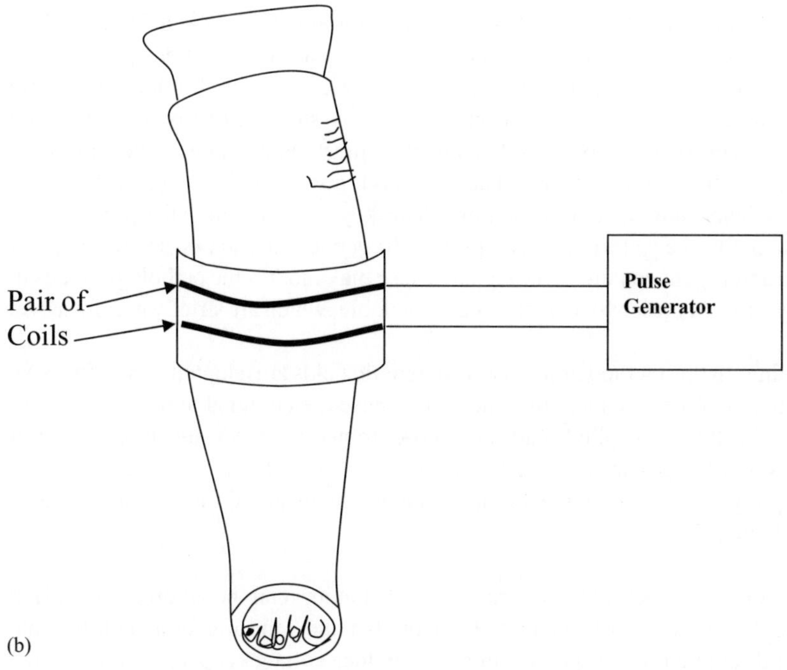


(b)

**Figure 5.10** *(Continued)*

potentials arising in nerves, skeletal or cardiac muscle and are quite different from sine waves (e.g., power lines). Signal parameters of PEMF affect signal efficacy. PEMF-induced tissue waveforms range from 50 to 100 μs in duration. The rise time in response to PEMFs varies between fractions of a microsecond to several milliseconds, which is equivalent to frequencies ranging from several hundred Hz to over 1 MHz (Bassett, 1989). Gaussian white-noise generators, centered on the dominant frequency range found in PEMFs, have demonstrated significant effects in fresh fracture repair as judged by technetium polyphosphate uptake (Wahlstrom, 1984). Antonson and Mann (1985) have found that the activities like walking can induce mechanical strains and endogenous currents primarily below 10 Hz. Consequently, the frequency of the PEMF is set at around 7.5 Hz, as used by several authors (Chyun and Raisz, 1984; Nefussi and Baron, 1985).

Goodman and Henderson (1986) have confirmed that PEMF effects are athermal, as they could not detect heat shock proteins. The inductively coupled fields expose tissues to an energy level of $10^{-10}$ W mm$^{-3}$, which will result in a change of 0.001 °C at nominal tissue impedances (Iannacone *et al.*, 1988; Bassett, 1989). These authors (Goodman and Henderson, 1987) also showed that pulse burst and single pulse triggered characteristically different translational indications. They found a continuous 72 Hz sine wave field, with energy matched to the 72 Hz single pulse, produced yet another "signature." Their studies strongly indicate that gene-specific activation and de-activation by selected PEMFs and other time varying magnetic fields (i.e., continuous fields) occurs.

As compared to the continuous wave frequency, Pulsed electromagnetic fields have been used successfully in several orthopedic pathologies and more detailed studies of the interaction between PEMFs and bone pathologies have been carried out in patients and animals. Pulsed electric fields have been found to stimulate bone growth, particularly after spinal fusion surgery. This is to distinguish other part of the electromagnetic spectrum. It implies therefore that continuous wave signals have no effect on biological systems. PEMFs show a highly selective biological effect when the living system and the time varying magnetic field to which these are exposed are appropriately matched. These methods now provide a valid alternative to bone grafting for delayed unions and non-unions. The key to their rational use lies in an increasing ability to define the patterns (e.g., amplitude, frequency, orientation and timing) that must be encoded as intelligent content in the signal. These may modify the pathologic states in the musculatetal system. At present two methods are available, which are strictly non-invasive:

1. Inductive coupling: In this method time varying magnetic fields are adopted as non-invasive means to alter the microenvironment of bone cells. These are confined to low frequencies only ($<1$ kHz). At the low applied field, induced currents in tissues are in the range of $\mu A\ cm^{-2}$ or less and the possibility of heating does not occur (Figure 5.10b).
2. Capacitive coupling of pulse modulated radiofrequency fields at the fracture site has also been adopted (Figure 5.11).

These methods make use of the fact that bone behavior can be altered by pulsed electromagnetic fields to produce osteogenesis. In methods 1 and 2 the affects are non-thermal in nature. These fields applied across the extremity will induce voltages capable of altering the pattern of bone healing. The inductive method, mainly introduced by Bassett, Pawluk and Pilla (1974), uses pulsating electromagnetic field to produce electrical currents in tissues without the placement of electrodes in physical contact. The method is based on the appropriate choice of electrical waveforms, which can selectively increase or decrease the calcium content of chondrocytes but not of fibroblasts. By a proper choice of pulse parameters (e.g., amplitude, rise time, shape, duration and repetition rates) it is possible to produce frequency matching

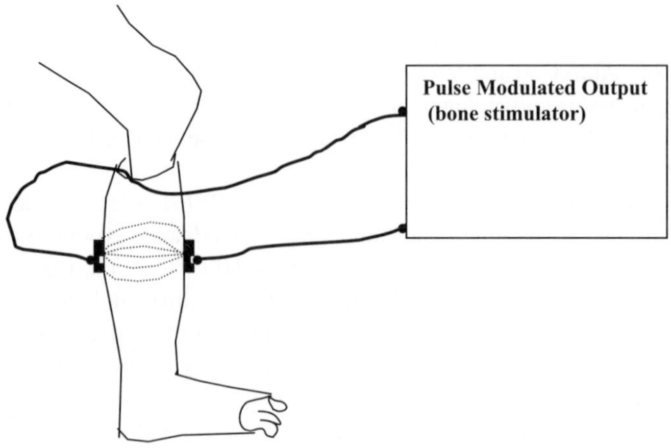


**Figure 5.11** Placement of capacitor plates on skin/cast for pulsed radio frequency field coupling at the fractured site

resonance effects. Selective action is not possible with direct currents. Furthermore, the biological effects (osteogenesis) of pulsed electromagnetic fields are not spatially limited as they are with DC electrodes.

## 5.4.4 Inductive Coupling

The exposure of cultured bone marrow cells to 60 Hz electric fields at 9.6 $\mu V\ cm^{-1}$ over 8 days suppressed the formation of osteoclast-like cells in the marrow culture (Rubin *et al.*, 1996; Shankar *et al.*, 1998). The results indicate that an approximate twofold increase in bone resorption in bone occurs when osteoclast–osteoblast co-cultures are exposed to the PEMF-generating coil for 18 h.

Earlier devices were developed so as to generate waveforms that would simulate stress-generated signal in bone. This would demand a multifrequency signal with a sizable component in the low frequency end. The chosen induced waveform parameters were based upon naturally occurring piezoelectric-like signals present in bone. This has also been used for tissue culture studies on fibroblasts (Bassett and Herrmann, 1961) and erythrocytes (Hassler *et al.*, 1977). Evidence has accumulated that an appropriate waveform can also increase or decrease the calcium content of chondrocytes (Bawin and Adey, 1976). Plastic immobilization and non weight bearing have been important clinical allies of PEMFs in producing the most consistent and rapid results in the treatment of non-union. (Bassett *et al.*, 1978). Bassett and his group adopted two type of waveforms: the first based upon the piezoelectric like approach and a peak induced voltage field at the osteotomy site of 2 $mV\ cm^{-1}$, a waveform duration of 1.5 ms, and a repetition rate of 1 Hz. The second, derived from the dynamic cellular model approach, had a peak induced voltage field of 20 $mV\ cm^{-1}$, a duration of 0.15 ms and a repetition rate of 65 Hz. The experiments were conducted on dogs. These authors discovered no evidence of malignant, premalignant or mitotic changes, including in the surrounding soft tissues.

Gonzalez-Riola *et al.* (1997) subjected growing female rats to a 3 mT, 100 Hz, Helmholtz-type electromagnetic field 24-hours a day for 30 days and compared the results with a corresponding control. These authors reported no significant difference in body weight, femur length and weight and growth cartilage thickness. These results indicate that the electromagnetic fields of this type reduced bone formation and increased bone resorption without affecting bone development in rats.

In this method the chamber is designed with the coils in a Helmholtz configuration which are connected to a signal generator and amplifier. In this configuration, the mean radius of the coils equals the distance of separation of the coils vertically and the magnetic and electric fields do not vary in the vertical plane within the partitioned rings (Figure 5.10a). In a typical configuration, the coils are made of copper wire 1/16 inch thick with 30 winds of each coil in the horizontal plane. The two coils are separated vertically by 14.3 cm with the mean diameter of the coils being 28.6 cm. An electric signal is pulsed through the coils, producing a PEMF with a repetitive train of 20 pulses for a total duration of 380 μs (positive going) and spaced 55 ms apart. The waveform is quasirectangular and repeated at a frequency of 72 Hz. The amplitude was adjusted to yield 15 mV on a standardized coil probe, placed midfield, parallel to the coil faces and connected directly to an oscilloscope. Figure 5.10b shows a general method of coil placement over the fracture site. There is abundant evidence that PEMF can stimulate the synthesis of the major components of bone matrix and cartilage (Murray and Farndale, 1985; Norton, 1982). Bassett, Pawluk and Pilla (1974) found that muscle and vessels are also normal. In another set of experiments on fibula fractures in beagles, with 65 Hz circuits, the stimulated legs demonstrated continuous enhanced bone repair with improved bone qualities 28 days after osteotomy. These authors found no evidence of neoplastic or preneoplastic processes in the stimulated bone. It is pointed out that bone growth involves continuous accretion on the surface and at the extremities and a continuous process of remodeling of its internal structure.

In another set of experiments, with a well-controlled calvarine unilateral delayed union model, pulsed PEMF stimulation for 1–4 h per day for 8 weeks significantly increased weight bearing on the affected limb, with higher mechanical strength of the healing osteotomy attributable to periosteal new bone formation. Another striking finding in this study is the effect of pulsed electromagnetic fields on reducing cortical porosity in the bone adjacent to the osteotomy when compared with the nontreated group (Inoue *et al.*, 2002). Double blind studies confirmed the clinical effectiveness of pulsed electromagnetic field stimulation on osteotomy healing (Borsalino *et al.*, 1988; Ilmeni *et al.*, 1996) and delayed union fractures (Sharrard, 1990).

However, other authors has reported opposite findings. The remodeling system of bone can be activated to respond maximally by a frequency periodic source that operates at approximately 0.7 Hz. Law *et al.* (1985) have described the effect of the non-invasive electromagnetic method operating at this frequency on the healing of experimental osteotomies in sheep tibia. They used a pair of coils each of 790 turns and 0.2 mm diameter with a rectangular winding cross section of $15 \times 17$ mm and a mean coil diameter of 52.5 mm. These authors concluded that electromagnetically induced current had no discernible effect on bone union as assessed clinically and radiographically (Rinaldi and Shamos, 1981). These authors have further concluded that the shape of the waveform is not so critical in fracture healing, which is contrary to many other findings.

A generally accepted approach emerges that the intensity of low frequency electromagnetic fields applied across an extremity will induce voltages of magnitude generally produced by deformation (0.1–50 $\mathrm{mV\,cm^{-1}}$) in bone. The induced voltage is apparently capable of altering the rate of bone healing in animals and can be extrapolated to humans.

Canine bone growth is electrically stimulated by a weak current between implanted electrodes and thereby causes accelerated fracture healing. Electric fields are reported to control tissue development. Pulsatile current is also found to stimulate bone growth. Electrical pulses are favored in bone therapy because they resemble electrical stimulation associated with physical activities, primarily walking. These signals are also credited with carrying a multiplicity of messages that can stimulate bone growth processes. Pulsed currents, hence, become important when they can be induced non-invasively. Magnetic or electromagnetic field pulses also stimulate other tissues, for example, collagen synthesis by chondrocytes and glycosaminoglycans and cellular transcription in eukaryotic cells. Tissue apparently responds to extremely low frequency (ELF) and bioelectric DC fields in a qualitatively similar manner. Generation of electric fields and potential gradients when mammalian or amphibian skin is injured is another indication of the role of endogenous electric stimuli in tissues. The potential gradient slowly diminishes as the healing proceeds. Transcutaneous potentials of $103 \pm 5$ mV were found in guinea pigs and in humans. The cell membrane is recognized as the prime site of electrical field interaction in the stimulation of bone cell growth. Nonhealing wounds have often been implicated as carcinogens. The effect of pulsed electromagnetic fields on bone and their ability to activate bone metabolism have been studied in multicenter randomized and prospective clinical studies, which confirmed PEMFs efficiency for treating certain orthopedic conditions (Lavine and Grodzinsky, 1987). Electromagnetic fields have also have a beneficial effect on the repair of fracture calluses (Bassett, Pawluk and Pilla, 1974) and they may increase bone mass (Martin and Guttman, 1978). A combination of EMF and $PGE_2$ (prostaglandin $E_2$) exhibited a trend of additive effect on bone formation (15 μg per kg of $PGE_2$) (Tsai *et al.*, 1991).

Three clinical trials (prospective randomized and double blinded) involving cartilage have also been undertaken in which the effects of the PEMF (Trock *et al.*, 1993; Trock, Bollet and Markoll, 1994), or a pulsed low amplitude, monophasic, 100 Hz spiked signal delivered via surface electrodes (Zizic *et al.*, 1995), were evaluated in osteoarthritis (OA) of the knee. All the clinical studies were of comparatively short duration (4–6 weeks) but showed a decrease in pain and an improvement in various clinical variables. None of these clinical studies showed that the electrical stimulation had a direct effect on the joint cartilage and they showed only a symptomatic relief of pain.

The utility of pulsed electromagnetic fields (0–300 Hz) is now successfully established for clinical use; however, their mechanism of interactions in relation to bone metabolism that influence bone healing remains to be fully understood. In particular, when coils are used to administer ELF-EMF, it is not clear whether the response obtained is due to the magnetic field or the resultant induced electric field.

The inductive method has the advantage of being an outpatient method and is non-invasive. However, it has certain disadvantages: (i) it does not allow full weight bearing on the lower extremity in a cast and (ii) and the equipment is not portable for long bone reunions and demands patient compliance. In this method, the magnetic fields can be shaped to localize the currents that occur in a volume conductor.

## *5.4.5 Capacitive Coupling*

### 5.4.5.1 Radio Frequency Bone Response

The electrical properties of bone in the high frequency limit ($>1$ MHz) have received scant attention, probably because the frequencies of physiological interest fall in the extremely low frequency limit. These govern the biological process in bone and tissues in general. Kosterich, Foster and Pollack (1983, 1984) have used an impedance analyzer and a vector impedance meter to study the electrical properties of fluid saturated bone in the range 10 kHz–100 MHz. While analyzing the data they predicted a relaxation frequency for fluid saturated bone well above 100 MHz (Kosterich, Foster and Pollack, 1984). Ray and Behari (1986) have studied bone conduction in the range 400–1300 MHz and reported relaxation frequencies around 1.0 GHz for dry and wet bone. Exposure of bone samples to UV light substantially affects the magnitude of relaxation frequencies. Reddy and Saha (1984) have developed a differential technique to measure the electric and dielectric properties of wet bone from DC to 1.0 MHz. They found a significant dependence of specific resistivity on frequency.

Singh and Behari (1984a, 1984b) (Figures 4.21 and 4.22) have measured the impedance and phase angle of bone specimens using a vector impedance meter from 0.5 to 108 MHz and dielectric parameters at microwave frequencies (Behari, Kumar and Aruna, 1982; Ray and

Behari, 1986). These investigations suggest that bone (*in vitro*) shows relaxation phenomena at radio frequencies. On the basis of these findings a bone transducer has been designed (Ray and Behari, 1987). If kHz–MHz frequency signals are applied externally at the bone fracture site these may affect the healing rate. High frequency (radio frequency) bone stimulation using capacitor plates seems not to have been attempted, apart for some scanty efforts (Brighton, Pollack and Windsor, 1981; Brighton and Pollack, 1985; Brighton *et al.*, 1985a, 1985b). Evidently, to optimize the effects of electrostimulation, knowledge of both the characteristics of the naturally occurring electromechanically induced signals inside the tissue and its electric field configuration with and without electrical stimulation is necessary.

Behari *et al.* (1991b), experimenting on rabbits, has identified that a pulsating frequency of 14 MHz (modulated at ELF frequencies 16–76 Hz) and externally applied voltage of 10 V p-p is sufficient to induce a voltage of requisite magnitude to accelerate the healing process (Figure 5.12). Electrostimulation using such a method is shown for humans in Figure 5.11 and for animals in Figure 5.13.

Several studies have demonstrated the beneficial effects of PEMFs in the treatment of non-union of fracture, osteonecrosis and tendonitis (Aaron, Ciombor and Jolly, 1989; Bassett *et al.*, 1981a, 1981b; Binder *et al.*, 1984; Borsalino *et al.*, 1988; Brighton, 1984; Coliacicco and Pilla, 1984; Cruess, Kan and Bassett, 1983; Farndale and Murray, 1985; Fitzsimmons *et al.*, 1986). It is claimed that electrical fields produce results in the healing to non-union similar to those of bone grafting alone, depending on the anatomical site and degree of non-union (Boyde, Lipinski and Wiley, 1961). Improved success rates have also been reported when exposure to electromagnetic fields is combined with surgical intervention (Bassett, Caulo and Kort, 1981; Bassett, Mitchell and Gaston, 1981; Rosen, 1979). These reports are, however, difficult to interpret due to the influence of additional procedures such as bone grafting and/or fracture stabilization, the unpredictable natural history of the underlying condition being treated (Bassett, Caulo and Kort, 1981a; Bassett, Mitchell and Gaston, 1981b; Brighton *et al.*, 1988; Paterson, Lewis and Cass, 1980) and the lack of well-designed studies. In one double blind study of 45 patients (Sharrard, 1990) no effect of electromagnetic fields was discernable on the rate of regression of non-union of tibial fractures. One study reported using low energy, time varying magnetic fields to induce specifically-configured electrical currents in rats

| | |
|---|---|
| **Carrier Frequencey** | **14.0 MHz** |
| **Modulating Frequency** | **16.0 Hz** |
| **Voltage output (Peak to Peak)** | **10.0 V** |
| **Output** | **Pulsating Type** |

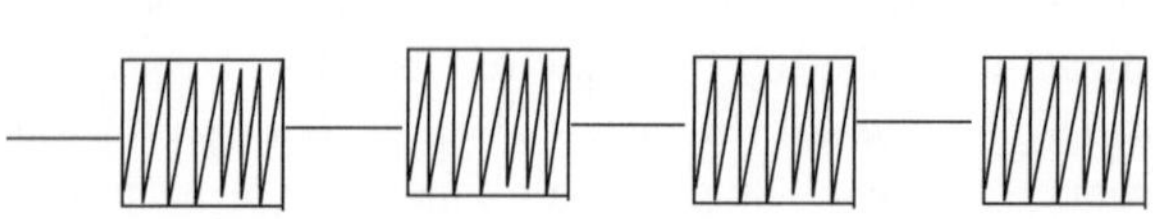

**Figure 5.12** Electromagnetic field stimulator used in the investigation has the following characteristics

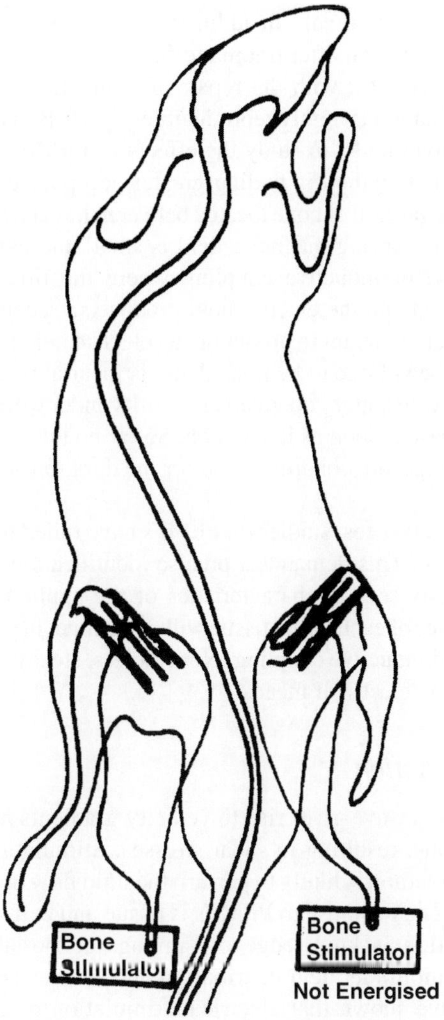


**Figure 5.13** Stimulating bone fracture healing using a pulsed electromagnetic field in a fractured rat bone

subjected to a model of disuse osteoporosis (Cruess, Kan and Bassett, 1983). Field exposure clearly stopped bone loss. Other reports also confirm the positive effects of capacitively coupled fields on bone density (Billota *et al.*, 1994).

Some authors chose to study the effects of PEMF on bone during limb lengthening. The surgical procedure involves an osteotomy, preserving where possible the periosteum. The proximal and distal segments are thereafter slowly distracted by means of external fixators. Regenerated bone is formed in the distraction gap (Ilizarov, Ledyasev and Shitin, 1969) at a rate dependent on the rate of distraction, stability of the fixator and on the underlying pathology. After distraction is complete, the affected site undergoes consolidation analogous to the changes that occur after fracture repair. In addition, marked osteoporosis occurs in the distal segment of the long bone, as seen after traumatic fractures (Eyres and Kanis, 1995; Kannus *et al.*, 1994a, 1994b). In contrast with the repair of traumatic fracture, calcification of the regenerate and its consolidation is highly reproducible (Eyres, Bell and Kanis, 1993a, 1993b), which makes this a suitable model to study the effects of PEMF.

The capacitively coupled method is finding greater acceptance. The method uses a time varying electric field in the portion of bone located between the capacitor plates. The magnitude of the accompanying time varying magnetic field is small and can be neglected. This is in contrast with the method of inductive coupling where the time varying magnetic fields produces the primary effect and the electric field produces a secondary effect in the tissues. Capacitive coupling is more efficient in producing an electric field in bone. It can cover a large area of the fracture and the volume to be treated and is invariably an outpatient method. The device is portable, can also be battery operated and works under weight bearing conditions and can be applied over the cast as soon as it is put on. So far no untoward side effects have been ascribed to any of these exposure conditions and for the durations and no safety measures are required.

Despite the significant advances, studies on PEMFs have failed to adequately deal with the dose–response relationship. This demands a precise identification of the targeted biological process and simultaneously several characteristics of the applied field. In fact, individual, passive electrical properties of each target tissue will determine, in part, the derivative electric-field patterns as a result of inductive coupling. Nonetheless, steady progress has been made in modifying cell function by electrical means.

### *5.4.6 Mechanism of Action*

The propagation of a stress wave gives rise to velocity gradients near the boundaries of the viscoelastic media of bone, resulting in shear stresses. Stimulation caused by the PEMFs adopting a non-invasive method is likely to perturb the fluid flow in bone and hence affect the bone healing processes. The response to PEMFs is tissue and cell specific. A prerequisite to effective electrostimulation is knowledge of, among the possible variables, the tissue structure, wave shape, amplitude, field distribution and species (Brighton and McCluskey, 1985). Investigations have shown that electrical stimulation of cartilage or bone cells in culture induces changes in the intracellular levels of cyclic AMP that are followed by a delayed increase in the synthesis of DNA (Rodan, Bourret and Norton, 1978; Korenstein *et al.*, 1984).

It is important to understand the molecular mechanisms linking cell exposure to PEMF in PEMF-biointeractions (Aaron *et al.*, 2004a; Aaron, Ciombor and Simon, 2004; Ibiwoye *et al.*, 2004). PEMFs have been demonstrated to regulate osteoclastogenesis, bone resorption, osteoproteglin, receptor activator of NFkB-ligand and macrophage colony-stimulating factor concentration in a murine marrow culture system (Chang *et al.*, 2005). One concern is the amount of PEMF energy that is absorbed by the tissue.

Heermeier *et al.* (1998) have concluded that human osteoclastic cells facilitates investigations on the effect of EMFs on human skeletal tissue under defined *in vitro* conditions. The EMF induced enhancement of collagen type I mRNA expression and substantial collagenous matrix synthesis, as observed in the normal human osteoblastic HO-197 cells as well as in myositis ossificans MO-192 cells, is comparable to the effects of TGF-β and TGF-I. This result may indicate the basis of EMFs observed treatments. Diniz, Soejima and Ito (2002) have studied osteoblasts of the MC3T3-E1 lineage, with or without an NO synthase inhibitor (L-NMMA), that were exposed to PEMF stimulation for 15 days. The electrical signal significantly increased NO synthesis, cellular proliferation and osteoblast differentiation. Consequently, NO production seems to be an essential parameter in the effect of electrical fields on bone cells.

Fracture repair is a complicated, multistage process that ends in a complete structural and functional restoration of the involved bone. The application of direct current would reduce local tissue that could transform polymorphic cells into bone (Brighton and Hunt, 1986). Such a mechanism also applies to mesenchymal cells associated with bone fracture hematoma. It has been postulated that the extracorporeal shock wave produces microtrauma or microstructure and induces neovascularization through hematoma formation, which increases osteoblast or fibroblast activity (Wang *et al.*, 2001). Bony continuity is re-established both by subperiosteal membranous bone formation adjacent to the fracture ends and by enchondral ossification of mesenchymal tissue that has replaced the hematoma in the fracture gap (Barnes *et al.*, 1999; Bostrom *et al.*, 1995; Ishidou *et al.*, 1995; Onishi *et al.*, 1998). The repair process includes inflammation and subsequent chemotaxis, proliferation and differentiation of cells, which is mediated by several growth factors (Barnes *et al.*, 1999; Einhorn, 1995; Schmitt *et al.*, 1999). Bone morphogenetic proteins (BMPs) belong to the transforming growth factor-β (TGF-β) superfamily and play a crucial role during the embryonic development of the skeleton (Sakou, 1998; Wozney and Rosen, 1998). Fracture healing is a close recapitulation of embryonic bone formation (Barnes *et al.*, 1999; Sakou, 1998; Schmitt *et al.*, 1999) and therefore it is suggested that the same growth factors (i.e., BMPs) are involved in both fracture repair and in embryonic skeletogenesis. In animal studies, clues have been found that indicate BMPs have a dominant role as regulating growth and differentiation factors during the healing process of fractures. PEMF stimulation is reported to induce osteogenesis through upregulating BMP-2 and BMP-4 in osteoblasts (Bodamyali *et al.*, 1998). Immunohistochemical analysis of closed fractures in rats has shown that BMP-2, BMP-4, osteogenic protein-1 (OP-1 or BMP-7) and BMP receptors are abundantly expressed in various cell types within the fracture callus (Bostrom *et al.*, 1995; Ishidou *et al.*, 1995; Onishi *et al.*, 1998). This finding indicates that OP-1 may be one of the key molecules in the regulation of normal fracture healing.

Brighton, Unger and Stambough (1984) have reported that bovine knee articular chondrocytes cultured in pellets and exposed to a capacitively coupled electric field showed an increase in glycosamino-glycan synthesis. Others, in articular cartilage chondrocytes or cartilage explants exposed to various electrical fields, found an increase in proteoglycan synthesis (Aaron, Ciombor and Jolly, 1989; Ciombor *et al.*, 2002; MacGinitie, Gluzband and Grodzinsky, 1994), a decrease in proteoglycan and an increase in cell proliferation (Grande *et al.*, 1991; Pezzetti *et al.*, 1999; Liu *et al.*, 1997), or an increase in protein synthesis (MacGinitie, Gluzband and Grodzinsky, 1994). It thus appears that the chondrocyte phenotype is susceptible to modulation by electrical stimuli.

Various PEMFs have been shown to promote bone formation *in vivo* (endochondral ossification) and skeletal homeostasis and in clinical studies of fracture non-unions. An important feature is the reduction in osteoclastic bone resorption *in vitro*. PEMFs may act to decrease bone resorption and maintain bone mass, thereby retarding the loss of structural integrity associated with resorption. Bassett, Schinik and Mitchell (1984) have presented data indicating up to five years of clinical benefit after treatment of the femoral head of osteonecrosis.

Recombinant human OP-1 (rhOP-1) induces enchondral bone formation after ectopic implantation in conjunction with a collagenous matrix in rats and baboons (Ripamonti *et al.*, 1997; Sampath *et al.*, 1992). *In vitro* studies have shown that rhOP-1 stimulates proliferation and chondroblastic and osteoblastic differentiation of mesenchymal and osteoblastic cell types (Asahina, Sampath and Hauschka, 1996; Asahina *et al.*, 1993; Sampath *et al.*, 1992). Bone marrow and periosteum contain mesenchymal stem cells with the potential to differentiate into chondrocytes or osteoblasts (Bruder and Fox, 1999; Einhorn, 1995; Nakahara *et al.*, 1990). Currently, it is supposed that, during fracture repair, BMPs, including OP-1, act on these cells, causing their proliferation and subsequent differentiation into cartilage and bone (Barnes *et al.*, 1999; Reddi, 1995; Schmitt *et al.*, 1999).

Because of its osteoinductive activity, rhOP-1 has great possibilities as a therapeutic tool for the enhancement of bone repair (Cook *et al.*, 1995). In segmental long bone defects in rabbits, dogs and monkeys, and in calvarial defects in baboons, local implantation of rhOP-1 in conjunction with a collagenous matrix induced healing of the defects (Cook *et al.*, 1994a, 1995; Ripamonti *et al.*, 1996). In contrast to these bone defect models, a closed fracture model has been used to investigate the efficacy of rhOP-1. In this study, the ability of rhOP-1 to accelerate undisturbed physiologic fracture healing was tested in a closed tibial fracture in goats. Bovine collagenous matrix is widely used as a carrier material of rhOP-1. However, bovine collagenous matrix complicates injection of the material and carries the risk of disease transmission.

It is now widely accepted that the primary site of PEMF and other extremely low frequency fields with tissue is at the cell membrane. Several groups of workers (Bawin *et al.*, 1978; Coliacicco and Pilla, 1984) have studied cell membrane associated activities such as $Ca^{2+}$ flux and cAMP metabolism in bone and bone cells. The calcium content of chondrocytes exposed to different pulse types is significantly increased by one and decreased by the another as compared with control values (Bassett *et al.*, 1979). Differential effects of various pulses of calcium uptake in chick limb have also been reported (Assailly *et al.*, 1981). One possible explanation is that an appropriate electrical signal to the bone and cartilage cells stimulates osteogenesis, or bone growth, by increasing cellular cyclic AMP or cytosolic calcium, both of which may act as intracellular second messengers.

It has been reported that PEMF inhibited the cAMP response to parathyroid hormone in an osteoblast-like cell line. Changes in cAMP metabolism correlated with changes in bone resorption. These data are consistent with the hypothesis that PEMF influences cell membrane activity, resulting in changes in the membrane-associated second messenger system such as $Ca^{2+}$, cAMP and subsequent cellular and organ responses. Cain, Adey and Luben (1981), basing their studies on cAMP metabolism in primary calvarial bone, proposed that PEMF inhibits PTH stimulated coupling of the adenylate cyclase system. They also pointed out that this inhibition does not affect the intrinsic activity of the G protein and the catalytic subunit. Coliacicco and Pilla (1984) have suggested that the electromagnetic field couples with specific processes such as bicarbonate dependent $Ca^{2+}$ adenosine triphosphatase and active ion transport at the plasmalemma. An alternating current creates a varying electrical and magnetic field, while DC produces a constant electric and magnetic field. Therefore, it is suggested that a DC creates a field that acts synergistically on the tissue, whereas AC fields effectively counteract each other.

Ishidou *et al.* (1994) have reported that TGF-β appeared in the stimulated area soon after the start of DC stimulation. Other investigators have suggested the involvement of such cell growth factors in electrically stimulated osteogenesis (Ryaby *et al.*, 1994). Some studies demonstrated that electrical stimulation altered the uptake of calcium ions by cells (Allen *et al.*, 1994). Adey (1986) hypothesized that changes in the $Ca^{2+}$ binding receptor (glycoprotein) took place.

Alkaline phosphatase (ALP) activity plays an important role in calcification. Its activity is useful as a marker of osteoblasts. The elevation in ALP observed after electrical stimulation (Yonemori *et al.*, 1996) reflects an increased number of osteoblasts and a greater degree of osteogenesis. The effect of electrical stimulation on ALP activity in tissue has also been reported (Matsunaga, 1986; Berg and Zhang, 1993). Such electrical stimulation elevates the transmembrane voltage and the resultant elevation in electrical conductivity of the cell membrane affects the functions of the cell membrane proteins and lipids, as well as the expression of genes, leading to cell proliferation.

In general, the repair process depends on the blood supply. Relating the pluripotential nature of these processes to the hemodynamic states suggests that in ischemic conditions the cells tend toward fibroblast and chondrocyte production, while hyper-vascularity simulates production of osteoblast and bone production (Hall, 1970). Ijiri *et al.* (1995) have reported that electrical stimulation increased local blood flow through the tissue as well as the permeability across blood vessels, and that the administration of indomethacin (to exert anti-inflammatory effects) during electrical stimulation suppressed osteogenesis. Based on these results Yonemori *et al.* (1996) have concluded that electrical stimulation in the presence of reactive cells leads to promotion of osteogenesis. Some *in vitro* experiments have yielded indirect evidence that cell proliferation is promoted by electromagnetic stimulation (Batista, Miklavac and Sersa, 1991; Becker and Esper, 1981; Cadossi *et al.*, 1985; Korenstein *et al.*, 1984). In contrast, some *in vitro* experiments have suggested that cell proliferation is suppressed by electromagnetic stimulation (Jones, Pedley and Ryaby, 1986; Revoltella *et al.*, 1993).

Using immunostaining, it has been demonstrated that CTGF/Hcs24 is localized in hypertrophic chondrocytes and proliferating chondrocytes in the regions of regenerating cartilage (Nakata *et al.*, 2002). These authors have reported that CTGF/Hcs24 is detected in hypertrophic chondrocytes on days 8 and 14; less immunostaining is detected in these cells on day 20. CTGF/Hcs24 is also detected in active osteoblasts in the regions of intramembranous ossification. In addition, it is detected in newly formed osteoid matrix, which indicates the presence of the secreted form of CTGF/Hcs24. As in chondrocytes, CTGF/Hcs24 is strongly detected in active osteoblasts on days 8 and 14, with less immunostaining detected on day 20. Because CTGF/Hcs24 stimulates the proliferation and differentiation of chondrocytes (Nakanishi *et al.*, 2000) and osteoblasts (Nishida *et al.*, 2000) *in vitro*, it is speculated that CTGF/Hcs24 could promote chondrogenesis and osteogenesis *in vivo* through its direct action on these cells.

CTGF/Hs24 is also detected in cells in fibrous tissue (on days 8 and 14). This finding supports data indicating that CTGF/Hcs24 is localized in fibroblasts (Bradham *et al.*, 1991; Igarashi *et al.*, 1993) and suggests that it is involved in the migration of fibroblasts and acts as a regulator of wound repair during fracture healing. In addition, CTGF/Hcs24 is detected in vascular endothelial cells in the callus from days 8 to 20. It has been shown that CTGF/Hcs24 promotes the proliferation, migration and tube formation of vascular endothelial cells in culture and angiogenesis *in vivo* (Shimo *et al.*, 1998, 1999). These observations may explain why CTGF/Hcs24 plays a role in angiogenesis during fracture healing.

CTGF/Hcs24 and CTGF/Hcs24 mRNA are detected (Nakata *et al.*, 2002) in various cells at the fracture sites, particularly proliferating and differentiating cells. These observations can be added to the findings of *in vitro* data that CTGF/Hcs24 stimulates proliferation and differentiation of chondrocytes (Nakanishi *et al.*, 2000) and osteoblasts (Nishida *et al.*, 2000) and proliferation and migration of fibroblasts (Frazier *et al.*, 1996) and vascular endothelial cells (Shimo *et al.*, 1998, 1999). It is therefore, suggested that CTGF/Hcs 24 is an important regulator in chondrogenesis, osteogenesis and angiogenesis *in vivo* and promotes intramembranous ossification and endochondral ossification in the process of fracture healing.

Rosenberg *et al.* (1965) have suggested that the different responses of cells to electrical stimulation may be attributed to differences in the culture systems employed in the experiments. To assess the effects of electrical stimulation on cells, it is essential to examine the effect of electrical stimulation in promoting osteoblast proliferation *in vivo*.

Some investigators (Croker and Nar, 1987; Woosley, 1991) have used NOR (nucleolar organizer region), a region of the cell nucleus associated with ribosomes, as an index of the proliferative activity of cells *in vivo*. The NOR is a DNA loop of the ribosomal RNA gene within the cell nucleolus. When stained with silver, this area is visible as a dot within the nucleus, representing the orgyrophilic proteins of the nucleolar organizer region (Ag NOR). Because ribosome is the site of protein synthesis, AgNOR is believed to correlate with cell proliferation and maturation, and the number of AgNOR dots correlates with the proliferative activity of cells (Grotto, Lorand-Metze and Metze, 1991; Lischwe *et al.*, 1979). AgNOR has been used as an indicator of the malignancy level of tumor cells because it is related to the proliferative activity of cells (Kuratsu *et al.*, 1991).

Yonemori *et al.* (1996) have reported that the proliferative activity of osteoblasts in bone marrow increased significantly at 7 and 14 days after electrical stimulation. This finding *in vivo* demonstrates that electrical stimulation promises osteoblast proliferation. Horii *et al.* (1993) have also shown that electrical stimulation promotes not only cell proliferation but also the differentiation of mesenchymal cells into osteoblasts. The effects of electrical stimulation on the cells of the osteogenic system might play a role in the promotion of osteogenesis.

Yonemori *et al.* (1996) have concluded that the electrical stimulation of osteogenesis is influenced by the local tissue environment and that osteogenesis is promoted markedly when electrical stimulation is provided in the presence of reactive cells. These observations then explain the clinical experience in which electrical stimulation for the treatment of non-union is effective in some cells, but is totally ineffective in others.

An indication of fracture healing process is its metabolic activity. This can be accomplished through measurement of isotope labeled glucon analogues as a function of fracture configuration and time since fracture (Severns *et al.*, 2004). The metabolic activity of a normally healing fracture is higher than that of a normal bone. Further, a bone defect destined not to heal will stabilize to a metabolic activity level that remains elevated relative to a normal bone.

### *5.4.7 Mechanism of PEMF Interaction at the Cellular Level*

As mentioned above, the induction of a change in calcium flux across the plasma membrane is a primary event following a short stimulation of bone cells by capacitively coupled electric fields.

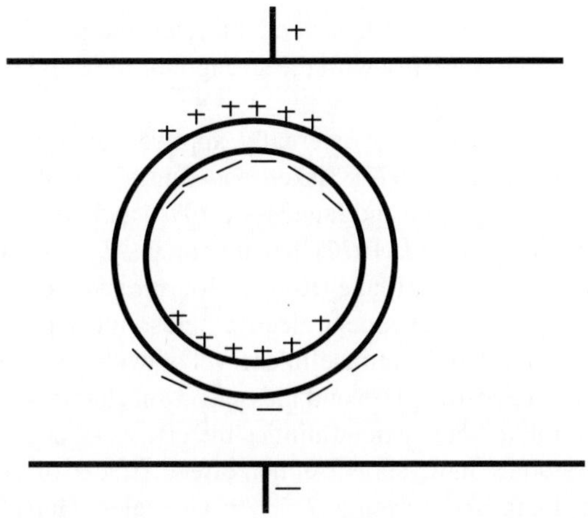

**Figure 5.14** Electrical polarization of a membrane vesicle (with an internal and an external conductive medium) by an external electric field. As is apparent, the polarization is of opposite sign in the lower and the upper hemispheres

The mechanism underlying this induced process involves interaction of the applied electric field mainly with the cellular plasma membrane. Bone cells in tissue culture form a monolayer of cells and are electrically uncoupled. The cells can be modeled by an array of membrane vesicles. Exposure of such a suspension of vesicles to an external electric field causes polarization of the membrane (Figure 5.14). The membrane polarization ($\Delta\psi$) induced by an external electric field ($E_{\text{ext}}$) is given by:

$$\Delta\psi = 1.5RE_{\text{ext}}\cos\theta \tag{5.7}$$

where $R$ is the radius of the membrane vesicle and $\theta$ is the angle between the external electric field direction and the radius vector to the membranal plane. Evidently, from Equation 5.7, membrane polarization is maximal at the poles of the two hemispheres and vanishes towards the equator. Furthermore, the polarizations at the lower and upper parts of the vesicle are of equal amplitude but of opposite sign (phase difference of $\pi$). It has been shown previously (Korenstein *et al.*, 1984) that when employing capacitive stimulation the electric fields prevailing in the medium in which the cells are in the range 12–54 V cm$^{-1}$. If we assume that the vesicles are 4 μm in diameter, the induced polarization for the above electric fields turns out to be 4–16 mV, respectively.

The cells are assumed to be attach to the substratum any rotational mobility of the cells is prohibited. Each electric field pulse applied will influence precisely the same region in the plasma membrane as indicated by Equation 5.7. Therefore, increasing the frequency of stimulation will always affect the same membranal region. The monotonic increase of calcium uptake with the frequency of stimulation suggests that the frequency region studied (0–100 Hz) is still far from the refractory region. Capacitively coupled electrical stimulation of bone cells in culture for several minutes induces an increase in calcium uptake, which increases monotonically with the strength of the applied electric field (Danon and Korenstein, 1984).

Modulation of cellular processes by externally applied electric fields accelerates limb regeneration in amphibia (Becker, 1972; Sisken and Smith, 1975; Jaffe, 1979) as well as peripheral nerve regeneration (Wilson and Jagadeesh, 1976) and bone fracture repair (Bassett, Pilla and Pawluk, 1977; Brighton *et al.*, 1977) in mammalians. *In vitro* studies of bone-derived cells have shown that exposure of cartilage (Rodan, Bourret and Norton, 1978) or bone cells (Korenstein *et al.*, 1984) to time dependent electric fields enhanced DNA synthesis. These effects are preceded by changes in the intracellular level of cyclic AMP as well as $Ca^{2+}$ influx (Korenstein *et al.*, 1984), suggesting a second messenger mechanism for mitogenesis. In this respect, electric stimulation to some extent mimics the effects of peptide hormones (Somjen *et al.*, 1983). Various studies have shown either direct effects of such hormones on the cytoskeleton (Freidkin, Legg and Rozenguri, 1979; Gonzales, Garrido and Vial, 1981) or effects on cell processes, which seem to be cytoskeleton mediated (Miller, Wolf and Arnaud, 1976; Brunk, Schellens and Westermark, 1976). It has been shown that several agents interacting with the plasma membrane induce actin polymerization (Pribluda, Laub and Rotman, 1981; Laub, Kaplan and Gitler, 1981). In these studies, ligand binding to receptor, receptor aggregation and possibly internalization must be considered as plausible initiators of the cascade of events, this includes cytoskeletal changes and eventually leads to proliferation. Since electric stimulation circumvents these initial steps, one would expect to obtain some clues as to the direct role of receptors in cytoskeleton changes.

There is a report (Laub and Korenstein, 1984) of a direct electric field effect on the cytoskeleton. Under optimal conditions we obtain polymerization of 28% of the G-actin pool. Similar results have been obtained with chemical stimulation (Pribluda, Laub and Rotman, 1981; Laub, Kaplan and Gitler, 1981). Some of these studies suggest that receptor rearrangements (aggregation) may be important in nucleating actin polymerization (Laub, Kaplan and Gitler, 1981; Bourgignon and Singer, 1977). It is indicative that ligand-induced receptor changes may not be essential, it is unlikely that short electric pulses directly induce the aggregation of specific receptors.

In parallel with actin polymerization, a transient rise in intracellular cyclic AMP (Korenstein *et al.*, 1984) and increased $Ca^{2+}$ uptake has been found within 1–2 min after electrical stimulation. This correlation suggests that second messengers could mediate the effect on action. Cyclic AMP and $Ca^{2+}$ are themselves interrelated by a complex feedback system (Berridge, 1975). Each of them could also affect the microfilament system in several ways (Dedman, Brinkley and Means, 1979).

Pulsed electric stimulation, coupled capacitively to bone cells isolated from rat embryo calvaria, caused changes in the intracellular level of cyclic AMP and enhanced DNA synthesis. The capacitive method of electrical stimulation is characterized in terms of displacement currents (0.7–4.0 mA) and voltages (10–54 V $cm^{-1}$) prevailing in the stimulation chamber. Changes, both in cyclic AMP and in incorporation of [$^3$H]thymidine into DNA, are correlated with the strength of the applied electric field. Unlike the mechanical stimulation of bone cells, the electrical stimulus is not medicated by de novo synthesis of prostaglandins. The findings suggest that cyclic-AMP changes, induced by the capacitive electrical stimulation of bone cells, trigger DNA synthesis (Korenstein *et al.*, 1984).

### 5.4.8 Spatial Coherence

Although the mechanism of EMF–cell interaction is not completely known, hypotheses can be constructed to account for thermal field (which may cause fluctuations in temperature) being ignored by cells. The data imply that cells distinguish between applied noise and thermal noise. Although both are (by definition) temporally incoherent, they do exhibit a distinct difference in their spatial behavior. At a given instant, the value (magnitude and direction) of the thermal noise field at any point is uncorrelated with its value at other locations more than a few nanometers or so apart. This property follows because the Debye screening length (roughly the range over which an ion is not shielded from other ions) in the extracellular fluid is only about 1 nm. Thus, thermally driven localized charge density fluctuations (and consequently, endogenous thermal noise fields) are spatially incoherent over distances greater than a few nanometers. By contrast, exogenously impressed fields (either temporally coherent or incoherent) always exhibit spatial coherence because their wavelengths are much larger than cellular dimensions. It is hypothesized that the spatial incoherence of the thermal noise field allows cells to ignore it. On the other hand, because externally imposed electromagnetic noise is spatially coherent cells respond to this. Two possibilities that should be examined are (Weaver and Astumian, 1990): that rather more cells must be simultaneously stimulated or (Adair, 1991) that there is a collection of receptors on each cell that must be excited to produce the observed biochemical responses. For signal transduction, in either case some cooperativity among the entities involving inter- or intracellular signaling, respectively, must be operative.

First the multicell stimulation hypothesis was developed by Weaver and Astumian (1992) and further elaborated by Gailey (1993). The basis for this is that communication among a group of cells is required for an alteration of cell functioning to occur. Gap junctions between adjacent cells have been proposed as a mechanism by which cells are coupled so that there is an "amplification" of the stimulating field energy. This explanation is obviously restricted to cellular systems in which gap junctions exist. For example, in *in vivo* studies, field-induced changes in the frequency of developmental abnormalities and in ODC activity levels could be described in terms of such a mechanism. However, it is not appropriate as a explanation of the EMF-induced effects in L929 cell cultures, because these cells do not form gap junctions (Matthews and Neale, 1989). Further, EMF effects have been observed with transformed or cancer-derived, attachment-independent cell lines, including HL-60 (Blank *et al.*, 1992; Goodman *et al.*, 1992), Daudi (Mullins, Krause and Litovitz, 1993) and CCM-CM3 (Phillips *et al.*, 1992), which should exhibit few, if any, gap junction couplings hence it is suggestive that only this possibility is unlikely. However, other forms of intercellular signaling may exist that would allow a multicell (as opposed to a single-cell, multireceptor) hypothesis to be preserved.

Other explanation relies on the spatial coherence of the field (Litovitz *et al.*, 1994a; Litovitz *et al.*, 1994b). Cellular detection of EMFs is assumed to result from the impressed fields modifying either the probability of ligand–receptor binding (Weaver and Astumian, 1990) or an increase in the number of receptor binding sites available on the cell surface (Adair, 1991). Either case would result in an increase in ligand binding and, therefore, a change in cell function. The first explanation is thermodynamically not viable, in that it does not explain how individual receptor binding would be modified by the EMF in the presence of large thermal noise. The probability of the second explanation is, therefore, greater. The requirement that many receptors over the cells surface be simultaneously activated (known as "cooperativity") prevents random activation of individual receptors from triggering an erroneous cellular response. It would also demand that EMF activation occur only if the stimulating field is essentially uniform over several receptor sites on the cell membrane. This requirement leads to spatial coherence in the EMF if a cellular response is to be evoked. Liburdy and Eckert (1994) have presented initial data indicating that the number of CD3 proteins available on the surface of Jurkat E6.1 cells is increased in the presence of an applied 60 Hz field.

One can determine the extent of cooperativity required by observing the threshold for biological response as a function of the concentration of the ligand near the cell. If $n$ receptors must be stimulated to produce a response, then it can be shown (Alberts *et al.*, 1983) that the response varies with concentration as:

$$\text{response} \propto 1 - \frac{1}{1+(C/C_{\mathrm{T}})^n} \tag{5.8}$$

where $C_{\mathrm{T}}$ is the threshold concentration. As $n$ increases, $C_{\mathrm{T}}$ becomes rather large. Extending this concept to EMF stimulation requires either a modification of the ligand–receptor binding or an assumption that the number of receptors available at the cell surface is proportional to the local field strength. The generalization of the equation for the response would be:

$$\text{response} \propto 1 - \frac{1}{1+(E/E_{\mathrm{T}})^n} \tag{5.9}$$

where $E$ is the field strength and $E_{\mathrm{T}}$ is the threshold field. A satisfactory fit to these data can be obtained for essentially any value of $n$ larger than about 10. These data are suggestive of biological cooperativity involving activation of a significant number of membrane integral protein receptors.

### *5.4.9 Effects of EMFs on Signal Transduction in Bone*

Although the effects of ELF fields at the tissue level have been clarified somewhat by research over the past two decades, the primary biochemical and biophysical effects at the molecular or ionic level remain obscure. One clear likelihood for the effects of ELF fields on bone is that the plasma membrane of osteoblasts is likely to be the major site of action. Figure 5.15 shows a possible mode of field amplification.

### *5.4.10 The Biophysical Interaction Concept of Window*

For a given signal type, it is interesting to know the range of waveform parameters over which the waveform may exhibit meaningful bioeffects. This is achieved by assuming a target pathway and spectral matching the frequency components of the input signal to the target's frequency response. The pathway chosen to illustrate this approach is $Ca^{2+}$ binding to calmodulin (CaM). Some studies suggest the EMF-sensitive pathway for bone repair may involve $Ca^{2+}$/CaM-dependent growth factor release (Zhuang *et al.*, 1997).

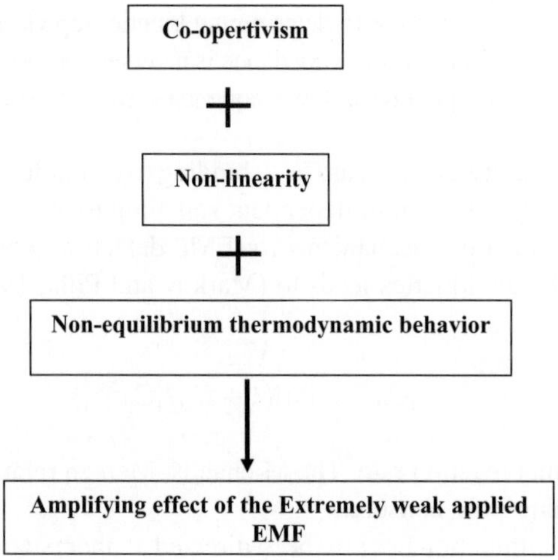


**Figure 5.15** Model for low frequency field amplification at the membrane site

Assuming first order binding kinetics:

$$\frac{\mathrm{d}Q}{\mathrm{d}t} = k_{\mathrm{b}} Q(t) \tag{5.10}$$

where $Q(t)$ is the concentration of the bound species (in this case $Ca^{2+}$) and $k_{\mathrm{b}}$ is the rate constant of $Ca^{2+}$ binding. The rate of binding resulting from the application of a time varying electric field $E(t)$ may be described in the frequency domain by Equation 5.11:

$$\Gamma(w) = \frac{k_{\mathrm{b}}}{jw}[-\Gamma(w) + \delta E(w)] \tag{5.11}$$

where $\Gamma(w)$ represents the frequency dependence of the surface concentration of the bound ion, and $\delta$ is an appropriate constant yielding the change in bound charge to first order in $E(w)$. The kinetics described by Equation 5.10 may be translated into a binding impedance $Z_{\mathrm{b}}(w)$ (Pilla, Nasser and Kaufman, 1994; Oin, Rubin and McLeod, 1998):

$$Z_{\mathrm{b}}(w) = \frac{1}{\delta}\left(\frac{1}{k_{\mathrm{b}}} + \frac{1}{\mathrm{i}w}\right) \tag{5.12}$$

which has the form of a series resistance–capacitance electrical equivalent circuit:

$$Z_{\mathrm{b}}(w) = R_{\mathrm{ion}} + \frac{1}{\mathrm{i}w\, C_{\mathrm{ion}}} \tag{5.13}$$

where $R_{\mathrm{ion}} = (1/\delta k_{\mathrm{b}})$ and $C_{\mathrm{ion}} = \delta$ are the equivalent resistance and capacitance, respectively, of the ion-binding pathway (Pilla, Nasser and Kaufman, 1994). The series circuit time constant,

$\tau_{ion} = R_{ion} \times C_{ion}$, is thus related to the rate-determining kinetic step via $\tau_{ion} = R_{ion} \times C_{ion} = 1/k_b$. The frequency response of the ion-binding impedance is inversely proportional to the applied field frequency, so that frequency components at low frequencies may be expected to have the greatest effect on the system.

Ion, or ligand, binding pathways such as $Ca^{2+}$ binding to calmodulin (CaM) are considered for which the ensuring steps of calcium-dependent signaling to intracellular enzyme systems may act as primary transduction mechanisms for EMF detection. For $Ca^{2+}$/CaM binding, a linearized Michaelis–Menten kinetics leads to (Markov and Pilla, 1997):

$$K_b = \frac{V_{max}}{[Ca^{2+}CaM](1 + K_D/[Ca^{2+}])} \tag{5.14}$$

where $V_{max}$ is the maximal reaction rate. The Michaelis–Menten relation thus determines $K_b$, that is, defines the binding time constant.

Research into PEMF stimulation has also been directed at understanding the mechanisms of response using tissue culture (Norton, 1982; Norton, Hanley and Turkewicz, 1984; Norton, 1985; Norton, Witt and Rovetti, 1988; Jannacone *et al.*, 1988; Aaron, Ciombor and Jolly, 1989). Grodzinsky (1983) conjectured that direct currents do not penetrate the cellular membranes and that control is achieved via extracellular matrix differentiation (Nods and Sato, 1985). Others have found changes in the second messenger, CAMP and cytoskeletal rearrangements due to electrical perturbations (Norton, Rodan and Bourret, 1977; Ryaby, Jones and Pilla, 1986; Jones, Pedley and Ryaby, 1986), as well as effects on glycosamino-glycan sulfation, hyaluronic acid, lysozyme activity and polypeptide sequences (Norton, 1982, 1985; Norton, Witt and Rovetti, 1988; Goodman and Henderson, 1988).

Luben (1991) has shown that the PEMF used in the most widespread clinical fracture treatment devices (Bassett, Pilla and Pawluk, 1977) produces activation of mouse osteoblasts *in vitro* by means of a strong inhibition of PTH responsiveness in the cells (Luben *et al.*, 1982). This leads to increased synthesis of collagen (Rosen and Luben, 1983), decreased levels of bone resorption and accelerated differentiation of osteoblasts from stem cells (Cain and Luben, 1987). The effects of EMFs on PTH responses are consistent with decreased coupling of receptors to adenylyl cyclase via the stimulatory G protein (Cain, Adey and Luben, 1987). Related studies (Luben, 1989) using flow cytometry and immunohistochemistry indicate that exposure to EMFs produces a change in the accessibility of some PTH receptor epitopes to monoclonal antibodies. This indicates that the fields induce a relatively persistent conformational change in the medium accessible domains of the receptor and a corresponding decrease in signal transduction capacity. Luben (1991) has proposed that some or all of the fracture healing effects of electromagnetic devices *in vivo* may be due to localized desensitization of osteoblasts to endogenous PTH at the site of application.

The probability that the effect of EMFs on PTH responsiveness is specific to the membrane is increased by the observation (Rosen and Luben, 1983) that the effects of 1,25-dihydroxyvitamin $D_3$ and dexamethasone on collagen synthesis by bone cells are not modified by EMF exposure. Both of these agents act primarily by means of cytosolic nuclear receptors rather than membrane receptors. Moreover, the effect appears to be receptor specific because EMF exposure does not change the responses of the cells to insulin or prostaglandin $E_2$ (Rosen and Luben, 1983; Cain and Luben, 1987). However, exposure of the bone cells to EMF did modify the responsiveness of the β-adrenergic receptor (Luben, 1993). These data are thus consistent with the hypothesis that EMFs may act at the membrane surface to cause a change in conformation of G-protein-linked receptors.

A 60 Hz, a 1.0-G sine wave field (encountered in environmental and occupational settings) causes a significant effect in osteoblasts in response to PTH (Luben, 1993). The effects of the 1-G, 60 Hz sinusoidly varying magnetic field on the activities of the signal transducing enzymes PKC and PKA has also been investigated. Exposure to the 1-G field has a very rapid and profound effect on PK activity in these cells – indeed a much stronger effect than on the AMP responses. Luben *et al.* (1982) have reported that exposure to the field for times ranging from 1 min to 24 h caused a transient activation, followed by progressive down-regulation of PKC activity associated with the particulate fraction of the cell. When combined with the observations of the effects of EMFs on PTH receptor function, the results support the hypothesis that EMFs may produce cellular responses by specifically modulating the conformation of G-protein-linked membrane receptors and thus perturbing associated signal transduction processes. PKC is thus taken to be the key enzyme involved in desensitization of the PTH receptor in bone cells (Ikeda *et al.*, 1991; Pernalete *et al.*, 1990). These data suggest that the effects of EMFs on PKC could be an important step in the signal transduction cascade that results in enhanced bone formation. PKC is also modulated by electromagnetic energy in developing rat brain is also now established (Paulraj and Behari, 2004, 2006). These findings have led to the suggestion (Luben, 1991) that G-protein-linked receptors may be a particular target for the interaction of EMFs with cell membranes in responsive cells. Overexpression of the serotonin receptor, for example, has been shown (Julius *et al.*, 1989) to cause neoplastic transformation of cells. G-protein-linked receptors can also be influential in controlling cell growth and development (Gupta, Gallego and Johnson, 1992). This may be by controlling gene expression via transcriptional regulators (e.g., *fos* and CREB) that are phosphorylated by PKC, PKA or both. Very small changes caused by EMFs in the surface charge distribution of cells bearing G-protein-linked receptors (Smith *et al.*, 1991) could possibly lead to significant changes in the signal transduction properties of those receptors. In studies using monoclonal antibodies specific for the PTH receptor, treatment with PEMFs produces changes in the expression of certain receptor determinants that are similar to changes observed in cells desensitized by treatment with hormone analogs. The receptor determinants whose expression is modified by ELF EMFs appear to be homologous to the signal transduction domains of other G-protein-linked receptors. Desensitization of other G-protein-linked receptors is known to be associated with changes in the configuration of the transmembrane domains adjacent to intracellular phosphorylation sites. These configurational changes lead to phosphorylation of the receptor by intracellular enzymes.

A wide range of ELF field effects on hormonal, neural and neurohormonal processes may become more amenable to direct molecular investigations. This hypothesis also seems to suggest that low energy fields may produce biological effects because the energy needed to produce small (but functionally significant) conformational changes in the surface domains of membrane proteins may be correspondingly low. This can also be evidenced by examining the effect of modulated radio frequency (RF) fields.

### *5.4.11 Mechanisms for EMF Effects on Bone Signal Transduction*

One key observation is that osteoblasts exposed to PEMFs for as little as 10 min exhibit a persistent desensitization to the effects of PTH on adenylyl cyclase. This desensitization may

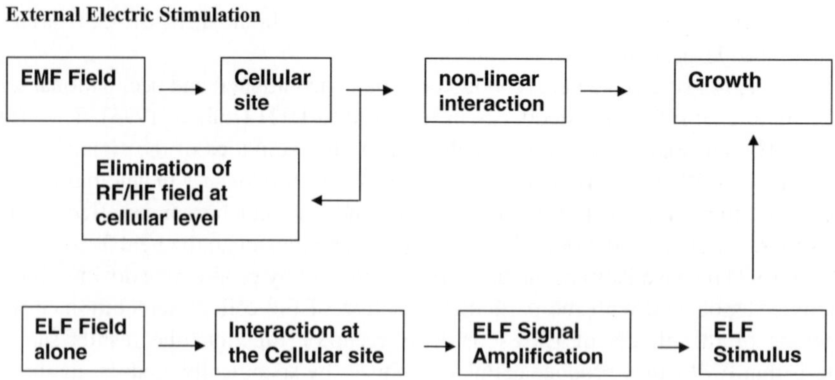


**Figure 5.16** Modulated RF/MW interaction and the demodulation at the interaction site (cell) and possible mechanism for tumor promotion

occur variously: in the absence of a decrease in the total amount of adenyl cyclase enzyme, in the number of hormone receptor sites, in the affinity of the hormone for the receptor, or in the fractional occupancy of the PTH receptor by hormone. Studies using biochemical probes of G protein coupling indicate that the ability of bound hormone receptor complex to activate G protein $\alpha$-subunits is impaired by treatment of the osteoblast with PEMFs. Desensitization of the PTH receptor results in an elevated rate of synthesis of collagen by the osteoblast and a decreased rate of bone resorption by osteoclasts. Both of these effects would tend to increase the amount of bone in a localized area exposed to PEMFs *in vivo* and these effects are in fact observed in healing bone in clinical situations.

RF/MW (microwave) fields modulated at ELF frequencies produce ELF-like effects. No single model yet explains the observed bioeffects across the spectrum from ELF to millimeter waves. Many effects are independent of RF but relate closely to modulation frequency (ELF) of the RF field. There is wide speculation that the detection of certain modulation components of RF/MW fields may be related to intrinsic property of tissue organization. The available data indicate similarities between certain cell ionic and biochemical responses to ELF fields, and to RF/MW fields' amplitude modulated at these same ELF frequencies. This suggests that demodulation of RF/MW fields may be a critical parameter in controlling the biological processes. Figure 5.16 presents a model for their cellular interaction with the modulated field. These findings suggest that ELF field treatment may change the conditions at the cell membrane surface in some as-yet-unknown way, so that the configuration of key residues of the PTH receptor is changed, leading to desensitization (possibly by phosphorylation) of the receptor and thus a shift in the balance of osteoblast activities toward increased bone formation.

## 5.5 Venous Pressure and Bone Formation

Although measurements in microgravity provide an important facility for research, logistical reasons limit the number of experiments that can be performed, or the number of subjects available for investigation. It is, therefore, useful to combine, experiments in microgravity with supporting ground based experiments. There is evidence from several independent sources that venous pressure and intraosseous pressure can modify bone mass. Both laboratory and clinical studies have shown that increasing venous pressure by application of an external tourniquet stimulates bone formation at the distal site. Studies have shown the application of a venous tourniquet proximal to the knee of puppies increased the fluid spaces in bone and the amount of periosteal new bone formation by significant amounts after a period of 7 days (Kelly and Bronk, 1990). These authors also demonstrated that application of a venous tourniquet increased significantly intraosseous pressure in the tibia. Work from the same laboratory has also shown the acceleration of fracture repair in tibia distal to a venous tourniquet (Kruse and Kelly, 1974).

A direct increase of intramedullary pressure in immature goats has also been shown to initiate periosteal bone apposition (Welch *et al.*, 1992; Welch *et al.*, 1993). Intraosseous pressures of 30–45 mmHg increased bone mineral significantly after 5 days. There was also a significant increase in cancellous new bone formation and osteoclast covered bone surface.

## 5.6 Ultrasound and Bone Repair

Wave propagation and vibration tests can be used to determine the *in vivo* properties of bone (Saha and Lakes, 1977; Behari and Singh, 1981a; Nokes *et al.*, 1984) by detecting stress waves in bone using the piezoelectric effect. Behari and Singh (1981a) have examined ultrasound (US) propagation (1.2 MHz) *in vivo* in bone and thereby measured the attenuation and relaxation time. The resonant frequency of the stress wave signal can be used to characterize physical properties of human long bones (Diherty, Bovill and Wilson, 1974; Lowet *et al.*, 1993) and in bone repair studies (Singh *et al.*, 1990). The elastic wave propagation technique has been used in estimating the mechanical properties of human tibia (Cheng, Timonen and Suominen, 1995). The resonant frequency of a long bone measured using the impulse response method can reflect the bending rigidity of the bone (Nakatsuchi, Tsuchikane and Nomura, 1996).

Several workers studying the properties of ultrasound in bone (Anast, Fields and Siegal, 1958; Rich *et al.*, 1966; Floriani, Debevoise and Hyatt, 1967; Brown, 1977; Brown and Mayor, 1974, 1976; Abendschein and Hyatt, 1970; Saha *et al.*, 1982; Rao *et al.*, 1982; Saulgozis *et al.*, 1996) have attempted to differentiate the various stages of a healing fracture by measuring the velocity change in the acoustic wave crossing a fracture gap. The small difference in this velocity can easily be overshadowed by a small difference in path length of the wave. However, the magnitude of the signal output (attenuation) seemed to be a more reliable measure of the different stages of the healing of the fracture (Saha *et al.*, 1982). Those studying the impact properties of bone (Clemedson and Johnson, 1961; Bird *et al.*, 1968; Kenner *et al.*, 1975; Lewis and Goldsmith, 1975; Wong, Goldsmith and Sackman, 1976) have used similar approaches, including bonded strain gages, to measure some of the dynamic properties of bone. From these observations it seems that it is feasible to use wave propagation characteristics of bone to measure some of the parameters of bone that are of clinical interest.

Mechanical vibrations in the form of ultrasound (Polyakov *et al.*, 1974; Dyson *et al.*, 1968; Forest *et al.*, 1953) produce bonding of bones and give rise to the phenomenon of osteogenesis. No pathway is involved in the stimulation of angiogenesis, mechanical stress and bone turnover. This also covers the three essential steps for bone repair. These observations may be crucial in helping to explain the valuable clinical roles of the therapeutic range of ultrasound in promoting angiogenesis and healing in soft tissue and bone. Uglow *et al.* (2003) did not find a significant improvement in the parameters of bone healing that were evaluated using low intensity pulsed ultrasound in a stress-free model. These authors have further concluded that different intensities may be required for different mechanical situations.

Changes in the mechanical and/or geometrical properties of cortical bone can be used to assess fracture stiffness (Barbieri *et al.*, 2006; Protopappas *et al.*, 2005; Protopappas, Fotiadis and Malozos, 2006). This relationship between velocity and elastic modulus indicates the onset of osteoporosis (Chen, Zhao and Mundy, 2004; Haiat *et al.*, 2005; Tatarinov *et al.*, 2005). Fracture healing of long bone is assessed by measuring the time delay brought about by the differences in velocity between cortical bone and the material filling the fracture site. The time delay is also dependent upon the fracture gap size. Therefore, a large change in velocity compared to the intact bone velocity for a particular transducer separation is desirable for the initial inflammatory stages of the healing process, which can then be compared with the velocity measured as the bone heals. Similarly, a large change compared to a baseline measurement on the intact bone is desirable when using the signal amplitude as a potential parameter to monitor healing in long bones (Saulgozis *et al.*, 1996). The ultrasound velocity is given by (Heaney *et al.*, 1989):

$$v = \sqrt{E/\rho}$$

where $E$ is the Young's modulus and $\rho$ is the density.

Dodd *et al.* (2007) have measured ultrasound velocity and attenuation. They employed an axial transmission technique using two 20 mm diameter, 200 kHz, unfocussed transducers, one acting as the source and the other as the receiver. The assembly was immersed in a tank containing distilled water. These authors plotted attenuation as sound pressure level (SPL) (db) versus transducer separation ($x$) using the relationship:

$$\text{SPL}(\chi) = 20\log[V(\chi)/V(\chi_0)] \tag{5.15}$$

where $V(\chi_0)$ is the arrival signal voltage at $\chi = \chi_0$ and $V(\chi)$ is the voltage at each measurement position $\chi$.

As parameter is defined using Equation 5.15 to quantify the effect on amplitude [SPL($\chi$)] of introducing a discontinuity to a specimen, that is, simulating a fracture. The difference in sound pressure levels, expressed as a fracture transmission loss (FTL), at a given measurement position $\chi$, is calculated using:

$$\text{FTL} = -[\text{SPL}_\text{F}(\chi) - \text{SPL}_\text{L}(\chi)] \tag{5.16}$$

where the subscript "L" denotes an intact specimen and "F" a fractured specimen. From Equation 5.16 it is apparent that a reduction in signal as a result of the fracture will give rise to positive FTL.

The velocity of an ultrasonic wave across a discontinuity in a structure, such as a fracture site, will be reduced, and a subsequent increase can be used as a measure of fracture healing. It can be compared to the wave velocity in the continuous structure (the unfractured contralateral bone) as the bone regains its structural integrity. Hence age and activity related changes in bone mechanical properties (i.e., osteoporosis) can also be detected using ultrasound techniques when compared to a normal population (Heaney *et al.*, 1989; Rubin *et al.*, 1987).

With uncomplicated healing, despite some early variation in the results for the control tibia, the velocity subsequently stabilizes to a constant value of about 2800 m $s^{-1}$. The results for fractured tibia approach those of the control tibia as healing progresses but by the time of clinical union (16 weeks) there is still a marked difference between the results for the control and fractured tibiae.

In the case of an undisplaced distal third tibial fracture, the ultrasound velocities in the control and fractured tibiae are coincident at the time of initial measurement and remain so throughout the course of treatment. These results suggest that the fracture is either incomplete or so closely aligned to minimally affect the transmission of the ultrasound across the fracture site. For the case of delayed healing of a commuted distal third fracture, the results for the fractured tibia show a steady, although slow, increase towards those of the control tibia. However, by 45 weeks post fracture the results are still markedly different and showed no signs of further convergence.

Low intensity pulsed ultrasound can show osteogenic effects relating to fracture repair (Maclman *et al.*, 1998; Mayr *et al.*, 2002; Leung *et al.*, 2004), treatment of non-union (Mayr *et al.*, 1999; Nolte *et al.*, 2001), or bone lengthening (Sato, Matsushita and Nakamura, 1999; Shimazaki *et al.*, 2000). The beneficial effects of low-intensity pulsed ultrasound has also been seen even in an *in vitro* study and in complex fracture healing (Leung *et al.*, 2004). The nonthermal effects of ultrasound have been shown in control trials to result in stimulation of tissue regeneration (Dyson, 1997; Young *et al.*, 1996), angiogenesis (Young and Dyson, 1990), healing of ulcers (Nussbaum, Biemann and Mustard, 1994), increased blood flow in chronically ischemic muscles (Fabrizio *et al.*, 1996), increased protein synthesis in fibroblasts (Doan *et al.*, 1999; Ramirez *et al.*, 1997) and tendon repair (Enwemeka, Rodriguez and Mendosa, 1990). Ultrasound effects in bone include the induction of bone formation *in vitro* (Reher *et al.*, 1997) and bone repair in animals (Azuma *et al.*, 2001; Nolte *et al.*, 2001; Takikawa *et al.*, 2001; Yang *et al.*, 1996). In an uncontrolled series of 22 cases of mandibular osteoradionecrosis, 20 of these healed in response to ultrasound (Harris, 1992). It may be inferred that the increase in resonant frequency with decreasing fracture gap is an indication of fracture healing (Singh *et al.*, 1990). The bone stiffness $K$, which is a measure of bone quality, is of greater clinical significance than an estimation of bone mineral quantity in fracture assessment and is estimated from the resonant frequency.

Ultrasound has also been reported to accelerate bone repair in various animal models. Corradi and Cozzolino (1953) showed that US treatment for 5 min $day^{-1}$ for 15 days at an intensity of 1.5 W $cm^{-2}$ accelerated healing compared to contralateral controls. One study showed that 200 mW $cm^{-2}$ enhances the healing of fibular fractures in rabbits (Klug, Franke and Knoch, 1986). A daily exposure of 5 min enhances fracture healing in rat fibula (Dyson and Brookes, 1983). A study on rabbits using intensities of 50 and 57 mW $cm^{-2}$ applied for 15 min daily (Duarte, 1983) suggest that US had stimulated fractured repair.

The results of a large study on fracture healing using rabbits has been reported by Pilla *et al.* (1990). In this study, a 30 mW $cm^{-2}$, 1.5 MHz pulsed US signal was applied for 20 min daily to a highly reproducible fibular osteotomy model. This study utilized biomechanical torsion testing to assess the effects of US. The results demonstrated a statistically significant acceleration of healing by a factor of 1.7. A power intensity dosimetry study in the same animal model (Pilla *et al.*, 1991) varied the US intensity from 1 to 45 mW $cm^{-2}$ for 20 min daily; the biomechanical data showed a statistically significant acceleration of healing at all doses except at 1 mW $cm^{-2}$. An independent study using the same power intensity (30 mW $cm^{-2}$) as used by Pilla *et al.* (1990) has been carried out using the standard rat femoral fracture model (Bonnarens and Einhorn, 1984) by Wang *et al.* (1994). The results showed a 22%increase in maximum torque versus the contralateral limb 21 days post-fracture. Signals with both 0.5 and 1.5 MHz carrier frequencies were effective. In another investigation, 50 mW $cm^{-2}$ was employed with resultant increases in maximum torque and torsional stiffness of 29% and 37%, respectively, at 21 days post-fracture (Yang *et al.*, 1996).

PEMF waveforms are typically employed for the repair of recalcitrant and fresh fractures, consisting of a 25 ms burst of 200 μs pulses. The peak induced electric field is 1–10 mV $cm^{-1}$ in a 2.5 cm radius target. The US signal consists of a 500 μs burst of 1.5 MHz sinusoidal waves repeated at 200 $s^{-1}$, inducing 1–10 mV due to the nonlinearities of the propagation of sound wave along the surfaces of the bone cells in the canaliculi (microstreaming). The SGP generates 0.1–1 mV $cm^{-1}$ peak amplitudes (for 1000 microstrain) at frequencies below 100 Hz and exponentially decays at higher frequencies (Pilla, 2002).

In bone repair, both osteogenesis and angiogenesis are essential. It has been shown that the therapeutic range of ultrasound stimulates bone formation (Reher *et al.*, 1997) and osteoblast proliferation (Daon *et al.*, 1999) and stimulates the synthesis of angiogenic factors, vascular endothelial growth factor (VEGF), basic fibroblast growth factor (FGF) and interleukin-8 (IL-8) (Reher *et al.*, 1999). This has been confirmed by animal studies (Hadjiargyrou *et al.*, 1998; Hantes *et al.*, 2004; Rubin *et al.*, 2001; Wang *et al.*, 1994; Yang *et al.*, 1996). Clinical trials (Kristiansen *et al.*, 1997; Leung *et al.*, 2004a, 2004b) have been carried out using ultrasonic techniques to accelerate the process of bone fracture healing. It is not clear, however, how the therapeutic range of ultrasound induces bone formation and angiogenesis. One possibility is that the ultrasound may act like mechanical stress, by which mechanical stimulation is transduced and leads to new bone formation. The latter may hold in the therapeutic range of interest. Prostaglandins (PGs) are necessary for the induction of bone formation and remodeling by mechanical stimuli (Chow and Chambers, 1994; Forwood, 1996; Meghji, 1992; Pead and Lanyon, 1989). In addition, nitric oxide (NO) production is required for mechanically induced bone formation (Fox, Chambers and Chow, 1996; Turner *et al.*, 1996). NO, like PGs, may also play an important role in bone remodeling because it also has potent affects on both osteoclast (Collin-Osdoby, Nickols and Osdoby, 1995; Kasten *et al.*, 1994; Macintyre *et al.*, 1991) and osteoblast (Riancho *et al.*, 1995a, Riancho *et al.*, 1995b) activity.

The results reported show that the therapeutic range of ultrasound can stimulate osteoblasts to produce NO and $PGE_2$. The synthesis of NO appears to be via the inducible form of NOS, which may explain the presence of NO at 18 and 24 h. The iNOS pathway is thought to be capable of generating much larger quantities of NO over a more prolonged period than the constitutive NOS (Moncada and Higgs, 1993; Palmer, Ferrige and Moncada, 1987). Of the NOS isoforms, eNOS has been the most widely reported in bone (Brandi *et al.*, 1995; Helfrich *et al.*, 1997; MacPherson, Noble and Ralston, 1999), but iNOS does not appear to be constitutively expressed in normal adult bone (Fox and Chow, 1998; Helfrich *et al.*, 1997) although it can be induced by proinflammatory cytokines and endotoxin (Armour *et al.*, 1999; Ralston *et al.*, 1995; Ralston *et al.*, 1994; VantHof and Ralston, 1997).

Bone adaptation to mechanical stimuli is dependent on the bone strain magnitude and the loading frequency (Fig. 1.10). Some stimulation of bone growth appears to occur with other wave patterns, including sine wave and the use of ultrasound (Li 1992; Bachner *et al.*, 1987; Pilla *et al.*, 1990; Heckman *et al.*, 1994). Low intensity ultrasound at a spatial averaged temporal averaged intensity of 30 mw/cm$^2$ may attenuate gradually in the soft tissue and posterior distal radius particularly in the cortical shell. Its averaged force is reported to be substantially less than the threshold for bone formation, and ineffective in stimulating remodeling (Turner *et al.*, 1994). Apart from the effects of low intensity pulsed ultrasound attenuation from the intact bone matrix, the bone in late adult life becomes less responsive to mechanical stimuli. This may be considered as another reason for the absence of a beneficial effect of low intensity pulsed ultrasound (Leung *et al.*, 2004). These finding seem to suggest that therapeutic effects are related through mechanical to electrical output. This in turn controls cellular function.

Low intensity pulsed ultrasound is suggested to enhance fracture healing by stimulating earlier synthesis of extracellular matrix protein, the aggrecan in cartilage, possibly altering chondrocyte maturation through the endochondral bone formation pathway (Yang *et al.*, 1996).

## *5.6.1 Ultrasonic Attenuation*

The attenuating properties of the oscalcis diminish with decreasing density (McCloskey *et al.*, 1990; McKelvie and Palmer, 1987), suggesting that density and porosity are inversely proportional. Hence the attenuation in the oscalcis decreases with increasing porosity. If, therefore, the attenuation coefficients of various porosities of trabecular bone display a similar peak to that given by the phantom, then the porosity of the oscalcis should be greater than 50%. Although several studies of the properties of the oscalcis have been undertaken, no direct information on its porosity is available. Density values are obtained by means of Archimedes' principle. However, these measurements can be unreliable if great care is not taken to exclude all air from the sample or if there is a degree of uncertainty about the nature of the pore filling material. For example, McKelvie and Palmer (1987) measured the density of samples of oscalcis to lie in the range 940–1170 kg m$^{-3}$, despite the density of marrow being $970 \pm 60$ kg m$^3$ and that of cortical bone being $1990 \pm 27$ kg m$^{-3}$ (ICRP, 1975). The existence of a peak in attenuation as the porosity is changed means that a lower than normal attenuation measurement at a particular skeletal site cannot be interpreted as an indication of osteoporotic bone without knowledge of the porosity of that bone. The nature of the ultrasonic attenuating mechanism in porous materials is the subject of debate. Two models have been suggested, one based purely on multiple scatter and the other taking care of viscoelastic mechanisms (Biot 1956, 1962). Biot theory treats attenuation as an effect of the viscous forces that are created as fluid is moved though the porous matrix.

In pursuance of the above, for the transmission of an ultrasound pulse across a fracture the amount of attenuation of a vibration pulse across the fracture site has also been used as a measure of fracture healing (Figures 5.17 and 5.18) (Pelker and Saha, 1985; Behari, Arya and Alex, 1991). In another set of investigations an impulse has been applied to the bone and the output from accelerometers mounted either on the skin (Nokes *et al.*, 1985), or directly on to the bone via hypodermic needles (Sonstegard and Matthews, 1976), at the proximal and distal of

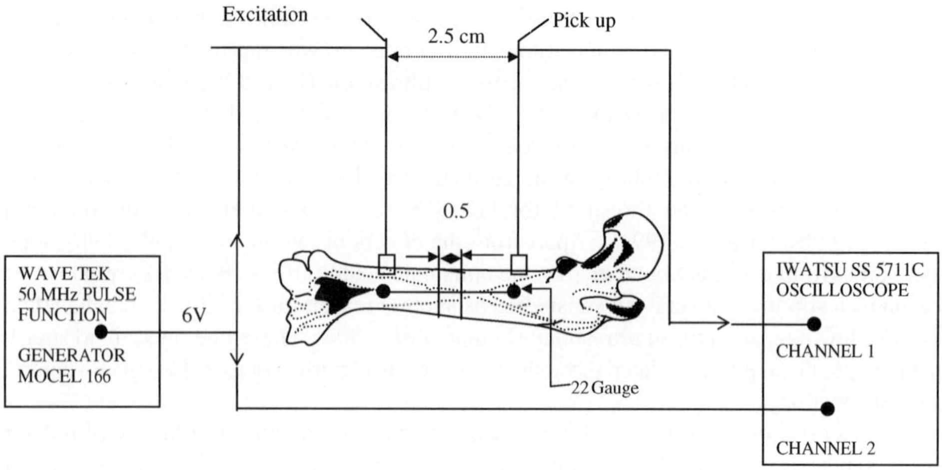


**Figure 5.17** Circuit diagram for measurement of ultrasonic attenuation in rat femur

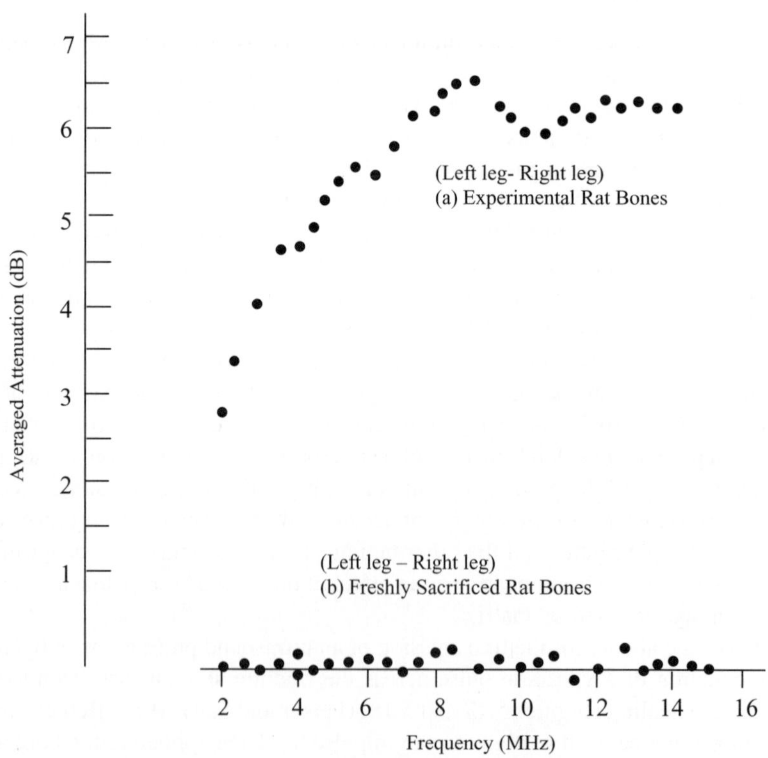


**Figure 5.18** Averaged difference in attenuation between a control and an experimental leg versus frequency in rat femur

positions the fracture site, has been measured. The output from the proximal and distal accelerometers was compared and the amount of attenuation of the signal across the fracture site was used as a measure of healing.

## *5.6.2 Measurements on Human Tibiae*

The transmitter (T) is pressed against the medial side of the tibia near the ankle and kept fixed at this position. The receiver (R) is then placed on the medial face at different known distances ($X$) from the transmitter (Figure 5.19a). For each position of the receiver, the traveling time of the US pulse is noted and a graph of distance between transducers against propagation time can be obtained, which is found to be linear. From these measurements the US propagation speed can be derived. By using the slope to calculate the velocity one excludes time delays due to soft tissues.

## *5.6.3 Measurements on Models*

To assess the influence of soft tissue thickness overlying the bone, a model consisting of a metal plate placed in a water tank (Figure 5.19b) can be used. The water height above the plate ($h$) can be changed, as can the distance ($X$) between the transducers.

The velocity of US in water is 1500 m s$^{-1}$, similar to that of soft tissue, while in metal it is 5570 m s$^{-1}$. The transmitter and receiver are kept a fixed distance apart, touching the water

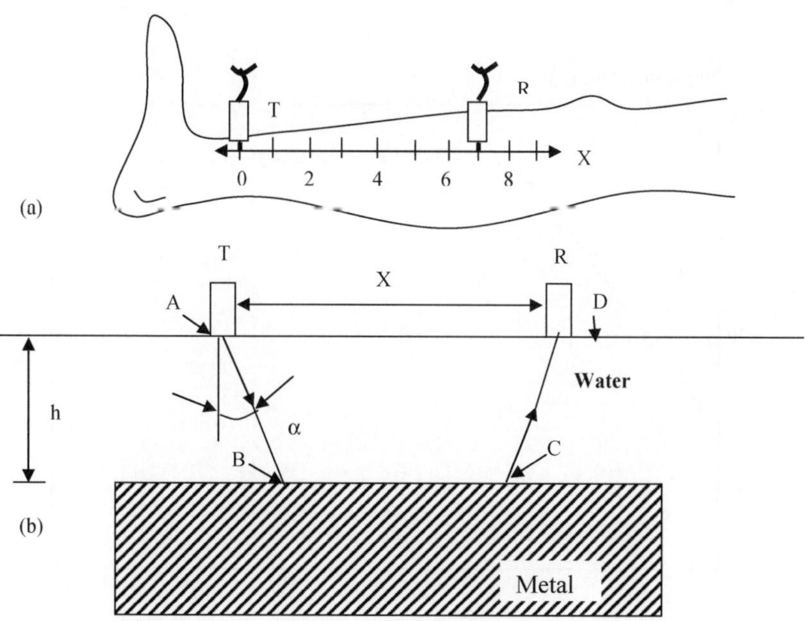


**Figure 5.19** Propagation and measurement of ultrasound parameters in a simulated phantom media (Van der Perre *et al.*, 1991)

surface. By changing the water height above the metal plate, it is intended to simulate the influence of soft tissue thickness on US transmission measurements. From this figure, a linear relationship between propagation time and water height is observed, until a certain water height, above which the propagation time appears to be constant, indicating that the wave is traveling only through water.

The behavior of the first part of the curve is linear. The propagation time ($t$) of the ultrasonic signal is the sum of the time needed to travel from A to B, from B to C and from C to D. This is:

$$t = 2\frac{h}{\cos(\alpha)}\frac{1}{v_{\mathrm{w}}} + \frac{X - 2.h.\tan(\alpha)}{v_{\mathrm{m}}} \tag{5.17}$$

where $v_{\mathrm{w}}$ and $v_{\mathrm{m}}$ are the US velocities in water and metal, respectively. This expression is minimal for $\alpha = \arcsin(v_{\mathrm{w}}/v_{\mathrm{m}})$.

Derivative of equation 5.17 fields

$$\frac{\mathrm{d}t}{\mathrm{d}h} = \frac{2}{\cos(\alpha)}.\frac{1}{v_{\mathrm{w}}} - \frac{2\tan(\alpha)}{v_{\mathrm{m}}} \tag{5.18}$$

For $v_{\mathrm{w}} = 1500$, $v_{\mathrm{m}} = 5570$ and $\alpha = 16.6°$, one obtains $\mathrm{d}h/\mathrm{d}t = 778\,\mathrm{m\,s^{-1}}$.

A linear regression on the first part of the curve of Figure 5.20 gives a slope $\mathrm{d}h/\mathrm{d}t$ of $754\,\mathrm{m\,s^{-1}}$, which is very close to these theoretical results.

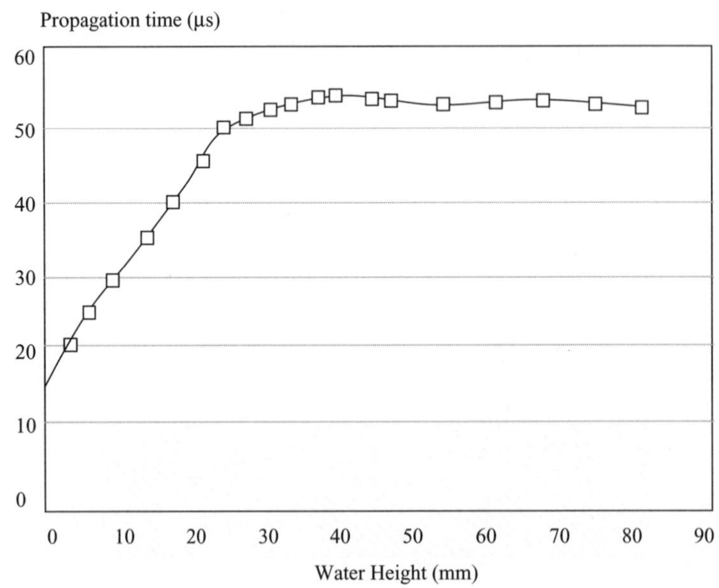


**Figure 5.20** Influence of water height on propagation time

To allow for transmission of the ultrasonic wave through the water (or soft tissue), the time elapsed is measured ($t$) and corrected. To obtain the time the wave would need to pass a distance in metal that is equal to the distance between the transducers:

$$t' = t - \frac{X}{v_m} = 2h\left(\frac{1}{\cos(\alpha)} \cdot \frac{1}{v_w} - \frac{\tan(\alpha)}{v_m}\right) \tag{5.19}$$

For *in vivo* measurements on bone, $v_{bone} = 3500\,m\,s^{-1}$. Bone mass measurement has shown a high degree of accuracy. It is well established that the measurement of bone mineral content (BMC) in the lumbar spine is highly reproducible in normal population, even if this diagnostic modality is greatly affected by the quantity of fat tissue and other physiological and pathological situations (Parrini, Gallazzi and Parrini, 1987). Forearm mineral density appears to be influenced by the site of measurement (distal or ultradistal), and since it measures the cortical bone compartment it does not provide a way to detect the early stages of bone deterioration (Mazess, 1983).

Notably, the heal BMC is possibly the most accurate measurement of bone mineral content (Wasnich *et al.*, 1987), because it is far less affected by obesity, it does not appear to be influenced by dominance of the site and it measures prevalently cancellous bone (more than 90% of the oscalcis is represented by trabecular bone). Furthermore, decreased mass is only one of several factors known to influence bone fragility.

Paradis and Kelley (1975) have suggested that the principal mechanism of control of mineral deposition in fracture healing is blood flow. Sympathectomy enhances blood flow to bone (Davis, Jones and Hungerford, 1987) and enhances the healing process, though this demands further confirmation. Some data suggest that vasoconstrictive responses that are induced by diverse mechanisms in the vascular bed of bone may be blocked by selective blocking agents. Pharmacological manipulations of the blood flow to the vascular bed of bone offer the potential to enhance the local supply of nutrients and oxygen and would probably promote improved fracture healing (Brinker *et al.*, 1990). An increase in blood flow to bone by prevention of local constriction may also be useful in the treatment of patients who have an infection of bone. Selective vasodilation of the vascular bed of bone may have important clinical applications.

## 5.7 SNR Analysis for EMF, US and SGP Signals

Clearly, from the foregoing discussions, both EMF and US signals have a clinically significant effect on bone repair, though the extent of their effect remains to be determined. The intensity of the signal (EMF and US) employed does not cause a physiologically significant temperature rise in the target cell/tissue area. The same is true for SGP signals. Bioeffects from weak EMF signals are caused by the time varying electric field [$E(t)$], induced from either an applied time varying magnetic field, $B(t)$, in the case of the therapeutic devices, or from streaming potentials in the case of mechanical signals. Both modalities are now a common part of the treatment for delayed union, non-union and fresh fractures. It is also clear that controlled weight bearing can also enhance bone repair. Such studies consider the dosimetry of both electromagnetic and mechanical signal modalities and suggest a possible unifying mechanism for the bioeffects from EMF, US and SGP signals. The $E(t)$ field is directly induced with EMF devices and is also indirectly induced via the streaming potentials associated with the mechanical movement of ionic fluids within bone canaliculi or directly past cell surfaces from US and SGP signals. It is shown here that both electrically and mechanically induced $E(t)$ have common waveform characteristics at the treatment site and thus can deliver similar doses of electrical stimulation. These are induced either by a time varying magnetic field, $B(t)$, in the case of therapeutic devices, or from streaming potentials in the case of mechanical signals. Most EMF clinical devices induce a peak $E$ of 1–10 mV cm$^{-1}$ at the treatment site (Otter, Goheen and Williams, 1988; Pilla, 1993), except for the combined magnetic field device, which induces a much lower $E$ and are designed to have a magnetic field component (McLeod and Liboff, 1987). It is then proposed that the time varying endogenous electric field, $E(t)$, from a time varying change in the mechanical environment of healing or remodeling bone, can act as a dose-dependent growth stimulus (Pilla, 2002). Ultrasound SGP signals induce similar electrical field amplitudes from the movement of charged fluids in contact with cell and molecular surfaces but with substantially different waveforms. An aspect of this is whether US and SGP signals induce similar electric field amplitudes from the movement of charged fluids in contact with cell and molecular surfaces but with substantially different waveforms or if at some stage they can have the equivalent bioeffects. An associated aspect is whether the signals can be detected at the molecular/cellular/tissue target in the presence of thermal noise, that is, the signal to thermal noise ratio (SNR) is important. The mechanistic pathway most often considered for EMF bioeffects involves the cell membrane, across which a time varying magnetic field can induce a transient voltage change.

A bioeffect from EMF, US or SGP signals requires that the induced $E$ has frequency components corresponding to the bandpass of the target and that the resulting transmembrane voltage ($V_M$) is detectable above thermal and other baseline voltage noise. To assess this, each of these waveforms has been applied to the cell array tissue model using $Ca^{2+}$/CaM binding at a bone cell as the target pathway. The effect of array length on the induced electric field from a typical SGP signal is around (1–10 Hz). Here it is seen that $E$ increases significantly for longer array lengths, but over a limited frequency range, corresponding to the bandpass imposed by the kinetic parameters of the $Ca^{2+}$ binding process. The mean transmembrane thermal noise is in the range 1–10 μV, depending the on membrane impedance and array lengths (Pilla, Nasser and Kaufman, 1994). The SNR was clearly satisfied only at the 100 μm and 1 mm array lengths. These lengths are typical for cells in gap junction contact for normal remodeling bone. Interestingly, for the 1 mm array, $E(w)$ exhibits a peak in the 10 Hz range. Remarkably, this is within the ideal frequency range for pulse mechanical modulation of disuse osteopenia (Oin, Rubin and McLeod, 1998).

The question to be settled is whether $E(t)$ from US and EMF signals can be similar enough to produce equivalent bioeffects. The EMF bioeffects involve the cell membrane, across which a time varying magnetic field can induce a transient voltage change (Foster and Schwan, 1986; Pilla, 1972). The argument then is that the induced transmembrane voltage change is buried in thermal and other voltage noise across the membrane, rendering the applied signal relatively insignificant. The simple model generally employed to justify this conclusion assumes that the EMF target is a spherical cell of approximately 10 μm radius. In addition, the dielectric characteristics of the cell membrane are limited to a simple membrane capacitance. These calculations often lead to an unfavorable SNR for many

EMF signals that have, however, demonstrated biological effects (Adair, 1991). However, the spherical cell model for SNR is certainly oversimplified and does not represent the geometric complexity of EMF cellular and tissue targets. All organized tissue is developed and maintained by an ensemble of complex geometry cells that probably have coordinated activity (Caveney, 1985). The most prevalent cell shape in living system tissue is elliptical and flattened, with processes extending in at least two directions. For example, human fibroblasts can typically exceed 100 μm when attached to a substrate (connective tissue). In addition, nerve axons can be tens of centimeters long. Furthermore, most cells are anisotropic in shape and function and can detect significantly more of the applied field than an isotropic shape can (Pilla *et al.*, 1992). In this connection, the role of cooperative organization in the EMF sensitivity of biological systems has been quantitatively considered (Adey, 1988). Gap junctions provide pathways for ionic and molecular intercellular communication (Lowenstein, 1981) and are present in all tissues, including bone (Doty, 1981). Gap junctions provide ionic coupling and metabolic cooperation, without which disorders in growth control and tissue repair, as well as neoplastic transformations, could occur (Sheridan and Atkinson, 1985).

In this connection, the functional modification of gap junctions by modulated MW fields as well as EMF signals has been reported (Fletcher *et al.*, 1986; Hu *et al.*, 2001). It has been shown that the potential effectiveness of any given EMF waveform on molecular, cellular or tissue targets may be first assessed by evaluation of the SNR. It is shown that, for connected cells in real tissues, sensitivity to EMF or mechanical signals is determined by both the geometry and the electrical characteristics of the target and can be several orders of magnitude higher than that for single isolated (e.g., circulating) cells or molecules. A distributed parameter can be an accurate representation of the electrical behavior of the cell array typical in repairing and remodeling bone tissue (Pilla, Nasser and Kaufman, 1993, 1994).

### *5.7.1 Ununited Fractures*

Occasionally, bone fractures fail to heal in a normal period of time and require some form of intervention to complete the process of healing. This condition is commonly referred to as non-union fracture. Many factors, such as smoking, diabetes and age, may contribute to the development of a non-union. However, in many cases the cause is unknown. The most common method of treating a non-union in bone is graft surgery. In this procedure, bone is taken from another location in the patients body and surgically implanted in the non-union site. Metal rods or plates and bone screws also may be attached to the broken bone to secure the ends in place while it heals. However, there is an element of serious health risk associated with bone graft surgery.

About 5% of long bone fractures fail to follow the normal repair pattern and the healing process stalls in the intermediate phase. Fibrocartilage may persist for months or years because it is never calcified. Various factors are responsible for this outcome, such as severe devascularization at the time of initial injury, uncontrolled motion, poor reduction of the fragments and infection and so on. Intraarticular fractures of the scaphoid and high femoral neck are typical sites for non-union, largely because of their tenuous blood supply. The end result is persistence of soft tissues, motion, pain and tenderness at the old fracture site along with disability; operative methods that are successful in establishing bony union result in calcification of this persistent soft tissue (fibrocartilage) in the gap (Adey, 1984).

The main application of PEMFs has been in orthopedics in healing non-union bone fractures, with an 80% success rate for the tibia, as reported by Bassett (1982a, 1982b, 1989) and Brighton *et al.* (1986). Using a capacitively coupled RF field such as used for bone fracture healing a high success rate in animals (rats) and humans for healing of bone fractures has also been reported (Behari, Arya and Alex, 1991). Yang, Chang and Liu (1994) have reported that the following settings can heal fracture non-union: frequency 7.5 Hz, pulse width of 300 μs and induced electric field intensity of 1–2 mV $cm^{-1}$. Borsalino *et al.* (1988) have reported results on the treatment of delayed and non-union of fractures using PEMFs. Their success when PEMF therapy was used in conjunction with surgery has been 100%, but for fractures treated with PEMF alone it was 86%.

Ultrasound technique can be easily used and is not time consuming. It compares the results of the fractured and the non-fractured bone. This has the advantage that the results are self-contained for a patient, but the disadvantage is that, if the results are not coincident on clinical healing, treatment may be prolonged for longer than necessary. The velocity of ultrasound depends on the mechanical properties, $E$ and $\rho$ of the bone, rather than the stiffness of the fracture. The results for the control and fractured bone are unlikely to be coincident until significant remodeling of the bone has taken place. When the results are coincident, it would be expected that the properties of the remodeling fracture zone are similar to those of the unfractured bone. This has the advantage that it would enable the later stages of healing, when treatment has usually been discontinued. However, should asymmetrical bridging of the fracture have taken place, then the ultrasound results would be the same as if complete bridging had occurred.

The positive effect of low intensity pulsed ultrasound on simple fractures, like distal radial fractures and diaphyseal fractures of the tibia, have been reported (Busse *et al.*, 2002; Duarte, 1983; Heckman *et al.*, 1994; Nolte *et al.*, 2001). The effects of low intensity pulsed ultrasound on more complicated fractures have also been reported. The physical property of low intensity ultrasound has limited penetration characteristics, the effect of which may be best shown on the stimulation of more superficial structures. However, tibial fractures were shown to be healed with large amount of callus that developed on the anterolateral surface, where the ultrasound probe is applied and on the posterolateral surface. The large amount of callus shows that the stimulatory effect of ultrasound is not localized but is propagated in the biologic tissues in the fracture site (Leung *et al.*, 2004). Leung *et al.* (2004) have investigated the effect of low intensity pulsed ultrasound on the human periosteal cells (proliferation and differentiation) and the biology of fracture healing [expression of alkaline phosphatase (ALP), osteocalcin, vascular endothelial growth factor and calcium deposition]. These authors used a low intensity ultrasound signal of a 200 μs burst of a sine wave, repeating at 1.0 kHz, with 30 mW $cm^{-2}$ spatial averaged, temporal averaged intensity. $5 \times 10^4$ periosteal cells per well were seeded into a six-well plate for 2 days. The ultrasound did not affect the total number of viable cells. However, it did stimulate cell proliferation at the early phase of cell culture. The activity of ALP increased significantly in the culture at day 4. A similar effect was seen for osteocalcin secretion and the response is dose dependent. Vascular endothelial growth factor secretion increased in day 2 and day 4 cultures, with a dose-dependent effect. The formation of calcium nodules found to be significantly higher with ultrasound treatment. Leung *et al.* (2004) have concluded that the low intensity pulsed ultrasound stimulated periosteal cell proliferation and differentiation toward oseogenic lineage. It has been suggested that ultrasound treatment be started from the beginning of fracture healing.

However, Emami, Petren-Mallmin and Larsson (1999) indicated that low intensity ultrasound had no effect on tibial fractures treated with intramedullary nailing.

Although differences may exist between the healing of the gap osteotomy and typical fractures, the results of Chen *et al.* (2003) support the potential beneficial effects of PTH in bone regeneration during fracture healing. Andreassen, Ejersted and Oxlund (1999) used 60 or 200 $\mu g\,kg^{-1}\,day^{-1}$ PTH (1–34) injections to promote callus formation and, thereby, increased the mechanical strength of tibia fractures in rats. They found that the high PTH dose (200 $\mu g\,kg^{-1}\,day^{-1}$) led to more beneficial effects on fracture healing than the low dose (60 $\mu g\,kg^{-1}\,day^{-1}$).

PTH delivered through daily injection and gene therapy resulted in a significant anabolic effect in femoral osteotomy sites despite the lack of significant results when applied individually (Chen *et al.*, 2003). The combination treatment group, when compared with other groups, showed significantly higher mineral bone density, bone mineral content and bone area using DEXA analysis, and a trend for greater bone area in microradiographic analysis. These authors have reported that an additive anabolic effect exists between intermittent PTH subcutaneous injection and local PTH gene therapy. This is at pointer toward the concept of a combination of local and systematic treatments.

## 5.8 Low Energy He-Ne Laser Irradiation and Bone Repair

Figure 5.21 shows the general features of laser–tissue interaction. A low energy laser beam is reported to control bone fracture healing (Figure 5.22), though this has yet to find a clinical acceptance. Low energy laser irradiation has been found to modulate various processes in different biological systems (Gallent, Bolognani and Ussia, 1992; Karu, 1989). Laser energy is photochemical in nature and the energy is absorbed in intracellular chromophores and converted into metabolic energy, most likely by involving the respiratory (cytochrome) chain

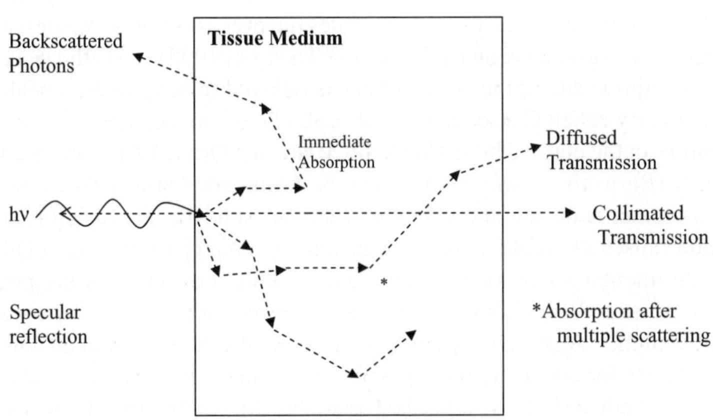


**Figure 5.21** General features of laser–tissue interaction

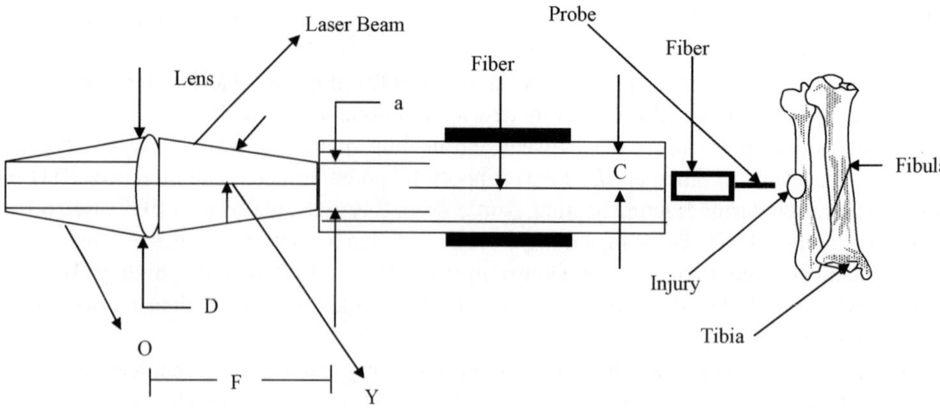


| | |
|---|---|
| Wavelength | 632 nm |
| Power | 5.3 mW (931 J/cm$^2$) |
| Beam Diameter | 1.9 mm |
| Duration | 120-180 Seconds/day |

**Figure 5.22** Treatment of bone injury by low level laser irradiation

(Karu, 1989). The sequential events of the repair process of the cortical bone of the tibia appear to be an elevation of alkaline phosphatase (ALP) activity in the injured site and $^{45}$Ca incorporation, which is an indication of mineralization. Calcium accumulation at the site of injury reaches a maximum after the peak of ALP and $^{45}$Ca incorporation in control rats (Yaakobi, Maltz and Oron, 1996). This phenomena has also been reported in other studies dealing with bone growth in rats (Reddi and Sullivan, 1980). In the study carried out by Yaakobi, Maltz and Oron (1996) peak ALP activity in the injured site was found on day six of post injury as compared with those of other workers (Barushika *et al.*, 1995). This finding supports the notion that the endosteal bone beneath it, as if a local response to injury following ablation of the medullary canal, precedes the process of bone repair in the cortical bone, as revealed by histological analysis (Barushika *et al.*, 1995). Thus the peak activity of osteoblast at the injured site of the cortical bone is delayed relative to their peak activity in the injured medullary canal. Consequently, the peak of calcium accumulation was found on day 13 post injury in the cortical bone (Yaakobi, Maltz and Oron, 1996), while it was day 11 in another study (Barushika *et al.*, 1995). Yaakobi, Maltz and Oron (1996) concluded that, by direct measurements of rate and extent of calculation at the site of injury, low energy laser irradiation causes a twofold increase in the rate of bone repair in the cortical part of the tibia in an experimental model of rats. This phenomenon corroborates the presence of a significantly higher number (2.2-fold) of active osteoblasts at the injured site – as reflected by significantly higher ALP activity. Interestingly, in the laser irradiated rats, the peak activity of ALP lasts for about day 9–15 post injury, whereas in the control rats the peak is observed on the ninth day. It is suggested that the differentiation of the osteoblast to osteocyte is delayed by the laser irradiation and it remains active for longer time interval in the injured zone. Kusakari, Orikasa and Tani (1992) have reported stimulation of DNA, protein synthesis and ALP activity in UMR 106 osteoblast like cells in culture by low energy laser irradiation at 780 nm. It can be deduced, therefore, that simulation of proliferation and/or bone matrix secretion by the laser irradiated osteoblast *in vivo* may also have caused the elevated rate of bone repair in the rat. There may be a possibility that the laser irradiation also promoted the new formation of blood vessels at the site of injury, essential for bone repair. This phenomenon also occurs during skeletal muscle regeneration (Bibikova, Belkin and Oron, 1994). Some other studies have also indicated a faster healing of bone in induced fractures in mice (Trelles *et al.*, 1990).

Various biostimulatory effects of low energy laser irradiation have been reported that involve wound healing (Mesler, Mesler and Mesler, 1985; Mesler, Nagylueskav Tisza and Mesler, 1978), fibroblast (Boulton and Marshal, 1986; Van Breugel and Bar, 1992) and collagen synthesis (Kana *et al.*, 1981).

The stimulatory mechanism of low energy laser irradiation on bone is not fully explored and understood. Osteoblast like cells may simultaneously secrete osteoblastic differentiation factor *in vitro* (Harris *et al.*, 1994) and local regulation of bone cell function is regulated by cytokine growth factors and prostaglandins (Marks, 1988). Exogenous application of factors can either increase or inhibit the number of nodules formed in cultures in a reproducible manner. These factors are bone morphogenetic proteins (Hughes *et al.*, 1995) and prostaglandins (Flanagan and Chambers, 1992; Nagata *et al.*, 1994).These investigators have pointed out that the increase in bone nodules is the result of the action of the factor on an earlier progenitor cell. This action may be comparable to the effect of laser irradiation (Ozawa *et al.*, 1998). Ozawa *et al.* (1998) have used osteoblast like cells isolated from fetal rat of calvarial, which were irradiated once with a low energy Ga-Al-As laser (830 nm 500 mW) at various cell culture stages (day 1–16). Laser irradiation at the early stages of culture significantly stimulated cellular proliferation, ALP activity and osteocalcin gene expression. Furthermore, laser irradiation at earlier stages of culture significantly stimulated a greater number (1.7-fold) and larger area (3.4-fold) of bone nodules that had developed in the culture dish on day 21. However, these effects could not be found by irradiation later. This suggests that the role of laser irradiation is twofold: one is stimulation of cellular proliferation, especially proliferation of nodule forming cells of osteoblast lineage, and the other is stimulation of cellular differentiation, especially to committed precursors. This results in an increase in the number of more differentiated osteoblastic cells and an increase in bone formation. Both bone formation stimulating roles may be exhibited by laser irradiation of immature cells only.

## 5.9 Electrostimulation of Osteoporosis

EMF stimulation of hard tissue healing and regeneration is now a well-recognized therapeutic area of medicine. Most of the data available are on animals though some success on humans is reported. This is because osteoporosis is in general a systematic disorder and hence the clinical trials and realization of EMF exposure is not easy. Nonetheless pulsed electromagnetic field exposure is an attractive proposition. Few studies have been performed on disuse osteoporosis in animal models. There are two standard procedures for inducing osteoporosis: (i) neuroectomy and (ii) ovariectomy. Experiments show that these exhibit osteoporosis one month after treatment. Tabrah *et al.* (1990) experimented with the prevention of systematic osteoporosis induced by bilateral ovariectomy with PEMF in studies using Helmholtz coils to generate uniform extremely low frequency electromagnetic fields on osteopenia. These experiments demonstrated that extremely low frequency single pulse electromagnetic fields significantly suppressed the trabecular bone loss and restored the trabecular bone structure in bilateral ovariectomized rats.

PEMFs have been used to deal with different kinds of osteoporosis in both animal and clinical experiments. Experimental results show that PEMF exposure stopped bone loss (Bassett *et al.*, 1979). Another kind of osteoporosis, which resulted from both bilateral ovariectomy and right sciatic neurectomy, has also been used to test the effects of PEMF. These results suggest that PEMF stimulation exerts a preventive effect against bone loss on osteoporotic hind-legs (Mishima, 1988). Other studies carried out by Cruess, Kan and Bassett (1983), McLeod and Rubin 1990, 1992) and Simske, Wachtel and Luttges (1991) have confirmed inhibition of bone loss by PEMF. On the basis of these studies, PEMF effects on osteoporosis in humans were carried out. Single photon absorptiometry of the distal and midradius was used to quantify changes in bone mineral density in a group of postmenopausal osteoporosis. The results show that the bone mineral density of the treated radii increased significantly as compared to controls (Tabrah *et al.*, 1990).

The ovariectomized (OVX) of rat has been widely used as a model of postmenopausal osteoporosis and has been validated as a clinically relevant model of human postmenopausal bone loss (Turner, Colvard and Spelsberg, 1990; Kalu, 1991; Wimalawansa, 1993; Wronski *et al.*, 1989, 1985). The similarity between the two include increased bone turnover, unequal bone resorption and formation, a greater loss of trabecular than cortical bone and a rapid initial phase of bone loss followed by a slower phase (Kalu, 1991).There is an age-related reduction of bone mass and strength in humans as well in rats (Martin, 1993; Turner, 1993). Miller *et al.* (1991) have reported that ovariectomy decreases the total calcium content and calcium percentage of ash in rat femur without affecting total ash weight of the femur. This suggests that the mechanical properties of bone could be partly influenced by the changes in chemical composition of bone matrix. Bone geometry is altered after ovariectomy as a result of increased periosteal apposition and endocortical resorption in the long bones. The marrow cavity expands in the femoral shaft in the OVX rats. Cortical bone thickness and trans sectional area did not change, however, indicating that both endocortical resorption and periosteal bone formation increased in parallel in the femur. This increase in the outer diameter and following increase in the polar moment of inertia is probably the most important factor in the increased maximal torque of the humerus in OVX rats (Peng *et al.*, 1997). Estrogen deficiency in rats leads to a significant and permanent loss of trabecular bone.

Although the model is useful for experimental design in studies of postmenopausal bone loss, its relevance in assessing postmenopausal women with osteoporosis needs further investigations. When OVX rats were fed a low calcium diet (LCD) the decrease in calcium absorption due to ovariectomy became significant (Kalu, 1991) and bone loss was significantly enhanced (Jiang *et al.*, 1997; Nordin, Horstnan and Marshall, 1979). One study using a OVX rat model has suggested delayed fracture healing and impaired biomechanical strength (Miwa *et al.*, 1999; Walsh *et al.*, 1997). The study demonstrated the influence of bone loss on the early phase of fracture healing in a rat osteoporotic model induced by ovariectomy and LCD (Jiang *et al.*, 1997; Wronski *et al.*, 1985).

Li (1992) has tested a Capacitively Coupled Electric Field (CCEF) signal (60 Hz symmetrical sine wave) for osteoporotic treatment. After treatment with CCEF the animals were tested for restoration of bone mass and strength. The author found a statistically significant enhancement of wet weight, dry weight, ashed weight, ultimate strength and cortical area. In a neurectomy model for localized disuse osteoporosis, single PEMFs were found to prevent the loss of tibia trabecular bone mass and mineral strength by evaluation of histomorphometry and mechanical testing system methods (Bassett *et al.*, 1979). Electrical or mechanical stimulation can be equally effective in maintaining or improving bone mass (Mcleod and Rubin, 1992). Cruess, Kan and Bassett (1983) used low energy time varying magnetic fields inducing specifically configured electrical currents in rats subjected to a model of disuse osteoporosis. In another study carried out by Rubin, McLeod and Lanyon (1989) functionally isolated ulna of an adult turkey was developed as a localized osteoporosis model after eight weeks of disuse by proximal and distal epiphyseal osteomis. The results revealed that short daily periods of PEMF stimulation (induced electric power between 0.01 and $0.04\,T^2\,s^{-1}$, an effective intensity window) had the ability to inhibit bone loss. Some other reports also confirm positive reports about the effectiveness of capacitively coupled fields on bone density (Billota *et al.*, 1994). Experimental results suggest that PEMF stimulation exerted a preventive effect against bone loss of osteoporotic hind legs (Mishima, 1988). Studies carried out by other investigators (Cruess, Kan and Bassett, 1983; McLeod and Rubin, 1990, 1992; Simske, Wachtel and Luttges, 1991) found the same experimental results of bone loss inhibition by PEMF.

Brighton *et al.* (1985a, 1985b) and Brighton, Nichols and Arangio, (1985) have determined the dose–response relationship of a low voltage, 60 kHz CCEF signal on an established osteoporosis in a rat tibial model with sciatic neuroctomy. In the first part of the study, rats subjected to different intensities of CCEF [0.25, 0.50, 1.0, 2.5, and 5.0 V (p-p)] for 12 days beginning on the 28th day after sciatic neuroctomy all showed mean losses of tibial ash weight that were significantly less than those of the controls. The rats that had a 0.5-volt p-p signal showed the least mean loss of tibial ash weight (only 6%). In the second part of the study, the duty cycles of a sine wave, 60 kHz, 0.5-V p-p signals, were varied (12.5, 50 and 100% on) and the wet, dry and ash weights were determined and compared with those of unstimulated osteoporotic controls. Only the 100% duty cycle was effective in reversing the loss of bone mass in the neurectomized tibiae.

Chang and Chang (2003) have demonstrated that reduction of trabecular bone mass, destruction of trabecular bone structure, and an increase of serum $PGE_2$ concentration resulting from estrogen deficiency in ovariectomized female rats were inhibited by exposure to PEMF stimulation. These workers used a stimulating waveform having repetitive single pulses with a pulse duration of 0.3 ms and frequency of 7.5 Hz. The intensity of magnetic field was 4–8 G. Since this PEMF stimulation had no effect on the bone formation rate, it seems reasonable to infer that the prevention effect of this PEMF stimulation on trabecular bone loss might be due to the suppression of the bone resorption rate.

Using Helmholtz coils and self-developed PEMF stimulations to generate uniform time varying electromagnetic fields, the effects of extremely low frequency electromagnetic fields on osteopenia have been investigated in bilateral OVX rats. Some investigators (Bassett *et al.*,

1979; Cruess, Kan and Bassett, 1983) have found that the a particular field (frequency 72 Hz, pulse width 380 μs and induced electric field intensity of 1.5 mV $cm^{-1}$) was as effective as pulse bursts in preventing disuse osteoporosis in rats. They applied fields by whole body coils for 24 h a day. These experiments demonstrated that extremely low intensity, low frequency, single pulse electromagnetic fields significantly suppressed the trabecular bone loss and restored the trabecular bone structure in bilateral OVX rats (Tabrah *et al.*, 1990). These investigators also examined bone mineral density changes of the radii of osteoporosis prone women during 12 weeks of daily PEMF exposure and for the subsequent months follow up. Increased bone density in the PEMF treated limb was also reported during treatment, with a return to the contralateral control values after treatment stopped.

Mishima (1988) found that pulse burst waves at a positive amplitude of 25 mV, burst width of 4.2 ms, pulse width of 230 μs and repetition rate of 12 Hz are effective in preventing osteoporosis caused by bilateral ovariectomy and right sciatic neuroectomy in rats. Pulse bursts with a pulse width of 30 ms, repetitive rate of 1.5 Hz and positive amplitude of 2.5 mV could inhibit bone loss in disuse osteoporosis.

A repetition rate of 1.5 Hz, with each burst containing 120 repetitions of an asymmetrical pulse, was effective in preventing bone loss by regulating cortical bone remodeling. Simske, Wachtel and Luttges (1991) found that twin pulses with a pulse width of 25 μs, twin pulse separation of 875 μs, and pulses in the pairs that were separated by 75 μs were capable of strengthening the biomechanical properties of tibiae of disuse osteoporosis caused by tail suspension in mice. However, Takayama *et al.* (1990) found that PEMF (square wave with a frequency of 15 Hz and a peak magnetic field strength of 15 G) did not show any protective effects against bone loss in estrogen deficient osteoporosis in rats. $PGE_2$ concentrations between $10^{-5}$ and $10^{-7}$ M were effective in promoting both bone formation and resorption. However, experiments performed by Mano *et al.* (2000) found that $PGE_2$ at a concentration of $10^{-6}$ or $10^{-7}$ M directly inhibits bone-resorbing activity of functionally mature osteoclasts. These results show that $PGE_2$ exerts diverse influences on bone remodeling under various concentrations and experimental models.

Jayanand and Behari (2005) induced osteoporosis in female Wistar female rats by bilateral ovariectomy. Pulsed electric stimulation of a 16 Hz modulated square wave (carrier frequency 14 MHz) and 10 V (p-p) was applied in one leg, while the other was treated as control (sham exposed). The treatment was applied for 2 h per day for 60 days. In a mineralogical analysis the bone mineral content, bone mineral density, calcium and phosphorous content were analyzed. These parameters showed a significant increase in the exposed group as compared to OVX group but differed insignificantly from the untreated (control) group (Table 5.1). Further to this, DNA single-strand break analysis performed by comet assay showed no genotoxic effect of exposure in bone marrow cells. Similar results were reported earlier by using sciatic neuroectomy (Jayanand *et al.*, 2003). The above-mentioned electric field is capacitively coupled to a portion of the long bone. Ovariectomy is associated with similar histologic changes in rat bone as in human bone, and it may thus provide beneficial information related to human post menopausal bone loss (Turner, Colvard and Spelsberg, 1990; Kalu, 1991).

Another way of looking at the management of osteoporosis by electromagnetic fields is to consider the role of osteoclasts. For instance, experiments in which cultured bone marrow cells were exposed to 60 Hz electric fields at 9.6 μV $cm^{-1}$ over 8 days resulted in the suppression of osteoclast like cells (Rubin *et al.*, 1996).

**Table 5.1** Effect of ovariectomy and pulsed electric field treatment on bone mineral content (BMC), bone mineral density (BMD), total calcium, total phosphorous and total carbon content in femur and tibia bones of rats (Jayanand *et al.*, 2003)

| | Control rat bones | | OVX rat bone | | OVX + PEMF rat bones | |
|---|---|---|---|---|---|---|
| | Femur | Tibia | Femur | Tibia | Femur | Tibia |
| BMC (mg) | 418 ± 27.71 | 284.15 ± 22.07 | 313 ± 54.77 | 220.77 ± 33.79 | 388.19 ± 70.33 | 273.71 ± 16.93 |
| BMD ($mg\,cm^{-1}$) | 1016.35 ± 148.28 | 751.41 ± 133.72 | 738.24 ± 133.72 | 540.29 ± 83.62 | 918.59 ± 180.68 | 714.29 ± 80.53 |
| Total Ca (mg) | 128.44 ± 18.51 | 86.69 ± 2.86 | 106.13 ± 4.59 | 69.21 ± 7.65 | 115.59 ± 12.84 | 84.57 ± 3.26 |
| Total P (mg) | 110.25 ± 5.16 | 72.95 ± 5.67 | 82.76 ± 3.13 | 48.62 ± 7.93 | 98.91 ± 5.42 | 70.07 ± 3.80 |
| Total C (%) | 22.40 ± 2.35 | 19.31 ± 1.63 | 19.02 ± 1.28 | 15.00 ± 1.82 | 20.51 ± 1.54 | 17.48 ± 2.86 |

## 5.10 Other Techniques: Use of Nanoparticles

Various barriers exist to the use of pharmaceutical agents to stimulate new bone formation. First, the agents can cause non-specific bone formation in areas not desirable. This is because these agents are often delivered in non-specific ways (such as through the mouth, directly into the blood stream, etc.). Second, if delivered locally to tissue around the area of low bone density, they may rapidly diffuse in the surrounding tissues, which is likely to limit their potential to promote prolonged bone formation in targeted areas of interest (weak osteoporotic bone). It is because of these limitations that even the best strategies to sufficiently increase bone mass (although, to date, in the experimental stage) require at least one year to see any change; a time period too long for the elderly. In addition to these the side effects of such pharmaceutical agents are always a target of close observation and often of suspicion.

For these reasons, nanotechnology (or the design of materials with $10^{-9}$ m dimensions) has been deployed to develop novel drug-carrying systems that will specifically attach to osteoporotic (not healthy) bone. Nanoparticle induction is essentially a procedure for affecting small volumes and small microstructural features. The method can be used to examine variations in the individual lamellar properties within osteons, as a function of distance from osteonal center. Characterization of lamellar bone properties by nano-indentation methods would thus aid in further theoretical developments of the mechanisms of fracture in bone. Moreover, some of these novel drug carrying systems will then distribute pharmaceutical agents locally to quickly increase bone mass. Efforts are therefore underway to obtain prolonged release of bioactive agents to efficiently regenerate enough bone for the patient to return to a normal active lifestyle. Specifically, inorganic biodegradable nanomaterials (including ceramics like hydroxyapatite) will be functionalized with bioactive chemicals such as bone morphogenetic protein-2 (BMP-2) that bind to bone of low mass. Such bioactive groups can be placed on the outer surface of nanoparticle systems using various techniques (such as covalent attachment). After bonding specifically to osteoporotic bone, excluding healthy bone, nanoparticle systems will deliver bioactive compounds to locally increase bone mass. Lastly, the outer coating of the embedded nanoparticle systems will be created to have different biodegradation rates for the release of bone building agents over various time spans. This will allow for not only quick bone formation but also long-term sustained bone regeneration. The method has inherent shortcomings in distinguishing osteoporotic bone from the healthy bone and hence the growth rate process will remain uncertain.

HA (hydroxy Apatite) is chemically very similar to the mineral component of bones and hard tissues in mammals. Nanosized HA consists of needle-like crystals approximately 5–20 nm wide by 60 nm long. They have properties such as surface grain size, pore size and can control protein interaction and related biological functions. Utilizing nanotechnology, calcium and phosphorous are programmed at the molecular level and then assembled to produce the desired structural and functional properties. Crystalline material is strong and osteoinductive. It is one of few materials that are classified as biocompatible and bioactive with respect to bone cells and tissues. Osteoblastic cells on HA exhibited unique attachment and subsequent behavior *in vitro*. Divalent cations ($Mg^{2+}$) are known to be active in cell adhesion mechanisms (Fitton, 1995; Hynes, 1992). It will, thus support bone in growth and osseointegration when used in orthopedic, dental and maxillofacial applications. This is because bone itself is composed of hydroxyapatite and other calcium phosphates. When loaded with bioactive compounds, such systems can also release the drug at therapeutic concentrations directly to the needed area.

Implants of this type can release the bioactive agents over the entire period of resorption. To achieve a fast resorption rate, amorphous calcium phosphate and nanocrystalline HA drug delivery carriers are excellent candidates.

It is suggestive that the ability to functionalize amino groups on nanophase HA and calcium phosphate materials are the critical criteria to allowing attachment of drug molecules, particularly BMP-2 and agents that will direct nanoparticles to osteoporotic bone. The next steps are to the determine attachment efficiency to osteoporotic compared to healthy bone, imbed bone building agents into the nanoparticles, determine release profiles once attached to weak bone and, finally, to determine bone regeneration. Nanophase coatings enhance bone integration and better device fixation mechanical testing reveals that the lightly adherent nanoporous coating on titanium has a mechanical property similar to that of bone (Paul and Sharma, 2006).

Du *et al.* (1998) have studied the tissue response of nano-HA-collagen implants in marrow cavities and concluded that nanoparticles allowed for a quicker implant surface turnover. The process of implant resorption and bone substitution is similar to bone remodeling.

Muller-Mai *et al.* (1995) have tested nanoapatite and nanoapatite/organic implants *in vivo*. From their results, it was seen that both materials were suitable for bone replacement and for drug release such as antibiotics, growth factors or other substances. In addition, the organic component can be used to control the physical properties in the bone implantation bed.

The following applications can be considered:

1. HA coating based on sol–gel technology (Liu *et al.*, 2002) or electrodeposition (Shirkhanzadeh, 1998; Manso *et al.*, 2000), allowing for the formation of thin adherent films. This do not affect the substrate morphology and topography.
2. Composite preparations with other materials, like chitosan (Yamaguchi *et al.*, 2001a; Yamaguchi *et al.*, 2001b), collagen (Du *et al.*, 1998; Itoh *et al.*, 2002; Kikuchi *et al.*, 2001; Itoh *et al.*, 2001; Du *et al.*, 1999; Ten Husen *et al.*, 1955) and other polymers (Cho *et al.*, 2001), that can reinforce the matrix while promoting osteoconduction, therefore providing scaffolding properties required in tissue engineering applications.
3. Nanosized HA can be used in drug delivery systems like intestinal delivery of insulin (Paul and Sharma, 2001) or other drugs such as antibiotics (Kano *et al.*, 1994). This in turn can also be controlled by the application of pulsed electromagnetic field applied across the treated site.
4. Further examples studied include its use in genetic therapy for certain types of tumors (Kano *et al.*, 1994).

Presently, it can be concluded that new HA-based materials are certainly among the most promising challenges in bioactive ceramics for the near future and, consequently, the research effort put in their development will continue to increase.

## 5.11 Possible Mechanism Involved in Osteoporosis

The relationship between fracture healing and osteoporosis is complex. The underlying etiology (which may include aging, hypogonadism, rheumatism, thyroid and parathyroid disorders, malignancy and mastocytosis) and the therapies commonly used for osteoporosis (estrogens, vitamin D and bisphosphonates) may all potentially affect fracture healing.

Fracture healing may also be complicated by accelerated local bone loss and regional osteoporosis (Raisz, 1988). Owing to these complexities, animal osteoporotic models, such as the rat, rabbit, or dog, may be more appropriate to study the effects of osteoporosis and to test drugs on fracture repair.

Some studies suggest that prostaglandin $E_2$ ($PGE_2$) might play a crucial role in mediating the responses of bone cells to external physical stimuli (Davidovitch *et al.*, 1984; Lorich *et al.*, 1998). In addition, other studies give direct proof that $PGE_2$ can stimulate bone formation (Chyun and Raisz, 1984; Jee *et al.*, 1985; Mori, Jee and Li, 1992). Chang and Chang (2003) have demonstrated that reduction of trabecular bone mass, destruction of trabecular bone structure and an increase of serum $PGE_2$ concentration resulting from estrogen deficiency in the ovariectomized female rats were inhibited upon exposure to PEMF stimulation.

Large pores, such as vascular, intertrabecular and resorption spaces in healing, remodeling and osteoporotic bone appear to play a role in stress generated potential (SP) generation. Bone

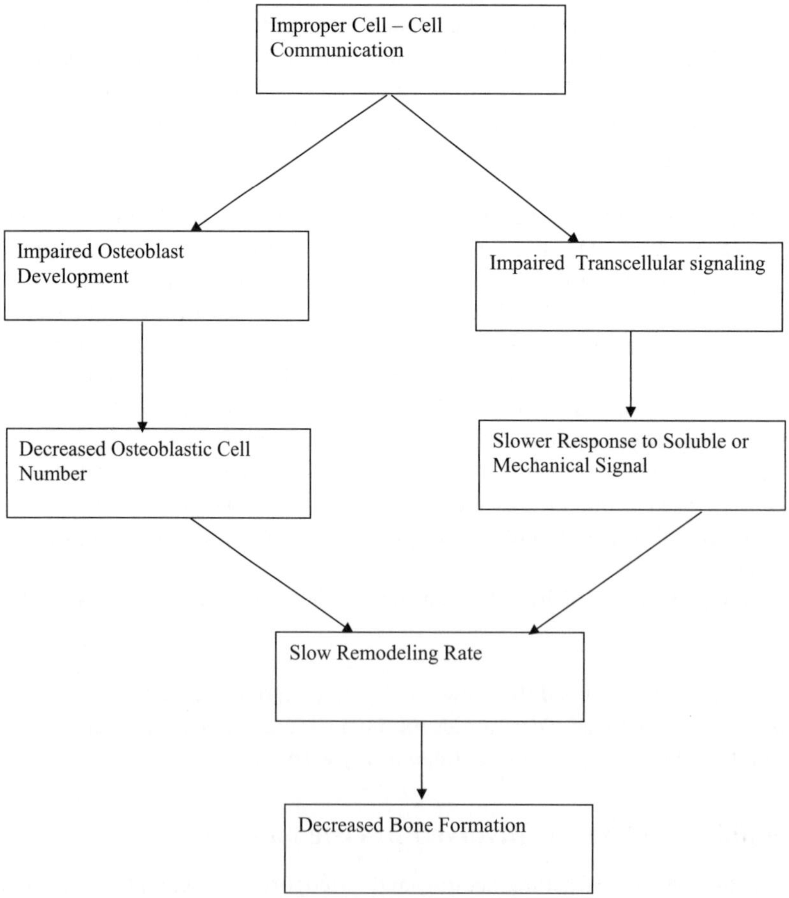


**Figure 5.23** Schematic representation of establishing that osteoporosis in bone is mainly due to defect remodeling

having more porosis (lower impedance) has generally smaller SPs and faster SP kinetics (MacGinitie *et al.*, 1993a; 1993b, Otter *et al.*, 1994a, 1994b, 1995).

In analogy to organ development and tissue differentiation, a changing pattern of cell–cell adhesion molecule (CAM) expression may also characterize the aging process at tissue level. It is suggested that age-related changes in the cell in the type and/or amount of CAM expressed may impair cell–cell communication, either directly or through abnormal expression of connexins (Figure 5.23). In theory, the impact of a deranged intercellular communication on bone can be multiple. An impaired transcellular signaling may lead to more difficult integration of input signals with a consequent slower system response to chemical or mechanical stimuli. In addition, if cell–cell communication is necessary to maintain the work of osteoblasts synchronous within multicellular units then inefficient signal exchanges may cause the remodeling units to become asynchronous, which may disrupt or slow down bone turnover. Ultimately, age-related abnormalities in the expression of CAMs and/or connexins may impair the ability of osteoblasts precursor to differentiate, with a consequent decrease in osteoblast number – a frequent finding in osteoporotic conditions. All these potential effects of ageing in the skeletal tissue may lead to decreased bone formation (Civitelli, 1995). Additionally, because cell–cell contact and communication may be required for osteoclast development (Suda, Takahashi and Martin, 1992), abnormal expression of CAMs may also lead to decreased osteoclast activity and translate into reduced activation frequency of the remodeling cycle at the tissue level. Because abnormalities in bone remodeling occur in most metabolic bone diseases such as osteoporosis, an impaired intercellular signaling may contribute to the pathogenetic mechanism and/or functions of CAMs. This may help in elucidating the role of cell–cell communications in healthy and pathologic conditions.

The ability to accurately assess bone quality *in vivo* is essential for reaching the diagnostic and therapeutic goals for bone loss from such varied etiologies as osteoporosis, microgravity, bed rest or stress shielding from an implant. Early diagnostic ability is important because the effectiveness of treatment diminishes with disease progression, yet patients are rarely symptomatic before considerable bone loss has occurred and sometimes not until after the first fracture (Davidson, 2003; Homminga *et al.*, 2004).

# 6

# Biophysical Parameters Affecting Osteoporosis

## 6.1 Introduction

Bone is a living tissue and normally there is an equilibrium between the amounts of old bone being removed and the new bone replacing it. In humans about 80% of the skeleton consists of cortical bone while the rest is cancellous bone, though its surface to volume ratio is much larger. Osteoporosis results when the rate of bone resorption exceeds that of new bone formation (Meunier *et al.*, 1980; Avioli, 1983) coupled with a defective osteoblast function. This causes an imbalance between bone resorption and bone formation. Osteoporosis is probably the most common metabolic disorder of bone and may be drug induced, idiopathic and postmenopausal or oophorectomy, but is mostly a result of aging. As a person ages, bone becomes more brittle, fracturing more easily and healing more slowly. Many elderly persons also are less active and have poor diets that are deficient in protein. This is a multifactorial and polygenic disease (Gennari and Brandi, 2001). Bone fragility may be the result of reduced bone size, reduced bone mineral density (BMD) or architectural changes (Duan, Parfitt and Seeman, 1999; Duboeuf *et al.*, 1997; Hans, Schott and Meunier, 1993; Karlsson *et al.*, 1993; Karlsson *et al.*, 1998; Vega *et al.*, 1998). There is evidence that the change of microstructure of trabecular bone such as loss of connectivity and removal of trabeculae, leaving those that remain more widely separated, can significantly decrease bone strength in osteoporosis (Parfitt, 1987; Aaron, Ciombor and Jolly, 1989; Parfitt, 1983; Riggs and Melton, 1986; Escher, Gambert and Rothschild, 1987; Consensus Development Conference, 1993; Nguyen *et al.*, 1996). In contrast to women, male osteoporosis is secondary in most cases (Orwoll and Klein, 1995). It has been proposed that postmenopausal osteoporosis is the consequence of impaired bone formation due to estrogen deficiency (Albright *et al.*, 1940). The other type of osteoporosis is due to calcium deficiency and aging of the skeleton (Riggs *et al.*, 1982). It is likely that multiple pathogenetic mechanisms converge to cause loss of bone mass and microarchitectural deterioration of skeletal structure. These factors seem to couple to contribute to a high incidence of fragility fractures in osteoporotic patients. The exact mechanism of osteoporosis at the cellular level remains to be fully understood but it is thought to be mainly due to the uncoupling of osteoclastic–osteoblastic activity, where the osteoclastic activity dominates (Avioli, 1983).

Osteoporosis is essentially a clinical syndrome in which there is a decrease in cancellous bone and to a variable extent also a decrease in cortical bone to levels below those required for mechanical support (Parfitt, 1983; Riggs and Melton, 1986; Escher, Gambert and Rothschild, 1987; Cann *et al.*, 1980). In normal women, there seems to be an accelerated decline in bone mass after menopause (Riggs *et al.*, 1981). Nevertheless, there is a considerable overlap in the amount of bone between age and sex-matched men and women with and without fractures (Aitken, 1984; Cummings, 1985). Rico *et al.* (1992) have concluded that total bone mass acquisition takes place earlier in women than in men, leading to more reduced bone mass value, which in turn may be an osteoporosis predisposing factor. The association between low body mass and fracture risk is now well established. Correspondingly, an inverse relation between bone size and fracture risk is also suggestive. When subjected to mechanical stress, a small bone is more likely to break early, than a large one of similar volumetric density, although it might be affected by bone geometry (Faulkner *et al.*, 1993). Studies of differential changes in bone mineral content (BMC) and bone mineral density (BMD) should provide more information about the relative contributions of bone size and bone density to skeletal strength in adults, although in the growing skeleton changes in bone size will dominate these parameters (Compston, Cooper and Kanis, 1995). Studies have reported that the assessment of trabecular architecture can give additional information for the prediction of bone strength (Chevalier *et al.*, 1992; Link *et al.*, 1998; Wallach *et al.*, 1992). Densitometry at most provides a measure of an average BMD but not architecture. While the first major pathogenetic mechanism, failure to achieve optimal peak bone mass, is determined largely by human genes, nutrition and lifestyle have an added effect during growth and developmental processes.

It has become increasingly clear that osteoporosis is a serious medical problem and a major cause of morbidity in the elderly. As of now, there is no effective treatment of established osteoporosis, hence the efforts are directed toward the prevention of bone loss and fragility (Kleerekoper and Krane, 1989). Increased bone mass has a protective effect against the effects of osteoporosis. A person with a large skeletal mass will withstand the disease better than a person with a small skeletal mass who suffers the same rate of bone loss. The demonstrated relationship between areal bone density and both bone strength (Dempster *et al.*, 1993) and fracture risk (Wasnich *et al.*, 1985; Cummings, Browner and Ettinger, 1990; Black *et al.*, 1992; Cummings *et al.*, 1993) provides justification for the use of areal density in clinical and epidemiological practice.

Osteoporosis (porous bone) is a disease characterized in addition to low bone mass and structural degradation of bone matrix but has normal composition. This is a clinically silent disease until it manifests in fracture. Accelerated bone loss occurs with limited weight bearing activity, as in the case of space flight, immobilization and long-term bed rest (Donaldson *et al.*, 1970; Lang *et al.*, 2004; Le Blanc *et al.*, 1990). Half of all white women suffer an osteoporotic fracture after the age of 50 (Chrischilles *et al.*, 1991). Such fractures not only entail high health risk costs (Chrischilles *et al.*, 1991; Ray *et al.*, 1997), but are also associated with pain and loss of quality of life, very often resulting from insufficient healing (Johnell, 1996). This leads to an increased morbidity and mortality risk (Center *et al.*, 1999). Risk factors for osteoporosis can be classified into two categories: modifiable and nonmodifiable (Heinemann, 2000). Nonmodifiable risk factors include the personal history of fracture. Potentially modifiable risk factors include prolonged glucocorticoid therapy, cigarette smoking, low body weight

(<127 lb), estrogen deficiency, early menopause or bilateral ovariectomies, prolonged pre-menopausal amenorrhea, impaired vision despite correction, alcoholism, recurrent falls, inadequate physical activity, low life long calcium intake and poor health (Lin and Lane, 2004). The lifetime risk of fractures of the hip, wrist and spine is 40%. It is estimated that at 80 years, 20% of women will have a hip fracture and at 90 years this will increase to approximately 50%. Women older than 85 years are approximately eight times more likely than 65–74 year-old women to fracture a hip (Lin and Lane, 2004).

Increased intracortical porosity has been observed as a consequence of aging, disease (hyperparathyroidism, osteoporosis) and pharmacologic intervention (thyroid hormone, fluoride, parathyroid hormone, prostaglandins) (Sietsema, 1995), and has been considered to be of importance for the mechanical strength of cortical bone (Barth, Williams and Kaplan, 1992; Martin and Ishida, 1989). The mechanical incompetence of bone would reveal impaired adaptations of bone strength and mechanical usage and hence a consequent mechanostat disorder.

Normal osteoporosis is primarily a decrease in BMD and predisposes to mechanical failure of the skeleton (Consensus Development Conference, 1987). The World Health Organization (Genant *et al.*, 1999) has defined osteoporosis on the basis of BMD measurements. This being obtained by dual energy X-ray absorptiometry (DEXA) as a $T$-score of greater than 2.5 standard deviations below the mean for normals, who are young healthy individuals at their peak bone mass. As such the importance of mechanical loading in maintaining skeletal health has been widely recognized. Clinical trials (Aloia *et al.*, 1978; Dalsky, 1989; Rikli and McManis, 1990) and animal studies (Barengolts *et al.*, 1993; Chow *et al.*, 1998; Westerland *et al.*, 1998) have described the significance of weight bearing activity in the maintenance of bone mass and strength. However, clinical investigations and biological tests are normal in about 50% of middle-aged men with osteoporosis. Some studies have shown that family factors and probably genetics factors contribute to bone loss in men (Cohen-Solal *et al.*, 1998). Despite our better understanding of the pathogenesis of fractures, the exact mechanism of bone loss remains to be understood in primary male osteoporosis.

Evidence is also accumulating that osteoporosis may be reprogrammed by environmental influences during critical periods of early development. Studies have shown that growth in infancy correlates with adult bone mass independently of adult life style (Cooper *et al.*, 1995, 1997; Jones, Riley and Dwyer, 1999). Dennison *et al.* (2003) have demonstrated a positive association between measures of the circulating GH (growth hormone) profile and bone mass at the lumbar spine in cohort British women (60–75 years). The relationship is more pronounced for basal GH concentration and for measures of BMC and areal density than for volumetric bone density. The effects remained significant after adjustment for body mass index, life style determinants of bone loss – including calcium intake, physical activity, cigarette smoking, alcohol consumption – and age at menopause. It also remained after adjustment for severity of osteoarthritis. These authors did not find a relationship between any measure of the circulating GH profile and bone loss over a four-year-period in this female cohost.

The ramifications of such bone loss continue to be a major health concern, emphasizing the need for effective countermeasures that can prevent or reduce the rate of bone loss. In postmenopausal women, BMD simplistically reflects peak bone mass, which is related to the age of menopause and rate of postmenopausal bone loss. Bone density at any stage in later life depends on the peak level attained during growth and consolidation of the skeleton and the subsequent rate of bone loss. Past and current regulation of these factors determines the bone mass in each individual although traditionally osteoporosis has been associated with fractures of the hip, spine and forearm. Evidence suggests that many other fractures are associated with reduced bone density (Seeley *et al.*, 1991; Editorial, 1990).

Osteoporosis arises in adults as a result of skeletal gains during the process of growth and losses after the attainment of peak bone mass, from the fourth decade of human life (Figure 6.1a and b). Peak bone mass is a major determinant of bone mass in later life and therefore of the risk

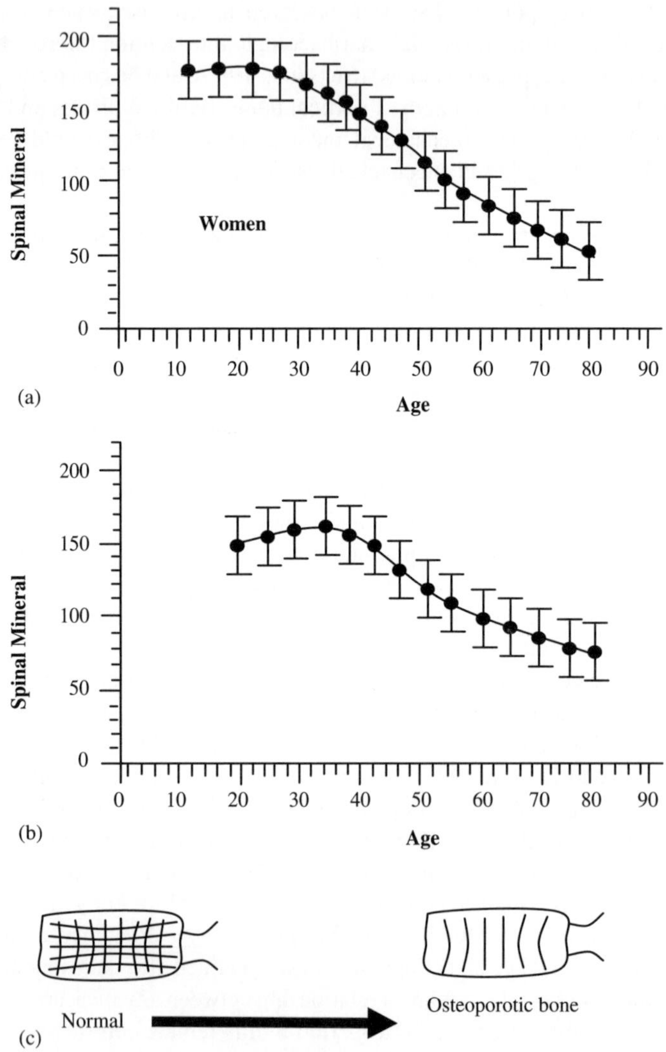


**Figure 6.1** (a) Women lose about 0.75–1.0% of their skeletal mass per year after age 35 – this may accelerate to as much as 2–3% per year following the menopause; (b) men lose about 0.4% of their skeletal mass per year after age 50; (c) patterns of trabecular bone loss in the spine

of developing osteoporosis at that time. Men reach a peak bone mass that is 10–15% higher than peak bone mass in women. This difference may be explained by gender differences in body size, mass as well as possibly in bone remodeling during the critical period for bone mass increment, that is, puberty. Indeed more than one half of peak bone mass is accepted during this critical period (Cidogan *et al.*, 1998). During puberty, the spurt is accompanied by a large increase in bone remodeling (Mora *et al.*, 1999; Szule *et al.*, 2000). Known determinants of bone remodeling are genetic factors, gender, puberty and environmental factors such as nutrition and exercise (Soyka, Fairfield and Klibanski, 2000).

It therefore seems imperative to examine the mechanism by which the skeletal mass is maintained and bone loss occurs in the adult and the implications for calcium nutrition. In adults in whom longitudinal growth of bone ceases, skeletal turnover is maintained by remodeling of bone, which consists a series of well-characterized morphological events. Bone turnover is important for the self-repair of skeletal tissue and when remodeling does not take place spontaneous fracture may arise, which is presumably related to the inability of the skeleton to repair fatigue damage.

### *6.1.1 Osteoporosis in Women*

Compared to men, women experience excessive bone loss during the time period surrounding menopause (Eriksen, Mosekilde and Melsen, 1985). Although the start and the length of menopause varies in women, the activation frequency of BMUs and the remodeling rate of both cortical and trabecular bone always increase due to a response to the decline in estrogen during menopause. The increased remodeling rate eventually returns to a rate parallel to that in men, but during the menopause period there is a net imbalance in resorption compared to formation and, consequently, a net bone loss. This net increased bone loss is thought to be a significant factor in the development of postmenopausal osteoporosis in women, because it relates to an increased probability of trabecular perforation during menopause (Parfitt *et al.*, 1983; Reeve, 1986). Also, the mineral apposition rate in postmenopausal women decreases markedly with age (Recker *et al.*, 1988).

A study by Lauretani *et al.* (2006) confirms the observational studies that muscle mass primarily affects bone size and not volumetric density in the loaded skeletal sites (Heinonen *et al.*, 1996, 1998).

### *6.1.2 Osteoporosis in Men*

Until recently, the focus of age-related bone loss has been on postmenopausal women, mainly because women start losing bone earlier than men. However, the male skeleton also undergoes predictable age-related changes in bone mass (Kassem, Melton and Riggs, 1996). Testosterone positively affects bone mineral density (Composton, 2001). In fact, androgen deprivation therapy in adult men with advanced prostate cancer is followed by rapid bone loss, similar to the loss of skeletal integrity in women after surgical ovariectomy (OVX) or during early menopause. Significant bone loss, though to a lesser extent, has also been observed at appendicular skeletal sites, including the hip and the radius, after the cessation of androgen therapy (Van der Schueren *et al.*, 2004).

The prevalence of osteoporosis is less in men than in women for several reasons: a greater evolution of skeletal mass during growth, a greater bone size, an absent midlife menopause, a slower rate of bone loss and a shorter male life expectancy (Amin and Felson, 2001). Nevertheless, osteoporosis in men is an increasing public health concern. The most commonly occurring hip fractures in men are associated with a greater morbidity and mortality than in women. The first two decades of life are characterized by rapid bone growth followed by peak bone mass at about 35 years of age. The peak mass persists for some time and is later followed by a gradual but progressive loss of bone with aging. Some epidemiologic studies show that approximately 30% of hip fractures occur in men (Legrand *et al.*, 1999, 2001; Seeman, 1999; Seeman, 2001), while the number of vertebral fractures is half of that reported in women (Cooper *et al.*, 1992). In men, osteoporosis causes vertebral compression fractures and there is an increase in the incidence of vertebral fractures in men in many countries (Anderson, Hodges and Moylan, 1999). Fractures related to osteoporosis occur mainly in the hip vertebral and the wrist (Melton, 1993). Currently, there are no established animal model for studying age-related male bone loss.

The cause of senile osteoporosis in men is unknown. The logical speculation that a decline in testicular or adrenal androgen production increases bone resorption is offset by the observation that most osteoporotic men have normal levels of plasma testosterone (Nordin, 1983). According to Jones *et al.* (1987), osteoporosis in men has different aspects than that observed in postmenopausal osteoporotic women. In men, it is more severe and is related to corticosteroid treatment and to idiopathic causes. These authors' results indicate a dominant decrease of bone mineral density of the anterior spongious portion of the vertebral body in osteoporosis. Both demineralization and increase of marrow fat occur in aging and osteoporosis (Minaire *et al.*, 1984a).

Some clinical surveys have indicated that adult bone mass diminishes at a mean rate of 0.5% per year and it can reach a loss of 2% per year after menopause. In man it is 0.5–0.7% (Matkovic *et al.*, 1979). One study (Mou *et al.*, 2001) found a 2.1% decrease at the lumbar spine while another (Black *et al.*, 2000) of 10 women found a 2.0% decrease only. The results of these two studies are in broad agreement with the finding of Kaur *et al.* (2003), who reported a small (statistically non-significant) decline in BMD at all sites.

Changes of bone density for the humerus, the third lumbar vertebra and the eight caudal vertebra of the rhesus monkey show differences due to age and sex (Pope *et al.*, 1989). In general, bone density decreases with age and then reaches a plateau at approximately 3–4 years in all bones measured. These results indicate that the rhesus monkey shows a natural pattern of change in bone mineralization that parallels that seen in humans. Age-related osteoporosis shows itself in altered urine biochemistry (Harma *et al.*, 1987). As the probability of fracture is closely related to a person's effective bone mass, a modality that could prevent or retard loss of bone might provide a substantial reduction in the incidence of skeletal morbidity. Hence the ability of measurements to capture elements of both bone size and density may prove advantageous to clinicians for making decisions.

Some workers (Cummings, 1985; Hall, Davis and Baran, 1987; Davis, 1987) have, however, concluded that bone mass may not provide an adequate indication of fracture risk. Using reports of fracture prevalence and incidence, a logistic regression procedure was used to evaluate the relationship between bone mass and fracture recurrence (Walker and Duncan, 1967). The logistic transformation of fracture risk, expressed as a multivariate model, is represented mathematically as:

$$\log\frac{(P)}{1-P} = a + \sum_{i}^{k} b_i X_i \tag{6.1}$$

where $P$ = the probability of fracture, $a$ = intercept, $X$ = covariates (e.g., BMC, age) and $b_i$ = coefficients for the corresponding covariates ($i$ = 1, 2, 3 ... $k$).

Vertebral crush fractures are often an indication of an accelerated case of osteoporosis (Ross *et al.*, 1992). Although wedging of vertebral is present in 60% of elderly women, only 5–10% develop two or more crush fractures (Nordin, 1983). Each vertebral crush fracture shortens height by approximately 1 cm, which confirms its occurrence. Vertebral crush fractures result from a combination of the osteoporotic process and minor trauma. The most commonly affected vertebrae are generally $T_{12}$, $T_{11}$ and $T_4$. Treatment of vertebrae fractures consists in having bed rest and most of these fractures heal without treatment.

## *6.1.3 Osteoporosis Types*

Riggs and Melton (1983, 1986) have suggested that evolutional osteoporosis can be divided into two categories: postmenopausal (type I) and age related (type II). This division is based on differences in patterns of fractures in different age groups: type I osteoporosis (Colles fracture) and type II osteoporosis (hip fracture). There are also differences in parathyroid function and mechanisms of fracture between these two groups (Riggs and Melton, 1990). In postmenopausal osteoporosis, increased bone resorption related to estrogen deficiency is a major predictive factor. Age-related factors such as decreased calcium absorption, vitamin D deficiency and overall inhibition of bone formation are important factors in type II osteoporosis.

Type I osteoporosis is present during the first 15–20 years after menopause and is characterized by an excessive and disproportionate loss of cancellous bone over cortical bone, leading to acute vertebral compression fractures and distal forearm (Colles) fractures. Although there are some exceptions, most studies in patients with type I osteoporosis have found values for serum estrogen and related hormones that are not different from age-matched controls (Davidson *et al.*, 1983). It is hypothesized that type I osteoporosis results from estrogen deficiency plus some additional factors, which produces an exaggeration of the rate and duration of the rapid postmenopausal phase of bone loss (Riggs and Melton, 1983). It has been reported (Pacifici *et al.*, 1989; Manolagas and Jilka, 1995; Kawaguchi *et al.*, 1995) that cytokines, such as interleukin-1, interleukin-6, tumor necrosis factor and prostaglandin $E_2$, may act as paracrine mediators of estrogen action in bone. In the presence of estrogen deficiency, perhaps, paracrine effectors may predispose some postmenopausal women, but not others, to excessive cancellous bone loss and thus to type I osteoporosis. Genetic polymorphism resulting in differences in the number or function of ER (endoplasmic reticulum) could also augment the effect of estrogen on bone cells. Another potential predisposing factor may be the presence of impaired renal tubular calcium transport leading to chronic renal losses in women in type I osteoporosis (Heshmati *et al.*, 1998). These abnormalities are not mutually exclusive and both could result from a local increase of the same cytokine(s) in bone and kidney.

The process causing type II osteoporosis in women and evolutional osteoporosis in men corresponds to the slow phase of age-related bone loss and involves comparable losses of both cancellous and cortical bone. This process appears to involve virtually the entire population of aging women and men and as more and more bone is lost those in the lower part will be at the greatest risk for fracture. The major fractures associated with this process are those of the proximal femur and wedge fracture of vertebrae, although various fractures occur also at sites with mixtures of cancellous and cortical bone (Riggs and Melton, 1986). Nordin *et al.* (1990) have estimated that over half of the bone loss after menopause is age related. At the tissue and cellular levels menopause-related bone mass is characterized by increased bone resorption. Correspondingly, decreased bone formation is probably the main cause of age-related bone loss. Loss of mineral salts causes osteoporotic bone to become more lucent than normal. This may be difficult to detect, since about 30% of the bone density must be lost before it be demonstrated as a lucent area on routine radiographs.

### *6.1.4 Spinal Cord Injury (SCI)*

Spinal cord injury (SCI) results in substantial and rapid osteoporosis and occurs primarily in sublesional sites rich in cancellous bone, where the BMC decreases by about 70% (Dauty *et al.*, 2000). Increased urinary calcium and hydroxyproline excretion during the first six months after injury is indicative of accelerated bone resorption (Claus-Walker *et al.*, 1975). Most bone loss occurs within first sixth months following injury (Biering-Sorensen *et al.*, 1990). This bone loss results in increased bone fragility (Lee *et al.*, 1997) and a subsequent increase in the risk for low-trauma fractures (Vestergaard *et al.*, 1998). It is therefore suggestive that the first six months are important if some intervention is to be applied.

Bassett, Schinik and Mitchell (1984) have presented successful data that indicate an up to five year clinical benefit after treatment of femoral head osteonecrosis.

### *6.1.5 Effect of Microgravity*

The two main physical factors that lead to physiological responses are (i) weightlessness, so that muscles are required to work at much lower load, resulting in smaller strains in bone; and (ii) redistribution of tissue fluid pressure. Under normal gravity there is a gradient of tissue fluid pressures, whereas in microgravity this disappears. The redistribution of tissue pressures has a significant impact on the function of the cardiovascular system, but the most easily visible manifestation of this is swelling of the face and thinning of the legs within a short time of exposure to microgravity.

Descriptions of bone density changes associated with microgravity have come from measurements on space stations (Lang *et al.*, 2004; McCarthy *et al.*, 2000) and models to simulate microgravity: prolonged bed rest with head down tilt (Leblanc *et al.*, 1990) and animal experiments involving tail suspension (Morey-Holton and Globus, 2002). Both these models simulate the fluid shifts observed in microgravity as well as unloading of parts of the skeleton. Data from spaceflight show that bone density changes are not uniform throughout the skeleton (McCarthy *et al.*, 2000). Bone is lost at around 1% per month from the lumbar spine and about 1.5% per month from the hip, with more trabecular than cortical bone loss from the hip. Similar distributions of bone changes have also been observed in some studies of tail suspended rats (Roer and Dillaman, 1990). It is normally assumed that loss of bone in microgravity is due to reduced loading and also produces a cranial fluid shift (Morey-Holton and Globus, 2002). These data suggest that bone density changes may be associated with fluid shifts occurring in microgravity. Experimental countermeasure studies have been performed based on this assumption, both in space (Goodship *et al.*, 1998) and on experimental animals (Fluckey *et al.*, 2002). During upright posture on earth there is a gravitational gradient of capillary pressures from head to foot, but during microgravity it has been postulated that there is a uniform capillary pressure of 30 mmHg. The change in capillary pressure is, therefore, positive in capillaries above the heart and negative for capillaries below the heart. Bone measurements in models of microgravity conditions have shown that the change in bone density is correlated with the change in capillary pressure (Le blanc *et al.*, 1990; Roer and Dillaman, 1990; Roer and Dillaman, 1994). In studies of bone density changes during a 180-day mission, the overall skeletal mass did not change, though there were significant decreases in bone density within the lumbar spine and proximal femur (McCarthy *et al.*, 2000). It has, therefore, been proposed that there is a causal relationship between tissue fluid pressures and bone density changes.

## *6.1.6 Bone Loss*

Bone mass starts to decline gradually from its peak during the forth decade of human life (Figure 6.1a and b). At the tissue and cellular levels, menopause-related bone loss is characterized by increased bone resorption. This is in agreement that fracture at most sites is associated with low BMD. Much of our understanding about age-related changes in bone is based on histomorphometric studies of bone and bone mineral density measurements. Studies have shown that age-related changes in cortical bone and in cancellous bone are different. There are also differences in the way the male and female skeletons lose bone during aging (Eriksen *et al.*, 1989). During aging, cortical bone begins to thin because of endosteal resorption. However, because the periosteum is still actively laying down new bone, there is also a continuous increase in bone width. In trabecular bone, the thickness of bone structural units decreases with advancing age. Trabecular thickness may not necessarily be reduced, it can even be increased, because net bone loss is mainly due to a result of perforative resorption and the complete removal of this perforated trabeculae (Reeve, 1986) (Figure 6.1c). Rodin *et al.* (1990) have demonstrated that significant premenopausal bone loss occurs from both the lumbar spine and neck femur. The rate of loss in neck femur is slower than that seen in the spine, but by the late forties the cumulative loss from peak density at both sites is comparable.

The WHO have established diagnostic criterion for the detection of osteoporosis based on BMD values attained in women (Amin and Felson, 2001). However, the applicability of these criteria for men has yet to be accepted. Some studies have suggested that fracture prevention depends not only on BMD values but is also influenced by the decrease in bone turnover (Bjarnason *et al.*, 2000; Garnero *et al.*, 1996). BMD and bone marker changes are associated with the ultimate clinical target, that is, fracture risk reduction. There is also evidence of an effect on fracture risk independent of bone density and bone turnover, which might be due to an alteration in the frequency of falls (Garnero, Darte and Delmas, 1999). Figure 6.1c shows the appearance of spine after bone loss.

Although bone loss in osteoporosis is generally related to bone density, biomechanical studies have demonstrated that other properties of bone, such as the architecture and chemical composition, contribute to its strength. For example, both the elastic modulus and the fracture stress increase with the amount of mineral deposited within collagen fibrils (Burstein *et al.*, 1975). Also, the size, orientation and local arrangement of the mineral particles are important determinants of bone strength (Martin and Ishida, 1989; Sasaki, Ikawa and Fukuda, 1991; Wagner and Weiner, 1992).

Bone loss, evaluated by iliac crest biopsies, proceeds mainly by trabecular thinning in men and correspondingly by loss of connectivity in women. Patients with fractures matched by iliac crest trabecular bone volume have a greater loss of connectivity than controls (Han *et al.*, 1996; Kleerekoper *et al.*, 1985). Karlsson *et al.* (2000) have suggested that BMD reduces trabecular connectivity, as connectivity is more greatly diminished in elderly women than in young women or elderly men, in persons with low BMD but not in persons with high BMD and in fracture cases more so than in nonfracture cases, as evaluated in the iliac bone by crest biopsies (not calcaneus) (Hans, Schott and Meunier, 1993; Kaufman and Einhorn, 1993; Salamone *et al.*, 1994). In two histomorphometric studies of men and women with spine fractures with matched controls by age, gender, body mass index and bone density, a greater reduction of trabecular connectivity was observed in fracture cases (Kleerekoper *et al.*, 1985; Legrand *et al.*, 2000). However, differences in other structural features, such as microarchitecture, geometry, bone size, ultrastructure and quality of bone matrix, and non-skeletal factors, such as soft tissue composition and thickness, may also contribute.

Women seem to experience more marked perforative resorption than men. This leads to more rapid disintegration of the trabecular structure than the simple thinning of the trabeculae seen in males. According to Aaron *et al.*, (1987), the main defect in males is decreased bone formation, but increased bone resorption predominates in women. Rico *et al.* (1993a) have indicated that measurements of specific body areas may not be extrapolated to others, due to different losses in different body areas, and that there is a marked bone loss rate in the axial skeleton in the first 5 years following menopause.

Age-related pathogenesis of male osteoporosis has, in part, been linked to gonadal hormone deficiency (Seeman, 1999; Jackson and Kleerekoper, 1990). For instance, serum total and free testosterone levels have been reported to be significantly lower in elderly men with hip fracture (Jackson, Riggs and Spiekerman, 1992). In addition to compromised testosterone levels, evidence suggests that a decrease in bioavailable estrogens may play a pivotal role in age-associated bone loss in men (Khosla *et al.*, 1998, 2001; Szule *et al.*, 2001). Additionally, in male animals, estrogens have been seen to be a predictor of bone density and metabolism (Erben *et al.*, 2000; Vandenput *et al.*, 2002).

The number of osteoblasts present in bone tissue depends not only on their proliferation rate but also on the rate of apoptosis. Several observations demonstrate that regulation of the osteoblasts life span is critical for maintenance of the skeleton. Osteoblast apoptosis increases in association with inflammation-mediated osteoporosis (Armour *et al.*, 2001) and glucocorticoid-induced osteoporosis (Weinstein *et al.*, 1998).

Mass reduction of the osteoporotic bone causes structural changes. Structural changes arrow the tolerable loading directions, which in turn may increase fracture risk. This is interspersed with individual life style stimulus to the body. Different types of non-invasive techniques provide different information on the structural changes in osteoporosis. To interpret the information correctly, it is important to understand the structural level of the bone and underlying assumptions on which the individual measurement is performed. Figure 5.2 summarizes a sequence of events leading to fracture. A large body of evidence suggests that bone mass is not adequate to completely explain either the skeletal fragility of osteoporosis or the effects of bone active agents (Heaney, 2003). This author has suggested that excess remodeling is a major source of osteoporotic bone fragility.

Several investigators have studied the mechanical properties of osteoporotic bones and reported a decrease in elastic modulus and strength in the transverse direction compared with those of the longitudinal direction (Kabel *et al.*, 1999). These findings correspond to an increase in intensity of the trabecular orientation in the loading direction in the osteoporotic bone. The ratio of the elastic modulus in a major principal axis (longitudinal direction) to that in the minor axis (transverse direction) has been reported as 1.00–1.51 in osteoporotic vertebral body (Hagiwara *et al.*, 2000) and 1.05–7.04 in the bovine proximal tibia (Inoue, 1987).

In Sprague–Dawley (SD) rats, from an age of about one month, cancellous and cortical bone parameters decreased with aging. The decrease with aging in cancellous bone in the proximal tibia of male SD rats in this study and in male Fisher 344 (F 344) rats was larger than in the lumbar vertebra and femoral neck (Erben *et al.*, 2000).

There are two reasons for selecting age-related bone loss in men (Wang *et al.*, 2001): (i) first, men do not normally experience the abrupt loss of sex hormones with aging that occurs in women following menopause, although gonadal deficiency can cause some osteopenia in men (Daniell, 1997; Katznelson *et al.*, 1996) and in male rats (Erben *et al.*, 2000; Vander Schulren *et al.*, 1992; Wink and Felts, 1980); (ii) secondly, men lose bone with aging despite clinically evident hypogonadism occurring only infrequently (Riggs, Khosla and Melton, 1998). Therefore, it is suggested that animal models of age-related bone loss in men be based on gonad-intact, aging, male animals and not on gonadectomized male animals, as has been done by some workers (Wink and Felts, 1980).

The lifetime risk of a forearm fracture in a postmenopausal woman is 16% (Riggs and Melton, 1986) – similar to the risk of hip and vertebral fractures. The occurrence of distal forearm fractures in Eastern Europe is similar to that observed in Scandinavia (Ismail *et al.*, 2002). Several studies have clearly established the relationship between osteoporosis and fractures of the distal radius, especially in women up to 65 years of age (Eastell, 1996; Kelsey *et al.*, 1992). An epidemiological association (Cleghorn *et al.*, 1991) has also been demonstrated. From cross-sectional data, the difference in BMD measured at the forearm among individuals with a fracture of the distal forearm and controls was about one half a standard deviation (SD) (Mallmin and Ljunghall, 1994; Wong and Pun, 1993), whereas the difference in BMD measured at the femoral neck or trochanter in individuals with a hip fracture turns out to be between 0.9 and 1 (SD) relative to the controls (Marshall, Johnell and Wedel, 1996). Similar results were obtained in two prospective studies (Kelsey *et al.*, 1992; Gardsell *et al.*, 1993).

Unlike hip and vertebral fractures, the risk of forearm fracture is not associated with smoking (Vestergaard and Mosekilde, 2003), and the variation in fall frequency could not explain the incidence of distal forearm fractures in women (O'Neill *et al.*, 1995). These studies indicate that fracture of the distal forearm is less osteoporotic than hip fracture. Important conditions in addition to BMD are involved in pathogenesis of distal forearm fragility fracture. Besides BMD measurements as a surrogate measure of bone strength, bone material and structural properties are also important components that determine bone's ability to withstand loading (Wehrli *et al.*, 2002). In addition, several extra skeletal factors play a substantial role (Ismail *et al.*, 2002; Honkanen *et al.*, 2000; Miller *et al.*, 2002).

Talmage *et al.* (1986) have found that athletic women did not undergo the normal postmenopausal bone loss and instead maintained higher bone mass than age-matched peers. Other studies have also shown that exercise can reduce the bone loss occurring after menopause (Smith *et al.*, 1989; Chow, Harrison and Notarius, 1987; Dalsky *et al.*, 1988; Berverly *et al.*, 1989).

### *6.1.7 Secondary Osteoporosis*

The causes of generalized secondary osteoporosis in adults may by one of the following types: (i) hormonal, (ii) nutritional, (iii) drugs related to mineral metabolism and (iv) inherited metabolic disorders and others. Factors identifying the occurrence of localized secondary osteoporosis are:

- A prolonged immobilization of a limb, for example in a plaster cast or high dorsal paraplegia/ low cervical quadriplegia and neurectomy rats. This form of osteoporosis is known as disuse osteoporosis.
- Monoarticular rheumatoid arthritis.

A practical method of the study of osteoporosis in population groups is to define or derive a fracture threshold, which is related to variations in the BMD. Based on this, fracture risk at the level of the neck of the femur, the proximal third of femur, fracture of the lumbar vertebrae with anterior wedging and fracture of the lower forearm, at the level of the distal third of the radius, are indications of reduction of bone mineral density and the onset of the osteoporosis.

The factors that determine the occurrence of a fracture in an osteoporotic individual depend on skeletal and extraskeletal causes. Sketal causes are bone mass per unit volume, cortical thickness at sites of maximum loading, the trabecular pattern of the bone matrix and the turnover of the bone mass per unit volume. The precipitating factor might be a fall.

## 6.2 Senile and Postmenopausal Osteoporosis

These two entities are the most common causes of generalized osteoporosis. Senile osteoporosis simply refers to the gradual loss of skeletal mass that is seen with advancing age. Postmenopausal osteoporosis refers to the increased bone loss seen in women following menopause. Both of these processes are very common and both commonly occur in the same individuals, so the two entities are often lumped together. In general, the gradual loss of skeletal mass begins in women in the fourth decade and in men in the fifth or sixth decade of life. This bone loss accelerates during women's menopause. The pathogenesis of both of these states is not clear, but probably involves a combination of decreased bone production and increased resorption.

Bone loss occurs in both cortical and trabecular bone. In the cortex, bone may be lost along the endosteal or subperiosteal surfaces, or may be lost along the surface of the Haversian canals (cortical tunneling). Trabecular bone loss occurs mostly among the trabeculae that are subject to the least stress. Conversely, the trabeculae that are unaffected the most are those that receive the most stress, generally the primary, vertical weight bearing trabeculae. In the femur, one sees gradual loss of both the compressive and tensile groups of trabeculae and the compressive group is usually the last to go. Similar patterns of trabecular bone loss are observed in the spine.

With loss of skeletal mass, one also loses bone strength. This loss of bone strength may be asymptomatic for many years, although significant loss may be accompanied by fractures and bone pain. In the spine, these fractures may take several forms, leaving the vertebral bodies wedged anteriorly, symmetrically flattened (vertebra plana) or with biconcave endplates. Increased thoracic hyphosis leads to a hump in elderly females. Fractures may occur in other areas as well, such as the proximal femur and humerus, distal radius and ribs.

Bone pain is a very common feature of many metabolic bone diseases such as metastatic bone disease, osteoporosis and Paget's disease. The common factor in many metabolic bone diseases is increased bone resorption, a pathophysiological process that often results in pain by various direct and indirect mechanisms. The majority of painful episodes of osteoporosis are not due to the osteoporotic process itself but are associated with fractures, particularly vertebral fractures. The mechanism of the pain produced is still not fully understood, but may be due to distortion of surrounding structures that contain nociceptors. Chronic back pain is often associated with osteoporosis, but may not be due to the osteoporotic process itself and is probably caused by either resulting structural changes or unrelated degenerative osteoarthritis of the spine.

Possibilities have been explored whereby effective treatments could be developed by manipulating the primary elements in bone biology that control the size and shape of bone structures. In the spirit of Wolff's law, changes in bone structure are brought about by a feedback system in which changes in peak mechanical strain drive bone cells to change bone structure (Frost, 1964a, 1964b, 1983, 1987; Currey, 1984; Hart and Davy, 1989; Beanpere, Orr and Carter, 1990). According to "mechanostat" theory due to Frost (1983, 1987, 1990a, 1990b) a feedback control system exists whereby bone structure is maintained such that ordinary mechanical strains do not exceed a minimum effective strain (MES) (1500–2500 microstrain). If the local strain within the bone surpass the MES, bone will undergo modeling and change its structure to reduce the local strains to below the MES. This theory has been expanded to incorporate another effective strain level of 50–200 microstrain (Martin and Burr, 1989). If local strains within the bone fall below the lower effective strain level, bone tissue will be resorbed until the local strains are increased.

The adaptation model, as presented in Figure 1.13, suggests a "physiological window" for bone (Frost, 1989; Martin and Burr, 1989). It is suggested that, if mechanical usage causes strain levels to fall outside the physiological window, the bone will be in a state of disuse or overuse. This leads to an adaptation response and an imbalance between bone formation and bone resorption (Bassett, 1971). Frost (1989) has further suggested that certain hormones and biochemical agents may alter the boundaries of the physiological window. Normal mechanical usage would then increase bone mass and bone strength significantly.

Bone remodeling takes place in several multicellular bone remodeling units (Frost, 1964). The first event (activation phase) is uncovering of the bone surface by osteoblasts to allow attachment of osteoclasts. A bone resorption phase lasts for about 10 days and is followed by invasion of the resorption cavity by mononuclear cells (reversal phase). Osteoblasts are then activated to form bone matrix. They first synthesize unmineralized osteoid, which is subsequently calcified. In osteoporosis, this tightly regulated system becomes unbalanced, leading to postmenopausal bone resorption, resulting in net bone loss. Rubin, Bain and Mcleod (1990) showed that a bone modeling response to a mechanical stimulus, which was exuberant in young animals, was almost nonexistent in old animals. The age-related decrease in the modeling capacity of bone may be due to decreased efficiency of the cellular effectors. For instance, an electron microscopic study of the lining cells on the bone's surface shows the presence of gaps between these cells that increase with age (Tonna, 1978).

At the molecular level, the various inorganic components of bone, including calcium ($Ca^{2+}$), phosphate ($PO_4^{3-}$), carbonate ($CO_3^{2-}$) and acid phosphate ($HPO_4^{3-}$), can affect the formation, maturation and dissolution of bone mineral (Posner, 1970). The organic component of bone is predominately collagen and is bonded to the inorganic mineral component to provide strength. Mutations of type I collagen and defects in post-translational processing can alter the strength and fatigue behavior of bone tissue (Japsen *et al.*, 1997). In addition, they indirectly affect the mechanical properties of bone by influencing the mineralization process. For example, in osteogenesis imperfecta, mutations in collagen disrupt the ability of collagen fibrils to incorporate crystalline mineral, leading to a higher fracture risk (Culbert *et al.*, 1995).

## *6.2.1 Type of Bone Pathogenesis*

Pathological disorders of bone may develop when osteoclast production or its function is impaired. A deficiency in osteoclast production or action results in deficient bone resorption and the development of osteoporosis. Osteoporosis is a relatively rare inherited disorder that may be due to either reduced numbers of osteoclasts or their decreased functional ability. The inability of the osteoclast precursor cells to differentiate and the inability to form a ruffled border have been associated with defects in the transcription factors fos and src, respectively. Another transcription factor is associated with deficient pycnodysostosis. In this case dissolution of the organic bone matrix following demineralization is defective.

In contrast to the above, excessive bone resorption results in osteoporosis. Excessive resorption can be due to a decrease in sex hormone levels, as occurs in post-menopausal women (type I osteoporosis) or castrated males. Excessive bone resorption also occurs in hyperparathyroidism due to increased release of PTH (parathyroid hormone) and with the prolonged use of glucocorticoid drugs. Acute calcium deficiency concernment with immobilization effectively ensures a high turnover state in bone due to elevated serum (PTH) (Cynthia *et al.*, 1999), though this is age dependent in rats. These authors further concluded that mechanical loading imposed on immobilized cortical bone experiencing high turnover effectively maintained BMD and tibial cross sectional area in that region experiencing bending forces during loading. Uncontrolled bone remodeling prevents the production of normal lamellar bone architecture and gives the bone a haphazard mosaic appearance in Paget's disease.

Osteogenesis imperfecta (OI) is a rare disease of the connective tissue characterized by osteopenia – progressive skeletal deformities cause severe growth retardation. OI is identified as a heterogeneous group of inherited disorders associated with bone fragility in the genes coding for type I collagen (Byers, 1989, 1990; Kuivaniemi, tromp and Prockop, 1991; Prockop, 1992). Blue scerae, dental and cardiac abnormalities, joint laxity and maturity onset deafness can be additional manifestations of this clinically variable disease (Kusick, 1972). Although OI is a disease characterized by increased bone fragility, the role of the mineral phase of bone in this disease is still a matter of speculation.

The expanding genetic diversity in OI as determined at the level of molecular genetics and biochemistry (Chu *et al.*, 1984; Pentinnen *et al.*, 1975; Williams and Prockop, 1983) has not been matched by an in-depth study to show possible difference in the quaternary structure of type I collagen that would be visualized by electron microscopy. Some electron microscopic studies have suggested (Doty and Mathews, 1971) that OI can be unequivocally differentiated from other osteopenias on the basis of characteristic changes involving the osteoblasts, osteoid and mineral formation. Some studies have also demonstrated that the structure of collagen fibrils in the osteoid is not uniform (Haebare, Yamasaki and Kyogokuni, 1969; Stoss and Freisinger, 1990).

Several workers (Sillence, Senn and Dnks, 1979; Sillence, 1988; Byers, 1989) have identified four major subgroups of OI patients (types I–IV). Type I OI is a mild, nondeforming type, while type II OI is the most severe form, resulting in perinetal death of several workers. Type III leads to progressive skeletal deformity and type IV is associated with a mild to moderate frequency of fracturing. A histomorphometric analysis of the diameters of collagen fibrils photographed by transmission electron microscopy indicated that type I collagen in OI bone was larger in diameter than normal bone. This increase in diameter of type I collagen fibrils may represent an alteration in the quaternary structure of the collagen fibril as a consequence of poorly packed collagen molecule. Such alterations in the collagen fibril may affect the formation and stability of bone mineral associated with it.

Most cases of type I result from mutations that decrease the amount of normal type I collagen synthesized (Rowe *et al.*, 1985; Genovese and Rowe, 1987). These mutations can produce nonfunctional proa (I) chains that cannot associate with their normal counterpart. However, most remaining OI cases are due to mutations that affect the structure of the triple helical domain of al (I) chain (Byers, 1990; Kuivaniemi, tromp and Prockop, 1991). Most of these mutations result in the synthesis of procollagen with decreased thermal stability, poor secretion and an increased rate of intracellular degradation (Bonadio and Byers, 1985; Steinmann *et al.*, 1984).

Despite the identification of mutations in a large number of OI patients, relatively few of these cases have been examined at the level of bone organization. Recent histological studies of bone from type II OI patients with known molecular defects have demonstrated severe deficiencies in trabecular and cortical bone, as well as the absence of lamellar bone structures and mature Haversian systems (Cole *et al.*, 1990a, 1990b, 1992).

The calcium-to-phosphorus ratio is lower in OI than in normal bone (Cassella and Ali, 1992). There has been speculation (Vetter *et al.*, 1991) that because crystal growth is tightly associated with collagen fibrils in bone (Weiner and Traub, 1989), defects in severely affected patients that lead to distributed fibril formation (Byers, 1988) also lead to retarded crystal growth. Haebare, Yamasaki and Kyogokuni (1969) have stated that the presence of pathologically immature collagen in the bone matrix causes irregular and granular calcium deposits and thus brittle bones build up.

The cellular mechanisms of cancellous bone loss contribute to the pathogenesis of vertebral fracture (Parfitt, 1988). The total amount of cancellous bone, expressed as bone volume/tissue volume (BV/TV), has been partitioned into two independent components – trabecular number (Tb.N) and thickness (Tb.Th) (Parfitt *et al.*, 1983). This gave rise to the concept that rapid bone loss is the result of complete removal of trabeculae, leading to a fall in Tb.N, whereas slow bone loss is the result of thinning of residual trabeculae, leading to a fall in Tb.Th. (Parfitt, 1988). There is considerable indirect evidence that a decline in Tb.N is mainly the result of increased resorption depth and, therefore, is osteoclast mediated. In contrast, a fall in Tb.Th is mainly the result of decreased wall thickness and is osteoblast mediated (Parfitt *et al.*, 1983; Parfitt, 1988). Possible etiologic agents include calcium malabsorption (Baylink *et al.*, 1980; Orwoll *et al.*, 1989), impaired protein synthesis and hypoproteinemai (Parfitt *et al.*, 1985; Parfitt, 1990a) and possibly the deficiency of one or more nutrients (Heaney, 1988; Avioli and Lindsay, 1990). Unfolding the pathogenesis of interstitial bone diseases may eventually shed light on the various more common forms of osteoporosis, in which similar abnormalities of osteoblast recruitment and function are often found (Parfitt, 1988, 1990b). Parfitt and Folds (1991) have concluded that when both (interstitial bone thickness) are reduced, trabecular thinning can still be accounted for mainly by a reduction in W.Th. (Croucher *et al.*, 1989).

The observation that reduced wall thickness is the major mechanism of trabecular thinning after intestinal shunt surgery is consistent with the significant correlation between W.Th and Tb.Th and with the other evidence of disordered osteoblast function (Parfitt *et al.*, 1985). Nonosteomalacic osteopenia with low bone turnover is common in a wide variety of gastrointestinal and hepatobilary values for osteoid apposition rate, as well as reduced wall thickness. These abnormalities can be accounted for by (i) reduced lifetime capacity for matrix synthesis by individual osteoblasts or (ii) by too few osteoblasts recruited during each cycle of remodeling (Parfitt, 1990b).

Pagel *et al.* (2003) have shown that thrombin inhibits apoptosis induced by serum deprivation in osteoblasts and osteoblast like cells. Osteoblasts express two of the three known thrombin receptors, but thrombin's inhibition of osteoblast apoptosis is independent of these receptors. The mechanism by which thrombin does inhibit osteoblast apoptosis would seem to result from the synthesis and/or secretion of a survival factor(s) by the osteoblasts, which then acts in an autocrine fashion to inhibit apoptosis.

Several studies have shown that short-term changes in the biochemical markers of bone turnover (bone markers) are related to the long-term response in bone mass during anti-resorptive therapy (Bjarnason and Christiansen, 2000; Christgau *et al.*, 1997; Garnero, Darte and Delmas, 1999; Gonnelli, Cepollaro and Pondrelli, 1997; Rosen, Chesnut and Mallinak, 1997; Delmas *et al.*, 2000). In clinical trials, the bone mass response within a treatment group can be estimated from the group wise degree of suppression in the bone markers (Ravn *et al.*, 1999a, 1999b; Riis, Overgaard and Christiansen, 1995). The practical consequences are that resources and time used for product development might be reduced if the bone markers are used for early screening and treatment effect in phase I and for interim analysis of phase II trials with the purpose of dose selection for phase III trials.

### *6.2.2 Risk Factors for Fractures*

In both children and adults, most extremity bone fractures from falls begin in diaphyseal and metaphyseal cortical bone. They rarely begin in trabecular bone, which can help to support metaphyseal cortices (Rockwood and Green, 1991; Ferrethi *et al.*, 1995).

Risk factors are a result of both trauma and decreased bone strength. All fractures are not related to osteoporosis. Trauma depends on factors related to falling and to the force of the impact. For example, bones will break more easily when subjected to torque than compression. Bone strength depends on both the density (quantity) and on the quality of the bone (structure of the trabecule, size of the crystals, visibility of bone cells, etc.). Clinically, factors can be described for low bone density or for risk of falling. Some risk factors, such as age, are in both categories. Older people are more likely to fall and they also have low bone density. Persons with illness also may fall more easily and have lower bone density. Figure 5.2 summarizes the compound effect of various factors leading to fracture as a result of a fall.

It is generally accepted that it is particularly important to increase trabecular bone mass in the treatment of osteoporosis. However, cortical bone represents 80% of bone mass. There are many differences between trabecular and cortical bone that are suggestive that bone compartments are distinctive structures with characteristic behavior and responses. Cortical bone contains much more osteocalcin and less osteonectin than trabecular bone and therefore the two may be subjected to different regulatory mechanisms (Ninomiya *et al.*, 1990). Eriksen *et al.* (1993) reported that bone remodeling has a different duration in the cortical and trabecular compartments, which indicates basic differences in metabolism pattern. Bone remodeling requires 200 days in trabecular bone, but only 120 days in cortical bone. In the early year of menopause, only a trabecular bone mass shows significant bone loss. Later, trabecular and cortical bone also behave differently. After about 15 years of menopause and coinciding roughly with the presentation of vertebral osteoporosis, only the cortical compartment shows significant bone loss. It has been reported that cortical thickness was present in 95% of cases regardless of whether the hip fracture was trochanteric or cervical, but trabecular osteoporosis was present only in 78% (Rico, 1997).

### *6.2.3 Fracture Risk Models*

Risk is defined as the probability of a mishap (fracture) occurring. Typically, the measures of risk are the frequency and rate of fractures (Ross *et al.*, 1989). The fracture threshold is often defined as a BMC cut-off, below which the large majority of fracture cases (80–95%) are observed (Ross *et al.*, 1987a, 1987b). However, this approach ignores the gradient that exists in fracture risk, which in turn is determined not only by present BMC but may also be influenced by age and other factors. Because it does not quantitate the level of risk, the fracture threshold approach cannot be used for cost–benefit analysis. The fracture threshold approach also does not permit the use of multiple treatment categories.

It is now documented with some certainty that 11–18% of the bone mass is lost in the first 10 years following menopause (Nordin *et al.*, 1990; Gotfredsen *et al.*, 1987). Menopause is preceded by a transitional phase that is characterized by rising gonadotropin levels and falling total estrogen output from the ovaries (Sherman, West and Korenman, 1976). Several studies have modeled bone loss immediately following menopause with an exponential function (Nordin *et al.*, 1990; Gotfredsen *et al.*, 1987; Gallagher, Goldgar and Moy, 1987; Heaney, 1989). According to these authors, menopause may be modeled as many small changes in the set point that converge to a new set point with the cessation of menses. On average 5 years pass from the first sign of menopausal transition to the cessation of menses (Treloar, 1981). During the menopausal transition, some bone loss may also occur. The logarithmic rate of bone loss associated with oophorectomy (Lindsay *et al.*, 1980) is about three times greater than that associated with menopause (Gotfredsen *et al.*, 1987). Therefore, it seems likely that the patterns of endocrine changes will vary between individuals, and it is assumed that genetic and environmental factors interact to influence the timing and profile of the hormone fluxes associated with menopause.

Studies have shown that bone loss due to disuse (Li *et al.*, 1990) and bone loss due to oophorectomy (Wronski, Clintron and Dunn, 1988a; Wronski *et al.*, 1988b) follow a similar pattern. Published data on bone mineral density measurements suggest that bone losses in men and women are around 0.8 and 1% per year, respectively. An acceleration to 2–3% per year occurs in women during menopause. There is sufficient evidence that bone density is a major determinant of vertebral fracture risk (Krolner, Berthelsen and Pors Nielsen, 1982; Firooznia *et al.*, 1984; Cann *et al.*, 1985; Hayes, Piazza and Zysset, 1991).

Riggs *et al.* (1981) found that bone loss from the vertebrae begins in young adulthood and is linear. However, the bulk of evidence now conflicts with this early model. Krolner and

Pors-Nielsen (1982) examined the lumbar spine bone mineral content in 70 women in relation to age. By fitting the data to gamma variate functions they showed a maximum spinal density at 34 years; however, this method of statistical handling has been subjected to criticism (Tothill, Smith and Sutton, 1983) – the authors results support the view that there is an increased loss of spinal bone mineral after menopause. Nilas and Christiansen (1987) carried out a cross sectional study of 178 healthy women and found no evidence that substantial premenopausal bone loss occurs from any site. Several authors (Hansson and Roos, 1978; Rodin *et al.*, 1990) noted a continuous age-related decrease of spinal bone density after the age of 35. Mazess *et al.* (1987), with a heterogeneous study population of 892 women, found a 10% decrease in spinal density in the decade preceding menopause. A similar pattern was shown for femoral bone density. Schaadt and Bohr (1988) found that the bone mineral content of the femoral neck decreased linearly after 30. Premenopausal bone loss is also suggested by histomorphometric studies (Meunier *et al.*, 1973; Marcus *et al.*, 1983) and by post-mortem investigations (Weaver and Chalmers, 1966; Arnold, 1973). The differing descriptions of age-related changes in trabecular bone are explained by technological differences in the sample sizes between the groups. Sambrook *et al.* (1987) pointed out the pitfalls of using cross sectional data to draw conclusions about longitudinal changes in population and emphasized the importance of an adequate sample size.

The above results provide compelling evidence that a significant loss of trabecular bone occurs before menopause. The significant increase in spinal bone density between the 20s and mid-30s may have important implications for the prevention of osteoporosis in later life. Also, the results of Rodin *et al.* (1990) provide further evidence that trabecular bone at different anatomical sites exhibits different patterns of age-related density changes, but the reason for this variation is not fully known.

More tenable models that have been constructed consider fracture risk as a continuum, using the results of either case control (cross sectional) (Melton, Eddy and Johnston, 1990) or prospective studies of fracture risk (Ross *et al.*, 1987a, 1987b, 1988). Because the progressive loss of bone and the risk of fracture are both cumulative over time, these models also consider the probability of survival when projecting fracture risk into the future. Models based on prospective data have not been developed yet that can predict vertebral fracture risk from spine BMC. It is desirable that separate models should be developed for various types of measurements (e.g., spine, radius, calcaneus, hip, etc.). Combinations of BMC measurements (including ultrasound) and combinations of BMC with other risk factors also need to be investigated (Wasnich *et al.*, 1989). Fracture risk prediction models for men may demand separate consideration.

In as much as all cost–benefit analysis depends directly on fracture costs, this is also one of the most important areas for improvements. Furthermore, because of comorbid conditions, it is difficult to determine which direct costs are associated with fractures. One approach for dealing with the problem of comorbid disease is that followed by Phillips *et al.* (1988). They examined the excess costs among patients with both osteoporosis related and comorbid conditions compared with hospitalized patients matched by a non-osteoporosis diagnostic related group. Two potential approaches are considered: (i) spinal bone density increases in young women until the mid-30s, and if this can be enhanced the peak spinal bone density can be maximized. (ii) If bone loss in the climacteric can be prevented or reduced, women would again reach menopause with a better reserve. Exercise and calcium supplementation may be suitable strategies, but before any measures can be recommended they must be fully evaluated and must be shown to reduce fracture risk.

## 6.3 Theoretical Analysis of Fracture Prediction by Distant BMD Measurement Sites

The mathematical model for describing the BMD distribution of a fracture population when there is just one measurement site is simplified if BMD values are expressed in terms of Z-scores that are calculated by taking the difference between the measured BMD and the mean BMD for a healthy population. Subjects are matched for age, gender and ethnic group and divided by the respective population SD. In the same patient the prediction at distant sites by the knowledge of fracture BMD at a particular site is discussed in chapter VII. (page 345). For a group of subject from the a group of subject from the general population with a narrow range of ages the distribution of Z-score values approximates to a Gaussian curve with its peak at $Z = 0$ and $SD = 1.0$:

$$\rho(Z) = \frac{1}{\sqrt{2\pi}} \exp(-Z^2/2) \tag{6.2}$$

Based on the proportional hazards model, the fracture risk can be modeled by an exponential curve that scales with $Z$ score as $\exp(-\beta Z)$, where $\beta$ is the logarithm of the relative risk [$\beta = \ln$ (RR)]. The Z-score distribution for the fracture population is found by multiplying Equation 6.2 by the fracture risk curve. It is found to be a Gaussian curve with its peak at $Z = -\beta$ and $SD = 1.0$ (Blake and Fogelman, 2001):

$$\rho_f(Z) = \frac{1}{\sqrt{\pi}} \exp\left[-(Z-\beta)^3/2\right] \tag{6.3}$$

Equations 6.2 and 6.3 introduce the relationship that the value of $\beta$ is equal to the separation between the two population distributions expressed in Z-score units.

The single measurement model described by Equations 6.2 and 6.3 can be generalized to two BMD measurement sites by using a bivariate Gaussian model (Blake *et al.*, 2003). Let the Z-score at the fracture site be represented by $Z_{frac}$ and that at the distant site by $Z_{dist}$.

Now with the correlation of $Z_{frac}$ to the fracture prediction by the $Z_{dist}$ measurement we can write $Z_{dist}$ in terms of its regression line with $Z_{frac}$ and $Z_{res}$:

$$Z_{dist} = rZ_{frac} + Z_{res} \tag{6.4}$$

Using the above relation in a bivariate Gaussian model and rearranging, one obtains the following equation for the distribution function of the fracture population:

$$\rho_f(Z_{frac}, Z_{dist}) = \frac{1}{\sqrt{2\pi}} \exp - \frac{(Z_{frac} + \beta_{frac})^2}{2} \frac{1}{\sqrt{2\pi(1-r^2)}} \exp - \frac{(Z_{res} + \beta_{dist} - r\beta_{frac})^2}{2(1-r^2)} \tag{6.5}$$

where $r$ is the correlation coefficient between $Z_{frac}$ and $Z_{dist}$.

The above equation is the product of two Gaussians, the first expressed in terms of the fracture site Z-score and its $\beta$-value $\beta_{frac}$ and the second in terms of the residual Z-score of the distant site and an associated $\beta$-value $\beta_{res} = (\beta_{dist} - r\beta_{frac})/\sqrt{(1-r^2)}$. The latter term quantifies the independent fracture discrimination of the measurement at the distant site after allowing for the effects of the correlation of $Z_{dist}$ with $Z_{frac}$. It follows that if the correlation hypothesis is true then $\beta_{res} = 0$ and $\beta_{dist} = r\beta_{frac}$ (Figure 6.2).

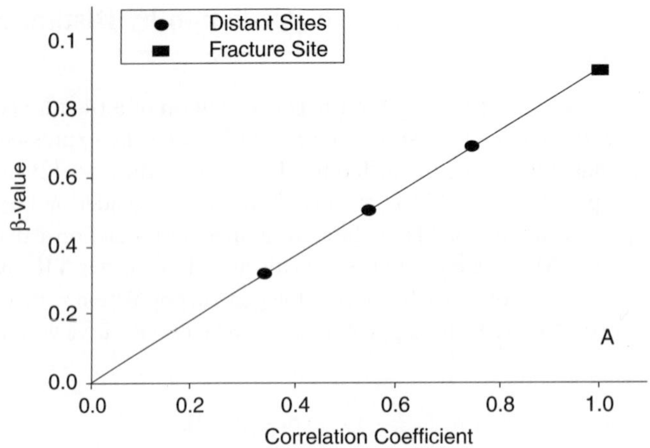


**Figure 6.2** Theoretical relationship between the bone mineral density correlation coefficient $\gamma$ and the gradient $\beta$ (BMD measurement errors are ignored)

All types of DEXA examination are affected by systematic and random measurement errors caused by inhomogeneity in the composition of soft tissue and bone marrow (Bolotin *et al.*, 2001). The magnitude of these errors can be assessed from cadaver studies (Svendsen *et al.*, 1995). The random measurements errors directly affect the *T*-score and *Z*-score figures used to interpret BMD results.

One consider now the effect of the random measurement errors on $\beta_{\text{dist}}$ and $\beta_{\text{frac}}$. Comparison of Equations 6.1 and 6.2 shows that the $\beta$-value is equal to the difference in mean BMD between the fracture population and the general population ($\delta_{\text{BMD}}$) divided by the population SD ($\sigma$):

$$\beta = \delta_{\text{BMD}}/\sigma \tag{6.6}$$

Since the error affects both populations equally they in fact do not change the value of $\delta_{\text{BMD}}$. However, they will lead to a larger population SD, which will be related to the original value of $\sigma$ by $\sigma^* = \sqrt{(\sigma^2 + \varepsilon^2)}$, where $\varepsilon$ is the root mean square error associated with the random measurement error. It follows that the effect of these errors is to reduce $\beta$ by the factor $\sigma/\sigma^*$:

$$\beta^* = \delta_{\text{BMD}}/\sigma^* = \beta(\sigma/\sigma^*) \tag{6.7}$$

We now consider the effect of the random errors on the value of the correlation coefficient between BMD measurements at the two sites. The value of $r$ is related to the covariance between the two BMD measurements and their individual SD values:

$$r = \text{cov}(\text{BMD}_{\text{frac}}, \text{BMD}_{\text{dist}})/\sigma_{\text{frac}}\sigma_{\text{dist}}$$

It follows that the effect of measurements is to reduce the value of the correlation coefficient to:

$$r^* = \text{cov}(\text{BMD}_{\text{frac}}, \text{BMD}_{\text{dist}}) \tag{6.8}$$

The bivariate Gaussian can be represented as:

$$\rho_{\mathrm{f}}(Z_{\mathrm{frac}}, Z_{\mathrm{dist}}) = \frac{1}{2\pi\sqrt{(1-r^2)}} \exp - \frac{\{(Z_{\mathrm{frac}} + \beta_{\mathrm{frac}})2 + (Z_{\mathrm{dist}} + \beta_{\mathrm{dist}})2 - 2r(Z_{\mathrm{frac}} + \beta_{\mathrm{frac}})(Z_{\mathrm{dist}} + \beta_{\mathrm{dist}})\}}{2(1-r^2)} \tag{6.9}$$

where $r$ is the correlation coefficient between $Z_{\mathrm{frac}}$ and $Z_{\mathrm{dist}}$.

The above equation has found to be the correct form to represent the fracture population and can be verified by projecting the distribution onto the $Z_{\mathrm{frac}}$ axis when it reduces to the Gaussian in Equation 6.9 with $\beta = \beta_{\mathrm{frac}}$. Clearly, therefore, the constant $\beta_{\mathrm{frac}}$ is related to the relative risk of fracture for the BMD measurement by $\beta_{\mathrm{frac}} = \ln(\mathrm{RR}_{\mathrm{frac}})$. In the same way it can be shown that the constant $\beta_{\mathrm{dist}}$ is related to the relative risk of the distant site BMD by $\beta_{\mathrm{dist}} = \ln(\mathrm{RR}_{\mathrm{dist}})$. This model along with experimental measurements of BMD can prove to be a powerful method for scanning large population for possibilities of fracture risk.

## 6.4 Markers of Osteoporosis

### *6.4.1 Structural Changes*

Postmenopausal osteoporosis is the result of increased bone resorption associated with menopause or oophorectomy. In the case of trabecular bone, it is not evident that postmenopausal osteoporosis is associated not only with a decrease in trabecular bone mass but also with a loss of bone trabeculae and a consequent decrease in their connectivity (Aaron, Ciombor and Jolly, 1989).

In cancellous bone, aging causes perforation of structural elements (trabeculae) and discontinuity of bone structure (Parfitt *et al.*, 1983, 1984; Vesterby *et al.*, 1989). Similar structural changes in cancellous bone also occur after long periods of disuse (Jaworski and Uhthoff, 1986). Increased mechanical usage tends to thicken existing trabeculae, but once trabeculae are lost they cannot be regenerated by increased mechanical usage. There is evidence that these irreversible structural changes in cancellous bone increase bone fragility, as osteoporotic patients have a lower trabeculae number and greater trabecular spacing than age-matched normal subjects (Kimmel and Jees, 1980). Both the above cases are likely to impair the resistance of trabecular bone to the bone loss. Changes of three-dimensional bone structure are, however, difficult to demonstrate by conventional bone histomorphometry. Osteoporotic changes generally are initiated at cancellous bone. Fractures associated with osteoporosis occur more frequently in the cancellous bone-rich regions, such as vertebral bodies of the spine, distal radius and proximal femur.

The introduction of vertical sections (Baddeley, Gundersen and Crees Orive, 1986) in bone histomorphometry (Vesterby *et al.*, 1987) has made it possible to obtain unbiased steroligical estimates of bone structure, as well as of changes in structure with age and in disease. The most promising of these parameters is the marrow space star volume. This parameter shows a steep and significant increase with age in lumbar vertebra, more pronounced in women than in men (Vesterby *et al.*, 1989; Vesterby, 1990). The only possible explanation for this finding is perforation or loss of trabeculae. The marrow space star volume has been estimated in an iliac crest specimen to evaluate sampling problems in such small bone specimens and the results indicate roughly parallel changes of trabecular bone in vertebrae and in iliac crest (Vesterby, 1990). These observations have important implications for the treatment of established osteoporosis in that all agents currently used affect bone remodeling. Inasmuch as remodeling is an activity associated with a surface, treatment may thicken the trabecular remnants without increasing their number or restoring their connectivity. There are two histological variants of postmenopausal osteoporosis: high turnover and low/normal turnover (also known as active and inactive forms, respectively) (Whyte *et al.*, 1982). In high turnover or active osteoporosis, a bone biopsy specimen demonstrates an abundance of bone cells (osteoblasts and osteoclasts).

In osteoporosis, there is a decrease in the reducible collagen crosslinks without an alteration in collagen concentration (Oxlund, Li and Ortoft, 1996); this would tend to increase bone fragility (Bailey *et al.*, 1993). Both animal (Knott *et al.*, 1995) and human (Bailey and Knott, 1999) studies suggest that newly synthesized collagen in osteoporotic bone matrix has increased lysine hydroxylation by 35–40%. This appears to be correlated with decreased strength in three-point bending tests (Bailey and Knott, 1999; Knott and Bailey, 1998; Knott *et al.*, 1995). Because of the interactions between matrix collagens, integrin expression and cell function (Bennett, Moffatt and Horton, 2001), this may have implications for the regulation of the osteoblast phenotype (Masi *et al.*, 1992; Shi, Kirk and Kahn, 1996) and cell signaling process (Green *et al.*, 1995) that finally underlie bone remodeling in osteoporosis.

Changes in collagen structure underlie an age-associated reduction in bone toughness, independent of mineral density or mean tissue age. This reduction in toughness increases fracture risk independent of bone mineral density (Langdahl *et al.*, 1998; Mann *et al.*, 2001). Apart from changes in mineral density, the inherent risk of fracture can change if the quality of the organic component of the bone is altered. This may contribute to the fracture risk (Hui, Slemenda and Johnston, 1988). The work of Wang *et al.* (2002) regarding this issue is important in showing that changes in the mechanical properties in bone collagen occur with age and that these are correlated to age-related changes in whole bone strength and toughness. They have gone beyond studies of simple association between biochemical changes in collagen and mechanical properties to identify specific changes in the mechanical properties of collagen with age. This means that it is not just changes in mineral content or mineral density that one must concern with in evaluating the propensity of bone to fracture but also changes in the organic nature of the bone matrix. This component of fracture risk may also be considered when risk assessments are made based on bone mineral density and geometry.

While fracture healing is well characterized in long bones, there is a scarcity of data in acute vertebral body fractures. Vertebral fractures commonly occur in both elderly men and women with osteoporosis.

## *6.4.2 Biophysical Parameters*

It is suggestive that exercise is beneficial to the skeleton, probably because it is associated with a substantial increase in mechanical stimulation of bone both directly and indirectly through the enhancement of muscle strength (Heinonen *et al.*, 1996).

Rai and Behari (1986) have studied the biophysical characteristics of osteoporotic rats by providing them food deficient in calcium and phosphorus. They reported changes in various biochemical parameters, for example, $P_{CO2}$, $P_{O2}$, Hb, RBC, WBC (mg per 100 mL, and $p$ (mg per100 mL), and so on. Infrared spectra of the two sets of bone samples differed significantly and so also did their X-ray diffraction patterns. Further, differences in histological parameters and mechanical strength of long bones were also observed. A decrease in cortical thickness and load carrying capacity were obtained in osteoporotic bones (Rai, Behari and Saha, 1986; Behari *et al.*, 1989).

Increased intracortical porosity has been observed as a consequence of aging, disease (hyperparathyroidism, osteoporosis) and pharmacologic intervention (thyroid hormone, fluoride, parathyroid hormone prostaglandins) (Sietsema, 1995) and has been considered to be important for the mechanical strength of cortical bone (Barth, Williams and Kaplan, 1992; Martin and Ishida, 1989). Small changes in porosity or density of compact bone exert a more pronounced influence on its strength than similar changes in trabecular bone would create (Schaffler and Burr, 1988; Snyder and Schneider, 1991). This observation is supported by the strong correlations of compact bone elastic modulus with bone volume fraction, a measure of intracortical porosity, found by various authors (Schaffler and Burr, 1988; Carter and Hayes, 1977; Currey, 1988). Shaffkey *et al.* (1988) have measured ultrasound attenuation and velocity in osteoporotic human femur bones. They found a lower velocity in osteoporotic bones but no observable difference in bone density. Singh and Behari (1994b) have reported that velocity rather than attenuation is more sensitive to a change in density.

Pharmacological treatment improves bone density and fracture risk at the spine, but the effects at the femoral neck are much less prominent (Gourlay, Richy and Reginster, 2003). A non-pharmacological intervention that would reduce fractures would be of value. Load bearing exercise influences bone mass (Lanyon, 1996). Vibration has been found to improve bone density in some very preliminary studies (Verschueren *et al.*, 2004) but the search continues for mechanical treatments for osteoporosis. They can be either used alone or in conjunction with pharmacological treatment to enhance its effect. The positive effect of physical exercise is highest at the femoral neck, which may be mediated through an improvement of muscle mass (Lanyon, 1996).

Type 1 collagen accounts for most of the organic matrix of bones, but it is also an important constituent of soft connective tissue. Assessment of the turnover of such collagen is particularly relevant to bone metabolism. Determination of the carboxy terminal propeptide of type 1 procollagen (PICP) is a means of estimating the rate of type 1 collagen synthesis in the body. Most circulating antigen relating to the carboxy terminal propeptide of type 1 procollagen is derived from bone metabolism. In osteoporosis studies this marker could be used to estimate the rate of bone turnover and to verify the effects of treatment (Risteli *et al.*, 1991).

Because bone matrix synthesis is performed by cells of osteoplastic lineage, matrix proteins synthesized by osteoblasts reflect the osteoblastic activity and bone formation. They may be used as biochemical markers in osteoporosis studies. The major macromolecular components synthesized by osteoblasts are type 1 collagen and bone carboxyglutamic acid (Gla). Osteoblasts also synthesize osteoclasts activating factors, prostaglandins, growth factors and several enzymes such as procollagenase and alkaline phosphatase. PTH and calcitriol are major controllers of osteoblast activity, but other factors, including cytokinins and growth factors, play important roles (Huffer, 1988). The level of vitamin D has been found to be below the average in aged people (Tasi *et al.*, 1984), especially in those who live in an institution or in a hospital (Lamberg-Allardt, 1984).

Parvianen *et al.* (1988) found that levels of the bone-specific isoform of alkaline phosphatase were age related: they were lower in patients 20–40 years-old than in those 40–60 years-old. These findings agree with those of Duda *et al.* (1988). Kuwana, Sugita and Yakata (1988) found an age-related decline in bone alkaline phosphatase in both sexes. In women, however, bone alkaline phosphatase increased from the sixth decade on. The decline in bone alkaline phosphatase up to a certain age obviously indicates decreased bone formation and turnover. Assays of bone-derived alkaline phosphatase and of osteocalcin are often used as noncollagenous markers of bone-cell metabolism.

In view of the above, different types of interventions, for example, exercise regiments (Baldwin *et al.*, 1996; Cavanagh, Davis and Miller, 1992; Nicogossian *et al.*, 1992; Norman *et al.*, 2000) drug/hormonal therapies (Bikle *et al.*, 1994, 1995; Turner *et al.*, 1998) and dietary modifications (Globus *et al.*, 1986; Navidi *et al.*, 1995), have been considered as countermeasures for bone loss associated with skeletal unloading. In addition, many types of pharmacological interventions are now in use.

## 6.5 Osteoporosis Interventions

The clinical significance of osteoporosis lies in fractures that occur mainly at the hip, spine and wrist (Boeley *et al.*, 1991). Given the cost of osteoporotic fractures and other consequences in terms of disability (Barrett-Connor, 1995), preventive strategies are clearly warranted. Fracture healing may also be complicated by accelerated local bone loss and osteoporosis (Raisz, 1988).

Systematic approaches for the prevention and treatment of postmenopausal bone loss may be categorized into pharmaceutical and nonpharmaceutical. The former may include antiresportive and stimulating agents (Dambacher *et al.*, 1998) and the later may include exercise programs (Qin *et al.*, 2002; Wolff *et al.*, 1999) and non-invasive biophysical interventions, such as vibration (Rubin *et al.*, 2001) magnetic fields (Eyres, Salch and Kanis, 1996; Mishima, 1988) and therapeutic low intensity pulsed ultrasound (Arai, Yokoyama and Tokashiki, 1993; Warden *et al.*, 2001a, 2001b).

There are many challenges to be faced in determining intervention thresholds for osteoporosis. There is a growing opinion that interventions should be based on absolute risk of fracture rather than on diagnostic thresholds provided by the *T*-score (Kanis and Gluer, 2000). The determination of absolute fracture risk probability at any one site depends on knowledge of the incidence of first fracture and mortality, and for long-term predictions on knowledge of the trends. The differential effects of determinants of peak bone mass on bone size and true bone density and the potential of life-style, dietary and pharmacological interventions to increase bone size in the adult skeleton demand serious consideration.

Nevitt *et al.* (2000) have developed a regression model based on the effects of antiresorptive therapies on vertebral fractures and nonvertebral fractures reported in 15 placebo-controlled trials of osteoporotic women. This model predicted that a reduction of vertebral fractures is associated with a reduction of non-vertebral fracture, but the magnitude of reduction is significantly less ($P < 0.01$) for the latter group.

## 6.6 Role of Estrogen

### 6.6.1 Steroid-Induced Osteoporosis

Hassager *et al.* (1991) have studied the effects of 1 year of sex-hormone therapy on serum PICP levels in osteoporotic women. It has been shown in a group of patients with various metabolic bone diseases that serum concentrations of PICP are correlated with cancellous bone formation rates (Parfitt, 1991).

The role of estrogen in bone pathogenesis stems from the fact that postmenopausal women whose estrogen levels are declining are at the risk of developing osteoporosis. Moreover, the biochemical markers indicate that bone remodeling is accelerated at the menopause as markers of both the resorption and formation are increased (Parfitt *et al.*, 1995; Ebeling *et al.*, 1996). Hence an increase in bone resorption is probably the set point for estrogen deficiency (Raisz, 2005). An increased bone formation that normally occurs in response to mechanical loading is diminished in estrogen deficiency, suggesting estrogen is both anticatabolic and anabolic (Lee *et al.*, 2003). Estrogen deficiency also causes bone loss in elderly men (Riggs, 2000). Bone cells contain receptors elucidating the molecular basis of estrogen action. Estrogen may also be important in the acquisition of peak bone mass in men (Khosla *et al.*, 2001). In fact estrogen has a greater effect than androgen in inhibiting bone resorption in men although androgen may still be important (Falahati-Nini *et al.*, 2000; Van Pottellbergh *et al.*, 2004). It is now widely accepted that women not taking estrogen therapy and taking less physical exercise are more prone to bone loss and subsequent fracture risk.

Steroids produce osteoporosis in a high percentage of individuals when they are used continuously over a long period. There is a significant correlation between the dose of the steroid and the extent of osteoporosis. Steroids affect bone metabolism by direct inhibition of osteoblast function, direct enhancement of bone resorption, inhibition of calcium absorption in the intestine, hypercalciuria and inhibition of gonadal hormonal functions. The immediate target site is the trabecular bone and, hence, vertebral compression fractures are very common.

Estrogen is critical for epiphyseal closure in puberty in both sexes and regulates bone turnover in men and women. It is well known that estrogen deficiency causes bone loss in rodents and humans (Frost, 1973; Riggs, 1991; Manolagas, Kousteni and Jilka, 2002; Eastell, 2003). It has also been established (Jeong *et al.*, 2005) that estrogen deficiency significantly reduces the proportion of newly synthesized bone matrix as well as the total amount of bone matrix. Other investigations (Dempster *et al.*, 1995; Kalu *et al.*, 1989; Wronski *et al.*, 1985; Wronski, Walsh and Ignaszewski, 1986; Swanberg, Mattila and Knuuttila, 1997) also support the findings that ovariectomy leads to thinning of the trabecular and an increased number of osteoclasts. Estrogen deficiency leads to an increase of osteclastogenesis – in a high turnover state there is a greater proportion of immature young bone than old. This increases the rate of the bone remodeling site depositing new tissue (Jeong *et al.*, 2005). These authors have pointed out that osteoclasts do not discriminate matrix age. When activated they breakdown old and fairly new matrix; in fact, immature young bone matrix is more labile to the attack of osteoclast. Furthermore, bone resorption is much faster than formation; it takes at least three months to rebuild bone resorbed in 2–3 weeks (Manolagas, Kousteni and Jilka, 2002; Eastell, 2003). This increased resorption, even when accompanied by a coupled increased formation, can cause bone loss owing to these kinetic differences. Jeong *et al.* (2005) have further concluded that a high proportion of new matrix in ovariectomized (OVX) rats is presumably degraded by increased osteoclasts.

A study has shown that the femoral head of the chickens treated with steroids showed that fat cell hypertrophy and proliferation in the subchondral area are evident as early as 1 week after initiating steroid administration (Cui *et al.*, 1997a, b).

Conversely, estrogen is also known to prevent bone loss and has been used for osteoporosis treatment and prevention of osteoporotic fragility fractures for three decades (Jensen *et al.*, 1982; Lindsay *et al.*, 1980). Estrogen treatment rapidly reduces bone breakdown in older women (Prestwood *et al.*, 1994). Fracture risk is inversely related to the estrogen level in postmenopausal women (Prestwood *et al.*, 2003). The estrogen sensitivity may be age related (Modder *et al.*, 2004). Presently estrogen is described as an antiresportive agent that slows bone turnover by depressing resorption, protecting the bone matrix, by reducing the bone turnover rate (Jeong *et al.*, 2005). Estrogen supplement after ovariectomy inhibits bone resorption by reducing the number of osteoclasts and decreasing the number of bone remodeling cycles by reducing the birth rate of osteoclasts and osteoblast (Manolagas, Kousteni and Jilka, 2002; Eastell, 2003). This model is consistent with the observation that estrogen deprivation occurring after either menopause or oophorectomy causes increased bone turnover and accelerates bone loss (Gotfredsen *et al.*, 1987; Gallagher, Goldgar and Moy, 1987; Slemenda *et al.*, 1987; Wronski, Cintron and Dann, 1988; Longcope *et al.*, 1989; Christiansen and Lindsay, 1990; Nordin *et al.*, 1990). The relationship between estrogen status and premenopausal bone density in women is not very clear. Studies of young women with severe estrogen deficiency resulting from illness (Rigotti *et al.*, 1984) or excess physical activity (Drinkwater *et al.*, 1984) have documented a marked reduction in axial and appendicular bone density. There is little information, however, on the association between bone density and circulating estrogen concentration within the normal range.

Several studies have demonstrated that estrogen therapy prevents bone loss in postmenopausal or oophorectomized women (Aitken, Hart and Lindsay, 1973; Lindsay *et al.*, 1976, 1980; Nachtingall, Nachtingall and Nachtingall, 1979; Genant *et al.*, 1982). It is argued that estrogen decreases bone formation; estrogen deprivation also leads to an increase in bone formation. However, there is considerable evidence that would suggest estrogen enhances bone formation as well. Ovariectomy-induced bone loss is observed in trabecular bone in the diaphyseal region of the femur. Trabecular bone areas are significantly reduced in OVX rats as compared with sham and estradiol treated rats. Changes in cortical bone are not observed (Jeong *et al.*, 2005). These authors have pointed out that ovariectomy increased the vacuoles in the bone marrow as well as the number of osteoclasts. Estradiol treatment restores all histopathological features induced by ovariectomy back to the control level, which is characterized by increased trabecular bone area and a decreased number of osteoclasts. The ratio of femur weight to body weight is lower in the OVX group than the sham and estradiol treated group though no significant difference is reported. Trabecular bone density increases by 15% and cortical bone density by 9% in the vertebra of girls during puberty; the increased mass is sustained throughout childbearing years. The excess bone mass is then lost after menopause (Gilsanz *et al.*, 1988). Also, young women who become amenorrheic lose bone mass (Cann *et al.*, 1984; Drinkwater *et al.*, 1984), and when normal menses are resumed the bone mass increases to a point near normal (Drinkwater *et al.*, 1986). A decline in plasma estrogen levels causes an increase in bone sensitivity to the resorptive action of parathyroid hormone (Orimo, Fujita and Yoshikawa, 1972; Gallagher *et al.*, 1980). The estrogen effect may be mediated by calcitonin, a hormone secreted by C cells of the thyroid, which lowers plasma calcium and phosphate and inhibits bone resorption (Greenberg *et al.*, 1986). It has been demonstrated that estrogen therapy increases 1,25-$(OH)_2$-vitamin $D_3$ serum levels (Stock, Ciders and Mallette, 1985) and calcium absorption across the intestine (Heaney, Recker and Saville, 1978). Other authors have also reported that calcitonin levels in postmenopausal patients increase with estrogen therapy (Morimoto *et al.*, 1980; Stevenson *et al.*, 1981). However, it has also been stated that estrogen does not affect basal calcitonin levels on a long-term basis and they question the role of calcitonin in the etiology of postmenopausal osteoporosis (Leggate *et al.*, 1984). Also, some strong evidence has been provided that bone cells do not contain estrogen receptors (Eriksen *et al.*, 1987; Komm *et al.*, 1987). These studies conclude that the increase in bone mass is due to estrogen effects on bone resorption. However, the evidence obtained from animal studies point to the contrary. Goulding and Gold (1990) have shown that estrogen deprivation associated with administration of a luteinizing hormone releasing hormone antagonist caused a significant decrease in total body calcium in rats. The lost calcium returns to normal after the normal estrus level has been reestablished. This result is significant because the majority of a rat's skeleton is made up of cortical bone and this does not remodel. In another study, estrogen treatment has been shown, in osteopenic adult rats, to increase cortical bone mass in the femur by 25% (Barengolts *et al.*, 1990). Therefore, the antiresportive properties of estrogen by themselves are not sufficient to increase cortical bone mass in rats. Therefore, it is concluded that a significant increase in cortical bone mass is probably caused by increased bone formation. These studies have two common features:

1. Estrogen increases the bone mass of an osteopenic skeleton. This is consistent with the observation that high dose estrogen therapy will increase the bone mass of a normal skeleton (Tobias *et al.*, 1990). This evidence also supports the postulate that the estrogen affects the set point for bone adaptation (Frost, 1987; Turner, 1991).
2. Tobias *et al.* (1990) have provided histomorphometric evidence that a high dose of estrogen treatment increases bone formation in trabecular bone as well, although the peak serum estradiol levels reached in their experiments are about 30× normal.

With estrogen replacement therapy, serum estradiol levels are in the neighborhood of 50–100 pg $mL^{-1}$ (Faguer de Mousteir *et al.*, 1989). These levels are only a fraction of the peaks seen in the normal menstrual cycle (600 pg $mL^{-1}$). Several studies (Goulding and Gold, 1990; Barengolts *et al.*, 1990; Tobias *et al.*, 1990) support the notion that high levels of estrogen can increase bone formation and restore lost bone mass. Thus, one might speculate that the levels of estrogen during estrogen replacements are insufficient either to increase bone formation or to restore lost bone mass. In support of this Studd *et al.* (1990) have shown that higher estrogen doses (serum estradiol levels of 56–241 pg $mL^{-1}$) increase spinal bone density in postmenopausal women by an average of 8.3% after 1 year of therapy. The increase in spinal bone density is correlated to serum estradiol levels, suggesting a dependence of bone mass on estrogen dose.

Turner (1991) has shown that estrogen given to growing female rats reduces periosteal bone formation and suppresses endosteal bone resorption. In the remodeling skeleton, estrogen deficiency following oophorectomy is known to increase bone formation as well as skeletal turnover. Thus, even though bone formation indices are increased, estrogen deprivation leads to significant net bone loss. Estrogen replacement reverses this trend by reducing skeletal turnover (Wronski *et al.*, 1988b).

One inconsistency between the effects of estrogen loss and disuse is their effects on bone formation. While both cause an eventual decrease in bone formation, estrogen loss causes transient increases in bone formation (Wronski, Cintron and Dann, 1988), whereas disuse appears to reduce bone formation (Young *et al.*, 1986; Sessions *et al.*, 1989). Yet, there is some evidence of a transient increase in bone formation following disuse (Heaney *et al.*, 1972).

One plausible theory for the estrogen effect on bone mass is that estrogen changes the set point of mechanical feedback. Furthermore, bone loss in rats caused by a 4-week-period of estrogen deprivation associated with administration of buserelin is totally reversible by withdrawing it for 4 weeks (Goulding and Gold, 1990). These findings are identical to the patterns of bone mass seen in animals subjected to long-term "immobilization and subsequent remobilization" (Jaworski and Uhthoff, 1986). Also, cellular events associated with loss of estrogen and immobilization are similar. Both cause (i) increased activation of remodeling (Young *et al.*, 1986; Wronski *et al.*, 1988b; Slemenda *et al.*, 1987), (ii) increased bone resorption (Young *et al.*, 1986; Jaworski and Uhthoff, 1986; Wronski *et al.*, 1988b; Lindsay *et al.*, 1980) and (iii) negative bone balance, leading to a net loss of bone mass.

Some *in vitro* studies (Ernst, Health and Rodan, 1989) suggest that the effects of estrogen are mediated through decreased responsiveness to PTH. In fact, estrogens have been used successfully to treat primary hyperparathrolidism in postmenopausal women (Gallagher and Nordin, 1972). However, other studies showed that estrogen effects on bone are pervasive in the presence or absence of PTH (Goulding and Gold, 1988). Some studies have shown that the ability of estrogen to induce bone resorption is not caused by indirect effects of plasma calcitonin levels (Hurley *et al.*, 1989; Goulding and Gold, 1988). It has also been suggested that estrogen effects on bone are mediated by osteoblasts, which have estrogen receptors (Komm *et al.*, 1988; Eriksen *et al.*, 1988). However, estrogen receptors on osteoblasts are present only in very low concentrations (Colston *et al.*, 1989). Direct effects of estrogen on bone cells in culture are not conclusive (Ernst, Health and Rodan, 1989; Colvard *et al.*, 1989). At present, the beneficial effects of estrogen on the skeleton cannot be completely explained by the direct effects of estrogen on bone forming cells.

An increase in bone turnover associated with estrogen deprivation occurs in the iliac crest (Steiniche *et al.*, 1989). However, observed changes in bone remodeling in the iliac crest are not nearly as dramatic as changes seen in the weight-bearing bones of rats (Wronski, Clintron and Dunn, 1988a; Wronski *et al.*, 1988b). It is possible that estrogen related changes measured in the iliac crest under represent the intensity of estrogen related effects in weight bearing portions of the skeleton. There is a strong possibility that the effects of estrogen are dose dependent.

The adipocyte has long been recognized as an estrogen producing cell, particularly in postmenopausal women. Thus, early postmenopausal women who lose bone rapidly have lower levels of both estrone and estradiol than the corresponding slow losers, and this may be accounted for by their lower fat mass (Riis, Rodbro and Christiansen, 1986). Reid *et al.* (1992) have confirmed a relationship between circulating estrone levels and bone density, but have shown this to be independent of the effects of fat mass. This implies that estrogen is not the only pathway by which fat influences bone density. This is supported by the finding of a fat–bone relationship in premenopausal women, in whom the adipocyte is a relatively minor source of estrogens. It is suggested that estrogen acts through two receptors ($\alpha$ and $\beta$); $\alpha$ appears to be the primary mediator of estrogen in actions on the skeleton (Lee *et al.*, 2003). Osteoblasts do express $\beta$, but the action of $\alpha$ and $\beta$ on bone are often in contradiction. However, other authors suggest that activation of these two receptors has similar effects on bone (Windhahl *et al.*, 2001; Sims *et al.*, 2002).

Evidently, from the above, there is strong evidence that estrogens are important for the maintenance of bone mineralization over the life span. However, because of their adverse effects on cardiovascular outcomes, estrogens are no longer considered the first choice intervention for the prevention and treatment of osteoporosis in menopausal women (McNagny *et al.*, 2002; Orwoll and Nelson, 1999).

Testosterone has conventionally been assumed to be the key sex steroid in bone metabolism in men. However, as a result of some investigations (Bilezikian, Morishima and Bell, 1998; Smith *et al.*, 1994) other factors have been pointed out. In fact the reason for the age-related decline in BMD in men seems to be a corresponding decline in levels of bioavailable estrogen (Gennari and Brandi, 2001). In an effort to elucidate how free testosterone and estrogen act on aging male skeleton, Falahati-Nini *et al.* (2000) eliminated the endogenous production of these steroids in elderly men and then studied these subjects under conditions of selective physiologic replacement. They point out that testosterone and estrogen are equally effective in preventing a decrease in the bone formation marker osteocalcin, suggesting a shared role in bone formation in the aging male skeleton.

It has been suggested that estrogen may play a role in regulating bone turnover in men as in women. Smith *et al.* (1994) have reported that men with homozygous mutations in the estrogen receptor gene in the presence of a normal testosterone level had marked osteopenia with an elevated index of bone turnover. Several authors (Morishima *et al.*, 1995; Carani *et al.*, 1997) have described two cases with mutations in the aromatase gene who had low BMD and high bone turnover despite normal testosterone levels (Carani *et al.*, 1997; Morishima *et al.*, 1995). Treatment with estrogen markedly increased BMD in both patients. An epidemiological study found that serum estradiol and sex hormone binding globulin (SHBG), but not testosterone levels, are associated with BMD in healthy men over 65 (Slemenda *et al.*, 1997). Also, Khosla *et al.* (1998) showed that serum bioavailable estrogen declined with age in a population based age-stratified sample of men and women and that it is an independent predictor of BMD in both genders. However, the role of estrogen deficiency is not clearly understood in young and middle-aged men with osteoporosis. Raloxifene has been shown to decrease the risk of vertebral fracture and does not appear to increase the risk of breast cancer (Modder *et al.*, 2004).

Legrand *et al.* (2001) have reported a marked increase of serum SHBG concentration and, consequently, a reduction of free androgen and free estrogen index in men with osteoporosis. This result suggests that bioavailable testosterone and bioavailable estradiol are reduced in men with osteoporosis and could partially explain bone loss. Several physiological and pathological conditions are known to be associated with an increase in SHBG levels, including aging, hypogonadism, hyperthyroidism, diabetes and severe alcoholic liver disease (Selby, 1990). Legrand *et al.* (2001) have suggested that the higher the SHBG serum concentration and the lower the bioavailable testosterone and estradiol level, the higher the bone remodeling and resorption. This mechanism could explain the accelerated bone loss in middle-aged men previously described by other investigators studying healthy but older men (Khosla *et al.*, 1998; Slemenda *et al.*, 1997). Legrand et al. (1999, 2000) have also shown that there is a strong association between hip BMD and vertebral fracture in middle-aged men. Further, for each SD increase in SHBG the risk of vertebral fracture increases twofold. Androgen receptors are present in osteoblasts. Like estrogen, androgens can regulate the interleukin (IL)-6 promoter, and, in experimental animals, orchidectomy has been associated with increased IL-6 production and bone loss. In men, a chronically low androgen level has been associated with low bone mass and testosterone has been associated with that. In contrast, testosterone can enhance bone mineral density. However, the role of androgen in maintaining bone mass in men and women is not clear. Legrand *et al.* (2001) have suggested that the serum SHBG concentration is increased in most middle-aged men with idiopathic or secondary osteoporosis and is correlated with bone remodeling markers, hip bone mineral density and the presence of vertebral fracture. It has been shown that serum SHBG level is an independent predictor of hip BMD in primary and secondary osteoporosis, even after adjusting for age, BMI and steroid levels. In contrast, there is no clear association between hip BMD and SHBG in age-matched controls, suggesting that an elevated serum SHBG level is a viable pathological finding in osteoporotic patients. It has been well documented that testosterone positively affects bone mineral density (Lauretani *et al.*, 2006; Composton, 2001). In fact, androgen therapy in adult men with advanced prostate cancer is followed by rapid bone loss, similar to the loss of skeletal integrity in women after surgical ovariectomy or during early menopause. Significant loss, though to a lesser extent, has been observed at appendicular skeletal sites, including the hip and the radius, after cessation of androgen therapy (Vander Schueren *et al.*, 2004).

## 6.6.2 Impact of HRT on Osteoporotic Fractures

Hormone replacement therapy (HRT) has been introduced, mainly given to suppress menopausal symptoms, but now commonly used for the prevention of postmenopausal osteoporosis (Rossouw *et al.*, 2002). HRT reduces postmenopausal bone loss and decreases the incidence of osteoporotic fractures by about 50% (Weiss *et al.*, 1980). Randomized intervention trials have demonstrated that hormone replacement (estrogen with and without progesterone) reduces bone turnover and increases bone mass (Chesnut *et al.*, 1997).

Indeed, HRT is now the most widely accepted and used intervention for the prevention of postmenopausal bone loss (Consensus Development Conference, 1991). Women who received 2–3 years of HRT had a significantly lower risk of vertebral fracture at the time of follow up compared with the women on placebo ($P = 0.03$). The number of vertebral fractures is also significantly reduced ($P = 0.025$). The risk of single vertebral fractures is reduced by 32% in HRT treated patients. When all fracture types are combined, HRT treatment for 2–3 years is associated with significantly reduced fracture risk.

Several studies have indicated that the time HRT was initiated had no influence on the optimal prevention of bone loss (Rossouw *et al.*, 2002; Cauley *et al.*, 2001; Hulley *et al.*, 1998; Schneider, Barrett-Connor and Morton, 1997). Some studies suggest that, for optimal prevention of fracture, HRT should be initiated within 5 years of the menopause and used for more than 5 years (Weiss *et al.*, 1980; Cauley *et al.*, 1995; Randell *et al.*, 2002). These data imply that to prevent osteoporosis women need to start HRT soon after the menopause and continue for the rest of their lives, which raises concern about breast cancer and other risks associated with extended use of HRT (Grady *et al.*, 1992; Rossouw *et al.*, 2002). To date, few studies have investigated the changes in bone mass following cessation of HRT (Cauley *et al.*, 1995; Christiansen, Christensen and Transbol, 1981; Jemollieres, Pouilles and Ribot, 2001; Greendale *et al.*, 2002; Riis, Johansen and Christiansen, 1988). All studies agree that once HRT is stopped, BMD decreases.

Bagger *et al.* (2004) concluded that even limited HRT given in the postmenopausal years can be of long-lasting benefit in terms of preventing bone loss and related fractures. Current recommendations suggest that prolonged HRT for at least 10 years or so is optimal, with relatively high (bone-sparing) doses of estrogens. There are few data from randomized controlled studies concerning the effects of HRT on fracture, but, supported by epidemiological evidence, it is widely believed that the preservation of bone mass results in a decrease in fracture risk. Longer-term treatments may increase this benefit but may also increase the risks. These extraskeletal risks and benefits of HRT treatment have not been quantified accurately because of the uncertain magnitude of the variability in the population data.

There also appears to be a differential response to HRT, depending on the skeletal site selected, with the response being greatest at the lumbar spine (Bagur *et al.*, 1996; Eiken, Nielsen and Kolthoff, 1997; Young *et al.*, 1996).

### *6.6.3 Role of Estrogen–Progesterone Combination*

Estrogen therapy has some built-in advantages. However, long-term estrogen therapy must be administered cautiously, because it has been associated with increased risk of endomentrial carcinoma (Jick *et al.*, 1979; Hulka *et al.*, 1980). Combined estrogen and progesterone therapy reduces that risk and is therefore recommended as the most suitable long-term therapy for postmenopausal bone loss prevention (Christiansen, Christensen and Transbol, 1981; Riis *et al.*, 1987).

The effect of progesterone on climacteric women has been studied and symptom relief has been reported (Appleby, 1962; Nordin, 1979). Progesterone may also have a role in preventing postmenopausal bone loss (Horsman *et al.*, 1977; Christiansen *et al.*, 1980; Lindsay *et al.*, 1987b; Lindsay *et al.*, 1978; Dequeker *et al.*, 1987; Dequeker and De Muylder, 1982; Abdalla, Hard and Lindsay, 1985; Nordin, 1979) by increasing bone formation, as suggested by elevated levels of serum alkaline phosphatase and sodium GLA protein (osteocalcin) (Christiansen *et al.*, 1985). Progestogens, administered to postmenopausal women, decrease urinary calcium and hydroxyproline excretion (Gallagher and Nordin, 1975; Selby *et al.*, 1985). Additionally, progesterone may exert a direct action on bone because progesterone receptors have been identified in this tissue (Yoshioka *et al.*, 1980).

The effects of experimental surgical castration on bone mass are usually studied in rats, as it has been demonstrated that oophorectomy in these animals led to rapid bone loss (Saville, 1969; Wronski *et al.*, 1985). Barbagallo *et al.* (1989) have concluded that the long-term consequence of oophorectomy in rats causes a significant decrease in bone mass and the administration of progesterone is very effective in preventing the subsequent development of osteoporosis. These authors have suggested that 17$\beta$-estradiol and progesterone treatment leads to similar results in oophorectomized rats. It is possible that not all progestins exert the same action on bone. Norethisterone has been proved to be able to prevent bone mass loss after 2 years of treatment (Nordin, 1979). Barbagallo *et al.* (1989) have suggested that, compared with progesterone alone, the combination of estrogen and progesterone is less effective in preventing post-oophorectomy bone loss in rats, in agreement with the findings of Lindsay *et al.* (1978). The results of Verharr *et al.* (1994) support the notion that estrogen and progestins are able to directly stimulate osteoblastic cells. This may also account for one of the mechanisms by which these steroids preserve bone mass by changing the balance between bone formation and resorption in a positive direction (Riggs *et al.*, 1969). It is suggested that progesterone may antagonize the action of estradiol in the vagina and uterus by decreasing the levels of cytoplasmic estrogen receptors (Hsueh, Peck and Clark, 1976), or by decreasing the tissue levels of estradiol subsequent to increasing the activities of the catabolic enzymes (Tseng and Gurpide, 1975).

Moreover, progesterone inhibits endometrial estrogen reception synthesis (Okulicz, Evans and Leavitt, 1991; Dicarlo *et al.*, 1983). Estro-progestin preparations currently available contain extremely low doses of progestins with very low molar progestin/estrogen ratios. Consequently, it is reasonable to assume that the dose of estradiol obscures any antagonistic effect of progesterone. Several studies have examined the relationship between pill use and bone density at various skeletal sites and have yielded discrepant results (Goldsmith and Johnston, 1975; Lindsay, Tohme and Kanders, 1986; Johnell and Nilsson, 1984; Stevenson *et al.*, 1989; Lloyd *et al.*, 1989; Rodin, Chapman and Fogelman, 1991; Dequeker *et al.*, 1987; Hreschychyn, Hopkins and Zylstia, 1988; Enzelsberger *et al.*, 1988; Mazess and Barden, 1991).

Studies by Cooper *et al.* (1993) suggest that pill use does not protect women against the occurrence of osteoporotic fractures in later life. Estrogen used alone, or in combination with progesterone and/or calcium, has a protective effect against bone loss at the forearm, spine and hip (Compston, 1997; Delmas, 1999; Hailey *et al.*, 1998; Holzherr *et al.*, 2000). Some data indicate that the estrogen, particularly when combined with progestins, increased the risk of breast cancer and cardiovascular disease in women. This then outweighs the benefits to the skeleton.

## 6.7 Glucocorticoid

Glucocorticoids have a multitude of effects on the immune response at several sites and are both anti-inflammatory and immunosuppressive when administered therapeutically. They are used from very low doses for basal immunosuppressive treatment in autoimmune diseases to extremely high doses for the treatment of acute situations such as spinal cord injury or acute exacerbations of life-threatening immunologically mediated disorders. However, bone loss is a frequent side effect of long-term glucocorticoid therapy. If the diagnosis, treatment and prevention of steroid induced osteopenia are to be improved, the pathogenetic mechanisms involved need to be understood in greater detail. Estrogen prevents glucocorticoid-induced osteoblast apoptosis *in vivo* and *in vitro* (Gohel, McCarthy and Gronowicz, 1999). Another agent capable of increasing bone mass *in vivo*, parathyroid hormone (PTH), appears to exert this effect in part through the inhibition of apoptosis (Jilka *et al.*, 1999). Present knowledge of these mechanisms makes special reference to glucocorticoid mediated effects on osteoblasts and osteoclasts. The interrelationship between glucocorticoids and the metabolism of parathyroid hormone and vitamin D and the ability of glucocorticoids to suppress gonadal function and the effects of glucocorticoids on intestinal and renal calcium transport is one of continuing clinical interest. There are at present evidence available of underlying the possible mechanisms. One question is how these substances act at the molecular level. Steroids were long thought to act exclusively by inducing changes in the transcriptional activity of cells by interacting with cytosolic receptors. In addition to the existence of specific and nonspecific nongenomic effects the existence of such genomic effects has since been confirmed (Buttgereit *et al.*, 1994; Lipworth, 2000). Genomic effects (at $>10^{-12}\,\mathrm{mol\,L^{-1}}$) occur within a few minutes and are mediated by membrane-bound glucocorticoid receptors. Nonspecific nongenomic effects (at $>10^{4}\,\mathrm{mol\,L^{-1}}$) occur within seconds. They appear to result from direct interactions with biological membranes (Buttgereit *et al.*, 2000).

For several years it has been widely accepted that glucocorticoids increase renal calcium elimination and decrease intestinal calcium absorption (Figure 6.3). A negative calcium balance resulting from changes in renal and intestinal calcium transport has been considered to be responsible for secondary hyperparathyroidism in glucocorticoid treated patients (Bikle *et al.*, 1993; Klein *et al.*, 1977; Lukert and Raisz, 1990; Reid, 1997). Although the precise mechanisms remain unclear, the increased urinary elimination of calcium that occurs in response to glucocorticoids may be due to diminished tubular reabsorption (Reid, 1997;

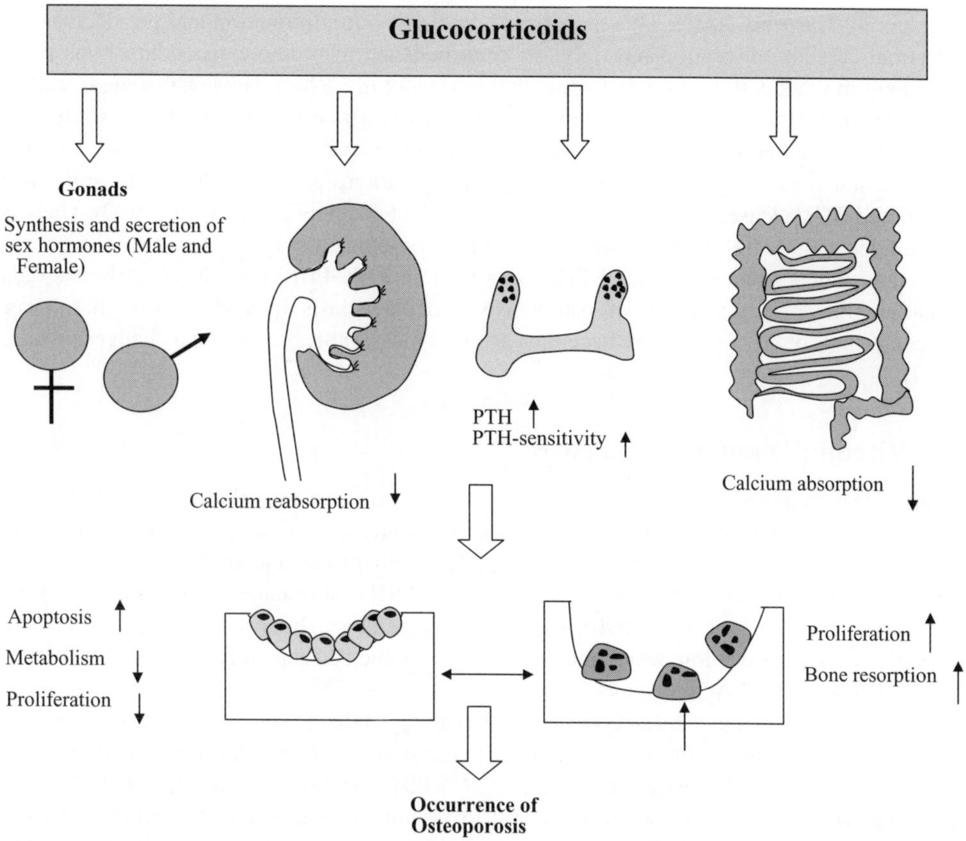


**Figure 6.3** Mechanisms of glucocorticoid-induced osteoporosis. In addition to their direct effects on bone cells, glucocorticoids have effects on the intestine, kidneys, gonads and probably the parathyroid glands, which indirectly lead to osteoporosis (Reproduced with permission from D. Patschan *et al.*, Molecular mechanisms of glucorticoid-induced osteoporosis, *Bone*, **29**(6), 498–505 © 2001, Elsevier B.V.)

Reid and Ibbertson, 1987). Findings on the effect of glucocorticoids on intestinal calcium transport are less consistent. As reported by Lukert and Raisz (1990), studies in humans and rats have found that intestinal calcium absorption remains unchanged, in the presence of glucocorticoids. Such variability in findings has been assigned to differences in steroid mediated effects on the upper and lower level. It has also been hypothesized that differences in intestinal calcium absorption are determined by the duration and intensity of glucocorticoid administration. A decrease in active transcellular transport and in normal brush border vesicle uptake of calcium, as well as diminished synthesis of calcium binding proteins, have been considered as mechanisms involved in diminished intestinal calcium absorption (Feher and Wasserman, 1979; Kimberg *et al.*, 1971; Shultz, Bollman and Kumar, 1982). These are time and dose dependent differences in intestinal calcium transport in various species and have relation to nongenomic glucocorticoid effects on the intestinal epithelium. Notwithstanding these differences, several studies have found net intestinal calcium absorption in humans to be decreased following glucocorticoid administration (Colette *et al.*, 1987; Klein *et al.*, 1977; Reid, 1997;

Ziegler and Kasperk, 1998). This increase in renal calcium elimination and the decrease in intestinal calcium absorption have always been believed to promote secondary hyperparathroidism in glucocorticoid treated patients (Gennari *et al.*, 1984). However, several studies using intact PTH assays in humans have failed to demonstrate this phenomenon (Hattersley *et al.*, 1994; Luengo *et al.*, 1991; Shultz, Bollman and Kumar, 1982; Slovik *et al.*, 1980).

In human parathyroid cells from hyperplastic parathyroid glands, dexamethasone also stimulated PTH release, both in time and dose dependent manner (Peraldi *et al.*, 1990). There is also evidence that PTH mediated effects are more pronounced in the presence of glucocorticoids and independent of higher PTH levels (Figure 6.3). In isolated bone cells, steroids modulate the PTH sensitivity of osteoblasts and osteoclasts in such a way that lower concentrations of bovine PTH are necessary to elicit measurable biochemical changes (Wong, 1979).

## 6.8 Vitamin D and Osteoporosis

Vitamin D deficiency is deleterious to bone and other body tissues (Plotinikoff and Quigley, 2003; Grant, 2002). Indeed, since hypovitaminosis D influences calcium-regulating hormones, insufficient levels of vitamin D has negative effects on skeletal metabolism. A decrease in serum concentrations of 25-hydroxyvitamin D (25-OHD) stimulates parathyroid hormone (PTH) secretion (Chapuy *et al.*, 1996), leading to an increment in bone remodeling, particularly osteoclastic bone resorption, which increases bone loss and consequently fracture risk (Parfitt, 1990c; Ooms *et al.*, 1995).

Vitamin D deficiency and secondary hyperparathyroidism can contribute not only to accelerated bone loss and increasing fragility, but also to neuromuscular impairment that can increase the risk of falls (Bischoff-Ferrari *et al.*, 2004; Sambrook *et al.*, 2004). The active hormonal form 1,25-dihydroxyvitamin (calcitriol) is not only necessary for optimal intestinal absorption of calcium and phosphorous, but also exerts a tonic inhibitory effect on PTH synthesis. Consequently, it appears that there are dual pathways that can lead to secondary hyperparathyroidism. However, calcium and vitamin D supplementation does not reduce fracture incidence if there is a greater deficiency of vitamin D (Grant *et al.*, 2005).

Two main types of vitamin D deficiency have been described: (i) primary vitamin D deficiency, which is caused by an inadequate supply of the parent compound and can be treated with plain vitamin D supplementation; (ii) 1,25-$(OH)_2D_3$ deficiency (mainly caused by insufficient renal production) or resistance (mainly caused by impaired tissue response), which requires the administration of active vitamin D analogs [1,25-$(OH)_2D_3$, calcitrol] or one of its precursors 1$\alpha$-$(OH)D_3$ (alfacalcidol) for correction (Tsai *et al.*, 1984). Both of these can occur with aging and both have been implicated as possible causes of intestinal calcium malabsorption, secondary hyperparathyroidism and involutional osteoporosis, leading to possible fracture (Lau and Baylink, 1999; Scopacasa, Wishart and Horowitz, 2004; Pattanaungkul, Riggs and Yergey, 2000). Figure 6.4 shows a sequential occurrence of such events.

Vitamin D is essential for mineralization of bone and its deficiency is thought to be related to rickets/osteomalacia. Several modifiable risk factors for osteoporotic fractures have been proposed (Cummings *et al.*, 1997; Cooper and Melton, 1992; Cumming *et al.*, 1997). One obvious factor is the diet. The influence and importance of dietary calcium intake, in particular, and the intake of vitamin D, in primary prevention of osteoporotic fractures, have however been highly debated (Cummings *et al.*, 1997; Kanis, 1999). In several randomized trials in

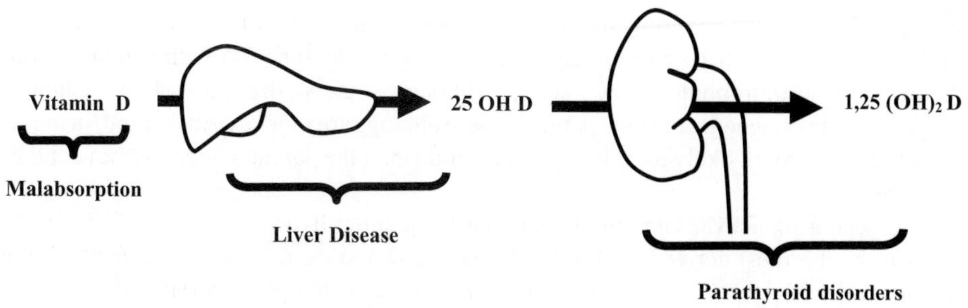


**Figure 6.4** Metabolism and abnormalities of vitamin D causing osteomalacia

postmenopausal women, supplemental calcium has slightly reduced loss of bone mineral density, but its impact on fracture is uncertain (Cummings and Nevitt, 1997; Kanis, 1999). Reviews (Gillespie *et al.*, 2001; Houselmann and Rizzoli, 2003) and meta-analyses (Cranney *et al.*, 2002; Papadimitropoulos *et al.*, 2002) of randomized controlled trials have shown that vitamin D treatment decreases the risk of osteoporotic fracture in postmenopausal women. An adequate serum level of vitamin D is, however, mainly maintained by endogenous production, but with increasing age this production is impaired and the exogenous component of vitamin D becomes important.

In osteoporosis, administration of 1,25-$(OH)_2D_3$ increases serum osteocalcin (Zerwekh, Sakhaee and Pak, 1985; Duda *et al.*, 1988). 1,25-$(OH)_2D_3$ is also a potent stimulator of bone resorption *in vitro* (Raisz *et al.*, 1972; Reynolds, Pavlovitch and Valsan, 1976) and in rats (Holtrop, Raisz and Simmons, 1974). This is therefore proposed as a potential activator in the activate, depress, free, repeat (ADFR) treatment of osteoporosis (Frost, 1981; Parfitt, 1988). It is suggestive of a large number of new remodeling episodes can be initiated in a coherent manner in a short time (Parfitt, 1990b).

Vitamin D supplementation is suggested to increase intestinal calcium absorption efficiency (Ikeda and Ogatta, 2000), synthesis of osteocalcin (Lian *et al.*, 1989) and bone mineralization (Van Leeuwen *et al.*, 2001). In particular, intestinal calcium regulation is regulated by the endocrine system, especially 1,25-dihydroxyvitamin $D_3$. On the other hand, an antiresorptive or anabolic effect of vitamin $K_3$ on the skeleton *in vivo* has been demonstrated in OVX, sciatic neurectomized or prednisolone treated rats (Akiyama *et al.*, 1999; Iwamoto, Yeh and Takeda, 2003; Hara, Kobayashi and Akiyama, 2002), suggesting that this may affect bone metabolism. In particular, vitamin $K_3$ improves poor calcium balance in OVX rats and also abnormal decline in serum calcium. This also improves magnesium levels and abnormal increase in serum PTH level and renal calcium concentration in calcium–magnesium deficient rats (Kobayashi, Hara and Akiyama, 2002). Synergistic effects of vitamin K and vitamin D supplementation on reducing bone loss in adult OVX rats and on increasing peak bone mass in young rats, and the efficiency of this combined treatment for increasing lumbar BMD in postmenopausal women with osteoporosis, have been demonstrated (Matsunaga, Ito and Sakou, 1999). It would thus be expected that vitamin K and vitamin D supplementation might act on bone synergistically. However, the effect of such supplementation on intestinal calcium absorption, renal calcium reabsorption and cortical and cancellous bone growth under pure mild calcium deficiency remains to be fully understood.

The question remains regarding the optimal form of vitamin D for this purpose and whether vitamin D treatment can be considered alone or only in combination with calcium supplementation (Papadimitropoulos *et al.*, 2002). Although some studies have shown vitamin D analogs to be advantageous for some patients, these analogs are more expensive and, in the case of clairol, may be more likely to cause hypercalcemia than the parent compound (Lau and Bay link, 1999).

Before becoming active, vitamin D undergoes sequential hydroxylation. 1,25-$(OH)_2D_3$ is thought to be the most active metabolite of vitamin D and exerts opposing effects on bone metabolism (Baylink *et al.*, 1980; Bordier *et al.*, 1969). Although 1,25-$(HO)_2D_3$ appears to correct many of the abnormalities of vitamin D deficiency in bone (Parfitt *et al.*, 1984; Gallagher *et al.*, 1986) it is as yet unclear whether this has a direct stimulator effect on bone mineralization process. Weinstein *et al.* (1984) have demonstrated that histological abnormalities in bone of vitamin D-deficient rats are almost completely reversed by the infusion of calcium (Ca) and phosphate ($P_i$). These authors have suggested that 1,25-$(OH)_2D_3$ exerts its effect only through elevation of the ionic product of serum and $P_i$ by enhancing their intestinal absorption. However, some reports show that the infusion of Ca or $P_i$ cannot completely eliminate the abnormalities in bone (Bordier *et al.*, 1969; Lee *et al.*, 1975). Vitamin D also contributes mineral homeostasis by increased release of Ca and $P_i$ from bone in blood (Norman, 1979). According to current concepts, the intestinal calcium absorption can be described as being the sum of a saturable (hyperbolic) and a nonsaturable (linear or diffusional) process (Broner, 1984; Heaney, Saville and Recker, 1975; Pansu, Bellaton and Bronner, 1979). The saturable process occurs along the proximal part of the intestine and it requires the metabolism of vitamin D.

In addition, studies using cultured osteoblasts or osteoblast-like cells have demonstrated that these cells process 1,25-$(OH)_2D_3$ receptor and that 1,25-$(OH)_2D_3$ alters alkaline phosphate activity as well as the synthesis of matrix proteins such as collagen, osteocalcin and matrix Gla protein in these cells (Price and Bankol, 1980; Manolagas, Burton and Deftos, 1981; Rowe and Kream, 1982; Walters *et al.*, 1982; Fraser, Otawara and Price, 1988). Therefore, there is a possibility that 1,25-$(OH)_2D_3$ may have a direct stimulatory effect on osteoblasts to enhance the bone mineralization process.

Geusens *et al.* (1991) compared the effect of short term 1,25-$(OH)_2D_3$ treatment of bone formation and resorption and the vitamin D endocrine system in osteoporosis. Repeated superphysiologic doses of 1,25-$(OH)_2D_3$ in the presence of high levels of dietary calcium may affect an increase in metaphysical osseous tissue and metaphysical osteoid by several mechanisms, including increased bone matrix formation, impaired bone matrix mineralization and decreased osteoclasis.

*In vitro* studies have demonstrated that fetal bone responds to parathyroid and 1,25-$(OH)_2D_3$ in serum free culture with an initial rapid elevation in bone resorption followed by a stimulation of bone precursor cells proliferation and an increase in osteoblastic activity with resultant new bone formation (Drivadahl, Liu and Baylink, 1984). Hypervitaminosis D has been reported in dogs to increase bone appositional rates and osteoid width subsequent to an increase in resorptive indices (Takahashi *et al.*, 1973). The findings of an increased number of osteoblasts and increased active osteoid surface associated with increased metaphysical osteoid and osteoid surface in this study also suggests that 1,25-$(OH)_2D_3$ increased bone matrix synthesis. The effect of 1,25-$(OH)_2D_3$ on osteoblasts could be followed by the changes in osteocalcin, a specific marker of osteoblast function. A significant increase in serum osteocalcin levels is observed after administration of 1,25-$(OH)_2D_3$, whereas the alkaline phosphate levels decreased (Brixen *et al.*, 1990).

Decreased osteoclasis also effects an increase in metaphysical bone mass. Osteoblasts per millimeter total trabecular surface perimeter and active resorbing surface decreased significantly in 1,25-$(OH)_2D_3$ treated rats on days 8 and 10. The decrease in osteoclasts by day 10 also confirms the findings of Boyce and Weisbrode (1983). These results are similar to those of Lindholm, Wilsson and Lindholm (1982), who demonstrated a progressive decrease in resorption indices in rats chronically treated with l$\alpha$-hydroxyvitamin $D_3$. Nakamura *et al.* (1992) have reported that 24R,25-$(OH)_2D_3$ reduces bone turnover and stimulates osteoblastic functions.

The response of bone to 1,25-$(OH)_2D_3$ may be catabolic, because 1,25-$(OH)_2D_3$ is a potent inducer of bone resorption, both *in vivo* (Tanaka and Deluca, 1971; Reynolds, Pavlovitch and Valsan, 1976) and *in vitro* (Raisz *et al.*, 1972) and has been shown to inhibit collagen synthesis (Raisz *et al.*, 1978, 1980). Ultrastructural studies of thyroparathyroidictomized rats have shown 1,25-$(OH)_2D_3$ to increase both osteoblastic and osteoclastic activity in rats that were fed a 0.05% calcium diet (Weisbrode, Capen and Norman, 1978), in contrast to increasing only osteoblastic activity in rats fed a 1.0% calcium diet (Weisbroade *et al.*, 1979). The combination of elevated dietary calcium and 1-hydroxyvitamin $D_3$, which is rapidly converted *in vivo* into 1$\alpha$,25-$(OH)_2D_3$ (Zerwekh *et al.*, 1974), has been reported to be successful in promoting a net increase in bone in human osteoporotic patients (Lindholm, Sevastikiglore and Lindgren, 1977, 1978; Lindholm, 1979). The increased skeletal mass induced by experimental intoxication with *Solanum malacoxylon*, a calcinogenic plant that contains a glycoside of 1,25-$(OH)_2D_3$ (Wasserman *et al.*, 1976), is enhanced by elevated dietary calcium (Dos Santos *et al.*, 1976). In addition, long-term studies dealing with the effects of 1,25-$(OH)_2D_3$ on bone have demonstrated that a low calcium diet abolishes enhanced bone formation mediated by 1,25-$(OH)_2D_3$ (Larsson, Ahlgren and Lorentzen, 1980).

Ramp *et al.* (1986) have reported that the a-release system in rat bone is dependent on the vitamin D status of the animals. The difference in net release of calcium between +D and –D tibial does not typically represent a difference in the rate of bone resorption during the 4-h incubations. These results suggest that separate mechanisms regulate the net release of Ca and $P_i$ from bone, but do not exclude the possibility that in severely D-deficient animals decreased bone resorption may also be a factor contributing to depressed Ca release.

Combined calcium and plain vitamin D have been proposed as a standard treatment for elderly people at risk for osteoporosis, but the therapeutic role of D analogs continuous to be debated.

Matsumoto *et al.* (1991) demonstrated that osteoblastic clone (MC3T3-E1) cells accumulate $^{45}Ca$ into cell and matrix layers after 2 weeks of culture. Also, most Ca appears to be deposited in the extracellular matrix. 1,25-$(OH)_2D_3$ stimulates the accumulation of $^{45}Ca$ in a dose-dependent manner without affecting alkaline phosphatase activity or collagen synthesis, in agreement with the findings of Kurihara *et al.* 1984, 1986). Also, neither hPTH (1–34) nor 24,25$(OH)_2D_3$ has any effect on *in vitro* mineralization by MC3T3-E1 cells in itself or in combination with 1,25-$(OH)_2D_3$. By using a cell culture system these authors were able to demonstrate that 1,25-$(OH)_2D_3$ has a direct stimulatory effect on the mineralization process by differentiated osteoblasts, although its mechanism of action is yet to be understood. It has also been demonstrated (Matsumoto *et al.*, 1985) that 1,25-$(OH)_2D_3$ stimulates the synthesis and increases the content of Phosphatidyl Serine (PS) in osteoblast. The stimulatory effect of 1,25-$(OH)_2D_3$ on PS synthesis may be associated with the stimulation of bone mineralization.

Bone inductive activity of extracellular matrix extracts is reported to be markedly reduced when bone matrix proteins are extracted from vitamin-D-deficient rats (Sampath, Weintroub and Reddi, 1984). Therefore, it is also possible that alterations in matrix protein metabolism is involved in the stimulation of matrix mineralization by 1,25-$(OH)_2D_3$. It is observed that 1,25-$(OH)_2D_3$ reduces the synthesis and enhances the degradation of proteoglycans in MC 3T3-E1 cells (Takeuchi *et al.*, 1989). Because the gap zone of collagen fibril is thought to be the site of mineralization and because gap zone associated proteoglycans are absent in mineralizing tissues, it has been postulated that proteoglycans inhibit the mineralization of collagen matrix (Scott 1988), Scott and Oxford (1981).

It was reported that not only 1,25-$(OH)_2D_3$ but also the presence of PTH and 24,25-$(OH)_2D_3$ is important for the stimulation of bone mineralization (Endo *et al.*, 1980; Tam *et al.*, 1986; Gunness-Hey and Hock, 1984). The latter effect was demonstrated when organ cultures of embryonic chick long bones were used *in vitro* or when these agents were administrated *in vivo*. Therefore, the enhancement of mineralization by these agents may also require the presence of other types of cells in addition to osteoblasts.

Serum 1,25-dihydroxyvitamin D concentrations are usually decreased in patients with malignant hypercalcemia due to the suppression of the renal 1-hydroxylase by hypercalcemia and low serum PTH concentrations. However, in patients with renal cell carcinoma, serum 1,25-dihydroxyvitamin D values may be within the normal range despite hypercalcemia (Yamamoto *et al.*, 1987). In some patients with B cell lymphomas, T-cell lymphoma and Hodgkin's' disease serum 1,25-dihdroxyvitamin D is increased and appears to cause the development and maintenance of hypercalcemia (Breslau *et al.*, 1984).

Richy *et al.* (2005) have suggested that vitamin D analogs alfacalcidol and calcitriol, compared to native vitamin D, may exhibit better efficacy in preventing spinal bone loss and spinal and/or nonspinal fracture's primary osteoporosis. D analogs are found to be active in preventing hip and spinal bone loss in corticosteroid induced osteoporosis whereas native vitamin D provides similar efficacy but is restricted to the hip level. Controversial conclusions surround the indirect or direct comparisons regarding fracture prevention in patients receiving glucocorticoid therapy. Trends favoring D analogs are found in indirect comparisons, whereas significantly better efficacy is found for alfacalcidol in the single superiority trial that is available. Michaelsson *et al.* (2003) concluded that there is no association between dietary vitamin D intake and fracture risk. Clinical trials on elderly individuals at high risk for calcium and vitamin D deficiency indicate that supplementation of both can reverse secondary hyperparathyroidism, decrease bone resorption, increase bone mass and decrease falling and fracture rates (Lips, 2001). In another study calcium and vitamin D supplementation did not reduce fracture incidence significantly, which may be because this population was less deficient in vitamin D (Grant *et al.*, 2005).

Although estrogen depletion is known to affect bone mass and calciotropic hormones, it has no effect on 25-OHD levels. Mastaglia *et al.* (2006) found that serum 25-OHD levels of OVX rats receiving dietary vitamin D were within the range of normal sham animals ($34.3 \pm 4.3$ $mg\,mL^{-1}$). This is similar to the normal range in human. Moreover, exposure or lack of exposure to fluorescent tube light does not modify 25-OHD levels in the OVX rats.

Vitamin K supplementation in calcium deficient rats is found to increase calcium balance and calcium retention, without increasing calcium intake, and retards the increase in serum PTH level (Iwamoto, Yeh and Takeda, 2003). However, vitamin K supplementation does not influence intestinal calcium absorption efficiency because of there was significant influence on serum 1,25-$(OH)_2D_3$ levels. Taking the findings of various groups of workers, it appears that the effect of vitamin K supplementation in calcium deficient rats is to stimulate renal calcium reabsorption and subsequent retardation of the increase in serum PTH level despite an effect on hypocalcemia. The role of vitamin K on bone metabolism remains to be fully explored and understood.

Further to this, vitamin E (i.e., $\alpha$-tocopherol) supplementation is of interest due to its ability to protect against oxidative damage by scavenging free radicals (Azzi *et al.*, 2000; Ricciarelli, Zingg and Azzi, 2001) and to inhibit certain pro-inflammatory mediators such as prostaglandin $E_2$ ($PGE_2$), interleukin (IL)-1 and tumor necrosis factor-alpha (TNF-$\alpha$) (Meydani *et al.*, 1990; Pathania *et al.*, 1999; Wu, Hayek and Meydani, 2001).

## 6.9 Role of Calcitonin

Calcitoin is now widely used in the treatment of several metabolic bone diseases and its clinical efficacy and safety has been well investigated (Ellerington *et al.*, 1996). In addition to its well-known antiosteoclastic effect (Chambers *et al.*, 1985; Murrils *et al.*, 1989) it is also suggested that this hormone exerts a direct action on osteoblasts (Farley *et al.*, 1988). This effect of calcitonin on bone formation has been documented to occur in experimental animal models during its initial phase (Weiss *et al.*, 1981).

Calcitonin acts primarily to produce anabolic effects in bone by suppressing osteoclastic activity (Chambers, 1985). This leads one to speculate that individuals with active forms of the osteoporotic process could respond much more rapidly to calcitonin therapy than those with less active disturbances, for example, in bone remodeling processes. This hypothesis has been substantiated with reports of rapid sustained responses of postmenopausal and immobilized patients to calcitonin treatment (Schoutens *et al.*, 1988; Minaire *et al.*, 1984b).

Reginster *et al.* (1987) have provided evidence that nasal spray calcitonin effectively counteracts the propensity to bone loss in women who had been menopausal for no more than 36 months and are most likely to have active bone remodeling because of estrogen loss. These findings, demonstrating a protective effect of calcitonin following ovarian failure when active bone turnover dominates, is also confirmed by other workers (Overgaard *et al.*, 1989 and Mazzuoli *et al.*, 1990). Civitelli *et al.* (1988) have reported that those patients with active osteoporotic syndromes responded to therapy with an average gain of 22% of vertebral bone mass after 12 months of therapy. The correlation obtained between blood BGP (bone Gla protein) and urinary hydroxyproline and the retention of $^{99m}$Tc-methylene diphosphonate in this study demonstrates the utility of these biochemical markers as a means of identifying patients with active osteoporosis or those who would respond to calcitonin therapy. The findings of Overgaard and coworkers (1989, 1990) not only confirms above results, but also revealed that the results could be achieved with 200 IU of nasal spray salmon calcitonin per day. These investigators also reported that discontinuous therapeutic regimens with nasal spray calcitonin preparation over a 3-year interval could produce a net gain of bone in both the vertebrae and axial skeleton. Gruben *et al.* (1984) have monitored the effect of calcitonin treatment for osteoporotic patients over a 26 month period. They found an initial increase followed by a decline in bone mass. The increase of bone mass associated with calcitonin treatment is attributed to the decreased bone resorption coupled with normal bone formation. When the bone mass increases to a point where the error signal falls below a threshold value, bone formation is decreased and bone resorption increases. However, bone resorption cannot increase because of the inhibitory effects of calcitonin. Therefore, a new steady state develops with decreased bone turnover. This phenomenon has been demonstrated clinically by Civitelli and coworkers (1988). Other studies demonstrate a striking sensitivity of patients with high turnover osteoporosis to calcitonin, with as much as 22% increments in vertebral bone mass recorded during a 12-month therapeutic level. These promising results may be compared with other forms of therapy in which an increase in vertebral bone mass of only 7–8% is sufficient to cause a significant decrease in the incidence of vertebral fracture rates.

One study (Arinoviche *et al.*, 1987) involving 32 patients with acute pain due to vertebral fractures demonstrated significant benefits in pain and mobility after 14 days of calcitonin therapy compared with placebo. In this study, salmon calcitonin produced a 70% reduction in pain compared with a 49% reduction in the placebo group. In an Italian study (Pontiroli *et al.*, 1986), 100 IU per day of human calcitonin given internasally was more effective in relieving bone pain that 100 IU given parenterally. Although the precise mechanism of calcitonin action remains to be fully understood, there is evidence that the analgesic effect of calcium may be mediated through the endogenous opioid system. The intranasal administration of calcitonin seems to be more effective in producing analgesia than parenteral administration.

Salmon calcitonin is more potent than procine calcitonin. One possible reason for this is a decreased metabolic clearance rate, as has been shown in normal volunteers (Huwyler *et al.*, 1979). In addition, there are also data to indicate that the calcitonin receptor has a higher affinity for salmon calcitonin: this would offer a second mechanism for increased potency of salmon calcitonin (Marx, Wopopdard and Aurbach, 1972). Salmon calcitonin at doses of 100 and 50 IU given subcutaneously inhibit osteoclastic activity (Pazzaglia *et al.*, 1993). Salmon calcitonin is a highly effective therapeutic agent in the treatment of Paget's disease, metastatic bone disease and osteoporosis. This peptide has an excellent safety profile and produces only mild side effects in a small percentage of patients. Radiologic healing of osteolytic lesions is particularly striking with calcitonin treatment.

Calcitonin has a role in preventing bone loss in the prophylaxis and treatment of osteoporosis from various causes (Christiansen, 1991). In patients with bone pain, the analgesic effect of calcitonin was first associated with an improvement of bone lesions. Numerous clinical studies have shown calcitonin to be effective in producing analgesia in patients with osteoporosis that is unrelated to its effects on bone. Treatment with salmon calcitonin nasal spray in patients with established osteoporosis prevents bone loss, normalizes bone turnover and has no side effects.

Calcitonin also appears to stimulate osteoblasts (Azria, 1989) and promotes internal calcium absorption (Chesnut *et al.*, 1984). Its effects match the pathogenetic factors involved in steroid osteoporosis and suppression of endogenous calcitonin secretion has been described during glucocorticoid therapy (Locasico *et al.*, 1982). Despite reduced intestinal calcium absorption, 24-h urine calcium excretion is increased during the first month of glucocorticoid therapy (Reid, 1989; Suzuki *et al.*, 1983) due to reduced tubular reabsorption of calcium (Reid and Ibbertson, 1987).

The principal target cell of calcitonin is the osteoclast, which possess numerous receptors for calcitonin (Marx, Wopopdard and Aurbach, 1972). However, calcitonin deficiency is not found in osteoporotic patients. Calcitonin therapy is less effective than other antiresorptive agents, possibly because osteoclasts can escape calcitonin inhibition (Rosen *et al.*, 2005; Cranney *et al.*, 2002). In bone biopsies obtained from patients with Paget's disease, the most dramatic histologic effect of intravenous administration of salmon or human calcitonin on osteoclasts is an alteration in the ruffled border. Within 30 min of calcitonin administration, osteoclasts exhibit disruption of the ruffles border and are less likely to be attached to bone, an effect that illustrates the potent and rapid anatomical changes resulting from interaction of calcitonin with receptors on the osteoclasts (Singer, Melvin and Miles, 1976).

When compared with controls, calcitonin treatment results in a decrease in urinary calcium and hydroxyproline excretion and a preservation of trabecular volume as quantitated from iliac crest biopsy specimen (Avioli, 1991). Because of these changes, it has been concluded that calcitonin therapy should be administered as soon as possible after the onset of paraplegia (Minaire *et al.*, 1984a) or as soon as possible following immobilization (Minaire *et al.*, 1984b).

## 6.10 Calcitonin and Glucocorticoids

Nasal calcitonin is helpful in the treatment of bony pain secondary to fracture and also works to increase BMD and decrease vertebral fracture risk. Typically, analgesia for bone pain is achieved as early as the first to second week of use. It is speculated that the endogenous opiate system may play a role in mediating the analgesic effects (Geonari, 2002). Chesnut *et al.* (2002) have carried out a 5 year, double blind, randomized, placebo controlled study in 1255 women with established osteoporosis and showed that the 200-IU dose reduces the risk of new vertebral fractures by 33% compared with placebo. There were gains of 1–5% in the lumbar spine BMD in patients receiving 100, 200 and 400 IU of calcitonin daily. However, calcitonin does not provide any fracture protection to the hip (Martens, 2003; Silverman, 2003).

Calcitonin can reduce bone turnover not only in osteoporosis but also in Paget's disease. In patients with malignant hypercalcemia, bone resorption is inhibited for relatively short periods – 24–72 h (Mundy, Wilkinson and Health, 1983), but despite continuous treatment serum calcium returns to pretreatment levels. This can be prevented by administration of glucocorticoids. Thus, the use of calcitonin in combination with glucocorticoids produces a more rapid, profound and persistent hypercalcemia activity (Mundy and Martin, 1982). However, this may be due to the addition of the calciuric effect of the two agents rather than to their activity on bone cells (Hosking, Stone and Foote, 1990).

Long-term glucocorticoid administration may result in osteopenia principally through two mechanisms. The first is direct inhibition of bone formation. This hypothesis is supported by histomorphometric studies (Bressot *et al.*, 1979; Dempster, Arlot and Meunier, 1983; Frost and Villaneuva, 1961), which show a marked decrease in the number of osteoid seams, a low mineral apposition rate and reduced mean wall thickness. Osteoblast-like cells have glucocorticoid receptors and glucocorticoids appear to have a direct inhibiting effect on osteoblastic replication and differentiation (Lukert and Raisz, 1990). The second mechanism is indirect stimulation of bone resorption (Hohn, 1978). Intestinal calcium absorption is inhibited by steroids, although the mechanism responsible for this inhibition of active transcellular transport of calcium in the duodenum is not clearly understood.

## 6.11 Parathyroid Hormone (PTH)

Although estrogen is used for preventing bone loss, a better option would be to reduce osteoporotic bone fracture risk by increasing bone formation to a level greater than bone resorption. This concurrently increases bone mass and its strength. Human parathyroid hormone (hPTH) possesses the anabolic property and may prove to be an effective therapeutic agent for fulfilling this need in the treatment of osteoporosis in humans (Finkelstein *et al.*, 1994; Hodsman *et al.*, 1991; Lindsay *et al.*, 1997). Cessation of hPTH is followed by rapid loss of cancellous bone mass (Cheng, Timonen and Suominen, 1995; Gunness-Hey and Hock, 1989).

PTH is a new anabolic agent that promises to treat osteoporosis efficaciously. PTH stimulates osteoblasts and provokes the remodeling sequence, including the elaboration of several cytokines that accelerate bone resorption. PTH, when administered as a pharmacologic agent in a pulsatile manner, augments bone mass in normal, intact animals (Hefti *et al.*, 1982; Podbesek *et al.*, 1983; Gunness-Hey and Hock, 1984; Mosekilde *et al.*, 1991; Hock and Gera, 1992). Similarly, PTH prevents bone loss in OVX rats (Hock *et al.*, 1988; Hori *et al.*, 1988; Liu and Kalu, 1990; Takahashi *et al.*, 1991) and restores lost bone in osteopenic OVX rats (Liu *et al.*, 1991; Ibbotson *et al.*, 1992; Shen *et al.*, 1992; Kimmel *et al.*, 1993). PTH works by increasing bone resorption and bone formation, increasing BMD and enhancing bone architecture and integrity (High *et al.*, 2002). Daily subcutaneous injections of the synthetic N-terminal portion of PTH, teriparatide, can increase bone mass and strength by markedly increasing bone formation while only modestly increase bone resorption (Neer *et al.*, 2001). A broad consensus of these studies is that the observed augmentation of bone mass in PTH-treated animals is due primarily to a powerful stimulatory effect of the hormone on bone formation. Clinical trials in patients with established osteoporosis have also detected PTH-induced increase in bone formation and cancellous bone mass (Reeve *et al.*, 1976, 1990, 1991; Slovik *et al.*, 1986).

Although the anabolic effects of PTH on cancellous bone in animal models and osteoporotic patients are well accepted, the effects of the hormone in cortical bone are controversial. Several clinical investigations have reported decreased cortical bone in hyperparathyroid and PTH-treated patients despite the observed increase in cancellous bone in the same patients (Hesp *et al.*, 1981; Reeve *et al.*, 1991; Parisien *et al.*, 1990). These authors suggest that PTH augments cancellous bone at the expense of cortical bone. However, other studies in hyperparathyroid and PTH-treated patients have failed to support this concept (Slovik *et al.*, 1986; Reeve *et al.*, 1990; Christiansen *et al.*, 1993). Several treatments have been tested clinically in which PTH has been combined with an antiresorptive agent or 1,25-hydroxyvitamin D to prevent an eventual cortical steal phenomenon (Slovik *et al.*, 1986; Hesch *et al.*, 1988, 1989). It is suggested that the action of PTH, causing an increase in spinal density, is not achieved at the expense of cortical bone from either the axial or appendicular skeleton. However, none of these studies with PTH monotherapy or co-therapy have included measurement of spinal deformity index or other measurements of bone strength.

Cortical bone loss has not been observed in PTH-treated animals (Gunness-Hey and Hock, 1984; Liu and Kalu, 1990; Liu *et al.*, 1991), but these studies were of relatively short duration (2–5 weeks). Wronski and Yen (1994) have described the long-term effect of PTH. They have concluded that, in contrast to estrogen and bisphosphonate risedronate, PTH stimulates cortical bone formation and augments cortical bone mass in the tibial diaphysis of OVX rats. Samnegard, Akhter and Recker (2001) have reported that treatment with hPTH ($75\,\mu g\,kg^{-1}$, three times per week) for 12 weeks restored lumbar vertebral BMD and strength in rats with established osteopenia up to the level of intact rats. They have further concluded that, upon withdrawal of hPTH, risedronate can maintain vertebral BMD and partially maintain bone strength for up to 36 weeks. These findings are potentially important as they may indicate that a hPTH pretreatment and bisphosphonate combination may be a possible option in clinical treatment studies in osteoporosis. hPTH pretreatment followed by estrogen in the given dose had no lasting effects on the vertebral body during the 36 weeks maintenance period. The data for the first 12 weeks period are consistent with another study that compared the efficacy of estradiol and risedronate for maintaining hPTH (1–34) induced tibial cancellous bone in rats with immobilization induced osteopenia (Ma *et al.*, 1995). These results agree with similar studies that have shown increased vertebral bone mass and bone strength after 5–24 weeks of hPTH treatment (Ejersted *et al.*, 1995; Mosekilde *et al.*, 1991; Mosekilde, Danielsen and Gasser, 1994; Mosekilde *et al.*, 1994; Kishi *et al.*, 1998) and OVX rats (Li *et al.*, 1995; Sogaard *et al.*, 1997).

Mosekilde *et al.* (1994) have examined the effect of PTH, estrogen and bisphosphonate (risedronate) and the combination of PTH with the antiresorptive drugs on vertebral bone mass and biomechanical competence in a rat model. These authors have concluded that sexually mature but still growing OVX rats treated with PTH alone show a pronounced anabolic effect on vertebral bone mass and strength. This is not further augmented by co-therapy with estrogen. However, long-term co-therapy with risedronate (Ris) showed a significant sustained increase in bone mass and strength throughout the whole treatment period. These authors have suggested that PTH alone or in combination with antiresorptive agents are a promising therapeutic regimen for postmenopausal osteoporosis.

Systematic intermittent PTH (1–34) administration once per day by subcutaneous injection consistently increases bone mass in animals (Jerome *et al.*, 1999; Jilka *et al.*, 1999) or humans (Lane *et al.*, 1998; Lindsay *et al.*, 1997) and exerts anabolic effects on trabecular (Gunness-Hey and Hock, 1984; Huo *et al.*, 1991) and cortical bone (Ejersted *et al.*, 1993; Wronski and Yen, 1994). Beyond its promising effects in treating osteoporosis in humans (Lane *et al.*, 1998; Neer *et al.*, 2001), systematic PTH administration also has beneficial effects in fracture healing (Andreassen, Ejersted and Oxlund, 1999; Andreassen *et al.*, 2001; Holzer *et al.*, 1999). This is indicted by increased callus formation and subsequent gain in mechanical strength during fracture healing in young (Andreassen, Ejersted and Oxlund, 1999) and old (Adreassen *et al.*, 2001) rats.

PTH administered locally also has been found to be beneficial in the treatment of bony defects using a gene therapy approach (Bonadio *et al.*, 1999; Fang *et al.*, 1996). The gene activated matrix (GAM) technology is a direct gene transfer strategy (Goldstein, 2000). A GAM consists of a three-dimensional carrier and associated DNA that expresses the gene of interest. During early osteotomy healing, granulation tissue fibroblast proliferates and migrates into the GAM, takes up and transiently expresses the osteoinductive plasmid DNA and promotes bone formation. Gene activated matrices made from collagen with a plasmid containing the complementary deoxyribonucleic acid (cDNA) sequences for PTH (1–34) have been used successfully to release PTH (1–34) locally and enhance osseous healing in osteotomy sites of rat femurs (Fang *et al.*, 1996) and dog tibias (Bonadio *et al.*, 1999).

Intramembranous low dose treatment with rhPTH (1–34) increased the recruitment of mesenchymal cells into the chondrocyte lineage, proliferation of mesenchymal (chondroprogenitor) cells and synthesis of type II collagen, yielding a larger cartilaginous callus. The PTH-induced cartilaginous callus is then rapidly replaced with bone (Nakazawa *et al.*, 2005). In another study intermittent low dose treatment with rhPTH (1–34) has been reported to increase hard callus formation (callus formed by intremembranous bone formation in periosteal osteoblasts and osteoprogenitor cells) and its mechanical strength without producing visible side effects (Nakajima *et al.*, 2002). These authors have demonstrated that treatment with PTH stimulates proliferation and differentiation of osteoprogenitor cells and production of bone matrix proteins early in the fracture healing process. Based on its osteogenic effects, treatment of fractures with intermittent low dose rhPTH (1–34) may represent an effective strategy for the enhancement of fracture healing and could be the first systematic intervention for the repair of skeletal injuries.

In men, PTH has been reported to be associated with volumetric bone mineral density ($\nu_{\text{BMD}}$). In women, serum PTH and $\nu_{\text{BMD}}$ are inversely correlated (Lauretani *et al.*, 2006). Estrogen enhances the osteoclast-promoting activity of PTH and the inverse relationship between PTH and $\nu_{\text{BMD}}$ is mediated by estrogen deficiency. These results support the finding of Riggs, Khosla and Melton (2002) that estrogen deficiency is indirectly responsible for the secondary hyperparathyroidism observed in late postmenopausal women. These data also agree with the findings of Neer *et al.* (1998) that PTH prevents a proximal femur and total body bone loss in young women with induced estrogen deficiency.

## 6.12 Role of Prostaglandins

Prostaglandins (PGs) are identified as a group of lipid mediators involved in the maintenance of local homeostasis. Of these, $PGE_2$ (prostaglandin $E_2$) has the best known effect on bone metabolism (Yoshida *et al.*, 2002; Sasaoka *et al.*, 2004; Tanaka, Sun and Yokota, 2004) by induction of both osteoblastogenesis and osteoclastogenesis and the integration of the two. Agonists of these receptors have been suggested for clinical use. $PGE_2$ exerts its influence through interaction with G-protein-coupled cell surface receptors (Narumiya, Sugimoto and Ushikubi, 1999), including EP1, EP2, EP3 and EP4, and is considered to be the most potent and specific receptor for mediating anabolic action in bone. Induction of the osteogenic effects by $PGE_2$ requires a concentration 100 times greater than that for specific $PGE_2$–EP4 agonist stimulation (Ke *et al.*, 1998; Rubin *et al.*, 2001).

Prostaglandins are potent stimulators of bone formation and resorption. Indomethacin inhibits ultrasound induced collagen synthesis by osteoblasts. Of the isoforms of COX, COX-1 is constitutively expressed in most cells and tissue, whereas COX-2 is induced in response to various stimuli and is thought to contribute to the generation of PGs in certain stages of cell proliferation and differentiation in bone turnover (Higashi, Ohishi and Kudo, 2000). There is large body of work showing the synthesis of NO and prostaglandins by bone cells (osteoblasts) when stimulated by mechanical stress.

$PGE_2$ is classified not only as resorptive agent (Klein and Raisz, 1970; Franklin and Tasjian, 1975; Dietrich, Goodson and Raisz, 1975; Raisz *et al.*, 1977; Katz *et al.*, 1983; Vargas and Raisz, 1990) but also stimulates DNA and collagen synthesis in cultured bones (Blumenkrantz and Sondergaard, 1972; Chyun and Raisz, 1984; Raisz and Fall, 1990; Vargas and Raisz, 1990). $PGE_2$ is produced in large amounts by solid tumors (Seyberth *et al.*, 1975) and has been reported to enhance bone formation (Chyun and Raisz, 1984). A systematic infusion of $PGE_2$ *in vivo* increased the calcified area of rat bone (Jee *et al.*, 1985). *In vivo* it appears to activate bone remodeling, which begins with either bone formation or bone resorption (Shen *et al.*, 1986). The initial rapid bone resorption occurring after oophorectomy and disuse can be retarded by cyclooxygenase inhibitors (inhibitors of prostaglandin synthesis) (Lane, Kimmel and Coble, 1989; Thompson and Rodan, 1988).

$PGE_2$ administered in rats has been shown to (i) increase trabecular bone mass by forming new woven trabeculae (Mori *et al.*, 1990), (ii) increase critical bone mass (Jee *et al.*, 1990) and

(iii) increase Haversian remodeling, probably by increasing activation (Jee *et al.*, 1990). Furthermore, infants treated for cyanotic heart diseases with long-term infusion of $PGE_2$ have shown periosteal bone formation (Jorgensen, Savnholm and Host, 1988). $PGE_2$ acts as a first messenger in the transduction of mechanical usage to bone cells. Increased mechanical strain causes $PGE_2$ production, which results in osteoblastic proliferation (Somjen *et al.*, 1980; Yen and Rodan, 1984; Binderman, Shimshoni and Somfjin, 1984). Therefore, $PGE_2$ has the ability to create a positive bone balance, where net bone formation results and thus $PGE_2$ may raise bone mass to a new steady state. $PGE_2$ indeed may have the ability to override or augment the effects of mechanical feedback signals and thus alter the set point of the bone structure control system.

Miller and Marks (1993) have reported that E-series prostaglandins *in vivo* have anabolic effects on bone. There are dose related increases in the periosteal surface with new bone formation, average new bone thickness, maximum thickness and new bone area. These authors have suggested that locally delivered $PGE_1$ may be useful in generating new bone in clinical situations of local bone deficit such as the local bone defect of the periodorial diseases and in situations where increased osteogenesis may be a desirable feature.

The effects of $PGE_2$ and Ris both separately and in combination ($PGE_2$ + Ris) have been studied on the intact aged female rat skeleton (Lin *et al.*, 1994). These authors have concluded that the combination of $PGE_2$ and Ris is more anabolic than $PGE_2$ or Ris alone when endochondrial ossification is active. However, $PGE_2$ + Ris is no more anabolic than $PGE_2$ alone in old bone without endochondral ossification. This is suggestive that $PGE_2$ alone and in co-treatment with Ris is accumulated in bone by different tissue mechanisms. $PGE_2$ alone created new bone by increasing the activation frequency 8.3-fold and the formation to resorption ratio 1.3-fold compared with controls. The combination of $PGE_2$ and Ris depressed the activation frequency and bone formation rate and eroded the surface so as to increase the formation to resorption ratio (by 3–4-fold) over $PGE_2$ alone.

It can be concluded that $PGE_2$ is needed to initiate changes in activity of the modeling/remodeling system associated with fracture healing, oophorectomy and disuse.

## 6.13 Thiazide Diuretics (TD)

Thiazide diuretics (TD) lower renal calcium excretion and may therefore affect bone metabolism (Rejnmark *et al.*, 2003). Histomorphometric analyses have shown a reduced extent of eroded surfaces, reduced bone formation rate, and decreased osteoid thickness. Consistent with a decreased bone turnover (Steiniche *et al.*, 1989), intake of TD has been associated with a 2–5% higher bone mineral density (BMD) at the hip and a 4–14% higher BMD at the lumbar spine (Dawson and Harris, 1993; Morton, Barrett Connor and Edelstein, 1994; Wasnich *et al.*, 1986). However, some other studies have indicated either no effects or a very small increase of approximately 1% in BMD in response to 2–3 years of TD treatment (Lacroix *et al.*, 2000; Reid *et al.*, 2000).

In a large population based case-control study Rejnmark, Vestergaard and Mosekilde (2005) found that treatment with TD was associated with a significant reduced overall fracture risk as well as a reduced risk of forearm fractures. Moreover, a dose–effect relationship was revealed for the risk of any fracture as well as the risk of forearm and hip fractures. The reason for the reduced fracture risk in users of TD may need to be investigated.

## 6.14 Effects of Fluoride

The concept that the consequences of osteopenia could be less severe in individuals with a history of fluoride exposure was first reported by Leone *et al.* (1955). Richards, Mosekilde and Sogaard (1994) revealed that there is an age-related increase in fluoride content of vertebral trabecular bone in normal individuals. However, these results suggest that it may not have an effect on the strength of human vertebral trabecular bone, supporting the earlier studies of Stein and Granik (1980). Fluoride stimulates osteoblast proliferation (Farley, Wergedal and Baylink, 1983). Initially, primitive woven-like bone is formed but later the bone is remodeled again in a normal way, creating lamellar bone organized in so-called packets. The remodeling process in trabecular bone is characterized by a slightly reduced activation frequency, a prolonged remodeling period and a slightly increased mineralization lag time. The final resorption depth is unchanged, but the completed wall thickness is increased, leading to a positive bone balance per remodeling cycle (Briancon and Meunier, 1981; Eriksen, Mosekilde and Melsen, 1985). Hence, treatment with fluoride may lead to an increase in bone mass and, consequently, decreases fracture rates (Mamelle, Meunier and Dusan, 1988; Pak *et al.*, 1989; Pak, Sakhaee and Zerweph, 1990; Farley *et al.*, 1992).

The initiation of osteogenesis and mineralization in embryonic chick mandibles maintained in organ culture has been found to be dependent on the presence of fluoride (Hall, 1987). However, other *in vitro* studies have not shown any evidence of direct effects of fluoride on osteoblasts (Kopp and Robey, 1990a, 1990b; Oreffo, Wells and Johnstone, 1991). This suggests that fluoride may act through an indirect mechanism on osteoblasts and probably requires local factors to increase bone cell proliferation and bone formation *in vivo* (Farley *et al.*, 1990). Chavasseiux *et al.* (1993a) have concluded that the function of osteoblasts is not modified after *in vitro* exposure to fluoride. In contrast, *in vivo* exposure of fluoride to rats for 1 month has a mitogenic effect on osteoblasts and stimulates bone activity. These data emphasize the hypothesis that fluoride may act either on osteoprogenitor cells or through an indirect mechanism mediated by a cofactor.

NaF has been identified as a useful therapeutic agent (Heaney *et al.*, 1989). There is general agreement that NaF has an anabolic effect on trabecular bone mass, particularly in the spine (Charles, Mosekilde and Taagehoj, 1985; Aaron, De Vernejoul and Kanis, 1991; Kanis and Meunier, 1984; Hansson and Roos, 1987; Raymakers *et al.*, 1987; Farley *et al.*, 1989; Hodsman and Drost, 1989; Riggs *et al.*, 1982, 1990; Pouilles, Tremolleres and Ribot, 1991). These authors have found that the effect of sodium fluoride is to stimulate bone formation. They have suggested a considerable reserve in the capacity of bone forming cells or their progenitors, even a severe idiopathic osteoporosis. This suggests that fluoride treatment may have potential for reestablishing or rebuilding trabecular connectivity. Fluoride also replaces the hydroxyl groups on the hydroxyapatite crystal, which is believed to increase resistance to resorption. However, some investigators have found that NaF increases the appendicular trabecular bone mass (Dambacher, Ittner and Ruegsegger, 1986; Resch *et al.*, 1993), whereas others have found no effect or even a decrease in predominantly cortical bone mass (Ringe, Kruse and Kuhlencordt, 1978; Hodsman and Drost, 1989; Kleerekoper *et al.*, 1991; Pouilles, Tremolleres and Ribot, 1991; Kleerekoper and Mendlovic, 1993). They have suggested that the improvement in axial bone density may be at the expense of cortical bone.

Several studies have shown structural abnormalities or mineralization defects in bone formed during fluoride administration (Vigorita and Suda, 1982; Eriksen, Mosekilde and

Melsen, 1985; Kragstrup *et al.*, 1989; Lundy *et al.*, 1989; Boivin *et al.*, 1993), thus indicating that the increase in bone mass is not necessarily followed by an improvement in bone quality. Also, three controlled trials (Dambacher, Ittner and Ruegsegger, 1986; Riggs *et al.*, 1990; Kleerekoper *et al.*, 1991) have failed to demonstrate any therapeutic advantage of NaF (75–80 mg) over placebo with respect to vertebral fracture rate. Sowers *et al.* (1991) have reported a significantly higher risk of wrist, spine or hip fractures in a community with a high fluoride water content compared with the control community. These findings, along with those of Turner, Akhter and Heaney (1992), suggest that there is a concentration threshold above which fluoride may become detrimental to bone strength and debate has arisen as to determining the level of toxic dose. There have also been reports of an increased incidence of nonvertebral fractures during fluoride administration (Hedlund and Gallagher, 1989; Melton, 1990; Schnitzler *et al.*, 1990). Some of these studies included incomplete stress fractures of the lower extremities in their measurement of peripheral rate (Riggs *et al.*, 1990; Schnitzler *et al.*, 1990).

One parameter for ascertaining fluoride treatment is to examine the marrow space star volume. Vesterby *et al.* (1991) have shown that marrow space star volume is significantly reduced after 5 years of continuous treatment with fluoride without changes in trabecular thickness or in the mean fractional amount of trabecular bone. The reduction in marrow space by a factor of two to three has been explained by assuming that the large marrow "cavities" have been split into two or more cavities by newly generated trabecular "structures." The small increase in bone mass in close relation to the reduced marrow space and the small decrease in trabecular thickness also point to the generation of new, but thin, walls.

Mackie *et al.* (1989) found no evidence of improved trabecular continuity and the creation of new trabeculae as a result of fluoride treatment. Grynpas and Rey (1992) have concluded that the changes induced by fluoride reduce the solubility of bone crystals by direct incorporation of fluoride ions in the apatite lattice and by decreasing the labile phosphate environments. Negative data (Riggs *et al.*, 1990) on vertebral fracture prevention during treatment with fluoride and calcium suggest that more attention should be drawn to the influence of the cortical component of vertebral bodies on bone strength. Although the marrow space star volume appears to be the most efficient and potent parameter for describing changes in bone connectivity, it has many sampling problems. An essential precondition for unbiased estimates is the use of vertical sections (Vesterby *et al.*, 1987). Sogaard *et al.* (1994) found that after five years of fluoride treatment of osteoporotic patients the iliac crest trabecular bone strength is reduced by 46–58% compared with pretreatment biopsies. Also, 1-year of fluoride administration seemed to reduce bone strength by 17–30%, although this is not significant. Budden *et al.* (1988) and Laurent *et al.* (1989) have suggested that an increase in bone volume can be reliably predicted in any patient who has accumulated a level of bone fluoride exceeding 0.2%.

It is suggested that fluoride causes resorption to be transferred from the thickened and increasingly metabolically inert trabeculae to the less fluoride loaded cortex (Faccini, 1967; Riggs *et al.*, 1980), an event that may predispose to fatigue damage at trabecular sites (Duffy *et al.*, 1986) and increased cortical resorption (Rune and Gallagher, 1989). However, it has been pointed out that the amount of fluoride accumulated in the cortical and trabecular regions of the ileum was the same and it is not higher in those patients (17%) in whom fracture recurred during treatment (Laurent *et al.*, 1989). In a study on rats, Turner, Akhter and Heaney (1992) showed that incorporation of fluoride into bone mineral has a dose-specific effect on bone strength, so that bone strength follows a biphasic relationship with bone fluorides.

Aaron, De Vernejoul and Kanis (1991) have concluded that the doses of fluoride conventionally used in the treatment of established osteoporosis are incapable of restoring abnormal skeletal architecture of idiopathic osteoporosis. In a study on experimental flurosis in pigs (Mosekilde, Krogstrup and Richards, 1987) the vertebral trabecular bone compressed strength decreased despite a concurrent increase in bone mass. The increase in bone mass is likely, therefore, to reduce the risk of fracture compared to that of the non-osteoporotic population and as such is not an indication of bone quality. However, high doses of fluoride cause osteosclerosis and osteomalasia. These conditions can be attributed to a change in bone cells sensitivity to mechanical stimuli. Furthermore, fluorosis often results in large, misshapen bones in which the new bone structure is disorganized. Exostosis is also common to chronic fluorosis (Weatherell and Weidmann, 1959). There is considerable controversy regarding the optimum dosage of NaF and several investigators have suggested that 50 mg $d^{-1}$ is appropriate (Riggs, 1991; Kleerekoper and Balena, 1991; Meunier, 1992). Dosage used in the study carried out by Sogaard *et al.* (1994) falls in the range 40–60 mg $d^{-1}$, but nevertheless failed to prove beneficial for the strength of iliac crest trabecular bone. Thus, the efficacy of fluoride therapy is still not well established.

## 6.15 Role of Growth Hormone (GH)

GH (growth hormone) stimulates linear growth in childhood and bone remodeling throughout life. It stimulates chondrocytes in the growth plate to secrete IGF-1 which in turn signals differentiation of these cells, leading to cartilage formation and bone growth. This is consistent with the observation that both GH and IGF-1 receptor co-exist on bone cell. Deficiency of GH is associated with low adult BMD (De Boer *et al.*, 1994; Hyer *et al.*, 1992; Kaufman *et al.*, 1992). It has been reported that lower circulating IGF-1 levels exist in patients with osteoporosis than in normal controls (Kassem *et al.*, 1994). It is possible that variations in the circulating GH profile, even within the physiologically normal range, may be associated with differences in body build (Denninson *et al.*, 2003). Growth hormone has similar effects on bone to $PGE_2$: it (i) increases Haversian remodeling, presumably by increasing activation, and (ii) increases bone formation (Aloia *et al.*, 1976; Kruse and Kuhlencordt, 1975; Haas *et al.*, 1976; Harris *et al.*, 1972). Also, growth hormone increases the efficiency of calcium absorption and decreases calcium excretion. A GH-induced increase in cortical bone mass has also been demonstrated (Harris *et al.*, 1972). The anabolic effects of GH on the skeleton are thought to be mediated by insulin-like growth factor I (IGF-I) (Isgaard *et al.*, 1998), and positive effects of IGF-I on bone cell proliferation and protein synthesis are documented (Canalis *et al.*, 1989).

Ghrelin is a gut peptide involved in growth (GH) secretion and energy homeostasis. It is reported that the adipocyte-derived hormone leptin, which also regulates energy homeostasis, plays a significant role in bone metabolism. This evidence implies that ghrelin may modulate bone metabolism. To study the role of ghrelin in skeletal integrity, Fukushima *et al.* (2005) have examined its effects on bone metabolism *in vitro* and *vivo*. These authors measured the growth hormone secretagogue receptor 1a (GHS –R1a) in rat osteoblasts using RT-PCR and immunohistochemistry. The effect of ghrelin on primary osteoblast-like cell proliferation is examined by recording changes in cell number and the level of DNA synthesis. Ghrelin and GHZ-R1a are identified in osteoblast-like cells. Ghrelin significantly increases osteoblast-like cell numbers and DNA synthesis in a dose-dependent manner. They have concluded that ghrelin increased BMD of both SD rats and SDRS, suggesting that it directly stimulates bone formation.

$PGE_2$, growth hormone and PTH tend to activate and accelerate remodeling, while estrogen tends to slow it down. This accelerated remodeling is commonly associated with disuse. Yet, each of these agents has the ability to cause anabolic effects on bone.

It may be stated that estrogen, $PGE_2$, growth hormone and PTH are classified as set point altering agents, whereas calcitonin and fluoride are agents that do not alter the set point. Alteration of the set point is invoked by adjusting the sensitivity of bone cells to mechanical stimuli. The bone cells most responsive to mechanical stimuli are probably the osteoblasts and osteoblast precursors. Therefore, it is speculated that changes in set point are communicated through these cells.

Age-related changes are an attribute of the bone tissue. The content of collagens and glycosaminoglycans in bone matrix decreases progressively with age (Vogel, 1979). The number of secondary osteons and resorption spaces in bone increases with age, leading to increased porosity (Martin and Burr, 1989; Atkinson, 1965). Age-related changes in bone tissue can also reduce its elasticity (Burstein, Reilly and Martens, 1976). Furthermore, the amounts of various growth factors and BMPS in the bone matrix decrease with age (Syftestad and Urist, 1980; Reddi, 1985), suggesting a decreased ability of osteoblasts to produce these growth factors. The administration of growth hormone can renew the osteoblasts ability to form growth factors (Syftestad and Urist, 1980).

Various age-related changes cause a reduced ability of the modeling/remodeling system to respond to mechanical feedback and the transduction of mechanical usage into a cellular signal will be altered somewhat by the reduction of stiffness of the bone tissue and changes in amounts of growth factors on the bone matrix. The net effect is a less effective response to changes in mechanical usage or biochemical agents.

## 6.16 Cholesterol

The hypothesis that increased total cholesterol may adversely affect BMD indicates a link between atherosclerosis and osteoporosis. Postmenopausal women with low bone mineral density (Barengolts *et al.*, 1998; Uyama *et al.*, 1997) or with greater amounts of bone loss (Hak *et al.*, 2000; Kiel *et al.*, 2001) have greater prevalence and severity of aortic calcification. Bone matrix proteins, including osteopontin (Giachelli *et al.*, 1995), osteocalcin (Fleet and Hock, 1994) and bone morphogenetic protein (Bostrom, 2001), have also been found in atherosclerotic plagues. Vitamin D (Price *et al.*, 2001) and osteoproteglun (Bucay *et al.*, 1998) regulators of bone resorption have been associated with vascular calcification, and interleukin-6 and homocysteine, as well as statin drugs used to treat hyperlipoproteinemia (Chung *et al.*, 2000; Edwards, Hart and Spector, 2000; Watanabe *et al.*, 2001), have links to osteoporosis (Jilka *et al.*, 1992).

## 6.17 Interleukin 1 (IL-1)

Besides its crucial role in triggering the immune response, IL-1 stimulates bone resorption *in vitro* and can induce hypercalcemia *in vivo* (Sabatini *et al.*, 1988). Some solid tumors associated with hypercalcemia have been shown to produce IL-1 (Sato *et al.*, 1987).

Jilka (1998) and Hentunen *et al.* (2000) pointed out that bone remodeling, which is coupled with biological activity of osteoclasts and osteoblasts, is controlled by cytokines such as interleukin-6 and tumor necrosis factor-$\alpha$ (TNF-$\alpha$). They also demonstrated that the synthesis of cytokines is regulated by estrogen and osteoresorption factors.

## 6.18 Bisphosphonates (BPs)

Bisphosphonates (BPs), like estrogen, possess antiresorptive properties that prevent bone loss mainly through the inhibition of osteoclast activity (Most *et al.*, 1995; Sato *et al.*, 1991; Shinoda *et al.*, 1983). Bisphosphonates and calcitonin inhibit glucocorticoid-induced osteoblast apoptosis (Plotkin *et al.*, 1999). BPs are analogs of pyrophosphonate and have powerful bone-resorption-inhibitory activities (BRIAS). They possess a non-hydrolysable P-C-P structure and bind strongly to the hydroxyapatite in bone and tend to accumulate when given repeatedly. Among BP derivatives, the aminobisphosphonates (aminoBPS) have much more powerful BRIAS (100 times or more) than the non-amino BPs (Bonjour *et al.*, 1994; Geddes *et al.*, 1994). However, clinical studies have found that amino BPs have inflammatory side effects, such as fever (in 10–50% treated patients), an increase in acute-phase proteins, gastrointestinal disturbances and ophthalmologic inflammation (Adami *et al.*, 1987; Sauty *et al.*, 1996; Thiebaud *et al.*, 1997; Siris, 1993; Macarol and Frauenfelder, 1994; Fleisch, 1997; Blank *et al.*, 1997). Endo *et al.* (1999) reported that various inflammatory reactions induced by a single intraperitoneal injection of an amino BP were strongly inhibited by chlodronate. In another study Monma *et al.* (2004) showed that a single injection in mice of alendronate (amino BP) or chlodronate increased bone density at a selective site in the tibia. This was detectable as a clear sclerotic line (BP line). In fact, the BP line is visible at the site at which the growth plate would have been located at the time of the injection of alendronate or chlodronate.

Bisphosphonates act as bone resorption inhibitors and are now used as therapeutic agents. All antiresorptive agents (estrogens, calcitonin or bisphosphonates) have in clinical studies proven themselves capable of increasing spinal bone mass by 2–5% in postmenopausal or elderly individuals (Gallagher, 1988; Chesnut, 1988; Harris *et al.*, 1991; Thamsborg *et al.*, 1991; Reginster, 1992). They prevent fractures despite the limited increase in spinal bone density, indicating that a higher bone quality might have been achieved in therapy (Overgaard *et al.*, 1989; Reginster *et al.*, 1989; Storm *et al.*, 1990; Watts *et al.*, 1990).

In man, resorption and formation are coupled in such a way that bone formation occurs at sites of previous resorption (Frost, 1964a, 1964b). Therefore, it is generally presumed that resorption inhibitors such as 3-amino-1 hydroxyprophylidine-1-bisphosphonate (AHPr BP) suppress bone formation through a reduction in the number of resorption sites available for bone formation. Similarly, in the rat at least some sequential remodeling occurs (Baron, Tross and Vignery, 1984). Shellhart *et al.* (1992), experimenting with rats, have demonstrated that dichloromethylene bisphosphonate ($10\,\text{mg}\,\text{kg}^{-1}$) induced a greater increase in periosteal apposition rate in loaded limbs and is proportional to the relative amount of body weight supported. However, it has not been confirmed that the suppression of formation by AHPr BP in the rat is due to reduced availability of resorption sites for induction of bone formation or to an effect of resorption inhibition on surfaces already forming bone. Tobias, Chow and Chambers (1993) concluded that AHPr BP is a resorption inhibitor, that is, it inhibits bone formation in the rat in a manner that appears inconsistent with the site-specific coupling seen in the adult human skeleton.

The effect of BP compounds in preventing bone loss has been reported by several authors using various immobilization models (Fleisch *et al.*, 1969; Cabnela and Jowsey, 1971; Michael, King and Francis, 1971; Muhlbauer *et al.*, 1971; Schoutens *et al.*, 1988; Thompson *et al.*, 1990), but the mechanical properties of the bone were not tested. In their experiments, clodronate, with a higher dose retarded the development of osteoporosis. Martino and Seideman (1980) also reported that the immobilization induced bone resorption was lowered by clodronate.

The results of Lepola, Jalovaara and Vaananen (1994) show that the decrease in bone mass during cast immobilization in rats can be prevented by clodronate, which also increases the mechanical strength of both immobilized and contralateral bones of immobilized animals. In non-immobilized rats, clodronate increases the mass but not the breaking strength of rat bone.

Clodronate has been used in some clinical trials with paraplegic patients (Minaire *et al.*, 1981, 1987) in which positive effects of serum and urine biochemistry and bone histomorphometry are found. Giammini *et al.* (1993) have concluded that cyclical low dose clodronate therapy induces a gain in lumbar spine bone mass in patients with postmenopausal osteoporosis and has no side effects.

Ibandronate has been reported (Epstein and Zaidi, 2005) to be a oral nitrogen-containing BP. It possesses (i) a strong affinity for bone, (ii) potent antiresorptive action, (iii) a high concentration, (iv) persistence in bone tissue and (v) excellent tolerability that enable simpler and less frequent dosing compared with other available drugs. This has been found to be effective in improving the probability of risk of fractures.

Adami *et al.* (1994) treated patients with mild Paget's disease of bone with 20 or 40 mg alendronate daily for 6 months. They concluded that the former is insufficient and the latter is associated with a high incidence of side effects. Further, the suppression of disease activity depends on the dose of bisphosphonate given daily or over a short period of time and lower doses cannot be compensated by a longer duration of the treatment course. So far, however, it has not been possible to demonstrate that beneficial structural bone changes follow treatment modalities of osteoporotic patients.

Two BPs (alendronate and risedronate) have shown a 30–50% reduction of vertebral and nonvertebral fractures in postmenopausal women with osteoporosis (Black *et al.*, 1996; Cummings *et al.*, 1998; Harris *et al.*, 1999; McClung *et al.*, 2001; Pols *et al.*, 1999; Reginster *et al.*, 2000). Bisphosphonates provide effective and relatively safe antiresportive therapy and have an enormous impact on the management of osteoporosis. The BPs in use (alendronate, risedronate and ibandronate) reduce the incidence of fractures, including both vertebral and non-vertebral fractures, particularly hip fractures (Rosen *et al.*, 2005). These agents bind to the bone surface and then are taken up by osteoclasts, leading to their inactivation and programmed cell death. However, some concern has been raised that their prolonged use may cause excessive inhibition of bone remodeling and slow the repair of fractures and microdamage.

## 6.19 Adipocyte Hormones

The adipocyte has long been recognized as an estrogen producing cell, particularly in postmenopausal women. Thus, early postmenopausal women who lose bone rapidly have lower levels of both estrone and estradiol than graded and this may be accounted for by their lower fat mass (Riis, Rodbro and Christiansen, 1986). A relationship has been developed between circulating estrone levels and bone density and has been shown to be independent of the effects of fat mass (Reid *et al.*, 1992). This implies that estrogen is not the only pathway by which fat influences bone density. This suggestion is supported by the finding of a fat–bone relationship in premenopausal women, in whom the adipocyte is a relatively minor source of estrogens.

The discovery of leptin offers the possibility of another link between these two tissues, fat and bone. Leptin, a 16-kDa protein encoded by the obese (ob) gene, hormonally regulates food intake and energy expenditure by negative feedback at the hypothalamic nuclei (Zhang *et al.*, 1994). Leptin is expressed predominantly in adipose tissue of normal mice and is absent in ob/ob mice, which are homozygous for the ob mutation. Leptin is thought to act primarily at the hypothalamus, where it has effects on appetite, energy expenditure and neuroendocrine axes (Compfield *et al.*, 1995). Leptin has an important function in the growth plate and plays a key role in endochondral ossification (Kishida *et al.*, 2005). Being produced in adipose tissue, leptin is produced in bone marrow adipocytes (Laharrague *et al.*, 1998) and its receptor is found on osteoblast (Steppan *et al.*, 2000; Sufang *et al.*, 1993; Thomas *et al.*, 1999). Osteoblasts exposed to leptin demonstrate increased proliferation and differentiation (Reseland *et al.*, 2001; Thomas *et al.*, 1999), as do chondrocytes (Cornish *et al.*, 2001). In addition, there is evidence that leptin regulates production of osteoprotein and the receptor activator of NF-kB ligand (RANKL) to result in diminished recruitment of osteoclasts and reduced bone resorption (Cornish *et al.*, 2001; Holloway *et al.*, 2002). Leptin also increases secretion of IL-1 receptor antagonist in monocytes and circulating levels of this antiosteolytic factor are substantially increased in obesity (Meier *et al.*, 2002). Thus, systemic administration of leptin in leptin-deficient mice (Steppan *et al.*, 2000) humans (Farooqi *et al.*, 1999) and wild-type mice (Cornish *et al.*, 2001) results in increased bone growth, increased skeletal mass and is related to bone density (Goulding and Taylor, 1998; Pasco *et al.*, 2001). Leptin levels are reduced in women with vertebral fractures (Yamauchi *et al.*, 2001).

The work of Ducy *et al.* (2000) has demonstrated another mode of action of leptin on bone. They found that the infusion of leptin into the third ventricle of either leptin deficient or wild-type mice is associated with bone loss. These experiments indicate that a mechanism exists for the regulation of bone mass via the central nervous system (CNS). One possible mechanism for this would be by way of leptin's potent effects on insulin secretion. The central or systemic administration of leptin to wild-type or leptin-deficient mice is found to reduce circulating insulin concentrations by up to 85% (Jang *et al.*, 2000; Pelleymounter *et al.*, 1995). Leptin regulates insulin secretion, in part, through the sympathetic nervous system (SNS) (Mizuno *et al.*, 1998). This effect, acting through the mechanisms outlined in Figure 6.5, contributes to the bone loss observed in these experiments. Leptin also influences several other neuropeptides that could themselves act directly or indirectly to produce bone loss. For instance, $\alpha$-MSH, which is secreted from the pituitary in response to leptin, causes profound bone loss when administered systemically to normal mice (Cornish *et al.*, 2001). Neuropeptide Y release in the hypothalamus is inhibited by leptin, yet central administration of this factor has effects on bone mass similar to those of leptin (i.e., decrease in trabecular bone volume) (Ducy *et al.*, 2000). Y2 is associated with increased bone mass (Baldock *et al.*, 2002). How these effects of leptin and neuropeptide Y fit together in normal physiology remains to be understood. However, experiments with systemic leptin administration have clearly demonstrated that the net effect of leptin is to cause bone gain.

The role of the CNS in the control of bone formation has been suggested by the inhibition of bone formation after intracerebroventricular injection of leptin (Ducy *et al.*, 2000; Harada and Rodan, 2003). It acts through the receptor expressed in the arcuate nuclei for the control of body weight and food intake and in the ventral hypothalamus for the control of bone formation. In

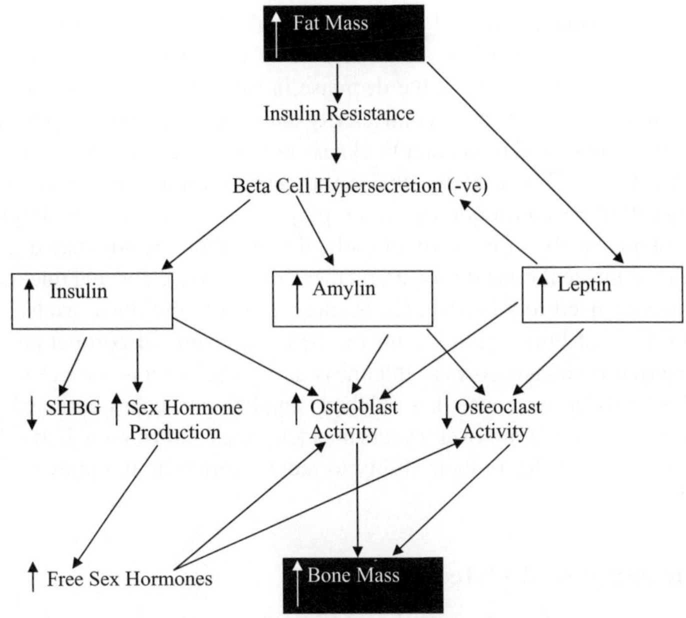


**Figure 6.5** Possible mechanisms by which fat mass may influence bone cell function and, thus, bone mass. Leptin not only decreases circulating insulin concentrations via its effects on the central nervous system but also has direct effects on osteoblasts; SHBG stands for sex hormone-binding globulin (Reid, 2005). (Reproduced with permission from I.R. Reid, Relationship among body mass, its components and bone, *Bone*, **31**(5), 547–555 © 2002, Elsevier B.V.)

mice the inhibition of bone formation in trabecular bone (at least in the spine) by leptin appears to be mediated by the SNS and $\beta 2$ adrenergic receptors expressed by osteoblasts (Takeda *et al.*, 2002). However, the role of leptin on cortical and trabecular bone still remains to be understood. Leptin receptors are expressed on bone cells and leptin affects directly osteoblast proliferation and/or differentiation (Reid, 2004; Thomas *et al.*, 1999). Therefore, it appears to be a possibility that leptin has control on bone formation.

Body weight affects both bone turnover and bone density, and is therefore a prominent risk factor for vertebral and hip fractures and is important alongside that of age. The effect of body weight is probably contributed by both fat mass and lean mass, although in postmenopausal women fat mass has been more consistently demonstrated to be important. Several mechanisms for the fat–bone relationship exist. These include the effect of soft tissue mass on skeletal loading, the association of fat mass with the secretion of bone active hormones from the pancreatic $\beta$ cell (including insulin, amylin and preptin) and the secretion of bone active hormones (e.g., estrogens and leptin) from the adipocyte.

## 6.20 Mechanism of Action of Antiresorptive Agents

The fact that raloxifene and nasal calcitonin reduce vertebral fracture incidence despite a small effect on BMD suggests that other mechanisms are also involved (Delmas, 2002).

Antiresorptive drugs reduce vertebral fractures through the inhibition of bone remodeling. Although the increase in cancellous bone volume is marginal, even with potent bisphosphonates (Chavasseiux *et al.*, 1997), the decrease in osteoclastic activity may reduce plate perforation and the disruption of plate connectivity that characterizes osteoporotic bone. High doses of alendronate increased trabecular thickness but did not change trabecular number and spacing in OVX baboons (Balena *et al.*, 1993). Antiresorptive drugs may have a positive effect on the cortical shell of vertebral bodies, which play a major role in their strength (Rockoff, Sweet and Blenstein, 1969). At the level of cortical bone, these agents may decrease cortical porosity, decrease endosteal bone turnover and therefore increase cortical thickness, but these possible mechanisms need to be confirmed in more exhaustive clinical trials.

The clinical effects of bone active agents on the components of cortical parameters (e.g., external and internal diameter, cortical thickness and porosity) has not yet been evaluated. Peacock *et al.* (2000) have shown that calcium supplements can significantly reduce the medullary expansion of the femur over 4 years in elderly men and women. It may be concluded that antiresorptive drugs differ in their ability to reduce vertebral and nonvertebral fractures (Delmas, 2002).

## 6.21 Genetic Studies of Osteoporosis

Fracture is a phenotype with obvious direct clinical relevance but its possible heritability has not been understood. Genetic factors are crucial in determining peak bone mass, as studies of sex difference, racial differences and twins show. Comparative studies between osteoporotic patients and their relatives confirm this observation (Evans *et al.*, 1988).

Deng *et al.* (2002) have demonstrated no significant heritability of wrist and spinal fractures and marginal heritability of hip fracture ($P = 0.048$). These authors also suggest that the genetic correlation between hip fracture and BMD is low. This appears to contradict several studies demonstrating that daughters of mothers who have themselves fractures have low BMD (Keen *et al.*, 1999; Seeman *et al.*, 1989; Seeman *et al.*, 1994). A twin study reports heritability of Colles fracture (Andrew *et al.*, 2005), while another twin study in Scandinavians found no significant increase in twin concordance for fracture (Kannus *et al.*, 1999), but its findings have been debated (MacGregor, Snieder and Spector, 2000). Thus, current evidence suggests that Colles and hip fracture may be heritable, but the magnitude of the genetic effect is not large enough to be widely confirmed and could cause alarm. Further studies of the genetic epidemiology of fracture are required before it is accepted as the principal focus of future genetic studies.

With regard to bone microarchitecture, it is not known if it is heritable, though it may be. Still to be determined are the number of genes involved and their likely magnitude of effect and whether there are gender or site-specific genetic effects. The correlation of microarchitecture with fracture risk is another aspect to be explored. These factors need to be established.

Two successful mappings of osteoporosis associated genes highlight the potential of the method: the identification of the LRP5 gene and demonstration of its role in BMD variation in the general population (Ferrari *et al.*, 2004; Gong *et al.*, 2001; Little *et al.*, 2002) and the demonstration of association of the BMP2 gene with the fracture BMD in the Icelandic population (Styrkarsdottir *et al.*, 2003). These studies demonstrate that with the use of crude phenotypes such as BMD and fracture (though holistic), gene mapping methods can succeed. The BMP2 gene is important in bone development. Its association is identified by studies in large families drawn from a relatively homogenous population. The LRP5 role is not identified in most general population linkage studies to date and its discovery is critically dependent on the availability of families with clear monogenic phenotypes.

## 6.22 Nutritional Aspects in Osteoporosis

Various types of interventions, including exercise regimens (Baldwin *et al.*, 1996; Cavanagh, Davis and Miller, 1992; Nicogossian *et al.*, 1992; Norman *et al.*, 2000), drug/hormonal therapies (Bikle *et al.*, 1995; Turner *et al.*, 1998) and dietary modifications (Globus *et al.*, 1986; Navidi *et al.*, 1995), have been considered as countermeasures for bone loss associated with skeletal unloading. The appeal of dietary interventions is that they may provide a practical and safe component to an osteoprotective regimen through the incorporation of specific nutrient rich foods or the consumption of supplements. However, the evidence that nutrition is a major determinant of osteoporosis is discordant and changes in the role of nutritional states are not clearly established (Schaafsma *et al.*, 1987). Physical inactivity is a major plausible contributory factor but it is not the only one. High levels of weight bearing activity are known to prevent osteoporotic fractures (Cooper, Barker and Wickman, 1988; Lau *et al.*, 1988).

### *6.22.1 Biochemical Markers*

Rico *et al.* (1993b) have identified several biochemical markers of nutrition – transferrins, prealbumin, retinal binding protein and fibronectin – that are significantly lower in women with vertebral osteoporosis. This suggests that postmenopausal osteoporosis may be associated with a nutritional deficiency. Here we discuss some of the observed effects of some important nutrients.

### *6.22.2 Salt Intake*

High salt intake may accelerate bone loss in individuals with a propensity to osteoporosis (Breslau *et al.*, 1982; Goulding and Lim, 1983; Parfitt, 1983). Increased ingestion of sodium chloride elevates the urinary excretion of calcium in animals (Goulding, 1980; Goulding and Campbell, 1983) and in man (Breslau *et al.*, 1982; Conney, Spears and Goulding, 1984; Castenmiller *et al.*, 1985; Sabto *et al.*, 1984) and also elevates blood levels of parathyroid hormone (Coe *et al.*, 1973, 1975). Saline and frusemide are administered as a standard treatment in hypercalcemia (Potts, 1980) to lower blood calcium levels rapidly by stimulating sodium mediated calciuresis (Suki *et al.*, 1970). Urinary calcium also rises when natriuresis is induced with arterial natriuretic peptide (Richards *et al.*, 1985) and diuretics other than thiazide (Suki, 1979; Brickman, Massr and Coburn, 1972). Furthermore, urinary hydroxyproline excretion rises in the presence but not in the absence of the parathyroids in salt loaded animals (Goulding, 1980), suggesting that salt consumption stimulates PTH-mediated bone collagen catabolism. These findings support the view that salt-mediated elevation of PTH increases bone resorption.

Salt-mediated hypercalciuria leads to an increased secretion of PTH (Castenmiller *et al.*, 1985; Coe *et al.*, 1973; Goulding, McIntosh and Campbell, 1984), whereas calciuria induced by high protein diets and metabolic acidosis does not (Adams, Gray and Lemann, 1979).

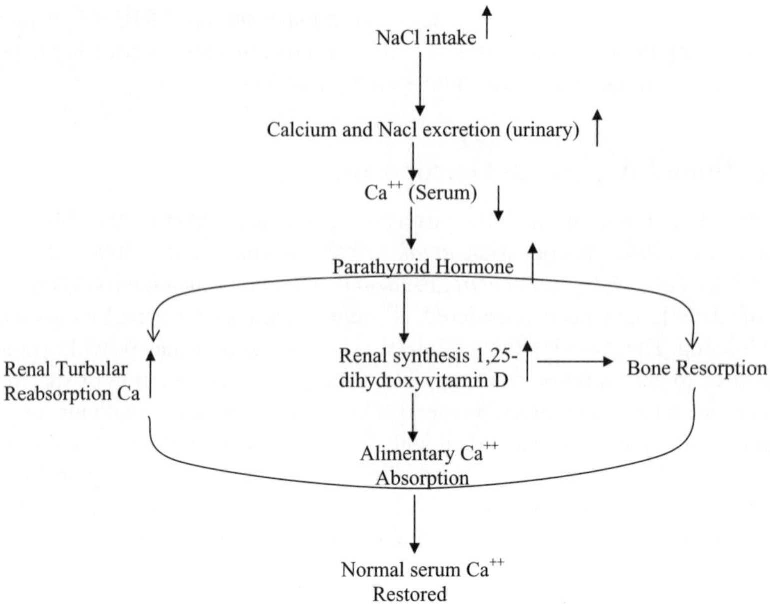


**Figure 6.6** The effect of an increase in sodium chloride intake

Coe *et al.* (1975) first demonstrated a rise in immunoassayable PTH (PTHi) by eliciting hypercalciuria with salt supplements and frusemide. In animals, saline infusions induce histological evidence of increased PTH synthesis (Sporonitz and Frick, 1973). Total and nephrogenous urinary cyclic AMP values also increase after salt supplementation in a manner consistent with increased PTH secretion (Muldowney, Freaney and Moloney, 1982; Fujita, Chan and Barter, 1984).

PTH hormone restores normocalcemia by increasing the input of calcium into the blood from the kidney, gut, via renal 1,25-dehydroxycholcalciferol [1,25-$(OH)_2D_3$] synthesis, and bone. Blood levels of PTHi and 1,25-$(OH)_2D_3$ rise in healthy volunteers but not in hypoparathyroid patients when salt intake is raised from 10 to 250 mmol $d^{-1}$ (Breslau *et al.*, 1982). Increasing salt also augments $^{47}Ca$ intestinal absorption (Breslau *et al.*, 1982; Meyer *et al.*, 1976) and may reduce fecal Ca loss (Fujita, Chan and Barter, 1984). Figure 6.6 summarizes the effect of sodium chloride intake on osteoporosis (Goulding *et al.*, 1986).

However, PTH, bicarbonate and thiazide diuretics lower the urinary excretion of calcium relative to sodium. Consequently, when PTH is elevated, or bicarbonate or thiazides are consumed, urinary calcium will be lower than usual for a given sodium output (Goulding and Campbell, 1984; Goulding, McIntosh and Campbell, 1984).

### 6.22.3 Calcium

The notion of early work that osteoporosis is primarily because of calcium deficiency, particularly in elderly, has been a counterpart to estrogen deficiency theory. The pathogenesis of the process of estrogen loss coupled with increase of fracture risk is in fact multifactorial.

Calcium deficiency leads to secondary hyperparathyroidism. The average calcium intake of women in their eighth and ninth decades of life is between 700 and 800 mg per day. A decrease in calcium intake, impaired intestinal absorption of calcium due to aging or disease, as well as vitamin D deficiency result in secondary hyperparathyroidism. Adequate calcium and vitamin D intake and appropriate physical activity may not only increase peak bone mass but also slow down bone loss and reduce fracture risk throughout life (Raisz, 2005).

Of the body's total calcium, 99% is located in bone and teeth. The remaining 1% is in the extracellular compartment. This is also very important, since calcium influences many physiological functions like the pace of the heart beat. Extracellular calcium levels are controlled by an equilibrium between absorption of calcium that occurs in the intestine, their excretion by the kidney and the deposition in bone. Intestinal absorption of calcium is regulated by a calcium pump that is controlled by an enzyme related to vitamin D.

### 6.22.3.1 Calcium Nutrition

Calcium absorption decreases with age (Allen, 1985). This decline is correlated with a decline in renal function as measured by the creatinine clearance (Nordin *et al.*, 1976). It is suggested that malabsorption of calcium in the elderly is predominantly due to vitamin D deficiency and renal impairment in the conversion of calcidiol into calcitriol (Francis *et al.*, 1983; Gallagher *et al.*, 1979). Low levels of calcidiol in plasma are often found in elderly people, particularly in those with hip fractures and this is attributed to decreased intestinal absorption of vitamin D (Lips, 1982; Parfitt *et al.*, 1982). In a cross sectional study of healthy elderly people, a significant correlation is found between the plasma calcidiol level and the bone mineral content of the radius, even though plasma calcidiol levels are within the normal range of young adults (Schaafsma, 1981). Vitamin D requirement of the elderly is 50–100% higher than that of a young adult and they recommend a daily oral intake of 15–20 μg for elderly people.

Calcium alters the bone related hormones and possibly the local hormones induced by mechanical stress. Calcium could alter the physical and chemical properties of the bone mineral. The major mechanism whereby calcium affects bone is probably through inhibition of PTH secretion.

In the healthy adult who is calcium deprived in the diet, the bone turnover is increased due to an increase in the activation frequency of new bone remodeling units. Conversely, when presented with high calcium challenges bone remodeling is reduced. A great deal of circumstantial evidence suggests that the mechanism for this relates to hormonal changes in calcium metabolism. During calcium depletion serum calcium tends to fall, which stimulates the secretion of parathyroid hormone and the synthesis of calcitriol, which increases the activation of bone remodeling. An increased intake of calcium increases the net intestinal absorption of calcium. The small rise in plasma calcium would reduce the secretion of parathyroid hormone and thus the rate of bone turnover, giving rise to a small but finite increase in reversible bone mass. The reduction in PTH secretion reduces the synthesis of calcitriol and offsets, to some extent, the dietary calcium challenge. Because of the long turnover time of bone these changes in bone mineral content may take several years to complete. It would be expected, therefore, that the balance for calcium intake minus total excretion would be negative for several years after a reduction in intake before the new equilibrium is attained.

### 6.22.3.2 Use of Calcium in Osteoporosis

In animals, calcium deficiency produces hypocalcemia, hyperparathyroidism, increased plasma calcitriol concentrations, increased bone resorption and osteoporosis irrespective of age (Beanusa, Metkovic and Kotial, 1978; Rader *et al.*, 1979; Sevastikoglou, Thomaidis and Lindholm, 1977; Stauffer *et al.*, 1973; Volpin and Salomon, 1978).

Signs of calcium deficiency in humans are not common, because it cannot easily be detected and does not reduce longitudinal growth. Several groups have claimed that low calcium intake is more prevalent in osteoporotic patients than in control subjects, but others have reported negative results (Exton-Smith, 1972). Circumstantial evidence suggests that low calcium diets predispose to osteoporosis. Thus, the high prevalence of osteoporosis in lactose deficient people has been attributed to a low calcium intake consequent to the avoidance of dairy products (Birge *et al.*, 1967; Kocian, Skala and Berkos, 1973; Newcorner *et al.*, 1978). The results of Matkovic *et al.* (1979) indicate that an habitual high calcium intake may prevent the development of osteoporosis by increasing adult peak bone mass; however, it does not reduce the rate of aging bone loss as also reported by Nilas, Christiansen and Rodro (1984).

In postmenopausal women, urine calcium excretion increases markedly after starting treatment with additional calcium. The effect is, however, attenuated with time and after two years reaches pretreatment values. This suggest that postmenopausal women can adapt to increments in dietary intake. The difference in adaptation between pre-and postmenopausal women is that in the latter the bone loss persists, though at a slower rate, because the balance of resorption and formation at remodeling sites remains unchanged. This imbalance, due to estrogen deficiency, is not reversed by calcium. In contrast, bone loss is halted by estrogen completely by correcting the imbalance and thereby preserving bone mass, irrespective of the time it is given after surgical castration or menopause. For all these reasons, there may be some potential for calcium supplementation of postmenopausal women who are at the risk from osteoporosis.

An impairment of calcium absorption has been described frequently in postmenopausal osteoporosis (Gallagher *et al.*, 1973) and variously attributed to calcitriol deficiency (Gallagher *et al.*, 1979) or resistance to the action of calcitriol in the gastrointestinal tract (Nordin *et al.*, 1988). Morris *et al.* (1991) have confirmed the significant difference in calcium absorption between normal and osteoporotic postmenopausal women, as indicated by the parameter $\alpha$ (Gallagher *et al.*, 1979; Burnell *et al.*, 1986; Nordin *et al.*, 1988). Although the main determinant of $\alpha$ in both groups is the serum calcitriol level it accounts for only about 20% of the difference in $\alpha$ between the two groups. The remainder of the difference in $\alpha$ can be attributed to "resistance" to the action of calcitriol on the gastrointestinal tract, assuming that the total calcitriol levels reflect the free concentrations. The resistance to calcitriol is not related to age in either group. A comparable blunting of the gastrointestinal response to calcitriol has been reported in aged rats (Wood *et al.*, 1988). Morris *et al.* (1990) have observed a similar gastrointestinal resistance to serum calcitriol in osteoporotic postmenopausal women on corticosteroid therapy and have attributed it to a deficiency in receptor number or receptor affinity (Morris and Steurer, 1988).

Further, Morris *et al.* (1991) have suggested that obligatory urinary calcium loss be regarded as a primary event and that it activates bone resorption, which in turn is related to urinary hydroxyproline. This suggests that the higher the calcium loss and the lower the calcium input from the gut, the higher is the bone resorption.

The usefulness of calcium in osteoporosis appears to be its ability to decrease bone turnover and thereby decrease skeletal losses, but it might not prevent it entirely when the imbalance between formation and resorption at each remodeling site persists. It appears likely that this is related to the small increase in serum calcium and the resulting decrease in the activation of bone in much the same way as seen in childhood and in women before the menopause. In addition, as serum calcitriol decreases, PTH levels have been shown to decrease, but this is not invariably noted, which is perhaps related to reversal of the effect between doses of calcium. Many studies have also shown that treatment with calcium supplements may be associated with the maintenance or even an increase in skeletal mass, though this is modest (2–10% depending upon the site measured). The reason for the transient increase in the bone mass relates to the decreased activation of new remodeling sites. Early on during treatment bone formation will continue at previously occurring remodeling sites and bone mass will increase transiently to fill in the resorption space. Because turnover is a slow process, particularly the terminal phase of the mineralization, the transient state may persist up to 3 years, and studies on the use of calcium supplements suggest that this is so. Similar though more rapid effects are observed at sites of cancellous bone.

### 6.22.3.3 Calcium Intervention

There is evidence of an 8% decrease in the probability of fracture rate upon taking a glass of milk per day. Calcium carbonate is currently the most widely used source of calcium in supplements and fortified products (Aronson *et al.*, 1988). Other salts of calcium such as lactate, citrate, gluconate and malate are used because of their solubility in beverages and juices. This implies that improved solubility increases calcium bioavailability (Heaney, Recker and Wever, 1990). None of these sources contain phosphorous, which together with calcium forms hydroxyapatite, the main mineral of bone. Shapiro and Heaney (2003) have evaluated the relative effect of calcium carbonate and two calcium phosphate salts (dicalcium phosphate and tricalcium phosphate) on bone development in growing rats under conditions of varying calcium and phosphorous deficiency. Under these conditions, the results clearly show, as had been anticipated, that calcium alone is insufficient and that both phosphate salts were superior to the carbonate salt with respect to essentially all the bone variables measured. At equivalent levels of calcium supplementation, the calcium phosphate salts promoted significantly greater body weight gain, femur weight, tensile strength and phosphorous deposition and calcium utilization. These workers have recognized the interdependence of calcium and phosphorous in bone development and quantified the interrelationship of the two nutrients. In fact phosphorous is the more critical of the two nutrients. An increase in diet phosphorous increased bone ash much more dramatically than increasing dietary calcium. By contrast, adding calcium without the addition of phosphorous had very little effect on bone ash or calcium content.

Intervention studies have indicated that oral calcium supplements of 1 g $d^{-1}$ (Horsman *et al.*, 1977; Lutwak, 1974; Recker, Saville and Heaney, 1977) or with vitamin D (Albanese *et al.*, 1973; Riggs *et al.*, 1976) are effective in decreasing bone loss. This is in accordance with the results of Horowitz *et al.* (1984), who showed that an oral supplement of the above amount reduces bone resorption in postmenopausal women, as measured by the urinary excretion of hydroxyproline. It is also in agreement with a study by Recker and Heaney (1985) in postmenopausal women showing a decrease in bone resorption and an improvement of calcium balance following 2 years of milk supplementation (680 mL $d^{-1}$). That calcium supplements set and consequently prevent bone loss is rather controversial. Some studies have shown a positive result (Horsman *et al.*, 1977; Recker, Saville and Heaney, 1977; Heaney, Recker and Saville, 1978; Riggs *et al.*, 1982; Pacifici *et al.*, 1988), whereas others have reported negative results (Ettinger, Genant and Hand Cann, 1987; Riis, Thomsen and Christiansen, 1987). The cause of this discrepancy is not clear. Stepan *et al.* (1989) have concluded that ossein hydroxyapatite might have specific effects on bone metabolism in addition to being a dietary supplement (Dent and Davies, 1980; Epstein *et al.*, 1982; Dambacher and Ruegsegger, 1987).

Several trials in elderly women have shown that combined. Calcium and vitamin D supplementation can reduce the rate of bone turnover, slow bone loss and even reduce the incidence of fractures (Baeksgaard, Andersen and Hyldstrup, 1998; Dawson *et al.*, 1997; Meier *et al.*, 2004; Larsen *et al.*, 2004).

In another study conducted by the World Health Initiative, calcium plus vitamin D trials in the women aged 50–79 revealed that supplementation with 1000 mg of calcium and 400 IU of vitamin D per day had no significant effect on fracture reduction at any site (Jackson *et al.*, 2006). However, secondary analysis reveals that women who adhered to the supplementation (at least 80% compliance) experienced a significant (29%) reduction in hip fracture. Factors known to be associated with increased bone resorption [Bio-T, 25(OH)-vitamin D and PTH] affected mainly volumetric BMD (Chapuy *et al.*, 2002; Grados *et al.*, 2003; Lau *et al.*, 2002; Lau *et al.*, 2001; Chee *et al.*, 2003).

A combination of supplemental vitamin D along with supplemental calcium substantially reduces the fracture risk in institutionalized elderly women (Chapuy *et al.*, 1992), and also reduced non-vertebral fracture in a community of American women and men of 65 years of age and older (Dawson *et al.*, 1997). It has been reported that the combination of above two help in producing a higher peak bone mass (Raisz, 2005). Michaelsson *et al.* (2003) have found no significant interaction between age, body mass or vitamin D intake and dietary calcium intake concerning an effect on fracture risk. These authors also pointed out that increasing dairy food intake was not associated with a reduced hip fracture risk or overall risk of osteoporotic fracture despite a more than fourfold difference in the monthly median intake frequency between the lowest and the highest.

### *6.22.4 Protein*

An increase in dietary protein or amino acid intake in humans has been associated with an increased urinary excretion of calcium. This is not accompanied by a proportional increase in apparent calcium absorption, thereby resulting in a decreased calcium balance (Allen, Oddaye and Margen, 1979a; Allen, Bartless and Bock, 1979b; Kim and Linkswiler, 1979; Schuette, Zemel and Linkswiler, 1980; Schuette et al., 1982). However, this is not confirmed by other workers (Spencer *et al.*, 1978a, 1978b, 1983). It is suggested that the urinary loss of calcium depends on the amount of protein, sulfur-containing amino acids and phosphate in the diet (Hegsted *et al.*, 1981; Schutte and Linkswiler, 1982; Whiting and Draper, 1980; Zemel *et al.*, 1981).

The calciuretic effect of the protein is attributable to two factors: (i) protein increases the glomerular filtration rate and thus the filtered calcium load and (ii) the sulfate that originates from the oxidation of sulfur-containing amino acids counteracts the tubular reabsorption of calcium in the kidneys (Allen, Oddaye and Margen, 1979a; Allen, Bartless and Bock, 1979b, Allen *et al.*, 1981; Allen, 1985; Bosch *et al.*, 1983). It has been shown that urinary loss of calcium is higher with isolated proteins or mixtures of amino acids than when mixed proteins (meat) are added to the diet. This has been attributed to the phosphate content of the protein source. Phosphate decreases urinary calcium secretion and may counteract the calciuric effect induced by protein (Hegsted *et al.*, 1981; Spencer *et al.*, 1978a, 1978b). However, it has been demonstrated that the hypocalciuric effect of phosphorous is not large enough to neutralize all of the protein effects at the phosphorous/protein ratios found in most natural protein sources (Heaney and Recker, 1982).

It has also been demonstrated that by increasing the protein content of the diet the increase in urinary calcium excretion was greater when phosphorous was given in a more acidic from than when it was less acidic (Petito and Evans, 1984). Ingestion of sodium bicarbonate alkalinized the urine and reversed the increase in urinary calcium associated with a higher protein intake (Lutz, 1984). Plasma calcitriol and parathyroid hormone levels show a clear increase (Coe *et al.*, 1975): neither changed (Adams, Gray and Lemann, 1979; Gallagher *et al.*, 1980) nor decreased (Licate, 1981) in response to an increased protein intake. In animals, no effect of long-term high protein intake on bone mass was found (Calvo, Bell and Forbes, 1982; Whiting and Draper, 1981; Yuen and Draper, 1983).

Marsh *et al.* (1980) have reported a more rapid postmenopausal bone loss amount in omnivorous women than in age-matched vegetarians. This difference might be related to a difference in type or amount of dietary protein.

The contradictory results of different research groups could be explained by such differences in diet components as acid ash and phosphate; more results are certainly needed.

### 6.22.5 *Lactose*

It is reported that lactose stimulates the intestinal calcium absorption in both animals and humans (Ali and Evans, 1973; Cochet *et al.*, 1983). The effect is independent of the presence of vitamin D and is exerted on the diffusional component of the intestinal transport system (Pansu, Bellaton and Bronner, 1979). In vitamin D-deficient hypocalcemic rats, lactose raises the plasma calcium concentration, increases bone mineralization and almost normalizes bone histology (Schaafsma *et al.*, 1987). The mechanism underlying the effect of lactose on calcium absorption is not completely understood, but it is thought that after its ingestion the sugar interacts with the villus membrane, thereby increasing its permeability for calcium (Armbrecht and Wasserman, 1976). However, contrary to lactose, which is digested rather slowly, most dietary sugars are absorbed before reaching the ileum (Vaughan and Filer, 1960).

It is suggested that the positive effect of lactose on calcium absorption could at least partly compensate for the decreased calcium absorption in the elderly, which is the vitamin D-dependent component of the intestinal calcium transport system. However, there are no data indicating that lactose is beneficial in the prevention of osteoporosis.

### 6.22.6 *Phosphorous*

Most of the calcium and phosphorous in the body is to found in bone, as hydroxyapatite (a phosphate salt). However, few supplements or food fortifications presently contain phosphorous. Phosphorous is, though, found in beverages and food additivities (Bassett *et al.*, 1997). Most studies on phosphorous have focused on the effect of high phosphorous intakes on calcium balance and bone metabolism. The essentiality of inorganic phosphorous in osteogenic cells and its importance in the function of osteoblastic cells have been confirmed (Caverzasio and Bonjour, 1996).

Shapiro and Heaney (2003) have shown that deficient phosphorous intake retarded both bone development and soft tissue growth. Body weight gain on a calcium-only regimen (calcium carbonate) is about half that on a calcium phosphate regimen. In contrast to those fed calcium carbonate the group fed calcium phosphate salts exhibited an increase in body weight gain as supplementation level increased. However, while femur ash and calcium increased slightly on the highest calcium intake, body weight gain actually decreased. This divergence suggests that either (or both) the demands of bone mineralization may have competed with soft tissue for the limited phosphorous available or the higher diet content of calcium provided by calcium carbonate complexed with some of the phosphate in the gut, reducing its absorption and hence creating phosphorous deficiency.

Dietary phosphorous reduces the urinary excretion of calcium and counteracts, at least in part, the calciuric action of dietary proteins. It is not caused by a decrease in intestinal calcium absorption (Spencer, Kramer and Osis, 1982) and might be related to parathyroid stimulation (Schaafsma *et al.*, 1985; Yuen and Draper, 1983). Dietary phosphate deprivation increases the renal synthesis of calcitriol independently of parathyroid hormone (Gray, 1981; Rader *et al.*, 1979). A negative correlation between serum phosphorous and calcitriol has been demonstrated in women (Gray *et al.*, 1977) as well as in men (Maierhofer, Gray and Lemann, 1984). Experiments on adult animals have shown that high phosphorus diets produce secondary hyperparathyroidism and increase bone resorption, resulting in bone loss (Draper and Bell, 1979). It may be mentioned that women with low phosphorous intake also have low protein intake and so their needs would be better served with combining total nutrition.

The stimulating effect of oral phosphorous on parathyroid function has been confirmed in humans (Reiss *et al.*, 1970), but there is no direct evidence that high-phosphorous diets produce bone loss in man. At varying levels of dietary calcium, calcium balance was unaffected when dietary phosphorous was raised by a factor of three or more (Heaney and Recker, 1982; Spencer *et al.*, 1965, 1975, 1978a, 1978b). Therefore, it is assumed that the effect of dietary phosphorous on renal calcium handling and bone turnover does not affect the calcium balance in man (Spencer, Kramer and Osis, 1982).

### 6.22.7 Lymphotoxin

Lymphotoxin is a powerful stimulator of osteoclastic recruitment and proliferation and it induces hypercalcemia *in vivo* (Thompson, Mundy and Chambers, 1987a; Pfeilschifter, Mundy and Roodman, 1989). Myeloma cells produce lymphotoxin, which appears to be a major, but not the sole, mediator of increased osteoclastic bone resorption (Garnett *et al.*, 1987).

### 6.22.8 Dietary Fiber, Oxalic Acid and Phytic Acid

The intestinal availability of minerals, particularly calcium, magnesium, iron and zinc, results in the formation of insoluble metal complexes (Ali *et al.*, 1981). The effect may vary with the type of fiber and may lead to a decreased calcium absorption and a reduction in calcium balance.

### 6.22.9 Alcohol

Reports have shown that the high prevalence of non-traumatic osteonecrosis is associated with alcohol abuse (Arlet, 1992; Hirota *et al.*, 1993; Hungerford and Lennox, 1985; Mont and

Hungerford, 1995; Solomon, 1985). The pathologic changes in steroid and alcohol-induced osteonecrosis are similar (Glimcher and Kenzora, 1979; Mont and Hungerford, 1995; Solomon, 1985). When blood perfusion is diminished in the femoral head, osteonecrosis eventually develops (Cui *et al.*, 1997). Ethanol is a cytotoxic factor that may play a key role in cell injury or death either directly or indirectly (Glimcher and Kenzora, 1979; Rubin and Rottenberg, 1982). The results of Wang *et al.* (2003) indicate that fat cell hypertrophy and proliferation and intracellular lipid deposits in bone marrow in the subchondral area of the femoral head in rabbits may be a direct result of treatment with alcohol. There is a common precursor for several cell lines, including osteogenic, adipogenic and chondrogenic lineages (Beresford *et al.*, 1992; Cui, Wang and Baliar, 1997a). Alcohol may enhance the differentiation of preadipocyte or pluripotential bone marrow cells into adipocytes and this decrease osteogenesis, resulting in a deficit in bone remodeling or repair of necrotic bone, eventually promoting the onset of osteonecrosis (Cui *et al.*, 1997b). Therefore, alcohol-induced lipid deposits in osteocytes, increased adipogenesis and decreased osteogenesis by bone marrow stroma may be a major factor contributing to alcohol-related osteonecrosis (Wang *et al.*, 2003). Several observations indicate an effect of alcohol on calcium metabolism, although its mechanism of action is not well understood. Chronically orally ingested alcohol decreases intestinal calcium absorption in animals (Kakikana, Nobel and Butte, 1968; Krawitt, 1974; Mendelson, Ogata and Mello, 1971) and in man even in the absence of hepatic cirrhosis (Vodoz *et al.*, 1977). Ingestion of alcohol can be toxic for bone cells of rats (Baran *et al.*, 1980), can cause an acute increase of urinary calcium excretion in man (Linderman, Adler and Yiengst, 1967) and may lead to an inadequate nutritional intake of calcium and vitamin D. Impairment of pancreatic function and liver cirrhosis, often seen in alcoholics, may cause malabsorption of calcium and vitamin D or impaired hydroxylation of vitamin D (Henke *et al.*, 1961; Hepner, Roginsky and Moo, 1976).

Bouillon *et al.* (1984) have reported in patients with alcohol abuse low plasma concentrations of albumin, vitamin D binding protein, total calcium, phosphorous and total calcitriol. Normal concentrations of ionized calcium, calcidiol and free calcitriol have also been reported. It has been suggested that the abnormalities of calcium metabolism are mainly due to decreased protein synthesis.

Osteoporosis induced due to chronic alcoholism is often correlated with an increase of femoral neck fracture associated with decreased bone mass (Seeman *et al.*, 1983; Spencer *et al.*, 1986; Cooper and Wickham, 1990). This would exert a disproportionate influence on the risk of moderate trauma fractures in postmenopausal women (Cooper *et al.*, 1992). An increased propensity to fall as the result of excessive alcohol consumption is an additional factor in estimating fracture rate.

A related phenomena is that of smoking, as the two often go hand in hand. Cigarette smoking has been found to be a risk factor for fracture and a determinant of bone density in several studies. Cigarette smokers have an earlier menopause, lower circulating levels of estrogen metabolites and lower body weights than non-smokers. All these factors could contribute to the association of smoking with osteoporosis.

### *6.22.10 Caffeine*

Caffeine is associated with higher levels of both urinary calcium and intestinal calcium excretion but not with calcium absorption efficiency. The effects are proportional to intake levels (Heaney and Recker, 1982; Massey and Wise, 1984). A high caffeine intake was noted in osteoporotic subjects when compared with age-matched controls (Daniel, 1976).

Caffeine, one of the main constituents of coffee, has various pharmacological and cellular responses in a wide spectrum of biological systems (Dews, 1982). These include stimulation of the central nervous system and cardiae muscle, increased urinary output and relaxation of smooth muscle. In addition to caffeine, coffee also contains a relatively large amount of tannin (a tea catechin analog) producing biological effects such as antioxidation, antimutation, angiogenesis, antibiotic action, antihypercholesterolemia, antihypertension and anti-inflammatory actions (Sugiyama *et al.*, 1999).

Several studies have suggested that consumption of coffee is associated with a significant increase in risk of fracture, osteoporosis and periodontal disease (Kamagata-Kiyoura *et al.*, 1999; Kiel *et al.*, 1990). These are associated with nutrition, exercise, alcohol intake, smoking and several other lifestyle factors (disorders). In fact, several epidemiological studies have reported the influence of caffeine on osteoporosis, but the effects of coffee on bone metabolism remain controversial (Johnell *et al.*, 1995; Lloyd *et al.*, 1997). Hypotheses to explain these associations have centered on the caffeine content of coffee (Johnell *et al.*, 1995). In fact, caffeine has various pharmacological actions and cellular responses in bone metabolism, resulting in increased calcium excretion in urine (Heaney and Recker, 1982) and an inhibitory effect on proliferation of osteoblast-like cells *in vitro* (Kamagata-Kiyoura *et al.*, 1999). According to Cooper *et al.* (1992) the median daily caffeine consumption of the women in their study is 210 mg (range 0–849.6 mg) based on intake of coffee, tea and other caffeinated beverages in male Wistar rats.

By analyses of bone histomorphometry and serum and urinary biomarkers of bone turnover, Sakamoto *et al.* (2001) have demonstrated that coffee consumption does not lead to bone resorption. Compared with the control, diets with high and low intake caffeine did not affect the number of osteoclasts and levels of urinary deoxypyridinoline and serum osteocalcin, which reflect bone resorption and bone formation. This result agrees with the report by Bergman, Newbrey and Massey (1988), who showed that caffeine does not stimulate bone to release calcium at concentrations between 5 and 500 $\mu g\,mL^{-1}$ in an *in vitro* system. Coffee led to greater excretion of urinary phosphorus compared with the control diet. However, according to Bover *et al.* (1999), this seemed to reflect dietary phosphorus ingestion and not bone turnover and urinary phosphorus excretion increased in a dose-dependent manner for dietary phosphorus. In fact, the phosphorus content of the instant coffee used was 350 mg per 100 g, whereas calcium is 150 mg per 100 g (Viani *et al.*, 1991). On the other hand, several investigators have demonstrated that chronic inflammatory diseases, such as rheumatoid arthritis, aseptic loosening and periodontal diseases, lead to bone resorption and are accelerated by TNF-$\alpha$, IL-6 and interleukin-$\beta$ (IL-1$\beta$), (Hentunen *et al.*, 2000; Johnell *et al.*, 1995), which are released from macrophages and other cells stimulated with Lipopolysaccharides (LPS). TNF-$\alpha$ and IL-6 levels of serum markedly increase at 2 h after injection of LPS (Givalois *et al.*, 1994). These authors also noted that the maximum peak of serum cytokines takes place 1–2 h after injection of LPS into rats. However, coffee consumption has no effect on the production of bone resorbing cytokines, such as TNF-$\alpha$ and IL-6, either with no stimulation or with LPS injection, despite long-term (140 days) coffee uptake. Consistent with these results, Hannan *et al.* (2000) reported that caffeine did not affect longitudinal bone loss in elderly women.

### 6.22.11 Other Factors

Biochemical parameters, especially bone Gla protein (Delmas *et al.*, 1983), have some predictive value in osteoporosis studies, but serum calcium and phosphate, alkaline phosphatase, vitamin D concentrations, and urinary calcium and hydroxyproline levels are of minor importance for the prediction of histological variation in osteoporosis (Meunier, 1986). Bone Gla protein has also been shown to reflect bone turnover, especially bone formation in other metabolic bone formation diseases (Malluche, Arnala and Faugere, 1988).

#### 6.22.11.1 Management of Osteoporosis in Elderly Women

It is important to discuss with the patient the necessity of an adequate intake of calcium and vitamin D (about 1000 mg day$^{-1}$, vitamin D 400–800 IU day$^{-1}$) and emphasize the need for physical exercise. If the patient accepts estrogen therapy (0.625 mg day$^{-1}$) then it is given in the usual dosage. If the patient does not accept it then the dose of calcium is increased to 1500 mg day$^{-1}$. It is also important to know whether any fatigue fracture has occurred in the neck of femur, lumbar spine or distal radius. If estrogen therapy is not accepted, then calcitonin or bisphosphonates may be an alternative choice. Persons with osteoporosis may benefit from an improved diet, including supplementation with vitamin D and calcium and moderate exercise to help slow further bone loss.

## 6.23 Osteoporosis: Prevention and Treatment

The best long-term approach in any disease, particularly with osteoporosis, is its prevention. Current medical practice for preventing osteoporotic fractures is often inadequate and hence ineffective. If children and young adults, particularly women, have a good diet (with enough calcium and vitamin D) and get plenty of exercise then they will build up and maintain bone mass. This will provide a good reserve against bone loss later in life. Exercise places stress on bones that builds up bone mass, particularly, skeletal loading from muscle contraction with weight training exercises. However, any exercise of any type is better than none at all and exercise also provides benefits for prevention of cardiovascular diseases that are more common in the elderly. Athletes tend to have greater bone mass than non-athletes. Exercise in later life will help to retard the rate of bone loss. It is presumed treatment to prevent bone loss would maintain both bone mass and architecture.

This may be because of proper diagnosis. Hormone replacement therapy in the early premenopausal years is of proven efficacy (Lindsay *et al.*, 1976). The findings of Rodin *et al.* (1990) raise the possibility that prophylaxis in premenopausal women should also be considered. Some physicians recommend estrogen replacement for all postmenopausal women, as this retards bone loss (Coleman, 1987; Hall, 1987a; Grisso, Baum and Turner, 1990). However, the duration of treatment has not been optimized and can be considered a variable parameter among individuals. Some women may need more than estrogen alone. As a result, only a small percentage of women are currently being treated for the prevention or treatment of osteoporotic fractures (Hemminki *et al.*, 1988; Johnston *et al.*, 1989; Grisso, Baum and Turner, 1990; Melton, Eddy and Johnston, 1990). However, no drug therapy can restore bone mass to normal.

As a rule most drug therapies work by decreasing bone resorption. At any given time, there is bone that has been resorbed but not replaced and this accounts for about 5–10% of bone mass.

By decreasing resorption of bone, a gain in bone density of the same amount is possible, taking about 2–3 years.

One of the more common non-estrogen therapies is the use of bisphosphonates such as alendronate or risedronate that act as an inhibitor of osteoclastic activity. Bisphosphonates may be beneficial, particularly in women who cannot tolerate estrogen therapy. Bisphosphonates are effective in inhibiting bone loss after menopause. Risedronate has also shown effectiveness in reducing the risk of hip fracture among elderly women with osteoporosis. Yasko *et al.* (1992) have demonstrated that recombinant human bone morphogenetic protein-2 (rh BMP-2) as a growth factor available in recombinant form induces osteoinductive activity in rats. This had led to a successful union in an orthotopic site. This supports the potential application of rh BMP-2 as a bone graft substitute.

Prevention of osteoporosis involves modification of the multiple factors associated with the disease, such as inadequate calcium intake, lack of exercise, postmenopausal estrogen loss and smoking and excessive alcohol consumption, and it may be related to genetic effects.

Papapoulos *et al.* (1992) have suggested that intermittent administration of the first generation bisphosphonate etidronate increases trabecular bone density up to 3 years, stabilizes it after 2 years and appears to reduce the rate of new vertebral fractures in women with postmenopausal osteoporosis. Reginster *et al.* (1989) have shown arrest of bone loss in bisphosphonate treated women, which was sustained for 6 months after stopping treatment. The results of Tenenbaum *et al.* (1992) indicate that, in addition to resorption, bisphosphonates have a significant and direct effect on mineralization in bone-forming cultures.

Another drug is raloxifene. This is a selective estrogen receptor modulator (SERM), and may also replace estrogen therapy. Raloxifene can act in concert with estrogen in bone to inhibit resorption and decrease the risk for fractures. Though raloxifene inhibits bone resorption, it lacks an anabolic effect. Additional potential benefits from raloxifene therapy include a decreased risk for breast cancer, because raloxifene acts antagonistically to estrogen on the uterus.

Teriparatide is a recombinant human parathyroid hormone administered by subcutaneous injection that binds to specific high-affinity cell surface receptors in bone and kidney. Daily administration of teriparatide stimulates new bone formation by promoting osteoblastic activity over osteoclastic activity. It improves trabecular bone architectural remodeling and increases bone mass.

Another hormone, calcitonin, that decreases bone resorption may be taken by injection or by nasal spray. Sodium fluoride can increase the measured bone density in vertebra, but seems to have no overall effectiveness in reducing vertebral fracture. In addition fluoride helps reduce tooth decay.

A risk factor can provide an indication of the probability of future fractures. It is felt that BMC measurements are incapable of separating fracture cases from nonfracture cases. BMC represents the cumulative composite effects of the main risk factors, which may vary in magnitude and duration over a person's lifetime. One of the primary purposes of identifying risk factors is to help identify etiologic factors that might be controlled to prevent further fractures (Davis *et al.*, 1989; Ross *et al.*, 1989; Wasnich *et al.*, 1989). Falch, Sandvik and Van Beresteijn (1992) have identified an index to predict individual postmenopausal bone loss. They identified low body weight, reduced renal phosphate reabsorption and smoking as significant independent risk factors. They have further concluded that the index may be helpful in identifying healthy premenopausal women in whom bone mass measurements should be considered. It is convenient to classify fracture risk into two categories: risk factors for low bone mass and risk factors for falls.

As such, BMC measurements should provide a more accurate indication of fracture risk than attempting to assess individual risk factors that exert their effects primarily through bone mass. Only 20–50% of the variability in BMC between individuals cannot be accounted for by age and other variables (Yano *et al.*, 1984; Johnston *et al.*, 1989; Slemenda *et al.*, 1990). In comparison with other risk factors, bone mass measurement can categorize patients much more effectively on the basis of fracture risk (Ross *et al.*, 1990). In addition, early intervention and monitoring of disease progression is possible with bone mass, where there is little or no ability to change such risk factors as age, body size, gender and race. Thus, at present, BMC appears to be the best risk factor for categorizing patients for treatment. The potential exists, however, that fracture risk assessment could be improved by measuring initial BMC and combining that information with risk factors predictive of bone loss.

Often, fracture risk factors, such as psychoactive drugs and gait speed, may influence fracture risk independently of bone mass by affecting the frequency and types of falls (Ray *et al.*, 1987; Cummings *et al.*, 1989).

At present, there are two approaches to the treatment of osteoporosis. The first consists of estrogen therapy and calcium supplementation used together. For postmenopausal osteoporosis the use of estrogen is somewhat acceptable, while its use in senile osteoporosis is controversial (Smith, 1987, 1990; Francis and Selby, 1987; Walker and Walker, 1987). Similarly, exercise has been shown to decrease the bone turnover rate associated with $NH_4Cl$-induced osteoporosis (Myburgh *et al.*, 1989). Furthermore, the initial events leading to bone loss after both estrogen loss and immobilization are prostaglandin mediated (Lane, Kimmel and Coble, 1989; Thompson and Rodan, 1988). Estrogen and calcium are likely to arrest this otherwise progressive disease. Several studies (Riggs *et al.*, 1980, 1982) have shown that calcium treatment alone reduces fracture rates by 50% in patients with osteoporosis and the combined use of calcium and estrogen further lowers the rate.

In the second approach, sodium fluoride therapy is combined with calcium supplementation. However, the mode of treatment needs examination. Some authors have reported improvement in bone mass as an actual increase in bone density with fluoride therapy. Others have reported increased bone density but with formation of uncalcified osteoid resembling a form of osteomalacia (Riggs *et al.*, 1982; Compston, Chadha and Merrett, 1980). More such trials are necessary.

Fluoride increases trabeculae bone volume (Briancon and Meunier, 1981; Kanis and Meunier, 1984) and mineral content, reduces urinary calcium excretion, reduces the incidence of fractures (Mamelle, Meunier and Dusan, 1988) and is not shown to be carcinogenic. The recommended dosage of sodium fluoride is 40–100 mg $d^{-1}$, with the serum level maintained at 5–10 μM.

It is also suggested that regular moderate physical activity might reduce bone loss in women (Lau *et al.*, 1990; Jonsson *et al.*, 1992). The effect of vitamin D injection is most pronounced in the bones of the upper limb and in ribs (Heikinheimo *et al.*, 1992). These authors recommended the supplementation of vitamin D in aged people. Vitamin D is also important for the function of nerves and muscles, and therefore may prevent a fall in the aged. This may be due to the organism's ability to regulate the mineral content of different bones (Whedon *et al.*, 1974): an aged organism with a vitamin D shortage might regulate its supply to different bones according to their needs. No toxic effects of vitamin D injection, so far, have been reported (Drinka and Nolten, 1984; Coles *et al.*, 1985).

Other modes of treatment that require further study include androgens and calcitonin. These agents may prove useful in osteoporotic men or in patients in whom estrogen therapy is ineffective or unacceptable.

### *6.23.1 Gene Therapy*

While a comprehensive mechanistic understanding of fracture healing at the molecular level is still illusive, sufficient information is available to proceed with gene therapy while further pathophysiological information is being generated.

An increasing demand is to develop a gene therapy to promote fracture healing (Rundle *et al.*, 2003). A single injection of a tiny volume of retroviral vector can be applied. External fixation would be minimally invasive and thereby constitute a very convenient therapy. In addition, if needed, gene therapy can deliver a growth factor that would result in long-term expression. In this regard vectors are available that integrate into the genomics and can result in long-term expression. Flexibility is also available in the form of an on-off switch that can be incorporated into the viral vector, thereby providing temporal control (Moutsatsos *et al.*, 2001). The dose of the growth factors can be computed.

The first variable that is evaluated in gene therapy is the viral vector. Retroviral vector has been shown to be: (i) Safe in trials (Hoogerbrugge *et al.*, 1996). (ii) The viral proteins produce a poor image response, regardless of the number of injections. Retroviral incorporation into the host genome provides potential for prolonged transgene expression (Marx *et al.*, 1999). (iii) Retroviruses specifically infecally transduce cells in fracture tissues and quantify retroviral vector efficiency. These authors found an average transduction efficiency of 70%. Upon application of the $\beta$-galactosidase gene to fracture tissues *in vivo*, they found evidence of transduction of periosteal mesenchymal cells at the fracture site, while MLV-$\beta$-galactosidase application failed to transduce unfractured femurs. Both observations are consistent gene transduction following a burst of cell proliferation at the fracture site soon after injury. It is anticipated that an optimum healing of fractures will eventually incorporate more than one type of growth factor gene (Barnes *et al.*, 1999; Einhorn *et al.*, 1995; Tatsuyama *et al.*, 2000).

Members of several growth factor families are expressed during fracture healing, including the fibroblast growth factors (Jinguishi *et al.*, 1990), insulin-like growth factor-1 (Andrew *et al.*, 1993), platelet-derived growth factor (Andrew *et al.*, 1995) and members of the transforming growth factor (TGF) superfamily (Bostrom and Asnis, 1998; Cho, Gerstenfeld and Einhorn, 2002). None of these agents has been shown to consistently promote bone healing in animal models. However, bone morphogenetic proteins (BMPs), a subfamily of the TGF superfamily that were initially identified on the basis of their ability to stimulate bone formation *in vivo* (Wozney *et al.*, 1988), re expressed during fracture healing (Cho, Gerstenfeld and Einhorn, 2002; Nakase *et al.*, 1994; Onishi *et al.* 1998). BMP-4 is an osteogenic growth factor that has been shown to promote the osteogenic phenotype *in vitro* (Chen *et al.*, 1991) and is expressed during endochondral bone formation in normal fracture healing. Furthermore, several BMP proteins have been used successfully to promote the healing of bony defects *in vivo* (Bostrom *et al.*, 1996; Cook *et al.*, 1994a, 1994b, Stevenson *et al.*, 1994; Yasko *et al.*, 1992). Therapy is, however, limited by the half-life of the protein.

Gene therapy offers an excellent opportunity for the expression of growth factors within healing tissues. Until now, gene therapy for skeletal repair has often been performed *ex vivo*, a process in which cells transduced with a therapeutic gene are introduced into the target tissue. While this technique has successfully expressed osteoinductive growth factor genes from cell vehicles and augmented bone production in skeletal tissues (Lieberman *et al.*, 1999), it does exhibit some disadvantages. Most notably, the clinical applications of this therapy are delayed by the necessity to transduce and accumulate sufficient numbers of the cells selected as the vehicle. In situations where the cell vehicle must be autologous and derived from the recipient of the therapy prior to therapeutic intervention, employing this process can result in additional delays in treatment.

To provide for immediate therapy to the fracture site, Rundle *et al.* (2003) have developed a procedure of *in vivo* gene therapy wherein one simply injects an engineered viral vector directly into the fracture site. Among the more common viral vectors, retroviruses, adenoviruses (Hitt, Addison and Graham, 1997) and lentiviruses (Gasmi *et al.*, 1999) have all been modified to utilize their various assets for gene transfer and expression. The application of a retroviral vector for *in vivo* gene therapy is rated higher for fracture healing. Murine leukemia virus (MLV) is a retroviral vector that produces a robust expression of transduced genes from viral regulator elements.

To apply gene therapy to the fracture it has been pointed out that the advantage of the proliferating nature of the fracture pertains to express the BMP-4 gene following the direct injection of our MLV based vector to the fracture tissue. Peng *et al.* (2001) have found that the secretion of BMP-r *in vitro* was low but could be overcome by constituting the BMP-4 gene. Although BMP-4 gene has been chosen for this application of gene therapy, MLV potentially transfers and expresses any gene within proliferating tissues, whether proliferatin is induced naturally as a result of wounding or artificially through the application of exogenous factors. The study of Rundle *et al.* (2003) describes the results with the MLV vector carrying this BMP-2/4 transgenesis. One has to target the fracture site with *in vivo* gene therapy of the feature of retroviral vectors that limits their transduction to proliferating cells.

pQCT segments for both soft and hard callus tissues show an increase in the size of fracture tissues with the BMP-2/4 transgene. The large soft tissue observed at 7 days of healing is unexpressive since BMP-4 expression is normally associated with differentiation, rather than proliferation, of skeletal tissue precursors. However, this tissue must have resulted from proliferation induced, directly or indirectly, by BMP-4 expression, as it was absent in fractures expressing the gene. A dose-dependent BMP-2 gene-increased proliferation has been observed in C3H/$10^{1/2}$ cells transduction with an adenoviral vector (Lou *et al.*, 1999).

## 6.24 Non-invasive Techniques

### *6.24.1 Electrical Stimulation and Osteoporosis*

As mentioned above, several prophylactic measures to prevent bone loss are available. Although these regimes have been shown to be effective in the treatment of osteopenia, limitations and dangers are inherent in their extended use. Even exercise, which provides the only endogenous stimulus to maintain normal bone mass, is potentially dangerous as it may sometimes lead to fracture, which it is supposed to prevent. The clinical potential for increasing bone mass or simply preventing loss of bone by alternative non-invasive means is, therefore, a matter of continuing interest.

Martin and Guttman (1982), using the rat model for osteoporosis, have shown that the effect of a capacitively coupled 30 Hz field was not dependent on the duration of stimulation (exposure for $2\,h\,d^{-1}$ was as effective as exposure for $8\,h\,d^{-1}$). In contrast to these results, Brighton, Nichols and Arangio (1985) have reported that the capacitively coupled electrical signal delivered at 60 kHz could reverse the loss of bone due to disuse, but needed a daily 24 h signal of only 0.5 V peak to peak. This is extremely important, as it represents the first *in vivo* evidence of the influence of electrical fields on osteogenesis that describes a specific dose–response effect in the modeling skeleton. Rubin, McLeod and Lanyon (1989) identified specific aspects of the electrical signal that have the potential to regulate bone cell activity. These workers have attempted to correlate cortical osteogenesis and the intensity of the induced electrical power, the level of which could potentially be generated during normal physiological activity (Bassett and Becker, 1962). The osteogenic effect with the exposure to low level magnetic fields acts with, the inhibition of the osteoporotic process.

Lirani-Galvao, Silva and Lazaratti-Castro (2007) have used bone densitometry to study the effect of electrical stimulation on OVX Wistar rats. They used a low intensity pulsed electrical field (1.5 MHz, $30\,mW\,cm^{-2}$), starting on the seventh day after surgery, five times a week for eleven weeks. These authors measured global spine and limb bone mineral density by dual-energy X-ray absorptiometry (DEXA), both before surgery and at the end of the protocol. They demonstrated that the electrical fields stimulate osteogenesis in OVX rats. They further reported a positive effect of electrical field stimulation on BMD (bone mineral density) and BMC (bone mineral content) in preventing osteoporosis in female rats.

Several authors have observed the effects of PEMF stimulation in nonrandomized studies (Hinsenkamp *et al.*, 1984; Wahlstrom, 1984; Coloson *et al.* (1974). A current of $4\,\mu A$ at approximately 0.5 V was continuously applied to the bone defect area. It was found that the electrically stimulated defect area consistently incorporates more label than its corresponding control femur. This labeling indicates an increased DNA synthesis by the cell population in the electrically stimulated defect area. These authors reported that PEMF stimulation increased callus formation and trabecular bridging at the osteotomy level. A comparison with their control counterpart revealed more bone formation.

Pulsed magnetic fields have also been used to heal ununited congenital "pseudoarthosis" of the tibia and surgically resistant non-unions in adults. Pulsed radio frequency fields have been used for fracture treating in animals and non-union in man (Behari, Dave and Arya, 1990; Behari, 1991). These authors have concluded that there is an effective window of pulsed electromagnetic field in which bone mass can be controlled. This is achieved by comparatively short daily periods of exposure (1–2 h) generated within a range of physiological intensity and frequency. Because of the physiological characteristics of the signal, the electricity influences the behavior of cell populations that are responsible for bone remodeling (Brighton and McCluskey, 1986). It has also been shown that these fields can slow, inhibit, or even reverse the osteoporotic processes that abnormally accompany disuse in animal models. Such information can also be of value in developing models to explain the cause of disuse osteoporosis.

To understand the mechanism of occurrence and treatment of osteoporosis, a model study in animals is needed. Ovariectomy induced bone loss in the rat (postmenopausal bone loss) (Kalu, 1991) and bone resorption inhibitors suppress bone formation in the rat as in humans (Wronski *et al.*, 1989; Movsowitz *et al.*, 1990; Seedor, Quertuccio and Thompson, 1991;

Chou *et al.*, 1992). Another method is to adopt neurectomy model (disuse osteoporosis) (Brighton, Tadduni and Pollack, 1985; Jayanand *et al.*, 2003).

The electrical method of reversing osteoporosis has promise, but more such investigations are needed.

## 6.24.2 Ultrasonic Methods

Ultrasonic assessment of bone for managing osteoporosis and other metabolic bone diseases has been proposed (Kaufman and Einhorn, 1993; Hans, Schott and Meunier, 1993) as an alternative to ionizing radiation based bone densitometry technology, for example, single-photon absorptiometry or dual energy X-ray absorptiometry (DEXA). In contrast with these clinical bone densitometric techniques, ultrasound is a mechanical wave and thus interacts with bone in a fundamentally different manner. Ultrasound is viewed as having great potential for assessing bone since its propagation is affected by the structure, composition and mass of the bone tissue being interrogated. Since the risk of bone fracture is related to the combined interactions of these features (Kleerekoper *et al.*, 1985), ultrasonic measurements may be able to provide important information on bone quality and more accurate estimates of the risk of bone fracture.

Numerous clinical (Salamone *et al.*, 1994; Roux *et al.*, 1993; Herd *et al.*, 1993; Palacios *et al.*, 1993; Waud *et al.*, 1992; Gluer *et al.*, 1992; Zagzebski *et al.*, 1991; Agren *et al.*, 1991; Baran *et al.*, 1991; Baran *et al.*, 1988) and *in vitro* human (McCloskey *et al.*, 1990; McKelvie *et al.*, 1989; McKelvie and Palmer, 1991) and animal (Gluer, Wu and Genant, 1993; Tavakoli and Evans, 1992; Kaufman *et al.*, 1994) ultrasonic studies have also been reported. For example, Evans and Tavakoli (1990) measured the correlation between the velocity of ultrasound and broadband ultrasound attenuation (BUA) with physical density in 44 samples of cancellous bovine femora. The results showed a correlation of $r = 0.85$ and $r = 0.33$, respectively. An *in vitro* study on density, as measured by quantitative computed tomography, in terms of BUA reported a correlation coefficient of $r = 0.92$, in the frequency range 200–600 kHz. In a clinical study, both ultrasonic velocity and BUA were measured in the oscalcis in 64 subjects (Waud *et al.*, 1992). The authors found respective correlations of 0.66 ($P < 0.01$) and 0.74 ($P < 0.73$) with bone density at the same site using DEXA.

Ultrasonic studies can be categorized under three headings. First, to explore in a comprehensive and consistent manner the interrelationships between ultrasonic velocity, ultrasonic attenuation and bone density over a very broad range of density values. Both human and bovine cancellous bone samples are used to obtain a set of specimens with a wide range of bone densities. Second, to determine if the combined use of ultrasonic attenuation and velocity in multivariate regressions would lead to improved accuracy in the estimation of bone density, in comparison to using either ultrasonic attenuation or velocity alone. Finally, to examine the effect of different nominal ultrasonic transducer frequencies on the measured velocities and attenuations.

### 6.24.2.1 Measurement Method

Figure 6.7 shows a typical experimental set-up. Two ultrasonic transducers are used, one acting as transmitter and the other as a receiver of the ultrasound wave. In a typical experiment, a bone specimen of thickness $d$ is placed between the two transducers. The transducers and bone are then

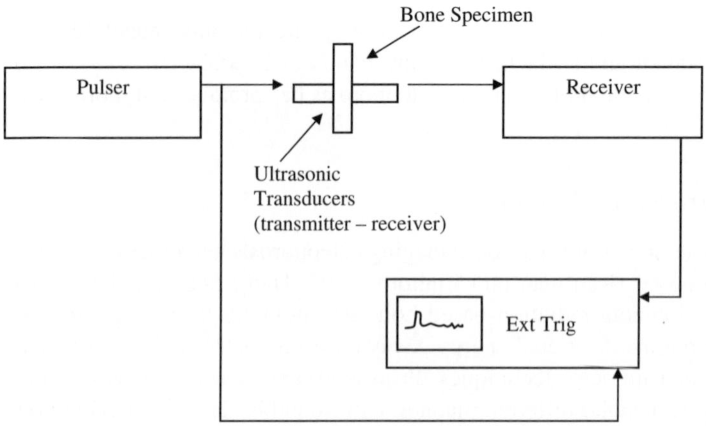


**Figure 6.7** Experimental set up for Ultrasonic Velocity and Attenuation measurements (Shaffkey *et al.*, 1988)

submerged in water. In this configuration, there are two water layers through which the ultrasonic wave propagates – one between the transmitter transducer and sample and the other between the receiver transducer and sample. The two transducers are coaxially positioned and separated by a fixed distance. The transmitting transducer is excited by an input signal, $v(t)$, so that an ultrasound pulse is transmitted through the water, into and through the sample and then through the water to the receiving transducer. The output voltage of the receiving transducer is denoted either by $v_s(t)$ or $v_r(t)$, depending on whether the measurement is made with a bone sample present or absent. In the latter case, the signal is termed a reference signal and corresponds to a measurement made through the water bath only. In the former case the signal, $v_s(t)$, corresponds to propagation through the water–sample water path. The techniques for estimating the ultrasonic attenuation and velocity are distinct and are carried out in the frequency and time domains, respectively. Figure 6.7 shows an arrangement using the pulse echo technique.

***Attenuation Estimation***

An expression for the Fourier transform magnitude, $|V_s(f)|$, associated with the bone sample signal, $v_s(t)$, is given by:

$$|V_s(f)| = T_t(f)T_r(f)|H(f)||U(f)| \tag{6.10}$$

In 6.10$T_t(f)$ and $T_r(f)$ are the magnitude of transfer functions of the transmitting and receiving transducers, respectively; $|U(f)|$ is the magnitude of Fourier transform of the input excitation signal, $v(t)$, and $|H(f)|$ is the acoustic magnitude transfer function of the bone specimen.

The magnitude transfer function $|H(f)|$ can be written in terms of an attenuation function $\alpha(f)$ as follows:

$$|H(f)| = \mathrm{e}^{-\alpha(f)} \tag{6.11}$$

The above analysis assumes that all reflection and transmission losses are included in the attenuation function $\alpha(f)$.

Using the specific attenuation function, $\mu(f) \equiv \alpha(f)/d$, (where $d$ is the distance traveled):

$$|H(f)| = \mathrm{e}^{-\alpha(f)/d} \quad (6.12)$$

If $\mu(f)$ is as an indication of attenuation, then:

$$\mu(f) = a + bf \quad (6.13)$$

then an estimate of the slope, $b$, of the specific attenuation can be obtained, over a specified frequency range (a is the intercept). The frequency range is chosen where the specific attenuation function is approximately linear, as assessed by the mean squared error in the least squares curve fit. The constant $b$ is known as the differential specific attenuation (DSA) and is expressed in units of dB cm$^{-1}$ MHz$^{-1}$; in terms of the more commonly used BUA it is given by DSA = BUA/$d$.

***Velocity Estimation***

The velocity is estimated in the time domain, according to the principle of the time of arrival of signal energy. It can also be estimated in the frequency domain, through the use of phase unwrapping and linear models, as described by Kaufman *et al.* (1995). However, most studies have relied on the pulse transit time technique.

Using $T_s$ and $T_r$ to denote the time of arrival of the sample and reference signals, respectively, the ultrasonic velocity ($v$) can be evaluated according to the following expression:

$$v = \frac{1}{\frac{1}{v_r} - \frac{(T_r - T_s)}{d}} \quad (6.14)$$

where $v_r$ is the velocity of ultrasound in the reference medium, in this case water.

The results obtained display the relationship between ultrasonic velocity and attenuation and bone mineral density of cancellous bone. Both bovine and human bone samples are often used to investigate the relationship over a relatively broad range of densities. For the wide range of bone densities studied, the velocity is a more consistent and accurate estimator of bone density. For both the bovine and human data, the ultrasonic velocity is highly correlated with density, having correlation coefficients in the range of 0.9. This is true for both the lower (500 kHz transducer pair) and higher (1 MHz transducer pair) frequency range. However, the same cannot be stated about attenuation. Although attenuation is highly correlated with density for human bone samples, with correlations greater than 0.8, the relationship for the higher density bovine samples is much weaker. The attenuation is strongly correlated with density for samples of relatively low density, but the correlation is significantly weakened in the case of higher densities. This indicates that a nonlinear relationship exists between the attenuation and bone density, at least when taking into account both the human and bovine sample data. The results of Serpe and Rho (1995) support these findings. A related observation arises from the nonlinear relationship between bone specific surface and porosity (Martin, 1984). If attenuation can be assumed to arise from the scattering of the ultrasonic wave in the bone structure, and if this scattering can be assumed to be proportional to the relative amount of specific surface area present, then a nonlinear dependence of attenuation on bone density is consistent with the above observations and the assumptions involved therein. Future investigations can be directed toward examining this phenomenon and the related frequency dependence of the differential specific attenuation. In particular, it may be possible to identify a transition frequency, that is a frequency at which the dependence of DSA on density changes from a positive to a negative slope, which could be useful for characterizing the mean pore size.

The range of ultrasonic velocity and attenuation values overlap for the human and bovine bone data. Thus it is not possible to analyze the human and bovine data together. The reasons for this may be related to two basic factors. First, the inherent assumed differences in trabecular architecture can lead to different absolute values in the acoustic parameters. Second, the experiments on the groups of samples may not have been carried out at the same time and, therefore, may be susceptible to some uncontrolled factors. Thus it is possible that a difference has arisen due to a change in the measurement conditions.

Clarke *et al.* (1994) have measured attenuation and velocity in a standard epoxy resin, fabricated to a predetermined porosity. They did this by the inclusion of a marrow-mimicking material, thereby introducing a known and controlled mean pore size. These authors performed measurements in the porosity range 10–80% in the frequency range 500–900 kHz. Also, broadband ultrasonic attenuation was measured using the relationship:

$$\alpha = \frac{20}{ds_1 - ds_2}\log_{10}\left(\frac{A_1}{A_2}\right) \tag{6.15}$$

where $ds_1$ and $ds_2$ are sample thicknesses, and $A_1$ and $A_2$ are the signal amplitudes.

These authors proposed that the velocity may change linearly with volume fraction and would, therefore, be more accurately predicted by the simpler expression:

$$V_P = V_a(1-\phi) + V_b\phi \tag{6.16}$$

where $V_P$ = predicted velocity in the material, $V_a$ = velocity in the matrix material, $V_b$ = velocity in the pore material and $\phi$ = porosity of the material.

Hasegawa *et al.* (1995) have used scanning acoustic microscopy to assess bone elasticity at a resolution of about 60 μm to estimate the bone quality. These authors compared the acoustic velocities and bone volumes of premenopausal (normal), osteoporotic and postmenopausal (normal) subjects. They found that the acoustic velocity of osteoporotic bone was 6.2% lower than that of the age-matched normal ($P < 0.02$) but 3.4% higher than that of bone from premenopausal normal women.

One question that remains is whether the combination of ultrasonic attenuation and velocity may be able to better estimate bone mineral density together than the two factors in isolation. The correlations of attenuation with bone density are so weak in the bovine data case that combining it with velocity does not lead to any significant improvement in the ability to estimate bone density. In contrast, however, for the human bone data the combination of both attenuation and velocity produces improvements, providing about an 11% improvement for the 500 kHz transducer pair data and about a 19% improvement for the 1 MHz transducer pair data.

This study does not consider the relationship of trabecular architecture to the ultrasonic measurements. This is an important aspect of the interaction of ultrasonic waves with trabecular bone, particularly with respect to the human bone samples (Kaufman and Einhorn, 1993). Additional data are required to elucidate the effects of age, architecture and bone quality on ultrasonic attenuation and velocity. In this regard, several reports have related ultrasonic measurements to osteoporotic fractures (Ross *et al.*, 1995), which suggests that ultrasound may be useful for the characterization and prediction of fracture risk as well.

The results obtained so far indicate that ultrasonic measurements are, in general, highly correlated with bone density in trabecular bone samples. This correlation is more consistent and stronger in relatively low density human samples than in higher density bovine samples. Whether this is related to the density value alone, or a combination of the density with differences in microstructure (architecture), remains to be determined. Additionally, the combination of both ultrasonic velocity and attenuation as measured in human cancellous bone appears to offer a significant improvement in accuracy of the bone density estimate. Studies are needed to investigate the application of more complex multivariate estimation techniques. Nonlinear neural network based methods can further enhance the accuracy of bone density estimates (Kaufman *et al.*, 1994). The neural network approach can also be possibly adopted to determine the ability of ultrasonic measurements to estimate bone architecture, strength and clinical bone fracture risk.

## 6.25 Conclusion

The ultimate objective of osteoporotic treatment (non invasive or otherwise) is to prevent the ultimate – occurrence of fractures. An obvious treatment is to adopt a package – weight bearing exercise, diet rich in calcim and vitamin and simultaneous reduction in smoking and excessive alcohol consumption. Inhibiting (or controlling) osteoclastic activity with hormone replacement. Therapy (oestrogen) bisphosphonates, selective estrogen receptor modulators along with promotion of osteoblastic activity (with agents such as PTH) show promise. Treatment of this type can be effectively coupled with early diagnosis of osteoporosis and to a significant extent decelerate (prevent) this by electromagnetic field stimulation.

# 7

# Non-Invasive Techniques used to Measure Osteoporosis

## 7.1 Introduction

As discussed in previous chapters, osteoporosis is essentially a multifactorial disease, with major contributions from genetic factors, lack of physical activity, inadequate nutritional intake and endocrine deficiencies. This causes a decrease in bone mass and secondary mechanical failure of the skeleton and has come to be identified as the commonest metabolic disorder of bone. Osteoporosis is not found to affect bone uniformly. It causes varying bone densities throughout the skeleton (Zamenhof, 1997). The decline in vitamin D status, dietary calcium intake and calcium absorption efficiency may account, in part, for the increase in serum parathyroid hormone (PTH) (Heaney, 1996). Chronically elevated serum PTH can contribute to increased rates of bone resorption and loss of bone mass at the femoral neck in the elderly (Meunier, 1996). Many older individuals also decrease voluntary physical activity or may even be subjected to prolonged bed rest, resulting in further exacerbating of bone loss (Le Blanc *et al.*, 1990). One in three women and one in eight men over the age of 50 are affected by enhanced bone fragility and consequently increased fracture risk. In osteoporosis, while new bone formation stays within the normal range in the elderly the rate of bone resorption is far greater. The decrease in mechanical loading incurred with limb immobilization results in localized bone loss that is similar in many respects to that observed with osteoporosis (Li *et al.*, 1990; Li and Jee, 1990; Maeda *et al.*, 1993; Shen *et al.*, 1997). Both types of bone tissue, that is, cortical and cancellous bone, are sensitive to age-related bone resorption and consequent increase in fragility. Cortical bone usually becomes more porous with advancing age. In addition, the cortices of long bones becomes thinner. Trabecular bone loss leads to increased porosity, thinning of trabecular elements and disruption of structure continuity. The relationship between intracortical porosity and cortical bone strength has been frequently examined. Strong correlations have been reported and power regression models have been established between these parameters (Barth, Williams and Kaplan, 1992; Carter and

*Biophysical Bone Behavior* Jitendra Behari

Hayes, 1977; Currey, 1988; Martin and Ishida, 1989; Schaffler and Burr, 1988; Yeni *et al.*, 1997).

McCalden *et al.* (1993) have reported that changes in porosity account for 76% of the reduction in the strength of cortical bone. Other microstructural parameters are also considered to be of importance for the mechanical strength of the cortical bone. The elastic modulus of bone decreases approximately with the square root of porosity. The increased porosity and the higher prevalence of porous structures have a markedly negative influence on the ability of the cortical shell to withstand stresses associated with a fall (Bell *et al.*, 1999). Both the average pore diameter and average pore area showed the highest correlation coefficients with yield stress as well as with elastic modulus, using a polynomial regression model ($r = 0.73$, $r = 0.68$ for yield stress, $p < 0.0005$; $r = 0.49$, $r = 0.51$ for elastic modulus, $P < 0.05$). These strong correlations between parameters describing porosity and porous structures and mechanical properties indicate high relevance for the prediction of bone strength.

Bone porosity can be defined as:

$$1 - \frac{\rho_{\text{apparent}}}{\rho_{\text{tissue}}}$$

where:

$$\rho_{\text{tissue}} = \frac{\text{mass of bone tissue}}{\text{bone tissue volume}}$$

$$\rho_{\text{apparent}} = \frac{\text{mass of bone tissue}}{\text{bulk bone volume}}$$

Additional variations in bone mineralization, collagen content and orientation, as well as the degree of collagen cross linking, may also influence the mechanical properties (Yeni, Brown and Norman, 1998; Yeni *et al.*, 1997). In addition to porosity, other microstructural parameters are important for the mechanical strength of cortical bone. Typical changes in cortical bone associated with advancing age are a decrease in osteonal area and a corresponding increase in Haversian canal area and an increasing number of osteons per unit area (Black *et al.*, 1974; Currey, 1964; Evans, 1976; Jowsey, 1960; Kelsey and Hoffman, 1987; Martin, Pickett and Zinaich, 1980; McCalden *et al.*, 1993).

Data reveal a significant positive correlation between osteon density and fraction of osteonal structures with the mechanical parameters of cortical bone. These findings are considered important for the estimation of fracture risk. Some authors have reported a significant predictive value of osteon density for fracture risk (Barth, Williams and Kaplan, 1992; Squillante and Williams, 1993) fracture resistance (Yeni *et al.*, 1997) and mechanical strength (Evans and bang, 1967). The high relevance of osteon morphology for mechanical properties of femoral cortical bone has also been confirmed. The results suggest that porosity and osteon density are the variables with the highest predictive value for strength. The morphological parameters together explained 49–68% of the variation in strength. It has been concluded that, although there must be other factors, such as biochemical components (Yeni, Brown and Norman, 1998), osteon morphology has an important influence on the fracture resistance of cortical bone.

Loss of mechanical strength is normally observed with disuse, which seem to reduce with BMD and is also significantly reduced at the proximal and distal tibial sites in immobilized subjects. In all patients with osteoporosis, vertebral body collapse and fractures of the proximal femur and distal forearm commonly occur with minimal trauma (Johnson and Epstein, 1981). In a clinical environment osteoporosis is often undetectable by conventional radiography until it reaches a relatively advanced stage (Avioli, 1977). Krolner and Pros Nielsen (1982) consider osteoporosis to be present when the bone mineral content (BMC) is below the .5% confidence limits for postmenopausal women. Riggs *et al.* (1981) have suggested the 10th percentile for vertebral BMD (measured by dual photon absorptiometry) for patients with non-traumatic vertebral fractures as the threshold below which the spine should be considered osteoporotic. Quantitative bone mineral analysis are widely used methods for assessing the skeleton in the diagnosis and treatment of metabolic bone disease. A reliable, non-invasive and readily available modality for *in vivo* measurement of BMC is necessary as an objective yardstick for the diagnosis of various skeletal disorders associated with bone loss.

Cortical bone contributes significantly to the mechanical strength of bone (Augat *et al.*, 1998; Bell *et al.*, 1999; Spadaro *et al.*, 1994). The assessment of cortical bone strength is therefore considered relevant for the prediction of fracture risk or the choice of suitable therapy strategies in orthopedic surgery. In conformity with this several approaches for the prediction of cortical bone strength have been described, most of them using cortical BMD or measures of cortical geometry (Augat, Reeb and Claes, 1996; Louis *et al.*, 1995). However, low correlation coefficients between cortical bone strength and BMD have often been reported (Snyder and Schneider, 1991; Stromsoe *et al.*, 1993).

Regarding the choice of animal models for studying age-related bone loss in men, clinically relevant bone sites, such as the vertebra and the femoral neck, are used because these are common fracture sites in men (Eastell *et al.*, 1998). SD F344 and Wistar rats are commonly-used laboratory animals. Clinical evidence suggests that osteoporotic fractures are expressed at weight bearing sites, where there was a significant fraction of trabecular bone, and therefore osteoporosis is a consequence of trabecular bone loss (Webber, 2006). It is also true that osteoporotic fractures occur at non-weight bearing sites such as the distal radius, where Colles fractures are expressed. Such fractures are now known to be an early warning of future, more serious, bone fractures, particularly in men (Haentijens *et al.*, 2004). Age-related bone loss has been found previously in the lumbar vertebra of aging male Wistar rats, which have been suggested to be a viable model of age-related bone loss in men (Schapira *et al.*, 1991). The vertebrae of male F344 rats appear not to lose substantial amounts of bone with aging (Erben *et al.*, 2000), whereas bones of aging male Sprague–Dawley (SD) rats appear to possess the characteristics of age-related bone loss in men (Wink and Felts, 1980). Changes in the skeletal morphology and mineral content are not identical. Hence, the diagnosis of osteoporosis may depend upon the method used as well as upon the part of the skeleton examined. Growing awareness of the impact of osteoporosis on the lives of elderly people (Cooper *et al.*, 1992), the costs of health care (Ray *et al.*, 1997) and the development of new treatments for preventing fractures (Black *et al.*, 1996; McClung *et al.*, 2001; Neer *et al.*, 2001) have all contributed to a growth in the demand for bone densitometry services. Today, scans to measure BMD play an essential role in the evaluation of patients at risk of osteoporosis (Baran *et al.*, 1997; Kanis *et al.*, 1997).

## 7.2 Measurement of the Mineral Content

Once a beam of electron enters the specimen its path length within the sample depends upon the depth from which a portion of it is scattered back (Figure 7.1a).

A two-component model of bone mineral and soft tissue is assumed (Goodman and Levin, 1970). For a beam of collimated, monoenergetic X-rays passing through bone mineral and soft tissue (Figure 7.1b), let:

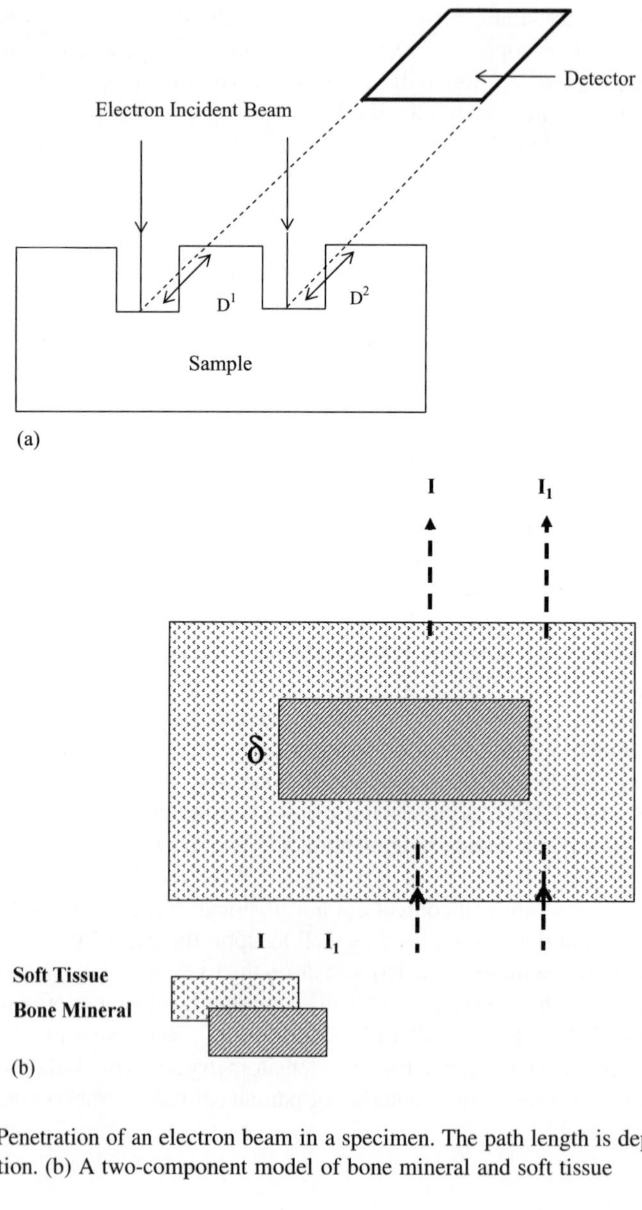


**Figure 7.1** (a) Penetration of an electron beam in a specimen. The path length is dependent upon the depth of penetration. (b) A two-component model of bone mineral and soft tissue

$$T = I_1/I = \exp[-\delta(P_m M_m - P_S M_S)] \\ = \exp\{-\delta P_m M_m[1 - (P_S M_S/P_m M_m)]\} \tag{7.1}$$

where $I$ is the intensity of X-rays through bone mineral and soft tissue; $I_1$ is the intensity through the same path length, but through soft tissue alone, $\delta$ is the path length through bone mineral and $P_m$ and $P_s$ are the densities of bone mineral and soft tissue, respectively.

We have $T = \exp(-\delta P_m M_m) = \exp(\sigma_m M_m)$, where:

$$\sigma_m = -\ln\frac{T}{M_m} = \delta P_m \tag{7.2}$$

is the surface density of mineral material in the path of the beam.

Figure 7.2 shows the schematic experimental set-up. To determine the amount of bone mineral in a section of the radius, a point by point measurement of $\sigma_m$ is made in uniform

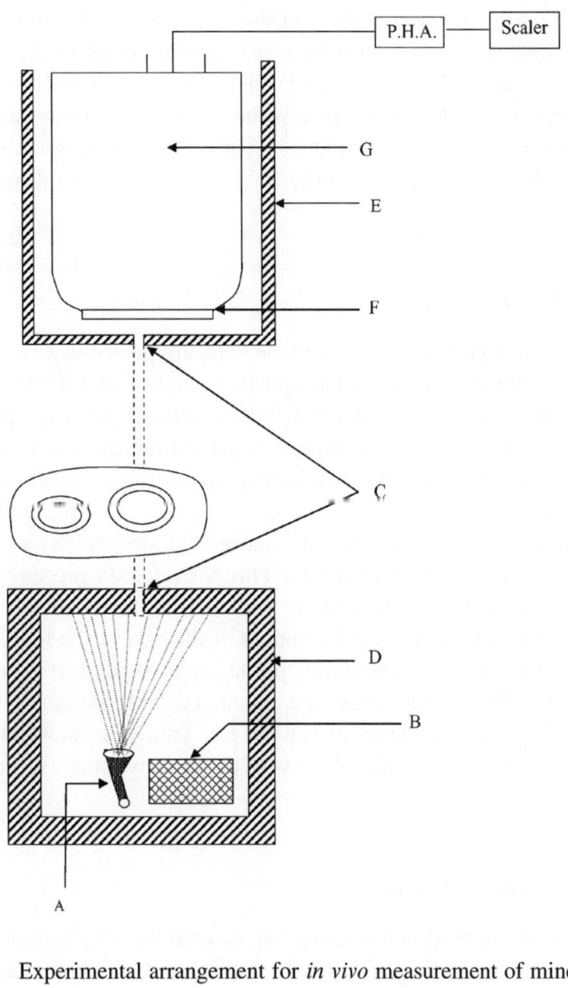


**Figure 7.2** Experimental arrangement for *in vivo* measurement of mineral content

steps across the radius. The total number of grams of bone mineral per cm length of bone is the sum:

$$M = \lambda \sum_{i-1}^{n} \sigma_{\mathrm{m}i} \tag{7.3}$$

in which it is assumed that the bone is crossed in steps.

The photons used in the measurement are barium-K X-rays from the fluorescence of a $BaSO_4$ target (A in Figure 7.2) that is irradiated with the bremsstrahlung from a $^{147}$Pm source (B) of high specific activity (Alffram and Bauer, 1962). The 32-keV X-rays so-obtained are practically monoenergetic and are of a convenient energy for transmission studies of the radius. The X-rays are collimated by slits (C) in the lead source holder (D) and the photomultiplier shield (E). The resulting ribbon-shaped beam, $1 \times 6$ mm in cross section, has its long dimensions parallel to the bone. The X-rays are detected and counted with a sodium iodide crystal (F), photomultiplier (G) single-channel, differential-analyzer assembly.

Changes in mineral metabolism are shown more readily in cancellous bone where metabolism is more rapid than in compact bone, and the distal part of the ulna is chosen to study mineralization. The shadow cast by a given mass of bone mineral above 1 cm$^2$ is independent of the distribution of the mineral along the beam. Furthermore, when the mineral content of a bone is inferred from comparison with a standard absorber, soft tissue within and outside the bone is to be the same as over the matching thickness of the comparable absorber. In this way the radiation transmitted by the matching thickness is a measure of mineral content of the bone in the direction of the beam.

### *7.2.1 Clinical Measurements*

To take two views of the ulna, the forearm and the hand are placed in a Perspex tank. First, the lateral view is taken without water, and then the postero-anterior view with water in the tank. The palm is placed against the side wall, close to the film, with the upper arm supported horizontally in an arm rest; a horizontal X-ray beam is used. The palm is turned so as to be flat on the floor of the tank and the upper arm is maintained in a horizontal position by means of the arm-rest. The beam is then turned through 90° so as to fall upon the top surface of the tank, keeping the same applied cone. A standard wedge of aluminium, attached to a strip of Perspex resting on the floor of the tank, is placed near and parallel to the ulna. The tank is filled with comfortably warm water to a level above the irradiated area. The system used provides automatically a sheet of constant thickness in the direction of the beam independent of the size of the limb.

To evaluate the postero-anterior radiograph, points on the shadow of the ulna are chosen at a particular distance from the styloid process and matched with the photographic densities on the standard wedge. On the densitometer an aperture of $8 \times 4$ mm is used, which is large enough to average out local unevenness within the shadow of the ulna, without exceeding the size of the wedge steps.

### *7.2.2 Calibration and Accuracy*

The problem of using the appropriate comparison standard has two aspects: the standard may be a bone of known content or a standard of near atomic number. Aluminium, which is close to

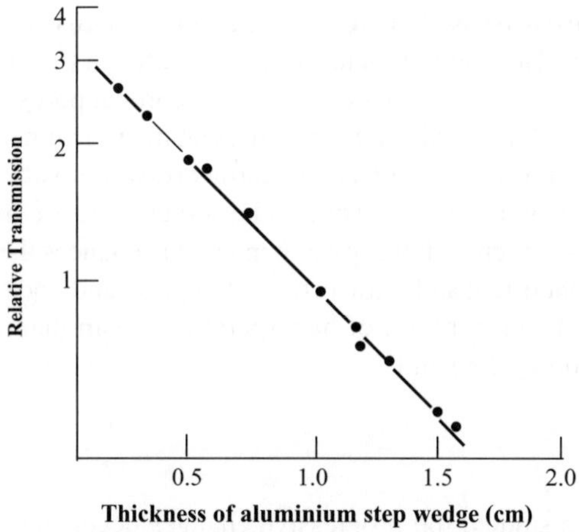


**Figure 7.3** Calibration plot of an aluminium step wedge with relative light transmission (Pelker and Saha, 1983)

bone mineral in terms of atomic number and specific gravity, can be used as a standard (Figure 7.3). An aluminium wedge can be calibrated by densitometric matching from the radiograph of the wedge placed in a tank filled with water and a slab of bone of known dimensions placed next to the wedge. The slab of bone is subsequently ashed and the ash weighed. The weight of the bone sample divided by the cross section perpendicular to the beam gives the mineral content per $cm^2$ on the same X-ray film.

Within the range of about 200–600 mg $cm^{-2}$ of bone mineral, the reproducibility of measurements on the ulna is of the order of within 8% average error, so that one would except to detect changes in the mineral content larger than 15–20%. The pathological changes in the mineral content due to an advanced degree of osteoporosis appear to be considerably greater (Omnell *et al.*, 1960).

## 7.2.3 Limitations

The limits to accuracy are the uncertainty in positioning of the patient forearm in the tank. Since the scan is made along the long axis of the bone and since the bone is not absolutely circular and uniform in cortical thickness the measurements are sensitive to rotation of the ulna about its long axis. The accuracy of transverse scans is, in any event, probably more sensitive to variations in amounts of muscle and subcutaneous fat than longitudinal scans and where the bone is rapidly changing in shape at the lower end of the ulna. Transverse scans are sensitive to minor errors in the position of the scanning pathway.

Technical difficulties in radiogrammetry can be largely overcome if the radiograph of the hand is taken on fine grain industrial rather than ordinary medical X-ray film. Although this requires a longer exposure time, the technique is otherwise similar to the regular X-ray technique. Viewing of this film at 6–8 magnification has been called microradioscopy, since a stereoscopic microscope can be used. Increased intracortical porosity can be clearly visualized by this simple procedure. Bone at the metacarpal midshaft can thus be characterized by two independent criteria: (i) cortical thickness and (ii) intracortical porosity. The measurement of the combined cortical thickness will be truly representative of mineralized bone mass only when there is no microradioscopic evidence of intracortical porosity.

Since X-rays of hands have always been one of the corner stones of the radiologic assessment of metabolic bone disease, microradioscopy of hand bone enhances their significance, permits separate assessment of endosteal and intracortical resorption and, incidentally, enables earlier recognition of subperiosteal resorption of hyperparathyroidism than is possible by ordinary radiographic examination of the hand.

### 7.2.4 Singh Index

Singh *et al.* (1970) have studied the progressive changes occurring in the trabeculae of the upper end of the femur. As normal bone deteriorates to severe osteoporosis, more and more of the trabeculae in the upper end of the femur would be resorbed or thinned to a degree where they would no longer be visible. It is confirmed that with increasing bone loss the trabeculae disappear from the roentgenograms in a definite order, which depends on the calculated mechanical stress of weight bearing. Thus, the roentanograms are analyzed on the basis of the presence or absence and the relative number and density of the trabeculae in the various patterns representing an increasing degree of bone loss (Figure 7.4). A highly significant correlation between this method and the histological grading of iliac crest biopsy specimens shows that the roentogenographic method also reflects bone loss elsewhere in the skeleton fairly accurately. It has been claimed that in cases where grading by the two methods differed

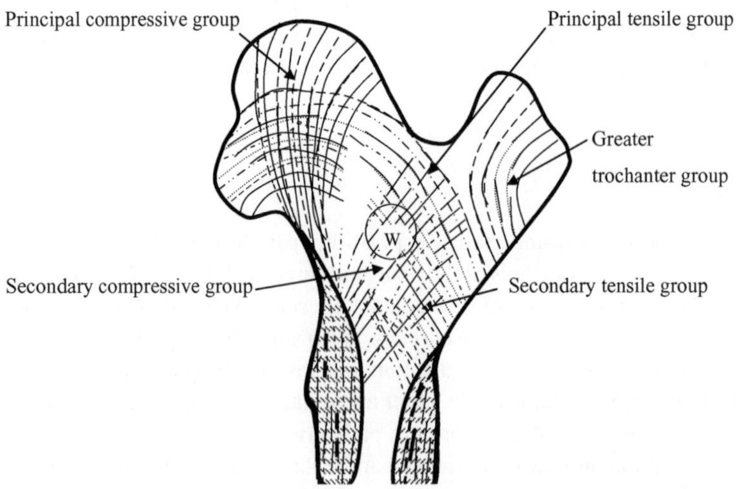


**Figure 7.4** Diagrammatic representation of the five normal groups of trabeculae in the upper end of femur. Ward's triangle is marked by "W" (Singh, Nagrath and Maiou, 1970)

the roentogenographic grading is a better index of the state of the general skeleton since it is related to the clinical picture.

Singh *et al.* (1972) have demonstrated a clear separation between persons with and without spinal crush fractures, using the femoral trabecular pattern index as a criterion. It has been used it as an index in an analysis of osteoporosis, which is a recognized factor in the etiology of these injuries. Theoretically this technique has several advantages over other methods (Barnett and Nordin, 1960; Delvin and Goldman, 1966; Smith and Rizek, 1966; Meema, 1963).

Although low bone mass is a strong predictor of fracture risk, other skeletal factors also contribute to it, including bone micro-architecture, bone material properties and bone microdamage:

1. Anatomically the upper end of the femur is a site of choice for the study of osteoporosis as this, along with spine, is almost invariably involved in the process (Collins, 1966). Since spinal roentgenograms are difficult to interpret, the upper end of the femur becomes the area of preferential choice. Hence the fracture itself may be used as an index of the underlying disease process.
2. The method is technically simple, requiring a roentgenogram of the hip of reasonable quality to delineate the internal architecture of the proximal end of the femur.
3. The method is universally applicable because the grading is based on bone structure rather than on bone quantity.

In recent years the Singh index has become accepted as a valid measure of osteoporosis not only in the femoral neck but also in general. However, the Singh index does not take into account absolute bone density and cortical thickness, both of which contribute to the integrity and strength of any bone and in particular to the femoral neck. The index also presupposes that in osteoporosis the geography of trabecular loss is time related and that the first trabecular group to be affected is always the secondary compressive group, progressing to the tensile group and finally to the principal compressive group.

## 7.3 Bone Densitometric Methods

Measurement of bone mineral density (BMD) is a reliable parameter for diagnosing osteoporosis and for monitoring patients. This is a measure of bone mineral content and a two-dimensional projected area in cm$^2$ of the bones. However, many clinical decisions are based on *T*-scores, which are defined as:

$$\frac{\text{(Patients' BMD-young normal mean)}}{\text{SD of young normal}}$$

The relationship between BMD and fracture is a continuous gradient, for which there is no fracture threshold. People with low bone mass (osteopenia) create problems because it includes people with a wide range of fracture risk. According to WHO criteria (Hussain *et al.*, 2003) a BDM of $\leq -2.5$ (standard deviation) below the mean for a heavy population group aged 20–40 years termed the T score of $\leq -2.5$, with or without pre-existing fragility fractures.

In general, bone status is difficult to evaluate because the bone changes are initially quite small, may not be highly related to apparent density and are effectively localized to a region (i.e., one femoral head but not the other) and type of bone (i.e., trabecular but not cortical bone). The definition of bone quality, thus, remains an ambiguous term in practice (Grimm and Williams, 1997; Pothuaud *et al.*, 2002). In essence, quality refers to a combination of apparent bone modulus and strength as well as some quantified measure of the trabecular geometry, stereology, microdamage, material (tissue) and chemical properties of the bone (Judex *et al.*, 2003).

To meet these increasing demands several techniques have been developed and several early comprehensive reviews (Whedon and Cameron, 1970; Dequeker and Johnston, 1982) are available on the subject. Here we review the available techniques in the light of the recent findings, and the relative merits of different methods are compared. Various non-invasive methods have also been developed to assess the degree of bone atrophy due to metabolic bone diseases. The age-related losses of both cortical and cancellous bone substantially increase the fragility and these alterations can be determined through a non-invasive method. Both cortical and cancellous bones are appropriate to assess the fracture risk. Specific theoretical developments and signal processing techniques are required to address each class of problems. Devices are currently available to measure accessible peripheral sites, such as the heel, finger phalanges and tibia, where the impact of a thin layer of surrounding soft tissue is less. These can be summarized as:

1. Radiographic methods:
   (a) picture contrast on X-ray film by visual reading, use of aluminium step wedge, and use of densitometer (Figure 7.5) (Anderson, Shimmins and Smith, 1966; Pelker and Saha, 1983; Wong, Pal and Saha, 1983);
   (b) change in trabeculae pattern (femoral neck, Singh, Nagrath and Maiou, 1970);
   (c) bone morphometry: metacarpal index, its modification (Garn, 1970; Dequeker and Roh, 1982), radial score, femur score, spine score and so on (Barnett and Nordin, 1960)
2. Photon absorptiometry (single or dual) (Cameron and Sorenson, 1963; Mazess, 1979)
3. Computed tomography (Cann and Genant, 1980)
4. Compton scattering (Odeblad and Norhagen, 1956; Huddleston *et al.*, 1978)
5. Neutron activation analysis (Chamberlin *et al.*, 1968; Kennedy, Eastell and Ferrington, 1982; Eastell *et al.*, 1983); radiations used in 1–5 are ionizing in nature and their effect is stochastic, and hence their exposure carries a risk
6. Infrasound and ultrasound (Angelo, 1971; Wishko, 1975)
7. Other techniques.

The same summary is made in Figure 7.6.

When a beam of X-rays or $\gamma$ rays passes through matter its intensity is reduced by an amount that depends upon the thickness, density and atomic number of the material through which the beam passes. Hence, for a given thickness of an absorbing material, the change in photon flux ($-\mathrm{d}I$) is equal to the thickness ($\mathrm{d}x$), initial flux $I_0$, and a constant ($\mu$) characteristic of the material (linear attenuation coefficient):

$$\mathrm{d}I = -\mu I, \mathrm{d}x$$

or:

$$I = I_0 \mathrm{e}^{-\mu x} \tag{7.3a}$$

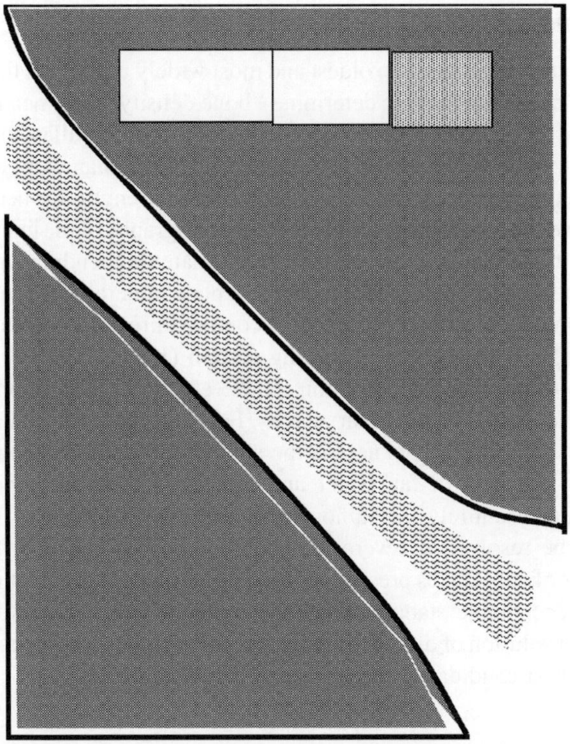

**Figure 7.5** Roentgenogram of a specimen aluminium step wedge (use of a densitometer) (Pelker and Saha, 1983)

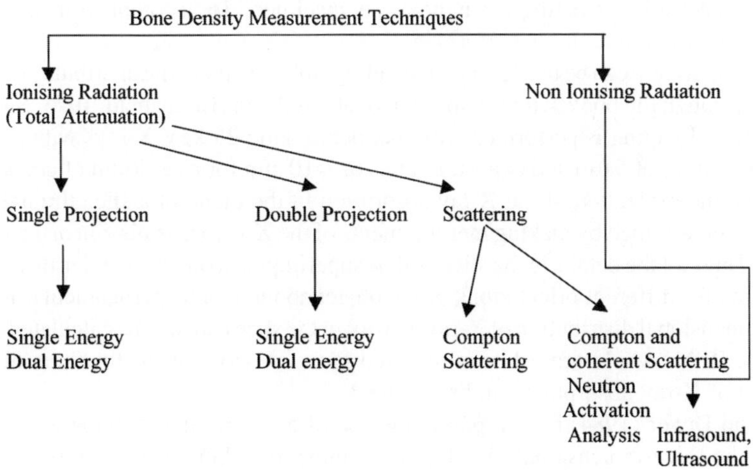


**Figure 7.6** Measurement of bone density

### 7.3.1 Radiographic Methods

Densitometry of X-rays is perhaps the oldest and most widely used of all the above-mentioned techniques that has been employed to determine "bone density" as a means of non-invasively assessing bone loss (Cameron, Mazess and Sorenson, 1968; Pullan and Roberts, 1978). Measurements of cortical thickness and the inner and outer diameters of long bones such as the metacarpal have been used. To directly assess bone strength, Klenerman, Swanson and Freeman (1967) have used such measurements from radiographs of the humerus, together with an elliptical model for the bone cross section, to estimate the bending strength.

A basic requirement of any method for measuring bone mineral that involves attenuation of an X-ray beam is a uniform primary beam. A second requirement is the elimination of non-uniform scatter before it reaches the recording system (Keane, Spiegler and Davis, 1959). A third requisite is to provide compensation for the effects of absorption in the soft tissues surrounding the bone, within its medullary cavity. For the peripheral skeleton, immersing the part in a water bath of uniform thickness is the simplest solution to this problem. However, because of varying amounts of fat, water and protein in different subjects it is virtually impossible to eliminate entirely the minor errors introduced by variability of soft tissues. The difficulty can be resolved if several monochromatic sources of different wavelengths equal to the number of substances present are used. The final requirement is calibration of the system against an acceptable standard for bone mineral. Pure hydroxyapatite (Heuck and Schmidt, 1960), or a solution of dipotassium hydrogen phosphate (Meema, Harris and Porrett, 1964), may be the best candidates for the role of a standard.

## 7.4 X-Ray Tomography

It can be imagined that the stress-state is quite different in the interior of a sample undergoing inelastic deformation as compared to its surface, and that this variation in the degree of stress has a profound effect on fracture properties (Knott, 1976). It is important to observe cracking behavior both at the surface and within the bulk of the sample to determine the sample toughening mechanism existing throughout the specimen. To accomplish this, synchrotron X-ray computed tomography is performed.

X-Ray tomography can be used in medical radiography to obtain the distribution of the X-ray absorption through the body, with a resolution of about 2 mm (Hounsfield, 1973; Swindell and Barrett, 1977). Imaging is performed with monochromatic 25-keV X-rays with a voxed size (spatial resolution) of 5 μm and exposure times of $\sim$10 min for over 1 mm of crack length. In this method, the projection of the X-ray absorption in the plane of a slice through the body onto a line is determined by making measurements of the X-ray transmission of a narrow beam of X-rays. Thus, all the details in the slice will be superimposed on this line. Further projections are measured for different orientations of the object about an axis perpendicular to the slice. The two-dimensional distribution of X-ray density in the slice can then be calculated from these projections. If this whole procedure is repeated for successive slices, the three-dimensional distribution of X-ray absorption can be obtained.

Elliott and Dover (1984) have reported the use of an X-ray microtomographic system to determine the three-dimensional distribution of mineral in bone without physically cutting sections. In this method X-rays are generated from a molybdenum target and passed through a zirconium filter and then through a 15 mm diameter aperture. The bone specimen is rotated about its long axis and the interval between successive projections is taken as 1/3°. Each projection consists of 128 points at intervals of 13 μm, each measured for 10 s. Six sections are made at intervals of 25 μm. The effective absorption coefficient of apatite has been determined from measurements on a flux grown slice of fluorapatite.

The upper limit of the linear absorption coefficient distribution for the $13 \times 13$ μm pixels was found to be about $11.2\,\text{cm}^{-1}$. If the assumption is made that all the absorption is due to stoichiometric hydroxyapatite, the maximum mineral content is:

$$\frac{11.2 \times (\text{density of hydroxyapatite})}{(\text{experimental linear absorption coefficient of hydroxyapatite})}$$

This gives $2.0\,\text{g cc}^{-3}$ if the experimental value for the linear absorption of fluorapatite is used, which will be very similar to that for hydroxyapatite. This value is higher than the $1.36\,\text{g cc}^{-3}$ reported by Sissons, Jowsey and Stewart (1960) and $1.3–1.5\,\text{g cc}^{-3}$ reported by Baud and Gossi (1980).

In L4 vertebra, no bone loss was observed until 19 months of age, which is past middle age in the life span of these rats, because the medial life span of male F344 rats is 31 months. All the cortical bone parameters studied gradually increased with age. The lack of age-related bone loss in the vertebra of male F344 rats has also been observed by Erben *et al.* (2000). Data obtained from the vertebra of male F444 rats and data available on the vertebrae of humans show two important differences. First, male F344 rats do not lose bone in the vertebra, unlike in men. Second, the decrease in BMD (cancellous) in the F344 male rats is seen only in the older age groups of 18, 24 and 27 months, whereas in men there is a high rate of vertebral deformation during middle age (Cummings *et al.*, 1985; O'Neil *et al.*, 1996).

In addition to the lumbar vertebra, cancellous bone has been analyzed in the proximal tibial metaphysis (PTM) and neck of the femur using pQCT. The PTM showed high bone loss when compared with any other site studied. This is in line with results on male F344 rats (Erben *et al.*, 2000). However, because fractures due to osteoporosis in humans do not normally occur in the PTM, this site may not be as relevant a site as the femoral neck and vertebra in studying age-related bone loss as occurs in men. In the neck of the femur of male F344 rats, the BMD decreased significantly in the 24- and 27-month-old rats, similar to men, in which the rate of bone loss is greater in the femoral neck (Fatayerji, Cooper and Eastell, 1999).

In another study, it has been demonstrated that the vertebra of Sprague–Dawley (SD) male rats lose bone from a much earlier age (9 months) (Wang *et al.*, 2000a; Wang, Banu and Kalu, 2000b) and to a higher extent than with F344. The intact male F344 rats lost an appreciable amount of bone only in very old animals in clinically relevant sites. The amount of bone lost is much less than with SD male rats. These authors concluded that male F344 rats may not make a good animal model to study age-related bone loss as it occurs in man.

## 7.5 Skeleton Roentgenology

Skeleton roentgenology makes it possible to determine a large number of morphologic parameters of osteoporosis, while bone scanning enables the amount of bone mineral to be measured in different parts of the skeleton.

In the method described here the roentgenologic classification is made in accordance with two principles: by examination of the trabecular structure and measurement of the cortical thickness of long bones. The inner and outer diameters of the cortex at the middle of the second metacarpal bone and the shaft of the femur (Barnett and Nordin, 1960) and of the proximal part of the radius (Meema, 1963) are measured. These authors determined the bone mineral content by roentgen spectrophotometry (Jacobson, 1964). The method is based on attenuation measurements employing a scintillation detector; the attenuation of the soft tissues in the radiation beam is compensated by using two radiation energies. The mineral content at a particular site is determined by scanning with a slit beam that is well collimated. Measurements are made at eight sites to obtain a representative estimate of the whole skeleton.

Certain trends are observed between the BMC for the various measuring sites. The closer the sites were to each other the greater the correlation. Moreover, the trabecular bone measurements agree with one another and the cortical bone measurements have a tendency to correlate internally. This is consistent with the results of Nordin (1968). The bone mineral content of the third lumbar vertebra, however, would seem to constitute an exception; but here the major part of the mineral content of the vertebra is accounted for by the transverse and spinous processes, which contain a greater proportion of cortical bone than the body.

The mineral content as in the trabecular parts of the skeleton had the greatest correlation with morphologic signs of osteoporosis, as indicated by sparse trabecular structure in the axial skeleton and wide marrow spaces in the appendicular bones.

## 7.6 Metacarpal Index

Here, dimensional measurements are taken across the narrowest portion of the shafts of the bones. Three measurements can be directly obtained:

1. Total width of the bone ($D$)
2. Width of the medullary cavity ($d$)
3. Total width of the cortical bone (i.e., left side + right side $C = D - d$.

$$\text{Cortical index(CI)} = D - d/D$$

CI is often referred to as the metacarpal cortical index (MCI).

Among the various radiographic assessment methods the metacarpal index method (MCI) is often preferred (Figure 7.7). It permits the measurement from an X-ray picture of the hand, which can be easily taken on ordinary equipment, and the position of the hand can be duplicated exactly by placing it on a cassette. However, Barnett's metacarpal indices, $(D–d)/D$, measured with a slide caliper may vary considerably. When the metacarpal index of the same X-ray picture is measured three times with a densitometer, it is found that the error decreased to about 1/7 of that for the value measured with a slide caliper. To circumvent the problem, Inoue *et al.* (1983) have assessed the degree of bone density based on the metacarpal index, using an aluminum step wedge, densitometer and computer. In this method, the hands, placed palm side down, are X-rayed from the back and an aluminium step wedge is placed between the hands. The optical density is measured at the middle portion of the second metacarpal on the X-ray picture using a microdensitometer. The densitometric pattern is

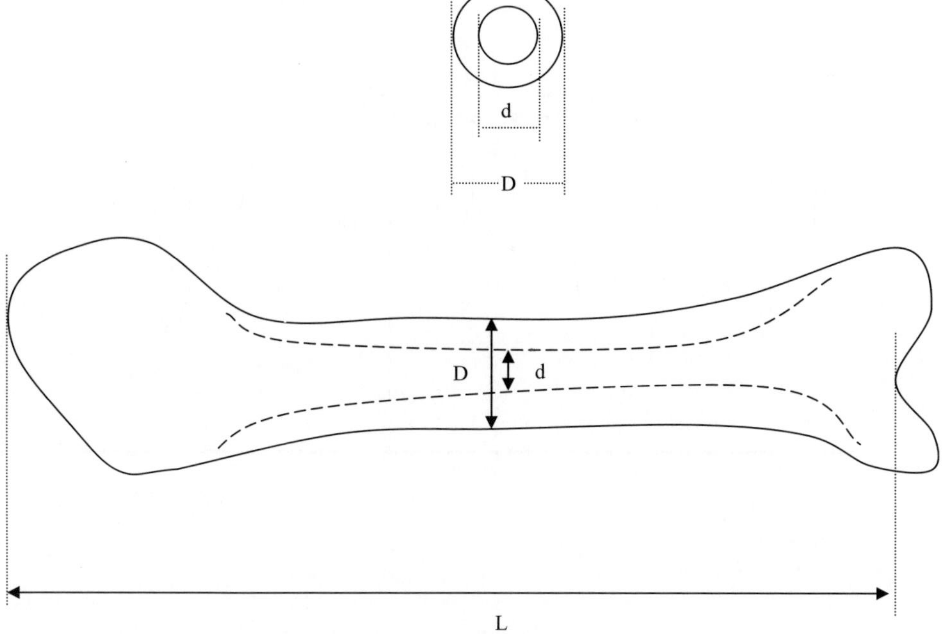


**Figure 7.7** Outline of metacarpal bone, showing measurements made from radiographs

recorded at 10 times magnification and simultaneously the optical density of each step of the aluminium wedge is measured. From the densitometric pattern shown in Figure 7.8, the bone width ($D$), marrow width ($d$), and cortical widths ($d_1$) (cortical width on the radial side), $h_2$ (peak of the cortex on the ulnar side) and $h_{min}$ (peak of the middle position of the bone marrow) are measured.

The errors in taking X-rays on the same day, errors in measurements of the X-ray picture by a densitometer, errors in chart analysis and the reproducibility in the X-ray pictures on different days are determined as coefficients of variation (CV) in normal subjects, patients with osteoporosis and dialyzed patients. The CV are extremely small, being less than 1% for errors in chart analysis and are is relatively small (0.4–2.8%) for errors in densitometer measurement. Regarding errors in taking the X-ray, including the errors in chart analysis and in densitometer measurement, the CV are relatively small (2%) for MCI but are 3–5% for the other indices.

When the CV are calculated for each index in the reproducibility test in normal subjects, patients with osteoporosis and dialyzed patients, it is found that the CV increased, that is, 2% for MCI and 5–7% for the other indices.

When a variation of more than +5% for MCI and +10% for the other indices is noted, the change can be regarded as significant. With a decrease in BMC, these authors have found a significant correlation between the changes in each index: MCI, marrow width ($d$), the peak heights converted into the number of steps of the aluminium step wedge ($G_{s\text{-}max}$, $G_{s\text{-}min}$), with $\sum G_s/D$ obtained by dividing the computer calculated densitometeric pattern area by the bone width $D$.

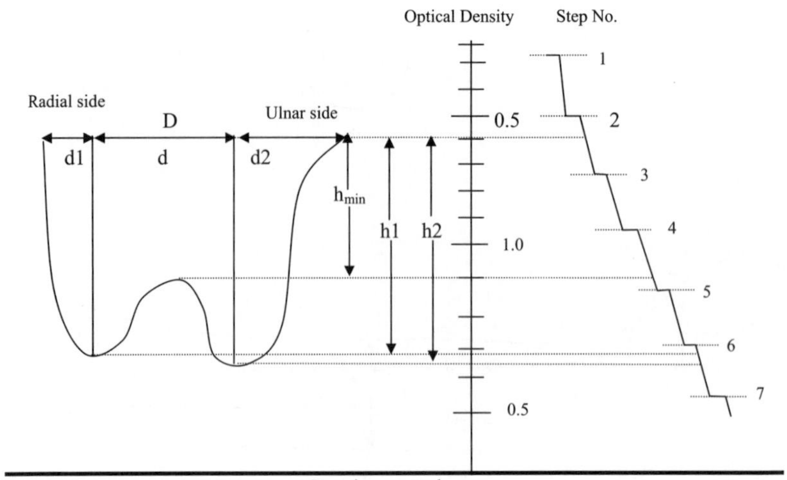


Densitometer chart

1: $MCI = \dfrac{d_1 + d_2}{D}$

2: d

3: hmin – Gsmin

4: $hmax = \dfrac{h_1 + h_2}{2} - \dfrac{GS_1 + GS_2}{2} = GSmax$

5: $\Sigma GS/D$

6: Densitometric pattern

**Figure 7.8** A densitometer chart (Inoue *et al.*, 1983). (Reproduced with permission from T. Inoue *et al.*, Quantitative assessment of bone density on x-ray picture, *Journal of the Japanese Orthopaedic Association*, **57**(12), 1923–1936 © 1983)

## 7.7 Analysis of Radiographic Methods

Most methods of bone density measurement are based on a comparison of the density produced by a standard step wedge with that produced by a bone simultaneously exposed on a radiograph. The photographic densities of the wedge measured by a densitometer are plotted against the wedge thickness and a calibration curve is obtained (Figure 7.3). The density of a given area of bone can be expressed in terms of the step wedge. There are distinct advantages when the effective atomic number of the standard and of hydroxyapatite are similar. Potassium hydrogen phosphate ($K_2HPO_4$) has an effective atomic number of 15.59 and can be obtained in pure form. At different monochromatic energies there is only a small variation in mass between $K_2HPO_4$ and HA (0.95–0.97).

When water, representing the combined thickness of water and plastic, is penetrated by monochromatic radiation of 35 keV, in the path of which two different absorbing bodies are placed, it is seen that in the presence of marrow fat 2–7% more hydroxyapatite is required to produce the same photographic density as is produced by a given amount of hydroxyapatite in water. However, at 35 keV, 5% less hydroxyapatite is required to produce the same photographic density as a given mass of $K_2HPO_4$. Therefore, the variable amount of fat compensates for the 5% difference by 2–7%. The experimental error resulting from these two factors combined would thus be approximately ±2–3% when a 100% weight by volume solution of $K_2HPO_4$ is employed. This is well within the total experimental error. Using this method the bone changes are obtained that are designated as cortical osteoporosis. The total amount of bone salt at a standardized site of measurement combines all three main factors: cortical thickness, its porosity and the degree of mineralization. It may, therefore, serve as an index of skeletal mineral reserves as a whole under conditions in which the mineral losses involve the entire skeleton at similar sites.

However, the method suffers from some limitations. The choice of using the soft tissue between the radius and the ulna for a measure of relevant "background" soft tissue is made for convenience and as a part of experimental investigation. Detailed examination of measurements on elderly people showed that in some cases this is unsatisfactory. The model neglects medullary calcium and is only a fair approximation to the geometry of a section of a real bone.

## 7.8 Direct Photon Absorption Method

Attempts to measure bone mineral using conventional X-ray film and optical densitometer encounter the following difficulties:

1. The heterogeneous X-ray beam changes its spectra while passing through an inhomogeneous medium, due to preferential absorption by softer components.
2. Some of the scattered radiation may also strike the film from the surrounding soft tissues.
3. X-Ray film as a detector has variability in response to radiation.

### *7.8.1 Theory*

If the X-ray beam is passing through a layer of $N$ materials, each having its own mass attenuation coefficient, density and thickness, then Equation 7.3a can be written as (Njeh and Shepherd, 2004):

$$I = I_0 \exp\left( - \sum_{Z_{i=1}}^{N} \mu_i \rho_i \chi_i \right) \tag{7.3b}$$

where $\rho_i$ is the density of the $i$th layer.

In single-photon absorptiometry, only two type of material [bone (b) and soft tissue (s)] contribute to the attenuation. Hence Equation 7.3b becomes:

$$I(\chi) = I_0 \exp[-(\mu_s \rho_s \chi_s(\chi) + \mu_b \rho_b \chi_b(\chi))]$$

Taking the total path length $W = \chi_b + \chi_s$, one gets, after taking the natural logs of both the sides:

$$\ln\left|\frac{I_0}{I(\chi)}\right| = (\mu_s\rho_s(W - \chi_b(\chi) + \mu_b\rho_b\chi_b(\chi)))$$

The bone path length ($W_b$) can be written as:

$$W_b(\chi) = \frac{\ln[I_0/I(\chi)] - \mu_s\rho_s W}{(\mu_b\rho_b - \mu_s\rho_s)}$$

$W$ can be related to the attenuation in soft tissue only:

$$\mu_s\rho_s W = \ln\left|\frac{I_s}{I_0}\right|$$

In this method the mineral content is normalized to give the amount of mineral present in a 1-cm length of the radius. It is assumed that the bone has a uniform composition in the cortical region (Dunn, 1974) and that the computed longitudinal modulus of elasticity of the bone is an average value over this region. The microscopic densities of bone mineral and soft tissue are assumed to be 3.2 and 1.0 g cm$^{-3}$, respectively.

In the experimental set-up (i) transmission of the photon beam is measured by means of a scintillation detector system, (ii) the photon beam used is essentially monochromatic, (iii) the photon beam and detector are well collimated and (iv) the effects of the tissue around the bone are also taken into account. These factors eliminate errors resulting from the variability of X-ray films and film development technique, reduce uncertainties in absorption coefficients, reduce the effects of scattered radiation and reduce errors arising from the presence of tissue.

An improved method for measuring the mineral content of bone *in vivo* by photon absorption techniques has been developed by Cameron and Sorenson (1963). The behavior of $\gamma$-rays and X-rays in matter is the same.

In single-photon absorptiometry (SPA) the radioactive photon source iodine-125 (5 mCi) is contained in a thin-walled small stainless steel tube. By filtering the emission with the K-absorption edge at 29.2 keV, a monoenergetic (95%) beam of 27.4 keV photons is produced by the preferential absorption of the higher energy photons. The tube is placed in a small lead cube that contains a hole. The source and detector system are rigidly coupled by mechanical means and are driven simultaneously by a motor driven system, in 1 mm steps, in a direction transverse to the bone. The transmission of the photon beam through the bone is measured for a 10-second interval after each step. During the scan, measurements of the transmission of the photon beam through the tissue alone, on either side of the bone, are also made. From the data obtained in these measurements, an equivalent thickness of bone mineral can be computed.

Figure 7.9 shows a schematic presentation of a system with the limb containing the bone surrounded by tissue-equivalent material to make the bone appear to be buried in a uniform thickness of "tissue." Therefore, the attenuation is measured through a constant thickness of equivalent soft tissue material. The mineral content and width of the bone can be calculated on the assumption that the mineral constituent of the bone is of uniformly standard composition. It can be further assumed that all the non-osseous tissue substances absorb radiation to the same degree as soft tissue, thus maintaining a two-component system.

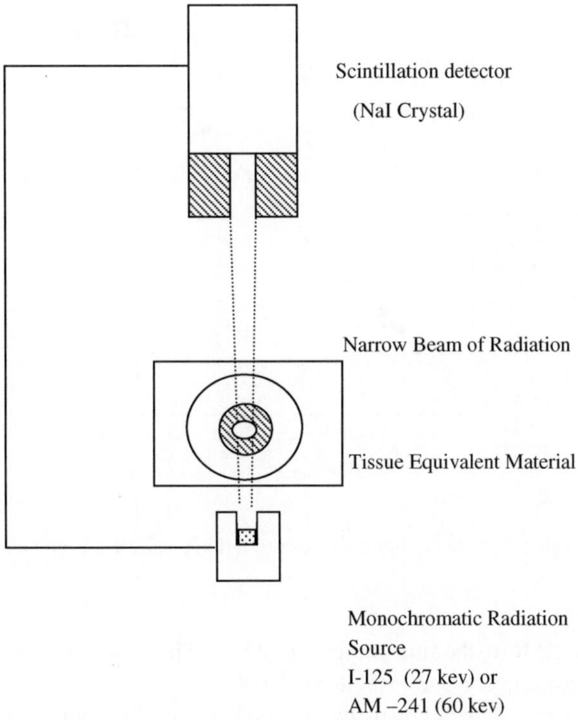


**Figure 7.9** Schematic diagram of the actual measuring arrangement in direct photon absorptiometry. (Cameron and Sorenson 1970)

The amount of bone mineral content ($M_b$, g cm$^{-2}$) in the path of the beam is given by:

$$M_b = [P_b/(\mu_b P_b - \mu_s P_s)]\ln(I_0/I) \tag{7.4}$$

where $\mu_b$ and $\mu_s$ are the mass absorption coefficients (cm$^2$ g$^{-1}$) for bone and soft tissue, respectively, and $P_b$ and $P_s$ are the corresponding microscopic densities (g cm$^{-3}$). $I_o$ and $I$ refer to the intensity of the radiation beam through a constant soft tissue thickness and through a combined thickness of soft tissue and bone, respectively.

To determine the total bone mineral content present across the bone, $M_b$ is integrated across the width of the bone. This is accomplished by synchronous movement of the $\gamma$-ray source and detector across the limb at a slow velocity (1 or 2 mm s$^{-1}$). The bone mineral content is usually determined by numerical integration using the expression:

$$\text{BMC} = k\left(Q\sum_{i=1}^{n} \text{In}(I_0/I)_i\right) \tag{7.5}$$

where BMC is the bone mineral content (g cm$^{-1}$), $Q$ is the distance traversed in an integration interval, $n$ is the number of integration intervals and $k$ is an empirically determined calibration

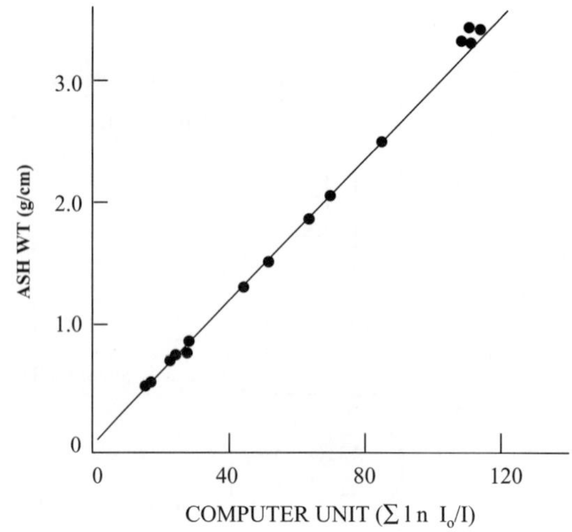


**Figure 7.10** Ash weight versus bone mineral (the correlation coefficient is greater than 0.99) (Cameron and Sorenson, 1970)

constant relating the BMC to the sum of the natural logs. The standard error of estimate in using this relation is approximately 1–2% (Wilson, 1972).

The technique can be evaluated by using excised bones of different sizes immersed in water (Cameron and Sorenson, 1970). Sections of the bones at each measuring site are cut into 1-cm lengths and ashed. The ash weight is then plotted against the measured BMC (Figure 7.10). The method gives the BMC and the bone width ($B_w$), or external bone diameter, of such accessible bones as the radius. It can also provide an estimate of the medullary canal diameter (Kramer, Pattersen and Smith, 1978). Wilson (1973) has shown that BMC of the midshaft of the femur correlates highly ($\gamma = 0.99$) with the compressive strength of the bone at the scan site. By studying cadaver bone, Martin and Burr (1982) have established that BMC decreases significantly with age in females, but torsional rigidity does not. Similarly, several authors have shown a significantly high correlation between mechanical strengths of both cortical and cancellous bone with their density and ash content (Saha and Gorman, 1981) (Figure 7.11).

### *7.8.2 Clinical Applications*

Several subjects have been studied (Sorenson, Cameron and Mazess, 1970) to determine the magnitude of bone mineral content change that might be measured by the direct photon absorption method. These investigators have measured the mineral content at a distance approximately one-third the length of the radius from the distal end while the bone width is taken to be the width of the scan. A high degree of correlation ($\gamma = 0.90$) has been reported (standard error $0.1\,\mathrm{g\,cm^{-1}}$).

This method has been used with minor modifications (Grubb *et al.*, 1984) to measure bone mass in the distal radius. The radius is scanned at two points, 3 cm from the distal end, an area primarily of cancellous bone, and 8 cm from the distal end, an area of cortical bone. Four scans are performed at each point. The base line count rate is first calculated and then the area

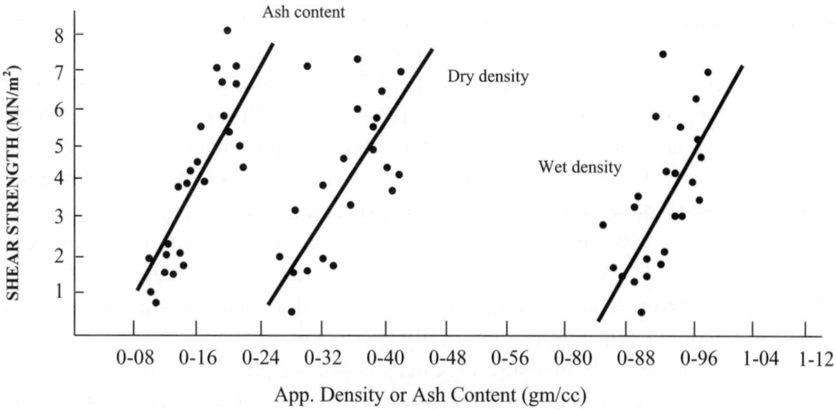


**Figure 7.11** Relationships between the shear strength of embalmed human cancellous bone (obtained from distal femur) and its apparent wet density (right), dry density (middle) and ash content (left) (Saha and Gorman, 1981)

between base line and bone is computed. The width of the radius, particularly at the 8 cm level, correlates well with the total area. The coefficient of variation for BMC is approximately 2.7% and for BMC/W it is 3.7%. The increased variation for BMC/W is due in part to the additional variability of bone width. Also, at 3 cm the bone is less regular and, thus, subject to more variation in position.

There have been two major problems with the use of measurements of the radius. First, these measurements have been accurate only for the appendicular skeleton and at sites that are largely cortical bone and regularly shaped. However, in practice most non-traumatic fractures occur in the axial skeleton and proximal femur, at sites containing significant amounts of cancellous (trabecular) bone a controversy exists as to the value of the radius for estimating the axial skeletal condition (Madsen, 1977; Mazess, 1979; Mazess *et al.*, 1984; Riggs *et al.*, 1981; Sandler and Herbert, 1981). Awbrey *et al.* (1984) have described an improved method of positioning the distal radius for single-photon absorptiometry, permitting rapid, accurate and reproducible measurements. For the single beam densitometer, the externally located "9/10" site on the ulna is used only for the purpose of orientation. The distal radial site is then determined by locating with the instrument a more distal position at which the ulna and the radius are separated by 5 mm. These authors confirmed earlier findings (Christansen *et al.*, 1968) that if the forearm is pronated >20° or supinated >10° the distance between the bones would be different and the "5 mm" site would be shifted. The design of the instrument used and the positioning of the subject maintained the arm in near neutral position. By the use of the new site, these authors overcame the two major problems of distal radial bone mass measurements: variability of measurement and the lack of trabecular bone.

The amount of trabecular bone present at the "9/10" site has often been a point of controversy (Madsen, 1977; Riggs *et al.*, 1981; Sandler and Herbert, 1981). Awbrey *et al.* (1984) have demonstrated that at the new site the cortical walls are the thinnest and the interior consists of a network of trabeculae. The percentage of trabecular bone is estimated to be 75%, which is supported by the work of earlier workers (Hayashi *et al.*, 1982; Schlenker and

Vonseggen, 1976). If distal radial measurements are to be correlated with bone loss in the spine, the site should contain a significant amount of trabecular bone.

Measurements performed (Christiansen *et al.*, 1975) on white women in a moderate to high socioeconomic group showed that bone mineral loss occurred proportionally in both appendicular and axial skeleton and in both compact and trabecular bone at rates that are in general agreement with those reported by Smith, Khairi and Johnston (1975). The correlation coefficient for rates of bone loss at all three measurement sites (lumbar spine, mid-radius and distal radius) are highly significant. Therefore, accurate assessment of bone mass status in "normal" women would be obtained at any of the three sites. Christiansen *et al.* (1975) also reported the use of a 5 mm separation between ulna and radius to locate the site for distal bone density measurements. Their data are internally consistent.

The distal radial bone density measurement in osteoporotic patients at the "2/3" and "9/10" radial measurement sites have been reported not to correlate well with vertebral densitometric changes. This has led to the conclusion that a radial density measurement is a poor indicator of spinal bone density (Mazess *et al.*, 1987; Riggs *et al.*, 1981).

## 7.9 Limitations of the Method

1. Using a 30 keV source there is a good contrast between bone and soft tissue, but for a relatively large section of the material a very active source would be needed to obtain sufficient intensity at the detector. If one wishes to make measurements on bone with much tissue cover, it is usually necessary to use a higher energy photon beam. Because of its 59.6 keV photon energy $^{241}$Am is more suitable for heavier bones, such as the humerus and femur. Its maximum activity, however, is limited by low specific activity and self-absorption in the source and a 100 mCi source is about 5 mm in diameter. One can reduce the diameter of the source by using a lead aperture, which in turn also reduces the intensity.
2. Equation 7.4 is valid only if the absorption in the bone and the surrounding tissue is exponential.
3. The physical dimensions of the source limit the size of the bone that can be accurately measured. Ideally, a point source with an infinitesimally small beam should be used to avoid "edge effects" as the beam enters or leaves the bone.
4. In real situations, the model of the bone as a two-component system cannot be assumed. Within the bone marrow and surrounding the total bone is a layer of adipose tissue. This fat tissue has a volumetric density that is 0.9 times that of muscle tissue, thereby attenuating the photon beam to a lesser extent than an equal thickness of soft tissue (Wishko, 1975). An error is caused by the continually varying thickness of the adipose tissue as the photon beam traverses the bone marrow cavity. The range of error has been calculated to be 1.5–5% (Wooten, Judy and Greenfield, 1973).

In single-photon absorptiometry (SPA) it is necessary to measure transmissions through adjacent tissue locations of equal thickness, one location consisting of both mineral and soft tissue and the other consisting of soft tissue alone. This requirement means that SPA could be applied only to peripheral locations where it is feasible to arrange for the thickness of the two transmission locations to be the same by the use of tissue equivalent material. The consequence of this limitation is that SPA can not be used at the relevant sites from the perspective of a physician.

Several secondary choices have to be made to optimize SPA for clinical applications. The energy of the radioisotopic source has to be chosen as a compromise between the need to maximize the difference between the photoelectric cross sections for bone mineral and soft tissue and the need to transmit photons through the object of interest. The requirement for uniform thickness of soft tissue and the low photon energy restricted SPA to peripheral sites such as the radius and the calcaneus. This constraint clashes with the clinical perspective which considers that the sites of interest to be those where bone fractures are most frequent (spine) are associated with significant morbidity and mortality (proximal femur). It appears that the essential features of a skeletal measurement site are (i) a significant fraction of trabecular bone, (ii) weight bearing these needs and (iii) a central location.

The shortcomings of SPA have been circumvented by the introduction of a low dose X-ray source. Here the radionuclide source has been replaced by an X-ray source (SXA). This improves precision and improves spatial resolution and reduces examination time (Kelley, Crane and Barner, 1994). SXA makes possible a quantitative assessment of bone mineral content at peripheral sites of the skeleton. It offers a valuable method for the diagnosis of osteoporosis, provides reasonable precision and less radiation exposure.

## 7.10 Dual-Photon Absorptiometry (DPA)

Single-photon absorptiometry has been improved considerably (Mazess, 1979; Mazess *et al.*, 1984) by making measurements at two distinct energies. This technique is referred to as dual-photon absorptiometry (DPA) and uses radioisotope sources such as $^{153}$Gd (Mazess, Ort and Judy, 1970; Roos, Rosengran and Skoldborn, 1970). This apparently minor advance opens up the possibility of measuring bone mass at central locations such as the lumbar spine and the proximal femur and encouraged the medical community into its clinical application. This helps eliminate the soft tissue problem around the bone and enables measurements at any body location. In addition, the influence of fat variations in the marrow on the measured bone mineral is reduced. Further, DPA techniques have been introduced to overcome the constant overall thickness of measurement sites.

In this method (Mazess, 1979; Krolner and Pors-Nielson, 1980) a strongly collimated beam of photons with two different $\gamma$-energies moves in a rectilinear scanning mode across the lumbar vertebrae. The radiation of the beam after passage through the body is registered by a collimated scintillation detector and a two-channel pulse height analyzer system (Figure 7.12). The radioactive source $^{153}$Gd has two well-defined energy peaks at 44 and 100 keV (Figure 7.13). The mean diameter of the radiation beam in the vertebral body is about 6 mm.

There is still the limitation imposed by the assumption that the object consists of only two radiologically distinct materials, each of which is characterized by a single mass attenuation coefficient of $\mu_b$ for bone mineral and $\mu_s$ for soft tissue. The expression of this can be appreciated from the equation for areal density measurement. For bone mineral, the areal density, $m_b$, is given by the expression:

$$\mu_b = \frac{\ln I_{0,l}/I_l - R_{ST} \ln I_{0,h}/I_h}{\mu_{b,l} - R_{ST}\mu_{b,h}} \tag{7.6}$$

where $I_0$ and $I$ are the incident and transmitted photon intensities and the subscripts l and h denote low and high energy photons respectively. $R_{ST}$ is the soft tissue ratio and is equal to $\mu_{s,l}/\mu_{s,h}$.

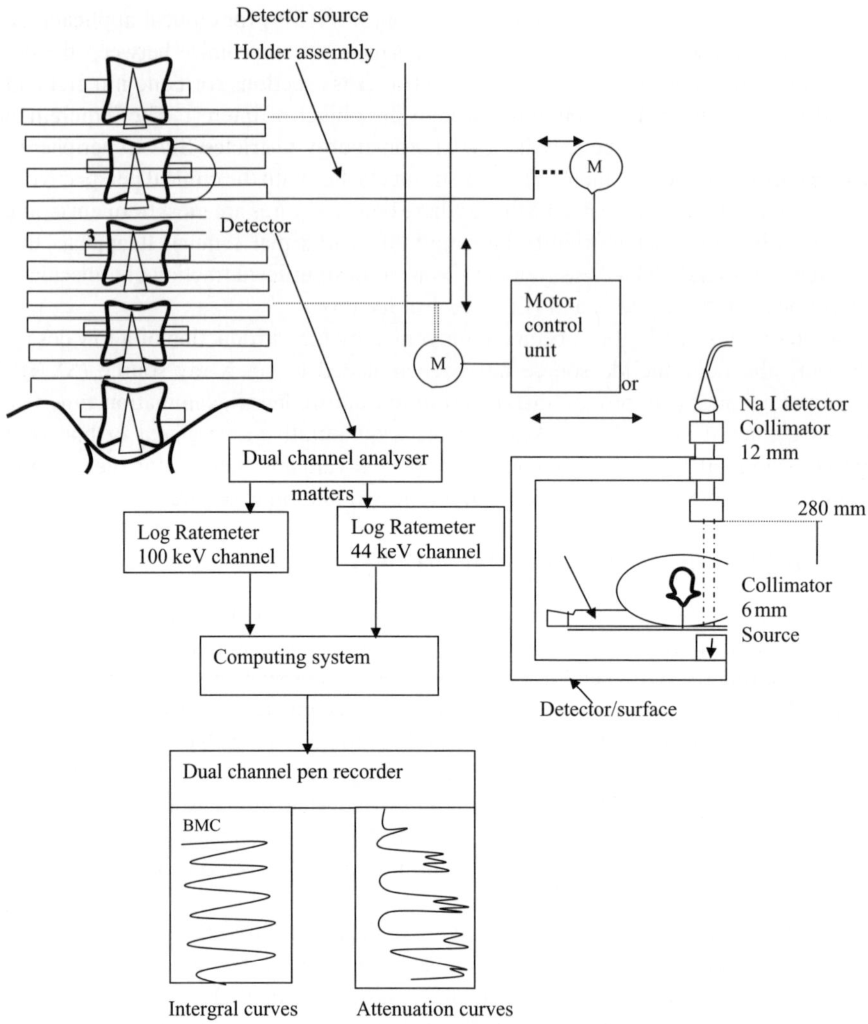


**Figure 7.12** Axial dual-photon absorptiometry using $^{153}$Gd, (Schaasdt and Bohr, 1982). (Reproduced with permission from O. Schaasdt and H. Bohr. Bone mineral by dual photon absorptiometry: Accuracy precision-sites for measurements. In: J. Dequeker and C.C. Johnston Jr, eds. *Non-Invasive Bone Measurements: Methodological Problems*. IRL Press, Oxford. © 1982, Oxford University Press)

If it is assumed that there is a universal composition of soft tissue then $R_{ST}$ is a constant and Equation 7.6 can be solved for area bone mass simply from measurements of the intersites of low and high energy photons to a single tissue location occupied by bone mineral and soft tissue. However, soft tissue composition varies among population and the difference between patients is encapsulated in $R_{ST}$, which is determined for each patient. This is achieved by application of Equation 7.6 to an anatomical location where bone mineral is not present (i.e., $m_b = 0$). Once $R_{ST}$ is determined for a specific patient, the assumption has to be made that

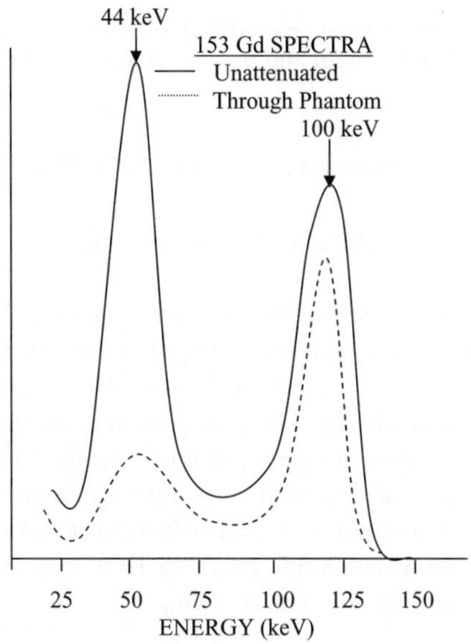


**Figure 7.13** The major source used for dual-photon absorptiometry is $^{153}$Gd, with peaks at 44 and 100 keV; in bone the lower energy peak is attenuated much more than the upper. (Mazess, 1983)

the soft tissue ratio measured at the non-mineral site applies to the soft tissue at the mineral/soft tissue site.

The deficiencies of an areal density have been recognized and efforts have been made to recover mass density from the attenuation measurements (Carter, Bouxsein and Marcus, 1992; Lu *et al.*, 1996).

It can be anticipated that the restrictions imposed by variable physical properties of soft tissue could be eliminated by increasing the number of photon energies in the incident beam. The feasibility of triple-photon absorptiometry (TPA) between lean and fat components of soft tissue results in considerable uncertainties in the mass of either component. A further increase in the number of incident photons is likely to produce an unacceptable radiation dose (Farrell and Webber, 1992).

### *7.10.1 Theoretical Background*

Two logarithmic absorption curves are generated during scanning, one for each energy peak. The influence of Compton cross talk is eliminated by an analog computer, which also eliminates the influence of soft tissue attenuation. The result of such computerization is a signal $M$ (mass per unit length), which for each scan is presented as a soft tissue corrected attenuation curve.

$M$ has the following characteristics:

1. It is constant despite varying amounts of soft tissue.
2. During scanning over bone tissue the change in $M$ ($\Delta M$) is proportional to the amounts of bone mineral in the path of the radioactive beam ($m_b$).

3. During scanning over adipose tissue $\Delta M$ is proportional to $m_f m_{f0}$, where $m_f$ is the amount of fat in the path of the radiation beam and $m_{f0}$ is the amount of fat at the point of calibration lateral to the vertebra.

The generated signal ($M$) is expressed by (Roos, 1974; Wilson and Madsen, 1977):

$$M = \ln\frac{I_0^{44}}{I^{44}} - 1.41\ln\frac{I_0^{100}}{I^{100}} \tag{7.6A}$$

where $I_0$ denotes the detected intensity of photons at the point of calibration and $I$ is the detected intensity of photons at a point during the scanning procedure. The numerals 44 and 100 refer to the corresponding isotopes, and $1.41 \approx \mu_s^{44}/\mu_s^{100}$ (Watt, 1975), where $\mu$ is the mass absorption coefficient and "s" denotes soft tissue. The dual-photon technique can be used for measurement of the mass of one component (mass per unit area) in a two-component system. The human body, however, must be regarded as a three-component system since fat has a significantly lower absorption coefficient than soft tissue at 44 keV. The influence of fat is eliminated if its thickness is constant during scanning. However, an erroneous result is obtained if the distribution of adipose tissue is non-uniform.

### 7.10.2 Procedure

The correct positioning of the patient over the source is ensured by radiological examination and palpation. The zero setting of the equipment is made during one longitudinal scan parallel with the lumbar spine to the right of the transverse processes of the vertebra. The scanning speed of the source–detector assembly is generally 4.0 mm s$^{-1}$ with scan steps of 4 mm between two scans. The length of one scan across the spine is 12.0 cm. A complete investigation consists of continuous scanning of all five lumbar vertebrae and takes about 25 min. Having identified the second, third and fourth lumbar vertebrae, one can calculate the sum of the BMC values of all scans over those vertebrae.

### 7.10.3 Nature of Attenuation

To clarify the nature of attenuation for bone, soft tissue and fat, the count rates at different absorber thickness are studied. Bone is simulated by an aqueous solution of KOH and $KH_2PO_4$ (Krolner and Pors-Nielson, 1980). Fat is simulated by triglyceride-3-oleic acid and triglyceride-3-palmitic acid in the ratio 2 : 1 (w/w). The two triglycerides constitute about 75% of human fat (Hirsch, 1965). Soft tissue is simulated by demineralized water. A Perspex layer, 11 cm thick, is placed between the radioactive source and the absorber in order to reduce the count rate to a magnitude similar to those used in human studies.

Figure 7.13 shows the absorption curves for the three different absorbers at the two energy levels. This figure clearly shows that the attenuation at 100 keV is mono-exponential for all three absorbers. The attenuation at 44 keV is mono-exponential for fat and water. For bone mineral the attenuation at 44 keV is practically mono-exponential for absorber thicknesses up to 19 cm. At higher thickness beam hardening occurs, which can be explained by (i) increasing Compton scatter at increasing absorber thickness, (ii) a change in geometry with increasing absorber thickness and (iii) the 44 keV radiation is not completely monoenergetic.

## 7.10.4 Reproducibility

### 7.10.4.1 *In Vitro*

The standard deviation of repeated measurements of the standards is enhanced with increasing BMC, but this enhancement is not significant.

### 7.10.4.2 *In Vivo*

*In vivo* precision has been investigated in ten women aged 35–70 (Krolner and Pors-Nielson, 1980). The BMCs of the women ranged from 20.7 to 53.0 U. Two measurements were performed in each subject with an interval of 14 days. The total variation of the method, including variation of equipment, subject variation and variation of curve reading, turned out to be 0.7 U. This corresponds to a mean coefficient of variation of 1.4% for normal premenopausal women and 2.6% for postmenopausal women with clinical osteoporosis.

The smallest significantly detectable difference in the ten women measured twice was 2.0 U, corresponding to a BMC change of 4% in normal premenopausal women and 7.4% in postmenopausal women with clinical osteoporosis (Christiansen and Rodbro, 1977).

Absorptiometry with $^{153}Gd$ has been used to measure spinal BMC (Wilson and Madsen, 1977; Price *et al.*, 1976; Dunn, Wahner and Riggs, 1980; Krolner and Pors-Nielson, 1980) and total body mineral (Mazess and Peppler, 1977; Peppler and Mazess, 1981). Total body bone mineral content (TBBM) is obtained by transmission scanning in a rectilinear raster pattern. It has been shown (Mazess *et al.*, 1981) that total body bone mineral is highly correlated with the actual weight of skeletons and with total body calcium as measured by neutron activity.

The mean BMC and BMD values show that males generally have larger values than females, while younger females have larger values than older females and osteoporotic patients had smaller values than controls. The osteoporotic females are about 15–20% below age-matched controls ($P < 0.001$) at appendicular sites, including the radial shaft site. The most notable difference, however, occurs in the axial skeleton, where the pelvis and spine are 31 and 38% below levels of young women for mass and 20 and 24% lower for areal density.

Measurements of lumbar BMC and dual-photon methods have been performed by several workers (Price *et al.*, 1976; Roos and Skoldborn, 1974; Roos, Hansson and Skoldborn, 1980). The method described here represents methodological improvements in the following respects: (i) The continuous two-dimensional scan mode in combination with a narrow radioactive beam makes anatomical localization simple. (ii) Crush fractures do not invalidate the measurements, since lumbar BMC is expressed as the sum of integral curves of the second, third and fourth lumbar vertebral.

Logically, the most reliable BMD measurement for predicting fracture risk at any given skeletal site is one made at the fracture site itself. This study examines the hypothesis that the ability of BMD measurements at other distant sites in the skeleton remote from the fracture site to predict fracture risk stems from their correlation with the fracture-site BMD measurements. This correlation predicts a linear relationship $\beta_{dist} = r\beta_{frac}$, where $\beta$ is the gradient of the exponential relationship between fracture risk and *Z*-score – $\beta_{frac}$ is the $\beta$-value for the fracture site and $\beta_{dist}$ is the $\beta$-value for the distant BMD site – and $r$ is the correlation coefficient between the *Z*-scores at the two sites. In practice it is important to consider the effect of BMD measurement errors on the $(r,\beta)$ relationship. The effect of errors at the distant site is to reduce $\beta_{dist}$ and $r$ in a way that preserves their original relationship. When errors at the fracture site are taken into account the effect on the $(r,\beta)$ plot is for the point representing the fracture site to fall below the extrapolation of the straight-line relationship predicted for the points representing the distant BMD sites. These authors tested the correlation hypothesis by using data from a study of osteoporotic fractures to examine the $(r,\beta)$ plots for hip, spine and forearm fractures. For the hip the data are consistent with the straight-line relationship and suggest that measurements made at other sites provide no additional information about fracture risk over and above that provided by hip BMD. For spine and forearm fractures the data show that distant sites do provide additional information about fracture risk in a way that is quantitatively consistent with the BMD measurement errors reported in cadaver studies. This method points to BMD correlation as an important factor in explaining the ability of measurements made at distant sites to predict the possibility of fracture risk. For the general population, this is discussed in chaper VI (Page 261).

## 7.11 Computed Tomography (CT)

Computed tomography (CT) (also called a CAT scan) is a widely adopted technique in a clinical environment. The method adopts ionizing radiation (X-rays) to scan the organ of the body under investigation (Figure 7.14).

### *7.11.1 Instrumentation and Clinical Procedure*

The CT scanner is a rotating gantry that has an X-ray tube mounted on one side and an arc-shaped detector mounted on the opposite side. An X-ray beam is emitted in a fan shape as the rotating frame spins the X-ray tube and detector around the patient. Each time the X-ray tube and detector make a 360° rotation and the X-ray passes through the patient's body, an image of a thin section is acquired. During each rotation the detector records about 1000 images (profiles) of the expanded X-ray beam. Each profile is then reconstructed by a computer into a two-dimensional image of the section that is scanned. When the image slices are reassembled by computer the result is a detailed, multidimensional view of the interior of the body. Very small,

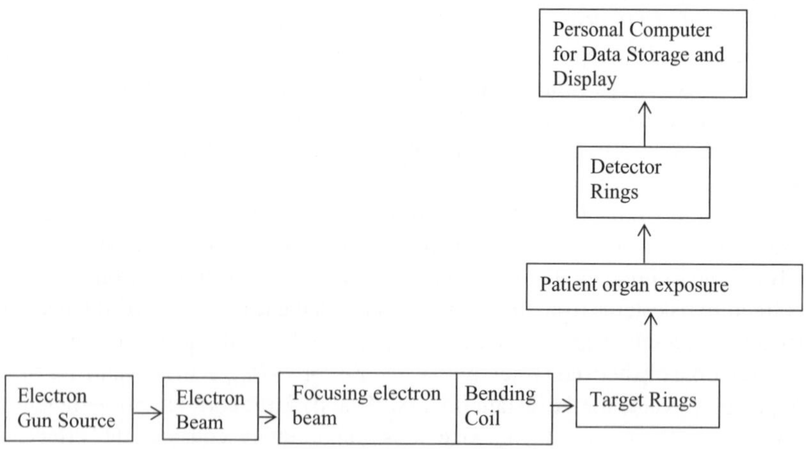


**Figure 7.14** Block diagram representing the arrangement in computer tomography (CT)

controlled amounts of X-ray radiation are passed through the body and different tissues absorb radiation at different rates. A CT examination often requires the use of different contrast materials to enhance the visibility of certain tissues or blood vessels. CT scanning causes no pain, and with the use of spiral CT the need to lie still for length of time is considerably reduced.

The mineral content in the compact bone of the tibia and femoral shaft has been measured accurately by using CT. In contrast, several studies have shown that determinations (Mazess, 1983) on trabecular bone samples (from the spine and elsewhere) *in vitro* involved large errors ($>30\%$). This accuracy reflects the influence of variations in marrow fat content with single-energy techniques at high Kilovolts peak (KVP). The normal variability of marrow fat ($\pm 20\%$) leads to uncertainties of 3–10% (dependent on beam energy and bone concentration) in the CT bone measurement. The high intrinsic accuracy of CT is also degraded by variations in the composition of bone substance. This uncertainty affects not only the diagnosis but also the monitoring of bone changes. By careful selection of images, a precision of 3% can be obtained for both spinal and radius but uncertainties in marrow composition can double the variance. The problem can be minimized by using either lower energy radiation on thinner body sections or dual-energy CT.

Lower energy measurements on the distal radius have been performed using specially constructed $^{125}$I-CT scanners. The long-term precision of bone mineral measurement is about 0.6%. In normal subjects a good correlation ($\gamma = 0.9$) is obtained with vertebral bone density and, consequently, there is fairly good sensitivity in the diagnosis of osteoporosis. The osteoporotic samples fell below the 95% confidence limit for age-matched controls. The low dose, high precision and good accuracy of these measurements makes this technique very useful. It has already been demonstrated that $^{125}$I-CT measurements on the distal radius can be used to follow immobilization, fracture healing and also various endocrine disorders. The density changes in radius due to immobilization are smaller and slower than those of the spine in immobilization (Mazess, 1983).

Assuming that bone trabeculae have the same composition as compact bone, then the trabecular bone density ($P_t$) is given by:

$$P_t = AP_c \tag{7.7}$$

where $P_c$ is the density of compact bone and $A$ is the fraction by volume of trabecular bone (Mullen *et al.*, 1985). Attenuation of the X-ray beam passing through the trabecular bone region is described by the linear attenuation coefficient $\mu$ of the mixture of trabecular bone and bone marrow:

$$\mu = A\mu_c + (I + A)\mu_m \tag{7.8}$$

where $\mu_c$ and $\mu_m$ are the linear attenuation coefficients of compact bone and bone marrow, respectively. The sensitivity to detecting bone changes is dependent on the amount of bone material present in the trabecular bone region.

Several workers have adopted a filtered back-projection technique to reconstruct CT images from the projection data (Shepp and Logan, 1974). The CT images are evaluated with regard to trabecular bone changes. For that purpose a "bone parameter" $\mu^I$ is defined as the average linear attenuation coefficient characterizing the material in the trabecular bone region. To avoid any contribution from cortical bone, a central disc limited to 50% of the total bone area is used for the evaluation. The precision of the bone parameter (dose dependent) $\mu^I$ is described by the ratio $\sigma/\mu^{I}$. The mean $\mu^{I}$ and the standard deviation $\sigma$ are calculated from a set of CT images (Brooks and DiChiro, 1976). The dose depends on the mass distribution of the model, the densities of the tissues and the slice thickness.

Carrying out measurements with the head phantom in the range 80–140 KVP, Keller *et al.* (1980) found a monotonically increasing detectability with decreasing energy when exposure was held constant – in line with the theoretical prediction of Mullen *et al.* (1985). For energies below 30 keV the sensitivity of the linear attenuation coefficient to changes of the trabecular bone density was higher, but the precision of the measured bone parameter $\mu^{I}$ is much smaller. For energies above 40 keV the precision is higher but the sensitivity is diminished. The soft tissue compartment in the trabecular bone region is assumed to have a constant composition during bone evaluations, though bone marrow might change its composition or density. To minimize this effect, the use of a high contrast between bone and soft tissue at low energies is often suggested. The accuracy of the bone parameter ($\mu^{I}$) depends, to a small extent, on the size of the X-ray beam.

One design of CT scanner, as developed by Hosie, Richardson and Gregory (1985), is a translation-rotation type in which the radiation source and detectors translate across the width of the arm. The translation is repeated at the new angle and the whole process is repeated through a total rotation of 180°. The radiation source is a 500 mCi 1-mm diameter bead of iodine-125. The source is collimated to produce four beams of radiation that have a narrow band of energies around 29 keV. Four photomultipliers operate in count mode and the energy-selected pulses are interfaced to a computer by means of specially built electronic counters.

To measure the length of the patient's ulna, the arm of the patient is secured in the arm holder of the scanner and positioned approximately by means of a light beam. The ends of the radius and ulna can be clearly defined and the scanner is automatically positioned at a site 30% from the end of the ulna. A set of three juxtaposed 2-mm thick slice measurements are made, starting at the 3% position, and a separate measurement is made at 25% of the ulna length, a site at which only cortical bone is present. The radiation dose to the patient is very low, typically about 40 rad, and is confined to a small region of the distal forearm.

### *7.11.2 Quantitative Computed Tomography (QCT)*

In this method BMD is quantified using a bone mineral reference phantom that is scanned simultaneously with the patient. To minimize the impact of beam hardening, the calibration phantom is placed as close as possible to vertebrae and is located under the lumbar spine of the patient, containing bone, red marrow and marrow fat. A single-energy QCT measurement can determine the mass of a bone in a volume consisting of two components (e.g., bone and red marrow), but not in a three-component system (Goodsitt *et al.*, 1994).

QCT is an accepted and long-used technique for the assessment of lumbar spine bone density but QCT of thoracic vertebrae is desirable because it measures bone mineral density in a skeletal region where compression fractures are prevalent (Zamenhof, 1997; Wasnich, 1996). A lateral protection is first made to localize the lumbar vertebral bodies and then scans of 8–10 mm thickness are acquired through 2–4 contiguous lumbar vertebral levels. The CT image is processed, along with analyses of the calibration phantom, to obtain BMD and so determine a region of interest in the trabecular bone of the vertebral body (Kalender, Klotz and Suss, 1987; Steiger *et al.*, 1990; Lang *et al.*, 1999). It is then possible to obtain the mean BMD of the vertebral region of interest as 80 μSv (value includes the dose of the lateral localizer scan) (Kalender, 1992). Again a significant advantage of thoracic QCT lies in the fact that the vertebrae necessary for densitometric measurements are in the field of view of chest scans. These scans, commonly acquired for the purposes of coronary calcium of lung assessment, serve as sources of valuable bone density information on patients and participants in research studies. In contrast to one-mineral density, the QCT density measurement is independent of bone size and thus is a more robust measure for comparisons of bone density between populations and, potentially, for growing children.

However, the use of thoracic QCT to measure BMD still needs more validation. To examine the validity of thoracic QCT as a method to measure BMD, Wong *et al.* (2005) have assessed the precision of thoracic QCT bone density values and their correlation to lumbar QCT density results.

In addition to producing precise results, thoracic QCT BMD compares with a high correlation to lumbar QCT results in both male and females. Although the two methods correlate well, the results are significantly different when analyzed with a paired *t*-test, with lumbar QCT producing a lower bone density mean. This variability between lumbar and thoracic bone mineral may be attributed to anatomic and physiologic differences or to the differential effect of osteoporosis on different anatomic sites. However, because lumbar and thoracic bone density are highly correlated, it suggests that measurement of the latter might be predictive of thoracic clinical features.

One inherent concern when using thoracic QCT to measure bone density is beam hardening, the distortion in attenuation coefficient caused by the presence of the lungs in the field of view. However, because the correlation between thoracic QCT and lumbar QCT bone density is high and the standard error of estimate of $18.98\,\mathrm{mg\,cm^{-3}}$ is large compared to the biologic variability of the populations, beam hardening does not seem to cause errors that are large enough to cause a significant risk of misclassification of fracture risk. If direct comparison of absolute bone density values are needed, lumbar QCT bone density ($x$) can be converted into the approximate thoracic QCT ($y$) value with the linear relationship:

$$Y = 0.87x + 22.97 \tag{7.9}$$

Carr *et al.* (2001) have validated thoracic QCT as a method of measuring BMD in an unembalmed cadaver by comparing *in situ* multidetector CT measurements of vertebral T8 to T11 to calcium ash measurements of the respective vertebral. These authors also performed *in vivo* bone density measurements in 15 study cases to compare bone density by lumbar QCT (T11-L3) to cardiac gated thoracic QCT (T6-T11). The cadaveric study found a high correlation ($P = 0.61$) between direct measurement of calcium ash and QCT of thoracic vertebrae and no significant difference in bone mineral density results (paired *t* test). The study presented an extremely high correlation between thoracic and lumbar QCT ($r = 0.99$). Lenchik *et al.* (2004) found a high correlation between thoracic and lumbar bone density with a higher bone density estimated from the thoracic vertebral.

Wong *et al.* (2005) have concluded that thoracic BMD is a good indictor of fracture risk, as shown by correlation to lumbar spine BMD, and also the typical age range of people having these scans makes them reasonable candidates for osteoporosis assessment. These authors have also shown that QCT of thoracic vertebral produces bone density measurements with rescan precision and regression errors that are relatively small compared to the biologic variable of BMD. Thoracic QCT bone density values are not identical to the results produced by lumbar

QCT, but its high correlation to the reference method and its small precision and regression errors validate the use of QCT in the thoracic spine as a reliable method of measuring BMD.

The presence of marrow fat means that single-energy QCT underestimates the mass of bone per unit tissue volume, an error that can be corrected using dual-energy acquisitions. In addition, the effect of marrow fat on QCT measurements is larger at the spine or peripheral skeletal sites. The conversion from red into fatty marrow tends to finish by the mid-20s, and this gradual age-related increase in the proportion of fat in the bone marrow that starts in youth continues through to old age.

The primary advantage of volumetric QCT of the spine is improved reliability in the measurement of trabecular BMD. The ability to examine both trabecular and cortical bone is an advantage for QCT of the hip, particularly in the case of therapies such as PTH. In this context Cann *et al.* (1999) reported that cortical bone maintained its density but significantly increased its mass and volume. In addition, a QCT image provides a mechanical model of the proximal femur based on the bone geometry and spatial distribution of material properties. Because the three-dimensional image can be connected for rotation of the vertebra in the coronal and sagittal plans, it is possible to make measurements with greater reliability in patients with scoliosis or lordotic curvature. In QCT the precision is better because the volumetric analysis can also correct for operator error in slice placement.

The major source of error in QCT bone measurement is the phenomenon of partial volume averaging. Because the voxel dimensions in QCT measurements (0.8–10 mm slice thickness) are larger than the dimensions and spacing of trabeculae, a QCT voxel includes both bone and marrow constituents.

Limited access to conventional all-purpose CT scanners has led to the development of peripheral QCT (PQCT). The basic PQCT unit consists of two major components, the scanner hardware and a computer system for hardware control and image data analysis. The scanner hardware contains a collimated X-ray source that emits a very narrow X-ray beam, a fixed set of solid-state detectors and a mechanical system allowing transverse, radial and axial displacements of the source detector assembly. The PQCT image is derived from a series of X-ray transmissions that represent the fraction of X-rays that pass through various lengths of the object. The PQCT image is composed of a discrete number of picture elements (pixels). However, since the tomographic bone slice has a predetermined thickness, the pixels that comprise the reconstruction image represent individual volume units referred to as "voxels." Each voxel in the PQCT represents the linear attenuation coefficient ($\mu$) of the tissue being studied. The attenuation coefficient is dependent on the energy of the X-ray beam as well as the amount of the absorbing material. The final density of the object imaged can be estimated from the linear attenuation coefficients. By summing the individual bone mineral content values assigned to each voxel, the PQCT allows us to determine the mineral mass [BMC (mg)] and the mineral density [BMD (mg $cm^{-3}$)] of the whole bone slice or of selected cortical and trabecular regions of interest.

PQCT allows the measurement of selective density calculation of cancellous bone (Genant, Steiger and Gluer, 1989; Muller, Ruegsegger and Ruegsegger, 1989; Ostlere and Gold, 1991). Wang *et al.* (2001) have defined the age-related characteristics of bones of male Sprague–Dawley (SD) rats by using PQCT densitometry and histomorphometry. PQCT densitometry is a relatively new, sensitive technique that has been developed for separately quantifying cancellous and cortical bones in humans (Augat *et al.*, 1998a, 1998b, Gasser, 1995). As in humans, the vertebrae and the femoral necks of rats contain both cancellous and cortical bones, which have been shown to be decreased to differing degrees by aging. Therefore, in male (SD) rats, it is appropriate to consider cortical bone and cancellous bone as separate compartments, as done by Wang *et al.* (2001). This enables us to monitor very small changes in a short period of time (Breen, Millest and Loveday, 1996; Ferrethi *et al.*, 1995). Banu, Wang and Kalu (2000) have analyzed both cortical and cancellous bones from vertebra, tibia and femur in intact male rats at different ages using PQCT densitometry. This is important because in large animals (Jerome, Johnson and Lees, 1995; Hotchkiss, 1999; Augat *et al.*, 1997) trabecular bone has an 8–10× larger surface area than cortical bone and therefore reacts much faster to metabolic changes brought on by disease or a pharmacological agent.

PQCT has been used in several studies and the results correlate well with histomorphometric measurements (Banu *et al.*, 1999; Gasser, 1995; Louis *et al.*, 1996; Rosen *et al.*, 1995; Takada *et al.*, 1996). Another advantage of PQCT is that it measures volumetric bone density – unlike other X-ray absorptiometric techniques, which only measure areal density – and it also uses a lower dose than the standard spinal QCT, because only the appendicular skeleton is irradiated.

Variables combining the density and geometry of the cortical shell are also derived from a PQCT image. The moment of inertia weighted by the cortical density of each voxel in the cortical shell has been shown to capture the strength of the bone. This so-called strength strain index (SSI) can be measured non-invasively and correlates strongly with bone strength (Schiessl *et al.*, 1998; Wilhelm *et al.*, 1998). A PQCT derived bone strength index (BSI), an index similar to the SSI, can predict the fracture load of bones in three-point pending tests. A comparison of BSI and fracture load in over 200 rat femurs revealed a correlation exceeding 0.9 (Feretti, Capozza and Zanchetta, 1996).

## 7.12 Modification of CT Methods

A relatively new technique, spiral (helical) CT, has improved the accuracy of CT in a clinical environment. A new vascular imaging technique, called spiral CT angiography, is non-invasive and less expensive than conventional angiography and allows us to see blood vessels. The term "spiral CT" comes from the shape of the path taken by the X-ray beam during scanning. The examination table advances at a constant rate through the scanner gantry while the X-ray tube rotates continuously around the patient, tracing a spiral path through the patient. This spiral path gathers continuous data with no gaps between images.

With spiral CT, refinements in detector technology support faster, higher-quality image acquisition with less radiation exposure. The current spiral CT scans are called multidetector CT and are most commonly 4-or 16-slice systems. CT scanners with 64 or more detectors are now available. These instruments can provide either faster scanning or higher resolution images. Using 16-slice scanner systems the radiologist can acquire 32 image slices per second. A spiral scan can usually be obtained during a single breath hold. This allows scanning of the chest or abdomen in 10 seconds or less. Such speed is beneficial in all patients but especially in elderly, pediatric or critically ill patients, populations in whom the length of scanning often becomes problematic. The multidetector CT also allows applications like CT angiography to be more successful.

With conventional CT, small lesions may go undetected when a patient breathes differently on consecutive scans because lesions may be missed by unequal spacing between scans. The speed of spiral scanning and a single breath hold increase the rate of lesion detection.

Dual energy CT scans (Alverez and Macovski, 1976) provide correction for a beam hardening. With this method the data obtained at two different scan energies are combined to produce two new sets of data that correspond to a " soft tissue" image and a "mineral" image. These reconstructed images are free of any beam hardening artifacts and can be used to obtain accurate CT numbers at any effective scan energy. In addition, dual energy CT provides more accurate values for the BMC when the fat content is significant (Firooznia *et al.*, 1984). However, this has not been used clinically to any great extent, for several reasons: the method demands additional dose and time and has an inherently poorer precision. In one study (Genant *et al.*, 1982) the precision on spinal samples *in vitro* is to reported to be 14%.

To develop a relationship between bone structure and bone geometry with the mechanical integrity of bone, micro-computed tomography (μCT) has been developed. μCT essentially consists of a micro-focused X-ray source, a detector array, an object stage, computing resource of reconstruction and visualization, and an optical bench to isolate environmental vibrations. The detector array records attenuated intensities of the X-ray beam from various orientations. Overall, the imaging system is contained in a lead shielded box that is interfaced to a data analysis system. This method allows us to understand the effect of various pharmacological agents or to examine various disease states. μCT gives insight into trabecular structure with a spatial resolution better than 50 μm (Gordon, 2004).

## *7.12.1 CT Methods: Benefits and Risks*

### 7.12.1.1 Benefits

- Unlike other imaging methods, CT scanning offers detailed views besides bone of many other types of tissue, for example, the lungs, soft tissues and blood vessels
- CT scanning is painless, non-invasive, fast, accurate and yet cost effective.
- Disease processes may be detected at an earlier stage than is possible with other techniques and lesions. It may be possible to detect in areas that are difficult to assess with conventional imaging. The most common technique is CT guided biopsy. CT guided drainage can also be used in the treatment of most deep seated abscesses and other pathological targets.
- Imaging at peak levels of contrast: arterial and venous phase.
- Besides, multiplanar and three-dimensional analysis, it offers CT angiography, facial bones and cancer straging.

### 7.12.1.2 Risks

- The effective X-ray radiation dose from this procedure is about 10 mSv, which is about the same as the average person receives from background radiation in three years. However, pregnant women need to inform the doctor of this before undergoing this procedure.
- Nursing mothers are advised to wait 24 hours after contrast injection before resuming breast feeding.
- The risk of serious allergic reaction to iodine-containing contrast material, though rare, still exists.

CT scanners have made a change in the speed at which images can be obtained. In the third generation a fan shaped beam of X-rays is directed to an array of detectors that are at fixed positions relative to the X-ray source. This eliminates the time consuming translation stage, allowing the scan time to be reduced, initially, to 10 seconds per slice. This advance dramatically improves the practicality of CT as scan times become shorter to image the lungs or abdomen. Previously scans were limited to the head, or to limbs. This design, introduced roughly simultaneously with third generation scanners, gives an approximately equal performance. Finally, instead of a row of detectors that moves with the X-ray source, new fourth generation scanners use a stationary 360° ring of detectors.

The precision of QCT for measurement of mineral content in humans is 1–3% for single energy (80 kvp) and 3–5% for dual energy (80 kvp/120 kvp) techniques (Genant *et al.*, 1982, 1983). The accuracy of single energy QCT is 1–2% for $K_2HPO_4$ solutions, 5% for monkey vertebrae specimens and 4–10% for human vertebrae specimens spanning a wide age range. The accuracy for QCT (and DPA), however, may be reduced to 15–20% in the very old osteoporotic population (Genant *et al.*, 1983; Laval-Jeantet *et al.*, 1984). With DPA, vertebrae compression with callous formation, angular and scoliotic deformity, and extraosseous calcification, and so on, which are widespread in the elderly, may result in inaccurate and poorly reproducible vertebral measurements. Highly significant correlations have been found between vertebral trabecular minerals determined by histomorphometry (De Vernegoul *et al.*, 1984).

### *7.12.2 Discussion*

Although the predictive value of the QCT for the dimensions of single osteons is low, the osteon density and osteonal fraction, parameters associated with the entire area of osteonal structures, may be accurately predicted. Conclusions about the reasons for high intracortical porosity and bone turnover require histological analysis. In addition, when applied *in vivo*, several factors, such as partial volume effect and the likely lower CT resolution, may influence the measurement of morphological cortical bone parameters considerably, whereas the influence on bone density measurement is considered to be lower.

CT methods have been useful in separating cortical from trabecular bone in the radius. This ability has enhanced the sensitivity of this technique to monitor metabolic changes by measuring trabecular bone with its high turnover rate independently of low turnover cortical bone. However, the measurement of radial mineral does not correlate well with mineral content in the axial skeleton in health or disease (Dalen and Jacobson, 1974; Wilson, 1974). Clinically, the most significant problems of bone loss occur in the vertebrae and femoral neck. The usefulness of CT for measuring bone mineral in the vertebrae lies in its ability to quantitatively image a thin transverse slice through the abdomen. With this technique, one can spatially separate cortical and cancellous bone in the spine. Thus, one can measure changes in bone mineral content: (i) non-invasively, (ii) in cancellous bone, where measurement of changes is more sensitive than in cortical bone, and (iii) at the site of clinical involvements, in the vertebrae. The diagnostic capability of vertebral CT in metabolic bone disease depends on the accuracy of measurement of the bone mineral contained within the vertebral body. This accuracy is influenced by (i) the accuracy of the calculated attenuation coefficients and (ii) the model assumed for vertebral composition.

Most scanners adopt correction factors due to hardening of the polychromatic X-ray spectrum as it passes through the body. The first-order correction for hardening is water and the second-order to remove artifacts due to bone. This may improve image quality but may also lead to other inaccuracies in the calculated CT numbers (Joseph and Spital, 1978).

The relevance of BMD measured by QCT for the mechanical properties of cortical bone has run into controversy, because low correlation coefficients have been reported between cortical bone strength and BMD (Snyder and Schneider, 1991; Stromsoe *et al.*, 1993). However, the relevance of BMD at the tissue level for the prediction of mechanical parameters of bone has also been reported for cancellous bone (Wachter *et al.*, 2001). Wachter *et al.* (2002) have revealed significant correlations between BMD and both mechanical parameters (yield stress $E$ modulus). BMD reveals a high predictive value for structural parameters strongly related to mechanical bone properties, for example, porosity, dimensions of pores and osteon density. For the correlation between elastic modulus and BMD ($r = 0.69$, $P < 0.0050$), the correlation coefficient is higher than previously reported (Snyder and Schneider, 1991). A possible explanation for this result could be that patients with different bone status (Singh index 3–6) are included. A wide range of BMD values of cortical bone might contribute to stronger correlations. However, patients with clinical symptoms of osteoporosis and a Singh index of <3 are excluded. By excluding patients with severe forms of osteoporosis and osteoarthritis one can reduce these effects on the measurements. However, this population of subjects can not be taken to be representative of the general population as far as the mechanical, densitometric and histological indices measurements are concerned. Although the effect of osteoarthritis on metaphyseal bone as well as on the femoral head and femoral neck may be considerable, the influence of these effects on the diaphyseal bone may be of minor relevance. Data reveal a significant correlation of BMD, assessed by QCT, with histologically determined porosity ($r = 0.51$ linear, $r = 0.85$ polynomial).

While assessing the bone loss, one has to distinguish between the effects of bone growth with the effects of aging (Kalu, 1991). Among the indices of vertebral cortical bone mass measured by PQCT densitometry, vertebral ct.Th is one that is affected appreciably by aging. Aging leads to a decrease of about 31% ($P < 0.0001$) in vertebral ct.Th (Wang *et al.*, 2001). From a peak at about 4 months of age, both vertebral cancellous BMC and vertebral cancellous BMD decreased linearly with aging as measured by PQCT densitometry, an observation consistent with the authors histomorphometric findings in the L-4 vertebral body of male SD rats. These authors also found that major structural features of vertebral cancellous bone loss in male SD rats are reductions in Tb.N and Tb.Th. In male SD rats a decrease with aging in osteoblastic activity is one of the possible mechanisms for the age-related loss of cancellous bone found in aging male SD rats, which is also believed to be the case in men (Clarke *et al.*, 1996).

## 7.13 Methods Based on Compton Scattering

Compton scattering has been explored as a technique that has the potential to yield a direct measurement of the mass of material within a defined volume of bone. In this method the incoming photon loses some of its energy to the electron and itself propagates in a new direction (undergoes scattering) but with reduced energy (increased wavelength). Figure 7.15a shows a schematic presentation of the physics of Compton scattering. The intensity of Compton scattered photons is proportional to the electron density of a substance. Electron densities are directly related to mass densities through the use of the effective ratio of the atomic number to the atomic weight of the material. The electrons are assumed to be free and at rest. Therefore, it is possible to determine the distribution of electron densities

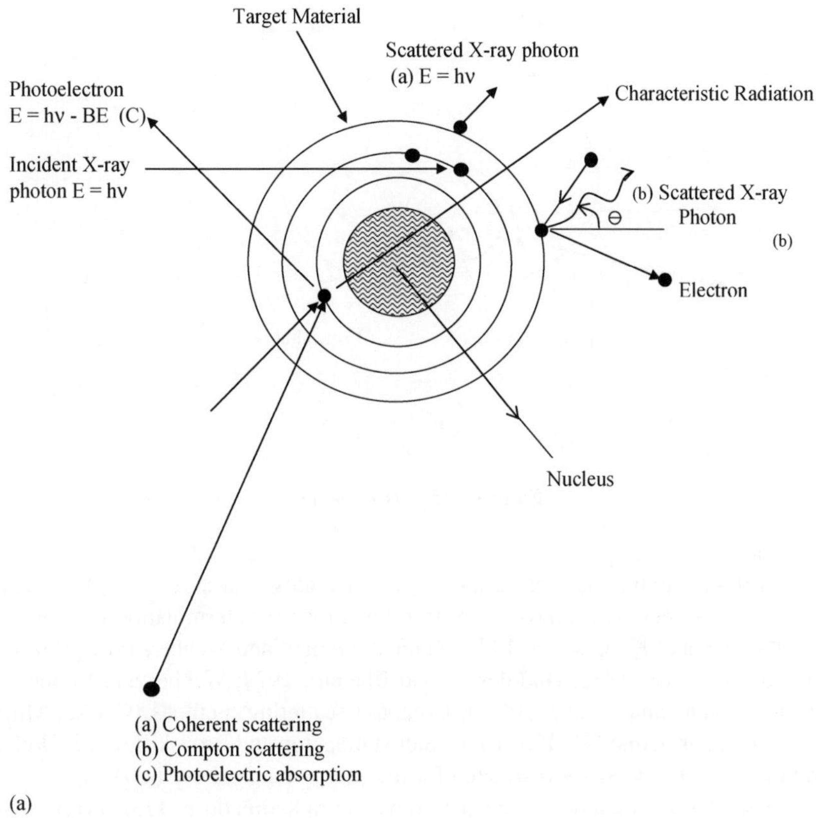


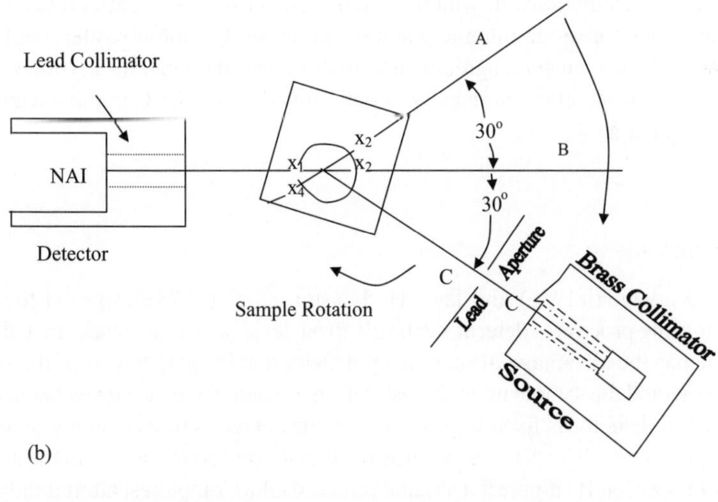


**Figure 7.15** (a) X-ray scattering and absorption process by an atom;(b) single-source, single-detector gamma scattering technique (Huddleston *et al.*, 1978);(c) simplified diagram of scattering geometry, showing source and detector and the paths in tissue of incident and scattered radiation

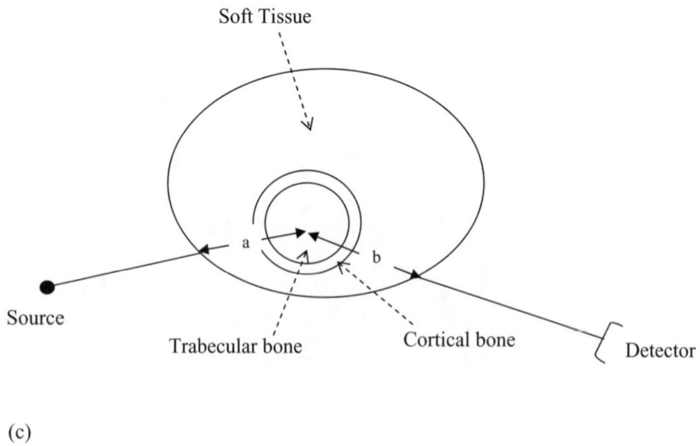


**Figure 7.15** *(Continued)*

throughout a tissue volume by measuring the scatter intensity at a given angle. Fixed point Compton scattering techniques have been developed for the determination of skeletal bone density (Olkkonen and Karjalainen, 1975; Kennett, Garnett and Webber, 1972; Hazan *et al.*, 1977; Huddleston *et al.*, 1978; Huddleston and Bhaduri, 1979; Webber and Kennett, 1976; Huddleston, Bhaduri and Weaver, 1979). Compton scattering methods (Clarke, Milne and Van Dyke, 1976; Farmer and Collins, 1974; Battista, Santon and Bronskill, 1977; Huddleston and Bhaduri, 1979) have been investigated for the energy range above 600 keV ($^{137}$Cs, $C_{60}$). However, these imaging methods utilize a high patient radiation dose (11.0 rad) (Battista and Bronskill, 1981). Assessment of the electron densities of body tissues is important in radiation therapy dosimetry calculations, in which the requirement for high spatial resolution may be relaxed. Corrections for tissue inhomogeneities, based on Compton scatter electron density measurements, have shown good agreement with those based upon transmission CT data. The accuracy with which the electron density is determined using the Compton scatter imaging method is estimated as 4.3%.

## *7.13.1 Technique*

Odeblad and Norhagen (1956) and later Huddleston *et al.* (1978) proposed the use of the Compton scattering process to determine the electron density of a substance. In a given region, this technique has the advantage of measuring density that is independent of the surrounding tissue. In this method the sample under investigation is irradiated by a narrow beam of photons. A collimated NaI detector is fixed in position and the source is moved about an axis through the center of the sample. The detector records the transmitted count rate from the sample when the source is in position B (Figure 7.15b) and it records the Compton scattered radiation when the source is moved to positions A and C. The sample is rotated clockwise by 30°. The scattering volume within the sample can be defined by the intersection of the projections of the source and the detector collimators.

The electron density ($P_e$) of the material within the scattering volume is given by:

$$P_e = k\left(\frac{S_1 S_2}{T_1 T_2}\right)^{1/2}$$

where $S_1$ and $S_2$ are the scattered count rates and $T_1$ and $T_2$ are the transmitted count rates and $k$ is a constant to be determined experimentally.

A single radioactive source is used for these measurements. Samarium-153 ($t_{1/2} = 46.7$ h) is chosen because its primary photon energy (103 keV) falls into the optimal energy range, as suggested by Garnett *et al.* (1973). An activity of 100–200 mCi has been used in these experiments. It is anticipated that higher activities will be required for *in vivo* measurements. The source and detector collimators are aligned using a small diameter laser beam. The samples to be tested are machined to fit inside an asymmetric plastic phantom that provides unequal absorption of the transmitted and scattered photon beams. A 30 keV symmetric energy window is used to count the scattered and the transmitted photopeaks in the multichannel analyzer. At the center of the sample, the diameter of the transmitter beam and scattering volume depends only on the detector/collimator arrangement. In all measurements, the scattering volume is completely enclosed within a homogeneous sample. Approximate agreement between independent methods for determining the scatter diameter in the sample is reported. The measured electron density depends on the diameter and the density range of the sample.

The single-source Compton scattering technique provides acceptable reproducibility and accuracy in standard samples. Changes in density of less than 3% have been claimed to be detected.

## 7.14 Coherent and Compton Scattering

Coherent scattering has received some attention as the most appropriate interaction mechanism upon which to base a bone measurement technique because of its strong dependence upon the atomic number of the initiated material (White, 1977). The difference in atomic number between soft tissue and bone would allow significant contrast between photons scattered coherently from bone and those scattered from soft tissue. Such a technique should offer a high sensitivity to time dependent changes in BMC and would have the advantage of a volume based measurement. The ratio of coherent to Compton scattering depends upon the effective atomic number of the material in which the scattering takes place. As the effective atomic number of bone mineral (cortical bone, $z = 13.8$) is considerably higher than that of soft tissue ($z = 6.4$) for yellow marrow (Guttmann and Goodsitt, 1995), this provides a more sensitive tool to measure bone mineral status than the method based on Compton scattering along. However, the coherent cross section decreases rapidly as photon energy increases.

When X-rays pass close to an atom they may cause the bound electrons to vibrate at a frequency corresponding to that of the X-ray photon. The electron re-radiates the energy in all directions (at the frequency of incident photons). This is the scattering of the energy without absorption. It occurs when the atomic number of the material is high (high binding energy) and the energy of the incident photon is comparatively low. This process is called elastic scattering.

Several workers (Puumalainen *et al.*, 1976; Puumalainen, Olkkonen and Sikanen, 1977, 1982; Kerr *et al.*, 1980; Stalp and Mazess, 1980; Olkkonen *et al.*, 1981; Ling *et al.*, 1982; Karellas *et al.*, 1983; Leichter *et al.*, 1984; Webster and Lillicrap, 1985) have used the measurement of the ratio of coherent to Compton scattered photons from irradiated bone to determine information on its composition. The technique offers the potential of monitoring bone mineral loss in diseased bone. Figure 7.15c shows the layout of this measurement technique. If the path length through the tissue is the same for coherent and Compton scattered radiations, the attenuation effects cancel out, since the energies of these two type of radiations are sufficiently close. Typically, with 600 keV incident $\gamma$ rays and a scattering angle of 90°, the energy difference between coherent and Compton scattering is 6.1 keV (Puumalainen *et al.*, 1976). Schatzler (1979) and Kerr *et al.* (1980) have shown that the coherent to Compton scatter ratio increases as the effective atomic number of the irradiated material increases.

The radiation source is either a 100 kvp X-ray beam filtered with 0.3 mm tungsten or a 500 MBq $^{241}$Am point source that provides $\gamma$-rays of 59.5 keV well separated in energy from other emissions. The tungsten filter suppresses the X-ray energy spectrum above 70 keV.

The energy loss $\Delta E$ of the Compton scattered photons is given by the equation:

$$\Delta E = \frac{E}{1 + k(1 - \cos\theta)} \tag{7.10}$$

where $\theta$ is the angle of scatter and $k$ is the incident energy in units of electron rest mass:

$$\left(\frac{E}{m_e c^2}\right)$$

Increasing the scattering angle therefore increases the separation of the coherent and Compton scattered photon energies.

For a broad X-ray spectrum, a value proportional to coherent Compton scattering ratio is obtained by sampling the scattered spectrum in two regions: (i) where the coherent scatter dominates and (ii) where the Compton scatter dominates. The higher energy region samples the coherently scattered spectrum close to the K-edge discontinuity, but at a higher energy than the position of the Compton shifted K-edge. The counts in this region are corrected for the presence of the small Compton distribution in the region from those X-rays in the primary beam at energies above the tungsten K absorption edge. The lower energy region samples an area where very little coherent scatter is present and where the Compton scatter dominates.

A heavily filtered X-ray beam can be used to determine the effective atomic number of bone *in vitro* by measuring the ratios of coherently to incoherently scattered X-rays. For a given scattering geometry and incident X-ray spectrum, the experimentally determined value of effective atomic number may be used as an indicator of bone composition. It is possible to separate out the coherent and incoherent components sufficiently to allow us to obtain a value similar to that found using a monoenergetic $\gamma$-ray beam. The obvious advantage of employing an X-ray unit over radionuclide sources is the difficulty in obtaining the latter of sufficient activity and specific gravity for clinical measurement over a prolonged period.

There is no general agreement concerning the optimal photon energy and scatter angle for this type of measurement. Small scatter angles are used (Schatzler, 1979; Kerr *et al.*, 1980) to increase the detection efficiency of coherent photons that scatter mostly in the forward direction. The use of low energy also favors coherent scattering since the differential cross section for this process decreases rapidly as the photon energy increases. However, the choice of small scatter angles and low energy photons results in poor spectral separation between the coherent and Compton peaks (Stalp and Mazess, 1980; Kerr *et al.*, 1980). A compromise must therefore be reached between adequate counting statistics and spectral resolution. Leichter *et al.* (1984) have examined the changes of sensitivity that occur as a result of a change in the incident photon energy. The combined effect of varying both the scatter angle and the photon energy can be investigated by studying the changes in the momentum transfer. These authors chose the range $8 \leq z \geq 11$ because it adequately covers the effective atomic numbers that also apply to trabecular bone. This method has been used to characterize this tissue since a change in its mineral density is associated with a significant change in its effective atomic number.

To investigate the effect of the incident photon energy and the momentum transfer on the sensitivity of measurement, experiments are carried out at three photon energies: 44 GHq of $^{241}$Am (60 keV), 1.18 GHq of $^{133}$Xe (81 keV) and 8.51 GHq of $^{99m}$Tc (140 keV). The coherent to Compton ratio ($R$) is measured by integrating the respective peaks in the pulse-height spectrum, after correcting for the contribution of the Compton counts to the coherent peak.

Figure 7.16 shows on a log–log scale the normalized measured ratio (R/RW) for each of the $K_2HPO_4$ solutions as a function if its effective atomic number ($z$). This figure displays a linear relationship to a very good approximation ($r = 0.99$) between $\log(R/R_w)$ and $\log z$ and the slope of the regression line increases with the photon energy. The slope represents the experimentally determined power dependence of the measured ratio on the effective atomic number for a scatter angle of 47°.

These findings seem agree with that of White (1977), who found that a small increase in the value when the photon energy increases in the range of 20–150 keV. An earlier study (Weber and Van den Berge, 1969) found an increase in $n$ (the power dependence) in the energy range of 20–60 keV, followed by a decrease of $n$ when the energy is further increased. Leichter *et al.* (1984) have explained the difference in their $n$ value from those of Weber and Van den Berge (1969) as due to the large differences in the cross sections reported by Grodstein (1957) and Hubbell and Overbo (1979). Differences in total coherent cross sections exceeding 30% are common between the above-mentioned sources. The results of Leichter *et al.* (1984), which are

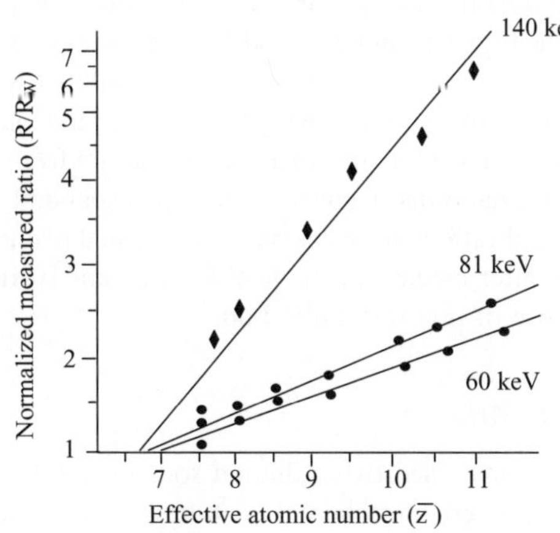


**Figure 7.16** The coherent to Compton measured ratio normalized to water ($R/R_W$) as a function of the effective atomic number of aqueous $K_2HPO_4$ solutions at various photon energies on a log–log scale. The scatter angle is 47° (Leichter *et al.*, 1984)

based on the tabulations of Hubbell and Overbo (1979), indicate that the effect of the photon energy on the power dependence of the total coherent cross sections on $z$ is quite small.

An increase in the photon energy or in the scatter angle results in a higher sensitivity, but it lowers the counting statistics. Since the number of detected coherent photons is much smaller than that of Compton scattered photons, it is the number of coherent photons that determines the statistical precision of the measurement. At large scatter angles, the coherent cross section decreases with the photon energy much more rapidly than in the case of small scatter angles. Therefore, if the energy is increased to improve the sensitivity, the percentage loss in the coherent count rate is less if a small scatter angle is used.

Several factors are affected by the variation of the photon energy, including the detector counting efficiency, the radiation dose to the patient, intensity of the coherently scattered photons and the intrinsic sensitivity of the measurement (Kerr *et al.*, 1980; Leichter *et al.*, 1984). To attain good sensitivity, the momentum transfer should be high; this can be achieved by increasing either the photon energy or the scatter angle.

Based on the above principle, Shukla *et al.* (1985) have developed a technique to measure the BMD of the calcaneus *in vivo*. In this technique the calcaneus is irradiated by a 60 keV photon beam from $^{241}$Am (44.4 GBq) sources and both the coherent and Compton scattered photons are detected at 71° by a high porosity Ge detector. The ratio of coherent to Compton scattered photons thus measured is converted into the BMD by use of a regression equation. A similarity in the scattering spectra of a calibration phantom and a normal volunteer is observed. The accuracy of the technique is approximately 5%, defined in terms of the standard error. Giganate and Sciuti (1985) have described a large angle ($\theta = 135°$) photon scattering mineralometer. The device operates with 59.5 keV line and is characterized by a very compact source–detector measuring head. These authors used two different experimental devices. In the first device a 7.4 GBq (200 mCi) $^{241}$Am point source, housed in a suitable lead collimator, is mounted on a rotating table of an X-ray spectrometer. A Ge detector and a multichannel analyzer are also part of the device.

The above experimental set-up provides a compact source–detector assembly that is characterized by (i) small overall dimensions, (ii) very short source–sample and sample–-detector lengths, to obtain the highest count rate and (iii) the use of two or more X-rays sources surrounding the detector.Leichter *et al.* (1985) have pointed out that $K_2HPO_4$ solution proved unsatisfactory for calibration purposes when using the coherent to Compton technique. These solutions differ markedly in their scatter spectra and composition from trabecular bone. These authors proposed the use of phantoms made of bone ash suspended in white petrolatum in varying concentrations. A calibration curve has been established using these phantoms with a range of BMD values. The scatter spectra and range of $R$ values and BMD of these phantoms are in good agreement with those of real trabecular bone.

### *7.14.1 Clinical Applications*

Clarke and Van Dyk (1973) have used two radiation sources and two detectors for *in vitro* measurements. A technique based on coherent scattering will be more sensitive, but the problem due to attenuation remains. Using two collimated sources around the detector allows variation in the shape of scattering volume. For good precision and accuracy for measurements *in vivo*, trabecular bone thickness should be no less than 2 cm. If the scattering volume includes a soft tissue layer, the precision and accuracy are reduced. The backscattering mode gives rise to an increase in relative sensitivity and to a high compactness of the measuring head, as is needed for an easy positioning of different bone peripheral sites. These advantages compensate for the decrease in counting rate in the coherent peak.

Henson and Fox (1984) have shown that a determination of the effective atomic number leads directly to the determination of percentage of calcium.

Regarding a material as equivalent to some hypothetical element with effective atomic number $Z^*$ and taking the ratio at two energies, $E_1$ and $E_2$, gives:

$$\frac{\mu_{\mathrm{m}}(E_1)}{\mu_{\mathrm{m}}(E_2)} = \frac{f(Z^*, E_1)}{f(Z^*, E_2)} \tag{7.11}$$

where $\mu_{\mathrm{m}}(E)$ is the linear attenuation coefficient of the material and $f(Z,E)$ describes the variation of the total atomic cross section with energy for the element of atomic number $Z$.

If the material contains $j$ elements and the mass fraction of the $i$th element is $a_i$, then the mass attenuation coefficient may be written as:

$$\frac{\mu_{\mathrm{m}}(E)}{\rho_{\mathrm{m}}} = \sum_j a_i \left(\frac{\mu(E)}{\rho}\right)_i \tag{7.12}$$

where:

$$\sum_j a_i = 1$$

Combining the above two equations gives:

$$\frac{\sum j\, a_i(\mu(E_1)/\rho)i}{\sum a_i(\mu(E_2)/\rho)i}\frac{f(Z^*, E_1)}{f(Z^*, E_2)} \tag{7.13}$$

Since the linear attenuation coefficient can be found from the CT numbers of the material at the two effective energies and the basic scanner relationship of CT number with the linear attenuation coefficient, $Z^*$ can be estimated. In bone, the relatively high atomic number of calcium, and consequently large photoelectric effect, is sufficient to allow realistic solutions to be found of Equation 7.13.

Experimental data (Hyuonen-Debek, Riihonen and Debek, 1979) on the Ca/P ratio for trabecular bone lie between 2.04 and 2.28, and the femoral data lies between 2.04 and 2.30. Henson and Fox (1984) used a range of 1.7–2.6 for Ca/P ratio in the computations to cover all reported values.

The means of the calcium distribution are plotted against $Z^*$ (Figure 7.17) together with their standard deviations. The results can be represented by an empirical relationship of the type:

$$\%\mathrm{Ca} = a(Z^*)^n + b\,Z^* + C \tag{7.14}$$

with a correlation coefficient exceeding 0.9999. The variation of calcium percentage between $Z^*$ values of 7.5 and 9.5 is approximately linear. This is in agreement with data of Adams *et al.* (1982). Errors in the experimental determination of $Z^*$ can arise from the estimation of the effective energies at the region of interest and from image noise.

The calcium percentage is calculable to within $\pm 13\%$ for $Z^*$ values from about 10 up to 13.5. The error increases with decreasing $Z^*$ and is almost 19% at $Z^* = 8$, corresponding to a calcium percentage of $2.2 \pm 0.4$. It is further pointed out (Henson and Fox, 1984) that these

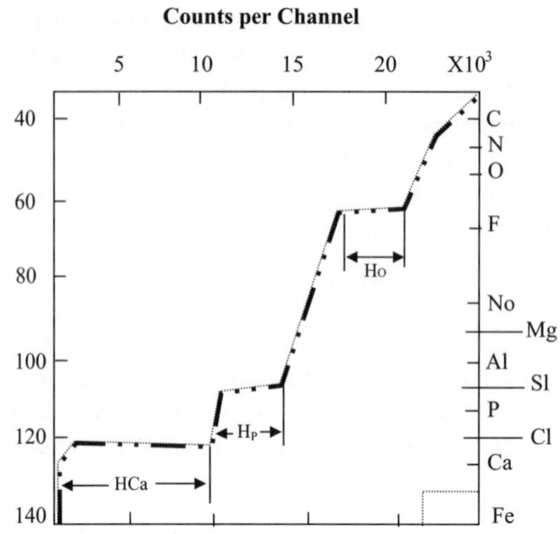


**Figure 7.17** Experimental setup for the dual energy Compton scattering method. (Huddleston and Scakler, 1985). (Reproduced with permission from A.L. Huddleston and J.P. Sackler, The determination of electronic density by the dual-energy Compton-scatter method, *Medical Physics*, **12**(1), 13–19 © 1985, American Association of Physicists in Medicine)

errors could be reduced by taking large trabecular areas to reduce the standard error of measurement or by selection of small cortical zones having low standard deviations of their CT numbers.

Phelps, Hoffman and Ter-Pogossian (1975) have shown that the linear attenuation coefficients (CT numbers) determined using computed tomography are linearly correlated with the electron density ($\rho_e$) and with the effective atomic number (Rutherford, Pullan and Isherwood, 1976).

## 7.15 Dual Energy Technique

### *7.15.1 Dual Energy X-ray Absorptiometry (DEXA)*

The assumption of constant thickness of the photon pathway restricts the use of SXA/SPA to peripheral sites. The assumption can not be applied to important fracture sites such as the spine, proximal hip and whole body. This is because of widely varying soft tissue thickness and soft tissue composition.

In dual energy X-ray absorptiomety (DEXA), dual energy X-ray beams are generated by an X-ray tube instead of an isotope source (Gustavson, Jacoson and Kusoffsk, 1974). This method has become the most widely used bone densitometric technique to assess risk of fragility fracture (Kanis and Gluer, 2000). Traditionally, the focus of clinical bone evaluation has been apparent bone mineral density (BMD) as measured by DEXA (Kanis *et al.*, 2002; Brunader and Shelton, 2002). With its advantages of high precision, short scan times and low radiation dose, DEXA is well suited to meet the needs of a non-invasive method of diagnosing osteoporosis (WHO, 1994; Royal College of Physicians, 2000). For the X-ray energies used in DEXA the dominant attenuation process in soft tissues is Compton scattering. Huddleston and Scakler

(1985) have investigated the dual energy technique using the principles of Compton scattering in the determination of the electron density of biological tissues. This is a technique to provide gradation among no risk, probable risk, to high risk categories with appropriate initiation of treatment. DEXA can usefully supplement SXA systems, eliminating the need for a water bath or water bolus.

Detailed issues concerning differences in the behavior of cortical and trabecular bone and at different anatomical sites have arisen. Consequently, several groups have devised methods for imposing investigator specified loading (e.g., Turner *et al.*, 1991; Chambers *et al.*, 1993; Torrance *et al.*, 1994; Hillam and Skerry, 1995; Forwood *et al.*, 1998; Yingling, Davies and Silva, 2001; Gross *et al.*, 2002; Guo *et al.*, 2002) and unloading (e.g., Yonezu *et al.*, 1999; Kodama *et al.*, 1999; Nakamura *et al.*, 2001; Yamashita *et al.*, 2001; Ito *et al.*, 2002) regimens on a single limb, using the contralateral limb as a control unit. Similarly, experimental investigations of fracture healing (e.g., Hiltunen, Vuoprio and Aro, 1993; Olmedo and Weiss, 1994; Chakkalakal *et al.*, 1999; Andreassen *et al.*, 2001; Kurth *et al.*, 2001; Meyer *et al.*, 2001; Uusitalo *et al.*, 2001) in small animal models routinely use contralateral limb controls. Such experiments require that site-specific data be generated and interpreted in proper perspective. Examination is feasible at the lumbar spine, hip and wrist, corresponding to common fracture sites. These features make DEXA an attractive proposition for diagnosis of osteoporosis. This allows patients to monitor the effectiveness of therapy for osteoporosis and as a endpoint for fracture in clinical investigations. Because of these advantages, DEXA has become a popular analytical technique in mice and other small animals, either as a substitute for more demanding techniques, such as biomechanical testing, histomorphometry and gravimetry, or as a supplement to them.

### *7.15.2 Theoretical Formulation and Instrumentation*

The method essentially utilizes a single source to provide the dual energy incident beam, a single detector and a fixed scattering angle. The energy ($E$) of the incident beam (photon energy) is shared with that of the scattered photon ($E^I$) and that of the recoil electron ($E$-$E^I$). These are related by the following relationship:

$$E^1 = E/[1 + (E/mc^2)(1 - \cos\phi)] \tag{7.15}$$

where $mc^2$ is the rest energy of the electron and $\phi$ is the scattering angle. The electron density determination is based upon the assumption that the tissue of interest (substance 2) is surrounded by a tissue of different density (substance 1). The electron density of substance 2 is calculated, using the measured scatter intensity, from:

$$P_{cs} = [I_s(E_1)/S(E_1)]\exp[\mu_{1\ \mathrm{eff}}(E_1)d_1 + \mu_{2\ \mathrm{eff}}(E_1)(D - d_1)] \tag{7.16}$$

where $D$ = the total path length in substances 1 and 2, $I_s(E_1)$ = intensity of scattered photons resulting from energy $E_1$, $P_{cs}$ = electron density of substance 2 determined by Compton scattering method (cs), $S(E_1)$ is the average scatter density per electron, which accounts for the scattering cross section, solid angle and detector efficiency corresponding to incident energy $E_1$; $\mu_{1\text{-eff}}(E_1)$ = effective linear attenuation coefficient of substance 1 at

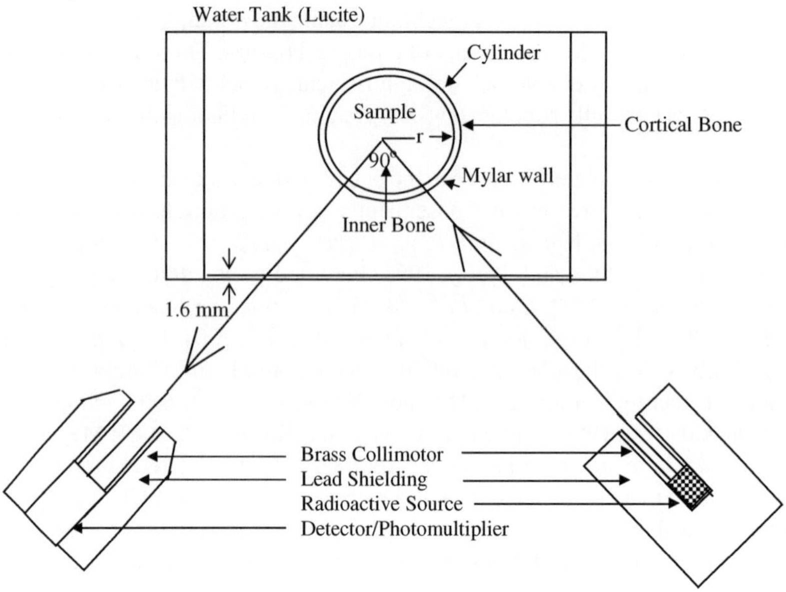


**Figure 7.18** A typical backscattering spectrum of a bone sample. All the elements present in the sample except hydrogen are shown. The locations of possible steps caused by different elements are indicated at the top. From the spectrum it is estimated that the sum of all elements heavier than calcium is at most 0.5% by weight of the bone ash. (Hyuonen-Debek, Riihonen and Debek, 1979)

incident photon energy $E_1$, $\mu_{2\text{-eff}}(E_1) =$ effective linear attenuation of substance 2 at incident photon energies $E_1$.

The method (Figure 7.18) utilizes a $7.4 \times 10^{10}$ Bq radionuclide source ($192_{\mathrm{Ir}}$) and a single NaI (TI) detector, in which the two scattered photo peaks, 195.6 and 244.3 keV, corresponding to the incident energies, 317.0 and 468.0 keV, respectively, are measured. The scattering angle (90°) was chosen because of interest in human vertebral electron density measurements. Here $\mu_{1\text{-eff}}$ and $\mu_{2\text{-eff}}$ represent all the attenuation interaction (photoelectric and scatter) that occurs along the incident beam path, as well as along the path from the scattering volume to the detector (Huddleston and Weaver, 1983). The scattered intensities $I_s$ (195.6 and 244.3 keV) are determined for three symmetric energy windows (20, 30 and 40 keV); while the selected counting time is maintained at 600 s. The effective linear attenuation coefficient ($\mu_{\text{cs-eff}}$) corresponding to each energy is determined by curve fitting techniques and expressed by the following expression (Huddleston and Scakler, 1985):

$$I_s(E) = \alpha \exp(-\beta d)$$

where $d$ is the path length in a sample and $\alpha$ and $\beta$ are the product $P_{\text{ec-S}}$ and $\mu_{\text{cs-eff}}$, respectively, where $P_{\text{ec}}$ is the calculated electron density based on elemental composition. Each scatter intensity obtained is the average of three measurements with the background subtracted for

**Table 7.1** Electron density based on elemental composition, and electron density determined by dual energy Compton scattering for known samples using geometries A, B and C

| Substance | Electron density based on elemental composition ($P_{ec}$) ($\times 10^{23}$ $e^-$ $cm^{-3}$) | Electron density from Compton scattering ($P_{cs}$) ($\times 10^{23}$ $e^-$ $cm^{-3}$) | | |
|---|---|---|---|---|
| | | Geometry A | Geometry B | Geometry C |
| 1. $H_2O$ | 3.344 | $3.282 \pm 0.067$ | $3.329 \pm 0.069$ | $3.233 \pm 0.045$ |
| 2. NaCl | 3.032 | $2.998 \pm 0.057$ | $3.025 \pm 0.034$ | $3.044 \pm 0.003$ |
| 3. KBr | 3.188 | $3.140 \pm 0.029$ | $3.239 \pm 0.013$ | $3.249 \pm 0.015$ |
| 4. $CuSO_4$ | 3.647 | $3.564 \pm 0.367$ | $3.598 \pm 0.022$ | $3.711 \pm 0.022$ |
| 5. $K_2HOP_4$ | 4.386 | $4.257 \pm 0.104$ | $4.375 \pm 0.035$ | $4.381 \pm 0.008$ |
| 6. $K_2HPO_4$ | 5.110 | $4.828 \pm 0.003$ | $4.913 \pm 0.024$ | $5.081 \pm 0.022$ |

each of the three selected energy windows. The substances are mixed just prior to the experiments and their electron densities are calculated using the relationship:

$$P_{\mathrm{ec}} = PN_{\mathrm{A}}\left(\frac{Z}{A}\right)^{\mathrm{eff}} \tag{7.17}$$

where $(Z/A)^{\mathrm{eff}}$ is calculated from the elementary composition for each substance, $N_{\mathrm{A}}$ is the Avogadro constant and $P$ is the mass density ($\mathrm{g\,cm^{-3}}$). The measurements are performed in three chosen geometries.

Table 7.1 compares the electron density values, $P_{cs}$, determined by dual-energy Compton scattering with electron density values calculated from elemental composition $P_{ec}$. The geometry in which the largest collimator is used gives the best agreement between $P_{cs}$ and $P_{es}$. When the $P_{cs}$ is compared with $P_{ec}$ the mean percent difference represented by the dual energy Compton scatter method is less than 1%.

### 7.15.3 Technical Details

DEXA provides an effective way to measure BMD in a specific region of interest (ROI) and is the most widely used diagnostic modality for assessing osteoporosis and osteopenia. A development in DEXA technology has been the introduction of a cone beam system in which the entire scan field is imaged in a single exposure using two-dimensional (2D) detector arrays. These are available for BMD measurements at peripheral sites such as the heel and hand (Fordham, Robert-Countant and Darbous, 2000; Bouxsein *et al.*, 1997). A cone beam DEXA system has also been introduced with a detector array large enough for performing BMD measurements at the spine and hip (Boudousq *et al.*, 2003). Cone beam technology for spine and hip BMD has the advantage of correcting for the effects of scattered radiation at the detector.

As in humans, site specificity of BMD has been reported in mice (Beamer *et al.*, 2001; Sheng *et al.*, 2002; Turner *et al.*, 2000; Masinde *et al.*, 2002). Regardless of species, site specificity has been sought primarily in comparisons of skeletal sites with marked differences in the proportion of cortical to trabecular bone. Therefore, little is known regarding the extent to which anatomically distant cortical bone sites resemble each other densitometrically. Existing rodent site specificity data are often insufficient to help guide interpretation of comparative densitometry of left and right bones. To bridge these gaps, Blake and Fogelman (2005) undertook a comparison of exercised mouse femora and humeri using DEXA.

The first DEXA scanners used a pencil-beam system (Cullum, Ell and Ryder, 1989; Mazess *et al.*, 1989). They used a pinhole collimator coupled to a single detector in the scanning arm and took 5–10 min to perform a BMD study of the patient's hip or spine. Subsequently, a fan-beam DEXA system was introduced that uses a slip-collimation coupled to a linear array of detectors in the scanning arm (Bouyoucef, Cullum and Ell, 1996; Mazess *et al.*, 2000). With faster image acquisition, the time to complete a spine or hip study is now only 10–30 s for the latest fan-beam models.

Lopez Franco *et al.* (2005) have reported that right–left correlations at each of these sites are high, as are intersite correlations using side averaged data. These findings demonstrate the impact of ROI analyses on measurement precision relative to whole body analyses. Moreover, these experiments reveal that left bones have significantly greater BMD than right ones. This observation has practical implications for mouse DEXA studies. The side disparities reflect a population wise pattern. These authors compared the number of mice in which the left values are greater and the number of mice in which the right values are greater for each measurement and each site and found significantly more left values to be greater for BMD at both sites. Left–right disparities of BMC mirrored the BMD findings, while bone area did not differ significantly between the left and right sides at either site.

These authors observed only a moderately high correlation of 0.78 between left and right BMDs at either the femur or the humerus – much lower than the correlation observed between right and left hips in humans (Faulkner, Genant and McClung, 1995; Bonnick *et al.*, 1996; Yang *et al.*, 1997; Mazess *et al.*, 2000; Rao, Reddy and Rao, 2000; Wong *et al.*, 2003). In addition, the intersite correlations are higher between side-averaged data than for either ipsilateral or contralateral comparisons. Lastly, they found that bone area is not normally distributed. Each of these findings reflect the relatively large pixel size ($0.18 \times 0.18$ mm) relative to the bones studied, as has been discussed by Lopez Franco *et al.* (2004). The average obtained from two-paired measurements are expected to be closer to the true values than either measurement taken individually. That side averaged data are more highly correlated between anatomic sites than unilateral data and that the correlations of these two intersite data demonstrates that the magnitude of individual measurement errors is large relative to the differences between right and left sided data. Comparison of unilateral and bilateral femur BMD measurement in humans has led to a similar interpretation (Mazess *et al.*, 2000). These intersite correlations Lopez Franco *et al.* (2005) are only slightly poorer than right–left correlations at a single anatomic site, suggesting that femoral BMD is highly predictive of humeral BMD and vice versa, supporting the notion that the various long bones possess similar biology. These authors have pointed out that these results may not be generalizable to other strains of mice, or other species. DEXA scanning is performed *ex vivo*, with bones positioned on a specimen tray provided by the instrument manufacturer, and are studied individually.

Multiple human investigations have compared densitometric features of bones and related the observations to known or hypothesized differences in mechanical loading. Some investigators did not relate side comparisons to handedness (Bonnick *et al.*, 1996; Mazess *et al.*, 2000), while others disagree about this relationship to BMD (Yang *et al.*, 1997;Rao, Reddy and Rao, 2000; Dane *et al.*, 2001; Akar *et al.*, 2002; Sadeghi *et al.*, 2000). Despite these limitations and adjustment for them, a large number of specimens do reveal a small but significant difference in right and left long bone BMD.

Investigations have revealed extensive clinical evidence (Bevier *et al.*, 1989; Chen *et al.*, 1997; Glauber *et al.*, 1995;Printchard, Nowson and Wark, 1996; Ravn *et al.*, 1999a, 1999b) of strong positive correlations between BMI and dual energy X-ray absorptiometry (DEXA) measurements of BMD. Yet, there is no insight into the underlying biological mechanism(s) responsible for the linkage(s) between these anthropometric parameters. The relevance of BMD measured by quantitative computed tomography for the mechanical properties of cortical bone has been controversial, because of low correlation coefficients between cortical bone strength and BMD (Snyder and Schneider, 1991; Stromsoe *et al.*, 1993).

DEXA delivers a very low radiation dose, thus also permitting its use in the pediatric population and allowing for repeated measurements. The effective dose of the DEXA scan is 1–3 μSv, depending on the scanning site. Although measurement of spine and hip bone mineral dominates in the investigations of osteoporosis in adults, total body measurements are more appropriate during growth. DEXA also has the advantage that measurements of soft tissue fat and lean masses can be determined at the same time, allowing bone mass to be related to these variables, as well as to age and weight. One factor that has been ignored completely is that the accuracy of DEXA is not absolute. X-Ray attenuation arises because of three main components: bone mineral, fat and lean soft tissue. As only two X-ray energies are used, complete solution of the equations is not possible and it is necessary to make assumptions about fat distribution. The results from different manufacturers and different software often give different results for three attenuating materials (Tothill, Avenell and Reid, 1994a; Tothill *et al.*, 1994b).

Overriding clinical evidence (Bevier *et al.*, 1989; Chen *et al.*, 1997; Glauber *et al.*, 1995; Hylstup *et al.*, 1993; Jensen, Nielsen and Skovsgaard, 1996; Liel *et al.*, 1988; Printchard, Nowson and Wark, 1996; Reid *et al.*, 1992; Reid, Evans and Ames, 1994) has provided strong, positive correlations between body mass index (BMI) and dual-energy X-ray absorptiometry (DEXA) measurements of BMD. Yet, no insight into the underlying biological mechanisms(s) responsible for the linkage(s) between these anthropometric parameters and observed changes in BMD has yet emerged. Nonetheless, these collective findings are generally considered of intrinsic importance, as a range of possible biologically causal connections and prognostic implications relative to osteopenia/osteoporosis and associated bone fragility have been inferred (Printchard, Nowson and Wark, 1996; Slemenda, 1995) from them. Ravn *et al.* (1999) have reported that the lowest BMI tertile of a carefully selected cohort of 417 recently scanned postmenopausal women had significantly lower measured BMD at spinal and proximal femoral sites than did those with higher body mass index {BMI = weight (kg)/ [height (m)]$^2$}. These authors confirmed the aggregation of clinical observations (Bevier *et al.*, 1989; Ravn *et al.*, 1999; Chen *et al.*, 1997; Compston *et al.*, 1992; Glauber *et al.*, 1995; Harris, 1992; Hylstup *et al.*, 1993; Jensen, Nielsen and Skovsgaard, 1996; Liel *et al.*, 1988; Printchard, Nowson and Wark, 1996; Reid *et al.*, 1992; Reid, Evans and Ames, 1994) of positive correlations between body mass/percent body fat/BMI and DEXA-measured *in vivo* BMD. Their findings also underline the relationship between patient soft tissue anthropometrics and measured prospective changes in BMD.

An extensive body of clinical evidence indicates that the use of DEXA *in vivo* BMD measurements is open to inherent inaccuracies (Antonacci, Hanson and Hegeness, 1996; Blake *et al.*, 1992; Bolotin, 1998a, 1998b; Farrell and Webber, 1989; Farrell and Webber, 1990;

Formica *et al.*, 1995; Gotfredsen *et al.*, 1988; Hangartner and Johnston, 1990; Hassager *et al.*, 1991; Ho *et al.*, 1990; Kuiper *et al.*, 1996; Sabin *et al.*, 1995; Sorenson, 1990; Sorenson, Doke and Smith, 1989; Svendsen *et al.*, 1995; Tothill, 1989; Tothill, Avenell and Reid, 1994a; Tothill *et al.*, 1994b; Tothill and Pye, 1992; Webber, 1987). These can have serious consequences, as they can influence final measurement outcomes and materially affect diagnostic interpretations and prognostic implications. These concerns are underpinned by reports of *in situ* and *in vitro* cadaveric investigations. These reports have presented data indicative of large DEXA BMD inaccuracies (exceeding ~20%), particularly for postmenopausal and elderly individuals displaying low BMD values. These findings are in agreement with more recent analytical and quantitative simulation studies of lumbar vertebral and proximal femoral sites (Bolotin, 1998a, 1998b). These investigations have shown that the trends and specific quantitative aspects of earlier clinical observations of strong positive correlations between soft tissue anthropometrics and BMD could be traced to manifestations of BMD inaccuracies inherent in DEXA methodology. Thus, the presence of sizeable DEXA *in vivo* BMD inaccuracies is based on a foundation of extensive and diverse clinical data, cadaveric studies and a fundamental understanding of the mechanisms and causes from which they originate.

The extent of such inaccuracies and their dependency on soft tissue anthropometrics are of salient importance for patient specific clinical DEXA measurements. This is due to the prospect that sizeable inherent systematic *in vivo* BMD inaccuracies could seriously compromise the integrity of DEXA measurements undertaken to evaluate predictive bone fragility. In addition, assessment of the efficacy and/or quantitative effectiveness of drug and other therapeutic regimens intended to ameliorate the osteoporotic condition may be confounded.

The results of Ravn *et al.* 1999a, 1999b) have introduced aspects of the relationship between BMD and soft tissue anthropometrics that demand consideration. Possible BMI–BMD correlations and observations of the BMI-independent effectiveness of remedial drug therapy that reflect actual changes in BMD indicate significant insight into bone resorption and remodeling processes. However, the possible link between these and the uncertainties prevailing in DEXA measurements needs to be ascertained. Clearly, it is important that these two conflicting interpretative views of the aggregation of confirmed clinical observations (Bevier *et al.*, 1989; Chen *et al.*, 1997; Glauber *et al.*, 1995; Harris and Dawson, 1996; Hylstup *et al.*, 1993; Jensen, Nielsen and Skovsgaard, 1996; Liel *et al.*, 1988; Printchard, Nowson and Wark, 1996; Ravn *et al.*, 1999a, 1999b; Reid *et al.*, 1992; Reid, Evans and Ames, 1994) be critically evaluated and compared before arriving at any conclusion.

### *7.15.4 Simulation Studies*

The mean photon energies of the low and high energy components of DEXA spectra ($E_L = 43$ keV, $E_H = 110$ keV: densitometers) (Dunn and Wahner, 1994) have been selected in simulation studies (Ravn *et al.*, 1999a, 1999b). Since beam hardening effects in DEXA are negligible (Dunn and Wahner, 1994; Sorenson, Doke and Smith, 1989), each component of the dual polychromatic spectrum is treated as monochromatic. In the absence of beam hardening the measured BMD can be treated as independent of overall torso thickness. Thus, the vertebra can be considered situated in a torso of fixed 20 cm overall thickness (the average anteroposterior thickness encountered clinically).

Since the procedure used in the dual energy Compton scattering method requires the determination of the effective linear attenuation coefficient ($\mu_{cs\text{-}eff}$), the accuracy of the dual energy Compton scatter method for use in electron density measurement is of the order of 1–2% for the 90° scatter geometry. Further, the precision of spine and hip BMD measurements from established pencil-beam and fan-beam DEXA systems are around 1% for the spine and 1–2.5% for different BMD sites in the hip (Patel *et al.*, 2000). Blake, Knapp and Fogelman (2005) have shown a strong linear relation between the cone beam and fan beam measurements with a correlation between $r = 0.95$ and 0.97. Although the mean BMD in the two systems agreed within 2% for spine and femoral neck BMD and 5% for total hip BMD, larger differences are found in some individual patients.

With pencil beam and fan beam systems, most scattered photons are likely to miss the detector and an accurate measurement of transmitted radiation is possible without any confusing signal from scatter. However, with cone beam systems measurement of the point-by-point transmission factor over the image field is complicated by the need to correct for the scatter contribution to the signal. Because of this inherent uncertainty, the possibility that a cone beam system can deliver accurate, reliable and precise measurements of spine and hip BMD is questionable.

Simulation studies are normally conducted on the vertebra. In this method, the transverse thickness of the vertebra is measured. The measured paths through the ROI matched fully the standard prescriptive geometry of anteroposterior vertebral DEXA X-ray direction (Wahner and Fogelman, 1994). All trabecular interstices are considered filled with the same bone marrow type (red/yellow mix). The vertebra are considered surrounded only by lean muscle tissue uniformly infused with adipose. The known total mass attenuation coefficients of each elemental constituent at the relevant X-ray energies are given in standards (Wapstra *et al.*, 1957; Report of the Task Group on Reference Man, 1975; Goodsitt, 1992; Goodsitt *et al.*, 1994). Using this method the overall X-ray attenuation along each given photon path through the bone ROI is obtained and the simulations of DEXA *in vivo* BMD scans are obtained (Bolotin, 1998a, 1998b). These simulation studies have been carried out for many combinations of different values of vertebral BMD, extraosseous fat-to-lean muscle tissue areal density ratios and red/yellow marrow mixes in the ranges clinically observed (Antonacci, Hanson and Hegeness, 1996; Bevier *et al.*, 1989; Levy, Lees and Stevenson, 1992; Meunier *et al.*, 1971; Wang *et al.*, 1989). In this context, the actual BMD ($BMD_{actual}$) is defined and used here to be the areal density of bone material ($g\,cm^{-1}$) that would result from a DEXA measurement with no BMD inaccuracy. The various simulated $BMD_{actual}$ values employed in this study are selected by changing the concentration of simulated trabecular bone material within the cortical enclosure, as also specified in earlier work (Bolotin, 1998a, 1998b). It is these BMD values that are defined as $BMD_{actual}$. These simulation studies are carried out as per the ideal DEXA prescribed posteroanterior scanning mode (Dunn and Wahner, 1994).

The expression used to formulate the simulated values of DEXA measured BMD is the standard DEXA equation:

$$BMD_{meas} = (\rho_B \chi_B)_{meas} = \frac{\left(\ln\left\{\frac{J_1}{J_0}\right\}\ln\left\{\frac{I_2}{I_0}\right\} - \ln\left\{\frac{J_2}{J_0}\right\}\ln\left\{\frac{I_1}{I_0}\right\}\right)}{\left(\lambda_{1B}\ln\left\{\frac{J_2}{J_0}\right\} - \lambda_{2B}\ln\left\{\frac{J_1}{J_0}\right\}\right)} \tag{7.18}$$

where $\rho_B$ and $\chi_B$ represent the volumetric density ($g\,cm^{-3}$) of bone material and the inner thickness of bone material present along any given DEXA X-ray path through the ROI, respectively; $\lambda_{1B}$ and $\lambda_{2B}$ represent the mass attenuation coefficients ($cm^2 g^{-1}$) of bone material at the lower (1) and higher (2) dual X-ray energies of the DEXA densitometer, respectively. Further, $\ln(J_{1,2}/J_0)$ and $\ln(I_{1,2}/I_0)$ are natural logarithms of the fraction of X-ray intensities transmitted at energies 1 and 2 through the soft tissue lateral to the bone and through bone material, respectively. The development and application of DEXA provides a means of categorizing patients into three categories: (i) low possibility of fracturing a bone, (ii) increased risk of fracture and (iii) those who have a higher risk of fracture (Webber, 2006). In addition DEXA permits the assessment of efficacy of various medications and treatment regimes considered to reduce the risk that patients may suffer a bone fracture.

To date, DEXA is the best accepted diagnostic tool for determination of BMD in a clinical environment. Bagi *et al.* (2006) have tested the ability of non-invasive methodologies (DEXA, pQCT and μCT) to asses and quantify cortical bone mass and geometry in a model in which bone mass is induced by surgical castration. These authors have concluded that these three techniques proved useful in describing cancellous and/or cortical bone parameters and positive correlations are demonstrated between data obtained by different methods. The cross sectional area of a bone structure is crucial for resisting loads in bending or torsion and is described as the "areal moment of inertia" for bending and as the "polar moment of inertia" in torsion. All three radiologic methods show a great deal of precision and ability to detect small changes in cortical bone morphology following drug treatment. Micro-CT has been reported to be sensitive even at medium resolution mode in detecting newly mineralized surfaces at the periosteal and endocortical bone envelops (not necessarily fully mineralized).

The technique shows an improvement over DPA in scan time, radiation dose, a finer collimation of the beam (1.5 mm), bone edge detection and cost of source renewal. This is free from a beam hardening effect due to tissue thickness (Sartoris and Resnick, 1989; Slosman *et al.*, 1990).

Non-invasive assessment of vertebral strength is critical for the effective treatment of osteoporosis. The current clinical standard, which is site-specific BMD assessment using DEXA (Bates, Black and Cummings, 2002; Cawte *et al.*, 1999), is only moderately effective in predicting the possibility of occurrence of vertebral fractures (Cummings, Bates and Black, 2002).

Several researchers have implicated that DEXA provides an inaccurate measure of vertebral fracture risk potential because it is structurally simple, that is, it indicates only volume averaged material properties. Consequently, they have focused their efforts on developing more sophisticated mechanical models of vertebral body fractures (Crawford, Cann and Keaveny, 2003; Crawford, Brouwers and Keaveny, 2004; Liebschner *et al.*, 2003). These models are patient specific and are generated from quantitative computed tomography (QCT) scans and can be classified as (i) simplistic models (Brinckmann, Biggemann and Hilweg, 1989; Crawford, Cann and Keaveny, 2003; Crawford, Brouwers and Keaveny, 2004; Eriksson, Isberg and Lindgren, 1989; Whealan *et al.*, 2000), making use of minimum axial rigidity and stiffness, and (ii) three-dimensional finite element models (Crawford, Cann and Keaveny, 2003; Crawford, Brouwers and Keaveny, 2004; Whealan *et al.*, 2000). However, as yet, there is no clear evidence that these methods have better outcomes than DEXA-BMD *in vivo* (Buckley, Loo and Motherway, 2007). These authors have further established that experimental strength and stiffness are modestly correlated, while quantitative computed tomography based finite element and strength are strongly correlated.

This technique provides only a 2D image of apparent density – its ability to assess early fracture risk is limited. While density (quantity) does positively correlate with strength (Keaveny *et al.*, 2001) about 30% of the variability in bone strength remains unexplained (National Institutes of Health, 2000; Hans *et al.*, 1997). DEXA also suffers from an inability to differentiate trabecular from cortical bone. This is because the effects of bone loss are more prevalent in trabecular bone due to its much higher surface area and because the greater amount of bone mineral content in cortical bone can conceal small changes in trabecular bone (Brismar, Budinsky and Majumdar, 2001).

## 7.16 Neutron Activation Analysis

Total body neutron activation analysis is now established as one of the most important methods of determining the elemental composition of the human body (Chamberlin *et al.*, 1968; Harrison, McNeil and Meema, 1974; Cohn *et al.*, 1976a; Palmer *et al.*, 1968; Spinks and Bewley, 1976; Kennedy, Boddy and Williams, 1979; Kennedy, Eastell and Ferrington, 1982; Eastell *et al.*, 1983). It can be used to determine both the absolute amount and changes with time of certain body elements, in particular calcium, sodium, chlorine, phosphorous and nitrogen (Cohn *et al.*, 1976b). Because of this, normalization of the results is necessary. Since 99% of total body calcium is present in bone (Heaney, 1963), the technique can be used to measure calcium directly. It can thus measure bone mass and in this way can avoid errors in sampling from different parts of the skeleton.

### *7.16.1 Technique*

Several techniques (Chamberlin *et al.*, 1968; Cohn, Dombrowski and Fairchild, 1970; Cohn *et al.*, 1972; Nelp *et al.*, 1970; Boddy, Hollowa and Elliott, 1973; Spinks and Bewley, 1976; Smith *et al.*, 1980) have been developed for total calcium measurement and used for patient examination. The subjects are irradiated by a uniform beam of partially moderated fast neutrons inducing several reactions, most notably $^{48}Ca(n/y)^{49}Ca$, $^{23}Na(n, y)^{24}Na$, and so on. The patient is either positioned at a distance from a single neutron source (Chamberlin *et al.*, 1968; Cohn, Dombrowski and Fairchild, 1970; Nelp *et al.*, 1970; Spinks and Bewley, 1976; Smith *et al.*, 1980) between an array of 15 radionuclide sources (Cohn *et al.*, 1972) or scanned between two opposed neutron generators (Boddy, Hollowa and Elliott, 1973). Alternatively, the patient is encased in a Perspex box, which moderates the neutrons and makes the flux through the patient uniform.

Anderson *et al.* (1964) described an experiment in which total body sodium and chlorine were estimated after irradiation with fast neutrons. The subjects were irradiated for ten minutes to a total dose of 0.10 rad. Fifteen minutes after irradiation, the subjects were taken and counted on a whole body counter. These counts are then compared with those obtained from an irradiated phantom, containing a known quantity of sodium and chlorine. The content in the subjects can then be estimated.

Palmer *et al.* (1968) studied $\gamma$-ray spectra from a human cadaver and from phantoms filled with tissue equivalent solution after irradiation with 14 MeV neutrons from the 3T (d, n) reaction. In both instances, a well-defined photopeak of induced $^{49}Ca$ is identified and separated from the higher energy portion of the spectrum. The characteristic 8.8 min half-life and 3.1 MeV energy of $^{49}Ca$ were also confirmed. The induced $^{49}Ca$ radioactivity is quantified with calcium standards that are irradiated simultaneously with the patient and can be counted immediately after counting the patient. Absolute calibration can be done by irradiating and counting several cadavers, after which the cadavers are ashed and the total body calcium (TBCa) determined by chemical analysis. Limitations in the geometry and uniformity of both irradiation and counting due to individual differences in body cavities are estimated to permit a measurement of total body calcium in a single individual with a accuracy of $\pm 8\%$. By observing continued changes in skeletal mass, a given patient can be compared with himself, thereby eliminating the problem of geometric differences in body habitus. These authors have estimated that it is possible to detect a change in total body calcium of $\pm 2\%$. The radiation dose of 0.1 rad that is necessary in this determination is measured with small tissue equivalent ionization chambers (Braby, 1966) placed at various locations within the phantom.

The accuracy of this method for determining the total amount of an element in man depends on a uniform irradiation of the body with the neutron beam.

Kennedy, Eastell and Ferrington (1982) have described a technique to measure TBCa with a high degree of precision. The method consists of two parts: The first to introduce corrections to eliminate the effects of variations in activation and detection efficiency due to varying body dimensions. Second, the normal predicted total body calcium is calculated using a formula derived from the results of a group of healthy normal volunteers (Nelp *et al.*, 1970; Cohn *et al.*, 1976).

The neutron source used for patient activation is a cyclotron producing neutrons of mean energy 6.5 MeV. During irradiation the subject is positioned in a moderator kiosk made of 3-cm thick polyethylene sheets mounted on a turntable. The total dose equivalent received by the patient is controlled by the neutron monitor chamber of the beam. The patient spectrum, covering the energy range 0–4 MeV, is recorded and analyzed.

The technique gives a precision of 1.8% on phantoms and is suitable for detecting small changes in total body calcium. The process of correcting for varying body dimensions is important for the accurate determination of the absolute amount of calcium in the body. These authors concluded that the percentage of calcium lost in normal women after the menopause (1.5%) is slightly higher (1.1%) than determined by Cohn *et al.* (1976). These authors further concluded that total body calcium was dependent partly on $(\text{weight})^{-00.34}$ for males and $(\text{weight})^{-0.18}$ for females, in broad agreement with the exponent (0.25) determined by Cohn *et al.* (1976).

In an improved experimentation described by Eastell *et al.* (1983) the induced radiation was measured using four sodium iodine detectors mounted in a whole body counter. Computer analysis of the spectrum and the value of a standard irradiated simultaneously with the subjects gave the total body Ca in $^{49}$Ca counts. The long-term precision based on anthropomorphic phantom measurements was 1.8% for a dose of 1.3 rem (Kennedy, Eastell and Ferrington, 1982). The data from $^{49}$Ca can be interpreted directly as an indicator of skeletal mass in the absence of ectopic calcification.

The total body calcium in normal women is $820 \pm 124$ g. This gives a combined variance (CV) of 15.1% without making an allowance for span or postmenopausal age. The greatest reduction in variance was obtained using the formula to obtain the predicted total body Ca postmenopause (TBCaP):

$$\text{TBCaP} = \alpha\, s\, 1.69\, \mathrm{e}^{-0.015y} (\gamma = 0.90, P < 0.001) \tag{7.19}$$

where TBCaP is in grams; $\alpha = 399$; $s =$ arm span in m; $y =$ years postmenopause.

The arm span not only related more closely to TBCa than height but also eliminated the error due to loss of height associated with vertebral fractures. The estimated rate of bone loss was 1.5% per year. The ratio TBCa/TBCaP has been defined as the calcium ratio ($Ca_R$) (Cohn, Ellis and Wallach, 1974) and for normal women it was found to be 1.0 with a CV of 6.6%.

The TBCa in women with vertebra and wrist fractures was $649 \pm 87$ g and $787 \pm 72$ g, respectively. In the women with wrist fractures the $Ca_R$s were $1.00 \pm 0.10$, which was normal. The formula for TBCaP cannot be extrapolated to cover women more than 22 years past the menopause.

When TBCa was normalized for span along (TBCa/$s$1.69), that is, $y = 0$ in Equation 7.19 for TBCaP, an estimate of osteopenia in patients with fractures can be made by comparing the values with the $Ca_R$ of normal women. The TBCa/$\alpha s$1.69 for women with vertebra and wrist fractures and for normal women were $0.69 \pm 0.06$, $0.84 \pm 0.08$, and $0.88 \pm 0.11$ respectively. It is likely that the precision of the measurements had at least 5% error (Mazess, 1983).

TBCa for the normal women ($820 \pm 104$ g) was similar to that reported by Cohn *et al.* (1976b), namely, $804 \pm 106$ g for normal women aged 50–59. The TBCa for women with vertebral fractures ($649 \pm 87$ g) resembled that described by Chesnut *et al.* (1977) ($688 \pm 94$ g) and Cohn, Ellis and Wallach (1974) ($590 \pm 108$ g). However, some of the patients in the latter two groups did not have vertebral compression fractures but only demineralized vertebrae.

Different formulae for TBCaP have been reported. Cohn *et al.* (1976b) use age from birth, height and total body potassium: the latter two also may be age dependent. Nelp *et al.* (1972) have predicted TBCa from the cube of the patient height. Eastell *et al.* (1983) noted that span was more effective than height in reducing variance and these authors have suggested the use of span, as it is unaffected by the height loss due to vertebral compression fractures.

Several workers (Smith and Tothill, 1979; McNeill *et al.*, 1973) have used radionuclide neutron sources ($^{252}$Cf or PU-Be) to measure calcium in the spine, while others (Smith *et al.*, 1980) have employed neutron generators for neutron sources. The method of trunk activation measures not only the spine but all the bone of the trunk. This trunk measurement encompasses about 30% of the total skeletal calcium and hence it is not specific to trabecular bone. As such it is exposed to the occurrence of ectopic calcification. Such extraosseous deposition, known to occur with aging and bone disease, changes the percentage of calcium outside the skeleton. Since much of the latter deposition occurs in the trunk area, the advantages of a trunk measurement (high percentage of trabecular bone) are offset by potential errors. Spinal calcium measurements involve fairly high doses (3 rem or more), however, which prohibits their use on normal subjects, as well as inhibiting repeat measurements on patients. This dose is particularly high considering that it may involve about 50% of the active marrow in the body.

### *7.16.2 Site Choice*

The larger the site to be measured the larger the neutron source and detection apparatus required. The hand was the first peripheral part body site to be examined, which requires a single and inexpensive radionuclide source (Catto, McIntosh and Maclead, 1973; Guey *et al.*, 1979). In certain clinical conditions the bone loss is initially most pronounced in the hand and the bone volume can be used to normalize the data for absolute measurements. The forearm has also been measured (Smith *et al.*, 1977) using two radionuclide sources. This, besides being easy to restrain, is more representative of calcium changes throughout the body.

In the study of osteoporosis, the spine is generally considered a most suitable site of measurements. Techniques have been developed to measure the lower spine (Al-Hiti *et al.*, 1976). McNeill *et al.* (1973) have developed a technique to measure the complete spine and pelvis, which contains about 40% of the total body calcium.

### *7.16.3 Dose*

The radiation dose to the patient is an important consideration with this method. The dose can be reduced by shorter activation times, which will reduce the precision. The dose to the hand given by Catto, McIntosh and Maclead (1973) was 15 rem, by Maziere *et al.* (1979) 8 rem and by Guey *et al.* (1979) 2 rem. The dose to the total body was less than 1 rem in the two techniques (Cohn, Dombrowski and Fairchild, 1970; Cohn *et al.*, 1972).

### *7.16.4 Limitations*

The particular problems encountered are:

1. High radiation dose ($>3$ rem), associated with the technique and the difficulties of housing neutron sources in a clinical environment.
2. Non-uniformity of neutron flux, and non-uniformity in counting geometry.

In addition, the technique is rather costly for a routine medical check up.

Consequently, despite significant advances, Compton scattering and neutron activation analysis have not so far been widely used in clinical settings.

## 7.17 Infrasound Method for Bone Mass Measurements

The above-mentioned techniques though accurate are beset with difficulties, such as exposure due to ionizing radiations and cost, which makes repetitive measurements somewhat difficult.

Studies of the application of acoustic energy for non-invasive skeletal diagnosis have shown its feasibility and demonstrated the advantages of utilizing acoustic (ultrasonic and subsonic) techniques for bone mass and strength measurements. Unlike conventional radiological techniques, acoustic techniques emit no radiation, are cost effective and utilize equipment that is portable and easy to operate.

Subsonic techniques utilize low frequency acoustic energy (hundreds of Hz) often called a vibration. Subsonic methods, known as impedance and resonance methods, involve excitation and measurements of the flexural mode of vibration of long bone and show some potential applicability for osteoporosis diagnosis. However, subsonic measurements are difficult to interpret and, to a great extent, depend upon a corresponding mathematical model of the tested object. The influence of surrounding soft tissue (its mass and damping effect) creates additional difficulties in the interpretation and use of these methods.

The infrasound method is different from the other methods. Instead of measuring flexural resonances, which involve both the mass and flexibility of a "rigid body" resonance of a long bone, such as tibia or ulna, it uses an artificial spring with known stiffness. The approach greatly simplifies the interpretation of measurements and is more accurate.

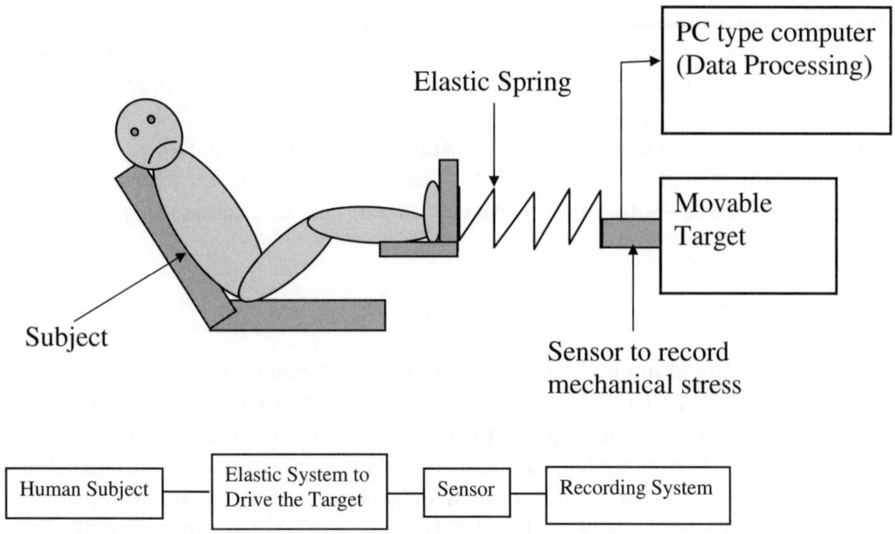


**Figure 7.19** A schematic presentation of a system to recording Infrasound measurement

The method employs very low frequency (below the audible range) longitudinal vibration excitation of human long bone through an artificial spring (Figure 7.19). The spring is much softer than the bone and its joints, but stiffer than surrounding soft tissue. Because of this, the flexibility of the bone and joints as well as mass of the surrounding soft tissue have no effect on longitudinal oscillation of the spring–bone system, where bone contribution is purely inertial. Thus, mechanically, the bone can be considered as a rigid mass. The spring mass system exhibits resonance behavior (amplification of the oscillation on the particular frequency, called the resonance frequency). The resonance frequency depends on the stiffness of the spring and the attached mass. Therefore, by measuring the frequency (with knowledge of the spring stiffness) the overall mass of bone can be determined. An important feature of this approach is that the resonance frequency can be shifted into the very low frequency range (infrasound range) by proper choice of the spring stiffness. This eliminates the damping effect of soft tissue raising the sharpness of the resonance and, hence, increases the accuracy of the measurements. Numerical analysis of the developed model has shown that the mass of bone and its loss can be accurately measured and that these measurements do not depend on variation in bone flexibility and soft tissue parameters.

## *7.17.1 The Ultrasonic Measurement: Concepts and Technique*

In the past few years, ultrasonic techniques have become very common and several commercial ultrasound measurements are available to measure osteoporosis. Quantitative ultrasonic sonography (QUS) is emerging as a powerful technique and is essentially used on the periphery and is suitable for use with soft tissue covering. Various sites, ranging from the calcaneus of the heel, the cortical bone of the tibia, the proximal phalanges of the fingers, the patella, to the distal radius, can by measured by currently available techniques. Other advantages of ultrasonic techniques are that the system is comparatively less expensive and is also portable. Ultrasonic characteristics of bone are directly related to the ultrasonic properties (Ashman *et al.*, 1984), which are in turn determined by the organic – as well as the inorganic – constituents of the tissue. QUS essentially measures two parameters: the speed of sound (SOS) and broadband ultrasound attenuation (BUA). The speed of a sound wave traveling in a given direction, with particle motion in a particular mode, can be expressed by a square root relationship of the form:

$$\nu = (c/\rho)^{1/2}$$

where $\nu$ is the velocity of ultrasound, $c =$ an appropriate combination of elastic constants and $\rho =$ the density of bone. This identifies a direct relationship with the structural integrity. Ultrasound measurements of the patella or calcaneus can predict fracture risk because sound transmission through bone is related to the density of the bone and its skeletal strength. For bone tissue, the SOS represents a combination of density and elastic modulus (Kaufman and Einhorn, 1993; Njeh, Boivin and Langton, 1997). Some workers have suggested that SOS is a better candidate parameter than BMD for evaluating osteoporosis (Kaufman and Einhorn, 1993; Njeh, Boivin and Langton, 1997; Miller, 2000; Nejh *et al.*, 2001; Wear, 2000; Drozdzowska and Pluskiewies, 2001; Javaid, Pitt and Moniz, 1990; Wuster *et al.*, 1992).

Accelerated bone loss in early postmenopause results not only in thinning of bone trabeculae but also in trabecular perforations and separations. Consequently, the modulus of elasticity of bone is to a certain extent determined not only by BMD but also by ultrasound transmission velocity in bone (Van Den Bergh *et al.*, 2000; Hans *et al.*, 1999). Ultrasonic waves are elastic waves that have the potential to probe multiple bone properties, such as bone density material properties, micro-architecture and even macro-structure. Several studies (Hans *et al.*, 1996; Thompson *et al.*, 1998) have demonstrated that ultrasound stiffness could be a predictor of fracture risk independent of BMD. This relationship is independent of BMD and differs with age. The elastic properties of bone may be affected by the quality of type 1 collagen, which is genetically determined. In addition, the COLIAI Spl polymorphism is also associated with lower values of ultrasound transmission and, therefore, with the modulus of elasticity of bone (Kann *et al.*, 2002), with femoral neck geometry (Qureshi *et al.*, 2001) and the rate of postmenopausal spinal bone loss (MacDonald *et al.*, 2001; Harrise *et al.*, 2000). These observations indicate that the COLIAI Spl genotype-dependent properties of bone contribute to fracture risk and hence act as a marker of reduced bone quality, namely the elasticity of the bone.

In view of its non-ionizing characteristics, and being easily accessible and low cost, ultrasound techniques have become an important tool in the assessment of osteoporosis and, hence, bone quality (Stewart and Reid, 2000; Alves *et al.*, 1996; Gnuidi *et al.*, 1985; Prins *et al.*, 1998). There is, therefore, an increasing need to correlate the ultrasound results with the associated mechanical properties. This method has a potential to estimate the elastic modulus in the bone, which in turn predicts the fracture risk of the osteoporotic bones. Ultrasound measurement of the cancellous bone shows the importance of trabecular orientation using the ultrasound measurement and is a significant contributor to the ultrasound measurement results (Chao *et al.*, 2004). Turner and Eich (1991) have concluded that only a fraction of the compressive strength of cancellous bone can be explained by apparent density without considering an anisotropy factor. Minakuchi *et al.* (1997) have investigated the correlation between the compressive stiffness or ultrasound velocity and the BMD and the intensity of the trabecular orientation using calcaneal cancellous bone from humans. This investigation suggests the importance of the trabecular orientation reflects the mechanical properties of cancellous bone with anisotropic behavior.

Ultrasound measures the density and anisotropy simultaneously and these parameters cannot be separated in an ultrasound measurement. Several investigators have reported that QUS predicts fractures after statistical adjustment for DEXA-derived BMD, implying that QUS is an indication of bone strength independent of BMD (Bauer *et al.*, 1997; Gonnelli *et al.*, 1995; Hans *et al.*, 1996). Several other studies support this notion (Gluer, Wu and Genant, 1993; Nicholson, Haddaway and Davie, 1994; Njeh, Boivin and Langton, 1997).

Most new clinical ultrasound (US) instruments that are designed for bone quality assessment measure broadband US attenuation (BUA) and/or speed-of-sound (SOS). BUA is defined as the slope of the frequency-dependent attenuation (Langton, Palmer and Porter, 1984) (Figure 7.20B). Previous *in vivo* and *in vitro* studies have not consistently demonstrated a relationship between BUA and bone mineral density ($BMD_{areal}$) in bovine and humane trabecular bone (Duquette *et al.*, 1997; Hodgskinson *et al.*, 1996; Serpe and Rho, 1996;Toyras, Kroger and Jurvelin, 1999). In clinical measurements, bone thickness, as well as variation in thickness and composition of soft tissue around the bone, can significantly affect BUA and especially the SOS (Gomez *et al.*, 1997; Johansen and Stone, 1997; Kotzki *et al.*, 1994; Serpe and Rho, 1996; Toyras, Kroger and Jurvelin, 1999). Differences in bone porosity have been proposed to affect US scattering and attenuation (Haire and Langton, 1999; Hodgskinson *et al.*, 1996; Langton *et al.*, 1998; Nicholson *et al.*, 2000; Serpe and Rho, 1996; Strelitzki, Evans and Clarke, 1997). Diffraction and interface losses may also affect US measurements (Strelitzki

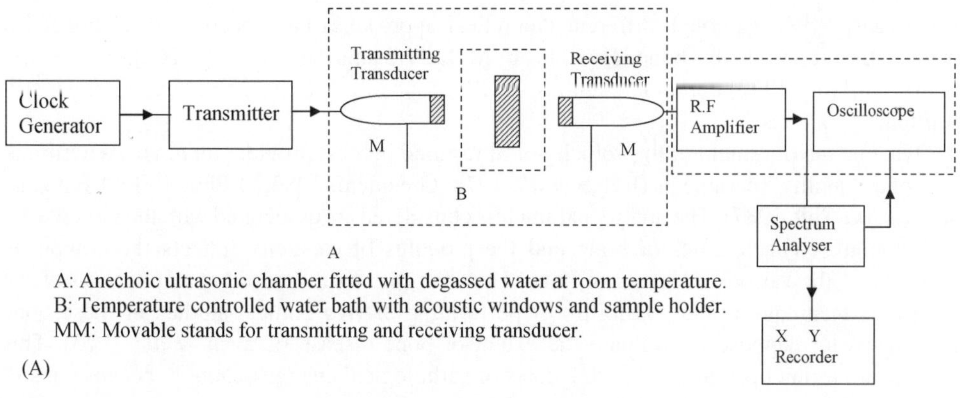


**Figure 7.20** Schematic presentation of ultrasonic measurements of bone sample trabecular bone. (A) The ultrasound pulse is transmitted from the transmitter to the receiver with and without the heel in a water bath; (B) determination of BUA as a slope of frequency dependent attenuation; (C) determination of SOS from a difference between time-of-flight measurements without and with the heel in the water bath (Chen, Zhao and Mundy, 2004)

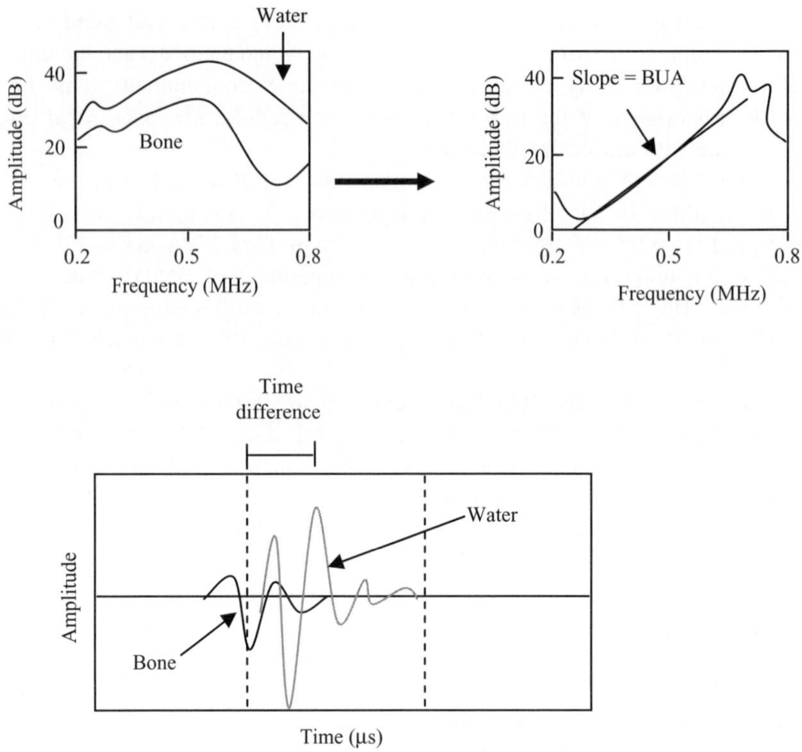


(B)

(C)

**Figure 7.20** *(Continued)*

and Evans, 1998). Although different theoretical approaches have been used to model US propagation in trabecular bone (Biot, 1956; McKelvie and Palmer, 1991; Nicholson *et al.*, 2000; Williams, 1992) our understanding of US–tissue interactions in trabecular bone is still limited.

Dynamic measurement of the velocity of ultrasound in bone provides an *in vivo* determination of a "quality" parameter (Reich *et al.*, 1976; Greenleaf, 1986; O'Phir, 1990; Kontonassions and O'Phir, 1987). The quality parameters considered are associated with the viscoelastic and strength characteristics of bone and the modulus of elasticity reflects the composite behavior of the various elements of bone. However, the more important implication of the ultrasonic technique is the possibility of performing *in vivo* clinical studies of these bone parameters to diagnose or evaluate the extent of bone disease (Craven *et al.*, 1973). This parameter has not been utilized in diagnosis of pathological changes, though, because in soft tissues the changes are only of the order of several percent.

### *7.17.2 Stress Wave Propagation in Bone and its Clinical Use*

The diagnosis of fractured bone, healing bone and bone in osteoporotic condition is still an important subject of investigation for orthopedic surgeons for the following reasons

(Kenwright, 1985):

1. The mechanical integrity of fractured bone, healing bone and osteoporotic bone may be very important in patients returning to normal work. Patients in these categories who have sustained tibial or femoral diaphyseal fractures may be at risk of re-fracture. Assessment and accurate estimation of bone fracture is also necessary to study the rate of fracture healing and for the correction procedures, in acute and severe fractures involving multiple bones and where the bone splits into two parts.
2. In certain long bone shaft fractures, the healing process is modified by the method of treatment so that clinical assessment of mechanical integrity is impossible and the interpretation of radiographs may be difficult.
3. Severe tibial fractures treated using external skeleton fixation may also cause difficulty in defining when healing is mechanically sound so as to remove fixation and allow free weight bearing.

A combination of these and the presence of osteoporosis make diagnosis of fracture union a difficult task.

Because of the wide variation in its structure and the variable cortical thickness surrounding it, cancellous bone may not be very suitable for ultrasonic diagnostic purposes. The voids in the cancellous bone result in a considerable attenuation of ultrasonic signal propagating through the specimen, making velocity measurements difficult.

A region of bone is required that can be assumed to be a circular cylinder (Wooten, Judy and Greenfield, 1973). Also, the procedure calls for an appendage that can be positioned with ease and cause minimum discomfort to the patient. In addition, SOS measurements require approximately parallel interfaces of cortical bone for adequate acoustical reflections (Figure 7.21). These requirements are met by using the proximal region of the radius. Also, there is a high correlation between the radial bone mineral content and the femoral neck mineral content (Wilson, 1972).

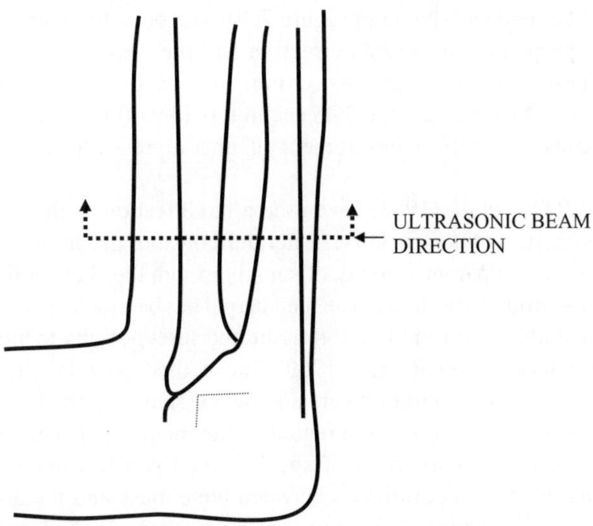


**Figure 7.21** Ultrasonic measurements in the proximal region of the radial bone

### 7.17.3 Measurement of Bone Parameters

Figure 7.20A shows a method of SOS measurement in cadaver bones using a pulse echo method. The time of flight (TOF) is derived from the transmitting pulse trigger and the front and rear boundaries of bone tissue as detected by the transducer pair. Transducer (1) transmits a broad band ultrasonic pulse that is normal to the test object and receives the echo pulse. Transducer (2) is adjusted to receive echoes only after an external trigger. The measurement sites may vary, ranging from the calcaneus of the heel, the cortical bone of the tibia, the proximal phalanges of the fingers, the patella, to the distal radius (Njeh, Boivin and Langton, 1997; Wear, 2000; Strelitzki and Truscott, 1998). The calcaneus is a major concern since, due to its high surface to volume ratio, it has approximately eight times the metabolic turnover rate of cortical bone and would consequently manifest bone metabolic changes before cortical bone (Vogel *et al.*, 1998). Moreover, the calcaneus is easily accessible and the media lateral surface is fairly flat and parallel, making it a favorite test site (Wasnich *et al.*, 1987; Black *et al.*, 1992). These authors have reported that the calcaneus appeared to be the optimal BMD measurement site for routine screening of perimenopausal women to predict fracture risk. The velocity on the calcaneus is highly correlated to the BMD on the femoral neck. However, the correlation coefficients between ultrasound velocity on the calcaneus and the BMD on the heel ranged widely, from 0.34 to 0.72. Several workers (Nejh *et al.*, 2001; Simkin, 1993) have claimed that the cortical bone of the tibia is a more predominant contributor to post-fracture resistance.

Since the shape of the long bone consists chiefly of compact bone, this has also been chosen for measurements by US methods (Singh *et al.*, 1984). Meema, Harris and Porrett (1964) and Meema, Bunker and Meema (1965) have used the proximal end of the radius for *in vivo* bone density measurements. The cortex in the region used is of relatively uniform thickness, making the site of the thickness measurement less critical than at other locations. Changes due to generalized skeletal demineralization are also seen early in the proximal end of the radius (Meema, Harris and Porrett, 1964). This site would therefore be satisfactory for making ultrasonic velocity measurements. The TOF through a test object can be measured from the interval of pulse peaks between two boundaries on pulse echo traces. The time delay measurements can be made as shown in Figure 7.20. The bone thickness can be determined by X-ray imaging perpendicular to this direction and the ultrasonic wave velocity can be calculated from these two measurements. O'Phir and Yazdi (1990) have used transaxial compression techniques to measure the SOS accurately for soft tissue *in vitro*. However, this method cannot be used for the measurement of uncompressible material, such as bone tissue.

Pratt (1980) and Rabin *et al.* (1983) have used a modification of the ultrasonic method to assess the elastic properties of bone in horse. Later, Jeffcott and McCartney (1985) attempted to determine the transverse apparent velocity of sound through the shaft of the third metacarpal bone and to assess its variation with age. The metacarpal has been selected because of its shape, accessible position, uniform surrounding soft tissue and susceptibility to biomechanical injury (Adams, Fee and Kenmore, 1966; Rooney, 1969). The method essentially measures the TOF of an ultrasonic beam between two transducers a known distance apart. It is possible to obtain consistent transmission of sound over the region of the metacarpal shaft that has reasonably parallel sides. The measurements made of specific gravity, ash, calcium and phosphorous content indicate that there is a correlation between bone mass and the apparent velocity of sound. The horse, which exhibited signs of senile osteoporosis, had normal ash although abnormal apparent velocity compared to normal adults. The results of the radiographic optical densitometry of the midcortical metacarpal region follow the same trend and showed increasing velocity of sound with decreasing optical density. This further confirms the trend of increasing velocity with advancing age to maturity.

Human arm is usually chosen as a site for investigation. Having obtained a clear reflection of the cortical bone interfaces, the time delay between the peaks is measured. The arm is then repositioned and the measurements repeated using the average of the two time-delay measurements. Cortical bone thickness measurements at the specific measurement site are made, and the SOS is determined. After this the arm is positioned on the bone mineral analyzer at the same site as the SOS measurement. A plot of the scan across the arm is also obtained.

### *7.17.4 Ultrasound System*

A typical system of this type has two unfocused ultrasound transducers, both having a 1.0 MHz normal center frequency and a $^1/_2$ inch diameter. Transducer (1) is excited by a pulser/receiver to transmit ultrasound pulses and then to receive the echoes bouncing back from the front and the rear boundaries of the bone tissues. Transducer (2) is connected to another pulser/receiver and is in synchronized by an external trigger from a function generator. The received pulse echoes are amplified and then digitized by the A/D conversion board. The data are then stored in a PC for further signal processing.

Transducer (1) is fixed and positioned normal to the tibia shaft. Transducer (2) is adjusted to ensure that it receives the echo signals. Both transducers are packaged in a Plexiglas box filled with coupling medium. To provide a good contact surface with the tibia shaft, a thin polyethylene film is placed at the bottom of the test module (Figure 6.7).

The dual transducer technique is a fairly simple and straightforward method and has the potential to be used in a clinical environment. The major advantage of this ultrasound method is that it acquires measurements of SOS in bone without bone thickness information, which is often not available in *in vivo* measurements. The test site, the tibia shaft, has a flat surface and is superficial to the skin, making it suitable for this *in vivo* approach. The acoustic speed obtained by Chen, Zhao and Mundy (2004) is highly correlated with the BMD from DEXA measurements ($r^2 = 0.93$).

The acoustic speed measured by Chen, Zhao and Mundy (2004) uses the mean velocity of the ultrasonic wave through both the cortical layer and the cancellous layer of the tibia shaft just as the BMD from DEXA measurements also includes information from the whole tibia shaft (cortical and cancellous bone). In this way these authors were able to demonstrate that the acoustic speed measured by this technique has good correlation with BMD. Simkin (1993) has reported that the measured acoustic speed correlates only moderately with BMD values ($r^2 = 0.56$).

The combination of these measurements with a bone model provides the necessary data to compute the compact bone density, bone mineral density, percent ash content and the modulus of elasticity.

### *7.17.5 Procedure for Obtaining Patient Data*

Chen, Zhao and Mundy (2004) have carried out studies to validate a newly developed dual transducer technique applied to human patients. In this method two transducers are utilized on the same side of the test object, one used as transmitter and the other as receiver. The SOS is based on the TOF from the signals received by both transducers and the separation distance between the transducer pair. In this approach the signals received from the two transducers provide the two variables needed to solve the acoustic speed without an assumption of object thickness. Acoustic speed measured by this technique can be compared to the measurement of BMD obtained from DEXA at the same site on the tibia shaft.

### 7.17.6 Analysis of Patient Data

The results of ultrasonic studies suggest that the method can be used in the diagnosis of structural changes in regard to bone health. Significant differences between males and females and patients and normals were observed (Wishko, 1975). The author has reported that when the parameters of elastic modulus and BMC are combined, a separation of approximately 3 : 1 between patients and normals is achieved for both males and females. Using the BMC parameter alone for diagnosing osteoporosis, there existed considerable overlap between normals and patients with clinically significant demineralization.

The same result is also obtained for the bone parameters of BMD and percent ash content. From this, it is concluded that the compact bone density, bone mineral density and percent ash content cannot differentiate between normals and patients and, therefore, these parameters may not be useful in diagnosing the health of the bone structure on a population basis. Wishko (1975) found that the SOS for most female patients less than 45 years of age fell within two standard deviations from the normal mean. Above this age, the speed of sound data deviate markedly from that obtained for the normal female population. It may be that factors associated with menopause are responsible for this dispersion.

A correction is included for the presence of adipose tissue in the marrow region of the bone. The factor effectively increases the observed bone mineral content, due to the inability of the fat to attenuate the photon beam as much as the soft tissue component. This can be calculated by the following equation (Wooten, Judy and Greenfield, 1973):

$$\text{Error} = \frac{0.9}{1.2}\left(\frac{r_i^2}{r_o^2 - r_i^2}\right) k \times 100 \qquad (7.20)$$

where $r_\mathrm{i}$ and $r_\mathrm{o}$ are the inner and outer radius of the hollow cylinder, respectively, and $k$ depends upon the energy of the radioactive source. This correction will amount to an increase of approximately 1.0–7.0% in the BMC in the proximal radius.

Several previous studies of human bone have suggested that the BUA–BMD correlation is linearly positive (Bouxsein and Radloff, 1997; Chappard *et al.*, 1997; Lees and Stevenson, 1993; Njeh, Boivin and Langton, 1997; Williams, 1992; Yeap *et al.*, 1998; Zagzebski *et al.*, 1991). In contrast, the results of Hodgskinson *et al.* (1996) and Serpe and Rho (1996) reveal that, in high-density cancellous bovine bone, the correlation between BUA and $\mathrm{BMD_{vol}}$ turns out to be negative.

US frequency dependent attenuation in viscoelastic and porous biological tissue, such as trabecular bone, is a highly complex physical problem. There are two essential mechanisms related to the attenuation of ultrasound in porous media: absorption, related to bulk composition, and scattering, related to spatial composition, such as pores (Haire and Langton, 1999). Several theoretical approaches have been developed to solve the problem (Biot, 1956;

McKelvie and Palmer, 1991; Williams, 1992). Biot theory considers attenuation to be due to absorption, that is, frictional and viscous forces opposing the motion of bone marrow, but ignores scattering (Biot, 1956; Haire and Langton, 1999; McKelvie and Palmer, 1991). The decrease in BUA with higher $BMD_{vol}$ values may be not due only to lower porosity and smaller pore size.

## *7.17.7 Verification of the In Vivo Bone Parameters*

The bone model employs two basic assumptions: (i) the density of the non-osseous content (collagen, cells and other organic tissues) is unity and (ii) the geometry of the proximal segment of the radius can be approximated as a circular cylinder. With these assumptions, the basic constitution of the cortical bone, the bone mineral content (BMC) and the non-osseous content are estimated. The three parameters can be expressed in terms of BMC/$A$ and the ratio $M_{coll}$/BMC as:

$$P_{cb} = \frac{\mathrm{BMC}}{A}\frac{\mathrm{BMC}}{(1 + M_{coll})}$$

$$\%\,\mathrm{Ash} = \frac{1}{(1 + M_{coll}/\mathrm{BMC})} \tag{7.21}$$

$$E = C^2\frac{\mathrm{BMC}}{A}(1 + M_{coll}/\mathrm{BMC}) \tag{7.22}$$

where $P_{cb}$ = compact bone density, $A$ is the total cross sectional area of the cortical bone, $M_{coll}$ is the mass of the collagen per running cm of cortical bone and $C$ is the speed of sound.

$M_{coll}$ is derived from the bone model and is computed by the relation:

$$M_{coll} = P_{coll}\frac{(A - \mathrm{BMC})}{3.2} \tag{7.23}$$

$$M_{coll} = P_{coll}\frac{[\pi(C_t)(B_w - T)\text{-}\mathrm{BMC}]}{3.2}$$

where $A$ = total cross sectional area of the cortical bone ($cm^2$); $P_{coll}$ = the density of collagen; $C_t$ = the cortical thickness; $B_w$ = the bone width, 3.2 = density of bone.

The major sources of error are the bone width and the cortical thickness. Because of the irregularities of the inner cortical surface, the ability to obtain an accurate measurement of the cortical thickness at the point of measurement is somewhat uncertain. The cortical thickness varies considerably with the azimuthal angle. This emphasizes the need for accurate positioning of the arm between the speed-of-sound apparatus and the radiograph so that the thickness measured from the radiograph accurately represents the distance traveled by the speed-of-sound pulse. The data indicate that the error produced in measuring the cortical thickness from the radiograph as compared to the actual cortical thickness involves an average positive bias of about 10% when the cortical thickness is measured using a magnifier.

Broadband ultrasound attenuation (BUA) may be more related to loss of connectivity as it may result in less frequency dependency of attenuation. Studies will be needed that match

BMD and trabecular thickness. Whether SOS is more dependent on bone mineral mass than connectivity demands studies of specimens not only differing in mineral density but also in connectivity (Parfitt, 1992; Salamone *et al.*, 1994; Ligrand *et al.*, 2000; Kaulsson *et al.*, 2001). However, it may not be easy to compare connectivity patterns with quantitative perfection.

Men lose trabecular bone mainly by trabecular thinning, whereas it occurs mainly by trabecular perforation and loss of connectivity in women (Aaron, Makins and Sagreiya, 1987; Parfitt *et al.*, 1983). Morphological parameters determined by the use of histomorphological techniques have proven to give a good prediction for the mechanical properties of cancellous bone (Hodgskinson and Currey, 1990; Parfitt, 1987). Nevertheless, other structural factors cannot be excluded.

Quantitative ultrasound (QUS) values are no different in young men and young women matched by femoral neck BMD, suggesting that there were no general differences in the structural property of bone at skeletal maturity. The data are consistent with the findings that peak trabecular number thickness and connectivity are similar in men and women (Aaron, Makins and Sagreiya, 1987). However, gender differences in QUS parameters were found in men and women with fractures and in elderly men and women with low BMD. Women with fractures had 10% lower BUA and an 18% lower stiffness index (SI) than men.

Several authors (Chaffai *et al.*, 2002; Laugier *et al.*, 1997; Roberjot *et al.*, 1996; Wear and Garra, 1998) have proposed ultrasound backscatter as a quantitative index for non-invasive assessment of microarchitecture within trabecular bone. Several groups have reported the anisotropy of ultrasonic parameters, reflecting the anisotropy of bone microstructure and biomechanical properties (Gluer, Wu and Genant, 1993; Hans *et al.*, 1999; Nicholson, Haddaway and Davie, 1994; Njeh, Boivin and Langton, 1997). Thus one would expect ultrasonic properties to be related to the orientation of the incident beam axis with respect to the trabecular network, as observed when measurements are performed along the orthogonal axis of the bone. Nicholson, Haddaway and Davie (1994) also observed that the degree of anisotropy (DA) is a significant independent predictor of ultrasonic properties in human vertebrae although they could not demonstrate the effect of DA in the transverse direction.

Backscatter is essentially a structure-related parameter and is correlated with density and microarchitecture of human calcaneal bone samples, but provided no strong incremental information on bone microstructure after adjustment for bone quality. Frediani *et al.* (2003) have found that the stiffness index, which is indicative of SOS and BUA, increased or is unchanged in a clodronate treated (bisphonate) group. This indicates that the quality of bone remains unmodified, whereas it declined significantly in the calcium plus vitamin D group (−8.06% in the os calcies).

The speed and attenuation of ultrasonic waves depend on density as well as on certain other properties of bone. According to one study (Waud, Lew and Baran, 1992) only 53% of the attenuation value and 44% of the speed value can be accounted for by density. Alves *et al.* (1996) showed that the acoustic speed is highly correlated with bone density measured by dual energy X-ray absorptiometry (DEXA) and might be able to provide information on bone quality for more accurate estimates of bone fracture risk. Quantitative ultrasound (QUS) predicts fragility fractures independent of BMD, whereas combined DEXA and QUS can improve fracture prediction (Bauer *et al.*, 1997; Gluer *et al.*, 1994; Gonnelli *et al.*, 1995; Hans *et al.*, 1996). These observations and several other studies (Gluer, Wu and Genant, 1993; Hans *et al.*, 1998; Nicholson *et al.*, 1998; Nicholson, Haddaway and Davie, 1994; Njeh, Boivin and

Langton, 1997; Njeh *et al.*, 1996; Strelitzki, Nicholson and Paech, 1998) have suggested that QUS may be a measure of bone fragility that also reflects structural properties of bone.

It is safe to conclude that ultrasonic techniques have considerable potential for the measurement of bone properties. QUS can be used in evaluation of the density and thickness of the cortex and this may assist in quantification of the degree of osteoporosis.

## 7.18 Other Techniques

Application of a vibrational technique to the problem of bone densitometry utilizes the ulna – an ideal bone for study because of its easy accessibility at each end (Jurist, 1970a). The product of resonant frequency ($F_o$) and length of long bone ($L$) is proportional to the speed of vibratory propagation: $F_oL = KC$, where $K$ depends on the mode of vibration and boundary conditions, and $C$ is the speed of sound. This relationship may be exploited in study of long bones, because $C = \sqrt{P/\rho}$ where $P$ is the average value of Young's modulus over the cross section of the bone and $\rho$ is its average density. Experimentally, $F_o$ may be obtained by recording the acceleration response of the head of the ulna as a function of driving frequency for excitation applied to the elbow (Olecranon process). $L$ is determined with a metric rule; small changes in cortical thickness may markedly affect bone elasticity and hence the vibratory properties.

Jurist (1970b) found that $F_oL$ increases from age 6 to about 15–20 and then decreases. The decrease after about 20 years of age occurs at a relatively constant rate in men until at least 80 years. The $F_oL$ of women decreases in the interval 25–55 years at roughly the same rate as for men. After 55 years, however, $F_oL$ decreases very rapidly (about 1% year) in women. It has been further shown that the mean $F_oL$ for osteoporotic women is significantly lower than the mean $F_oL$ of age-matched controls. The $FL$ distribution for clinically normal women more than 45 years of age suggests that about 35% of these women may be developing osteoporosis. Measurement of ulnar $F_oL$ has shown about 82% discrimination of women with symptomatic osteoporosis from their age-matched normal. The vibratory properties have also been correlated with material properties such as mineral density and porosity (Pelker and Saha, 1983). Measurement of ulnar $F_oL$ seems to be a promising approach to follow the development of osteoporosis in aging or immobilized population on a long-term basis.

### *7.18.1 Magnetic Resonance Imaging (MRI)*

Methods utilizing X-rays essentially measure bone density, while those based on ultrasound make use of velocity and attenuation measurements. Clearly, from this, BMD alone is not a good predictor of fracture risk. To provide mechanical stability, microhardness is another factor that defines patterns of bone alterations with aging and pathology. Microarchitecture is important in understanding the mechanisms of bone fragility as well as the action of drugs, which could possibly prevent osteoporotic fractures (Dalli Carbonare and Giannini, 2004).

In the recent past, magnetic resonance imaging (MRI) has emerged as a means of measuring trabecular bone structure *in vivo*. MRI applies principles of magnetism to create images reflecting the dynamic behavior of water and lipid protons in the tissues. Since bone (cortical and trabecular) does not contain appreciable protons, it provides low signal intensity, compared with soft tissues. The presence of a network in trabecular bone and its impact on marrow relaxation time is utilized in assessing trabecular bone properties using quantitative MR. The pore spaces of trabecular bone are filled with hematopoietic (red) or fatty (yellow) marrow. MRI produces a positive image of the interstitial space between trabeculae and a negative image of the bone matrix. This is directly correlated to the bone porosity (Hong *et al.*, 2000).

#### 7.18.1.1 Theory, Experimental Design and Results

The use of MRI is based on the fact that there is an abundance of hydrogen atoms: nearly two-third of the body's hydrogen atoms are found in water and fat molecules. In our bodies, the nuclei of hydrogen atoms (protons) normally are randomly oriented. In the presence of a magnetic field in a MRI chamber, the nuclei line up in parallel formation (rows of tiny magnets). With the application of a pulse of radio waves from the MRI machine the parallel alignment gets disturbed. As they fall back into alignment they absorb energy from the RF signal (generator radio frequency oscillator, RFO) and in the process of transition produce a detectable radio signal. A receiver coil (radio frequency detector, RFD) is used, which is tuned to the same RF frequency as the exciter coil (Figure 7.22). The signal is recorded and transferred to the computer. The difference in magnetic susceptibility between trabecular bone matrix and bone marrow results in a distortion in the magnetic lines of force in the main static magnetic field inhomogeneities in the tissue (at the interface of the bone and marrow) and results in an enhanced decay of signal intensity in gradient-echo images (Gugliemi *et al.*, 1995). The diffusion of water (protons) in these magnetic field inhomogeneities results in an

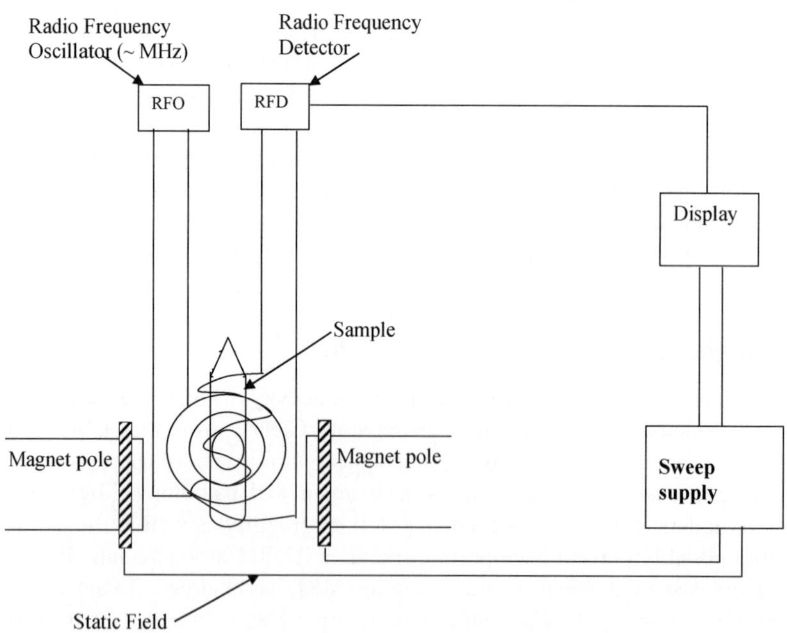


**Figure 7.22** Diagram of an NMR spectrometer. The display is a cathode ray oscilloscope; RFO, radio frequency oscillator; RFD, radio frequency detector. Protons (hydrogen atoms) absorb energy from the radio frequency field

irreversible loss of magnetization. This also causes a reduction in transverse relaxation type ($T_2$). The effect is dependent on magnetic field strength and is greater at higher magnetic fields. The reduction in the relaxation time is dependent on trabecular bone density and its spatial distribution. Thus, in a normal dense trabecular network $T_2$ shortening should be more pronounced than in rare field osteoporotic trabeculae. The nature and strength of the signal is of the tissue type. Since bone is deficient in hydrogen, it appears black. In contrast, tissues containing large amount of hydrogen (e.g., brain) produce a bright image. MRI produces high quality, cross sectional pictures of the part of the body being studied. Each picture represents a virtual slice through that part of the body.

The signal amplitude decays as the net magnetization gradually realigns with the magnetic field. The signal also decays as processing spins lose coherence, thus reducing net magnetization. There are two relaxation times: $T_1$ (time constant exponential decay of signal after excitation phase). $T_2$ measures how long a tissue maintains its signal.

A MRI scanning machine consists of a large donut-shaped magnet with a sliding scanning table. In a typical measurement cylindrical specimens of bone from the vertebral bodies of whales have been used (Hong *et al.*, 2000). MRI examinations were performed using a 1.5 T field and a scanner, with a quadrature head coil. To homogenize signal intensity differences between red and yellow marrow, a proton density sequence is used. Homogenous specimens (8 mm long segment, with marrow intact) are submerged in corn oil and degassed in a vacuum chamber. Bone density is measured, in a random pickup, using a MRI scan. These authors also performed experiments in *in vivo* (human volunteers, male and female) and reported a high correlation between MRI and QCT measurements of ash density for a 10 mm diameter trabecular region of the distal tibia of human volunteers.

Quantitative MRI-based measurements of bone density are facilitated by a proton density sequence that provides similar signal intensities from both red and yellow marrow. Conversion of red into yellow marrow within long bones of the appendicular skeleton reaches nearly 100% by age 25 (Vogler and Murphy, 1988). Although the proportionality of water, fat and protein differ in red and yellow marrow, the density of the protons is about the same.

Bone marrow is considered as a liquid and has an interface with the solid trabecular surface. The interaction between the two can be attributed to the bonding and dipole–dipole interactions relating to the molecular behavior of the liquid that adheres to or is close to the solid interfaces. The changes in molecular motion that are manifest in MR as a modification of $T$ (relaxation times) are particularly relevant at high field strengths. The modified $T_1$ depends on the surface area to volume ratio of this solid–liquid interface, and increases at high magnetic field strengths. This tends to increase when the number of bone and marrow interfaces increase, that is, as bone density increases. $T_2$ relaxation times are governed by a similar mechanism to $T_1$, but are also affected by processes such as diffusion.

Several authors (Fransson, Grampp and Imhof, 1999; Ford and Wehrli, 1991; Wehrli *et al.*, 1991; Majumdar and Genant, 1992; Sugimoto, Mimura and Ohasawa, 1993) have concluded (*in vivo*) that there is a significant correlation between BMD and the relaxation rate. The same study has also been undertaken *in vitro* (Kong *et al.*, 1999; Takahashi *et al.*, 2000; Brismar *et al.*, 1999). In this approach specimens of vertebral bodies have been immersed in saline, an immersion of peanut oil and water to replicate marrow (Majumdar *et al.*, 1991). To measure susceptibility mediated effects on marrow relaxation times $T$, $T_2$ and $T_2^*$ experiments have been conducted at 1.5 T. Standard single energy quantitative CT (QCT) measures of BMD are than correlated with measures of $T_1$, $T_2$ and $T_2^*$. Remy and Guillot (1998) have demonstrated a dependence of $T_2$ on bone mineral density. Field inhomogeneities, both RF and static, and duration pose the greatest obstacle in measuring bone properties with MRI.

Gugliemi *et al.* (1994) have shown regional variation in $1/T_2$ in the calcaneus, reflecting the known variations in $1/T_2$ in calcaneal BMD and structure. The use of MR microscopy may be an additional MR-based technique to study trabecular microarchitecture in a quantitative manner.

Inhomogeneities in the magnetic field may also cause a reduction in relaxation time ($T_2$). Because of this, it can emerge as a good supplement for standard BMD methods for assessing osteoporosis and evaluating longitudinal changes. Developments in high-resolution MR imaging techniques have opened up new perspectives for the characterization of trabecular bone architecture by non-invasive methods. In fact a fractal dimension (Dr) has been proposed and used as a suitable tool to measure complexity variation of most biomedical functions and structures during aging and disease (Lipstiz and Goldberger, 1992; Vaillancourt and Newell, 2002; Lipstiz, 2004). MR imaging is ideally suited for analyzing the texture of trabecular bone because bone marrow has a uniform high signal intensity while bone appears with background intensity.

MRI does not use ionizing radiations and is more sensitive to changes in tissue pathology. MRI can also directly make coronal, sagittal and transverse cross sectional slices at equal resolution with the patient in a single position, making it convenient to obtain comprehensive geometric information without repositioning the patient.

Lacunarity (lack or hole) is another fractal property, describing the texture of a fractal and can measure fractal space filling capacity (Zaia *et al.*, 2005). Lacunarity refers to gap distribution in a fractal (Mandelbrot, 1982, 1983). This has been introduced to describe random and fractal spatial patterns (Plotnick *et al.*, 1996; Smith, Lange and Marks, 1996). Lacunarity describes the most suitable tool to characterize trabecular bone texture. It is able to describe discontinuity of the bone network as well as sizes of bone marrow spaces, changes of which are an index of increased fracture risk. These studies (Zaia *et al.*, 2005) deal with the application of lacunarity analysis to trabecular bone in vertebra from MR image. This then provides a method to assess trabecular bone architecture. The method needs more quantification, probably with more human data. Clinical application on humans for lacunarity analysis has been reported for calcaneus CT images and micro-radiography images of bone biopsies (Dougherty and Henebry, 2002; Chappard *et al.*, 2001).

## 7.19 Relative Advantages and Disadvantages of the Various Techniques

The type of non-invasive measurement of bone mass to be selected for a specific study depends to a large extent on the nature and the type of data required. Measurements of bone mass generally are employed in two basic type of studies, longitudinal and cross sectional. Large-scale cross-sectional studies of osteopenia are easier and less expensive to conduct if bone mass can be ascertained by the relatively simple radiogrammetric and densitometric measurement of the metacarpals or the radius than by the more sophisticated neutron activation methods.

Each of the presently available non-invasive methods used to determine the nature and degrees of changes in skeletal metabolism has significant advantages and limitations. Radiographic techniques are applicable to all parts of the entire skeleton, but lack the sensitivity required for quantifying levels of changes associated with the development of pathological conditions. Radiogrammetry, which measures cortical thickness of the appendicular skeleton, is relatively precise, but not necessarily indicative of the status of the axial skeleton. Photon absorptiometric techniques, although highly quantitative and precise, also provide information on a highly localized portion of the appendicular skeleton. Extrapolation of the findings to the axial skeleton cannot be assumed to be valid for all individuals.

Computed tomography offers significant potential advantages. DPA offers measurements at a lower radiation dose than CT and is less affected by crush fractures, so it may be preferred, particularly in serial studies in young subjects (Sambrook *et al.*, 1985). However, CT may offer better diagnostic discrimination between osteoporotic and normal subjects. These two methods show a reasonable correlation in a cross sectional study but comparative longitudinal studies are required to assess their usefulness for predicting bone loss in clinical work.

Many other currently available diagnostic modalities (e.g., peripheral DEXA, single-energy X-ray absorptiometry, radiographic absorptiometry and single- and dual-photon absorptiometry) suffer from the same shortcomings as DEXA since they bear the mark of the same technology (Mittra, Rubin and Qin, 2004). Other modalities such as quantitative computed tomography (QCT) can differentiate between cortical and trabecular bone (which DEXA can not) but, because of its limitation on resolution, can not provide very detailed microstructural information (Brismar, Budinsky and Majumdar, 2001).

The advantage of the Compton scattering technique is that the volume can be located entirely within weight bearing trabecular bone. Repositioning of the scattering volume is relatively simple, hence the measurement is not a function of the thickness of the bone nor of its possible rotation. Thus, the precision of the method is very high. This technique has great potential for bone density measurements. However, problems to be resolved are the evaluation of the contribution of photons that have undergone multiple scattering. In this technique, the precision depends largely on the size of the volume measured. To date, it has not been possible to separate clearly osteoporotic individuals from normal on the basis of the density measurements of their trabecular bone.

With neutron activation analysis, calcium may be measured in a portion of the body (e.g., the hand, the torso) or the entire body. The low radiation dose to the subject allows the study of large normal populations as well as osteopenic populations. The precision of the technique also makes it useful as a sensitive indicator of overall therapeutic efficiency. An obvious disadvantage of this technique is that it is incapable of revealing whether a change is confined to a small volume of the body. Small changes in bone mass at a particular site are masked by the large amount of calcium in the total body. Also this method cannot distinguish between the skeletal calcium and ectopic calcification, a condition that frequently occurs in renal osteodystrophic patients.

Ultrasonic methods show some promise owing to their simplicity, though the information contents are limited and need more quantification. This technique needs to be developed and more work is required before it becomes a part of routine clinical practice. Wright, Glade and Gopal (1987) have shown that ultrasound velocity is a potential method for assessing the bone strength and mineralization in human newborns. It is apparent from Figure 7.23 that as the density of cadaver bone increases the ultrasonic velocity also increases linearly and hence it can be a reliable parameter.

Finally, screening for osteoporotic treatment is a major future task. The diagnostic procedures are expensive and repetitive measures are quite demanding: A certain population type may be targeted with some allowance for individual variation. As of now there are no known risk factors (or combination) that can be identified as likely parameters for the diagnosis

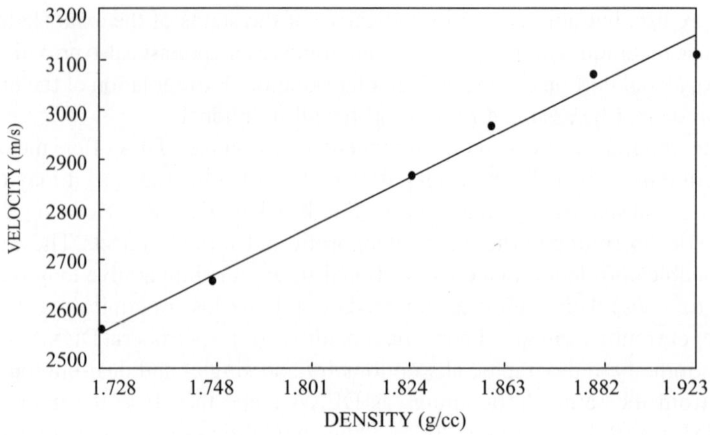


**Figure 7.23** Velocity of ultrasound as a function of density of cadaver bone of an aged human patient (Singh and Behari, 1994)

of osteoporosis. Various decision making indications are those of the National Osteoporosis Foundation (NOF) (1999), simple calculated osteoporosis risk assessment (SCORE) (Lydick *et al.*, 1998) and osteoporosis risk assessment instrument (ORAI) (Morley, Whitfield and Willick, 2001). SCORE and ORAI have been evaluated as better than NOF (White, 2002). With these markers available, the identification of subjects for osteoporotic screening may be made easier.

# References


Aaron, J.E., De Vernejoul, M.C. and Kanis, J.A. (1991) The effect of sodium on trabecular architecture. *Bone*, **12**, 307–310.

Aaron, J.E., Francis, R.M., Peacock, M. and Makins, N.B. (1989) Contrasting microanatomy of idiopathic and corticosteroid induced osteoporosis. *Clinical Orthopaedics and Related Research*, **243**, 294–305.

Aaron, J.E., Makins, N.B. and Sagreiya, K. (1987) The microanatomy of trabecular bone loss in normal aging men and women. *Clinical Orthopaedics*, **215**, 260–271.

Aaron, R.K., Boyan, B.D., Ciombor, D.M. *et al.* (2004) Stimulation of growth factor synthesis by electric and electromagnetic fields. *Clinical Orthopaedics*, **419**, 30–37.

Aaron, R.K. and Ciombor, D.M. (1997) Role of electrical stimulation in the treatment of a vascular necrosis of the femoral head. *Current Orthopaedics*, **11**, 187–192.

Aaron, R.K., Ciombor, D.M. and Jolly, G. (1989) Stimulation of experimental endochondral ossification by low energy pulsing electromagnetic fields. *Journal of Bone and Mineral Research*, **4**, 227–233.

Aaron, R.K., Ciombor, D.M. and Simon, B.J. (2004) Treatment of non unions with electric and electromagnetic fields. *Clinical Orthopaedics and Related Research*, **419**, 21–29.

Abdalla, H.I., Hard, D.M. and Lindsay, R. (1985) Prevention of bone mineral loss in postmenopausal women by norethysterone. *Obstetrics and Gynecology*, **66**, 789–792.

Abendschein, W.F. and Hyatt, G.W. (1970) Ultrasonics and selected physical properties of bone. *Clinical Orthopaedics and Related Research*, **69**, 294–301.

Abendschein, W.F. and Hyatt, G.W. (1971) Ultrasonics and the physical properties of healing bone. *Journal of Trauma*, **12**, 297–301.

Adair, R.K. (1991) Constraints on biological effects of extremely low frequency electromagnetic fields. *Physical Review A*, **43**, 1038–1049.

Adami, S., Bhalla, A.K., Dorizzi, R. *et al.* (1987) The acute phase response after bisphosphonate administration. *Calcified Tissue International*, **41**, 326–331.

Adami, S., Main, M., Gathi, P. *et al.* (1994) Effects of two oral doses of alendronate in the treatment of Paget's disease of bone. *Bone*, **15**, 415–417.

Adams, J.E., Chen, S.Z., Adams, P.H. and Isherwood, I. (1982) Measurement of trabecular bone mineral by dual energy computed tomography. *Journal of Computer Assisted Tomography*, **6**, 601–607.

Adams, J.P., Fee, N. and Kenmore, P.I. (1966) Tear-gas injuries. A clinical study of hand injuries and an experimental study of its effects on peripheral nerves and skeletal muscles in rabbits. *The Journal of Bone and Joint Surgery. American Volume*, **48** (3), 436–442.

Adams, N.D., Gray, R.W. and Lemann, J. (1979) The calciuria of increase fixed acid production in humans: evidence against a role for parathyroid hormone and 1,25 $(OH)_2$ vitamin D. *Calcified Tissue International*, **28**, 233–238.

Adey, W.R. (1984) Nonliner, nonequilibuium aspects of electromagnetic field interaction at cell membranes, in *Nonlinear Electrodynamics in Biological Systems* (eds W.R. Adey and A.F. Lawrence), Plenum Press, New York, p. 3.

Adey, W.R. (1986) The sequence and energetics of cell membrane transducing coupling to intracellular enzyme systems. *Bioelectrochemistry and Bioenergetics*, **15**, 447–456.

Adey, W.R. (1988) Cell membranes: The electrochemical environment and cancer promotion. *Neurochemical Research*, **13**, 671.

Agren, M., Karellas, A., Leashey, D. *et al.* (1991) Ultrasound attenuation of the calcaneus: A sensitive and specific discriminator of osteopenia in postmenopausal women. *Calcified Tissue International*, **48**, 240–244.

Aitken, J.M. (1984) Relevance of osteoporosis in women with fracture of the femoral neck. *British Medical Journal*, **288**, 287–601.

Aitken, J.M., Hart, D.M. and Lindsay, R. (1973) Oestrogen replacement therapy for prevention of osteoporosis after oophorectomy. *British Medical Journal*, **3**, 515–518.



*Biophysical Bone Behavior* Jitendra Behari

Akar, S., Sivrikaa, H., Canikli, A. and Varoglu, E. (2002) Lateralized mineral content and density in distal forearm bones in right handed men and women: relation of structure to function. *The International Journal of Neuroscience*, **112**, 301–311.

Akiyama, Y., Hara, K., Kobayashi, M. *et al.* (1999) Inhibitory effect of vitamin $K_2$ (menatetrenone) inhibits bone resorption in ovariectomized rats: A histomorphometric and dual energy X-ray absorptiometric study. *Japanese Journal of Pharmacology*, **80**, 67–74.

Albanese, A.A., Edelson, A.H., Woodhull, M.L. *et al.* (1973) Effect of a calcium supplement on serum cholesterol, calcium, phosphorous and bone density of "normal healthy"elderly females. *Nutrition Reports International*, **8**, 119–130.

Albert, J.L., Boyle, J.P., Roberts, J.A. *et al.* (1997) Regulation of brain capillary endothelial cells by P2Y receptors coupled to $Ca^{2+}$, phospholipase C and mitogen-activated protein kinase. *British Journal of Pharmacology*, **122**, 935–941.

Alberts, B., Bray, D., Lewis, J. *et al.* (1983) *Molecular Biology of the Cell*, Garland, New York, p. 752.

Albright, F., Bloomberg, E. and Smith, P. (1940) Postmenopausal osteoporosis. *Transactions of the Association of American Physicians*, **55**, 298–305.

Albright, F., Smith, P.H. and Richardson, A.M. (1941) Postmenopausal osteoporosis. *The Journal of the American Medical Association*, **116**, 2465.

Alffram, P.A. and Bauer, G.C.H. (1962) Epidemiology of fractures of the forearm. A biochemical investigation of bone strength. *The Journal of Bone and Joint Surgery*, **44**, 105–114.

Al-Hiti, K., Thomas, B.J., Al-Tikrity, S.A. *et al.* (1976) Spinal calcium: Its *in vivo* measurement in man. *International Journal of Applied Radiation and Isotopes*, **27**, 97–102.

Ali, R. and Evans, J.L. (1973) Lactose and calcium metabolism: a review. *Journal of Agriculture of the University of Puerto Rico*, **57**, 149–164.

Ali, R., Staub, H., Coccordrilli, Jr G. and Schaubacher, L. (1981) Nutritional significance of dietary fiber: effect on nutrient bioavailability and selected gastrointestinal functions. *Journal of Agricultural and Food Chemistry*, **29**, 465–472.

Alkhiary, Y.M., Gerstenfeld, L.C., Krall, E. *et al.* (2005) Enhancement of experimental fracture healing by systemic administration of recombinant human parathyroid hormone (PTH 1–34). *The Journal of Bone and Joint Surgery. American Volume*, **87**, 731–741.

Allen, F.D., Pollack, S.R., Gupta, R. and Hung, C.T. (1994) The intracellular calcium response of osteoblast – like cells to 60 KHz electric field stimulation from 10 V/m to 1 KV/m. Proceedings of the 14th Annual Meeting of the Society for Physical Regulation in Biology and Medicine, Washington D.C., October 13, p. P1.

Allen, L.H. (1985) Evidence for a role for insulin in the hypercalciuretic response to macronutrients, in *Nutritional Bioavailability of Calcium*, ACS Symposium Series, American Chemical Society, Washington D.C., No. 275, pp. 117–124.

Allen, L.H., Bartless, R.S. and Bock, G.D. (1979a) Reduction of renal calcium reabsorption in man by consumption of dietary protein. *The Journal of Nutrition*, **109**, 1345–1350.

Allen, L.H., Block, G.D., Wood, R.J. and Bryce, G.F. (1981) The role of insulin and parathyroid hormone in the protein induced calciuria of man. *Nutrition Research*, **1**, c–11.

Allen, L.H., Oddaye, E.A. and Margen, S. (1979b) Protein induced hypercalciuria: a longer term study. *The American Journal of Clinical Nutrition*, **32**, 741–749.

Aloia, J.F., Cohn, S., Ostuni, J. *et al.* (1978) Prevention of involutional bone loss by exercise. *Annals of Internal Medicine*, **89**, 356–358.

Aloia, J.F., Zanzi, T., Ellis, K. *et al.* (1976) Effects of growth hormone in osteoporosis. *The Journal of Clinical Endocrinology and Metabolism*, **43**, 992–999.

Alverez, R.E. and Macovski, A. (1976) Energy selective reconstructions in X-ray computerized tomography. *Physics in Medicine and Biology*, **21**, 733–744.

Alves, J.M., Xu, W., Lin, D. *et al.* (1996) Ultrasonic assessment of human and bovine trabecular bone: A comparison study. *IEEE Transactions on Biomedical Engineering*, **43**, 249–258.

Ament, Ch. and Hofer, E.P. (2000) A fuzzy logic model of fracture healing. *Journal of Biomechanics*, **33**, 961–968.

Amin, S. and Felson, D.T. (2001) Osteoporosis in men. *Rheumatic Disease Clinics of North America*, **27**, 19–47.

Amling, M. and Delling, G. (1996) Cell biology of osteoclasts and molecular mechanisims of bone resorption. *Pathologi*, **17**, 358–367.

Anast, G., Fields, T. and Siegal, I. (1958) Ultrasonic technique for the evaluation of bone fractures. *American Journal of Physical Medicine*, **37**, 157–159.

Anderson, F.H. and Cooper, C. (1999) Hip and vertebral fractures, in *Osteoporosis in Men: The Effects of Gender in Skeletal Health* (ed. E.S. Orwell), Academic Press, New York, pp. 29–50.

Anderson, H.C., Hodges, P.T. and Moylan, P.E. (1999) Bone morphogenetic protein expression by cells of human and rat growth plate, metaphysis and articular cartilage. *Journal of Bone and Mineral Research*, **14** (Suppl 1), 5309.

Anderson, J., Osborn, S.B., Tomlinson, R.W.S. *et al.* (1964) Neutron activation analysis in man *in vivo*. *Lancet*, **2** (7371), 1201–1205.

Anderson, J.B., Shimmins, J. and Smith, D.A. (1966) A new technique for the measurement of metacarpal density. *British Journal of Radiology*, **39**, 443–450.

Anderson, J.C. and Eriksson, C. (1968) Electrical properties of wet collagen. *Nature*, **218**, 166–168.

Anderson, J.C. and Eriksson, C. (1970) Piezoelectric properties of dry and wet bone. *Nature*, **227**, 491–492.

Andrabi, W.H. (1978) Electret Phenomena in Bone. M. Phil. Dissertation submitted to JNU, India.

Andrabi, W.H. (1980) Biophysical Parameters affecting Bone Hebavior Ph.D Thesis, Jawaharlal Nehru University, New Delhi, 1–158.

Andrabi, W.H. and Behari, J. (1981a) Formation of bone electrets and their charge decay characteristics. *Biomaterials*, **2** (22), 120–123.

Andrabi, W.H. and Behari, J. (1981b) Bone photovoltaic effect in Hall geometry. *Calcified Tissue International*, **33**, 239–242.

Andrabi, W.H., Rai, D.V. and Behari, J. (1980) Photoelectric effects in bone. *International Journal of Biological Macromolecules*, **2** (5), 306.

Andreassen, T.T., Ejersted, C. and Oxlund, H. (1999) Intermittent parathyroid hormone (1–34) treatment increases callus formation and mechanical strength of healing rat fractures. *Journal of Bone and Mineral Research*, **14**, 960–968.

Andreassen, T.T., Fledelius, C., Ejersted, C. and Oxlund, H. (2001) Increases in callus formation and mechanical strength of healing fractures in old rats treated with parathyroid hormone. *Acta Orthopaedica Scandinavica*, **72**, 304–307.

Andrew, J.G., Hoyland, J., Freemont, A.J. and Marsh, D.R. (1993) Insulin factor gene expression in human fracture callus. *Calcified Tissue*, **53**, 97–102.

Andrew, J.G., Hoyland, J.A., Freemont, A.J. and Marsh, D.R. (1995) Platelet growth factor expression in normally healing human fracture. *Bone*, **16**, 455–460.

Andrew, T., Anitoniades, L., Scurrah, K.J. *et al.* (2005) Risk of wrist fracture in women is heritable and is influenced by genes that are largely independent of those influencing BMD. *Journal of Bone and Mineral Research*, **20**, 67–74.

Angelo, C.M. (1971) The Investigation of an Ultrasonic Technique for Detecting Bone Mineral Loss, Ph.D. thesis, University of California, LA.

Antonacci, M.D., Hanson, D.S. and Hegeness, M.H. (1996) Pitfalls in the measurement of bone mineral density by dual energy X-ray absorptiometry. *Spine*, **21**, 87–91.

Antonson, E.K. and Mann, R.W. (1985) The frequency content of gait. *Journal of Biomechanics*, **18**, 39–47.

Appleby, B. (1962) Norethysterone in the control of postmenopausal symptoms. *Lancet*, **1**, 407–409.

Arai, T., Yokoyama, S. and Tokashiki, M. (1993) 50 L Scale perfusion culture of hybridoma cells by gravitational settling for cell separation, in *Animal Cell Technology: Basic and Applied Aspects*, (eds S. Kaminogawa, A. Ametani and S. Hachimura), Kluwer Academic Publishing, Dordrecht (The Netherlands), pp. 341–346.

Arinoviche, R. *et al.* (1987) Calcitonin in acute pain due to vertebral fractures in osteoporosis. *Revista Medica de Chile*, **115**, 1039–1043.

Arlet, J. (1992) Nontraumatic avascular necrosis of the femoral head: Past, present and future. *Clinical Orthopaedics*, **277**, 12–21.

Armbrecht, H.J. and Wasserman, R.H. (1976) Enhacement of $Ca^{2+}$ uptake by lactose in the rat small intestine. *The Journal of Nutrition*, **106**, 1265–1271.

Armour, K.E., Van't Hof, R.J., Grabowski, P.S. *et al.* (1999) Evidence for a pathogenic role of nitric oxide in inflammation induced osteoporosis. *Journal of Bone and Mineral Research*, **14**, 2137–2142.

Armour, K.J., Armour, K.E., Van't Hof, R.J. *et al.* (2001) Activation of the inducible nitric oxide synthase pathway contributes to inflammation-induced osteoporosis by suppressing bone formation and causing osteoblast apoptosis. *Arthritis and Rheumatism*, **44**, 2790–2796.

Arnett, T.R. and King, B.F. (1997) ATP as an osteoclast regulator? *The Journal of Physiology (London)*, **503**, 236.

Arnold, J.S. (1973) Amount and quality of trabecular bone in osteoporotic and vertebral fractures. *The Journal of Clinical Endocrinology and Metabolism*, **2**, 221–238.

Aro, H.T. and Chao, E.Y.S. (1991) Biomechanics of fracture fixation. In Mow, V.C., Heyes, W.C. (Eds.) Basic Orthopaedic Biomechanics. New York, Raven Press Ltd., 293–336.

Aro, H.T., Wahner, H.T. and Chao, E.Y.S. (1991) Healing patterns of transverse and oblique osteotomies in the cavine tibia under external fixation. *Journal of Orthopaedic Trauma*, **5**, 357–369.

Aro, H.T., Wippermann, B.W., Hodgson, S.F. *et al.* (1989) Prediction of properties of fracture callus by measurement of mineral density using micro-bone densitometry. *The Journal of Bone and Joint Surgery*, **71**, 1020–1030.

Aro, H.T., Wippermann, B.W., Hodgson, S.F. *et al.* (1990) Internal remodeling of periosteal new bone during fracture healing. *Journal of Orthopaedic Research*, **8**, 238–246

Aronson, J. and Harp, J.H. (1990) Factors influencing the choice of external fixation for distraction osteogenesis. *Instructional Course Lectures*, **39**, 175–183.

Aronson, J., Harrison, B., Boyel, C.M. *et al.* (1988) Mechanical induction of osteogenesis. The importance of pin rigidity. *Journal of Pediatric Orthopaedics*, **8**, 396–401.

Asahina, I., Sampath, T.K. and Hauschka, P.V. (1996) Human osteogenic protein-1 induces chondroblastic osteoblastic and/or adipocytic differentiation of clonal murine target cells. *Experimental Cell Research*, **222**, 38–47.

Asahina, I., Sampath, T.K., Nishimura, I. and Hauschka, P.V. (1993) Human osteogenic protein-1 induces both chondroblastic and osteoblastic differentiation of osteoprogenitor cells derived from newborn rat calvaria. *The Journal of Cell Biology*, **123**, 921–933.

Ascenzi, A. (1988) The micromechanics versus the macromechanics of cortical bone – a comprehensive presentation. *Journal of Biomechanical Engineering*, **110**, 357–363.

Ascenzi, A. and Bonucci, E. (1966) The oseon calcification as revealed by the electron microsope, in *Proceedings of the third European Symposium on Calcified Tissue*, (eds H. Fleisch *et al.*), Springer-Verlag, New York, p. 142.

Ascenzi, A. and Bonucci, E. (1968) The compressive properties of single osteons. *The Anatomical Record*, **172**, 499.

Ascenzi, A., Bonucci, E. and Bocciarelli, D.S. (1965) An electron microscope study of osteon clacification. *Journal of Ultrastructure Research*, **12**, 287–303.

Ashhurst, D.E. (1986) The influence of mechanical stability on the healing of experimental fractures in the rabbit: a microscopical study. *Philosophical Transactions of the Royal Society of London. Series B, Biological Sciences*, **313**, 271–302.

Ashman, R.B., Cowin, S.C., Van Buskirk, W.C. and Rice, J.C. (1984) A continuous wave technique for the measurement of the elastic properties of cortical bone. *Journal of Biomechanics*, **17**, 349–361.

Ashton, I.K. and Dekel, S. (1983) Fracture repair in the small dwarf mouse. *British Journal of Experimental Pathology*, **64**, 479–486.

Assailly, J., Mohit, J.D., Goureau, Y. *et al.* (1981) Effect of weak inductively coupled pulsating currents on calcium uptake in embryonic chick tibia explants. *Bioelectrochemistry and Bioenergetics*, **8**, 515.

Assima, C.O. (1968) Wound healing promotion by use of negative electrical current. *The American Surgeon*, **34**, 423–431.

Athenstaedt, H. (1968) Permanent longitudinal electric polarization and pyroelectric behaviour of collagenous structures and nervous tissue in man and other vertebrates. *Zeitschrift fuer Zellforschung und Mikroskopische Anatomie*, **91**, 135.

Athenstaedt, H. (1969) Permanent electric polarization and pyro-electric behavior of the vertebrate skeleton. *Zeitschrift fuer Zellforschung und Mikroskopische Anatomie*, **97**, 537–548.

Athenstaedt, H. (1970) Permanent longitudinal electric polarization and pyro-electric behavior of the collagenous structures and nervous tissues in man and other vertebrates. *Nature (London)*, **228**, 830–834.

Athenstaedt, H. (1979) Pyroelectricity and piezoelectricity properties of vertebrates. *Annals of the New York Academy of Sciences*, **238**, 68–99.

Atkinson, P.J. (1964) Quantitative analysis in cortical bone. *Nature*, **201**, 373.

Atkinson, P.J. (1965) Changes in resorption spaces in femoral cortical bone with age. *Journal of Pathology & Bacteriology*, **89**, 173–178.

Augat, P., Fan, B., Lane, N.E. *et al.* (1998a) Assessment of bone mineral at appendicular sites in females with fractures of the proximal femur. *Bone*, **22**, 395–402.

Augat, P., Gordon, C.L., Lang, T.F. *et al.* (1998b) Accurcy of cortical and trabecular bone measurements with peripheral quantitative computed tomography (PQCT). *Physics in Medicine and Biology*, **13**, 2873–2883.

Augat, P., Merk, J., Genant, H.K. and Claes, L. (1997) Quantative assessment of experimental fracture repair by peripheral computed tomography. *Calcified Tissue International*, **60**, 194–199.

Augat, P., Reeb, H. and Claes, L.E. (1996) Prediction of fracture load at different skeletal sites by geometric properties of the cortical shell. *Journal of Bone and Mineral Research*, **11**, 1356–1363.

Avioli, L.V. (1977) Osteoporosis problem. *Current Concepts in Nutrition*, **5**, 99–103.

Avioli, L.V. (1983) Osteoporosis, in *Bone and Mineral Research Annual* (ed. W.A. Peck), Exercepta Medica, Amsterdam, pp. 280–318.

Avioli, L.V. (1991) Use of calcitonin in the treatment of bone pain associated with osteoporosis. *Calcified Tissue International*, **49**, S9–S13.

Avioli, L.V. and Lindsay, R. (1990) The female osteoporosis syndrome(s), in *Metabolic Bone Disease and Clinically Related Disorders*, 2nd edn (eds L.V. Avioli and S.M. Krane), W.B. Saunders, Philadelphia, PA, pp. 397–451.

Awbrey, B.J., Jaobson, P.C., Grubb, S.A. *et al.* (1984) Bone density in women: A modified procedure for measurement of distal radial density. *Journal of Orthopaedic Research*, **2**, 314–321.

Azria, M. (1989) Endogenous calcitonin, in *The Calcitonin: Physiology and Pharmacology* (ed. M. Azia), Karger, Basel, p. 48.

Azuma, Y., Ito, M., Harada, Y. *et al.* (2001) Low intensity pulsed ultrasound accelerates rat femoral fracture healing by acting on the various cellular reactions in the fracture callus. *Journal of Bone and Mineral Research*, **16**, 671–680.

Azzi, A., Breyer, I., Feher, M. *et al.* (2000) Specific cellular responses to alpha-tocopherol. *The Journal of Nutrition*, **130**, 1649–1652.

Bacabac, R.G., Smit, T.H., Mullender, M.G. *et al.* (2004) Nitric oxide production by bone cells is fluid shear stress rate dependent. *Biochemical and Biophysical Research Communications*, **315**, 83–829.

Bacabac, R.G., Smit, T.H., Mullender, M.G. *et al.* (2005) Initial stress-kick is required for fluid shear stress induced rate dependent activation of bone cells. *Annals of Biomedical Engineering*, **33** (1), 104–110.

Bachman, C.H. and Ellis, E.H. (1965) Fluorescence of bone. *Nature*, **206**, 1328–1330.

Bachner, E.J., Klion, M. and Caruana, J.J. (1987) Ultrasonic effects on fresh fracture repair. A study of two rabbit models. *BRAGS Transactions*, **7**, P3.

Baddeley, A.J., Gundersen, H.J.G. and Crees Orive, L.M. (1986) Estimation of surface area from vertical sections. *Journal of Microscopy*, **142**, 259–276.

Baeksgaard, L., Andersen, K.P. and Hyldstrup, L. (1998) Calcium and Vitamin D supplementation increases spinal BMD in healthy postmenopausal women. *Osteoporosis International*, **8**, 255–260.

Bagger, Y.Z., Tanko, L.B., Alexandersen, P. *et al.* (2004) Two to three years of hormone replacement treatment in healthy women have long term preventive effects on bone mass and osteoporotic fractures: the PERF study. *Bone*, **34**, 728–735.

Bagi, C.M., Hanson, N., Andresen, C. *et al.* (2006) The use of micro-CT to evaluate cortical bone geometry and strength in nude rats: Correlation with mechanical testing, pQCT and DXA. *Bone*, **38**, 136–144.

Bagur, A., Wittich, A., Ghiringhelli, G. *et al.* (1996) Hormone replacement therapy increases trabecular and cortical bone density in osteoporotic women. *Medicina*, **56**, 247–251.

Bailey, A.J. and Knott, L. (1999) Molecular changes in bone collagen in osteoporosis and osteoarthritis in the elderly. *Experimental Gerontology*, **34**, 337–351.

Bailey, A.J., Wotton, S.F., Sims, T.J. and Thompson, P.W. (1993) Biochemical changes in the collagen of human osteoporotic bone matrix. *Connective Tissue Research*, **29**, 119–132.

Bakker, A.D., Soejima, K., Klein-Nulend, J. and Burger, E.H. (2001) The production of nitric oxide and prostaglandin $E_2$ by primary bone cells is shear stress dependent. *Journal of Biomechanics*, **34**, 671–677.

Baldock, P.A., Sainsbury, A., Couzens, M. *et al.* (2002) Hypothalamic Y2 receptors regulate bone formation. *The Journal of Clinical Investigation*, **109**, 915–921.

Baldwin, K.M., White, T.P., Arnaud, S.B. *et al.* (1996) Musculoskeletal adaptions to weightlessness and development of effective countermeasures. *Medicine and Science in Sports and Exercise*, **10**, 1247–1253.

Balena, R., Toolan, B.C., Shea, M. *et al.* (1993) The effect of 2 year treatment with the amino bisphonate alendronate on bone metabolism, bone histomorphometry and bone strength in ovariectomized non union primtes. *The Journal of Clinical Investigation*, **92**, 2577–2586.

Banu, J., Wang, L. and Kalu, D.N. (2000) Age-related changes in bone mineral content and density in intact male F344 rats. *Bone*, **30** (1), 125–130.

Banu, M.J., Orthii, P.B., Mejia, W. *et al.* (1999) Analysis of the effects of growth hormone, voluntary exercise and food restriction on diaphyseal bone in female F344 rats. *Bone*, **25**, 469–480.

Baran, D.T., Faulkner, K.G., Genant, H.K. *et al.* (1997) Diagnosis and management of osteoporosis: guidelines for the utilization of bone densitometry. *Calcified Tissue International*, **61**, 433–440.

Baran, D.T., Kelly, A.M., Karellas, A. *et al.* (1988) Ultrasound attenuation of the os calcis in women with osteoporosis and hip fractures. *Calcified Tissue International*, **43**, 138–142.

Baran, D.T., McCarthy, C.K., Leahey, D. and Lew, R. (1991) Broadband ultrasound attenuation of the calcaneus predicts lumbar and femoral neck density in caucasian women: A preliminary study. *Osteoporosis International*, **1**, 110–113.

Baran, D.T., Teitelbaum, S.L., Bergfeld, M.A. *et al.* (1980) Effect of alcohol ingestion on bone and mineral metabolism in rats. *American Journal of Physiology*, **238**, 507–510.

Barbagallo, M., Carbognani, A., Palummeri, E. *et al.* (1989) The comparative effect of ovarian hormone administration on bone mineral status in oophorectomized rats. *Bone*, **10**, 113–116.

Barbieri, G., Barbieri, C.H., De Matos, P.S. *et al.* (2006) Ultrasonometric evalue of bone healing: experimental study using a model of diaphyseal transverse osteotomy of sheep tibiae. *Ultrasound in Medicine and Biology*, **32**, 875–882.

Barengolts, E.I., Berman, M., Kukreja, S.C. *et al.* (1998) Osteoporosis and coronar atherosclerosis in asymptomatic postmenopausal women. *Calcified Tissue International*, **62**, 209–213.

Barengolts, E.I., Currey, D.J., Bapna, M.S. and Kukreja, S.C. (1993) Effects of endurance exercise on bone mass and mechanical properties in intact and overiectomized rats. *Journal of Bone and Mineral Research*, **8**, 937–942.

Barengolts, E.I., Rosol, T.J., Botsis, J. and Kukreja, S.C. (1990) Comparison of the effects of progesterone, estrogen and medoxyprogesterone on established bone loss in overiectomized aged rats. *Journal of Bone and Mineral Research*, **5**, S246.

Barnes, G.L., Kostenuik, P.J., Gerstenfeld, L.C. and Einhorn, T.A. (1999) Growth factor regulation of fracture repair. *Journal of Bone and Mineral Research*, **14**, 1805–1815.

Barnett, E. and Nordin, B.E.C. (1960) The radiological diagnosis of osteoporosis: A new approach. *Clinical Radiology*, **11**, 166–174.

Baron, R., Tross, R. and Vignery, A. (1984) Evidence of sequential remodeling in rat trabecular bone: morphology, histomorphometry and changes during skeletal mutations. *The Anatomical Record*, **208**, 137–145.

Barrett-Connor, E. (1995) The economic and human costs of osteoporotic nature. *The American Journal of Medicine*, **98**, 3S–8S.

Barth, R.W., Williams, J.L. and Kaplan, F.S. (1992) Osteon morphometry in females with femoral neck fractures. *Clinical Orthopaedics*, **283**, 178–186.

Barushika, A., Yaakobi, T. and Oron, U. (1995) Effect of low energy laser (He-Ne) irradiation on the process of bone repair in the rat tibia. *Bone*, **16**, 47–55.

Bassett, C.A.L. (1962) Current concepts of bone formation. *The Journal of Bone and Joint Surgery*, **44**, 1217.

Bassett, C.A.L. (1965) Electrical effects in bone. *Scientific American*, **213**, 18–25.

Bassett, C.A.L. (1968) Biological significance of piezoelectricity. *Calcified Tissue Research*, **1**, 252–272.

Bassett, C.A.L. (1971) Biophysical principles effecting bone structure, in *Biochemistry and Physiology of Bone*, 2nd edn, vol. **III** (ed. G.H. Bourne), Academic Press, New York, p. 1.

Bassett, C.A.L. (1978) Pulsating electromagnetic fields: A new approach to surgical problems, in *Metabolic Surgery* (eds H. Buchwald and L.V. Richard), Grune and Stratton, Inc, p. 225.

Bassett, C.A.L. (1982a) Pulsing electromagnetic field treatment in ununited fractures and failed arthrodeses. *The Journal of the American Medical Association*, **247**, 623–628.

Bassett, C.A.L. (1982b) Pulsing electromagnetic fields: a new method to modify cell behavior in calcified and noncalcified tissues. *Calcified Tissue International*, **34**, 1–8.

Bassett, C.A.L. (1984) The development and application of pulsed electromagnetic fields (PEMF's) for ununited fractures and arthrodeses. *The Orthopedic Clinics of North America*, **15**, 61–87.

Bassett, C.A.L. (1989) Fundamental and practical aspects of therapeutic uses of pulsed electromagnetic fields (PEMFs). *Critical Reviews in Biomedical Engineering*, **17**, 451–529.

Bassett, C.A.L. (1991) *Fundamental and Practical Aspect of Therapeutic uses of Pulsed Electromagnetic Fields (PEMFGs), Critical Reviews of Bioengineering*, CRC Press, p. 451.

Bassett, C.A.L. and Becker, R.O. (1961) Generation of electrical potentials by bone in response to mechanical stress. *Science*, **137**, 1063–1064.

Bassett, C.A.L. and Becker, R.O. (1962) Generation of electrical potentials in bone in response to mechanical stress. *Science*, **137**, 1063–1064.

Bassett, C.A.L., Caulo, N. and Kort, J. (1981) Congenital pseudoarthroses of the tibia: Treatment with pulsing electromagnetic fields. *Clinical Orthopaedics and Related Research*, **154**, 136–149.

Bassett, C.A.L. and Herrmann, I. (1961) Influence of oxygen concentration and mechanical factors on differentiation of connective tissues *in vivo*. *Nature*, **190**, 460.

Bassett, C.A.L., Mitchell, S.N. and Gaston, S.R. (1981) Treatment of ununited tibial diaphyseal fracture with pulsing electromagnetic fields. *The Journal of Bone and Joint Surgery*, **63**, 511–523.

Bassett, C.A.L., Mitchell, S.N., Norton, L. and Pilla, A.A. (1978) Repair of non-unions by pulsing electromagnetic fields. *Acta Orthopaedica Belgica*, **44**, 706–724.

Bassett, C.A.L., Mitchell, S.N. and Schink, M.M. (1982) Treatment of therapeutically resistant non-unions with bone grafts and pulsing electromagnetic fields. *American Journal of Bone and Joint Surgery*, **64**, 1214–1220.

Bassett, C.A.L. and Pawluk, R.J. (1972) The electrical behavior of cartilage during loading. *Science*, **178**, 982.

Bassett, C.A.L., Pawluk, R.J. and Becker, R.O. (1964) Effects of electric currents on bone *in vivo*. *Nature*, **204**, 652–654.

Bassett, C.A.L., Pawluk, R.J. and Pilla, A.A. (1974) Acceleration of fracture repair by electromagnetic fields. *Annals of the New York Academy of Sciences*, **238**, 242–262.

Bassett, C.A.L., Pilla, A.A. and Pawluk, R.J. (1977) A non-operative salvage of surgically resistant pseudarthroses and non-unions by pulsing electromagnetic fields: A preliminary report. *Clinical Orthopaedics and Related Research*, **124**, 128–143.

Bassett, C.A.L., Schink, M. and Mitchell, S.N. (1984) Treatment of osteonecrosis of the hip with specific pulsed electromagnetic fields: A preliminary report, in *Bone Circulation* (eds J. Arlet, R. Ficat and D.S. Hungerford), Williams and Wilkins, Baltimore, pp. 343–354.

Bassett, L.S., Tzitzikalkis, G., Pawluk, R.J. and Bassett, C.A.L. (1979) Prevention of disuse osteoporosis in the rat by means of pulsing electromagnetic fields, in *Electrical Properties of Bone and Cartilage: Experimental Effects and Clinical Applications* (eds C.T. Brighton, J. Black and S.R. Pollack), Grune and Stratton, New York, pp. 311–331.

Bassey, E.J. and Ramsdale, S.J. (1994) Increase in femoral bone density in young women following high impact exercise. *Osteoporosis International*, **4**, 72–75.

Bates, D.W., Black, D.M. and Cummings, S.R. (2002) Clinical use of bone densitometry-clinical applications. *The Journal of the American Medical Association*, **288**, 1898–1900.

Batge, B., Diebold, J., Bodo, M. *et al.* (1992) Evidence for matrix alterations in osteoporosis, in *Current Research in Osteoporosis and Bone Mineral Measurement II* (ed. E.F.J. Ring), British Institute of Radiology, London.

Batista, U., Miklavac, D. and Sersa, G. (1991) The effect of low-level direct current on V-79 cell line *in vitro*. *Periodicum Biologorum*, **93**, 225–226.

Batra, N.K., Li, Y.J., Yellowley, C.E. *et al.* (2005) Effects of short term recovery periods on fluid induced signaling in osteoblastic cells. *Journal of Biomechanics*, **38**, 1909–1917.

Battista, J.J. and Bronskill, M.J. (1981) Compton scatter imaging of transverse section: An overall appraisal and evaluation for radiotherapy planning. *Physics in Medicine and Biology*, **26**, 81–99.

Battista, J.J., Santon, L.W. and Bronskill, M.J. (1977) Compton scatter imaging of transverse sections: Corrections for multiple scatter and attenuations. *Physics in Medicine and Biology*, **22**, 229–244.

Baud, C.A. and Gossi, M. (1980) Degree of mineralization of bone tissue as determined by quantitative micrograph: effect of age, sex and pathological conditions, in *Proceedings of the 4th International Conference on Bone Measurement, Ontario*, (ed. R.B. Mazess), U.S. National Institutes of Health, Washington, D.C., Publication No. 80-1938, pp. 345–352.

Bauer, D.C., Gluer, C.C., Culey, J.A. *et al.* (1997) Bone ultrasound predicts fracture strongly and independently of densitometry in older women: A prospective study. *Archives of Internal Medicine*, **157**, 629–634.

Bawin, S.M. and Adey, W.R. (1976) Sensitivity of calcium binding in cerebral tissue to weak electromagnetic fields oscillating at low frequency. *Proceedings of the National Academy of Sciences of the United States of America*, **73**, 1199.

Bawin, S.M., Adey, W.R. and Sabbot, I.M. (1978) Ionic factors in release of $^{45}Ca^{2+}$ from cerebral tissue by electromagnetic fields. *Proceedings of the National Academy of Sciences of the United States of America*, **75**, 6314.

Baylink, D.J., Morey, E.R., Ivey, J.L. and Stauffer, M.E. (1980) Vitamin D and bone, in *Basic and Clinical Nutrition Vol. 2, Vitamin D Molecular Biology and Clinical Nutrition* (ed. A.W. Norman), Marcel Dekker, New York, pp. 387–453.

Beamer, W.G., Shultz, K.L., Donahue, L.R. *et al.* (2001) Quantitative trait loci for femoral and lumbar vertebral bone mineral density in C57BL/6J and C3H/HeJ Inbred strains of mice. *Journal of Bone and Mineral Research*, **16**, 1195–1206.

Beanpere, G.S., Orr, T.E. and Carter, D.R. (1990) An approach for time dependent bone modeling and remodeling theoretical development. *Journal of Orthopaedic Research*, **8**, 651–661.

Beanusa, M., Metkovic, V. and Kotial, K. (1978) Kinetic parameters of calcium metabolism and femur morphometry in rats. II. Influence of ovariectomy and dietary calcium. *Pflügers Archiv*, **375**, 239–244.

Beaupre, G.S., Orr, T.E. and Carter, D.R. (1990) An approach for time dependent modeling and remodeling - theoretical development. *Journal of Orthopaedic Research*, **8**, 651–661.

Becker, R.O. (1972) Stimulation of partial limb regeneration in rats. *Nature*, **235**, 109–111.

Becker, R.O., Bassett, C.A.L. and Bachman, C.H. (1964) Bioelectrical factors affecting bone structure, in *Bone Biodynamics* (ed. C.H. M. Frost), Little, Brown, Boston, pp. 209–232.

Becker, R.O. and Brown, F.M. (1965) Photoelectric effects in human bone. *Nature*, **206**, 1325–1328.

Becker, R.O. and Esper, C. (1981) Letters to the Editor: Electrostimulation and undetected malignant tumors. *Clinical Orthopaedics*, **161**, 336–339.

Becker, R.O. and Marino, A.A. (1966) Electron paramagnetic resonance spectra of bone and its major components. *Nature*, **210**, 583–588.

Becker, R.O. and Murray, D.G. (1967) A method for producing cellular differentiation by means of very small electrical currents. *Annals of the New York Academy of Sciences*, **29**, 606–616.

Becker, R.O. and Murray, D.G. (1970) The electrical control system regulating fracture healing in amphibians. *Clinical Orthopaedics*, **73**, 169–198.

Becker, R.O., Spadaro, J.A. and Marino, A.A. (1977) Clinical experiences with low intensity direct current stimulation of bone growth. *Clinical Orthopaedics*, **124**, 75.

Behari, J. (1991) Electrostimulation and bone fracture healing. *Critical Reviews in Biomedical Engineering*, **18** (4), 235–254.

Behari, J. (1991a) Solid bone behavior, progress in biophysics and molecular biology **56**, 1–41.

Behari, J. (1991b) Electrostimulation and bone fracture healing. *Critical review in biomedical engineering*, **18** (4) 235–253.

Behari, J. (1994) Biophysical parameters effecting osteoporosis. *Critical reviews in biomedical engineering*, **22** (2), 69–137.

Behari, J. and Andrabi, W.H. (1978) Generation of Hall voltage in bone. *Connective Tissue Research*, **6**, 181–184.

Behari, J. and Andrabi, W.H. (1981) PN junction characteristics and photoelectricmagnetic effect in bone. *Biomaterials*, **2**, 23–27.

Behari, J., Arya, V.P. and Alex, Z.C. (1991) Acclerated bone fracture healing using a capacitively coupled pulsed radio frequency electric field: An animal study. *Bioelectricity*, **10**, 231.

Behari, J., Dave, P.K. and Arya, V.P. (1990) Accelerated fracture healing using a capacitively coupled electric field: a case study of a non-union, in *Biomaterials, in Orthopaedics* (ed. R. Prakash), Golden Printers, pp. 109–115.

Behari, J. and Guha, S.K. (1976) Mechanism of charge transport and photoconduction in human bone. *Medical and Life Sciences Engineering*, **2**, 35.

Behari, J. and Guha, S.K. (1977) Absorption spectra of bone. *Calcified Tissue Research*, **23**, 113.

Behari, J., Guha, S.K. and Agarwal, P.N. (1974) Temperature dependence of the electrical conductivity of bone. *Connective Tissue Research*, **2**, 325–328.

Behari, J., Guha, S.K. and Agarwal, P.N. (1975) The effect of ultraviolet radiations on the electrical conductivity of human bone. *Calcified Tissue Research*, **19**, 223–227.

Behari, J., Kumar, H. and Aruna, R. (1982) Effect of ultraviolet light on the dielectric behavior of bone at microwave frequencies. *Annals of Biomedical Engineering*, **10**, 139–144.

Behari, J., Rai, D.V. and Jha, R. (1979) On the solid state of bone. *Calcified Tissue International*, **28**, 33–36.

Behari, J., Rai, D.V., Saha, S. and Mathur, M. (1989) Bone metabolism in calcium and phosphorous deficient rats. *Medical and Life Sciences Engineering*, **11**, 14–23.

Behari, J. and Singh, S. (1980) Ultrasound propagation in *in vivo* bone. *Ultrasonics*, **19** (21), 87–91.

Behari, J. and Singh, S. (1981a) Ultrasound propagation in '*in vivo*' bone. *Ultrasonics*, **19** (2), 87–90.

Behari, J. and Singh, S. (1981b) Bioelectric characteristics of unstressed *in vivo* bone. *Medical & Biological Engineering & Computing*, **19** (1), 49.

Beigi, R., Kobatake, E., Aizawa, M. and Dubyak, G.R. (1999) Detection of local ATP release from activated platelets using cell surface attached firefly luciferase. *The American Journal of Physiology*, **276**, C267–C278.

Bell, K.L., Loveridge, N., Power, J. *et al.* (1999) Regional differences in cortical porosity in the fractured femoral neck. *Bone*, **24**, 57–64.

Bellows, C.G., Aubin, J.E., Heersche, J.N.M. and Antoz, M.E. (1986) Mineralized bone nodules formed *in vitro* from enzymatically released rat calvaria cell populations. *Calcified Tissue International*, **38**, 143–154.

Bennett, J.H., Moffatt, S. and Horton, M. (2001) Cell adhesion molecules in human osteoblasts: Structure and function. *Histology and Histopathology*, **16**, 603–611.

Bennok, F.K. (1972) in *Biomechanics* (ed. Y.C. Fung), Prentice Hall, Inc., Englewood Cliffs, New Jersey.

Bentolila, V., Boyce, T.M., Fyhrie, D.P. *et al.* (1998) Intracortical bone remodeling in adult rat lung bones after fatigue loading. *Bone*, **23**, 275–281.

Benya, P.D., Brown, P.D. and Padilla, S.R. (1988) Microfilament modification by dihydrocytochalasin B causes retinoic acid-modulated chondrocytes to re-express the differentiated collagen phenotype without a change in shape. *The Journal of Cell Biology*, **106**, 161–170.

Ben-Ze'ev, A., Robinson, G.S., Bucher, N.L.R. and Farmer, S.R. (1988) Cell-cell and cell-matrix interactions differentially regulate the expression of hepatic and cytoskeletal genes in primary cultures of rat hepatocytes. *Proceedings of the National Academy of Sciences of the United States of America*, **85**, 2161–2165.

Berendsen, H.J.C. (1962) Nuclear magnetic resonance study of collagen hydration. *Journal of Chemical Physics*, **36**, 3297–3305.

Beresford, J.N., Bennett, J.H., Delvin, C. *et al.* (1992) Evidence for an inverse relationship between the differentiation of adipocytic and osteogenic cells in rat marrow stromal cell cultures. *Journal of Cell Science*, **102**, 341–351.

Beretta, D.A. and Pollack, S.R. (1986) Ion concentration and the zeta potential of bone. *Journal of Orthopaedic Research*, **4**, 337–345.

Berg, H. and Zhang, L. (1993) Electrostimulation in cell biology by low frequency electromagnetic fields. *Electro and Magnetobiology*, **12**, 147–163.

Bergman, E.A., Newbrey, J.W. and Massey, L.K. (1988) Caffeine does not cause *in vitro* calcium loss from neconatal mouse calvaria. *Calcified Tissue International*, **43**, 281–283.

Berkenbilt, S.I., Frank, E.H., Bombard, D.L. and Grodzinsk, A.J. (1995) Electrokinetic interactions in cartilage: relevance to chondrocyte metabolism and diagnostics for tissue degeneration. Proceedings of the 1995 Bioengineering Conference, Beaver Greek, Co.

Bernad, M., Martinex, M.W., Escalona, M. *et al.* (2002) Polymorphism in the type I collagen (COLOAO) gene and risk of fractures in postmenopausal women. *Bone*, **30**, 223–228.

Bernengo, J., Rous, B. and Herbage, U. (1974) Electrical birefringence study of monodisperse collagen solutions. *Biopolymers*, **13**, 641.

Berridge, M.J. (1975) Cyclic nucleotides and cellular calcium metabolism. *Advances in Cyclic Nucleotide Research*, **5**, 375–394.

Berridge, M.J. *et al.* (1998) Calcium-a life and death signal. *Nature*, **395**, 645–648.

Bertram, J.E.A. and Biewener, A.A. (1988) Bone curvature sacrificing strength for loading predictability? *Journal of Theoretical Biology*, **131**, 75–92.

Bertram, J.E.A. and Swartz, S.M. (1991) The "law of bone transformation": A case of crying Wolff? *Biological Reviews of the Cambridge Philosophical Society*, **66**, 245–273.

Bethe, H.A. and Schwinger, J. (1943) Perturbation theory of resonant cavities NDRC Report D1, Radiation Laboratory MIT.

Beverly, M.C., Rider, T.A., Evans, M.J. and Smith, R. (1989) Local bone mineral response in brief exercise that stresses the skeleton. *British Medical Journal*, **299**, 233–235.

Bevier, W.C., Wiswell, R.A., Pyka, G. *et al.* (1989) Muscle strength, physical fitness and weight but not age predict femoral neck bone mass. *Journal of Bone and Mineral Research*, **4**, 441–448.

Bibikova, A., Belkin, V. and Oron, U. (1994) Enhancement of angiogenesis in regenerating gastrocnemius muscle of the toad (Bufo yiridis) by low energy laser irradiation. *Anatomy and Embryology*, **190**, 597–602.

Biering-Sorensen, F., Bohr, H.H. and Schaadt, O.P. (1990) Longitudinal study of bone mineral content in the lumbar spine, the forearm and the lower extremities after spinal cord injury. *European Journal of Clinical Investigation*, **20**, 330–335.

Bijvoet, O.L.M., Froeling, P.G.A.M. and Sluys Veer, J.V.D. (1972) Calcitonin and diphosphonate in Paget's diseases. *The Journal of Bone and Joint Surgery*, **54**, 554.

Bikle, D.D., Halloran, B., Fong, L. *et al.* (1993) Elevated 1,25-dihydroxyviamin D levels in patients with chronic obstructive pulmonary disease treated with prednisone. *The Journal of Clinical Endocrinology and Metabolism*, **76** (2), 456–461.

Bikle, D.D., Harris, J., Hallora, B.P. *et al.* (1995) The molecular response of bone to growth hormone during skeletal unloading: regional differences. *Endocrinology*, **136**, 2099–2109.

Bikle, D.D., Morey-Holton, E.R., Doty, S.B. *et al.* (1994) Alendronate increases skeletal mass of growing rats during unloading by inhibiting resorption of calcified cartilage. *Journal of Bone and Mineral Research*, **9**, 1777–1787.

Bilezikian, J.P., Morishima, A. and Bell, J. (1998) Increased bone mass as a result of estrogen therapy in a man with aromatase deficiency. *The New England Journal of Medicine*, **339**, 599–603.

Billota, T.W., Zati, A., Gundi, S. *et al.* (1994) Electromagnetic fields in the treatment of postmenopausal osteoporosis: an experimental study conducted by densitometric, dry ash weight and metabolic analysis of bone tissue. *Chirurgia Degli Organi di Movimiento*, **79**, 309–313.

Binder, A., Parr, G., Hazleman, B. and Fitton-Jackson, S. (1984) Pulsed electromagnetic field therapy of persistent rotor cuff tendonitis. A double blind controlled assessment. *Lancet*, **11**, 695.

Binderman, I., Shimshoni, Z. and Somfjin, D. (1984) Biochemical pathways involved in the translation of physical stimulus into biological message. *Calcified Tissue International*, **36**, 582–585.

Binderman, I., Somjen, D. and Shimshoni, Z. (1986) Growth induction and cartilage cells by physical forces, in *Tissue Nutrition and Viability* (ed. A.R. Hargens), Springer-Verlag, New York, pp. 121–133.

Biot, M.A. (1956) Theory of propagation of elastic waves in fluid saturated porous solid. *The Journal of the Acoustical Society of America*, **28**, 168–191.

Biot, M.A. (1962) Generalised theory of acoustic propagation in porous dissipative media. *The Journal of the Acoustical Society of America*, **34**, 1254–1264.

Bird, F., Becker, H., Healer, J. and Messer, M. (1968) Experimental determination of the mechanical properties of bone. *Aerospace Medicine*, **39**, 44–48.

Birge, S.J., Kentmann, H.T., Cuatrecasas, P. and Whedon, G.D. (1967) Osteoporosis, intestinal lactase deficiency and low dietary calcium intake. *The New England Journal of Medicine*, **276**, 445–448.

Birkenhager-Frenkel, D.H. and Nigg, A.L. (1993) Age related bone loss as reflected by changes of interstitial bone thickness. *Calcified Tissue International*, **52**(Suppl. 1), S60 (Abstract).

Bischoff-Ferrari, H.A. *et al.* (2006) Effect of vitamin D on falls: A meta analysis. *The Journal of the American Medical Association*, **291** (16), 1999–2006.

Bizarov, G.A., Barabash, A.P., Imerlishvili, I.A. *et al.* (1984) Morphological characteristics of the formation and reconstruction of bone tissue in the replacement of extensive bone defects. *Ortopediia Travmatologiia i Protezirovanie*, **1**, 16–20.

Bjarnason, N.H. and Christiansen, C. (2000) Early response in biochemical markers predicts long-term response in bone mass during HRT in early postmenopasal women. *Bone*, **26**, 561–569.

Bjarnason, N.H., Sarkar, S., Duong, T. *et al.* (2000) Six and twelve months changes in bone turnover are related to reduction in vertebral fracture risk during 3 years of raloxifene treatment I postmenopausal osteoporosis. *Osteoporosis International*, **12**, 922–930.

Black, A.J., Topping, J., Durham, B. *et al.* (2000) A detailed assessment of alteration in bone turnover, calcium homeostatis, and bone density in normal pregnancy. *Journal of Bone and Mineral Research*, **15**, 557–563.

Black, D.M., Cummings, S.R., Genant, H.K. *et al.* (1992) Axial and appendicular bone density predicts fractures in older women. *Journal of Bone and Mineral Research*, **7**, 633–638.

Black, D.M., Cummings, S.R., Karpf, D.B. *et al.* (1996) Randomized trial of the effect of alendronate on risk of fracture in women with existing vertebral fractures. *Lancet*, **348**, 1535–1541.

Black, J., Baranowski, T.J. and Brighton, C.T. (1984) Electrochemical aspects of D.C. stimulation of osteogenesis. *Bioelectrochemistry and Bioenergetics*, **12**, 323–327.

Black, J. and Korostoff, E. (1973) Dynamic mechanical properties of viable human cortical bone. *Journal of Biomechanics*, **6**, 435.

Black, J. and Korostoff, E. (1974) Strain related potentials in living bone. *Annals of the New York Academy of Sciences*, **238**, 95–119.

Black, J., Mattson, R. and Korostoff, E. (1974) Haversian osteons: Size, distribution, internal structure and orientation. *Journal of Biomedical Materials Research*, **8**, 299.

Blake, G.M. and Fogelman, I. (2001) Peripheral or central densitometry: Does it matter which technique we use? *Journal of Clinical Densitometry*, **4**, 83–96.

Blake, G.M. and Fogelman, I. (2005) Fracture prediction by bone density measurements at sites other than the fracture site: the contribution of BMD correlation. *Calcified Tissue International*, **76**, 249–255.

Blake, G.M., Knapp, K.M. and Fogelman, I. (2005) Dual X-ray absorptiometry: clinical evaluation of a new cone-beam system. *Calcified Tissue International*, **76**, 113–120.

Blake, G.M., McKeeney, D.B., Chhaya, S.C. *et al.* (1992) Dual energy X-ray absorptiometry: The effects of beam hardening on bone density measurements. *Medical Physics*, **19**, 459–465.

Blake, G.M., Patel, R., Knapp, K.M. and Fogelman, I. (2003) Does the combination of two BMD measurements improve fracture discrimination? *Journal of Bone and Mineral Research*, **18**, 1955–1963.

Blank, M. (1987) The influence of surface charge on oligometric reaction as a basis for channel dynamics, in *Mechanistic Approaches to Interactions of Electric and Electromagnetic Fields with Living Systems* (eds M. Blank and E. Findl), Plenum Press, New York, pp. 151–160.

Blank, M.A., Ems, B.L., Gibson, G.W. *et al.* (1997) Nonclinical model for assessing gastric effects of bisphosphonates. *Digestive Diseases and Sciences*, **42**, 281–288.

Blank, M., Soo, L., Lin, H. *et al.* (1992) Changes in transcription in HL-60 cells following exposure to alternating currents from electric fields. *Bioelectrochemistry and Bioenergetics*, **28**, 301–309.

Blenman, P.R., Carter, D.R. and Beaupre, G.S. (1989) Role of mechanical loading in the progressive ossification of fracture callus. *Journal of Orthopaedic Research*, **7**, 398–407.

Block, M.S. and Brister, G.D. (1994) Use of distraction osteogenesis for maxillary advancement: preliminary results. *Journal of Oral and Maxillofacial Surgery*, **52**, 282–286.

Blumenkrantz, N. and Sondergaard, J. (1972) Effect of prostaglandin E1 and Fla on biosynthesis of collagen. *Nature (New Biology)*, **239**, 246.

Boarder, M.B., Turner, J.T., Erb, L. and Weisman, G.A. (1994) Classification of P2 purinoceptors. *Trends in Pharmacological Sciences*, **15**, 28.

Bockholt, S.M. and Burridge, K. (1993) Cell spreading on extracellular matrix proteins induces tyrosine phosphorylation of tensin. *The Journal of Biological Chemistry*, **268**, 14565–14567.

Bodamyali, T., Bhatt, B., Hughes, F.J. *et al.* (1998) Pulsed electromagnetic fields simultaneously induce osteogenesis and upregulate transcription of bone morphogenetic proteins 2 and 4 in rat osteoblasts *in vitro*. *Biochemical and Biophysical Research Communications*, **250**, 458–461.

Boddy, K., Hollowa, I. and Elliott, A. (1973) A simple facility for total body *in vivo* activation analysis. *International Journal of Applied Radiation and Isotopes*, **24**, 228–230.

Bodin, P., Bialy, D. and Burnstock, G. (1991) Increased flow induced ATP release from isolated vascular endothelial cells but not smooth muscle cells. *British Journal of Pharmacology*, **103**, 1203–1205.

Bodin, P. and Burnstock, G. (1996) ATP-stimulation release of ATP by human endothelial cell. *Journal of Cardiovascular Pharmacology*, **27**, 872–875.

Boeley, D.G., Browner, W.S., Nevitt, M.C. *et al.* (1991) Which fractures are associated with low appendicular bone mass in elderly women? The study of osteoporotic fractures research group. *Annals of Internal Medicine*, **115**, 837–842.

Boivin, G., Duriez, J., Chapuy, M.C. *et al.* (1993) Relationship between bone fluoride content and histological evidence of mineralization defects in osteoporotic women long-term treated with sodium fluoride. *Osteoporosis International*, **3**, 204–208.

Bolotin, H.H. (1998a) A new perspective on the causal influence of soft tissue composition on DXA measured *in vivo* bone mineral density. *Journal of Bone and Mineral Research*, **13**, 1739–1746.

Bolotin, H.H. (1998b) Analytic and quantitative exposition of patient specific systematic inaccuracies inherent in planar DXA derived *in vivo* BMD measurements. *Medical Physics*, **25**, 139–151.

Bolotin, H.H., Sievanen, H., Grashuis, J.L. *et al.* (2001) Inaccuracies inherent in patient-specific dual-energy X-ray absorptiometry bone mineral density measurements: comprehensive phantom-based evaluation. *Journal of Bone and Mineral Research*, **16**, 417–426.

Bonadio, J. and Byers, P.H. (1985) Subtle structural alterations in the chains of type I procollagen produce osteogenesis imperfects type II. *Nature*, **316**, 363–366.

Bonadio, J., Smiley, E., Patil, P. and Goldstein, S. (1999) Localized direct plasmid gene therapy *in vivo*: prolonged therapy results in reproducible tissue regeneration. *Nature Medicine*, **5**, 753–759.

Bonewald, L.F. (2002) Osteocytes: a proposed multifunctional bone cell. *Journal of Musculoskeletal and Neuronal Interactions*, **2**, 239–241.

Bonfield, W. and Clark, E.A. (1973) Elastic deformation of compact bone. *Journal of Materials Science*, **8**, 1590–1594.

Bonfield, W. and Li, C. (1967) Anisotropy of nonelastic flow in bone. *Journal of Applied Physiology*, **38**(6), 2450–2455.

Bonjour, J.P., Rizzoli, R., Ammann, P. and Chevalley, T. (1994) Bisphosphonates in clinical medicine, in *Bone and Mineral Research*, vol. **8** (eds N.M. Heershe and J.K. Kanis), Elsevier Science BV, Amsterdam, pp. 205–264.

Bonnarens, F. and Einhorn, T.A. (1984) Production of a standard closed fracture in laboratory animal bone. *Journal of Orthopaedic Research*, **2**, 97–101.

Bonnick, S.L., Nichols, D.L., Sanborn, C.F. *et al.* (1996) Right and left proximal femur analyses: Is there a need to do both? *Calcified Tissue International*, **58**, 307–310.

Bordier, P., Hioco, D., Rouquier, M. *et al.* (1969) Effects of intravenous vitamin D on bone and phosphate metabolism in osteomalacia. *Calcified Tissue Research*, **4**, 78–83.

Bordier, P., Matrajt, H., Miravet, L. and Hioco, D. (1964) Histological Measure of the Volume and Resorption of Bone joints. *Pathological Biology*, **12**, 1238–1243.

Borsalino, G., Bagnacani, M., Bettati, E. *et al.* (1988) Electrical treatment of delayed and nonunion of fractures using pulsed electromagnetic fields. *Journal of Biomedical Engineering*, **10**, 301.

Bosch, J.P., Saccage, A., Lauer, A. *et al.* (1983) Renal functional reserve in humans: effect of protein intake on glomerular filtration rate. *The American Journal of Medicine*, **75**, 943–950.

Bostrom, K. (2001) Insights into the mechanism of vascular calcification. *The American Journal of Cardiology*, **88**(2A), 20E–22E.

Bostrom, M.P. and Asnis, P. (1998) Transforming growth factor beta is repair. *Clinical Orthopaedics and Related Research*, **355** (Suppl.), S124–S131.

Bostrom, M.P.G., Lane, J.M., Berberian, W.S. *et al.* (1995) Immunolocalizaiton and expression of bone morphogenetic protein 2 and 4 fracture healing. *Journal of Orthopaedic Research*, **13**, 357–367.

Bostrom, M., Lane, J.M., Tomin, E. *et al.* (1996) Use of bone morphogenetic protein-2 in the rabbit ulnar nonunion model. *Clinical Orthopaedics and Related Research*, **327**, 272–282.

Boudousq, V., Kotki, P.O., Dinten, J.M. *et al.* (2003) Total dose incurred by patient's and staff from BMD measurement using a new 2D digital bone densitometer. *Osteoporosis International*, **14**, 263–269.

Bouillon, R., Auwrex, J., de Keyser, L. *et al.* (1984) Serum vitamin D metabolites and their binding protein in patients with liver cirrhosis. *The Journal of Clinical Endocrinology and Metabolism*, **59**, 86–89.

Boulton, M. and Marshal, J. (1986) He-Ne stimulation of human fibroblast stimulation and attachment *in vitro* lasers. *Life Sciences*, **1**, 125–134.

Bourgignon, L.Y.W. and Singer, S.J. (1977) Transmembrane interactions and the mechanism of capping of surface receptors by their specific ligands. *Proceedings of the National Academy of Sciences of the United States of America*, **74**, 5031–5035.

Bouxsein, M.L., Michaeli, D.A., Plass, D.B. *et al.* (1997) Precision and accuracy of computed digital absorptiometry for assessment of bone density of the hand. *Osteoporosis International*, **7**, 444–449.

Bouxsein, M.L. and Radloff, S.E. (1997) Quantitative ultrasound of the calcaneus reflects the mechanical properties of calcaneal trabecular bone. *Journal of Bone and Mineral Research*, **12**, 839–846.

Bouyoucef, S.E., Cullum, I.D. and Ell, P.J. (1996) Cross-calibration of a fan-beam X-ray densitometer with a pencil-beam system. *British Journal of Radiology*, **69**, 522–531.

Bover, J., Jara, A., Trinidad, P. *et al.* (1999) Dynamics of skeletal resistance to parathyroid hormone in rat: Effect of renal failure and dietary phosphorus. *Bone*, **25**, 279–285.

Bowler, W.B. *et al.* (2001) Extracellular nucleotide signaling: a mechanism for integrating local and systemic responses in the activation of bone remodeling. *Bone*, **28**, 507.

Bowler, W.B., Bilbe, G. and Gallagher, J.A. (1998) G-protein coupled receptors in bone. *Frontiers in Bioscience: A Journal and Virtual Library*, **3**, D769–780.

Bowler, W.B., Birch, M.A., Gallagher, J.A. and Bilbe, G. (1995) Identification and cloning of human P-2U purinoceptor present in osteoclastoma, bone and osteoblasts. *Journal of Bone and Mineral Research*, **10**, 1137–1145.

Bowler, W.B., Dixon, C.J., Halleux, C. *et al.* (1999) Signaling in human osteoblasts by extracellular nucleotides: Their weak induction of the c-fos protoncogene via $Ca^{2+}$ mobilization is strongly potentiated by a parathyroid hormone/camp-dependent protein kinase pathway independently of mitogen-activated protein kinase. *The Journal of Biological Chemistry*, **274**, 14315–14324.

Bowler, W.B., Tattersall, J.A., Hussein, R. *et al.* (1998a) Release of ATP by osteoblasts: Modulation by fluid shear forces. *Bone*, **22** (Suppl.), 3.

Bowler, W.B., Tattersall, J.A., Hussein, R. *et al.* (1998b) Real time measurement of ATP release from human osteoblasts. *Journal of Bone and Mineral Research*, **13**, 525.

Boyce, R.W. and Weisbrode, S.E. (1983) The effect of dietary calcium on the response of bone to 1,25 $(OH)_2D_3$. *Laboratory Investigation*, **48**, 683–689.

Boyde, H.B., Lipinski, S.W. and Wiley, J.H. (1961) Observations on nonunion of the shafts of long bones. *The Journal of Bone and Joint Surgery*, **43**, 159–168.

Boyden, L.M., Mao, J., Belsky, J. *et al.* (2002) High bone density due to a mutation in LDL-receptor-related protein 5. *The New England Journal of Medicine*, **346**, 1513–1521.

Boyle, W.J. *et al.* (2003) Osteoclast differentiation and activation. *Nature*, **423**, 337–342.

Braby, L.A. (1966) Pacific Northwest Laboratory Annual Report for 1965 in Physical Sciences, Doc No. BNWL-235, 2, 27.

Braddock, M., Schwachtgen, J.-L., Houston, P. *et al.* (1988) Fluid shear stress modulation of gene expression in endothelial cells. *News in Physiological Sciences*, **13**, 241–246.

Bradham, D.M., Igarashi, A., Potter, R.L. and Grotendorst, G.R. (1991) Connective tissue growth factor: A cysteine rich mitogen secreted by human vascular endothelial cells is related to the SRC induced immediate early gene product CEF-10. *The Journal of Cell Biology*, **114**, 1285–1294.

Brandi, M.L., Hukkanen, M., Umeda, T. *et al.* (1995) Bidirectional regulation of osteoclast function by nitric oxide synthase isoforms. *Proceedings of the National Academy of Sciences of the United States of America*, **92**, 2954–2958.

Breen, S.A., Millest, A. and Loveday, B. (1996) Regional analysis of bone mineral density in the distal femur and proximal tibia using peripheral quantitative computed tomography in the rat *in vivo*. *Calcified Tissue International*, **58**, 449.

Breger, L. (1972) Piezoelectricity of Bone and Tendon, PhD thesis, University of Illinois, Urbana, USA.

Breslau, N.A., McGuire, J.L., Zerwekh, J.E. and Pak, C.Y.C. (1982) The role of dietary sodium on renal excretion and intestinal absorption of calcium and on vitamin D metabolism. *The Journal of Clinical Endocrinology and Metabolism*, **55**, 369–373.

Breslau, N.A., McGuire, J.L., Zerwekh, J.E. *et al.* (1984) Hypercalcimia associated with increased serum calcitriol levels in three patients with lymphoma. *Annals of Internal Medicine*, **100**, 1–7.

Bressot, C., Meunier, P.G., Chapuy, M.C. *et al.* (1979) Histomorphometric profile, pathophysiology and reversibility of corticosteroid induced osteoporosis. *Metabolic Bone Disease & Related Research*, **1**, 303–309.

Briancon, D. and Meunier, P.J. (1981) Treatment of osteoporosis with fluoride, calcium and vitamin D. *The Orthopedic Clinics of North America*, **12**, 629–648.

Brickman, A.S., Massr, S.G. and Coburn, J.W. (1972) Changes in serum and urinary calcium during treatment with hydrocholorothiazide: studies on mechanisms. *The Journal of Clinical Investigation*, **51**, 945–954.

Brighton, C.T. (1984) The semi-invasive method of treating nonunion with direct current. *The Orthopedic Clinics of North America*, **15**, 33–45.

Brighton, C.T., Black, J. and Friedenberg, Z.B. (1981) A multicenter study of the treatment of non-union with constant direct current. *The Journal of Bone and Joint Surgery*, **63A**, 2–13.

Brighton, C.T., Friedenberg, Z.B., Black, J. *et al.* (1986) Direct current versus capacitive coupling in treating non unions. *Transactions of Bioelectrical Repair and Growth Society*, **6**, 84.

Brighton, C.T., Friedenberg, Z.B., Mitchell, E.I. and Booth, R.E. (1977) Clinical experiences for the treatment of non-union with constant DC. *Clinical Orthopaedics and Related Research*, **124**, 106–123.

Brighton, C.T., Friedenberg, Z.B., Zemsky, F.M. and Polis, R.R. (1975) Direct current stimulation of nonunion and congenital pseudosrthrosis. *The Journal of Bone and Joint Surgery*, **57A**, 368–377.

Brighton, C.T., Hozalk, N.J., Breger, M.D. *et al.* (1985a) Fracture healing in the rabbit fibula when subjected to various capacitively coupled electric fields. *Journal of Orthopaedic Research*, **3**, 331.

Brighton, C.T. and Hunt, R.M. (1986) Ultrastructure of electrically induced osteogenesis in the rabbit medullary canal. *Journal of Orthopaedic Research*, **4**, 27–36.

Brighton, C.T., Katz, E.J., Goll, S.R. *et al.* (1985b) Prevention and treatment of sciatic denervation of disuse osteoporosis in the rat tibia with capacitively coupled electrical stimulation. *Bone*, **6**, 87–97.

Brighton, C.T. and McCluskey, W.P. (1985) The early response of bone cells in culture to a capacitively coupled electric field. *The Journal of Bone and Joint Surgery*, **67**, 577.

Brighton, C.T. and McCluskey, W.P. (1986) Cellular response and mechanisms of action of electrically induced osteogenesis, in *Bone and Mineral Research*, vol. **4**(ed. W.A. Kek), Elsevier, New York, pp. 213–254.

Brighton, C.T., Nichols, C.E. III and Arangio, G.A. (1985) Amelioratin of oxygen induced osteoporosis in the *in vitro* fetal rat tibia with capacitively coupled electric field. *Journal of Orthopaedic Research*, **3**, 311–320.

Brighton, C.T. and Pollack, S.R. (1985) Treatment of recalcitrant non-union with a capacitively coupled electrical field. A preliminary report. *American Journal of Bone and Joint Surgery*, **67**, 577–585.

Brighton, C.T., Pollack, S.R. and Windsor, R.E. (1981) Stimulation of fracture healing by a capacitively coupled electric field in the rabbit fibula. *Transactions of the Orthopaedic Research Society*, **6**, 93.

Brighton, C.T., Tadduni, G.T., Goll, S.R. and Pollack, S.R. (1988) Treatment of sciatic denervation/disuse osteoporosis in the rat with a capacitatively coupled electrical signal: Effects on bone formation and bone resorption. *Journal of Orthopaedic Research*, **6**, 676–684.

Brighton, C.T., Tadduni, G.T. and Pollack, S.R. (1985) Treatment of sciatic devervation disuse osteoporosis in the rat tibia with capacitively coupled electrical stimulation: dose response and duty cycle. *The Journal of Bone and Joint Surgery*, **67**, 1022–1028.

Brighton, C.T., Unger, A.S. and Stambough, J.L. (1984) *In vitro* growth of boving articular cartilage chondrocytes in various capacitively coupled electrical fields. *Journal of Orthopaedic Research*, **2**, 15–22.

Brighton, C.T., Wang, W., Seldes, R. *et al.* (2001) Signal transduction in electrically stimulated bone cells. *The Journal of Bone and Joint Surgery. American Volume*, **83**, 1514–1523.

Brinckmann, P., Biggemann, M. and Hilweg, D. (1989) Prediction of the compressive strength of human lumbar vertebrae. *Spine*, **14**, 606–610.

Brinker, M.R., Lippton, H.L., Stephen, D.C. and Hyman, A.L. (1990) Pharmacological regulation of the circulation of bone. *The Journal of Bone and Joint Surgery*, **72**(7), 964–975.

Brismar, T.B., Budinsky, L. and Majumdar, S. (2001) Evaluation of trabecular bone orientation in wrists of young volunteers using MR relaxometry and high resolution MRI, in *Noninvasive Assessment of Trabecular Bone Architecture and the Competene of Bone* (eds S. Majumdar and B.K. Bay), Kluwer Academic/Plenum Publishers, New York, pp. 1–7.

Brismar, T.B., Karlsson, M., Li, T. and Ringertz, H. (1999) The correlation between $R_2$ and bone mineral measurements in human vertebrae: an *in vitro* study. *European Radiology*, **9**, 141–144.

Brixen, K., Nielsen, H.K., Mosekilde, L. and Flyvbjerg, A. (1990) A short course of recombinant human growth hormone treatment stimulates osteoblasts and activates bone remodeling in normal human volunteers. *Journal of Bone and Mineral Research*, **5**, 609–618.

Bronckers, A.L.J.J., Goei, W., Luo, G. *et al.* (1996) DNA fragmentation during bone formation in neonatal rodents assessed by transferase-mediated and labeling. *Journal of Bone and Mineral Research*, **11**, 1281–1291.

Broner, F. (1984) Role of intestinal calcium absorption in plasma regulation of the rat. *The American Journal of Physiology*, **246**, R680–R683.

Brooks, R.A. and DiChiro, G. (1976) Statistical limitations in X-ray reconstructive tomography. *Medical Physics*, **3**, 237–240.

Brown, G.P. and Aftergut, S. (1962) Electronic properties of organic compounds, in *Organic Semiconductors* (eds T.T. Brophy and J.W. Buttey), Macmillan, New York, pp. 89–99.

Brown, P.D. and Benya, P.D. (1988) Alterations in chondrocyte cytoskeletal architecture during phenotypic modulation by retionic acid and dihydrocytochalasin B-induced re-expression. *The Journal of Cell Biology*, **106**, 171–179.

Brown, S. (1977) Polymeric fixation and ultrasound assessment of fractures and healing. *Bulletin of the New York Academy of Medicine*, **53**, 710–724.

Brown, S.A. and Mayor, M.B. (1976) Ultrasonic assessment of early callus formation. *Biomedical Engineering*, **11**, 124–127.

Broz, J.J., Simske, S.J. and Greenberg, A.R. (1995) Material and compositional properties of selectively demineralized cortical bone. *Journal of Biomechanics*, **28**, 1357–1368.

Bruder, S.P. and Fox, B.S. (1999) Tissue engineering of bone. Cell based strategies. *Clinical Orthopaedics*, **367** (Suppl.), S68–S83.

Brunader, R. and Shelton, D.K. (2002) Radiologic bone assessment in the evaluation of osteoporosis. *American Family Physician*, **65**, 1357–1364.

Brunk, U., Schellens, Y. and Westermark, B. (1976) Influence of epidermal growth factor (EGF) on ruffling activity, pinocytosis and proliferation of cultivated human glia cells. *Experimental Cell Research*, **103**, 295–302.

Bucay, N., Sarosi, I., Dunstan, C.R. *et al.* (1998) Osteoprotegerin-deficient mice develop early onset early osteoporosis and arterial calcification. *Genes & Development*, **12**(9), 1260–1268.

Buckely, K.A., Hipskind, R.A., Gaw, A. *et al.* (2000) Effects of extracellular ATP on resorption and transcription factor activation in co-culture and RANK ligand derived human osteoclasts. *Calcified Tissue International*, **51**(Suppl.), S62.

Buckley, J.M., Loo, K. and Motherway, J. (2007) Comparison of quantitative computed tomography-based measures in predicting vertebral compressive strength. *Bone*, **40**, 767–774.

Buckley, M.J., Banes, A.J., Levin, L.G. *et al.* (1988) Osteoblasts increase their rate of division and align in response to cyclic, mechanical tension. *Bone & Mineral*, **4**, 225–236.

Buckwalter, J.A. and Mankin, H.J. (1997) Articular cartilage Part I Tissue design and chondrocyte matrix interactions. *The Journal of Bone and Joint Surgery*, **79**, 600–611.

Budden, F.H., Bayley, T.A., Harrison, J.E. *et al.* (1988) The effect of fluoride on bone histology in postmenopausal osteoporosis depends on adequate fluoride absorption and retention. *Journal of Bone and Mineral Research*, **3**, 127–133.

Bur, A.J. (1976) Measurement of the dynamic piezoelectric properties of bone as a function of temperature and humidity. *Journal of Biomechanics*, **9**, 495–507.

Burger, E. (2000) Experiments on cell mechanosensitivity: Bone cells as mechanical engineers, in *Bone Mechanics Handbook*, 2nd edn (ed. S.C. Cowin), CRC Press, Boca Raton, FL, pp. 28–1-28-16.

Burger, E.H. and Klein-Nulend, J. (1999) Mechanosensory transduction in bone role of the lacuno-canalicular network. *The FASEB Journal*, **13**, S101–S112.

Burnell, J.M., Baylink, D.J., Chesut, H. III and Teubner, E.J. (1986) The role of skeletal calcium deficiency on postmenopausal osteoporosis. *Calcified Tissue International*, **38**, 187–192.

Burnstock, G. (1972) Purinergic nerves. *Pharmacological Reviews*, **24**, 509–581.

Burnstock, G. (1993) Hypoxia, endothelium and purines. *Drug Development Research*, **28**, 301–306.

Burr, D.B. (2002) Targeted and non-targeted remodelling. *Bone*, **30**(1), 2–4.

Burr, D.B. and Martin, R.B. (1992) Mechanisms of bone adaptation to the mechanical environment. *Triangle (Sandoz)*, **31**, 59.

Burr, D.B., Martin, R.B., Schaffler, M.B. and Radin, E.L. (1985) Bone remodeling in response to *in vivo* fatigue microdamage. *Journal of Biomechanics*, **18**, 189–200.

Burr, D.B., Mori, S., Boyd, R.D. *et al.* (1993) Histomorphometric assessment of the mechanisms for rapid ingrowth of bone to HA/TCP coated implants. *Journal of Biomedical Materials Research*, **27**, 645–653.

Burr, D.B., Schaffler, M.B. and Frederickson, R.G. (1988) Composition of the cement line and its possible mechanical role as a local interface in human compact bone. *Journal of Biomechanics*, **21**, 939–945.

Burr, H.S. (1950) Bioelectricity: potential gradients. In Medical Physics, (Otto Glasser, Chicago, Publishing Co.). 2, 90–94.

Burstein, A.H., Reilly, D.T. and Martens, M. (1976) Aging of bone tissue: Mechanical properties. *The Journal of Bone and Joint Surgery*, **58**, 82–86.

Burstein, A.H., Zika, J.M., Heiple, K.G. and Klein, L. (1975) Contribution of collagen and mineral to the elastic properties of bone. *The Journal of Bone and Joint Surgery*, **57**, 959–961.

Burstein, A.H., Zika, J.C., Heiple, K.G. and Klein, L. (1977) Contribution of collagen and mineral to elastic plastic properties of bone. *The Journal of Bone and Joint Surgery*, **57a**, 956–961.

Busse, J.W., Bhandari, M., Kulkarni, A.V. *et al.* (2002) The effect of low intensity pulsed ultrasound therapy on time to fracture healing. A meta analysis. *Canadian Medical Association Journal*, **166**, 437–441.

Buttgereit, F., Brunester, G.R. and Brand, M.D. (2000) Bioenergetics of immune functions; fundamental and therapeutic aspects. *Immunology Today*, **21**(4), 189–196.

Buttgereit, F., Grant, A., Muller, M. and Brand, M.D. (1994) The effects of methylprednisolone on oxidative phosphorylation in concanavalin-A-stimulated thymocytes. Top-down elasticity analysis and control analysis. *European Journal of Biochemistry*, **223** (2), 513–519.

Byers, P.H. (1988) Osteogenesis imperfecta: a update growth. *Genetics and Hormones*, **4**, 1–5.

Byers, P.H. (1989) Disorders of collagen biosynthesis and structure, in *The Metabolic Basis of Inherited Disease*, 6th edn (eds C.R. Scriver, A.L. Beaudet, W.S. Aly and D. Valle), McGraw Hill, New York, pp. 2805–2842.

Byers, P.H. (1990) Brittle bones-fragile molecules: disorders of collagen gene structure and expression trends. *Genetics*, **6**, 293–300.

Cabnela, M.E. and Jowsey, J. (1971) The effects of phosphates on experimental osteoporosis. *Calcified Tissue Research*, **8**, 114–120.

Cadossi, R., Emilia, G., Torelli, G. *et al.* (1985) The effect of low frequency pulsing electromagnetic fields on the response of human normal lymphocytes to phtohaemagglutinin (PHA). *Bioelectrochemistry and Bioenergetics*, **14**, 115–119.

Cady, W.G. (1964) Piezoelectricity, in *An Introduction to the Theory and Applications of Electromechanical Phenomena in Crystals*, vol. 1, Dover, New York.

Cain, C.D., Adey, W.R. and Luben, R.A. (1981) Evidence that pulsed electromagnetic fields inhibit coupling of adenylate cyclase by parathyroid hormone in bone cells. *Journal of Bone and Mineral Research*, **2**(5), 437.

Cain, C.D., Adey, W.R. and Luben, R.A. (1987) Evidence that. pul: electromagnetic fields inhibit coupling of adenylate cyel, by parathyroid hormone in bone cells. *Journal of Bone and Mineral Research*, **2**, 437–441.

Cain, C.D. and Luben, R.A. (1987) *Interaction of Biological Systems with ELF* (eds L.E. Anderson, B.J. Kelman and R.J. Weigel), Conference Publication No. 24, Batelle Laboratories, Richland, WA, pp. 269–278.

Calvo, M.A., Bell, R.R. and Forbes, R.M. (1982) Effect of protein induced calciuria on calcium metabolism and bone status in adult rats. *Journal of Nutrition*, **112**, 1401–1413.

Cameron, J.R., Mazess, R.B. and Sorenson, J.A. (1968) Precision and accuracy of bone mineral determination by direct photo absorptiometry. *Investigative Radiology*, **3**, 141–150.

Cameron, J.R. and Sorenson, J.A. (1963) Measurement of bone mineral *in vivo*: An improved method. *Science*, **142**, 230–232.

Cameron, J.R. and Sorenson, J.A. (1970) Measurement of bone mineral by the direct photon absorption method: Principles and instrumentation, in *Progress in Methods of Bone Mineral Measurement*, U.S. Department of Health Education and Welfare, Public, Health Service, National Institute of Health pp. 253–68.

Campbell, J. and Jurist, J. (1971) Mechanical impedance of the femur: a preliminary report. *Journal of Biomechanics*, **4**, 319–322.

Canalis, E., Centrella, M., Burch, W. and McCarthy, T.L. (1989) Insulin like growth factor I mediates selective anabolic effects of parathyroid hormone in bone cultures. *The Journal of Clinical Investigation*, **83**, 60–65.

Cane, V., Zaffe, D., Cavani, F. *et al.* (1996) PEMFs modulate the enzymatic activity during the bone repair process. *Bone*, **19**, 1338.

Cann, C.E. and Adachi, R.R. (1983) Bone resorption and mineral excretion in rats during spaceflight. *The American Journal of Physiology*, **244**, R327–R331.

Cann, C.E. and Genant, H.K. (1980) Precise measurement of vertebral bone mineral content using computed tomography. *Journal of Computer Assisted Tomography*, **4**, 493–500.

Cann, C.E., Genant, H.K., Ettinger, B. and Gordon, G.S. (1980) Spinal mineral loss in oophorectomized women. *The Journal of the American Medical Association*, **244**, 2056–2059.

Cann, C.E., Genant, H.K., Kolb, F.O. and Ettinger, B. (1985) Quantitative computed tomography for prediction of vertebral fracture risk. *Bone*, **6**, 1–7.

Cann, C.E., Martin, M.C., Genant, H.K. and Jaffe, R.B. (1984) Decreased spinal mineral content in amenorrheic women. *The Journal of the American Medical Association*, **251**, 626–629.

Cann, C., Roe, E., Sanchez, S. and Arnud, C. (1999) PTH effects in the femur envelope specific responses by 3D QCT in postmenopausal women. *Journal of Bone and Mineral Research*, **140**, 5137.

Carani, C., Qin, K., Simoni, M. *et al.* (1997) Effect of testosterone and estradiol in a man with aromatase activity. *The New England Journal of Medicine*, **337**, 91–95.

Cardew, M.H. and Eley, D.D. (1959) The semiconductivity of organic semiconductors. *Discussions of the Faraday Society*, **27**, 115–128.

Carr, J., Shi, R., Lenchik, L. *et al.* (2001) Validation of quantitative computed tomography for measurement of bone mineral density in the thoracic spine during a cardiac gated protocol for coronary vascular calcium. *Radiology*, **221**(4 Suppl), 380.

Carter, D.R., Beaupre, G.S., Giori, N.J. and Helms, J.A. (1998) Mechanobiology of skeletal regeneration. *Clinical Orthopaedics and Related Research*, **S355**, S41–S55.

Carter, D.R., Blenman, P.R. and Beaupre, G.S. (1988) Correlations between mechanical stress history and tissue differentiation in initial fracture healing. *Journal of Orthopaedic Research*, **6**, 736–748.

Carter, D.R., Bouxsein, M.L. and Marcus, R. (1992) New approaches for interpreting projected bone densitometry data. *Journal of Bone and Mineral Research*, **7**, 137–145.

Carter, D.R. and Hayes, W.C. (1977) The compressive behavior of bone as a two-phase porous structure. *The Journal of Bone and Joint Surgery. American Volume*, **59**, 954–962.

Carter, D.R., Hayes, W.C. and Schurman, D. (1976) Fatigue life of compact bone–II. Effects of microstructure and density. *Journal of Biomechanics*, **9**, 211.

Carter, D.R. and Spengler, D.M. (1978) Mechanical properties and composition of cortical bone. *Clinical Orthopaedics*, **135**, 192–217.

Casciani, F.S. (1971) Identification of hydrate water in enamel, dentine, cementum and bone, in *Tooth Enamel II: Its Composition, Properties and Fundamental Structure* (eds R.W. Fearnhead and M.V. Stack), John Wright & Sons, Ltd, Bristol, pp. 14–23.

Cassella, J.P. and Ali, S. (1992) Abnormal collagen and mineral formation in osteogenesis imperfecta. *Bone and Mineral*, **17**, 123–128.

Castenmiller, J.J.M., Mensink, R.P., Van der Heijden, L. *et al.* (1985) The effect of dietary sodium on urinary calcium and potassium excretion in normotensive men with different calcium intake. *The American Journal of Clinical Nutrition*, **41**, 52–60.

Casuccio, C. (1962) Concerning osteoporosis. *The Journal of Bone and Joint Surgery*, **44**, 453.

Caswell, A.M. and Russell, R.G.G. (1988) Evidence that ectonucleoside triphosphate pyrophosphatase serves in the generation of inorganic pyrophosphate in human bone and articular cartilage. *Biochimica et Biophysica Acta*, **966**, 310–317.

Catto, G.D.R., McIntosh, J.A.R. and Maclead, M. (1973) Partial body neutron activation analysis *in vivo*: A new approval to the investigation of metabolic bone disease. *Physics in Medicine and Biology*, **18**, 508–517.

Cauley, J.A., Soeley, D.G., Ensrud, K. *et al.* (1995) Estrogen replacement therapy and fractures in older women. Study of osteoporotic fractures research group. *Annals of Internal Medicine*, **122**, 9–16.

Cauley, J.A., Zmud, J.M., Ensrud, K.E. *et al.* (2001) Timing of estrogen replacement therapy for optimal osteoporosis prevention. *The Journal of Clinical Endocrinology and Metabolism*, **86**, 5700–5705.

Cavanagh, P.R., Davis, B.L. and Miller, T.A. (1992) A biomechanical perspective on exercise countermeasures for long-term spaceflight. *Aviation, Space, and Environmental Medicine*, **63**, 482–485.

Caveney, S. (1985) The role of gap junction in development. *Annual Review of Physiology*, **47**, 319–335.

Caverzasio, J. and Bonjour, J.P. (1996) Characteristics and regulation of $P_i$ transport in osteogenic cells for bone metabolism. *Kidney International*, **49**, 975–980.

Cawte, S.A., Pearson, D., Green, D.J. *et al.* (1999) Cross-calibration, precision and patient dose measurements in preparation for clinical trials using dual energy X-ray absorptiometry of the lumbar spine. *British Journal of Radiology*, **72**, 354–362.

Center, J.R., Nguyen, T.V., Scheider, D. and Sambrook, P.N. (1999) Mortality after all major types of osteoporotic fracture in women: an observational study. *Lancet*, **353**, 878–882.

Cerquiglini, S., Cignitti, M., Marchetti, M. and Salleo, A. (1976) On the origin of electrical effects produced by stress in the hard tissues of living organisms. *Life Sciences*, **6**, 2651–2660.

Chaffai, S., Peyrin, F., Xiuzzo, S. *et al.* (2002) Ultrasonic characterization of human cancellous bone using transmission and backscatter measurements: Relationship to density microstructure. *Bone*, **30**(1), 229–237.

Chakkalakal, D.A. (1989) Mechanoelectric transduction in bone. *Journal of Materials Research*, **44**, 1034–1046.

Chakkalakal, D.A. and Johnson, M.W. (1981) Electrical properties of bone. *Clinical Orthopaedics*, **161**, 133.

Chakkalakal, D.A., Johnson, M.W., Harper, R.A. and Kartz, L.J. (1980) Dielectric properties of fluid saturated bone. *IEEE Transactions on Biomedical Engineering*, **BME-27**, 95–100.

Chakkalakal, D.A., Strates, B.S., Mashoof, A.A. *et al.* (1999) Repair of segmental bone defects in the rat: an experimental model of human fracture healing. *Bone*, **25**, 321–332.

Chalmers, J. and Weaver, J.K. (1966) Cancellous bone: its strength and changes with aging and an evaluation of some methods for measuring its mineral content. *The Journal of Bone and Joint Surgery. American Volume*, **48**, 299.

Chamberlin, M.J., Fremlin, J.H., Peters, D.K. and Philip, H. (1968) Total body calcium by whole body neutron activation: New technique for study of bone disease. *British Medical Journal*, **2**, 581–583.

Chambers, T.J. (1985) The pathobiolog of the osteoclast. *Journal of Clinical Pathology*, **38**, 241–252.

Chambers, T.J., Evans, M., Gardner, T.N. *et al.* (1993) Induction of bone formation in rat tail vertebrae by mechanical loading. *Bone & Mineral*, **20**, 167–178.

Chambers, T.J., McSheehy, P.M.J., Thompson, B.M. and Fuller, K. (1985) The effect of calcium regulating hormones and prostagladins on bone resorption by osteoclasts disaggregated from neonatal rabbit bones. *Endocrinology*, **60**, 234–239.

Chang, K. and Chang, W.H.-S. (2003) Pulsed electromagnetic fields prevent osteoporosis in an ovariectomized female rat model: A prostaglandin $E_2$-associated process. *Bioelectromagnetics*, **24**, 189–198.

Chang, K., Chang, W.H., Huang, S. *et al.* (2005) Pulsed electromagnetic fields stimulation affects osteoclast formation by modulation of osteoprotegerin. RANK ligand and macrophage colony-stimulating factor. *Journal of Orthopaedic Research*, **23**, 1308–1314.

Chao, E.Y.S., Inoue, N., Kou, T.K.K. and Kim, Y.H. (2004) Biomechanical considerations of fracture treatment and bone quality maintenance in elderly patients and patients with osteoporosis. *Clinical Orthopaedics and Related Research*, **425**, 12–25.

Chapman, G.E., Danyluk, S.S. and McLauchlan, K.A. (1971) A model for collagen hydration. Proceedings of the Royal Society of London. Series B, Biological Sciences, **178**, 465.

Chappard, C., Laugier, P., Fournier, B. *et al.* (1997) Assessment of the relationship between broadband ultrasound attenuation and bone mineral density at the calcaneus using BUA imaging and DXA. *Osteoporosis International*, **7**, 316–322.

Chappard, D., Legrand, E., Haethich, B. *et al.* (2001) Fractal dimensions of trabecular bone: Comparison of three histomorphometric computed techniques for measuring the architectural two-dimensional complexity. *The Journal of Pathology*, **195**, 515–521.

Chapuy, M.C., Arlog, M.E., Duboeuf, F. *et al.* (1992) Vitamin $D^3$ and calcium to prevent hip fractures in women. *The New England Journal of Medicine*, **327**, 1637–1642.

Chapuy, M.C., Pamphili, R., Paris, E. *et al.* (2002) Combined calcium and vitamin D supplementation in elderly women: confirmation of reversal of secondary hyperparathyroidism and hip fracture risk: the Decalyos II Study. *Osteoporosis International*, **13**, 257–264.

Chapuy, M.C., Schott, A.M., Garnero, P. *et al.* (1996) Healthy elderly French women living at home have secondary hyperparathyroidism and high bone turnover in winter. EPIDOS Study Group. *The Journal of Clinical Endocrinology and Metabolism*, **81**, 1129–1133.

Charles, P.L., Mosekilde, L. and Taagehoj, J.F. (1985) The effects of sodium fluoride, calcium phosphate and vitamin D for one to two years on calcium and phosphorous metabolism in postmenopausal women with spinal crush fracture osteoporosis. *Bone*, **6**, 201–206.

Chavasseiux, P.M., Arlot, M.E., Rede, C. *et al.* (1997) Histomorphometric assessment of the long term effects of alendronate on bone quality and remodeling in patients with osteoporosis. *The Journal of Clinical Investigation*, **100**(6), 1475–1480.

Chavasseiux, P., Chenu, C., Valentin-opran, A. *et al.* (1993a) *In vitro* exposure to sodium fluoride does not modify activity and proliferation of human osteoblastic cells in primary cultures. *Journal of Bone and Mineral Research*, **8**, 37–44.

Cheal, E.J., Mansmann, K.A., DiGioia, A.M. *et al.* (1991) Role of interfragmentary strain in fracture healing: bovine model of a healing osteotomy. *Journal of Orthopaedic Research*, **9**, 131–142.

Chee, W.S., Suriah, A.R., Chain, S.P. *et al.* (2003) The effect of milk supplementation on bone mineral density in postmenopausal Chinese women in Malaysia. *Osteoporosis International*, **14**, 828–834.

Chen, D., Zhao, M. and Mundy, G.R. (2004) Bone morphogenetic proteins. *Growth Factors*, **22**, 233–241.

Chen, H., Frankenburg, E.P., Goldstein, S.A. and McCauley, L.K. (2003) Combination of local and systematic parathyroid hormone enhances bone regeneration. *Clinical Orthopaedics and Related Research*, **416**, 291–302.

Chen, N.X., Ryder, K.D., Pavalko, F.M. *et al.* (2000) $Ca^{2+}$ regulates fluid shear induced cytoskeletal reorganization and gene expression in osteoblasts. *American Journal of Physiology. Cell Physiology*, **278**, C989–C997.

Chen, T. Chen, P.J., Fung, C.S. *et al.* (2004) Quantitative assessment of osteoporosis from the tibia shaft by ultrasound techniques. *Medical Engineering & Physics*, **26**, 141–145.

Chen, T.L., Bates, R.L., Dudley, A. *et al.* (1991) Bone morphogenetic protein-2B stimulation of growth and osteogenotypes in rat osteoblast like cells: comparison with TGF. *Journal of Bone and Mineral Research*, **6**, 1337–1393.

Chen, Z., Lohman, T.G., Stini, W.A. *et al.* (1997) Fat or lean tissue mass: Which one is the major determinant of bone mineral mass in healthy postmenopausal women? *Journal of Bone and Mineral Research*, **12**, 144–151.

Cheng, S., Timonen, J. and Suominen, H. (1995) Elastic wave propagation in bone *in vivo*: Methodology. *Journal of Biomechanics*, **28**, 471–478.

Chenu, C. (2002) Glutamatergic innervation in bone. *Microscopy Research and Technique*, **58**, 70–76.

Chesnut, C.H. (1988) Drug therapy: calcitonin biphosphonates, in *Anabolic Steroids and PTH* (eds B.L. Riggs and L.F. Belton), Raven Press, New York, pp. 404–418.

Chesnut, C.H. III (1993) Bone mass and exercise. *The American Journal of Medicine*, **95**, 34S–36S.

Chesnut, C.H., Baylink, D.J., Gruber, H.E. *et al.* (1984) Synthetic salmon calcitonin in postmenopausal osteoporosis: current concepts, in *Calciton: Das Therapeutische Potential be Introduce Osteoporose, Scharrauer* (eds L.V. Avioli M.A. Dambacher), Springer Verlag, Stuttgart.

Chesnut, C.H. III, Bell, N.H., Clark, G.S. *et al.* (1997) Hormone replacement therapy in postmenopausal women: urinary N-telopeptide of Type I collagen; monitoring therapeutic effect and predicting response of bone mineral density. *The American Journal of Medicine*, **102**, 29–37.

Chesnut, C.H. III, Help, W.B., Baylink, D.J. and Denney, J.D. (1977) Effect of methandrostenolone on postmenopausal bone wasting as assessed by changes in total bone mineral mass. *Metabolism: Clinical and Experimental*, **26**, 267–277.

Chesnut, C.H. III, Silverman, S., Andriano, K. *et al.* (2002) A randomized trial of nasal spray salmon clcitonin in postmenopausal women with established osteoporosis: The prevent recurrence of osteoporotic fracture study. Proof Study Group. *The American Journal of Medicine*, **109**, 267–276.

Chevalier, F., Laval Jeantet, A.M., Laval Jeantet, M. and Bergot, C. (1992) CT image analysis of the vertebral trabecular network *in vivo*. *Calcified Tissue International*, **51**, 8–13.

Cho, C., Kobayashi, A., Takei, R. *et al.* (2001) Receptor mediated cells modulator delivery to hepatocyte using nanoparticles coated with carbohydrate carrying polymers. *Biomaterials*, **22**, 45–51.

Cho, T.-J., Gerstenfeld, L.C. and Einhorn, T.A. (2002) Differential temporal members of the transforming growth factor B step during murine fracture healing. *Journal of Bone and Mineral Research*, **17**, 51.

Chou, J., Tobias, J.H., Colston, K.W. and Chambers, T.J. (1992) Estrogen maintains trabecular bone volume in rats not only by suppression of bone resorption but also by stimulation of bone formation. *The Journal of Clinical Investigation*, **89**, 74–78.

Chow, J.W.M. and Chambers, T.J. (1994) Indomethacin has distinct early and late actions on bone formation induced by mechanical stimulation. *The American Journal of Physiology*, **267**, E287–E292.

Chow, J.W.M., Jagger, C.J. and Chambers, T.J. (1993) Characterization of osteogenic response to mechanical stimulation in cancellous bone of rat caudal vertebrae. *The American Journal of Physiology*, **265**, E340–E347.

Chow, J.W.M., Wilson, A.J., Chambers, T.J. and Fox, S.W. (1998) Mechanical loading stimulates bone formation by reactivation of bone lining cells in 13-week-old rats. *Journal of Bone and Mineral Research*, **13**, 1760–1767.

Chow, R., Harrison, J.E. and Notarius, C. (1987) Effect of two randomized exercise programs on bone mass of healthy postmenopausal women. *British Medical Journal*, **295**, 1441–1444.

Chow, T.S. (1977) Elastic moduli of filled polymers: the effect of particle shape. *Journal of Applied Physiology*, **48**, 4072–4075.

Chow, T.S. (1978) Effect of particle shape at finite concentration on the elastic moduli of filled polymers. *Journal of Polymer Science*, **16**, 959–965.

Chow, T.S. (1980) Review. The effect of particle shape on the mechanical properties of filled polymers. *Journal of Materials Science*, **15**, 1873–1888.

Chow, T.S. (1982) Tensile strength of filled polymers. *Journal of Polymer Science*, **20**, 2103–2109.

Chrischilles, E.A., Butler, C.D., Davis, C.S. and Wallace, R.B. (1991) A lifetime osteoporosis impact. *Archives of Internal Medicine*, **151**, 202–206.

Christansen, J.B., Adams, J.P., Cho, K.O. and Miller, L. (1968) A study of the interosseous distance between the radius and ulna during rotation of the forearm. *The Anatomical Record*, **160**, 261–272.

Christgau, S., Rosenquist, C., Alexandersen, P. *et al.* (1998) Clinical evaluation of the serum Cross Laps One Step ELISA, a new assay measuring the serum concentration of bone derived degradation products of type I collagen C-telopeptides. *Clinical Chemistry*, **44**, 2290–2300.

Christgau, M., Schmalz, G., Wenzel, A. and Hiller, K.-A. (1997) Periodontal regeneration of intrabony defects with resorbable and nonresorbable membranes: 30-month results. *Journal of Clinical Periodontology*, **24**, 17–27.

Christiansen, C. (1991) Use of normally administered salmon calcitonin in preventing bone loss. *Calcified Tissue International*, **49**, S14–S15.

Christiansen, C., Christensen, M.C. and Transbol, I. (1981) Bone mass in postmenopausal women after withdrawal of estrogen/gestrogen replacement therapy. *Lancet*, **1**, 459–461.

Christiansen, C., Christensen, M.S., McNair, P. *et al.* (1980) Prevention of early postmenopausal bone loss: controlled 2 years study in 315 normal females. *European Journal of Clinical Investigation*, **10**, 273–279.

Christiansen, C. and Lindsay, R. (1990) Estrogen bone loss and preservation. *Osteoporosis International*, **1**, 7–13.

Christiansen, C., Riis, B.J., Nilas, L. *et al.* (1985) Uncoupling of bone formation and resorption by combined oestrogen and progestagen therapy in postmenopausal osteoporosis. *Lancet*, **11**, 800–801.

Christiansen, C. and Rodbro, P. (1977) Long term reproducibility of bone mineral content measurements. *Scandinavian Journal of Clinical and Laboratory Investigation*, **37**, 321–323.

Christiansen, C., Rodbro, P. and Jensen, H. (1975) Bone mineral content in the reararm measured by photon absorptiometry. *Scandinavian Journal of Clinical and Laboratory Investigation*, **35**, 323–330.

Christiansen, P., Steiniche, T., Brockstedt, H. *et al.* (1993) Primary hyperparathyroidism: Iliac crest cortical thickness, structure and remodeling evaluated by histomorphemetric methods. *Bone*, **14**, 403–408.

Chu, M.L., Williams, C.J., Pepe, G. *et al.* (1984) Internal deletion in a collagen gene in perinatal lethal form of osteogenesis imperfecta. *Nature*, **304**, 78–80.

Chung, Y.S., Lee, M.D., Lee, S.K. *et al.* (2000) HMG-CoA reductase inhibitors increase BMD in type 2 diabetes mellitus patients. *The Journal of Clinical Endocrinology and Metabolism*, **85**(3), 1137–1142.

Chyun, Y.S. and Raisz, L.G. (1984) Stimulation of bone formation by prostaglandin $E_2$. *Prostaglandins*, **27**, 97–103.

Cidogan, J., Blumsohn, A., Barker, M. and Eastell, R.A. (1998) Longitudinal study of bone gain in pubertal girls: anthropometric and biochemical cortical. *Journal of Bone and Mineral Research*, **13**, 1602–1612.

Cieszynski, T. (1964) Studies on the regeneration of osseal tissue. Influence of positive and negative electricity on callus formation in humans. *Archivum Immunologiae et Therapiae Experimentalis*, **12**, 269–297.

Cignitti, M., Figura, F., Marchetti, M. and Salleo, A. (1970) Electro-kinetic effects in mechanoelectrical phenomenology of the bone. *Archivio di Fisiologia*, **68**, 232–249.

Ciombor, D.M., Lester, G., Aaron, R.K. *et al.* (2002) Low frequency EMF regulates chondrocyte differentiation and expression of matrix proteins. *Journal of Orthopaedic Research*, **20**, 40–50.

Civitelli, R. (1995) Cell-cell communication in bone. *Calcified Tissue International*, **56**(S), S29–S31.

Civitelli, R., Gonnelli, S., Zaccheri, F. *et al.* (1988) Effect of calcitonin treatment. *The Journal of Clinical Investigation*, **82**, 1268–1274.

Claes, L.E., Augat, P., Suger, G. and Wilke, H.J. (1997) Influence of size and stability of the osteotomy gap on the success of fracture healing. *Journal of Orthopaedic Research*, **15**, 577–584.

Claes, L.E. and Heigele, C.A. (1999) Magnitudes of local stress and strain along bony surfaces predict the course and type of fracture healing. *Journal of Biomechanics*, **32**, 255–266.

Clarke, A.J., Evans, J.A., Truscott, J.G. *et al.* (1994) A phantom for quantitative ultrasound of trabecular bone. *Physics in Medicine and Biology*, **39**, 1677–1687.

Clarke, B.L., Ebeling, P.R., Jones, J.D. *et al.* (1996) Changes in quantitative bone histomorphometry in aging health men. *The Journal of Clinical Endocrinology and Metabolism*, **81**, 2264–2270.

Clarke, R.L., Milne, E.N.C. and Van Dyke, G. (1976) The use of Compton scattered gamma rays for tomography. *Investigative Radiology*, **11**, 225–235.

Clarke, R.L. and Van Dyk, G. (1973) A new method for measurement of bone mineral content using both transmitted and scattered beams of gamma-rays. *Physics in Medicine and Biology*, **18**, 532–539.

Claus-Walker, J., Spencer, W.A., Carter, R.E. *et al.* (1975) Bone metabolism in quadriplegia: dissociation between calciuria and hydroxyprolinuria. *Archives of Physical Medicine and Rehabilitation*, **56**, 327–332.

Cleghorn, D.B., Polley, K.J., Bellon, M.J. *et al.* (1991) Fracture rates as a function of forearm mineral density in normal postmenopausal women: retrospective and prospective data. *Calcified Tissue International*, **49**, 161–163.

Clemedson, C. and Johnson, A. (1961) Transmission and reflection of high explosive shock waves in bone. *Acta Physiologica Scandinavica*, **51**, 47–61.

Cobbold, P.H. and Lee, J.A.C. (1991) *Cellular-Calcium: A Practical Approach* (eds J.G. McCormack and P.H. Cobbold), Oxford University, Oxford, pp. 54–81.

Coble, R.L. and Kingery, W.D. (1956) Effect of porosity on physical properties of alumina. *Journal of the American Ceramic Society*, **39** (11), 377–385.

Cochet, B., Jung, A., Griessen, M. *et al.* (1983) Effects of lactose on intestinal calcium absorption in normal and lactase deficient subjects. *Gastroenterology*, **84**, 935–940.

Cochran, G.V.B. (1966) Electromechanical characteristics of moist bone. PhD thesis. Columbia University, New York.

Cochran, G.V.B., Otter, M.W., Bieber, W. and Wu, D. (1993) Streaming potentials associated with gap healing in canine tibia. Transactions of the Bioelectrical Repair and Growth Society 13th Annual Meeting Dana Point, CA.

Cochran, G.V.B., Pawluk, R.J. and Bassett, C.A.L. (1968) Electromechanical characteristics of bone under physiologic moisture conditions. *Clinical Orthopaedics*, **58**, 249–270.

Coe, F.L., Canterbury, J.M., Firpo, J.J. and Reiss, E. (1973) Evidence for secondary hyperparathyroidism in idiopathic hypercalciuria. *The Journal of Clinical Investigation*, **52**, 134–14.

Coe, F.L., Firpo, J.J., Hollandsworth, D.L. *et al.* (1975) Effect of acute and chronic metabolic acidosis on serum immunoreactive parathyroid hormone in man. *Kidney International*, **8**, 262–273.

Cohen-Solal, M.E., Baudoin, C., Omouri, M. *et al.* (1998) Bone mass in middle aged osteoporotic men and their relatives. Familial effect. *Journal of Bone and Mineral Research*, **13**, 1909–1914.

Cohn, S.H., Dombrowski, C.S. and Fairchild, R.G. (1970) *In vivo* neutron activation analysis of calcium in man. *International Journal of Applied Radiation and Isotopes*, **21**, 127–137.

Cohn, S.H., Ellis, K.J. and Wallach, S. (1974) Absolute and relative deficit in total skeletal calcium and radial bone mineral in osteoporosis. *Journal of Nuclear Medicine*, **15**, 428–435.

Cohn, S.H., Shukla, K.K., Dombrowski, C.S. and Fairchild, R.G. (1972) Design and calibration of a "broad beam" 238 PUBE neutron source for total body neutron activation analysis. *Journal of Nuclear Medicine*, **13**, 487–492.

Cohn, S.H., Vaswani, A., Zanzi, I. and Ellis, K.J. (1976a) Effect of aging on bone mass in adult women. *The American Journal of Physiology*, **230**, 143–148.

Cohn, S.H., Vaswani, A., Zanzi, I. *et al.* (1976b) Changes in body chemical composition with age measured by total body neutron activation. *Metabolism: Clinical and Experimental*, **25**, 85–95.

Cohn, S.M., Zabzi, I., Vaswani, S. *et al.* (1976c) Quantitation of the degree of osteoporosis by measure of total-body calcium employing neutron activation. *International Journal of Calcif Tissue*, **21**, Suppl., 375–379.

Cole, W.G., Campbell, P.E., Rogers, J.G. and Baterman, J.F. (1990a) The clinical features of osteogenesis imperfecta resulting from a nonfunction carboxy terminal prox (I) propeptide of type I prolollagen and a severe deficiency of normal type I collagen in tissues. *Journal of Medical Genetics*, **27**, 545–551.

Cole, W.G., Chow, C.G., Rogers, J.G. and Baterman, J.F. (1990b) The clinical features of three babies with osteogenesis imperfecta resulting from the substitution of glycine by arginine in the Prox (I) chains of type I prolollagen and a severe deficiency of normal type I collagen in tissues. *Journal of Medical Genetics*, **27**, 228–235.

Cole, W.G., Patterson, E., Bonadio, J. *et al.* (1992) The clinicopathological features of three babies with osteogenesis imperfecta resulting from the substitution of glycine by valine in the Prox (I) chain of the type I procollagen. *Journal of Medical Genetics*, **29**, 112–118.

Coleman, T.H. (1987) Bone mineral screening for osteoporosis (letter). *The New England Journal of Medicine*, **317**, 315–316.

Coles, J., Hamdy, R., Bocuquet, J. *et al.* (1985) The treatment of biochemical osteomalacia abstract. XIII International Congress Gerontology, New York, p. 147.

Colette, C., Monnier, L., Pares-Herbute, N. *et al.* (1987) Calcium absorption in corticoid treated subjects: effects of a single oral dose of calcitriol. *Hormone and Metabolic Research*, **19**(7), 335–338.

Coliacicco, G. and Pilla, A.A. (1984) Electromagnetic modulation of biological processes: influence of culture medium and significance of methodology in the calcium uptake of embryonal chick tibia *in vitro*. *Calcified Tissue International*, **36**, 167–174.

Collin-Osdoby, P., Nickols, G.A. and Osdoby, P. (1995) Bone cell function, regulation and communication-A role for nitric oxide. *Journal of Cellular Biochemistry*, **57**, 399–408.

Collins, D.H. (1966) *Pathology of Bone*, Butlerworths, London.

Coloson, D.J., Browett, J.P., Fiddian, N.J. *et al.* (1974) Uptake of tritiated thymidine during electrical stimulation of induced cortical bone defects. *Annals of the New York Academy of Sciences*, **38**, 307.

Colston, K.W., King, R.J.B., Hayward, J. *et al.* (1989) Estrogen receptors and human bone cells. Immunocytochemical studies. *Journal of Bone and Mineral Research*, **4**, 625–631.

Colvard, D., Keeting, P., Scott, R. *et al.* (1989) Effects of estrogen on growth and differentiation of normal human osteoblast like cells. *Journal of Bone and Mineral Research*, **4**, S118.

Compfield, L.A., Smith, F.J., Guise, Z.Y. *et al.* (1995) Recombinant mouse OB protein: evidence for a peripheral signal linking adiposity and central neural net works. *Science*, **269**, 546–549.

Compston, J.E. (1997) Hormone replacement therapy. *Baillière's Clinical Rheumatology*, **11**, 583–596.

Compston, J.E. (2001) Sex steroids and bone. *Physiological Reviews*, **81**, 419–447.

Compston, J.E., Chadha, S. and Merrett, A.L. (1980) Osteomalacia developing during treatment of osteoporosis with sodium fluoride and vitamin D. *British Medical Journal*, **281**(6245), 910–911.

Compston, J.E., Cooper, C. and Kanis, J.A. (1995) Bone densitometry in clinical practise. *British Medical Journal*, **310**, 1507–1510.

Compston, J.E., Laskey, M.A., Croucher, P.I. *et al.* (1992) Effect of diet-induced weight loss on total body bone mass. *Clinical Science*, **82**, 429–432.

Conney, J.M., Spears, G.F.S. and Goulding, A. (1984) Effects of varying dietary NaCl intake on 24 hour urinary sodium, calcium and hydroxyproline values in women consuming low and high dietary calcium intakes. Proceedings 3rd International Symposium: Bone Structure Function and Disease, Adelaide, p. 40.

Connolly, J.F., Hahn, H. and Jardon, O.M. (1977) The electrical enhancement of periosteal proliferation in normal and delayed fracture healing. *Clinical Orthopaedics*, **124**, 97.

Connolly, J.F., Ortiz, J., Price, R.R. and Bayuzick, R.J. (1974) The effect of electrical stimulation on the biophysical properties of fracture healing. *Annals of the New York Academy of Sciences*, **238**, 519–529.

Cook, S.D., Baffes, G.C., Wolfe, M.W. *et al.* (1994a) Recombinant human bone morphogenetic protein-7 induces healing in a canine long bone segmental defect model. *Clinical Orthopaedics*, **301**, 302–312.

Cook, S.D., Baffes, G.C., Wolfe, M.W. *et al.* (1994b) The effect of recombinant human osteogenic on healing of large segmental bone defects. *The Journal of Bone and Joint Surgery*, **76**, 827–838.

Cook, S.D., Wolfe, M.W., Salkeld, S.L. and Rueger, D.C. (1995) Effect of recombinant human osteogenic protein-1 on healing of segmental defects in non-human primates. *The Journal of Bone and Joint Surgery*, **77-A**, 734–750.

Cooper, C., Atkinson, J., Wahner, H.W. *et al.* (1992) Is caffeine consumption a risk factor for osteoporosis? *Journal of Bone and Mineral Research*, **7**, 465–471.

Cooper, C., Barker, D.J.P. and Wickman, C. (1988) Physical activity, muscle strength and calcium intake in fracture of the proximal femur in Britain. *British Medical Journal*, **279**, 1443–1446.

Cooper, C., Cawley, M., Bhalla, A. *et al.* (1995) Childhood growth, physical activity and peak bone mass in women. *Journal of Bone and Mineral Research*, **10**, 940–947.

Cooper, C., Fall, C., Egger, P. *et al.* (1997) Growth in infancy and bone mass in later life. *Annals of the Rheumatic Diseases*, **56**, 17–21.

Cooper, C., Hannaford, P., Croft, P. and Kay, C.R. (1993) Oral contraceptive pill use and fracture in women. A prospective study. *Bone*, **14**, 41–45.

Cooper, C. and Melton, L.J.I. (1992) Epidemiology of osteoporosis. *Trends in Endocrinology and Metabolism*, **3**, 224–229.

Cooper, C. and Wickham, C. (1990) Cigarette smoking and the risk of age related fractures, in *Smoking and Hromone Related Disorders* (eds N. Wald and P. Froggatt), Oxford University Press, Oxford, pp. 93–100.

Cooper, M.S. and Keller, R.E. (1984) Perpendicular orientation and directional migration of amphibian neural crest cells in dc electrical field. *Proceedings of the National Academy of Sciences of the United States of America*, **81**(1), 160–164.

Cornish, J., Callon, K.E., Bava, U. *et al.* (2001) The direct actions of leptin on bone cells increase bone strength *in vivo* – an explanation of low fracture rates in obesity. *Bone*, **28**(Suppl.), S88.

Cornish, J., Callon, K.E., Mountjoy, K.G. *et al.* (2001) Alpha-melanocyte stimulating hormone – A novel factor affecting bone turnover. *Journal of Bone and Mineral Research*, **16**(Suppl. 1), S201.

Corradi, C. and Cozzolino, A. (1953) Gli ultrasuoni e Tevoluzione delle fratture sperimentalli dei conigli. *Archives of Orthopaedic*, **66**, 77–98.

Courtney, A.C., Hayes, W.C. and Gibson, L.J. (1996) Age related differences in post yield damage in human cortical bone: Experiment and model. *Journal of Biomechanics*, **29**, 1463–1471.

Cowin, S.C. (1999) Bone poroelasticity. *Journal of Biomechanics*, **32**, 217–238.

Cowin, S.C., Moss-Salentijn, L. and Moss, M.L. (1991) Candidates for the mechanosensory system in bone. *Journal of Biomechanical Engineering*, **113**(2), 191–197.

Cowin, S.C., Weinbaum, S. *et al.* (1995) A case for bone canaliculi as the anatomical site of strain generated potentials. *Journal of Biomechanics*, **28**(11), 1281–1297.

Cranney, A., Guyatt, G., Griffith, L. and Wells, G. (2002) IX: Summary of meta-analyses of therapies for postmenopausal osteoporosis. *Endocrine Reviews*, **23**, 570–578.

Craven, J.D., Costantine, M.A., Greenfield, M.A. and Stern, R. (1973) Measurement of the velocity of ultrasound in human cortical bone and its potential clinical importance. *Investigative Radiology*, **8**, 72–77.

Crawford, R.P., Brouwers, J. and Keaveny, T.M. (2004) Accurate prediction of vertebral strength using voxel-based non-linear finite element models. *Transactions of the Orthopaedic Research Society*, **29**, 1123.

Crawford, R.P., Cann, C.E. and Keaveny, T.M. (2003) Finite element models predict *in vitro* vertebrae body compressive strength better than quantitative computed tomography. *Bone*, **33**, 744–750.

Crawford, R.P. and Keaveny, T.M. (2004) Relationship between axial and bending behaviors of the human thoracolumbar vertebra. *Spine*, **29**, 2248–2255.

Croker, J. and Nar, P. (1987) Nucleolar organized regions in lymphomas. *The Journal of Pathology*, **152**, 111–118.

Croucher, P.I., Mellish, R.W.E., Vedi, S. *et al.* (1989) The relationship between resorption depth and mean interstitial bone thickness: age related changes in man. *Calcified Tissue International*, **45**, 119.

Crowninshield, R.D. and Pope, M.H. (1974) The response of compact bone in tension at various strain rates. *Annals of Biomedical Engineering*, **2**, 217–225.

Cruess, R.L. and Dumont, J. (1975) Fracture healing. *Canadian Journal of Surgery*, **18**, 403–413.

Cruess, R.L., Kan, K. and Bassett, C.A.L. (1983) The effect of pulsing electromagnetic fields upon bone metabolism in an experimental model of disuse osteoporosis. *Clinical Orthopaedics and Related Research*, **173**, 345.

Cui, Q., Wang, G.J. and Balian, G. (1997) Steroid-induced adipogenesis in a pluripotential cell line bone marrow. *The Journal of Bone and Joint Surgery*, **79**, 1054–1063.

Cui, Q., Wang, G.J., Su, C.C. and Balian, G. (1997a) The Otto Aufranc Award. Lovastatin prevents steroid-induced adipogenesis and osteonecrosis. *Clinical Orthopaedics and Related Research*, **344**, 8–19.

Cui, Q., Wang, G.J., Su, C.C. and Balian, G. (1997b) Lovastatin prevents steroid induced adipogenesis and osteonecrosis. *Clinical Orthopaedics*, **344**, 8–19.

Culbert, A.A., Lowe, M.P., Atkinson, M. *et al.* (1995) Substitutions of aspartic acid for glycine-220 and of arginine glycine-664 in the triple helix of the pro alpha 1 (I) chain of type I procollagen produce lethal osteogenesis imperfecta and disrupt of ability of collagen fibrils to incorporate crystalline hydroxyapatite. *The Biochemical Journal*, **311**, 815–820.

Cullum, I.D., Ell, P.J. and Ryder, J.P. (1989) X-ray dual photon absorptiometry: a new method for the measurement of bone density. *British Journal of Radiology*, **62**, 587–592.

Cummings, R.G. and Nevitt, M.C. (1997) Calcium for prevention of osteoporosis fractures in postemenopausal women. *Journal of Bone and Mineral Research*, **12**, 1321–1329.

Cummings, R.G., Nevitt, M.C. and Cummings, S.R. (1997) Epidemiology fractures. *Epidemiologic Reviews*, **19**, 244–257.

Cummings, S.R. (1985) Are patients with hip fractures more osteoporotic? *The American Journal of Medicine*, **78**, 487–494.

Cummings, S.R., Bates, D. and Black, D.M. (2002) Clinical use of bone densitometry – scientific review. *The Journal of the American Medical Association*, **288**, 1889–1897.

Cummings, S.R., Black, D.M., Nevitt, M.C. *et al.* (1993) Bone density at various sites for prediction of hip fractures. *Lancet*, **341**, 72–75.

Cummings, S.R., Black, D.M., Thompson, D.E. *et al.* (1998) Effect of alendronate on risk of fracture in women with low bone density but with low bone density but without vertebral fractures. Results from the fracture intervention trial. *The Journal of the American Medical Association*, **280**, 2077–2082.

Cummings, S.R., Browner, W.S. and Ettinger, B. (1990) Should prescription of postmenopausal hormone therapy be based on the results of bone densitometry? *Annals of Internal Medicine*, **113**, 565–567.

Cummings, S.R., Kelsey, J.L., Nevitt, M.C. and O'Dowd, K.J. (1985) Epidemiology of osteoporosis and osteoporotic fractures. *Epidemiological Reviews*, **7**, 178–208.

Cummings, S.R., Nevitt, M.C., Browner, W.S. *et al.* (1989) Hip and wrist fractures are due to different types of falls, not different types of osteoporosis. *Journal of Bone and Mineral Research*, **4**(Suppl.), S170.

Cummings, S.R., Nevitt, M.C., Browner, W.S. *et al.* (1995) Risk factors for hip fracture in white women. *The New England Journal of Medicine*, **332**, 767–773.

Currey, J.D. (1964a) Some effects of ageing in human haversian systems. *Journal of Anatomy*, **98**, 69.

Currey, J.D. (1964b) Three analogies to explain the mechanical properties of bone. *Biorheology*, **2**, 1–10.

Currey, J.D. (1965) An elasticity in bone and echinoderm skeletons. *The Journal of Experimental Biology*, **43**, 279–292.

Currey, J.D. (1968) The adaptation of bones to stress. *Journal of Theoretical Biology*, **20**, 91–106.

Currey, J.D. (1969) The relationship between the stiffness and the mineral content of bone. *Journal of Biomechanics*, **2**, 477–480.

Currey, J.D. (1975) The effects of strain rate, reconstruction, and mineral content on some mechanical properties of bovine bone. *Journal of Biomechanics*, **8**, 81–86.

Currey, J.D. (1979) Changes in impact energy absorption of bone with age. *Journal of Biomechanics*, **12**, 459–469.

Currey, J.D. (1984) Effects of differences in mineralization on the mechanical properties of bone. *Proceedings of The Royal Society of London. B, Biological Sciences*, **304**, 509–518.

Currey, J.D. (1988) The effect of porosity and mineral content on the Young's modulus of elasticity of compact bone. *Journal of Biomechanics*, **21**, 131–139.

Currey, J.D. (1989) Strain dependence of the mechanical properties of reindeer antler and the cumulative damage of bone fracture. *Journal of Biomechanics*, **22**, 469–476.

Currey, J.D. (2002) *Bones. Structure and Mechanics*, Princeton University Press, Princeton, NJ.

Currey, J.D. (2003) The many adaptations of bone. *Journal of Biomechanics*, **36**, 1487–1495.

Currey, J.D., Brear, K. and Zioupos, P. (1996) The effects of aging and changes in mineral content in degrading the toughness of human femora. *Journal of Biomechanics*, **29**, 257–260.

Currey, J.D. and Butler, G. (1975) The mechanical properties of bone tissue in children. *The Journal of Bone and Joint Surgery*, **57-A**, 810.

Currey, J.D., Foreman, J., Laketic, I. *et al.* (1997) Effect of ionizing radiation on the mechanical properties of human bone. *Journal of Orthopaedic Research*, **15**, 111–117.

Cynthia, L.I., Gordon, L.W., Harry, A.H. and Susan, A.B. (1999) Mechanical loading attenuates bone loss due to immobilization and calcium deficiency. *American Physiological Society*, **87**(1), 189–195.

da Silva, E.F. and Mephun, M.K. (1978) Calibration tech. for one part measurement. *Microwave Journal*, **21**, 97–100.

Dalen, N. and Jacobson, B. (1974) Bone mineral assay: Choice of measuring sites. *Investigative Radiology*, **3**, 174–185.

Dalli Carbonare, L. and Giannini, S. (2004) Bone microarchitecture as an important determinant of bone strength. *Endocrinological Investigation*, **27**, 99–105.

Dalsky, G.P. (1989) The role of exercise in the prevention of osteoporosis. *Comprehensive Therapy*, **15**, 30–37.

Dalsky, G.P., Stocke, P.S., Ehsani, A.A. *et al.* (1988) Weight bearing exercise training and lumbar bone mineral content in postmenopausal women. *Annals of Internal Medicine*, **108**, 824–828.

Damask, C. (1978) *Medical Physics*, Academic Press, New York, ch. 5, p. 121.

Dambacher, M.A., Ittner, J. and Ruegsegger, P. (1986) Long term fluoride therapy of postmenopausal osteoporosis. *Bone*, **7**, 199–205.

Dambacher, M.A., Neff, M., Kissling, R., Qin, L. *et al.* (1998) Highly precise peripheral quantitative computed tomography, bone density, loss of bone density and structure's: Consequences for prophylaxix and treatment. *Drugs Aging*, **12**, 15–24.

Dambacher, M.A. and Ruegsegger, P. (1987) Therapy of osteoporosis with an osseinhydroxyapatite compound, evaluated with quantitative computed tomography. *Journal of Bone and Mineral Research*, **2**(Suppl. 1), 325.

Dane, S., Akar, S., Hacibeyoglu, I. and Varoglu, E. (2001) Differences between right and left femoral bone mineral densities in right and left handed men and women. *The International Journal of Neuroscience*, **111**, 187–192.

Daniell, H.W. (1976) Osteoporosis and the stander smoker: vertebrae compression fractures and loss of metacarpal cortex in relation to postmenopausal cigarette smoking and lack of obesity. *Archives of Internal Medicine*, **136**, 298–304.

Daniell, H.W. (1997) Osteoporosis after ostectomy for prostate cancer. *The Journal of Urology*, **157**, 439–444.

Danielsen, C.C., Mosekilde, Li., Bollersley, J. and Mosekilde, Le. (1994) Thermal stability of cortical one collagen in relation to age in normal individuals and in individuals with osteoporosis. *Bone*, **15**, 91–96.

Danon, A. and Korenstein, R. (1984) 720-Capactive pulsed electrical stimulation of bone cells: Induction of calcium uptake. *Bioelectrochemistry and Bioenergetics*, **13**, 49–54.

Danz, A.M., Zittermann, A., Schiedermaier, U. *et al.* (1998) The effect of a specific strength development exercise on bone mineral density in perimenopausal and postmenopausal women. *Journal of Women's Health*, **7**, 701–709.

Daon, N., Reher, P., Meghji, S. and Harris, M. (1999) *In vitro* effects of therapeutic ultrasound on cell proliferation, protein synthesis and cytokine production by human fibroblasts, osteoblats and monocytes. *Journal of Oral and Maxillofacial Surgery*, **57**, 409–419.

Dartsch, P.C. and Betz, E. (1989) Response of cultured endothelial cells to mechanical stimulation. *Basic Research in Cardiology*, **84**, 268–281.

Date, M. (1972) Piezoelectric constant in dispersed systems. *Reports in Progress on Polymer Physics in Japan*, **15**, 385–387.

Dauty, M., Perrouin Verbe, B., Maugars, Y. *et al.* (2000) Supralesional and sublesional bone mineral density in spinal cordinjured patients. *Bone*, **27**, 305–309.

Davidovitch, Z., Shanfeld, J.L., Montgometry, P.C. *et al.* (1984) Biochemical mediators of the effects of mechanical forces and electric currents on mineralized tissues. *Calcified Tissue International*, **36**, 865–975.

Davidson, M.R. (2003) Pharmacotherapeutics for osteoporosis prevention and treatment. *Journal of Midwifery & Women's Health*, **48**, 39–52.

Davidson, S., Mansour, A., Gallily, R. *et al.* (1983) Mast cell differentiation depends on T cells and granule synthesis on fibroblasts. *Immunology*, **48**, 439–452.

Davis, J.W., Voge, J.M., Ross, P.D. and Wanich, R.D. (1989) Diseases versus etiology: the distinction should not be lost in the analysis. *Journal of Nuclear Medicine*, **30**, 1273–1276.

Davis, M.R. (1987) Screening for postmenopausal osteoporosis. *American Journal of Obstetrics and Gynecology*, **156**, 1–5.

Davis, R.F., Jones, L.C. and Hungerford, D.S. (1987) The effect of symphectomy on blood flow in bone. Regional distribution and effect over time. *The Journal of Bone and Joint Surgery*, **69**, 1384–1390.

Dawson, H.B. and Harris, S. (1993) Thiazides and seasonal bone change in healthy postmenopausal women. *Bone & Mineral*, **21**, 41–51.

Dawson, H.B., Harris, S.S., Krall, E.A. and Dallal, G.E. (1997) Effect of calcium and vitamin D supplementation on boe density in men and women 65-years of age or older. *The New England Journal of Medicine*, **337**, 670–676.

De Boer, H., Blok, G.J., Van Lingen, A. *et al.* (1994) Consequences of childhood-onset growth hormone deficiency for adult bone mass. *Journal of Bone and Mineral Research*, **9**, 1319–1326.

De Mercato, G. and Garcia-Sanchez, F.J. (1988) Dielectric properties of fluid saturated bone: a comparison between diaphysis and epiphysis. *Medical and Biological Engineering and Computing*, **26**, 313–316.

De Mercato, G. and Garcia-Sanchez, F. (1992) Correlation between low frequency electric conductivity and permittivity in the diaphysis of bovine femoral bone. *IEEE Transactions on Biomedical Engineering*, **39**, 523–526.

De Reggi, A.S., Roth, S.C., Kenny, J.M. and Edelman, S. (1981) Piezoelectric polymer probe for ultrasonic applications. *The Journal of the Acoustical Society of America*, **69**, 853–859.

De Vernegoul, M.C., Laval-Jeanter, A.M., Garcia-Carrasco, M. and Rckewaert, A. (1984) Bioconcavity of lumbar vertebrae as an index of osteopenia? A histomorphometric, CT densitometric and radiological study (Abstract). Copenhagen International Symposium on Osteoporosis.

Dedman, J.R., Brinkley, B.R. and Means, A.R. (1979) Regulation of microfilaments and microtubules by calcium and cyclic AMP. *Advances in Cyclic Nucleotide Research*, **11**, 131–174.

Delmas, P.D. (1999) HRT in the prevention and treatment of osteoporosis. *Journal of Epidemiology and Biostatistics*, **4**, 155–160.

Delmas, P.D. (2002) Different effects of antiresorptive therapies on vertebral and nonvertebral fractures in postmenopausal osteoporosis. *Bone*, **30**(1), 14–17.

Delmas, P.D., Hardy, P., Garnero, P. and Dain, M.-P. (2000) Monitoring individual response to hormone replacement therapy with bone markers. *Bone*, **26**, 553–560.

Delmas, P.D., Wahner, H.W., Mann, K.G. and Riggs, B.L. (1983) Assessment of bone turnover in postmenopausal osteoporosis by measurement of serum bone Gla protien. *The Journal of Laboratory and Clinical Medicine*, **102**, 470–476.

Delvin, H.B. and Goldman, M. (1966) Backache due to osteoporosis in an industrial population. A survey of 481 patients. *Irish Journal of Medical Science*, **6**, 141–148.

Dempster, D.W., Arlot, M.A. and Meunier, P.J. (1983) Mean wall thickness and formation periods of trabecular bone packets in corticosteroid induced osteoporosis. *Calcified Tissue International*, **35**, 410–417.

Dempster, D.W., Cosman, F., Parisien, M. and Shen, V. (1993) Anabolic actions of parathyroid hormone on bone. *Endocrine Reviews*, **14**, 690–709.

Dempster, D.W., Birchman, R., Lindsay, R. *et al.* (1995) Temporal changes in cancellous bone structure of rats immediately after overiectomy. *Bone*, **16**, 157–161.

Dempster, W.T. and Liddicoat, R.T. (1952) Compact bone as a non isotropic material. *The American Journal of Anatomy*, **91**, 331–362.

Deng, H.W., Chen, W.M., Recker, S. *et al.* (2000) Genetic determination of Colles' fracture and differential bone mass in women with and without Colles' fracture. *Journal of Bone and Mineral Research*, **15**, 1243–1252.

Deng, H.W., Mathaney, M.C., Williams, J.T. *et al.* (2002) Relevance of the genes for bone mass variation to susceptibility to osteoporotic fractures and its implications to gene search for complex human diseases. *Genetic Epidemiology*, **22**, 12–25.

Dennison, E.M., Hindmarsh, P.C., Kellingray, S. *et al.* (2003) Growth hormone predicts bone density in elderly women. *Bone*, **32**, 434–440.

Dent, C.E. and Davies, I.J.T. (1980) Calcium metabolism in bone disease: effect of treatment with microcrystalline calcium hydroxyapatic and dihydrotachysterol. *Journal of the Royal Society of Medicine*, **73**, 780–785.

Dequeker, J. and De Muylder, E. (1982) Long term progestogen treatment and bone remodeling in postmenopausal women: a longitudinal study. *Maturitas*, **4**, 309–312.

Dequeker, J. and Johnston, C.C. Jr (1982) *Non-Invasive Bone Measurements Methodological Problems*, IRL Press Ltd, Oxford, England and Washington D.C., p. 255.

Dequeker, J. and Roh, S. (1982) *Non-Invasive Bone Measurements: Methodological Problems* (eds J. Dequeker and C.C. Johnston Jr), IRL Press, pp. 33–42.

Dequeker, J., Rutten, V., Verstraeten, A. and Geusens, P. (1987) Effect of menarche, lactation and use of oral contraceptives on peripheral and axial bone mass, in *Osteoporosis* (eds C. Christiansen *et al.*), Osteo Press APS, Copenhagen, pp. 432–434.

Dews, P.B. (1982) Caffeine. *Annual Review of Nutrition*, **2**, 323–341.

Dicarlo, G., Gallo, W., Conti, G. and Racca, S. (1983) Changes in the binding of estradiol to uterine estrogen receptors induced by some progesterone and 19-nor testosterone derivatives. *Journal of Endocrinology*, **98**, 385–389.

Dickenson, R.P., Huton, W.C. and Stott, J.R.R. (1981) The mechanical properties of bone in osteoporosis. *The Journal of Bone and Joint Surgery*, **63-B**, 233–238.

Dickerson, J.W.T. (1962) The effect of development on the composition of a long bone of the pig, rat and fowl. *The Biochemical Journal*, **82**, 47–53.

Dietrich, J.W., Goodson, J.M. and Raisz, L.G. (1975) Stimulation of bone resorption by various prostaglandins in organ culture. *Prostaglandins*, **10**(2), 231–240.

Diherty, W.P., Bovill, E.G. and Wilson, E.L. (1974) Evaluation of the use of resonant frequencies to characterize physical properties of human long bones. *Journal of Biomechanics*, **7**, 559–561.

Diniz, P., Soejima, K. and Ito, G. (2002) Nitric oxide mediates the effects of pulsed electromagnetic field stimulation on the osteoblast proliferation and differentiation. *Nitric Oxide*, **7**, 18–23.

Dixon, S.J. and Sims, S.M. (2000) P2 purinergic receptors on osteoblasts and osteoclasts: Potential targets for drug development. *Drug Development Research*, **49**, 187–200.

Doan, N., Reher, P., Meghji, S. and Harris, M. (1999) In vitro effects of therapeutic ultrasound on cell proliferation, protein synthesis and cytokine production by human fibroblasts, osteoblasts and monocytes. *Journal of Oral and Maxillofacial Surgery*, **57**, 409–419.

Dobnig, H. and Turner, R.T. (1995) Evidence that intermittent treatment with parathyroid hormone increases bone formation in aged rats by activation of bone lining cells to osteoblasts. *Endocrinology*, **136**, 3632–3638.

Dodd, S.P., Cunningham, J.L., Milas, A.W. *et al.* (2007) An *in vitro* study of ultrasound signal loss across simple fractures in cortical bone mimies and bovine cortical bone samples. *Bone*, **40**, 656–661.

Doherty, W., Bovill, E. and Wilson, E. (1974) Evaluation of the use of resonant frequencies to characterize physical properties of human long bones. *Journal of Biomechanics*, **7**, 559–561.

Dolmetsch, R. (2003) Excitation transcription coupling: signaling by ion channels, to the nucleus. *Science's STKE: Signal Transduction Knowledge Environment*, (166), PE4.

Dolmetsch, R.E., Lewis, R.S. *et al.* (1997) Differential activation of transcription factors induced by $Ca^{2+}$ response amplitude and duration. *Nature*, **386**(6627), 855–858.

Donahue, S.W., Donahue, H.J. *et al.* (2003a) Osteoblastic cells have refractory periods for fluid flow-induced intracellular calcium oscillations for short bouts of flow and display multiple low magnitude oscillations during long term flow. *Journal of Biomechanics*, **36**(1), 35–43.

Donaldson, C.L., Hulley, S.B., Vogel, J.M. *et al.* (1970) Effect of prolonged bed rest on bone mineral. *Metabolism: Clinical and Experimental*, **19**, 1071–1077.

Dos Santos, M.N., Nunes, V.A., Nunes, I.J. *et al.* (1976) Solanum malacoxylon toxicity: inhibition of bone resorption. *Cornell Veterinarian*, **66**, 565–588.

Doty, S.B. (1981) Morphological evidence of gap junctions between bone cells. *Calcified Tissue International*, **33**, 509.

Doty, S.B. and Mathews, R.S. (1971) Electron microscopic and histochemical investigation of osteogenesis imperfecta. *Clinical Orthopaedics and Related Research*, **80**, 191–201.

Dougherty, G. and Henebry, G.M. (2002) Lacunarity analysis of spatial pattern in CT images of vertebrae trabecular bone for assessing osteoporosis. *Medical Engineering & Physics*, **24**, 129–138.

Draper, H.H. and Bell, R.R. (1979) Nutrition and osteoporosis, in *Advances in Nutritional Research*, vol. **2**(ed. H.H. Draper), Plenum Press, New York, pp. 70–106.

Drinka, P.J. and Nolten, W.E. (1984) Hazards of treating osteoporosis and hypertension concurrently with calcium, vitamin D and distal diuretics. *Journal of the American Geriatrics Society*, **32**, 405–407.

Drinkwater, B.L., Nilson, K., Chesnut, C.H. III *et al.* (1984) Bone mineral content of amenorrheic and eumenorrheic athletes. *The New England Journal of Medicine*, **31**, 277–281.

Drinkwater, R.H., Nilson, K., Ott, S. and Chesnut, C.H. III (1986) Bone mineral density after resumption of menses in amenorrheic women. *The Journal of the American Medical Association*, **256**, 380–382.

Drivadahl, R.H., Liu, C.C. and Baylink, D.J. (1984) Regulation of bone repletion in rats subjected to varying low calcium stress. *The American Journal of Physiology*, **246**, R190–R196.

Drozdzowska, B. and Pluskiewies, W. (2001) Qualitative ultrasound at calcaneus in premenopausal women and their postmenopausal mothers. *Bone*, **1**, 79–83.

Du, C., Cui, F.Z., Feng, Q.L. *et al.* (1998) Tissue response to nano-hydroxyapatite/collagen composite implants in marrow cavity. *Journal of Biomedical Materials Research*, **42**, 540–548.

Du, C., Cui, F.Z., Zhu, X.D. and de Groot, K. (1999) Three-dimensional nano-Hap/collange matrix loading with osteogenic cells in organ culture. *Journal of Biomedical Materials Research*, **44**, 407–415.

Duan, Y., Parfitt, A. and Seeman, E. (1999) Vertebral bone mass, size and volumetric density in women with spinal fractures. *Journal of Bone and Mineral Research*, **14**, 1796–1802.

Duarte, L.R. (1983) The stimulation of bone growth by ultrasound. *Archives of Orthopaedic and Traumatic Surgery*, **101**, 153–159.

Duboeuf, F., Hans, D., Schott, A.M. *et al.* (1997) Different morphometric and densitometric parameters predict cervical and trochanteric hip fracture: The EPIDOS Study. *Journal of Bone and Mineral Research*, **12**, 1895–1902.

Ducy, P., Amling, M., Takeda, S. *et al.* (2000) Lepting inhibits bone formation through a hypothalamic relay: A central control of bone mass. *Cell*, **100**, 197–207.

Ducy, P., Zhang, R., Geoffroy, V. *et al.* (1997) Osf2/Cbfa1: a transcriptional activator of osteoblast differentiation. *Cell*, **89**, 747–754.

Duda, R.J., O'Brien, J.F., Kartzmann, J.A. *et al.* (1988) Concurrent assays of circulating bone Gla protein and bone alkaline phosphatase: effects of sex, age and metabolic bone disease. *The Journal of Clinical Endocrinology and Metabolism*, **66**, 951–957.

Duffy, J.D., Wahner, H.W., O'Fallon, W.M. *et al.* (1986) Mechanisms of acute lower extremity pain syndrome in fluoride treated osteoporotic patients. *The American Journal of Medicine*, **80**, 561–566.

Dunlap, W.C. (1950) Some properties of high resistivity P-type germaninium. *Physical Review*, **79**(2), 286.

Dunn, W.L. and Wahner, H.W. (1994) Commercial instruments available for bone mineral measurements, in *The Evaluation of Osteoporosis: Dual Energy X-ray Absorptiometry in Clinical Practice* (eds H.W. Wahner and I. Fogelman), Dunitz, London, pp. 35–48.

Dunn, W.L., Wahner, H. and Riggs, B.L. (1980) Measurement of bone mineral content in human vertebrae and hip by dual photon absorptiometry. *Radiology*, **136**, 485–487.

Dunstan, C.R., Evans, R.A., Hills, E. *et al.* (1990) Bone death in hip fracture in the elderly. *Calcified Tissue International*, **47**, 270–275.

Dunstan, C.R., Somers, N.M. and Evans, R.A. (1993) Osteocyte death and hip fracture. *Calcified Tissue International*, **53**(5), 5113–5117.

Duquette, J., Honeyman, T., Hoffman, A. *et al.* (1997) Effect of bovine bone constituents on broadband ultrasound attenuation measurements. *Bone*, **21**, 289–294.

Durand, B., Christel, P. and Assailly, J. (1978). In vitro stury of electrical impedance of bone. In Electric Stimulation of Bone Growth and Repair (ed. F. Buny, E. Herbst and M. Hinsenkamp) Springer Verlag, New York, pp. 15–18.

Dymling, J.F. (1964) Calcium kinetics in osteopenia and parathyroid disease. *Acta Medica Scandinavica*, **175** (Suppl.), 1408, 1.

Dyson, M. (1997) Advances in wound healing physiology: the comparative perspective. *Veterinary Dermatology*, **8**, 227–233.

Dyson, M. and Brookes, M. (1983) Stimulation of bone repair by ultrasound. *Ultrasound in Medicine and Biology*, **8**, 61–66.

Dyson, M., Pond, J.B., Joseph, J. and Warwick, R. (1968) The stimulation of tissue regeneration by means of ultrasound. *Clinical Science*, **35**, 273–285.

Eanes, E.D. (1965) Effect of fluoride on human bone apatite crystals. *The New York Academy of Sciences*, **151**, 727.

Eanes, E.D. (1973) X-ray diffraction of vertebrate hard tissue, in *Biological Mineralization* (ed. I. Zipkin), John Wiley & Sons, Inc, New York, pp. 227–255.

Eanes, E.D. and Posner, A.S. (1965) Kinetics and mechanism of conversion of non-crystalline calcium phosphate to crystalline hydroxyapatite. *Transactions of the New York Academy of Sciences*, **28**, 233.

Eastell, R. (1996) Forearm fracture. *Bone*, **18**, 203s–207s.

Eastell, R. (2003) Management of osteoporosis due to overian failure. *Medical and Pediatric Oncology*, **41**, 222–227.

Eastell, R., Boyle, I.T., Compston, J. *et al.* (1998) Management of male osteoporosis: Report of the UK Consensus group. *Quarterly Journal of Medicine*, **91**, 71–92.

Eastell, R., Kennedy, N.S.J., Smith, M.A. *et al.* (1983) The assessment of postmenopausal osteoporosis by total body neutron activation analysis. *Metabolic Bone Disease & Related Research*, **5**, 65–67.

Eastoe, J.E. (1961) in "Biochemists' Handbook" (C. Long, ed.) p. 715. Spon, London.

Ebeling, P.R., Atley, L.M., Guthrie, J.R. *et al.* (1996) Bone turnover markers and bone density across the menopausal transition. *The Journal of Clinical Endocrinology and Metabolism*, **81**, 3366–3371.

Editorial (1990) Fracture patterns revised. *Lancet*, **2**, 1290–1291.

Edwards, C.J., Hart, D.J. and Spector, T.D. (2000) Oral statins and increased bone mineral densit I postmenopausal women. *Lancet*, **355**, 2218–2219.

Eiken, P., Nielsen, S.P. and Kolthoff, N. (1997) Effects on bone mass after eight years of hormonal replacement therapy. *British Journal of Obstetrics and Gynaecology*, **104**, 702–707.

Einhorn, T.A. (1995) Enhancement of fracture healing. *The Journal of Bone and Joint Surgery*, **77-A**, 940–956.

Einhorn, T.A. (1998) The cell and molecular biology of fracture healing. *Clinical Orthopaedics and Related Research*, **355**, S7–S21.

Einhorn, T.A., Majeska, R.J., Rush, E.B. *et al.* (1995) The expression of cytokine activity by fracture callus. *Journal of Bone and Mineral Research*, **10**, 1272–1281.

Einhorn, T.A., Simon, G., Devlin, V. *et al.* (1990) The osteogenetic response to distant skeletal injury. *The Journal of Bone and Joint Surgery*, **72**, 1374–1378.

Eisman, J.A., Kelly, P.J., Sambrook, P.N. *et al.* (1990) Physical activity and bone mass, in *Osteoporosis: Physiological Basis, Assessment and Treatment* (eds H.F. DeLuca R. Mazess), Elsevier Science, New York, pp. 277–283.

Ejersted, C., Andreassen, T.T., Hauge, E.M. *et al.* (1995) Parathyroid hormone (1–34) increases vertebral bone mass, compressive strength and quality in old rats. *Bone*, **17**, 507–511.

Ejersted, C., Andreassen, T.T., Oxlund, H. *et al.* (1993) Human parathyroid hormone (1–34) and (1–84) increase the mechanical strength and strength of cortical bone in rats. *Journal of Bone and Mineral Research*, **8**, 1097–1101.

Eley, D.D., Lockhart, N.C. and Richardson, C.N. (1977) Electrical properties and structural transitions in the mitochondrion. *Journal of Bioenergetics and Biomembranes*, **9**, 289–301.

Eley, D.D., Lockhart, M.C. and Richardson, C.N. (1979) Thermoelectric effects and dielectric polarization in biopolymers. *Transactions of the Faraday Society 1*, **79**, 323–334.

Eley, D.D. and Spirey, D.E. (1960) The semiconductivity of organic substances. *Transactions of the Faraday Society*, **56**, 1432–1442.

El-Hakim, I.E., Azim, A.M.A., El-Hassan, M.F.A. and Maree, S.M. (2004) Preliminary investigation into the effects of electrical stimulation on mandibular distraction osteogenesis in goats. *International Journal of Oral and Maxillofacial Surgery*, **33**, 42–47.

Ellerington, M.C. *et al.* (1996) Intranasal salmon calcitonin for the prevention and treatment of postmenopausal osteoporosis. *Calcified Tissue International*, **59**, 6–11.

Elliott, J.C. and Dover, S.D. (1984) Three-dimensional distribution of mineral in bone at a resolution of 15 mm determined by X-ray microtomography. *Metabolic Bone Disease & Related Research*, **5**, 219–221.

Emami, A., Petren-Mallmin, M. and Larsson, S. (1999) No effect of low intensity ultrasound on healing time of intramedullary fixed tibial fractures. *Journal of Orthopaedic Trauma*, **13**, 252–257.

Endo, H., Kiyoki, M., Kawashima, K. *et al.* (1980) Vitamin $D_3$ metabolites and PTH synergistically stimulate bone formation of chick embryonic femur *in vitro*. *Nature*, **286**, 262–264.

Endo, Y., Shibazaki, M., Nakamura, M. and Kosugi, H. (1999) Inhibition of inflammatory actions of aminobisphosphonates by dichloromethylene bisphosphonate, a non-aminobisphosphonate. *British Journal of Pharmacology*, **126**, 903–910.

Enemeto, S. and Krimm, S. (1962) Elastic moduli of helical polypeptide chain structures. *Biophysical Journal*, **2**, 317–325.

Engler, G.L., Neil, I. Spieholz, N.J. (1978) Somatosensory evoked potentials during Harrington instrumentation for scoliosis. *The Journal of Bone and Joint Surgery*, **60A/4**, 528–532.

Engstrom, A. and Zetterstrom, R. (1951) Experimental Cell Research, 2, 268 (1951. Relation between collagen and mineral salts in bone tissue. Cellular and Molecular Life Sciences (CMLS), 8(7), 259.

Enlow, D.H. (1991) A study of the post-natal growth and remodeling of bone. *The American Journal of Anatomy*, **110**, 79–101.

Enwemeka, C.S., Rodriguez, O. and Mendosa, S. (1990) The biomechanical effects of low intensity ultrasound on healing tendons. *Ultrasound in Medicine and Biology*, **16**, 801–807.

Enzelsberger, H., Metka, M., Heytmanck, G. *et al.* (1988) Influence of oral contraceptive use in bone density in climaeteric women. *Maturitas*, **9**, 375–378.

Epker, B.N. and Frost, H.M. (1965) Correlation of bone resportion and formation with the physical behavior of loaded bone. *Journal of Dental Research*, **44**, 33.

Epstein, O., Kato, Y., Dick, R. and Sherlock, S. (1982) Vitamin, D, hydroxyapatite and calcium gluconate in treatment of cortical bone in postmenopausal women with primary biliary cirrhosis. *The American Journal of Clinical Nutrition*, **36**, 426–430.

Epstein, S. and Zaidi, M. (2005) Biological properties and mechanism of action of ibandronate: Application to the treatment of osteoporosis. *Bone*, **37**, 433–440.

Erben, R.G., Eberie, J., Stahr, K. and Goldberg, M. (2000) Androgens deficiency induces high turnover osteopenia in aged male rats: A sequential histomorphometric study. *Journal of Bone and Mineral Research*, **15**, 1085–1098.

Eriksen, E.F. (1986) Normal and pathological remodeling of human trabecular bone: three dimensional reconstruction of the remodeling sequency in normals and in metabolic bone disease. *Endocrine Reviews*, **7**, 379–408.

Eriksen, E.F., Berg, N.J., Graham, M.L. *et al.* (1987) Evidence of estrogen receptors in human bone cells. *Journal of Bone and Mineral Research*, **2**(Suppl. 1), Abstr. 238.

Eriksen, E.F., Clavard, D.S., Berg, N.J. *et al.* (1988) Evidence of estrogen receptors in normal human osteoblast like cells. *Science*, **241**, 84.

Eriksen, E.F., Mosekilde, L. and Melsen, F. (1985a) Effect of sodium fluoride, calcium phosphate and vitamin $D_2$ on trabecular bone balance and remodeling in osteoporotics. *Bone*, **6**, 381–389.

Eriksen, E.F., Mosekilde, L. and Melsen, F. (1985b) Trabecular bone resorption depth decreases with age: Differences between normal males and females. *Bone*, **6**, 141–146.

Eriksen, E., Steiniche, T., Mosekilde, L. and Melsen, F. (1989) Histomorphometric analaysis of bone in metabolic bone disease. *Endocrinology and Metabolism Clinics of North America*, **18**, 919–954.

Eriksen, E.F., Vesterby, A., Kassem, M. *et al.* (1993) Bone remodeling and bone structure, in *Physiology and Pharmacology of Bone* (eds G.R. Undy and T.J. Martin), Springer-Verlag, Berlin, pp. 67–109.

Eriksson, C. (1974) Streaming potentials and other water dependental effects in mineralized tissue. *Annals of the New York Academy of Sciences*, **238**, 321.

Eriksson, C. (1976) Electrical properties of bone, in *Biochemisrry and Physiology of Bone IV* (ed. G.H. Bournej), Academic Press, London, pp. 330–382.

Eriksson, S.A., Isberg, B.O. and Lindgren, J.U. (1989) Prediction of vertebral strength by dual photon absorptiometry and quantitative computed tomography. *Calcified Tissue International*, **44**, 243–250.

Ernst, M., Health, J.K. and Rodan, G.A. (1989) Estradiol effects on proliferation, messenger ribonucleic acid for collagen and insulin like growth factor I and parathyroid hormone-stimulated adenylate cyclase activity in osteoblastic cells from calvaria and long bones. *Endocrinology*, **125**, 825–838.

Escher, J.E., Gambert, S. and Rothschild, B. (1987) Metabolic bone disease. *Age*, **10**, 62–69.

Eshelby, J.D. (1957) The determination of the elastic field of an ellipsoidal inclusion and related problems. *Proceedings of the Royal Society of London*, **241**, 376–396.

Ettinger, B. Genant, H. and Cann, C.E. (1987) Postmenopausal bone loss is prevented by treatment with low dosage estrogen with calcium. *Annals of Internal Medicine*, **106**, 40–45.

Evans, F.B. (1976) Mechanical properties and histology of cortical bone from younger and older men. *The Anatomical Record*, **185**, 1.

Evans, F.B. and Bang, S. (1967) Differences and relationships between the physical properties and the microscopic structure of human femoral, tibial and cortical bone. *The American Journal of Anatomy*, **120**, 79.

Evans, F.G. (1973) *Mechanical Properties of Bone*, Charles C Thomas, Springfield, IL.

Evans, F.G. and Lebow, M. (1951) Regional difference in some of the physical properties of the human femur. *Journal of Applied Physiology*, **3**, 563–572.

Evans, F.G. and Vincentelli, R. (1969) Relations of collagen fibre orientation to some mechanical properties of human cortical bone. *Journal of Biomechanics*, **2**, 63–71.

Evans, J.A. and Tavakoli, M.B. (1990) Ultrasonic attenuation and velocity in bone. *Physics in Medicine and Biology*, **35**(10), 1387–1396.

Evans, R.A., Marel, G.M., Lancester, E.K. *et al.* (1988) Bone mass is low in relations of osteoporotic patients. *Annals of Internal Medicine*, **109**, 870–873.

Exton-Smith, A.N. (1972) Physiological aspects of aging: relationship to nutrition. *The American Journal of Clinical Nutrition*, **25**, 853–859.

Eyre, D.R., Dickson, I.R. and Van Ness, K. (1988) Collagen cross-linking in human bone and articular cartilage. Age-related changes in the content of mature hydroxypyridinium residues. *Biochemical Journal*, **252**, 495–500.

Eyres, K.S., Bell, M.J. and Kanis, J.A. (1993a) New bone formation during leg lengthening. Evaluated by dual energy X-ray absorptiometry during limb lengthening. *The Journal of Bone and Joint Surgery*, **75**, 96–106.

Eyres, K.S., Bell, M.J. and Kanis, J.A. (1993b) Methods of assessing new bone formation during limb lengthening. *The Journal of Bone and Joint Surgery*, **75**, 358–364.

Eyres, K.S. and Kanis, I.A. (1995) Evaluation of bone losses after tibial fractures using dual energy X-ray absorptiometry. *The Journal of Bone and Joint Surgery*, **77**, 473–478.

Eyres, K.S., Salch, M. and Kanis, J.A. (1996) Effect of pulsed electromagnetic fields on bone formation and bone loss during limbs lengthening. *Bone*, **18**, 505–509.

Fabrizio, P.A., Schmidt, J.A., Clemente, F.R. *et al.* (1996) Acute effects of therapeutic ultrasound delivered at varying parameters on the blood flow velocity in a muscular distribution artery. *Journal of Orthopaedic & Sports Physical Therapy*, **24**, 294–302.

Faccini, J.M. (1967) Inhibition of bone resorption in the rabbit by fluoride. *Nature (London)*, **214**, 1269–1271.

Faguer de Mousteir, B., Conard, J., Guyene, T.T. *et al.* (1989) Comparative metabolic study of percutaneous versus oral microionized 17 $\beta$-estradiol in replacement therapy. *Maturitas*, **11**, 275–286.

Falahati-Nini, A, Riggs, B.L., Atkinson, E.J. *et al.* (2000) Relative circulation of testosterone and estrogen I regulating bone resorption and formation in normal elderly men. *The Journal of Clinical Investigation*, **106**, 1553–1560.

Falch, J.A., Sandvik, L. and Van Beresteijn, E.C.H. (1992) Development and evaluation of an index to predict early postmenopausal. *Bone*, **13**, 337–341.

Fang, J., Zhu, Y.Y., Smiley, E. *et al.* (1996) Stimulation of new bone formation by direct transfer of osteogenic plasmic genes. *Proceedings of the National Academy of Sciences of the United States of America*, **93**, 5753–5758.

Farley, J.R., Tarbaux, N.M., Hall, S.L. *et al.* (1988) The anti-bone resorptive agent calcitonin also acts *in vitro* to directly increase bone formation and bone cell proliferation. *Endocrinology*, **123**, 159–167.

Farley, J.R., Tarbaux, N., Susan, H. and Baylink, D.J. (1990) Mitogenic action of fluoride on osteoblast line cells. Determinants of the response *in vitro*. *Journal of Bone and Mineral Research*, **5**(Suppl.), S107–S113.

Farley, J.R., Wergedal, J.E. and Baylink, P.J. (1983) Fluoride directly stimulates proliferation and alkaline phosphatase activity of bone forming cells. *Science*, **222**, 330–332.

Farley, S.M., Libanati, C.R., Odvina, C.V. *et al.* (1989) Efficacy of long term fluoride and calcium therapy in correcting the deficit of spinal bone density in osteoporosis. *Journal of Clinical Epidemiology*, **42**, 1067–1074.

Farley, S.M., Wergedal, J.E., Farley, J.R. *et al.* (1992) Spinal fractures during fluroride therapy for osteoporosis: relationship to spinal bone density. *Osteoporosis International*, **2**, 213–218.

Farmer, F.T. and Collins, M.P. (1974) A further appraisal of the Compton scattering method for determining anatomical cross-sections of the body. *Physics in Medicine and Biology*, **19**, 808–818.

Farndale, R. and Murray, J.C. (1985) Pulsed electromagnetic fields promote collagen production in bone marrow fibroblasts via athermal mechanisms. *Calcified Tissue International*, **37**, 178–182.

Farooqi, I.S., Jebb, S.A., Langmack, G. *et al.* (1999) Effects of recombinant leptin therapy in a child with congenital leptin deficiency. *The New England Journal of Medicine*, **341**, 879–884.

Farrell, T.J. and Webber, C.E. (1989) The error due to fat inhomogeneity in lumbar spine bone mineral measurements. *Clinical Physics and Physiological Measurement*, **10**, 57–64.

Farrell, T.J. and Webber, C.E. (1990) Triple photon absorptiometry cannot correct for fat inhomogeneities in lumnar spine bone mineral measurements. *Clinical Physics and Physiological Measurement*, **11**, 77–84.

Farrell, T.J. and Webber, C.E. (1992) Phantom studies of triple photon absorptiometry and bone mineral measurement at a hip porosthesis. *Acta Radiologica*, **33**, 103–109.

Fatayerji, D., Cooper, A.M. and Eastell, R. (1999) Total body and regional bone mineral density in men: effect of age. *Osteoporosis International*, **10**, 59–65.

Faulkner, K.G., Cummings, S.R., Black, D. *et al.* (1993) Simple measurement of femoral geometry predicts hip fracture: the study of osteoporotic fractures. *Journal of Bone and Mineral Research*, **8**, 1211–1217.

Faulkner, K.G., Genant, H.K. and McClung, M. (1995) Bilateral comparison of femoral bone density and hip axis length from single and fan beam DXA scans. *Calcified Tissue International*, **56**, 26–31.

Faulkner, K.G., Gluer, C.-C., Palermo, L. *et al.* (1992) Geometric measurements from dual X-ray absorptiometry scans predict hip fracture (abstract). *Journal of Bone and Mineral Research*, **7**, S117.

Feher, J.J. and Wasserman, R.H. (1979) Intestinal calcium binding protein and calcium absorption in cortisol treated chicks: Effects of vitamin D3 and 1,25-dihydroxyvitamin D3. *Endocrinology*, **104**(2), 547–551.

Feretti, J.L., Capozza, R.F. and Zanchetta, J.R. (1996) Mechanical validation of a tomographic (PQCT) index for non invasive estimation of rat femur bending strength. *Bone*, **18**, 97–102.

Fernandez-Moran, H. and Engstrom, A. (1957) Electron microscopy and X-ray diffraction of bone. *Biochimica et Biophysica Acta*, **23**, 260–264.

Fernandez-Seara, M.A., Wehrli, S.L., Takahashi, M. and Wehrli, F.W. (2004) Water content measured by proton-deuteron exchange NMR predicts bone mineral density and mechanical properties. *Journal of Bone and Mineral Research*, **19**, 289–296.

Fernandez-Seara, M.A., Wehrli, S.L. and Wehrli, F.W. (2002) Diffusion of exchangeable water in cortical bone studied by nuclear magnetic resonance. *Biophysical Journal*, **82**, 522–529.

Ferrari, S.L., Deutsch, S., Choudhury, U. *et al.* (2004) Polymorphisms in the low density lipoprotein receptor related protein 5 (LRP5) gene are associated with variation in vertebral bone mass, vertebral bone size and stature in whites. *American Journal of Human Genetics*, **74**, 866–875.

Ferrethi, J.L., Gaffuri, O., Capozza, R. *et al.* (1995) Dexamethasone effects on mechanical, geometric and densitometric properties of rat femur diaphyses as described by peripheral quantitative computerized tomography and bending tests. *Bone*, **16**, 119–124.

Ferry, J.D. (1980) *Feeding Behavior of the Narrow-Striped Dwarf Siren Pseudobranchus Striatus Axanthus Viscoelastic Properties of Polymers*, John Wiley & Sons, Inc, New York.

Feyen, J.H., Van der Wilt, G. *et al.* (1984) Stimulation of arachidonic acid metabolism in primary cultures of osteoblast-like cells by hormones and drugs. *Prostaglandins*, **28**(6), 769–781.

Finkelstein, J.S., Klibanski, A., Schalfer, E.H. *et al.* (1994) Parathyroid hormone for the deficiency. *The New England Journal of Medicine*, **331**, 1618–1623.

Firooznia, H., Golimbu, C., Rafii, M. *et al.* (1984) Quantitative computed tomography assessment of spinal trabecular bone. II. Osteoporotic women with and without vertebral fractures. *The Journal of Computed Tomography*, **8**, 99–103.

Fitton, J.H. (1995) Divalent cation modulation of cell adhesion to surface. Proceedings 5th Annual Conference of Australian Society Biomaterials, p. PA22.

Fitton-Jackson, S. and Randall, J.T. (1956) The fine structure of bone. *Nature*, **178**(798), 409.

Fitzsimmons, R., Farley, J., Adey, W.R. and Baylink, D. (1986) Embryonic bone matrix formation is increased after exposure to a low amplitude capacitively coupled electrical field *in vitro*. *Biochimica et Biophysica Acta*, **882**, 51–56.

Flagg-Newton, J.L. and Lowenstein, W.R. (1980) Asymmetrically permeable membrane channels in cell junction. *Science*, **207**, 771.

Flanagan, A.M. and Chambers, T.J. (1992) Stimulation of bone nodule formation *in vitro* by prostaglandins $E_1$ and $E_2$. *Endocrinology*, **130**, 443–448.

Fleet, J. and Hock, J. (1994) Identification of osteocalcin in MRNA in nonosteotonal tissue of rats and human by reverse transcription-polymerase chain reaction. *Journal of Bone and Mineral Research*, **9**, 1565–1573.

Fleisch, H.A. (1997) Bisphosphonates: preclinical aspects and use in osteoporosis. *Annals of Medicine*, **29**, 56–62.

Fleisch, H., Russell, R.G.G., Simpson, B. and Muhlbauer, R.C. (1969) Prevention by a diphosphonate of immobilization "osteoporosis" in rats. *Nature*, **223**, 211–212.

Fletcher, W.H., Shiu, W.W., Havil, D.A. *et al.* (1986) Proceedings, 8th Annual Meeting of the Bioelectromagnetics Society, Madison, WI, p. 12.

Floriani, L., Debevoise, N. and Hyatt, G. (1967) Mechanical properties of healing bone by the use of ultrasound. *Surgical Forum*, **18**, 468–479.

Fluckey, J.D., Dupont-versteegden, E.E., Montague, D.C. *et al.* (2002) A rat resistance exercise regimen attenuates losses of musculoskeletal mass during hindlimb suspension. *Acta Physiologica Scandinavica*, **176**, 293–300.

Ford, J.C. and Wehrli, F.W. (1991) *In vitro* quantitative characterization of trabecular bone by NMR interferometry and localized proton spectroscopy. *Magnetic Resonance in Medicine*, **17**, 543–551.

Fordham, J.N., Robert-Countant, C. and Darbous, M. (2000) Identification of women with reduced bone density at the lumbar spine and femoral neck using BMD at the oscalcis. *Osteoporosis International*, **11**, 797–802.

Foresbergh, P.W. Jr (1956) Piezoelectricity, electrostriction and ferroelectricity, in *Encyclopedia of Physics*, vol. **17**(ed. S. Flugge), Springer, Berlin, pp. 264–392.

Forest, R.D., Herrick, J.F., James, J.M. and Krursen, F.H. (1953) Effect of ultrasound on growing bone. *Archives of Physical Medicine and Rehabilitation*, **34**(1), 21–31.

Formica, C., Loro, M.L., Gilsanz, V. and Seeman, E. (1995) Inhomogeneity in body fat distribution may result in inaccuracy I the measurement of vertebral bone mass. *Journal of Bone and Mineral Research*, **10**, 1504–1511.

Forwood, M.R. (1996) Inducible cyclo-oxygenase (COX-2) mediates the induction of bone formation by mechanical loading *in vivo*. *Journal of Bone and Mineral Research*, **11**, 1688–1693.

Forwood, M.R., Bennett, M.B., Blowers, A.R. and Nadorfi, R.L. (1998) Modification of the *in vivo* four point loading model for studying mechanically induced bone adaptation. *Bone*, **23**, 307–310.

Forwood, M.R., Owan, I. *et al.* (1996) Increased bone formation in rat tibia after a single short period of dynamic loading *in vivo*. *The American Journal of Physiology*, **270**(3Pt 1), E419–E423.

Forwood, M.R. and Turner, C.H. (1994) The response of rat tibiae bone incremental bouts of mechanical loading: a quantum concept for bone formation. *Bone*, **156**, 603–609.

Foster, K.R. and Schwan, H.P. (1986) Biophysics of electromagnetic field bioeffects, in *Handbook of Biological Effects of Electromagnetic Fields* (eds C. Polk and E. Postow), CRC Press, Boca Raton, FL, p. 9.83.

Foster, K.R. and Schwan, H.P. (1989) Dielectric properties of tissues and biological materials: a critical review. *Critical Reviews in Biomedical Engineering*, **17**, 25–104.

Fox, S.W., Chambers, T.J. and Chow, J.W.M. (1996) Nitric oxide is an early mediator of the induction of bone formation by mechanical stimulation. *Bone*, **19**, 687–687.

Fox, S.W. and Chow, J.W.M. (1998) Nitric oxide synthase expression in bone cells. *Bone*, **23**, 1–6.

Fraher, L. (1993) Biochemical markers of bone turnover. *Clinical Biochemistry*, **26**, 431–432.

Francis, R.M., Peacock, M., Storer, J.H. *et al.* (1983) Calcium malabsorption in the elderly: the effect of treatment with oral 25hydroxy vitamin $D_3$. *European Journal of Clinical Investigation*, **13**, 391–396.

Francis, R.M. and Selby, P.L. (1987) Osteoporosis: cause and management. *British Medical Journal*, **294**, 702.

Frangos, J.A. and Johnson, D. (1995) Fluid flow in bone: Stimulated release at remodeling mediators. *Biorheology*, **32**, 187–187(1).

Franklin, R.B. and Tasjian, A.H. (1975) Intravenous infusion of prostaglandin E2 raises plasma calcium concentration in the rat. *Endocrinology*, **98**, 240–243.

Fransson, A., Grampp, S. and Imhof, H. (1999) Effects of trabecular bone on marrow relaxation in the tibia. *Journal of Magnetic Resonance Imaging*, **17**, 69–82.

Fraser, J.D., Otawara, Y. and Price, P. (1988) 1,25-dihydroxy vitamin $D_3$ stimulates the synthesis of matrix Y-carboxyglutamic acid, protein by osteosarcoma cells. *The Journal of Biological Chemistry*, **263**, 911–916.

Fratzl, P., Fratzl-Zelman, N. and Klaushofer, K. (1993) Collagen packing and mineralization. *Biophysical Journal*, **54**, 260–266.

Frazier, K., Williams, S., Kothapalli, D. *et al.* (1996) Stimulation of fibroblast cell growth, matrix production and granulation tissue formation by connective tissue growth factor. *The Journal of Investigative Dermatology*, **107**, 404–411.

Fredholm, B.B., Abbracchiol, M.P., Brnstock, G. *et al.* (1994) Nomenclature and classification of purinoceptors. *Pharmacological Reviews*, **64**, 143–156.

Frediani, B., Falsetti, P., Baldi, F. *et al.* (2003) Effects of 4-year treatment with once-weekly clodronate on prevention of corticosteroid-induced bone loss and fractures in patients with arthritis: evaluation with dual-energy X-ray absorptiometry and quantitative ultrasound. *Bone*, **33**(4), 575–581.

Freeman, J.R. (1967) Dielectric properties of mineralized tissues. *Transactions of the New York Academy of Sciences*, **29**, 623–633.

Freidkin, M., Legg, A. and Rozenguri, E. (1979) Antitubulin agents enhance the stimulation of DNA synthesis by polypeptide growth factors in 3T3 mouse fibroblasts. *Proceedings of the National Academy of Sciences of the United States of America*, **76**, 3909–3912.

Fricke, H. (1924) A mathematical treatment of the electrical conductivity and capacity of disperse system. *Journal of Physical Review*, **24**, 575.

Friedenberg, Z.B., Andrews, E.T., Snolenski, B.I. *et al.* (1970) Bone reaction of varying amounts of direct current. *Surgery, Gynecology & Obstetrics*, **131**(5), 894–899.

Friedenberg, Z.B. and Brighton, C.T. (1966) Bioelectric potential in bone. *The Journal of Bone and Joint Surgery*, **48**, 915–923.

Friedenberg, Z.B. and Brighton, C.T. (1974) Electrical fracture healing. *The New York Academy of Sciences*, **238**, 564–574.

Friedenberg, Z.B., Harlow, M.C., Heppenstall, R.B. and Brighton, C.T. (1973) The cellular origin of bioelectric potentials in bone. *Calcified Tissue International*, **13**(1), 53–62.

Friedenberg, Z.B., Roberts, P.G., Didizian, N.H. and Brighton, C.T. (1971) Stimulation of fracture healing by direct current in the rabbit fibula. *The Journal of Bone and Joint Surgery*, **53**, 1400–1408.

Friedenberg, Z.B., Zemsky, L.M., Pollis, R.P. and Brighton, C.T. (1974) The response of non-traumatized bone to direct current. *The Journal of Bone and Joint Surgery*, **56**, 1023.

Fritton, S.P., McLeod, K.J. and Rubin, C.T. (2000) Quantifying the strain history of bone: spatial uniformity and self-similarity of low magnitude strains. *Journal of Biomechanics*, **33**, 317–325.

Frost, H.M. (1960a) Presence of microscopic cracks *in vivo* in bone. *Henry Ford Hospital Medical Bulletin*, **8**, 25–35.

Frost, H.M. (1960b) Some aspects of the mechanics and dynamics of blood-bone interchange. *Henry Ford Hospital Medical Bulletin*, **8**, 36.

Frost, H.M. (1963) *Bone Remodeling Dynamics*, Little Brown & Co, Boston.

Frost, H.M. (1964a) *Mathematical Elements of Lamellar*, Bone Remodeling, Charles C Thomas, Springfield, IL.

Frost, H.M. (1964b) *The Laws of Bone Structure*, Charles C Thomas, Springfield, IL.

Frost, H.M. (1964c) Dynamics of bone remodeling, in *Bone Biodynamics* (ed. H.M. Frost), Little Brown, Boston, p. 315.

Frost, H.M. (1973a) *Orthopaedic Biomechanics*, Charles C Thomas, Springfield, IL.

Frost, H.M. (1973b) *Bone Remodeling and its Relationship to Metabolic Bone Disease*, Charles C Thomas, Springfield, IL.

Frost, H.M. (1980) Skeletal physiology and bone remodeling. An overview, in *Fundamental and Clinical Bone Physiology* (ed. M.R. Urist), J.B. Lippincott Co., Philadelphia, p. 208.

Frost, H.M. (1981) Coherence treatment of osteoporoses. *Orthopedic Clinics of North America*, **12**, 649–669.

Frost, H.M. (1983) A determinant of bone architecture. The minimum effective strain. *Clinical Orthopaedics*, **200**, 198–225.

Frost, H.M. (1987) Bone "mass" and the "mechanostat" a proposal. *The Anatomical Record*, **219**, 1–9.

Frost, H.M. (1989) Mechanical usage, bone mass, bone fragility: a brief review, in *Clinical Disorders in Bone and Mineral Metabolism* (eds M. Kleerekoper and S. Krane), Mary Ann Liebow, New York, pp. 15–140.

Frost, H.M. (1990a) Skeletal structural adaption to mechanical usage (SATMU). I. Redefining Wolff's Law: the bone modeling problem. *The Anatomical Record*, **226**, 403–413.

Frost, H.M. (1990b) Skeletal structural adaptations to mechanical usage (SATMU). I. remodeling Wolff's Law: the remodeling problem. *The Anatomical Record*, **226**, 414–422.

Frost, H.M. (1997) Perspectives: On our age-related bone loss and muscle strength: Insights from a new paradigm. *Journal of Bone and Mineral Research*, **12**, 1529–1546.

Frost, H.M. and Villaneuva, A.R. (1961) The effect of cortisone on lamellar osteoblastic activity. *Henry Ford Hospital Medical Journal*, **9**, 97–99.

Fry, W.J. and Dunn, F. (1962) Ultrasound: analysis and experimental methods in biological research, in *Physical Techniques in Biological Research*, vol. **IV**(ed. W.L. Nastuk), Academic Press, New York, pp. 261–394.

Fujita, T., Chan, J.C.M. and Barter, F.C. (1984) Effects of oral furosemide and salt loading on parathyroid function in normal subjects. Physiological basis for renal hypercalciuria. *Nephron*, **38**, 109–114.

Fukada, E. (1968) Mechanical deformation and electrical polarization in biological substances. *Biorheology*, **5**, 199–208.

Fukada, E. (1974a) *Advances in Biophysics*, **6**(ed. M. Kotani), University Tokyo Press, Tokyo, p. 121.

Fukada, E. (1974b) Piezoelectric properties of organic polymers. *Annals of the New York Academy of Sciences*, **238**, 7–25.

Fukada, E. and Hara, K. (1969) Piezoelectric effect in blood vessel walls. *Journal of the Physical Society of Japan*, **26**, 777.

Fukada, E. and Yasuda, I. (1957) On the piezoelectric effect of bone. *Journal of the Physical Society of Japan*, **120**, 1158–1162.

Fukada, E. and Yasuda, I. (1964) Piezoelectric effects in collagen. *Japanese Journal of Applied Physics*, **3**(2), 117–121.

Fukushima, N., Hanada, R., Teranishi, H. *et al.* (2005) Ghrelin directly regulates bone formation. *Journal of Bone and Mineral Research*, **20**, 790–798.

Fuller, J.B. (1975) *Synergetics*, Macmillan, New York.

Fung, B.M. and Tautmann, P. (1971) Deuterium NMR and EPR of hydrated collagen fibers in the presence of salts. *Biopolymars*, **10**, 391.

Gabriel, C., Gabriel, S. and Corthout, E. (1996) The dielectric properties of biological tissues: I Literature Survey. *Physics in Medicine and Biology*, **41**, 2231–2249.

Gailey, P.C. (1993) Presented at the Fifteenth Annual Meeting of the Bioelectromagnetics Society, Los Angeles, CA June 13–17.

Gallagher, J.C. (1988) *Drug Therapy of Osteoporosis: Calciu, Estroge and Vitamin D* (eds B.L. Riggs and Lf. Melton), Raven Press, New York, pp. 389–401.

Gallagher, J.C., Aaron, J., Horsman, A. *et al.* (1973) The crush fracture syndrom in postmenopausal women. *The Journal of Clinical Endocrinology and Metabolism*, **1**, 293–315.

Gallagher, J.C., Beneton, M., Harvey, L. and Lawson, D.E.M. (1986) Response of rachitic rat bones to 1,25-dihydroxy vitamin $D_3$. Biphasic effects on mineralization and lack of effect on bone resorption. *Endocrinology*, **119**, 1603–1609.

Gallagher, J.C., Goldgar, D. and Moy, A. (1987) Total bone calcium in normal women: effect of age and menopause status. *Journal of Bone and Mineral Research*, **2**, 491–496.

Gallagher, J.C. and Nordin, B.E.C. (1972) Treatment with estrogen of primary hyperparathyroidism in postemenopausal women. *Lancet*, **1**(7749), 503–507.

Gallagher, J.C. and Nordin, B.E.C. (1975) Effect of estrogen and progestogen therapy on calcium metabolism in postmenopausal women. *Frontiers of Hormone Research*, **3**, 150–176.

Gallagher, J.C., Riggs, B.L., Eisman, J. *et al.* (1979) Intestinal calcium absorption and serum vitamin D metabolites in normal subjects and osteoporotic patients. Effects of age and dietary calcium. *The Journal of Clinical Investigation*, **64**, 729–736.

Gallagher, J.C., Riggs, B.L., Jerpbak, C.M. and Arnaud, C.D. (1980) The effect of age on serum immunoreactive parathyroid hormone in normal and osteoporotic women. *The Journal of Laboratory and Clinical Medicine*, **95**, 373–385.

Gallent, G., Bolognani, L. and Ussia, G. (1992) *Lasers Applications in Medicine and Surgery*, Monduzi/editore, Bologna.

Gallinaro, B.J., Reimer, W.J. and Dixon, S.J. (1995) Activation of protein kinase C inhibits ATP- induced $[Ca^{2+}]_i$ elevation in rat osteoblastic cells: Selective effects on $P_{2Y}$ and $P_{2U}$ signaling pathway. *Journal of Cellular Physiology*, **162**, 305–314.

Gallop, P.M. and Paz, M.A. (1975) Post translational protein modifications with special attention to collagen and elastin. *Physiological Reviews*, **55**, 418–487.

Gammaitoni, L. (1998) Stochastic response. *Reviews of Modern Physics*, **70**, 223–287.

Garcia-Sanchez, F.J. and De Mercato, G. (1995) A study of dielectric anisotropy in dehydrated cortical bone. *Medical Progress Through Technology*, **21**, 165–170.

Gardner, T.N., Stoll, T., Marks, L. *et al.* (2000) The influence of mechanical stimulus on the pattern of tissue differentiation in a long bone fracture an FEM study. *Journal of Biomechanics*, **33**, 415–425.

Gardsell, P., Johnell, O., Nilsso, B.E. and Gullberg, B. (1993) Predicting various fragility fractures in women by forearm bone densitometry: a follow-up study. *Calcified Tissue International*, **52**, 348–353.

Garn, S.M. (1970) *The Earlier Gain and Later Loss of Cortical Bone, in Nutritional Prospective*, Charles C Thomas, Springfield, IL, p. 146.

Garner, E., Lakes, R., Lee, T. *et al.* (2000) Viscoelastic dissipation in compact bone; implications for stress induced fluid flow in bone. *Journal of Biomechanical Engineering*, **122**(2), 166–172.

Garnero, P., Darte, C. and Delmas, P.D. (1999) A model to monitor the efficacy of alendronate treatment in women with osteoporosis using a biochemical marker of bone turnover. *Bone*, **24**, 603–609.

Garnero, P., Hausherr, E., Chapuy, M.C. *et al.* (1996) Markers of bone resorption predict hip fracture in elderly women: the EPIDOS prospective study. *Journal of Bone and Mineral Research*, **11**, 1531–1538.

Garnett, E.S., Kennett, T.J., Kenyon, D.B. and Webber, C.E. (1973) A photon scattering technique for the measurement of absolute bone density in man. *Radiology*, **106**, 209–212.

Garnett, O., Duriet, B.G.M., Nedwin, G.E. *et al.* (1987) Production of the bone resorbing cytokine lymphotoxin by cultured human melome cells. *The New England Journal of Medicine*, **317**, 526–532.

Gasmi, M., Glynn, J., Jin, M.J. *et al.* (1999) Requirements of efficient production and transduction of human immunodeficiency virus type 1-based vectors. *Journal of Virology*, **73**, 1828–1834.

Gasser, J.A. (1995) Assessing bone quantity by pQCT. *Bone*, **17**(Suppl.), 145S–154S.

Geddes, A.D., D'Souza, S.M., Ebentino, F.H. and Kenneth, J.I. (1994) Bisphosphonates: Structure-activity relationships and therapeutic implications, in *Bone and Mineral Research*, vol. **8**(eds N.M. Heershe and J.K. Kanis), Elsevier Science BV, Amsterdam, pp. 265–306.

Geddes, L.A. (1972) *Electrodes and the Measurements of Bioelectric Events*, Wiley Interscience, New York, London, p. 8.

Genant, H.K., Cann, C.E., Boyd, D.P. *et al.* (1983) Quantitative computed tomography for vertebrae mineral determination. Proceedings of Henry Ford Hospital Symposium on clinical Disorders of Bone and Mineral Metabolism. Excerpt Medica, pp. 40–47.

Genant, H.K., Cann, C.E., Ettinger, B. and Gordan, G.S. (1982) Quantitative computed tomography on vertebral spongiosa: A sensitive method for detecting early bone loss after oophorectomy. *Annals of Internal Medicine*, **97**, 699–705.

Genant, H.K., Cooper, C., Poor, G. *et al.* (1999) Interim report and recommendations of the World Health Organization Task- Force for osteoporosis. *Osteoporosis International*, **10**, 259–264.

Genant, H.K., Steiger, P. and Gluer, C.C. (1989) New developments in bone densitometry. *Postgraduate Medicine*, **4**, 18–22.

Gennari, C., Imbimbo, B., Montagnani, M. *et al.* (1984) Effects of prednisone and Deflazacort on mineral metabolism and parathyroid hormone activity in humans. *Calcified Tissue International*, **36**(3), 245–252.

Gennari, L. and Brandi, M.L. (2001) Genetics of male osteoporosis. *Calcified Tissue International*, **69**, 200–204.

Genovese, C. and Rowe, D.W. (1987) Analysis of cytophasmic and nuclear messenger RNA in fibroblasts from patients with type I osteogenesis imperfecta. *Methods in Enzymology*, **145**, 223–235.

Geoffroy, V., Kneissel, M., Fournier, B. *et al.* (2002) High bone resorption in adult aging transgenic mice overexpressing Cbfa1/Runx2 in cells of the osteoblastic lineage. *Molecular and Cellular Biology*, **22**, 6222–6233.

Geonari, C. (2002) Analgesic effect of calcitoin in osteoporosis. *Bone*, **30**(Supp5), 675–705.

Gerling, J.A., Sinclair, P.M. and Roa, R.L. (1985) The effect of PEMF on condylar growth in guinea pig. *American Journal of Orthodontics*, **87**, 211–223.

Geusens, P., Vanderschueren, D., Verstraeten, A. *et al.* (1991) Short term course of 1,25 $(OH)_2D_3$ stimulates osteoblasts but not osteoclasts in osteoporosis and osteoarthritis. *Calcified Tissue International*, **49**, 168–173.

Giachelli, C.M., Liaw, L., Murr, C.E. *et al.* (1995) Osteopontin expression in cardiovascular diseases. *Annals of the New York Academy of Sciences*, **760**, 109–126.

Giammini, S., D'Angelo, A., Malvasi, L. *et al.* (1993) Effects of one year cyclical treatment with clodronate on postmenopausal bone loss. *Bone*, **14**(2), 137–141.

Gibson, L.J. (1985) The mechanical behavior of cancellous bone. *Journal of Biomechanics*, **18**(5), 317–328.

Giganate, G.E. and Sciuti, S. (1985) A large angle coherent/Compton scattering method for measurement *in vitro* of trabecular bone mineral concentration. *Medical Physics*, **12**, 321–324.

Gillespie, W.J., Avenell, A., Henry, D.A. and O'Connell, D.L. (2001) Vitamin D and vitamin D analogues for preventing fractures associated with involutional and post-menopausal osteoporosis. *Cochrane Database of Systematic Reviews*, **1**, CD000227.

Gilmore, R.S. and Katz, J.L. (1968) Elastic properties of apatites. Proceedings International Symposium Structural properties of Hydroxyapatite and related Compounds. NBS, Washington, D.C.

Gilsanz, V., Gibbens, D.T., Roe, T.F. *et al.* (1988) Vertebral bone density in children effect of puberty. *Radiology*, **166**, 847–850.

Givalois, L., Dornand, J., Mckauche, M. *et al.* (1994) Temporal cascade of plasma level surges in ACTH, corticosterone and cytokines in endotoxin-challenged rats. *The American Journal of Physiology*, **267**, R167–R170.

Gizdulich, P. and Aschero, G. (1993) Day-to-day trend of dielectric properties of bones. Proceedings 15th Annual International Conference IEEE, San Diego 28–31 Oct. 1993, 1024–1025.

Gizdulich, P. and Calsolaro, A. (1993) Electrical impedance measurements of bones. *Physica Medica*, **9**, 87–89.

Gjelsvik, A. (1973) Bone remodelling and piezoelectricity. *Journal of Biomechanics*, **6**, 69–77.

Glauber, H.S., Vollmer, W.M., Nevitt, M.C. *et al.* (1995) Body weight versus body fat distribution, adiposity and frame size as predictors of bone density. *The Journal of Clinical Endocrinology and Metabolism*, **80**, 1118–1123.

Glimcher, M.J. (1959) Molecular biology of mineralized tissues with particular references to bone. *Reviews of Modern Physics*, **31**(2), 359–393.

Glimcher, M.J. (1976) Composition, structure and organization of bone and other mineralized tissues and the mechanism of calcification, In *Handbook of Physiology – Endocrinology*, vol VII, Williams and William Co., Baltimore, p. 25.

Glimcher, M.J. (1992) The nature of the mineral component of bone and the mechanisms of calcification, In. *Disorders of Bone and Mineral Metabolism* (eds F.L. Coe and M.J. Favus) New York, Raven Press, pp. 265–286.

Glimcher, M.J. and Kenzora, J.E. (1979) The biology of osteonecrosis of the human femoral head and its clinical implications: III Discussion of the etiology and genesis of the pathological sequelae: Comments on treatment. *Clinical Orthopaedics*, **140**, 273–312.

Glimcher, M.J. and Krane, M. (1968) *Treatise on Collagen*, vol. **2**(ed. B.S. Gould), Academic Press, London and New York, p. 99.

Globus, R.K., Bikle, D.D., Halloran, B.P. and Morey-Holton, E. (1986) Skeletal response to dietary calcium in a rat model simulating weightlessness. *Journal of Bone and Mineral Research*, **1**, 191–197.

Gluer, C.C., Vahlensieck, M., Faulkner, K.G. *et al.* (1992) Site matched calcaneal measurements of broadband ultrasound attenuation and single X-ray absorptiometry: Do they measure different skeletal properties? *Journal of Bone and Mineral Research*, **7**(9), 1071–1079.

Gluer, C.C., Wu, C.Y. and Genant, H.K. (1993) Broadband ultrasound attenuation signals depend on trabecular orientation: An *in vitro* study. *Osteoporosis International*, **3**, 185–191.

Gluer, C.C., Wu, C.Y., Jergan, M. *et al.* (1994) Three quantitative ultrasound parameters reflect bone structure. *Calcified Tissue International*, **55**, 46–52.

Gnuidi, S., Malavolta, N., Ripamonti, C. and Caudarella, R. (1985) Ultrasound in the evaluation of osteoporous: a comparison with bone mineral density at distal radius. *British Journal of Radiology*, **68**, 476–480.

Godfrey, L., Sutherland, M., Boy, D. and Gomberg, N. (1991) Scaling of limb joint surface areas in anthropoid primates and other mammals. *Journal of Zoology, London*, **223**, 603–625.

Gohel, A., McCrthy, M.B. and Gronowicz, G. (1999) Estrogen prevents glucocorticoid-induced apoptosis in osteoblasts *in vivo* and *in vitro*. *Endocrinology*, **140**, 5339–5347.

Goldsmith, F. and Johnston, O. (1975) Bone mineral: effects of oral contraceptives, pregnance and lactation. *The Journal of Bone and Joint Surgery*, **57**, 657–668.

Goldstein, S.A. (1987) The mechanical properties of trabecular bone: dependence of anatomic location and function. *Journal of Biomechanics*, **20**, 1055–1061.

Goldstein, S.A. (2000) *In vivo* nonvircal delivery factors to enhance bone repair. *Clinical Orthopaedics*, **379** (Suppl.), S113–S119.

Gomez, B. Jr, Ardakani, S., Ju, J. *et al.* (1995) Monoclonal anti-body assay for measuring bone-specific alkaline phos-phatase activity in serum. *Clinical Chemistry*, **41**, 1560–1566.

Gomez, M., Aguado, F., Manuel, J. *et al.* (1997) Influence of soft tissue (fat and fat-free mass) on ultrasound bone velocity: An *in vivo* study. *Investigative Radiology*, **32**, 609–612.

Gong, Y., Slee, R.B., Fukai, N. *et al.* (2001) LDL receptor-related protein 5 (LRP5) affects bone accrual and eye development. *Cell*, **107**, 513–523.

Gonnelli, S., Cepollaro, C., Agnusdei, D. *et al.* (1995) Diagnostic value of ultrasound analysis and bone densitometry as predictors of vertebral deformity in postmenopausal women. *Osteoporosis International*, **5**, 413–418.

Gonnelli, S., Cepollaro, C. and Pondrelli, C. (1997) The usefulness of bone turnover I predicting the response to transdermal estrogen therapy in postmenopausal osteoporosis. *Journal of Bone and Mineral Research*, **12**, 624–631

Gonzales, A., Garrido, J. and Vial, J.D. (1981) Epidermal growth factor inhibits cytoskeleton-related changes in the surface of parietal cells. *The Journal of Cell Biology*, **88**, 108–114.

Gonzalez-Riola, J., Pamies, J.A., Hernandez, E.R. *et al.* (1997) Influence of electromagnetic fields on bone mass and growth in developing rats: A morphometric. Densitometric and histomorphometric study. *Calcified Tissue International*, **60**, 533–537.

Goodman, L.S. and Levin, B. (1970) *In vivo* measurement of the mineral content of a section of the radius, in *Methods of Bone Mineral Measurement*, U.S. Department of Health, Education and Welfare. Public, Health Service pp. 204–215.

Goodman, R., Bumann, J., Wei, L.-X. and Henderson, A.S. (1992) Exposure of human cells to electromagnetic fields: effect of time and field strength on transcripts levels. *Electro and Magnetobiology*, **11**, 19–28.

Goodman, R. and Henderson, A.S. (1986) Some biological effects of electromagnetic fields. *Bioelectrochemistry and Bioenergetics*, **15**, 39.

Goodman, R. and Henderson, A.S. (1987) *Patterns of Transcription and Translation in Cells Exposed to EM fields: A review, in Mechanistic Approaches to Interactions of electric and Electromagnetic Fields and Living Systems* (eds M. Blank and E. Findl), Plenum Press, New York, p. 217.

Goodman, R. and Henderson, A.S. (1988) Exposure of salivary gland cells to low frequency electromagnetic fields alters polypeptide synthesis. *Proceedings of the National Academy of Sciences of the United States of America*, **85**, 3928–3932.

Goodship, A.E., Cunningham, J.l., Oganov, V. *et al.* (1998) Bone loss during long term space flight is prevented by the application of a short term impulsive mechanical stimulus. *Acta Astronaut*, **43**, 65–75.

Goodship, A.E. and Kenwright, J. (1985) The influence of induced measurement upon the healing of experimental tibial fracture bone. *The Journal of Bone and Joint Surgery*, **67**, 650–655.

Goodsitt, M.M. (1992) Evaluation of a new set of calibration standards for the measurement of fat content via DPA and DXA. *Medical Physics*, **19**, 35–44.

Goodsitt, M.M., Hoover, P., Veldee, M.S. and Hsueh, S.L. (1994) The composition of bone marrow for a dual energy quantitative computed tomography technique: A cadaver and computer simulation study. *Investigative Radiology*, **29**, 696–704.

Gordon, C.M., Bachrach, L.K., Carpenter, T.O. *et al.* (2004) Bone health in children and adolescents: a symposium at the annual meeeting of the Pediatric Academic Societies/Lawson Wilkins Pediatric Endocrine Society. *Current Problems in Pediatric Adolescent Health Care*, **34**, 226–242.

Gori, F., Hofbauer, L.C., Dunstan, C.R. *et al.* (2000) The expression of osteoprotegerin and RANK ligand and the support of osteoclast formation by stromal-osteoblast lineage cells is developmentally regulated. *Endocrinology*, **141**, 4768–4776.

Goss, S.A., Johnston, R.L. and Dunn, F. (1978) Comprehensive compilation of empirical ultrasonic properties of mammalia tissue. *The Journal of the Acoustical Society of America*, **67**, 423–457.

Gotfredsen, A., Hadberg, A., Nilas, L. and Christiansen, C. (1987) Total body bone mineral in healthy adults. *The Journal of Laboratory and Clinical Medicine*, **110**, 362–368.

Gotfredsen, A., Podenphant, J., Norgaard, H. *et al.* (1988) Accuracy of lumbar spine bone mineral content by dual photon absorptiometry. *Journal of Nuclear Medicine*, **29**, 248–254.

Goulding, A. (1980) Effects of dietary NaCl supplement on parathyroid function, bone turnover and bone composition in rats taking restricted amount of calcium. *Mineral and Electrolyte Metabolism*, **4**, 203–208.

Goulding, A. and Campbell, D.R. (1983) Dietary NaCl loss promote calciuria and bone loss in adult oophorectomized rats consuming a low calcium diet. *The Journal of Nutrition*, **113**, 1409–1414.

Goulding, A. and Campbell, D.R. (1984) Hypocalciuric effects of hydrochlorothiazide in the rat during $NaHCO_33$, NaCl and $NH_4Cl$ loading. *Renal Physiology*, **7**, 185–191.

Goulding, A., Everitt, H.E., Cooney, J.M. and Spears, G.F.S. (1986) Sodium and osteoporosis, in *Recent Advances in Clinical Nutrition* (eds M.L. Wahlqvest and A.S. Truswell) Food & Nutrition Press.

Goulding, A. and Gold, E. (1988) Effects of chronic prednisolone treatment on bone resorption and bone composition in intact and ovariectomized rats and in ovariectomized rats receiving â-estradiol. *Endocrinology*, **122**, 482–487.

Goulding, A. and Gold, E. (1990) Buserelin-mediated osteoporosis: effects of restoring estrogen on bone resorption and whole body calcium content in the rat. *Calcified Tissue International*, **46**, 14–19.

Goulding, A. and Lim, P.E. (1983) Effects of varying dietary salt intake on the fasting urinary excretion of sodium, calcium and hydroxyproline in young women. *The New Zealand Medical Journal*, **96**, 853–854.

Goulding, A., McIntosh, J. and Campbell, D.R. (1984) Effects of $NaHCO_3$ and 1,25 $(OH)_2D_3$ on calcium and phosphorous balances in the rat. *The Journal of Nutrition*, **114**, 653–659.

Goulding, A. and Taylor, R.W. (1998) Plasma leptin values in relation to bone mass and density and to dynamic biochemical markers of bone resorption and formation in postmenopausal women. *Calcified Tissue International*, **63**, 456–458.

Gourlay, M., Richy, F. and Reginster, J.Y. (2003) Strategies for the prevention of hip fracture. *The American Journal of Medicine*, **115**, 309–317.

Grados, F., Brazier, M., Kamel, S. *et al.* (2003) Effects on bone mineral density of calcium and vitamin D supplementation in elderly women with vitamin D deficiency. *Joint Bone Spine*, **70**, 203–208.

Grady, D., Rubin, S.M., Petitti, D.B. *et al.* (1992) Hormone therapy to prevent disease and prolong life in postmenopausal women. *Annals of Internal Medicine*, **117**, 1016–1037.

Grande, D.A., Magee, F.P., Weinstein, A.M. and McLeod, B.R. (1991) The effect of low-energy combined AC and DC magnetic fields on articular cartilage metabolism. *Annals of the New York Academy of Sciences*, **635**, 404–407.

Grant, A.M. *et al.* (2005) Oral vitamin $D^3$ and calcium for secondary prevention of low-trauma fractures in elderly people (randomised location of calcium or vitamin D, RECORD), a randomized placabo-controlled trial. *Lancet*, **365**, 161–1628.

Grant, E.H., Buchanan, T.J. and Cook, H.F. (1957) Dielectric behavior of water at microwave frequencies. *Journal of Chemical Physics*, **26**(1), 156–161.

Grant, E.H., Keefe, S.K. and Takashima, S. (1968) The dielectric behavior of aqueous solutions of brain serum albumin from radiowave to microwave frequencies. *The Journal of Physical Chemistry*, **72**(13), 4374–4380.

Grant, E.H., Sheppard, R.J. and South, G.P. (1978) *Dielectric Behavior of Biological Molecules in Solution*, Clarendon Press, Oxford, pp. 101–120.

Grant, W.B. (2002) An estimate of premature cancer mortality in the United States due to inadequate doses of solar ultraviolet B radiation. *Cancer*, **94**, 1867–1875.

Gray, R.W. (1981) Effects of age and sex on the regulation of plasma 1,25 $(OH)_2$-D by phosphorus in rat. *Calcified Tissue International*, **33**, 477–484.

Gray, R.W., Wilz, D.R., Caldas, A.E. and Lemann, J. Jr (1977) The importance of phosphate in regulating plasma 1,25 $(OH)_2$-vitamin D levels in humans: studies in healthy subjects, in calcium stone formers and in patients with primary hyperparathroidism. *The Journal of Clinical Endocrinology and Metabolism*, **45**, 299–306.

Green, A.E. and Taylor, G.L. (1945) Stress system in alolotropic planes. *III Proceedings of the Royal Society*, **184A**, 181–195.

Green, J., Schotland, S., Stauber, D.J. *et al.* (1995) Cell matrix interaction in bone: Type I collagen modulates signal transduction in osteoblast like cells. *The American Journal of Physiology*, **37**, C1090–C1103.

Greenberg, C., Krikreia, S.C., Bowser, E.N. and Hargis, A. (1986) Effect of estradiol on calcitonin secretion. *Endocrinology*, **118**, 2594–2598.

Greendale, G.A., FitzGerald, G., Huang, M.H. *et al.* (2002) Dietary soy isoflavones and bone mineral density: results from the study of women's health across the nation. *American Journal of Epidemiology*, **155**(8), 746–754.

Greenleaf, J.F. (1986) *Tissue Characterization with Ultrasound*, CRC Press, Boca Raton, FL.

Grimm, M.J. and Williams, J.L. (1997) Assessment of bone quantity and 'quality' by ultrasound attenuation and velocity in the heel. *Clinical Biomechanics*, **5**, 281–285.

Grisso, J.A., Baum, C.R. and Turner, B.J. (1990) What do physicians in practice do to prevent osteoporosis? *Journal of Bone and Mineral Research*, **5**, 213–219.

Grodstein, G.W. (1957) X-ray attenuation coefficients from 10 keV to 100 MeV. *NBS Circular*, **2**, 583.

Grodzinsky, A. (1983) Electromechanical and physiochemical properties of connective tissue. *Critical Reviews in Biomedical Engineering*, **9**, 133.

Gross, D. and Williams, W.S. (1982) Streaming potential and the electromechanical response of physiologically moist bone. *Journal of Biomechanics*, **15**, 277–295.

Gross, T., Edwards, J., McLeod, K. and Rubin, C. (1997) Strain gradients correlate with sites of periosteal bone formation. *Journal of Bone and Mineral Research*, **12**, 982–988.

Gross, T.S., Srinivasan, S., Liu, C.C. *et al.* (2002) Noninvasive loading of the murine tibia: an *in vivo* model for the study of mechanotransduction. *Journal of Bone and Mineral Research*, **17**, 493–501.

Grubb, S.A., Jacobson, P.C., Awbrey, B.J. *et al.* (1984) Bone density in osteopenic women: A modified distal radius density measurement procedure to develop an "At Risk" value for use in screening women. *Journal of Orthopaedic Research*, **2**, 322–327.

Gruben, H.E., Ivey, J.L., Baylink, D.J. *et al.* (1984) Long term calcitonin therapy in postmenopausal osteoporosis. *Metabolism: Clinical and Experimental*, **33**, 295–303.

Grynpas, M. (1993) Age and disease related changes in the mineral of bone. *Calcified Tissue International*, **53** (Suppl.), S57–S64.

Grynpas, M.D. and Rey, C. (1992) The effect of fluoride treatment on bone mineral crystals in the rat. *Bone*, **13**, 23–429.

Guey, A., Leitienne, P.H., Zech, P.Y. *et al.* (1979) *Utilisation d'un irradiateur neutronique charge en californium-252 sour l'analyse in vivo du contenu calcique osseux de la main, Nuclear Activation Techniques in the Life Sciences*. IAEA, Vienna, pp. 707–717.

Gugliemi, G., Gluer, C.C., Majumdar, S. *et al.* (1995) Current methods and advances in bone densitometry. *European Radiology*, **5**, 129–139.

Gugliemi, G., Majumdar, S., Jergas, M. *et al.* (1994) Quantitative magnetic resonance to assess the regional variation in trabecular bone of the calcaneus. *Bone & Mineral*, **25**, S18.

Guilak, F. and Mow, V.C. (2000) The mechanical environment of the chondrocyte: a biphasic finite element model of cell-matrix interactions in articular cartilage. *Journal of Biomechanics*, **33**, 1663–1673.

Guilak, F., Ratcliffe, A. and Mow, V.C. (1995) Chondrocyte deformation and local tissue strain in articular cartilage: a confocal microscopy study. *Journal of Orthopaedic Research*, **13**(3), 410–421.

Gundjian, A.A. and Chen, H.L. (1974) Determination of the bone crystallites distribution function by X-ray diffraction. *Medical and Biological Engineering*, **12**, 531–535.

Gunness-Hey, M. and Hock, J.M. (1984) Increased trabecular bone mass in treated with human synthetic parathyroid hormone. *Metabolic Bone Disease & Related Research*, **5**, 177–181.

Gunness-Hey, M. and Hock, J.M. (1989) Loss of anabolic effect of parathyroid hormone on bone after discontinuation of hormone in rats. *Bone*, **10**, 447–452.

Guo, J., Bringhurst, F.R. and Kronenberg, H.M. (2004) Collagen promoter directed overexpression of Dkk1 in mice causes dwarfism and very short limbs. *Journal of Bone and Mineral Research*, **19**, S6.

Guo, X.E., Eichler, M.J., Takai, E. and Kim, C.H. (2002) Quantification of a rat tail vertebra model for trabecular bone adaptation studies. *Journal of Biomechanics*, **35**, 363–368.

Gupta, S.K., Gallego, C. and Johnson, G.L. (1992) Mitogenic pathways regulated by G proteins. *Molecular Biology of the Cell*, **3**, 123–128.

Gupta, T.D., Jain, V.K. and Tandon, P.N. (1991) Comparative study of bone growth by pulsed electromagnetic fields. *Medical & Biological Engineering & Computing*, **29**, 113–120.

Gustavson, L., Jacoson, B. and Kusoffsk, L. (1974) X-ray spectophoometry for bone mineral determinations. *Medical and Biological Engineering and Computing*, **12**, 113–118.

Gutmann, F. and Lyons, L.E. (1967) in *Organic Semiconductors*, John Willey & Sons Inc, New York, pp. 566–569.

Guttmann, G.D. and Goodsitt, M.M. (1995) The effect of fact on the coherent to Compton scattering ratio in the calcaneuous: A computational study. *Medical Physics*, **22**, 1229–1234.

Guzelsu, N. (1978) A piezoelectric model for dry bone and tissue. *Journal of Biomechanics*, **11**, 257.

Guzelsu, N. and Demiray, H. (1979) Electromechanical properties and related models of bone tissues. *International Journal of Engineering Science*, **17**, 813–851.

Guzelsu, N. and Regimbal, R. (1990) The origin of the electro-kinetic properties of compact bone tissue: the organic phase. *Journal of Biomechanics*, **23**, 661–672.

Guzelsu, N. and Saha, S. (1981) Electromagnetic wave propagation in long bones. *Journal of Biomechanics*, **14**, 19–33.

Guzelsu, N. and Saha, S. (1984) Electro-mechanical behavior of wet bone. *Journal of Biomechanical Engineering*, **106**, 249–261.

Guzelsu, N. and Walsh, W.R. (1990) Streaming potential of intact wet bone. *Journal of Biomechanics*, **23**, 673–686.

Haas, H.G., Damgacher, M.A., Goschke, H. *et al.* (1976) Growth hormone in osteoporosis. *Calcified Tissue Research*, **21**, 467–468.

Hadjiargyrou, M., Ahrens, W. and Rubin, C. (2000) Temporal expression of the chondrogenic and angiogenic growth factor CYR61 during fracture repair. *Journal of Bone and Mineral Research*, **15**, 1014–1023.

Hadjiargyrou, M., McLeod, K., Ryaby, J.P. and Rubin, C. (1998) Enhancement of fracture healing by low intensity ultrasound. *Clinical Orthopaedics and Related Research*, (355 Suppl.), 216–224.

Haebare, H., Yamasaki, Y. and Kyogokuni, M. (1969) An autopsy case of osteogenesis imperfecta congetia histological and electron microcopical studies. *Acta Pathologica Japonica*, **19**, 377–394.

Haentijens, P., Johnell, O., Kanis, J.A. *et al.* (2004) Evidence from data searches and life table analysis for gender-related differences in absolute risk of hip fracture after colle's of spine fracture as an early and sensitive marker of skeletal fragility in white men. *Journal of Bone and Mineral Research*, **19**, 1933–1944.

Hagiwara, H., Inoue, N. Matsuzaki, H. *et al.* (2000) Relationship between structural anisotropy of the vertebral body and bone mineral density. *Transactions of the Orthopaedic Research Society*, **25**, 738.

Hagiwara, T. and Bell, W.H. (2000) Effect of electrical stimulation on madibular distraction osteogenesis. *Journal of Cranio-Maxillo-Facial Surgery*, **28**, 12–19.

Haiat, G., Padilla, F., Barkmann, R. *et al.* (2005) *In vitro* speed of sound measurement in intact human femur specimens. *Ultrasound in Medicine and Biology*, **31**, 987–996.

Hailey, D., Sampietro Colom, I., Marshall, D. *et al.* (1998) The effectiveness of bone density measurement and associated treatments for prevention of fractures. An international collaborative review. *International Journal of Technology Assessment in Health Care*, **14**, 237–254.

Haire, T.J. and Langton, C.M. (1999) Biot theory: A review of its application to ultrasound propagation through cancellous bone. *Bone*, **24**, 291–295.

Hak, A.E., Pols, H.A., Van Hemert, A.M. *et al.* (2000) Progression of aortic calcification is associated with metacarpal bone loss during menopause: a population-based longitudinal study. *Arteriosclerosis, Thrombosis, and Vascular Biology*, **20**(8), 1926–1931.

Hall, B.K. (1970) Cellular differentiation in skeletal tissue. *Biological Review*, **45**, 455–484.

Hall, B.K. (1987a) Sodium fluoride as an initiator of osteogenesis from embryonic mesenchyme *in vitro*. *Bone*, **8**, 111–116.

Hall, F.M. (1987b) Bone mineral screening for osteoporosis. *American Journal of Radiology*, **149**, 120–122.

Hall, F.M., Davis, M.A. and Baran, D.T. (1987) Bone mineral screening for osteoporosis. *The New England Journal of Medicine*, **316**, 212–214.

Hall, L. (1948) The origin of ultrasonic absorption in water. *Physical Review*, **73**, 775.

Hall, M.C. (1965) *The Locomotor System-Functional Histology*, Charles C Thomas, Springfield, IL.

Hamburg, N.J., Satson, J., Sivager, A. and Sashle, D.J.B. (1971) Effect of microampere electrical currents on bone '*in vivo*' and measurements using strontium-85. *Nature*, **231**, 190–191.

Hambury, H.J., Watson, J., Sivyer, A. and Ashley, D.J.B. (1971) Effect of microamp electrical currents on bone *in vivo* and its measurement using strontium-85 uptake. *Nature*, **231**, 190.

Hamilton, W.C. (1968) *Structural Chemistry and Molecular Biology*, WH Freeman & Co, Boston, pp. 466–483.

Han, M. and Mould, J.R. (1990) The Velocity Field in the Local Supercluster. *Astrophysical Journal*, **360**, 448–464.

Han, Z.H., Palnitkar, S., Rao, D.S. *et al.* (1996) Effect of ethnicity and age or menopause on the structure and geometry of iliac bone. *Journal of Bone and Mineral Research*, **11**, 1967–1975.

Hancox, N.M. (1972) *Biology of Bone*, Cambridge University Press, New York.

Hancox, N.M. and Boothroyd, B. (1965) Electron microscopy of the earl stages of osteogenesis. *Clinical Orthopaedics*, **40**, 153.

Hangartner, T.N. and Johnston, C.C. (1990) Influence of fat on bone measurements with dual energy absorptiometry. *Bone & Mineral*, **9**, 71–81.

Hannan, M.T., Felson, D.T., Dawson-Hughes, B. *et al.* (2000) Risk factors for longitudinal bone loss in elderly men and women: The Framingham osteoporosis study. *Journal of Bone and Mineral Research*, **15**, 710–720.

Hans, D., Dargent Molina, P., Schott, A.M. *et al.* (1996) Ultrasonographic heel measurements to predict hip fracture in elderly women: the EPIDOS prospective study. *Lancet*, **348**, 511–514.

Hans, D., Fuerst, T., Lang, T. *et al.* (1997) How can we measure bone quality? *Bailliere's Clinical Rheumatology*, **11**, 495–515.

Hans, D., Njeh, C.F., Genant, H.K. and Menunier, P.J. (1998) Quantitative ultrasound in bone status assessment. *Revue du Rhumatisme (English Edition)*, **65**, 489–498.

Hans, D., Schott, A.M. and Meunier, P.J. (1993) Ultrasonic assessment of bone: A review. *The European Journal of Medicine*, **2**, 157–163.

Hans, D., Wu, C., Njeh, C.F. *et al.* (1999) Ultrasound velocity of trabecular cubes reflects mainly bone density and elasticity. *Calcified Tissue International*, **64**, 18–23.

Hansson, R. and Roos, B. (1978) Effect of combined therapy with sodium fluoride, calcium and vitamin D on the lumbar spine in osteoporosis. *American Journal of Roentgenology*, **126**, 1294.

Hansson, T. and Roos, B. (1987) The effect of fluoride and calcium on spinal bone mineral content, a controlled perspective (3 years) study. *Calcified Tissue International*, **40**, 315–317.

Hantes, M.E., Mavrodontedis, A.N., Zalavras, C.G. *et al.* (2004) Low intensity transosseous ultrasound accelerates osteotomy healing in a sheep fracture model. *The Journal of Bone and Joint Surgery*, **86**(10), 2275–2282.

Hara, K., Kobayashi, M. and Akiyama, Y. (2002) Effects of vitamin K2 (menatetrenone) inhibits bone loss induced by predwisolone partly through enhancement of bone formation in rats. *Bone*, **31**, 575–581.

Harada, S. and Rodan, G.A. (2003) Control of osteoblast function and regulation of bone mass. *Nature*, **423**, 349–355.

Harlow, M.C., Friendenberg, Z.B. and Brighton, C.T. (1971) Electric currents hasten bone healing. *The Journal of the American Medical Association*, **218**, 1123.

Harma, M., Parviainen, M., Koskinen, T. *et al.* (1987) Bone density, histomorphometry and biochemistry in patients with fractures of the hip or spine. *Annals of Clinical Research*, **19**, 378–382.

Harper, R.A. and Posner, A.S. (1966) Measurement of non-crystalline calcium phosphate in bone mineral. *Proceedings of the Society for Experimental Biology and Medicine*, **123**, 137.

Harrington, D.B. and Becker, R.O. (1973) Electrical stimulation of RNA and protein synthesis in the frog erythrocyte. *Experimental Cell Research*, **76**, 95–98.

Harris, M. (1992) The conservative management of osteoradionecrosis of the mandible with ultrasound therapy. *British Journal of Oral and Maxillofacial Surgery*, **30**, 313–318.

Harris, S.E., Sabatem, M., Harris, M.A. *et al.* (1994) Expression of bone morphogenetic protein messenger RNA in prolonged cultures of fetal rat calvaranal cells. *Journal of Bone and Mineral Research*, **9**, 389–394.

Harris, S.S. and Dawson-Hughes, B. (1996) Weight body composition and bone density in postmenopausal women. *Calcified Tissue International*, **59**, 428–432.

Harris, S.S., Patel, M.S., Cole, D.E. and Dawson Hughes, B. (2000) Associations of the collagen type 1 alpha 1 Spl polymorphism with five-year rates of bone loss in older adults. *Calcified Tissue International*, **66**, 268–271.

Harris, S.T., Genant, H.K., Baylink, D.J. *et al.* (1991) The effects of estrogen on spinal bone density of postmenopausal women. *Archives of Internal Medicine*, **151**, 1980–1984.

Harris, S.T., Watts, N.B., Genant, H.K. *et al.* (1999) Effects of risedronate treatment on vertebral and nonvertebral fractures in women with postemenopausal osteoporosis. *The Journal of the American Medical Association*, **282**, 1344–1352.

Harris, W.H. (1960) A microscopic method of determining rates bone growth. *Nature (London)*, **188**, 1038.

Harris, W.H. and Heaney, R.P. (1969) Skeletal renewal and metabolic bone disease. *The New England Journal of Medicine*, **280**, 193–202.

Harris, W.H., Heaney, R.P., Jowsey, S. *et al.* (1972) Growth hormone. The effect on skeletal renewal in the adult dog. I. Morphometric studies. *Calcified Tissue Research*, **10**, 1–13.

Harrison, J.E., McNeil, K.G. and Meema, S. (1974) Partial body calcium measurement by *in vivo* neutron activation analysis. *Journal of Nuclear Medicine*, **15**, 929–934.

Hart, R.T. and Davy, D.T. (1989) Theories of bone modeling and remodeling, in *Bone Mechanics* (ed S.C. Cowin), CRC Press, Boca Raton, FL, pp. 253–277.

Harting, M., Joos, U. and Wiesmann, H.P. (2000) Capacitively coupled electric fields accelerate proliferation of osteoblast – like primary cells and increase bone extracellular matrix formation *in vitro*. *European Biophysics Journal*, **29**, 499–506.

Hasegawa, K., Turner, C.H., Recker, R.R. *et al.* (1995) Elastic properties of osteoporotic bone measured by scanning. *Acoustic Microscopy Bone*, **16**(1), 85–90.

Hashin, T. and Strikman, S. (1963) A variational approach to the elastic behavior of multiphase materials. *Journal of the Mechanics and Physics of Solids*, **11**, 127–140.

Hassager, C., Jensen, L.T., Podenphant, J. *et al.* (1994) The carboxy-terminal pyridinoline cross-linked telopeptide of type I collagen in serum as a marker of bone resorption: The effect of nandrolone decanoae and hormone replacement therapy. *Calcified Tissue International*, **54**, 30–33.

Hassager, C., Jensen, S.B., Gotfredsen, A. and Christiansen, C. (1991) The impact of measurement errors on the diagnostic value of bone mass measurements: Theoretical consideration. *Osteoporosis International*, **1**, 250–256.

Hassler, C.R., Rybicki, E.F., Diegle, R.B. and Clark, L.C. (1977) Studies of enhanced bone healing via electrical stimuli. *Clinical Orthopaedics and Related Research*, **124**, 9–19.

Hastings, G.W., El Messiery, M.A. and Rokowski, S. (1981) Mechano-electrical properties of bone. *Biomaterials*, **2**, 225–233.

Hastings, G.W. and Mahmud, F.A. (1991) The electromechanical properties of fluid-filled bone: a new dimension. *Journal of Materials Science: Materials in Medicine*, **2**, 118–124.

Hattersley, A.T., Meeran, K., Burrin, J. *et al.* (1994) The effect of long and short-term corticosteroids on plasma calcitonin and parathyroid hormone levels. *Calcified Tissue International*, **54**(3), 198–202.

Hayashi, Y., Igarashi, M., Masuda, M. *et al.* (1982) Measurement of bone mineral distribution by photon beam scanner. *Bone Metabolism (Japan)*, **15**, 227–235.

Hayes, W.C., Piazza, S.J. and Zysset, P.K. (1991) Biomechanics of fracture risk prediction of the hip and spine by quantitative computed tomography. *Radiologic Clinics of North America*, **29**, 1–8.

Hays, V.W. and Swenson, M.J. (1984) Minerals. In *Dukes' Physiology of Domestic Animals*, 10th edn (ed J.M. Swenson), Comstock Publishing Associates, Ithaca, NY, pp. 449–485.

Hazan, G., Leichter, I., Loewinger, E. *et al.* (1977) The early detection of osteoporosis by Compton gamma ray spectroscopy. *Physics in Medicine and Biology*, **22**, 1073–1084.

Heaney, R.P. (1963) Evaluation and interpretation of calcium kinetic data in man. *Clinical Orthopaedics*, **31**, 153–183.

Heaney, R.P. (1988) Nutritional factors in bone health, in *Osteoporosis: Etiology, Diagnosis and Management* (eds L. Riggs and L.J. Melton), Raven Press, New York, pp. 359–372.

Heaney, R.P. (1989) Nutritional factors in bone health in elderly subjects. Methodological and contextual problems. *The American Journal of Clinical Nutrition*, **50**, 1182–1189.

Heaney, R.P. (1996) Calcium parathyroid function bone and aging. *The Journal of Clinical Endocrinology and Metabolism*, **81**, 1697–1698.

Heaney, R.P. (2003) Is the paradigm shifting? *Bone*, **33**, 457–465.

Heaney, R.P., Avioli, L.V., Chesnut, C.H. *et al.* (1989) Osteoporotic bone fragility. Detection by ultrasound transmission velocity. *The Journal of the American Medical Association*, **261**, 2986–2990.

Heaney, R.P., Harris, W.H., Cockin, J. and Weinberg, E.H. (1972) Growth hormone: the effect on skeletal renewal in the adult dog. II. Mineral kinetic studies. *Calcified Tissue Research*, **10**, 14–22.

Heaney, R.P. and Recker, R.R. (1982) Effect of nitrogen phosphorus and caffeine on calcium balance in women. *The Journal of Laboratory and Clinical Medicine*, **99**, 46–55.

Heaney, R.P., Recker, R.R. and Saville, P.O. (1978) Menopausal changes in calcium balance performance. *The Journal of Laboratory and Clinical Medicine*, **92**, 953–956.

Heaney, R.P., Recker, R.R. and Wever, C.M. (1990) Absorbability of calcium sources: The limited role of solubility. *Calcified Tissue International*, **46**, 300–304.

Heaney, R.P., Saville, P.D. and Recker, R.R. (1975) Calcium absorption as a function of calcium intake. *The Journal of Laboratory and Clinical Medicine*, **85**, 881–890.

Heaney, R.P. and Whedon, G.D. (1958) The effect of aluminum hydroxide in serum calcium, serum phosphorus and calc1um turnover in uraemic patients. *Journal of Clinical Endocrinology*, **18**, 1248.

Heckman, J.D., Ryaby, J.P., McCabe, J. *et al.* (1994) Acceleration of tibial fracture healing by non-invasive, low intensity ultrasound. *The Journal of Bone and Joint Surgery*, **76**, 26–34.

Hedlund, L.R. and Gallagher, J.C. (1989) Increased incidence of hip fracture I osteoporotic women treated with sodium fluoride. *Journal of Bone and Mineral Research*, **4**, 223–225.

Heermeier, K., Spanner, M., Trager, J. *et al.* (1998) Effects of extremely low frequency electromagnetic field (EMF) on collagen Type I mRNA expression and extracellular matrix synthesis of human osteoblastic cells. *Bioelectromagnetics*, **19**, 222–231.

Hefti, E., Trechsel, U., Bonjour, J.P. *et al.* (1982) Increase of whole body calcium and skeletal mass in normal and osteoporotic adult rats treated with parathyroid hormone. *Clinical Science*, **62**, 389–396.

Hegsted, M., Schuette, S.H., Zemel, M.B. and Linkswiler, H.M. (1981) Urinary calcium and calcium balance in young men as affected by level of protein and phosphorous intake. *The Journal of Nutrition*, **111**, 553–562.

Heikinheimo, R.J., Inkovaara, J.A., Harju, E.J. *et al.* (1992) Annual injection of vitamin D and fractures of aged bones. *Calcified Tissue International*, **51**, 105–110.

Heinemann, D.F. (2000) Osteoporosis: An overview of the National Osteoporosis Foundation Clinical Practice Guide. *Geriatries*, **55**, 31–36.

Heinonen, A., Kannus, P., Sievanen, H. *et al.* (1996) Randomized controlled trial of effect of high impact exercise on selected risk factors for osteoporotic fractures. *Lancet*, **348**, 1343–1347.

Helfrich, M.H., Evans, D.E., Grabowski, P.S. *et al.* (1997) Expression of nitric oxide synthase isoforms in bone and bone cells cultures. *Journal of Bone and Mineral Research*, **12**, 1108–1115.

Hemminki, E., Kennedy, D.L., Baum, C. and McKinlay, S.M. (1988) Prescribing of noncontraceptive estrogens and progestins in the United States, 1974 to 1986. *American Journal of Public Health*, **78**, 1478–1481.

Henke, W.J., Vacca, J.B., Goidsenhoven Van, G.V. and Knight, W.A. Jr (1961) Evaluation of pancreatic function tests in confirmed pancreatic disease. *Gastroenterology*, **41**, 233–241.

Henson, P.W. and Fox, R.A. (1984) A relationship between the percentage of calcium by mass and the effective atomic number of regions containing bone. *Physics in Medicine and Biology*, **29**, 979–984.

Hentunen, T.A., Choi, S.J., Boyce, B.E. *et al.* (2000) A murine model of inflammatory bone disease. *Bone*, **26**, 183–188.

Hepner, G.W., Roginsky, M. and Moo, H.F. (1976) Abnormal vitamin D metabolism in patients with cirrhosis. *The American Journal of Digestive Diseases*, **21**, 527–532.

Herd, R.J.M., Blake, G.M., Ramalingam, T. *et al.* (1993) Measurement of postmenopausal bone loss with a new contact ultrasound system. *Calcified Tissue International*, **53**, 153–157.

Hernandez, C.J., Beaupre, G.S., Keller, T.S. and Carter, D.R. (2001) The influence of bone volume fraction and ash fraction on bone strength and modulus. *Bone*, **29**(1), 74–78.

Herring, G.M. (1968) The chemical structure of tendon, cartilage, dentin and bone matrix. *Clinical Orthopaedics*, **60**, 261.

Hesch, R.D., Busch, U., Prokop, M. *et al.* (1989) Increase of vertebral density by combination therapy with pulsatile 1–38 hPH and sequential addition of calcitonin nasal spray in osteoporotic patients. *Calcified Tissue International*, **44**, 176–180.

Hesch, R.D., Hock, J.M., Delling, G. *et al.* (1988) Results of a stimulatory therapy of low bone metabolism osteoporosis with (1–38) hPTH and disphosphonate EHDP. *Klin Wochensehr*, **66**, 976–984.

Heshmati, H.M., Khosla, S., Burritt, M.F. *et al.* (1998) A defect in renal calcium conservation may contribute to the pathogenesis of postmenopausal osteoporosis. *The Journal of Clinical Endocrinology and Metabolism*, **83**(6), 1916–1920.

Hesp, R., Hulme, P., Williams, D. and Reeve, J. (1981) The relationship between changes in femoral bone density and calcium balance in patients with involutional osteoporosis treated with human parathyroid hormone fragment (hPTH 1–34). *Metabolic Bone Disease & Related Research*, **2**, 331–334.

Heuck, F. and Schmidt, E. (1960a) Quantitative determination of the mineral content in bone from the X-ray picture. *Fortschritte auf dem Gebiete der Röntgenstrahlen und der Nuklearmedizin*, **93**, 523–554.

Heuck, F. and Schmidt, E. (1960b) The practical use of a method for quantitative determination of the content of calcium salt in healthy and diseased bone. *Fortschritte auf dem Gebiete der Röntgenstrahlen und der Nuklearmedizin*, **93**, 761–783.

Heywood, B.R., Sparks, N.H.C., Shellis, R.P., Weiner, S. and Mann, S. (1990) Ultrastructure, morphology and crystal growth of biogenic and synthetic apatites. *Connective Tissue Research*, **25**, 1–17.

Higashi, S., Ohishi, H. and Kudo, I. (2000) Augmented prostaglandin E-2 generation resulting from increased activities of cytosolic and secretory phospholipase A (2) and induction of cyclooxygenase 2 in interleukin 1 beta stimulated rat calvarial cells during the mineralizing phase. *Inflammation Research*, **49**, 102–111.

High, J.J., Gatch, G.A., Seheele, W.H. *et al.* (2002) A randomized double blind study to compare the efficacy of teriparatide (recombinant human–hormone (1–34) with alendronate postmenopausal women with osteoporosis. *The Journal of Clinical Endocrinology and Metabolism*, **87**, 4528–4535.

Hill, C.R.(ed) (1986) *Physical Principles of Medical Ultrasonics*, Ellis Horwood, p. 147.

Hill, R. (1952) The elastic behavior of a crystalline aggregate. *Proceedings of the Physical Society of London A*, **65**, 349–354.

Hillam, R.A. and Skerry, T.M. (1995) Inhibition of bone resorption and stimulation of formation by mechanical loading of the modeling rat ulna *in vivo*. *Journal of Bone and Mineral Research*, **10**, 683–689.

Hiltunen, A., Vuoprio, E. and Aro, H.T. (1993) A standardized experimental fracture in the mouse tibia. *Journal of Orthopaedic Research*, **11**, 305–312.

Hinsenkamp, M., Burny, F., Donkerwolcke, M. and Coussaert, E. (1984) Electromagnetic simulation of fresh fractures treated with Hoffmann external fixation. *Orthopaedics*, **7**, 411.

Hipskind, R.A. and Bilbe, G. (1998) MAP kinase signaling cascades and gene expression in osteoblasts. *Frontiers in Bioscience: A Journal and Virtual Library*, **3**, D804–D816.

Hirota, Y., Hirohata, T., Fukuda, K. *et al.* (1993) Association of alcohol intake, cigarette smoking and occupational status with the risk of osteopathic osteonecrosis of the femoral head. *American Journal of Epidemiology*, **137**, 530–538.

Hirsch, T.J. (1962) Modulus of elasticity of concrete as affected by elastic moduli of cement paste matrix and aggregate. *Proceedings of the American Concrete Institute*, **59**, 427.

Hirsch, J. (1965) Falty acid patterns in human adipose tissue. P. 181, in *Handbook of Physiology, Section 5: Adipose Tissue* (eds A.E. Renold and G.F. Cahill), American Physiological Society, Washington.

Hitt, M.M., Addison, C.L. and Graham, F.L. (1997) Human adenovirus vectors for gene transfer into mammalian cells, in *Advances in Pharmacology*, vol. **40** (Gene Therapy), Academic Press, New York, pp. 137–296.

Ho, C.P., Kim, R.W., Schaffler, M.B. and Sartoris, D.J. (1990) Accuracy of dual-energy radiographic absorptiometry of the lumbar spine: Cadaver Study. *Radiology*, **176**, 171–173.

Hock, J.M. and Gera, I. (1992) Effects of continuous and intermittent administration and inhibition of resorption on the anabolic response of bone to parathyroid hormone. *Journal of Bone and Mineral Research*, **7**, 65–72.

Hock, J.M., Gera, I., Fonseca, J. and Raisz, L.J. (1988) Human parathyroid hormone (1–34) increases bone mass in ovariectomized and orchidectomized rats. *Endocrinology*, **122**, 2899–2904.

Hodge, A.J. (1967) *Treatise on Collagen* (ed. G.N. Ramachandran), Academic Press, p. 185.

Hodgskinson, R. and Currey, J.D. (1990) The effect of variation in structure on the Young's modulus of cancellous bone: A comparison of human and non-human material. *Proceedings of the Institution of Mechanical Engineers. Part H*, **204**, 115–121.

Hodgskinson, R., Njeh, C.F., Whitehead, M.A. and Langton, C.M. (1996) The non-linear relationship between BUA and porosity in cancellous bone. *Physics in Medicine and Biology*, **41**, 2411–2420.

Hodsman, A.B. and Drost, D.H. (1989) The response of vertebral bone mineral density during the treatment of osteoporosis with sodium fluoride. *The Journal of Clinical Endocrinology and Metabolism*, **69**, 932–938.

Hodsman, A.B., Steer, B.M., Fraher, L.J. and Drost, D.J. (1991) Bone densitometric and histomorphometric response to sequential human parathyroid hormone (1–38) and salmon calcitonin in osteoporotic patients. *Bone & Mineral*, **14**, 67–83.

Hohn, T.J. (1978) Corticosteroid induced osteopenia. *Archives of Internal Medicine*, **138**, 882–885.

Holloway, W.R., Collier, F.M., Aitken, C.J. *et al.* (2002) Leptin inhibits osteoclast generation. *Journal of Bone and Mineral Research*, **17**, 200–209.

Holt, D.J., Martin, R.B. and Advani, S. (1977) Measurements of bone's 18 piezoelectric constants. Transactions, 23rd Annual Meeting of the Orthopaedic Research Society, Las Vegas, Nevada, 1–3 Feb.

Holtrop, M.E., Raisz, L.G. and Simmons, H.A. (1974) The effect of parathyroid hormone, colchicines and calcitonin on the ultrastructural and the activity of osteoclasts in organ culture. *The Journal of Cell Biology*, **60**, 346–355.

Holzer, G., Majeska, R.J., Lundy, M.W. *et al.* (1999) Parathyroid hormone enhances fracture healing: a preliminary report. *Clinical Orthopaedics*, **366**, 2158–263.

Holzherr, M.L., Retallack, R.W., Gutteridge, D.H. *et al.* (2000) Calcium absorption in postmenopausal osteoporosis: Benefit of HRT plus calcitriol, but not HRT alone in bone matabsorbers and normal absorbers. *Osteoporosis International*, **11**, 43–51.

Homminga, J., Van Riethergen, B., Lohmuller, E.-M. *et al.* (2004) The osteoporotic vertebral structure is well adapted to the loads of daily life, but not to infrequent "error" loads. *Bone*, **34**, 510–516.

Hong, J., Hipp, J.A., Mulkern, R.V. *et al.* (2000) Magnetic resonance imaging measurements of bone density and cross sectional geometry. *Calcified Tissue International*, **66**, 74–78.

Honkanen, R.J., Honkanen, K., Kroger, H. *et al.* (2000) Risk factors for perimenopausal distal forearm fracture. *Osteoporosis International*, **11**, 265–270.

Hoogerbrugge, P.M., Van Beusechem, V.W., Fischer, A. *et al.* (1996) Bone marrow gene transfer in three patients with adenosine deaminase deficiency. *Gene Therapy*, **3**, 179–183.

Hope, E., Johnston, S.D., Hegstad, R.L. *et al.* (1993) Effects of sample collection and handling on concentration of osteocalcin in equine serum. *American Journal of Veterinary Research*, **54**(7), 1017–1019.

Hori, M., Uzawa, T., Morita, K. *et al.* (1988) Effect of human parathyroid hormone (PTH 1–34) on experimental osteopenia of rats induced by overiectomy. *Bone and Mineral*, **3**, 193–199.

Horii, A., Takeda, N., Matsunaga, T. *et al.* (1993) Effect of unilateral vestibular stimulation on histamine release from the hypothalamus of rats *in vivo*. *Journal of Neurophysiology*, **70**(5), 1822–1826.

Horowitz, M., Need, A.G., Philcox, J.C. and Nordin, B.E.C. (1984) Effect of calcium supplementation on urinary hydroxyproline in osteoporotic postmenopausal women. *The American Journal of Clinical Nutrition*, **39**, 957–959.

Horsman, A., Gallagher, J.C., Simpson, M. and Nordin, B.E.C. (1977) Prospective trial of estrogen and calcium in postmenopausal women. *British Medical Journal*, **ii**, 789–792.

Hosie, C.J., Richardson, W. and Gregory, N.L. (1985) A gamma ray computed tomography scanner for the quantitative measurement of bone density. *Journal of Biomedical Engineering*, **7**, 30–34.

Hosking, D.J., Stone, M.D. and Foote, J.W. (1990) Potentiation of calcitonin by glucocorticoid during the treatment of the hypercalcaemia of malignancy. *European Journal of Clinical Pharmacology*, **38**, 37–41.

Hotchkiss, C.E. (1999) Use of quantitave computed tomography for densitometry of the femoral neck and spine in cynomolgus monkeys. *Bone*, **24**, 101–107.

Hou, J.C.H., Salem, G.J., Zernicke, R.F. and Barnard, R.J. (1990) Structural and mechanical adaptions of immature trabecular bone to strenuous exercise. *Journal of Applied Physiology*, **69**, 1309–1314.

Hounsfield, G.N. (1973) Computerized transverse axial scanning (Tomography): Part I: Description System. *British Journal of Radiology*, **46**, 1016–1022.

Houselmann, H.J. and Rizzoli, R. (2003) A comprehensive review of treatments for postmenopausal osteoporosis. *Osteoporosis International*, **14**, 2–12.

Hreschychyn, M.M., Hopkins, A. and Zylstia, S. (1988) Association of parity, breast-feeding and birth control pills with lumbar spine and femoral neck bone densities. *American Journal of Obstetrics and Gynecology*, **159**, 318–322.

Hsueh, A.J.W., Peck, E.J. and Clark, J.H. (1976) Control of uterine estrogen receptor levels by progesterone. *Endocrinology*, **98**, 438–444.

Hu, G.L., Chiang, H., Zeng, Q.L. *et al.* (2001) ELF magnetic field inhibits gap junctional inter cellular communication and induces hyperphosphorylation of connexin 43 in NIH 3T3 cells. *Bioelectromagnetics*, **22**, 568–573.

Hubbell, J.H. and Overbo, I. (1979) Relativistic atomic form-factors and photon coherent scattering cross-sections. *Journal of Physical and Chemical Reference Data*, **8**(1), 69–105.

Huddleston, A.L., Agarwal, S.K., Friesen, E.J. and Bhaduri, D. (1978) Assessment of bone density using the single source Compton scattering technique. *Applied Radiology*, **7**, 232–237.

Huddleston, A.L. and Bhaduri, D. (1979) Compton scatter densitometry in cancellous bone. *Physics in Medicine and Biology*, **24**, 310–318.

Huddleston, A.L., Bhaduri, D. and Weaver, J.B. (1979) Geometrical considerations for Compton scatter densitometry. *Medical Physics*, **6**, 519–525.

Huddleston, A.L. and Scakler, J.P. (1985) The determination of electron density by the dual-energy Compton scatter method. *Medical Physics*, **12**, 13–19.

Huddleston, A.L. and Weaver, J.B. (1983) Dual energy Compton scatter densitometry. *International Journal of Applied Radiation and Isotopes*, **34**, 997–1002.

Huffer, W.E. (1988) Morphology and biochemistry of bone remodeling: possible control by vitamin D parathyroid hormone and other substances. *Laboratory Investigation*, **59**, 418–442.

Huggin, C. and Young, N.C. (1962) Induction and extinction of mannary cancer. *Science*, **137**, 257–262.

Hughes, F.J., Collyer, J., Stantield, M. and Goodman, S.A. (1995) The effect of bone morphogenetic proteins 2,4 and 6 on differentiation of rat osteoblast cells *in vitro*. *Endocrinology*, **136**, 2671–2677.

Hughes, S., Davies, R., Khan, R. and Kelly, P. (1978) Fluid spaces in bone. *Clinical Orthopaedics and Related Research*, **134**, 332–341.

Hui, S., Slemenda, C.W. and Johnston, C.C. (1988) Age and bone mass as predictors of fracture in a prospective study. *The Journal of Clinical Investigation*, **81**, 1804–1809.

Huiskes, R., Ruimerman, R., Van Lenthe, G.H. and Janssen, J.D. (2000) Effects of mechanical forces on maintenance and adaptation of form in trabecular bone. *Nature*, **405**, 704–706.

Hukkanen, M.V.J., Polak, J.M. and Hughes, S.P.F. (1998) *Nitric Oxide in Bone and Joint Disease*, Cambridge University Press, Cambridge.

Hulka, B.S., Fowler, W.C. Jr, Kaufman, D.G. *et al.* (1980) Estrogen and endometrial cancer: cases and two control groups from North Corlina. *American Journal of Obstetrics and Gynecology*, **137**, 92–101.

Hulley, S., Grady, D., Bush, T. *et al.* (1998) Randomized trial of estrogen plus progestin for secondary prevention of coronary heart disease in postmenopausal women. Heart and Estrogen/Progestin Replacement Study (HERS) Research Group. *The Journal of the American Medical Association*, **280**, 605–613.

Hulmes, D.J.S., Miller, A., Parry, D.A.D., Piez, K.A. and Woodhead-Galloway, J.W. (1973) Analysis of the primary structure of collagen for the origin of molecular packing. *Journal of Molecular Biology*, **79**, 137–148.

Hung, C. T., Allen, F.D., Pollack, S.R. and Brighton, C.T. (1996) What is the role of the convective current density in the real time calcium response of cultured bone cells to fluid flow? *Journal of Biomechanics*, **29**, 1403–1409.

Hung, C.T., Pollack, S.R., Reilly, T.M. and Brighton, C.T. (1995) Real time calcium response of cultured bone cells to fluid flow. *Clinical Orthopaedics and Related Research*, **313**, 256–269.

Hungerford, D.S. and Lennox, D.W. (1985) The importance of increase intraosseous pressure in the development of osteonecrosis of the femoral head: Implication for treatment. *The Orthopedic Clinics of North America*, **16**, 635–654.

Hunter, M.J. (1966) A method for the determination of protein partial specific volumes. *The Journal of Physical Chemistry*, **70**, 3285–3291.

Hunter, R.J. (1981) *Zeta Potential in Colloid Science*, Academic Press, New York.

Huo, M.H., Troiano, N.W., Pelker, R.R. *et al.* (1991) The influence of ibuprofer on fracture repair: Biomechanical, biochemical, histologic and histomorphometric parameters in rats. *Journal of Orthopaedic Research*, **9**, 383–390.

Hurley, D.L., Tiegs, E.D., Barta, J. *et al.* (1989) Effects of oral contraceptive and estrogen administration on plasma calcitonin in pre- and postmenopausal women. *Journal of Bone and Mineral Research*, **4**, 89–95.

Hussain, S.A., Weston, R., Stephenson, R.N. *et al.* (2003) Immediate dual energy X-ray absorptiometry reveals a high incidence of osteoporosis in patients with advanced prostate cancer before hormonal manipulation. *Osteoporosis International*, **92**, 690–694.

Huwyler, R., Born, W., Ohnhaus, E. and Fischer, J.A. (1979) Plasma kinetics and urinary excretion of exogenous human and salmon calcitonin in man. *The American Journal of Physiology*, **236**, E15–E19.

Hyer, S.L., Rodin, D.A., Tobias, J.H. *et al.* (1992) Growth hormone deficiency during puberty reduces adult bone mineral density. *Archives of Disease in Childhood*, **67**, 1472–1474.

Hylstup, L., Andersen, T., McNair, P. *et al.* (1993) Bone metabolism in obesity: Changes to severe overweight and dietary weight reduction. *Acta Endocrinologica*, **129**, 393–398.

Hynes, R.U. (1992) Integrins: Versatility, modulation and signaling in cell adhesion. *Cell*, **69**, 11.

Hyuonen-Debek, M., Riihonen, M. and Debek, J.T. (1979) Elemental analysis of bone mineral by back scattering of alpha particles. *Physics in Medicine and Biology*, **24**, 988–998.

Iannacone, W., Korostoff, W. and Pollack, S.R. (1979) Microelectrode study of stress-generated potentials obtained from uniform and nonuniform compression of human bone. *Journal of Biomedical Materials Research*, **13**, 753–763.

Iannacone, W.M., Pienkowski, D., Pollack, S.R. and Brighton, C.T. (1988) Pulsing electromagnetic field stimulation of the *in vitro* growth plate. *Journal of Orthopaedic Research*, **6**, 239.

Ibbotson, K.J., Orcutt, C.M., D'Souza, S.M. *et al.* (1992) Contrasting effects of parathyroid hormone and insulin like growth factor I in an aged overiectomized rat model of postmenopausal osteoporosis. *Journal of Bone and Mineral Research*, **7**, 425–432.

Ibiwoye, M.O., Powell, K.A., Garbiner, M.D. *et al.* (2004) Bone mass is preserved in a critical sized osteotomy by low energy pulsed electromagnetic fields as quantitated by *in vivo* microcomputed tomography. *Journal of Orthopaedic Research*, **22**, 1086–1093.

ICRP (1975) Report of the Task Group of Reference Man. ICRP Report 23, 97.

Igarashi, A., Okochi, H., Bradham, D.M. and Grotendorst, G.R. (1993) Regulation of connective tissue growth factor gene expression in human skin fibroblasts and during wound repair. *Molecular Biology of the Cell*, **4**, 637–645.

Iida, H. Ko, S., Miyashita, Y. *et al.* (1956) On electric callus produced by an alternating current. *Journal of the Kyoto Prefect Medical University*, **60**, 561–564.

Ijiri, K. Matsunagos, S., Fukada, T. and Shimzu, T. (1995) Indomethacin inhibition of ossification induced by direct current stimulation. *Journal of Orthopaedic Research*, **13**, 123–131.

Ikeda, K. and Ogatta, E. (2000) Modulation of bone remodeling by active vitamin D. Its role in the treatment of osteoporosis. *Mechanisms of Ageing and Development*, **116**, 103–111.

Ikeda, K., Sugimoto, T., Fukase, M. and Fujita, T. (1991) Protein kinase C is involved in PTH-induced homologous desensitization by directly affecting PTH receptor in the osteoblastic osteosarcoma cells. *Endocrinology*, **128**, 2901–2906.

Ikeda, T. (1990) *Fundamentals of Piezoelectricity*, Oxford University Press.

Ilfeld, F.W., Weinberg, G., Rosen, V. and August, W. (1974) Direct current induced mosaic bone architecture. *Clinical Orthopaedics*, **99**, 298.

Ilizarov, G.A., Ledyasev, V.I. and Shitin, V.P. (1969) Experimental studies of bone lengthening. *Eksperimentalnaia Khirurgiia i Anesteziologiia*, **14**, 3–6.

Ilmeni, G.I., Rocchi, R., Cadossi, R. *et al.* (1996) The electromagnetic stimulation of tibial osteotomies. *Clinical Orthopaedics*, **288**, 246–253.

Imamura, K., Ozawa, H. *et al.* (1990) Continuously applied compressive pressure induces bone resorption by a mechanism involving $PGE_2$ synthesis. *Journal of Cellular Physiology*, **144**(2), 222–228.

Ingber, D.E. (1993) Cellular tensegrity: defining new rules of biological design that govern the cytoskeleton. *Journal of Cell Science*, **104**, 613–627.

Ingber, D.E. (2003) Tensegrity II. How structural networks influence cellular information processing networks. *Journal of Cell Science*, **116**, 1397–1408.

Ingber, D.E. and Folkman, J. (1989) Mechanochemical switching between growth and differentiation during fibroblast growth factor-stimulated angiogenesis *in vitro*: role of extracellular matrix. *The Journal of Cell Biology*, **109**, 317–330.

Ingber, D., Karp, S., Plopper, G. *et al.* (1993) Mechanochemical transduction across extracellular matrix and through the cytoskeleton, in *Physical Forces and the Mammalian Cell* (ed. J.A. Frangos), Academic Press, New York, pp. 61–79.

Inoue, N. (1987) Mechanical properties of cancellous bone. Dependential strength and elastic modulus on trabecular orientation. *Journal of Kyoto Prefectural University of Medicine*, **96**, 847–865.

Inoue, N., Ohmishi, I., Chen, D.A. *et al.* (2002) Effect of pulsed electromagnetic fields on osteotomy of healing in a canine tibial ends. *Journal of Orthopaedic Research*, **20**, 957–966.

Inoue, T., Kusida, K., Miyamoto, S. *et al.* (1983) Quantitative assessment of bone density on X-ray picture. *Journal of the Japanese Orthopaedic Association*, **57**, 1923–1936.

International Commission on Radiological Protection (1975) *Report of the Task Group on Reference Man*, Pergamon, Oxford, Publication No. 23.

Isgaard, J., Carlsson, L., Isakssen, O.G.P. and Jansson, J.O. (1998) Pulsatile intravenous growth hornone (GH) infusion to hyprphysectomized rats increases insulin like growth factor I messenger RNA in skeletal tissues more effectively than continuous GH infusion. *Endocrinology*, **123**, 2605–2610.

Ishidou, Y., Ijiri, K., Matsunaga, S. *et al.* (1994) Transforming growth factor-$\beta$ in osteogenesis by direct current stimulation. Proceedings of the 4th Conference of the International Society for Fracture Repair. Kobe, Japan, September 27, p. 91.

Ishidou, Y., Kitajima, I., Obama, H. *et al.* (1995) Enhanced expression of type 1 receptors for bone morphogenetic proteins during bone formation. *Journal of Bone and Mineral Research*, **10**, 1651–1659.

Ismail, A.A., Pye, S.R., Cockerill, W.C. *et al.* (2002) Incidence of limb fracture across Europe: results from the European Prospective Osteoporosis Study (EPOS) *Osteoporosis International*, **13**, 565–571.

Ito, M., Nishida, A., Nakamura, T. *et al.* (2002) Differences of three dimensional trabecular microstructure in osteopenic rat models caused by overiectomy and neurectomy. *Bone*, **30**, 594–598.

Itoh, S., Kikuchi, M., Koyama, Y. *et al.* (2002) Development of an artificial vertebral body using a novel biomaterial, hydroxyapatite/collagen composite. *Biomaterials*, **23**, 3919–3926.

Itoh, S., Kikuchi, M., Takakuda, K. *et al.* (2001) The biocompatibility and osteoconductive activity of a novel hydroxyapatite/collagen composite biomaterial and its function as a carrier of rhBMP-2. *Journal of Biomedical Materials Research*, **54**, 445–453.

Iwaki, A., Jinguishi, S., Oda, Y. *et al.* (1997) Localization and quantification of proliferating cells during rat fracture repair: detection of proliferating cell nuclear antigen by immunochemistry. *Journal of Bone and Mineral Research*, **12**, 96–102.

Iwamoto, J., Yeh, J.K. and Takeda, T. (2003) Effect of vitamin $K_2$ on cortical and cancellous bones in orchidectomized and/or sciatic neurectomized rats. *Journal of Bone and Mineral Research*, **18**, 776–783.

Jackson, J.A. and Kleerekoper, M. (1990) Osteoporosis in men diagnosis, pathophysiology and prevention. *Medicine (Baltimore)*, **69**, 137–152.

Jackson, J.A., Riggs, M.W. and Spiekerman, A.M. (1992) Testosterone deficiencies as a risk factor for hip fractures in men: a case control study. *The American Journal of the Medical Sciences*, **304**, 4–8.

Jackson, R.D., La Croix, A.Z., Gass, M. *et al.* (2006) Calcium plus vitamin D supplementation and the risk of fractures. *The New England Journal of Medicine*, **354**, 669–683.

Jacobs, C.R., Yellowley, C.E., Davis, B.R. *et al.* (1998) Differential effect of steady versus oscillating flow on bone cells. *Journal of Biomechanics*, **31**, 969–976.

Jacobson, B. (1964) X-ray spectrophotometry *in vivo. American Journal of Roentgenology*, **91**, 202–210.

Jacobson, M.D., Weil, M. and Raff, M.C. (1997) Programmed cell death in animal development. *Cell*, **88**, 347–354.

Jaffe, L.F. (1979) *Membrane Transduction Mechanisms* (eds R.A. Cone and J.E. Dowling), Raven, New York, pp. 199–231.

Jameson, M.W., Hood, J.A. and Tidmarsh, B.G. (1993) The effects of dehydration and rehydration on some mechanical properties of human dentine. *Journal of Biomechanics*, **26**, 1055–1065.

Jancar, J. and Kucera, J. (1990) Yield behavior of pp/$CaCo_3$ and pp/$Mg(OH)_2$ composites II: Enhanced interfacial adhesion. *Polymer Engineering*, **30**, 714–719.

Jang, M., Mistry, A., Swick, A.G. and Romsos, D.R. (2000) Leptin rapidly inhibits hypothalamic neuropeptide Y and secretion and stimulates corticotropin-releasing hormone secretion in adrenalectomized mice. *The Journal of Nutrition*, **130**, 2813–2820.

Jankovich, J.P. (1970) Structural Development of bone in the Rat under Earth Gravity, Simulated Weight Lossness, Hypergravity and Mechanical Vibration, PhD Dissertation, University of Kentucky.

Jannacone, W.M., Pienkowski, D., Pollack, S.R. and Brighton, C.T. (1988) Pulsing electromagnetic field stimulation of the *in vitro* growth plate. *Journal of Orthopaedic Research*, **6**, 239–247.

Japsen, K.J., Schaffler, M.B., Kuhn, J. *et al.* (1997) Type I collagen mutation alters the strength and fatigue behavior of MOV 13 cortical tissue. *Journal of Biomechanics*, **30**, 1141–1147.

Javaid, P.P., Pitt, P. and Moniz, C. (1990) Bone mineral density of patella and its relation to lumbar spine. *Osteoporosis*, **2**, 662–663.

Jaworski, Z.F.G. and Uhthoff, H.K. (1986) Reversibility of nontraumatic disuse osteoporosis during its active phase. *Bone*, **7**, 431–439.

Jayanand and Behari, J. (2005) Proceedings of International Union of Radio Science (URSI) held in October 13–16, New Delhi.

Jayanand, Behari, J. and Rajeev, L. (2003) Effects of low level pulsed radio frequency fields on induced osteoporosis in rat bone. *Indian Journal of Experimental Biology*, **41**, 587–591.

Jee, S.H., Shen, S.C., Tseng, C.R. *et al.* (1998) Curcumin induces a p53-dependent apoptosis in human basal cell carcinoma cells. *The Journal of Investigative Dermatology*, **111**(4), 656–661.

Jee, W.S. (1988) The skeletal tissues, in *Cell and Tissue Biology: A Textbook of Histology* (ed. L. Weiss), Urban & Schwarzenberg, Baltimore, MD, pp. 218–254.

Jee, W.S., Ke, H.Z. *et al.* (1991) Long term anabolic effects of prostablandin-$E_2$ on tibiae diaphyseal bone in male rats. *Bone and Mineral*, **15**(1), 33–55.

Jee, W.S.S. and Li, X.J. (1990) Adaptation of cancellous bone to overloading in the adult rat: A single photon absorptiometry and histomorphometry study. *The Anatomical Record*, **227**, 418–426.

Jee, W.S.S., Mori, S., Li, X.J. and Chan, S. (1990) $PGE_2$ enhances cortical bone mass and activates intracortical bone remodeling in intact and overiectomzed female rats. *Bone*, **11**, 253–266.

Jee, W.S.S., Ueno, K., Deng, Y.P. and Woodbury, D.M. (1985) The effects of prostaglandin E2 in growing rats increased metaphysical hard tissue and corticoendosteal bone formation. *Calcified Tissue International*, **37**, 148–157.

Jeffcott, L.B. and McCartney, R.N. (1985) Ultrasound as a tool for assessment of bone quality in the horse. *The Veterinary Record*, **116** (13), 337–342.

Jemollieres, F.A., Pouilles, J.M. and Ribot, C. (2001) Withdrawal of hormone replacement therapy is associated with significant vertebral bone loss in postmenopausal women. *Osteoporosis International*, **12**, 385–390.

Jensen, B.V., Nielsen, S.L. and Skovsgaard, T. (1996) Treatment with angiotensin-converting-engyme inhibitor for epirubicin-induced dilated cardiomyopathy. *Lancet*, **347**, 297–299.

Jensen, G.F., Christiansen, C. and Transbol, I. (1982) Fracture frequency and bone preservation in postmenopausal women treated with estrogen. *Obstetrics and Gynecology*, **60**, 493–496.

Jeong, K.S., Lee, J., Jeong, W. *et al.* (2005) Measurement of estrogen effect on bone turnover by $^2H_2O$ labeling. *Calcified Tissue International*, **76**, 365–370.

Jepsen, K.J., Goldstein, S.A., Kuhn, J.L. *et al.* (1996) Type-I collagen mutation compromises the post-yield behavior of Mov13 long bone. *Journal of Orthopaedic Research*, **14**, 493–499.

Jepsen, K.J., Mansoura, M.K., Kuhn, J.L. *et al.* (1992) An *in vivo* assessment of the contribution of type I collagen to the mechanical properties of cortical bone. *Transactions of the Orthopaedic Research Society*, **17**, 93.

Jepsen, K.J., Hu, B., Tommasini, S.M. *et al.* (2007) Genetic randomization reveals functional relationships among morphologic and tissue-quality traits that contribute to bone strength and fragility. *Mammalian Genome*, **18**(6–7), 492–507.

Jerome, C.P., Johnson, C.S. and Lees, C.J. (1995) Effect of treatment for 3 months with human parathoroid hormone 1–34 peptide in overiectomized cynomolgus monkeys (Macaca Fascicularis). *Bone*, **17**, S415–S421.

Jerome, C.P., Johnson, C.S., Vafai, H.T. *et al.* (1999) Effect of treatment for 6 months with human parathyroid hormone (1-34) peptide in ovariectomized cynomolgus monkeys (Macaca fasciculairies). *Bone*, **25**, 301–309.

Jha, S.C. and Behari, J. (1993) Stress wave propagation bone. *Biomedicine*, **1**, 13–20.

Jiang, Y., Zhao, J., Augat, P. *et al.* (1998) Trabecular bone mineral and calculated structure of human bone specimens scanned by peripheral quantitative computed tomography: relation to biomechanical properties. *Journal of Bone and Mineral Research*, **13**, 1783–1790.

Jiang, Y., Zhao, J., Genant, H.K., Dequeker, J. and Geusens, P. (1997) Long-term changes I bone mineral and biomechanical properties of vertebrae and femur in aging, dietary calcium restricted, and/or estrogen-deprived/replaced rats. *Journal of Bone and Mineral Research*, **12**, 820–831.

Jick, H., Watkins, R.N., Hunter, J.R. *et al.* (1979) Replacement estrogen and endometrial cancer. *The New England Journal of Medicine*, **300**, 218–222.

Jilka, R.L. (1998) Cytokines, bone remodeling and estrogen deficiency. A 1998 update. *Bone*, **23**, 75–81.

Jilka, R.L., Hangoc, G., Girascole, G. *et al.* (1992) Increased osteoclast development after estrogen loss: Mediation by interleukin-6. *Science*, **257**, 88–91.

Jilka, R.L., Weinstein, R.S., Bellido, T. *et al.* (1999) Increased bone formation by prevention of osteoblast apoptosis with parathyroid hormone. *The Journal of Clinical Investigation*, **104**, 439–446.

Jinguishi, S., Hedemann, A., Kana, S.K. and Macey, L.R. (1990) Acidic fibroblast growth factor (aFGF) injection stimulates enlargement and inhibits cartilage gene expression in rat fracture healing. *Journal of Orthopaedic Research*, **8**, 364–371.

Jingushi, S., Iwaki, A., Higuchi, O. *et al.* (1998) Serum 1$\alpha$,25-dihydroxyvitamin $D_3$ accumulates into the fracture callus during rat femoral fracture healing. *Endocrinology*, **139**, 1467–1473.

Johansen, A. and Stone, M.D. (1997) The effect of ankle oedema on bone ultrasound assessment at the heel. *Osteoporosis International*, **7**, 44–47.

Johnell, O. (1996) Advances in osteoporosis: better identification of risk fracture reduces morbidity and mortality. *Journal of Internal Medicine*, **239**, 299–304.

Johnell, O., Gullberg, B., Kanis, J.A. *et al.* (1995) Risk factors for hip fracture in European women: The MEDOS study. *Journal of Bone and Mineral Research*, **10**, 1802–1815.

Johnell, O. and Nilsson, B.E. (1984) Lifestyle and bone mineral mass in perimenopausal women. *Calcified Tissue International*, **36**, 354–356.

Johnson, C.C. and Epstein, S. (1981) Clinical biochemical, radiographic, epidemiologic and economic features of osteoporosis. *The Orthopedic Clinics of North America*, **12**, 559–569.

Johnston, C.C., Melton, L.J., Linkday, R. and Eddy, D.M. (1989) Clinical implications for bone mass measurements. *Journal of Bone and Mineral Research*, **4**(Suppl 2), 1028.

Johnson, C.C., Smith, D.M., Yu, P.L. and Deiss, W.P. Jr (1968) *In vivo* measurement of bone mass in the radius. *Metabolism: Clinical and Experimental*, **17**, 1140.

Johnson, D.L., McAllister, T.N. and Frangos, J.A. (1996) Fluid flow stimulates rapid and continuous release of nitric oxide in osteoblasts. *The American Journal of Physiology*, **271**, E205–E208.

Johnson, L.C. (1964) Morphologic analysis in pathology: The kinetics of disease and general biology of bone, in *Bone Biodynamics* (ed. W.M. Frost), Little, Brown, Boston, MA, p. 543.

Johnson, M.W., Chakkalakal, D.A., Harper, R.A. *et al.* (1982) Fluid flow in bone *in vitro*. *Journal of Biomechanics*, **15**, 881–885.

Johnson, M.W., Chakkalakal, R.A., Harper, R. and Katz, J.L. (1980) Comparison of the electromechanical effects in wet and dry bone. *Journal of Biomechanics*, **13**, 437–442.

Johnson, R.S., Spiegelman, B.M. and Papaioannou, V. (1992) Pleiotropic effects of a null mutation in the c-fox protooncogene. *Cell*, **71**, 577–586.

Jones, C.D., Laval Jeanlet, A.M., Laval Jeantet, M.H. and Genant, H.K. (1987) Importance of measurement of spongious vertebral bone mineral density in the assessment of osteoporosis. *Bone*, **8**, 201–206.

Jones, D.B., Pedley, R.B. and Ryaby, J.T. (1986) The effects of pulsating electromagnetic fields on differentiation and growth in cloudman S91 murine melanoma cells *in vitro*. *Journal of Bioelectricity*, **5**, 145–169.

Jones, G., Riley, M. and Dwyer, T. (1999) Maternal smoking during pregnancy growth and bone mass in prepubertal children. *Journal of Bone and Mineral Research*, **14**, 146–151.

Jones, S.J., Gray, C., Boyde, A. and Burnstock, G. (1997) Purinergic transmitters inhibit bone formation by cultured osteoblasts. *Bone*, **21**, 393–399.

Jonsson, B., Riongsber, K., Josefsson, P.O. *et al.* (1992) Effects of physical activity on bone mineral content and muscle strength in women: a cross sectional study. *Bone*, **13**, 191–195.

Jonsson, U., Ranta, H. and Stromberg, L. (1985) Growth changes of collagen cross linking, calcium and water content in bone. *Archives of Orthopaedic and Traumatic Surgery*, **104**, 89–93.

Jorgensen, H.R.I., Savnholm, H. and Host, A. (1988) Bone formation induced in an infant by systematic prostaglandin $E_2$ administration. *Acta Orthopaedica Scandinavica*, **59**, 464–466.

Jorgensen, N.R., Geist, S.T., Civitelli, R. and Steinberg, T.H. (1997) ATP-and gap junction-dependent intracellular calcium signaling in osteoblastic cells. *The Journal of Cell Biology*, **139**, 497–506.

Jorgensen, T.E. (1972) Measurements of stability of crural fractures treated with Hoffmann osteotaxis. II. Measurements of crural fractures. *Acta Orthopaedica Scandinavica*, **43**, 207–218.

Joseph, P.M. and Spital, R.D. (1978) A method for correcting bone induced artifacts in computed tomography scanners. *Journal of Computer Assisted Tomography*, **2**, 100–108.

Jowsey, J. (1960) Age changes in human cortical bone. *Clinical Orthopaedics*, **17**, 210.

Joyce, M.E., Roberts, A.B., Sporn, M.B. and Bolander, M.E. (1990) Transforming growth factor-$\beta$ and the initiation of chondrogenesis and osteogenesis in the rat femur. *The Journal of Cell Biology*, **110**, 2195–2207.

Judex, S., Boyd, S., Qin, Y.-X. *et al.* (2003) Combining high-resolution micro-computed tomography with material composition to define the quality of bone tissue. *Current Osteoporosis Reports*, **1**, 11–19.

Judex, S. and Zernicke, R.F. (2000) High impact exercise and growing bone: Relation between high strain rates and enhanced bone formation. *Journal of Applied Physiology*, **88**, 2183–2219.

Julius, D., Livelli, T.J., Jessell, T.M. and Axel, R. (1989) Ectopic expression of the serotonin 1c receptor and the triggering of malignant transformation. *Science (Washington, DC)*, **244**, 1057–1062.

Jurist, J. (1970a) *In-vivo* determination of the elastic response of bone. I. Method of ulnar resonant frequency determination. *Physics in Medicine and Biology*, **15**(3), 417–426.

Jurist, J. (1970b) *In vivo* determination of the elastic response of bone. II. Ulnar resonant frequency in osteoporotic diabetic and normal subjects. *Physics in Medicine and Biology*, **15**(3), 427–434.

Jurist, J. (1972) Difficulties with measurement of ulnar resonant frequency. Correspondence. *Physics in Medicine and Biology*, **17**, 290–291.

Jurist, J. and Kianinan, K. (1973) Three models of the vibrating ulna. *Journal of Biomechanics*, **6**, 331–342.

Kabel, J., Odgaard, B., Von Riebergen, B. and Huisks, R. (1999) Connectivity and elastic properties of cancellous bone. *Bone*, **24**, 115–120.

Kakikana, R., Nobel, E.P. and Butte, J.C. (1968) Corticosterone response to ethanol in inbred strains of mice. *Nature (London)*, **218**, 360–361.

Kalender, W.A. (1992) Effective dose values in bone mineral measurements by photon absorptiometry and computed cosmography. *Osteoporosis International*, **2**, 82–87.

Kalender, W.A., Klotz, E. and Suss, C. (1987) Vertebral bone mineral analysis: an integrated approach. *Radiology*, **164**, 419–423.

Kaltakci, M.Y. (1995) Stress concentration and failure criteria in an isotropic plates with circular holes subjected to tension or compression. *Computers and Structers*, **61**, 67–78.

Kalu, D.N. (1991) The ovariectomized rat model of postmenopausal bone loss. *Bone & Mineral*, **15**, 175–192.

Kalu, D.N., Liu, C.C., Hardin, R.R. and Hollis, B.W. (1989) The aged rat model of overian hormone deficiency bone loss. *Endocrinology*, **124**, 7–16.

Kamagata-Kiyoura, Y., Ohta, M., Cheuk, G. *et al.* (1999) Combined effects of caffeine and prostaglandin $E_2$ on the proliferation of oseoblast like cells (UMR106-01). *Journal of Periodontology*, **70**, 283–288.

Kana, J., Hutsschenretter, G., Hama, D. and Wandehch, W. (1981) Effect of low power density laser radiation on healing of open skin wounds in rats. *Archives of Surgery*, **116**, 293–296.

Kanis, J.A. (1999) The use of calcium in the management of osteoporosis. *Bone*, **24**, 279–290.

Kanis, J.A., Black, D., Cooper, C. *et al.* (2002) A new approach to the development of assessment guidelines for osteoporosis. *Osteoporosis International*, **13**, 527–536.

Kanis, J.A., Delmas, P., Burckhardt, P. *et al.* on behalf of the European foundation for osteoporosis, Bone Disease (1997) Guidelines for diagnosis and treatment of osteoporosis. *Osteoporosis International*, **7**, 390406.

Kanis, J.A. and Gluer, C.C. (2000) Committee of scientific advisors. Internationational Osteoporosis Foundation. An update on the diagnosis and assessmet of osteoporosis with densitometry. *Osteoporosis International*, **11**, 192–202.

Kanis, J.A., Melton, L.J., Christiansen, C. *et al.* (1994) The diagnosis of osteoporosis. *Journal of Bone and Mineral Research*, **9**, 1137–1141.

Kanis, J.A. and Meunier, P.J. (1984) Should we use fluoride to treat osteoporosis? A review. *The Quarterly Journal of Medicine: New Series*, **210**, 145–164.

Kann, P., Bergink, A.P., Fang, Y. *et al.* (2002) The collagen IaI spl polymorphism is associated with differences in ultrasound transmission velocity in the calcaneus in postmenopausal women. *Calcified Tissue International*, **70**, 450–456.

Kannus, P., Jarvinen, M., Sievanen, H. *et al.* (1994a) Reduced bone mineral density in man with a previous femur fracture. *Journal of Bone and Mineral Research*, **11**, 1729–1736.

Kannus, P., Jarvinen, M., Sievanen, H. *et al.* (1994b) Osteoporosis in men with a history of tibial fracture. *Journal of Bone and Mineral Research*, **9**, 423–429.

Kannus, P., Jozsa, L., Kvist, M. *et al.* (1996) Expression of osteocalcin in the patella of experimentally immobilized and remobilized rats. *Journal of Bone and Mineral Research*, **11**(1), 79–87.

Kannus, P., Palvanen, M., Kaprio, J. *et al.* (1999) Genetic factors and osteoporotic fractures elderly people: prospective 25 year follow up of a nationwide cohort of elderly Finnish twins. *British Medical Journal*, **319**, 1334–1337.

Kano, S., Yamazaki, A., Otsuka, R. *et al.* (1994) Application of hydroxyapatite-sol as drug carrier. *Bio-Medical Materials and Engineering*, **4**, 283–290.

Karachalios, T., Lyritis, G.P., Giannarakos, D.G. *et al.* (1992) Calcitoin effects on rabbit bone. Bending tests on ulnar osteotomies. *Acta Orthopaedica Scandinavica*, **63**, 615–618.

Karellas, A., Leichter, I., Carven, J.D. and Greefield, M.A. (1983) Characterization of tissue via coherent-to-Compton scattering ratio: Sensitivity considerations. *Medical Physics*, **10**, 605–609.

Karlsson, M.K., Johnell, O., Nilsson, B.E. *et al.* (1993) Bone mineral mass in hip fracture patients. *Bone*, **14**, 161–165.

Karlsson, M.K., Obrant, K.J., Nilsson, B.E. and Johnell, O. (1998) Bone mineral density assessed by quantitative ultrasound and dual energy X-ray absorptiometry. Normative data Malmo, Sweden. *Acta Orthopaedica Scandinavica*, **69**, 189–193.

Karlsson, M.K., Weigall, S.J., Duan, Y. and Seeman, E.J. (2000) Bone size and volumetric density in women with anorexia nervosa receiving estrogen replacement therapy and in women recovered from anorexia nervosa. *The Journal of Clinical Endocrinology and Metabolism*, **85**, 3177–3182.

Karsenty, G. (2003) The complexities of skeletal biology. *Nature*, **423**, 316–318.

Karsenty, G. and Wagner, E.F. (2002) Reaching a genetic and molecular understanding of skeletal development. *Developmental Cell*, **2**, 389–406.

Karu, T. (1989) Photobiology of low power laser effects. *Health Physics*, **56**, 691–704.

Kassem, M., Brixen, K., Blum, W.F. *et al.* (1994) Normal osteoclastic and osteolastic responses to exogenous growth hormone in patients with postmenopausal spinal osteoporosis. *Journal of Bone and Mineral Research*, **9**, 1365–1370.

Kassem, M., Melton, I.J. and Riggs, B.L. (1996) The type I/type II model for involutional osteoporosis, in *Osteoporosis* (eds R. Marcus, D. Feldman and J. Kelsey), Academic Press, New York, pp. 691–702.

Kasten, T.P., Collinosdoby, P., Patel, N. *et al.* (1994) Potentiation of osteoclast bone resorption activity by inhibition of nitric-oxide synthase. *Proceedings of the National Academy of Sciences of the United States of America*, **91**, 3569–3573.

Katagiri, T. and Takahashi, N. (2002) Regulatory mechanisms of osteoblast and osteoclasts differentiation. *Oral Diseases*, **8**, 147–159.

Kato, M., Patel, M.S., Levasseur, R. *et al.* (2002) Cbfal-independent decrease in osteoblast proliferation, osteopenia and persistent embryonic eye vascularization in mice deficient in LRP5, a Wnt coreceptor. *The Journal of Cell Biology*, **157**, 303–314.

Katz, J.L. (1971) Hard tissue as a composite material – I. Bounds on the elastic behavior. *Journal of Biomechanics*, **4**, 455–473.

Katz, J.L. (1977) Anisotropy of Young's modulus of bone. *Nature*, **283**, 106–107.

Katz, J.L. (1980) The structure and biomechanics of bone, in *Mechanical Properties of Biological Materials* (eds J.F.V. Vincent and J.C. Currey), Cambridge University Press Cambridge, pp. 137–168.

Katz, J.L. (1981) Composite material models for cortical bone, in *Mechanical Properties of Bone*, vol. **45**(ed S.C. Cowin), AMD, American Society of Mechanical Engineers, New York, pp. 171–184.

Katz, J.L. and Ukraincik, K. (1971) On the anisotropic elastic properties of hydroxyapatite. *Journal of Biomechanics*, **4**, 221–227.

Katz, J.L. and Ukraincik, K. (1971) On the anisotropic properties of hydroxyapatite. *Journal of Biomechanics*, **4**, 221–227.

Katz, J.L. and Yoon, H.S. (1984) The structure and anisotropic mechanical properties of bone. *IEEE Transactions on Biomedical Engineering*, **31**(12), 878–883.

Katz, J.M., Skinner, S.J.M., Wilson, T. and Gray, D.H. (1983) The *in vitro* effect of indomethacin on basal bone resorption, on prostaglandin production and on the response to added prostaglandins. *Prostaglandins*, **26**, 545–555.

Katznelson, L., Finkelstein, J.S., Schoenfeld, D.A. *et al.* (1996) Increase in bone density and lean bone mass during testosterone administration in men with acquired hypogonadism. *The Journal of Clinical Endocrinology and Metabolism*, **81**, 4358–4365.

Kaufman, J.J., Alves, J.M., Siffert, R.S., Magee, F.P. and Ryaby, J.T. (1994) Ultrasound assessment of trabecular bone density using neural networks. *Journal of Bone and Mineral Research*, **9**(S1), 188.

Kaufman, J.J. and Einhorn, T.A. (1993) Perpectives: Ultrasound assessment of bone. *Journal of Bone and Mineral Research*, **8**(5), 517–525.

Kaufman, J.J., Xu, W., Chiabrera, A.E. and Siffert, R.S. (1995) Diffraction effects in insertion mode estimation of ultrasonic group velocity. *IEEE Transactions on Ultrasonics, Ferroelectrics, and Frequency Control*, **42**(2), 232–242.

Kaufman, J.M., Taelman, P., Vermeulen, A. and Van de Weghe, M. (1992) Bone mineral states I growth hormone-deficient males with isolated and multiple pituitary deficiencies of childhood onset. *The Journal of Clinical Endocrinology and Metabolism*, **74**, 118–123.

Kaulsson, M.K., Duan, Y., Ahlborg, H. *et al.* (2001) Age, gender and fragility fractures are associated with differences in qualitative ultrasound independent of bone mineral density. *Bone*, **28**, 118–122.

Kaur, R., Sharma, S. and Rampal, S. (2003) Effect of sub-chronic selenium toxicosis on lipid peroxidation, glutathione redox cycle and antioxidant enzymes in calves. *Veterinary and Human Toxicology*, **45**, 90.

Kawaguchi, H., Pilbeam, C.C., Harrusion, J.R. and Raisz, L.G. (1995) The role of postaglandins in the regulation of bone metabolism. *Clinical Orthopaedics and Related Research*, **313**, 36–46.

Kawai, H. (1969) The piezoelectricity of poly(vinylidene fluoride). *Japanese Journal of Applied Physics*, **8**, 975–976.

Ke, H.Z., Shen, V.W., Oi, H. *et al.* (1998) Prostaglandin E2 increases bone strength in intact rats and in ovariectomized rats with established osteopenia. *Bone*, **23**(3), 249–255.

Keane, B.E., Spiegler, G. and Davis, R. (1959) Quantitative evaluation of bone mineral by a radiographic method. *British Journal of Radiology*, **32**, 162–167.

Keaveny, T.M.K., Morgan, E.F., Biebur, G.L. and Yeh, O.C. (2001) Biomechanics of trabecular bone. *Annual Review of Biomedical Engineering*, **3**, 307–333.

Keen, R.W., Hart, D.J., Arden, N.K. *et al.* (1999a) Family history of appendicular fracture and risk of osteoporosis: a populaion based study. *Osteoporosis International*, **10**, 161–166.

Keen, R.W., Woodford-Richens, K.L., Grant, S.F. *et al.* (1999b) Association of polymorphism at the type I collagen (COLIAI) locus with reduced bone mineral density, increased fracture risk and increased collagen turnover. *Arthritis & Rheumatism*, **42**, 285–290.

Keller, M.R., Kessler, R.M., Brooks, R.A. and Kirkland, L.R. (1980) Optimum energy for performing CT iodinated contrast studies. *British Journal of Radiology*, **53**, 576–579.

Keller, T.S. (1994) Predicting the compressive mechanical behavior of bone. *Journal of Biomechanics*, **27**, 1159–1168.

Kelley, P.J. and Bronk, J.T. (1990) Venous pressure and bone formation. *Microvascular Research*, **39**, 364–375.

Kelley, T.L., Crane, G. and Barner, D.T. (1994) Single X-ray absorptiometry of the forearm precision and reference data. *Calcified Tissue International*, **54**(3), 212–218.

Kelly, A. (1973) *Strong Solids*, 2nd edn, Clarendon Press, Oxford.

Kelsey, J.L., Browner, W.S., Seeley, D.G. *et al.* (1992) Risk factors for fractures of the distal forearm and proximal humerus. *American Journal of Epidemiology*, **135**, 477–489.

Kelsey, J.L. and Hoffman, S. (1987) Risk factors for hip fractures. *The New England Journal of Medicine*, **316**, 404.

Kennedy, A.C., Boddy, K. and Williams, E.D. (1979) Whole body elemental composition during drug treatment of rheumatoid arthritis: A preliminary study. *Annals of the Rheumatic Diseases*, **38**, 137–140.

Kennedy, N.S.J., Eastell, R. and Ferrington, C. (1982) Total body neutron activation analysis of calcium: Calibration and normalization. *Physics in Medicine and Biology*, **27**, 697–707.

Kenner, G.H., Gabrielson, E.W., Lovelly, J.E. *et al.* (1975) Electrical digest osteoporosis. *Calcified Tissue Research*, **18**, 111–117.

Kennett, T.J., Garnett, E.S. and Webber, C.E. (1972) An *in vivo* measurement of absolute bone density. *Journal of the Canadian Association of Radiologists*, **23**, 168.

Kenwright, J. (1985) Biomechanical measurement of fracture repair, in *Biomechanical Measurement in Orthopaedic Practice* (eds W. Michael and H. Derek), Claredon Press, Oxford, pp. 3–9.

Kerner, E.H. (1956) Elastic and thermoelastic properties of composite media. *Proceedings of the Physical Society (London)*, **69**, 808–813.

Kerr, D., Morton, A., Dick, I. and Prince, R. (1996) Exercise effects on bone mass in postmenopausal women are site specific and load dependent. *Journal of Bone and Mineral Research*, **11**, 218–225.

Kerr, S.A., Kouris, K., Webber, C.E. and Kennett, T.J. (1980) Coherent scattering and the assessment of mineral concentration in trabecular bone. *Physics in Medicine and Biology*, **25**, 1037–1047.

Kessler, L.W. (1973) VHF ultrasonic attenuation in mammalian tissue. *The Journal of the Acoustical Society of America*, **53**, 1759–1760.

Keyak, J.H., Lee, I. and Skinner, H.B. (1994) Correlations between orthogonal mechanical properties and density of trabecular bone: use of different densitometric measures. *Journal of Biomedical Materials Research*, **28**, 1329–1336.

Khachigian, L.M., Lindner, V., Williams, A.J. and Collins, T. (1996) Egr-1 induced endothelial gene expression: a common theme in vascular injury. *Science*, **271**, 1427–1431.

Khosla, S., Melton, J., Atkinson, E.J. *et al.* (1998) Relation of serum sex steroid levels and bone turnover markers with bone mineral density in men and women. A key role for bioavailable estrogen. *The Journal of Clinical Endocrinology and Metabolism*, **83**, 2266–2274.

Khosla, S., Melton, L.J. III, Atkinson, L.J. and O'Fallon, W.M. (2001) Steroid levels to longitudinal changes in bone density in young versus elderly men. *The Journal of Clinical Endocrinology and Metabolism*, **86**, 3555–3561.

Kiel, D.P., Felson, D.T., Hannan, M.T. *et al.* (1990) Caffeine and the risk fracture. *American Journal of Epidemiology*, **132**, 675–684.

Kiel, D.P., Haunan, M.T., Broe, K.E., Felson, D.T., Cupples, L.A. *et al.* (2001) Bone metacarpal cortical area predict the occurrence of hip fracture in women and men over 3 decades of follow up? Results from the Farmingham Osteoporosis Study. *Journal of Bone and Mineral Research*, **16**(12), 2260–2266.

Kikuchi, M., Itoh, S., Ichinose, S. *et al.* (2001) Self organization mechanism in a bone like hydroxyapatite/collagen nanocomposite synthesized *in vitro* and its biological reaction *in vivo*. *Biomaterials*, **22**, 1705–1711.

Kim, Y. and Linkswiler, H.M. (1979) Effect of level of protein intake on calcium metabolism and on parathyroid hormone and renal function in the adult human male. *The Journal of Nutrition*, **109**, 1399–1404.

Kimberg, D.V., Baerg, R.D., Gershon, E. and Graudusius, R.T. (1971) Effect of cortisone treatment on the active transport of calcium by the small intestine. *The Journal of Clinical Investigation*, **50**(6), 1309–1321.

Kimmel, D.B., Bozzato, R.P., Kronis, K.A. *et al.* (1993) The effect of recombinant human (1–84) or synthetic human (1–34) parathyroid hormone on the skeleton of adult osteopenic ovariectomized rats. *The Journal of Endocrinology*, **132**, 1577–1584.

Kimmel, D.B. and Jees, W.S. (1980) A quantitative histologic analysis of the growing long bones mataphysis. *Calcified Tissue International*, **32**, 123–133.

Kingery, W.S., Offley, S.C., Guo, T.-Z. *et al.* (2003) A substance P receptor effects of sciatic nerve section. *Bone*, **33**, 927–936.

Kishi, T., Hagino, H., Kishimoto, H. and Nagashima, H. (1998) Bone responses at various skeletal sites to human parathyroid hormone in overiectomized rats: Effects of long term administration, withdrawal and readministration. *Bone*, **22**, 515–522.

Kishida, Y., Hirao, M., Tamai, N. *et al.* (2005) Leptin regulates chondrocyte differentiation and matrix maturation during endochondral ossification. *Bone*, **37**, 607–621.

Kittel, C. (1974), in *Introduction to Solid State Physics*, 4th edn, Willey Eastern Private Ltd, Delhi, pp. 269–289.

Klafter, J. and Shlesinger, M.F. (1986) On the relationship among three theories of relaxation in disordered systems. *Proceedings of the National Academy of Sciences of the United States of America*, **83**, 848–851.

Klapper, L. and Stellard, R.E. (1974) Mechanism of electrical stimulation of bone formation. *Annals of the New York Academy of Sciences*, **238**, 530.

Kleerekoper, M. and Balena, R. (1991) Fluorides and osteoporosis. *Annual Review of Nutrition*, **11**, 309–324.

Kleerekoper, M. and Krane, S.M. (1989) *Clinical Disorders of Bone and Mineral Metabolism*, May and Liebert, Inc, New York.

Kleerekoper, M. and Mendlovic, D.B. (1993) Sodium fluoride therapy of postmenopausal osteoporosis. *Endocrine Reviews*, **14**, 312–323.

Kleerekoper, M., Peterson, E.L., Nelson, D.A. *et al.* (1991) A randomized trial of sodium fluride as a treatment for postmenopausal osteoporosis. *Osteoporosis International*, **1**, 155–161.

Kleerekoper, M., Villanueva, A.R., Stanciu, J. *et al.* (1985) The role of three dimensional trabecular microstructure in the pathogenesis of vertebral compression fractures. *Calcified Tissue International*, **37**, 594–597.

Klein, D.C. and Raisz, L.G. (1970) Prostaglandin: Stimulation of bone resorption in tissue culture. *Endocrinology*, **86**, 1436–1440.

Klein, R.G., Arnaud, S.B., Gallagher, J.C. *et al.* (1977) Intestinal calcium absorption in exogenous hypercortisonism. Role of 25, hydroxyvitamin D and corticosteroid dose. *The Journal of Clinical Investigation*, **60**(1), 253–259.

Klein-Nulend, J., Roelofsen, J., Sterck, J.G. *et al.* (1995a) Mechanical loading stimulates the release of transforming growth factor beta activity by cultured mouse calvariae and periosteal cells. *Journal of Cellular Physiology*, **163**, 115–119.

Klein-Nulend, J., Semeins, C.M., Ajubi, N.E. *et al.* (1995c) Pulsating fluid flow increases nitric oxide (NO) synthesis by osteocytes but not periosteal fibroblasts-correlation with prostaglandin upregulation. *Biochemical and Biophysical Research Communications*, **217**, 640–648.

Klein-Nulend, J., Van Der, P.A., Semeins, C.M. *et al.* (1995b) Sensitivity of osteocytes to biomechanical stress *in vitro*. *The FASEB Journal*, **9**, 441–445.

Klenerman, L., Swanson, S.A.V. and Freeman, M.A.R. (1967) A method for the clinical estimation of the strength of a bone. *Proceedings of the Royal Society of Medicine*, **60**, 1–4.

Klug, W., Franke, W.G. and Knoch, H.G. (1986) Scintigraphic control of bone fracture healing under ultrasonic stimulation: an animal experimental study. *European Journal of Nuclear Medicine*, **11**, 494–497.

Knothe Tate, M.L. and Knothe, U. (2000) An *ex vivo* model to study transport processes and fluid flow in loaded bone. *Journal of Biomechanics*, **33**, 247–254.

Knothe Tate, M.L., Niederer, P. and Knothe, U. (1998) *In vivo* tracer transport through the lacuno canalicular system of rat bone in an environment devoid of mechanical loading. *Bone*, **22**, 107–117.

Knothe Tate, M.L. and Schaffler, M.B. (2002) Loss of osteocyte integrity colocolizes with bone resorption following disuse. Transactions 48th Annual Meeting of the Orthop. Research Society, Dallas, Texas.

Knothe Tate, M.L., Steck, R. *et al.* (2000) *In vivo* demonstration of load induced fluid flow in the rat tibia and its potential implications for processes associated with functional adaptation. *The Journal of Experimental Biology*, **203**(Pt 18), 2737–2745.

Knothe Tate, M.L., Tami, A., Bauer, T.W. and Knothe, U. (2002) Micropathoanatomy of osteoporosis: indications for a cellular basis of bone disease. *Advances in Osteoporotic. Fracture Management*, **2**(1), 9–14.

Knott, J.F. (1976) *Fundamentals of Fracture Mechanics*, Butterworth & Co. (Publishers) Ltd, London, UK.

Knott, L. and Bailey, A.J. (1998) Collagen crosses links in mineralizing tissues: A review of their chemistry, function and clinical relevance. *Bone*, **22**, 181–187.

Knott, L., Whitehead, C.C., Fleming, R.H. and Vailey, A.J. (1995) Biochemical changes in the collagenous matrix of osteoporotic avian bone. *The Biochemical Journal*, **310**, 1045–1051.

Ko, R. (1953) The tension test upon the compact substance in the long bones of cattle extremities. *Journal of the Kyoto Prefect Medical University*, **53**, 503–525.

Kobayashi, M., Hara, K. and Akiyama, Y. (2002) Effect of menatetrotone (vitamin $K_2$) on bone mineral density and bone strength in Ca/Mg deficient rats. *Nippon Yakurigaku Zasshi*, **120**, 195–204.

Kocian, J., Skala, I. and Berkos, K. (1973) Calcium absorption from milk and lactose free milk in health subjects and patients, with lactose intolerance. *Digesteon*, **9**, 317–324.

Kodama, Y., Dimai, H.P., Wergedal, J. *et al.* (1999) Cortical tibial bone volume in two strains of mice: effects of sciatic neurectomy and genetic regulation of bone response to mechanical loading. *Bone*, **25**, 183–190.

Kolsky, H. (1963) *Stress Waves in Solids*, Dover, New York.

Komm, B.S., Sheetz, L., Baker, M. *et al.* (1987) Bone related cells in culture express putative estrogen receptors mRNA and 125 1–17 beta estradiol binding. *Journal of Bone and Mineral Research*, **2**(Suppl. I), Abstr., 237.

Komm, B.S., Terpening, C.M., Benz, D.Z. *et al.* (1988) Estrogen binding receptor RNA and biologic response in osteoblast like osteosarcoma cells. *Science*, **241**, 81.

Kondepudi, D.K. (1991) Detection of gravity through nonequilibrium mechanisms. *American Society for Gravitational and Space Biology Bulletin*, **4**, 119–124.

Kong, C., Paley, M., Ordidge, R. and Speller, R. (1999) $R_2$ measured in trabecular bone *in vitro* relationship to trabecular separation. *Journal of Magnetic Resonance Imaging*, **17**, 989–995.

Kontonassions, T. and O'Phir, J.Y. (1987) Variance reduction of speed of sound estimation in tissue using beam-tracking methods. *IEEE Transactions on Ultrasonics, Ferroelectrics and Frequency Control*, **34**(2), 524–530.

Koop, A. and Cobbold, P.H. (1993) Continuous bioluminescent monitoring of cytoplasmic ATP in single isolated rate hepatocytes during metabolic poisoning. *The Biochemical Journal*, **295**, 165–170.

Kopman, C.R., Boskey, A.L., Lane, J.M. *et al.* (1987) Biochemical characterization of fracture callus proteoglycans. *Journal of Orthopaedic Research*, **5**, 7.

Kopp, J., Bonnet, M. and Renou, J.P. (1989) Effect of collagen cross linking on collagen – water interaction (a DSC investigation). *Matrix (Stuttgart, Germany)*, **9**, 443–450.

Kopp, J.B. and Robey, P.G. (1990a) Sodium fluoride does not increase human bone cell proliferation of protein synthesis *in vitro*. *Calcified Tissue International*, **47**, 221–229.

Kopp, J.B. and Robey, P.G. (1990b) Sodium fluoride lacks mitogenic activity for fetal human bone cells *in vitro*. *Journal of Bone and Mineral Research*, **5**(Suppl. 1), S137–S141.

Korenga, R., Ando, J., Tsubii, H. *et al.* (1994) Laminar flow stimulates ATP and shear stress-dependent nitric oxide production in cultured bovine endothelial cells. *Biochemical and Biophysical Research Communications*, **198**, 213–218.

Korenstein, R., Somjen, D., Fischler, H. and Binderman, I. (1984) Capacitative pulsed electric stimulation of bone cells induction of cyclic-AMP changes and DNA synthesis. *Biochimica et Biophysica Acta*, **803**, 302–307.

Korostoff, E. (1979) A linear piezoelectric model for characterizing stress generated potentials in bone. *Journal of Biomechanics*, **12**, 335–347.

Koskinen, E.V., Lindholm, R.V., Nieminen, R.A. *et al.* (1975) Human growth hormone in fractures of long bones with delayed fracture healing: fractures of tibial diaphysis and other injuries. *Die Medizinische Welt*, **26**, 1905–1910.

Kosterich, J.D., Foster, K.R. and Pollack, S.R. (1983) Dielectric permittivity and electrical conductivity of fluid saturated bone. *IEEE Transactions on Biomedical Engineering*, **30**, 81–86.

Kosterich, J.D., Foster, K.R. and Pollack, S.R. (1984) Dielectric properties of fluid saturated bone- the effect of variation in conductivity of immersion fluid. *IEEE Transactions on Biomedical Engineering*, **31**, 369–374.

Kotzki, P.O., Buyck, D., Hans, D. *et al.* (1994) Influence of fat on ultrasound measurements of the oscalcis. *Calcified Tissue International*, **54**, 91–95.

Kowalchuk, R.M., Pollack, S.R. and Corcoran, T.A. (1993) in *Electricity and Magnetism in Medicine and Biology* (ed. M. Blank), San Franscisco Press, New York, pp. 689–692.

Kragstrup, J., Shijie, Z., Mosekilde, Le. and Melsen, F. (1989) Effects of sodium fluoride, vitamin D and calcium on cortical remodeling in osteoporotic patients. *Calcified Tissue International*, **45**, 337–341.

Kramer, H.W., Pattersen, J.F. and Smith, J.C. (1978) Combined cortical thickness and bone density determination by photon absorptiometry. *Physics in Medicine and Biology*, **23**, 1101–1114.

Krawczyk, A., Kuropka, P., Kuryszko, J. *et al.* (2007) Experimental studies on the effect of osteotomy technique on the bone regeneration in distraction osteogenesis. *Bone*, **40**, 781–791.

Krawitt, E.L. (1974) Effect of acute ethanol administration on duodenal calcium transport. *Proceedings of the Society for Experimental Biology and Medicine*, **146**, 406–408.

Kristiansen, T.K., Ryaby, J.P., McCabe, J. *et al.* (1997) Accelerated healing of distal radial fractures with the use of specific, low intensity ultrasound a multicenter, prospective, randomized, double blind, placebo-controlled study. *The Journal of Bone and Joint Surgery*, **79**, 961–973.

Kroger, H., Jurvelin, J., Arnala, I. *et al.* (1995) Ultrasound attenuation of the calcaneous in normal subjects and in patients with wrist fracture. *Acta Orthopaedica Scandinavica*, **66**, 47–52.

Krolner, B., Berthelsen, B. and Pors Nielsen, S. (1982) Assessment of vertebral osteopenia: comparison of spinal radiography and dual-photon absorptiometry. *Acta Radiologica: Diagnosis (Stockholm)*, **23**, 517–520.

Krolner, B. and Pors-Nielson, S. (1980) Measurement of bone mineral content of the lumbar spine I. Theory and application of a new two dimensional dual photon attenuation method. *Scandinavian Journal of Clinical and Laboratory Investigation*, **6**, 653–663.

Krolner, B. and Pors Nielsen, S. (1982) Bone mineral content of the lumbar spine in normal and osteoporotic women. *Clinical Science*, **62**, 329–336.

Krolner, B., Toft, B., Nielsen, S.P. and Topndevold, E. (1983) Physical exercise as prophylaxis against involutional vertebral bone loss: a controlled tribal. *Clinical Science*, **64**, 541–546.

Kruse, H.P. and Kuhlencordt, F. (1975) On an attempt to treat primary and secondary osteoporosis with human growth hormone. *Hormone and Metabolic Research*, **7**, 488–491.

Kruse, R.L. and Kelly, P.J. (1974) Acceleration of fracture healing distal to a venous tourniquet. *The Journal of Bone and Joint Surgery*, **56-A**, 750–759.

Kufahl, R.H. and Saha, S. (1990) A theoretical model for stress-generated fluid flow in the canaliculi-lacunae network in bone tissue. *Journal of Biomechanics*, **23**, 171–180.

Kuhn, K. (1969) The structure of collagen. *Essays in Biochemistry*, **5**, 59.

Kuiper, J.W., Van Kuijk, C., Grashuis, J.L. *et al.* (1996) Accuracy and influence of marrow fat on quantitative CT and dual energy X-ray absorptiometry measurements of the femoral neck *in vitro*. *Osteoporosis International*, **6**, 25–30.

Kuivaniemi, H., Tromp, G. and Prockop, D.J. (1991) Mutations in collagen genes: causes of rare and some common disease in humans. *The FASEB Journal*, **5**, 2052–2060.

Kumagi, H., Sacktor, B. and Filburn, C.R. (1991) Purinergic regulation of cytosolic calcium and phosphoinositide metabolism in rat osteoblast like osteosarcome cells. *Journal of Bone and Mineral Research*, **6**, 667–708.

Kuratsu, S., Aozasa, K., Myoui, A. *et al.* (1991) Prognostic significance of argyrophilic nucleolar organizer staining in soft tissue sarcomas. *International Journal of Cancer*, **48**, 21–214.

Kurihara, N., Ikeda, K., Hakeda, Y. *et al.* (1984) Effects of 1,25-dihydroxy vitamin $D_3$ on alkaline phosphatase activity and collagen synthesis in osteoblastic cells, clone MC3T3-E1. *Biochemical and Biophysical Research Communications*, **119**, 761–767.

Kurihara, N., Ishizuka, S., Hakeda, Y. *et al.* (1986) Effects of 1,25- dihydroxy vitamin $D_3$ on osteoblastic MC3T3 cells. *Endocrinology*, **118**, 940–947.

Kurokouchi, K., Jacobs, C.R. *et al.* (2001) Oscillating fluid flow inhibits TNF-alpha-induced NF-Kappa B activation via an Ikappa B kinase pathway in osteoblast-like UMR 106 cells. *The Journal of Biological Chemistry*, **276**, 13499–13504.

Kurth, A.H., Wang, C., Hayes, W.C. and Shea, M. (2001) The evaluation of a rat model for the analysis of densitometric and biomechanical properties of tumor-induced osteolysis. *Journal of Orthopaedic Research*, **19**, 200–205.

Kusakari, H., Orikasa, N. and Tani, H. (1992) Effects of low-power laser on wound healing of gingiva and bone. in *Proceedings of the 3rd World Congress of the International Society for Low Power Laser Applications in Medicine*, (eds G. Galletti, L. Bolognani and G. Ussia), Monduzzi Editore, Bologna, pp. 49–56.

Kusick, V.A. (1972) Osteogenesis imperfecta, in *I: Heritable Disorders of the Conective Tissue*, C. V. Mosby, St. Louis.

Kuwana, T., Sugita, O. and Yakata, M. (1988) Reference limits of bone liver alkaline phosphatase isoenzymes in the serum of healthy subjects according to age and sex as determined by wheat germ lectin affinity electrophoresis. *Clinica Chimica Acta*, **173**, 273–280.

Lacroix, D. and Prendergast, P.J. (2002) A mechano-regulation model for tissue differentiation during fracture healing: analysis of gap size and loading. *Journal of Biomechanics*, **35**, 1163–1171.

Lacroix, D., Prendergast, P.J., Marsh, D. and Li, G. (2000) Simulation of tissue differentiation during fracture healing. *Transactions of the Orthopaedic Research Society*, **25**, 869.

Laharrague, P., Larrouy, D., Fontamilles, A.M. *et al.* (1998) High expression of leptin by human bone marrow adipocytes in primary culture. *The FASEB Journal*, **12**, 747–752.

Laird, G. and Kingsbury, H. (1973) Complex viscoelastic modulae of bovine bone. *Journal of Biomechanics*, **6**, 59–67.

Lakes, R.S., Harper, R.A. and Katz, J.L. (1977) Dielectric relaxation in cortical bone. *Journal of Applied Physics*, **48**, 808–811.

Lakes, Ft. and Katz, J.L. (1977) Viscoelastic and dielectric properties of bone. *Bulletin of the Hospital for Joint Diseases*, **37**(23), 23–25.

Lakes, R.S. and Katz, J.L. (1979) Viscoelastic properties of wet cortical bone –2. Relaxation mechanisims. *Journal of Biomechanics*, **12**, 679–687.

Lakes, R.S. and Saha, S. (1977) The effect of soft tissue on the non-invasive techniques of determining the mechanical properties of bone, in *Proceedings of the 5th New England Bioengineering Conference*, (ed. M.R. Cannon), Pergamon Press, Oxford, pp. 36–40.

Lakes, R.S. and Saha, S. (1978) A non-contracting electromagnetic device for the determination of *in vivo* properties of bone. *Medical Instrumentation*, **12**, 106–109.

Lakes, R.S. and Saha, S. (1979) Cement line motion in bone. *Science*, **204**, 501–503.

Lamberg-Allardt, C. (1984) Serum 25-hydroxy vitamin D intake (Dissert Helsinki). *Annals of Nutrition and Metabolism*, **28**, 144–150.

Lane, N., Kimmel, D. and Coble, T. (1989) Effect of Naprosyn on bone volume and formation in overectomized rats. *Journal of Bone and Mineral Research*, **4**, S (Abstr.) 193.

Lane, N.E., Sanchez, S., Modin, G.W. *et al.* (1998) Parathyroid hormone treatment can reverse corticosteroid induced osteoporosis: Results of a randomized controlled clinical trial. *The Journal of Clinical Investigation*, **102**, 1627–1633.

Lang, S.B. (1966) Pyroelectric effect in bone and tendon. *Nature (London)*, **212**, 704–705.

Lang, S.B. (1969) Thermal expansion coefficients and the primary and secondary pyroelectric coefficients of animal bone. *Nature (London)*, **224**, 798.

Lang, S.B. (1970) Ultrasonic method for-measuring elastic coefficients of bone and results on fresh and dried bovine bones. *IEEE Transactions on Biomedical Engineering*, **17**, 101.

Lang, T., Le Blanc, A., Evans, H. *et al.* (2004) Cortical and trabecular bone mineral loss from the spine and hip in long duration spaceflight. *Journal of Bone and Mineral Research*, **19**, 1006–1015.

Lang, T.F., Li, J., Harris, S.T. and Genant, H.K. (1999) Assessment of vertebral bone mineral density using volumetric quantitative CT. *Journal of Computer Assisted Tomography*, **23**(1), 130–137.

Langdahl, B.L., Ralston, S.H., Grant, S.F.A. and Eriksen, R.F. (1998) An Spl. Binding site polymorphism in the COLIAI gene predicts osteoporotic fracture in both men and women. *Journal of Bone and Mineral Research*, **13**, 1384–1389.

Langton, C.M., Palmer, S.B. and Porter, R.W. (1984) The measurement of broadband ultrasonic attenuation in cancellous bone. *Engineering in Medicine*, **13**, 89–91.

Langton, C.M., Whitehead, M.A., Haire, T.J. and Hodgskinson, R. (1998) Fractal dimension predicts broadband ultrasound attenuation in stereolithography models of cancellous bone. *Physics in Medicine and Biology*, **43**, 467–471.

Lanyon, L.E. (1992) The success and failure of the adaptive response to functional load-bearing in averting bone fracture. *Bone*, **13** (Suppl.), S15–S21.

Lanyon, L.E. (1993) Osteocytes, strain detection and bone modeling and remodeling. *Calcified Tissue International*, **53**, 102–107.

Lanyon, L.E. (1994) Mechanically sensitive cells in bone, in *Society for Experimental Biology Seminar Series 54: Biomechanics and Cells* (eds F. Lyall and A.J. El Haj), Cambridge University Press, Cambridge, pp. 178–186.

Lanyon, L.E. (1996) Using functional loading to influence bone mass and architecture: objective, mechanism and relationship with estrogen of the mechanically adaptive process in bone. *Bone*, **18**, 375–435.

Lanyon, L.E. and Bourn, S. (1979) The influence of mechanical function on the development and remodeling of the tibia. *The Journal of Bone and Joint Surgery*, **61**, 263–273.

Lanyon, L.E. and Rubin, C.T. (1985) Functional adaption in skeletal structure, in *Functional Vertebrate Morphology* (eds M. Hildebrand, Bramble, D.K. Liem and D. Wake), Harvard University Press, Cambridge, MA, pp. 1–25.

Larsen, E.R., Mosekilde, L. and Foldspang, A. (2004) Vitamin D and calcium supplementation prevents osteoporotic fractures in elderly community dwelling residents: a pragmatic population-based 3-year intervention study. *Journal of Bone and Mineral Research*, **19**, 370–378.

Larsson, S.E., Ahlgren, O. and Lorentzen, R. (1980) Experimental studies of the influence of 1,25-dihydroxy cholecalciferol on the regulation of mature bone mass. *Calcified Tissue International*, **30**(3), 250.

Last, J.A., Armstrong, L.G. and Reiser, K.M. (1990) Biosynthesis of collagen crosslinks. *The International Journal of Biochemistry*, **22** (6), 559–564.

Lau, E.M.C., Cooper, C., Wickham, C. *et al.* (1990) Hip fractures in Hong Kong and Britain. *International Journal of Epidemiology*, **19**, 199–121.

Lau, E., Donnanm, S., Barker, D.J.P. and Cooper, C. (1988) Physical activity and calcium intake on fracture of the proximal femur in Hong Kong. *British Medical Journal*, **297**, 1441–1443.

Lau, E.M., Lynn, H., Chen, Y.H. and Woo, J. (2002) Milk supplementation prevents bone loss in postmenopausal Chinese women over 3 years. *Bone*, **31**, 536–540.

Lau, E.M., Woo, J., Lam, V. and Hong, A. (2001) Milk supplementation of the diet of postmenopausal Chinese women on a low calcium intake retards bone loss. *Journal of Bone and Mineral Research*, **16**, 1704–1709.

Lau, K.H. and Baylink, D.J. (1999) Vitamin D therapy for osteoporosis; plain vitamin D therapy versus active vitamin D analog (D-hormone) therapy. *Calcified Tissue International*, **65**, 295–306.

Laub, F., Kaplan, M. and Gitler, C. (1981) Actin polymerization accompanies thy-1-capping on mouse thymocytes. *FEBS Letters*, **124**, 35–38.

Laub, F. and Korenstein, R. (1984) Acting polymerization induced by pulsed electric stimulation of bone cells *In Vitro*. *Biochemica Biophysica Acta*, **803**, 308–313.

Laugier, P., Giat, P., Chappard, C. *et al.* (1997) Clinical assessment of the backscatter coefficient in osteoporosis, in *Ultrasonics Symposium Proceedings*, (eds S.C. Schneider, M. Levy and B.R. McAvoy), Institute of Electrical and Electronic Engineers, New York, pp. 1101–1103.

Laurent, V., De Vernejoul, M.C., Brouard, A. *et al.* (1989) Interest du dosage de fluor oxxeux ehe2 1' osteoporotique traite pare le fluorure de sodium. *La Presse Médicale*, **18**, 679–682.

Lauretani, F., Bandinelli, S., Russo, C.R. *et al.* (2006) Correltes of bone quality in older persons. *Bone*, **39**, 915–921.

Laval-Jeantet, A.M., Laval-Jeantet, M., Roger, B. *et al.* (1984) Interêt et limites de la mesure tomodensitométrique de la minéralisation vertébrale. *Journal de Radiologie*, **65**, 151–157.

Lavine, L.S. and Grodzinsky, A.J. (1987) Electrical stimulation of repair of bone. *The Journal of Bone and Joint Surgery*, **69**, 626–630.

Lavine, L.S., Lustrin, I., Rinaldi, R.A. and Shamos, M.H. (1974) Clinical and ultrastructural investigations of electric enhancement of bone healing. *Annals of the New York Academy of Sciences*, **238**, 552.

Lavine, L.S., Lustrin, I., Shamos, M.H. *et al.* (1972) Electrical enhancement of bone healing. *Science*, **175**, 1118–1120.

Lavine, L.S., Lustrin, I., Shamos, M.H. and Moss, M.L. (1971) The influence of electric current on bone regeneration *in vivo*. *Acta Orthopaedica Scandinavica*, **42**, 305–314.

Law, H.T., Annan, I., Mc Carthy, I.D. *et al.* (1985) The effect of induced electric currents on bone after experimental osteotomy in sheep. *The Journal of Bone and Joint Surgery*, **67**, 463.

Lawrence, L.A., Ott, E.A., Miller, G.J. *et al.* (1994) The mechanical properties of equine third metacarpals as affected by age. *Journal of Animal Science*, **72**, 2617–2623.

Lawrence, T.L.J. and Fowler, V.R. (1997) Compensatory growth, in *Growth of Farm Animals*, CAB Int, pp. 219–246.

Lazarowski, E.R., Homolya, L., Boucher, R.C. and Harden, T.K. (1997) Direct demonstration of mechanically induced release of cellular UTP and its implication for uridine nucleotide receptor activation. *The Journal of Biological Chemistry*, **272**, 24348–24354.

Le Blanc, A., Schneider, V., Evans, H.J. *et al.* (1990) Bone mineral loss and missense recovery after 17 weeks of bed rest. *Journal of Bone and Mineral Research*, **5**, 843–850.

Le Geros, R.Z., Bonel, G. and Legros, R. (1978) Types of "$H_2O$" in human enamel and in precipitated apatites. *Calcified Tissue Research*, **26**, 111–118.

Lee, B., Thirunavukkarasu, K., Zhou, L. *et al.* (1997) Missense mutations abolishing DNA binding of the osteoblast-specific transcription factor OSF2/CBFA1. In cleidocranial dysplasia. *Nature Genetics*, **16**, 307–310.

Lee, D.B.N., Mills, B.G., Brickman, A.S. and Teitelbaum, S.L. (1975) Osteomalacia, a comparative study of the separate effects of intervenous calcium and phosphorous on parathyroid function, intestinal calcium absorption and bone morphology, in *Vitamin D and Problems to Uremic Bone Disease* (eds A.S. Norman, K. Schaeffer, H.G. Grigoleit *et al.*), Walter de Gruyter, New York, pp. 181–187.

Lee, H.M., Choi, J.H., Chae, S.W. *et al.* (2003) Expression of epidermal growth factor receptor and its ligands in chronic sinusitis. *The Annals of Otology, Rhinology, and Laryngology*, **112**, 132–138.

Lee, K., Jessop, H., Suswillo, R. *et al.* (2003) Endocrinology bone adptation requires estrogen receptor-alpha. *Nature*, **424**, 389.

Lee, W. (1964) Appositional bone formation in canine dog: A quantitative microscopic study using tetracycline markers. *Journal of Anatomy*, **48**, 665.

Lee, W., Marshall, J.H. and Sisson, H.A. (1965) Calcium accretion and bone formation in dogs. *The Journal of Bone and Joint Surgery American Volume*, **47**, 157.

Lees, B. and Stevenson, J.C. (1993) Preliminary evaluation of a new ultrasound bone densitometer. *Calcified Tissue International*, **53**, 149–152.

Lees, S. (1979) A model for the distribution of HAP crystallites in bone -an hypothesis. *Calcified Tissue International*, **27**, 53–56.

Lees, S. (1981) A mixed packing model for bone collagen. *Calcified Tissue International*, **33**, 591–602.

Lees, S. and Davidson, C.L. (1977) The role of collagen in the elastic properties of calcified tissues. *Journal of Biomechanics*, **10**, 473–486.

Lees, S. and Davidson, C.L. (1979) A theory relating sonic velocity to mineral content in bone. S I'525, pp. 179–187.

Leggate, J., Farish, E., Fletcher, C.D. *et al.* (1984) Calcitonin and postmenopausal osteoporosis. *Clinical Endocrinology*, **20**, 85–92.

Legrand, E., Chappard, D., Pascaretetti, C. *et al.* (1999) Bone mineral density and vertebral fractures in men. *Osteoporosis International*, **10**, 265–270.

Legrand, E., Chappard, D., Pascaretti, C. *et al.* (2000) Trabecular bone microarchitecture: bone mineral density and vertebral fractures in male osteoporosis. *Journal of Bone and Mineral Research*, **15**, 13–19.

Legrand, E., Hedde, C., Gallois, Y. *et al.* (2001) Osteoporosis in men: a potential role for the sex hormone binding globulin. *Bone*, **29**(1), 90–95.

Leichter, I., Karallas, A., Craven, J.D. and Greenfield, M.A. (1984) The effect of the momentum transfer on the sensitivity of a photon scattering method for the characterization of tissues. *Medical Physics*, **11**, 31–36.

Leichter, I., Karellas, D., Shukla, S.S. *et al.* (1985) Quantitative assessment of bone mineral by photon scattering: Calibration considerations. *Medical Physics*, **12**, 466–468.

Leidner, J. and Woodhams, R.T. (1974) Strength of polymeric composites containing spherical fillers. *Journal of Applied Polymer Science*, **18**, 1639–1654.

Lenchik, L., Shi, R., Register, T.C. *et al.* (2004) Measurement of trabecular bone mineral density in the thoracic spine using cardiac gated quantitative computed tomography. *Journal of Computer Assisted Tomography*, **28**, 134–139.

Leone, N.C., Stevenson, C.A., Hilbish, T.F. and Sosman, M.C. (1955) A roengenologic study of a human population exposed to high fluoride domestic water (a 10 year study). *American Journal of Roentgenology*, **74**, 874–885.

Leong, W.S., Russell, R.G.G. and Caswell, A.M. (1994) Stimulation of cartilage resorption by extracellular ATP acting at P2 purinoceptors. *Biochimica et Biophysica Acta*, **1201**, 298–304.

Lepage, O.M., Des Coteaux, L., Marcoux, M. and Tremblay, A. (1991) Circadian rhythms of osteocalcin in equine serum. Correlation with alkaline phosphatase, calcium, phosphate and total protein levels. *Canadian Journal of Veterinary Research*, **55**, 5–10.

Lepola, V., Jalovaara, P. and Vaananen, K. (1994) The influence of clodronate on the torional strength of the growing rat tibia in immobilization osteoporosis. *Bone*, **15**, 367–371.

Leung, K.S., Cheung, W.H., Zhang, C. *et al.* (2004) Low intensity pulsed ultrasound stimulates osteogenic activity of human periosteal cells. *Clinical Orthopaedics and Related Research*, **418**, 253–259.

Leung, K.S., Lee, W.S., Cheung, W.H. and Qin, L. (2004a) Lack of efficancy of low intensity pulsed ultrasound on prevention of postmenopausal bone loss evaluated at the distal radius in older Chinese women. *Clinical Orthopaedics and Related Research*, **427**, 234–240.

Leung, K.S., Lee, W.S., Tsui, H.P. *et al.* (2004b) Complex tibial fracture outcomes following treatment with low intensity pulsed ultrasound. *Ultrasound in Medicine and Biology*, **30**(3), 389–395.

Levy, C.J., Lees, B. and Stevenson, J.C. (1992) Sex and menopause associated changes in body fat distribution. *The American Journal of Clinical Nutrition*, **55**, 950–954.

Levy, D.D. (1974) A pulsed electrical stimulation technique for inducing bone growth. *Annals of the New York Academy of Sciences*, **238**, 457.

Levy, J.C., Ilisca, E. and Mochane, J.L. (1972) Pseudohelicoidal surface spin waves. *Physical Review B*, **5**, 187–190.

Lewis, J. and Goldsmith, W. (1975) The dynamic fracture and prefracture response of compact bone by split Hopkinson bar methods. *Journal of Biomechanics*, **8**, 27–40.

Li, M. (1992) Electrical stimulation in the treatment of osteoporosis in sciatic denervated rat tibia. *Chung-Hua–Wai-Ko-Tsa-Chi*, **30**, 458–460.

Li, M., Mosekilde, L., Sogaard, C.H., Thomsen, J.S. and Wronski, T.J. (1995) Parathyroid hormone monotherapy and cotherapy with antiresorptive agents restore vertebral bone mass and strength in aged ovariectomized rats. *Bone*, **16**(6), 629–635.

Li, X.J. and Jee, W.S.S. (1990) Adaptation of diaphyseal structure to aging and decreased mechanical loading in the adult rat: a densitometric and histomorphometric study. *The Anatomical Record*, **229**, 291–297.

Li, X.J., Jee, W.S.S., Chow, S. and Woddbury, D.M. (1990) Adaptation of cancellous bone to aging and immobilization in the rat. A single photon absorptiometry and histomorphometry study. *The Anatomical Record*, **227**, 12–24.

Lian, B.L. and Stein, G.S. (1999) The cells of bone, in *Principles of Bone and Cartilage Metabolism* (eds M.J. Seibel, S. Robbins and J.P. Bilezikian), Academic Press, San Diego, pp. 165–185.

Lian, J.B. and Gundberg, C.M. (1988) Osteocalc: Biochemical considerations and clinical applications. *Clinical Orthopaedics and Related Research*, **226**, 267–291.

Lian, J., Stewart, C., Pucha, C.Z.E. *et al.* (1989) Structure of the rat osteocalcin gene and regulation of vitamin D-dependent repression. *Proceedings of the National Academy of Sciences of the United States of America*, **86**, 1143–1147.

Liboff, A.R. and Furst, M. (1974) Pyroelectric effect in collagenous strucrures. *Annals of the New York Academy of Sciences*, **238**, 26.

Liboff, A.R., Rinaldi, R.A., Lavine, L.S. and Shamos, M.H. (1975) On electrical conduction in living bone. *Clinical Orthopaedics*, **106**, 330.

Liboff, A.R. and Shamos, M.H. (1973) *Biological Mineralization* (ed. I. Zipkin), John Wiley & Sons, Inc, New York.

Liburdy, R.P. and Eckert, V. (1994) *Abstracts of Papers, The Annual Review of Research on Biological Effects of Electric and Magnetic Fields from the Generation, Delivery, and use of Electricity*, U.S. Department of Energy, Washington, DC.

Licate, A.A. (1981) Acute effects of increased meat protein on urinary electrolytes and cyclic adenosine monophosphates and serum parathyroid hormone. *The American Journal of Clinical Nutrition*, **34**, 1779–1784.

Lieberman, D.E. and Crompton, A.W. (1998) Responses of bone to stress, in *Principles of Biological Design: The Optimization and Symmorphosis Debate* (eds E. Wiebel, C. Taylor and L. Bolis), Cambridge University Press, Cambridge, UK, pp. 78–86.

Lieberman, D.E., Devlin, M.J. and Pearson, O.M. (2001) Articular area responses to mechanical loading effects of exercise, age and skeletal location. *American Journal of Physical Anthropology*, **116**, 266–277.

Lieberman, J.R., Daluski, A., Stevenson, S. *et al.* (1999) The effect of gene therapy with one morphogenetic protein-2 producing marrow cells on the repair of segmental femoral defects. *The Journal of Bone and Joint Surgery*, **81-A**, 905–917.

Liebschner, M.A., Kopperdahl, D.L., Rosenberg, W.S. and Keaveny, T.M. (2003) Finite element modeling of the human thoracolumbar spine. *Spine*, **28**, 559–565.

Liel, Y., Edwards, J., Shar, J. *et al.* (1988) The effects of race and body habitus on bone mineral density of the radius, hip and spine in premenopausal women. *The Journal of Clinical Endocrinology and Metabolism*, **66**, 1247–1250.

Lin, B.Y., Jee, W.S.S., Ma, Y.F. *et al.* (1994) Effects of prostaglandin E2 and risedronate administration on cancellous bone in older female rats. *Bone*, **15**, 489–496.

Lin, C.T., Zhou, E.H. and Quek, S.T. (2006) Mechanical models for living cells-A review. *Journal of Biomechanics*, **39**, 195–216.

Lin, J.T. and Lane, J.M. (2004) Osteoporosis: A review. *Clinical Orthopaedics and Related Research*, **425**, 126–134.

Linderman, R.D., Adler, S. and Yiengst, M.J. (1967) Influence of various nutrients on urinary divalent cation excretion. *The Journal of Laboratory and Clinical Medicine*, **70**, 236–245.

Lindholm, T.S. (1979) Two years of experience of oral treatment with I, hydroxyvitamin $D^3$ and calcium in patients with senile or postmenopausal osteoporosis, in *Vitamin D Basic Research and Its Clinical Application* (eds A.W. Norman, K. Schaefer, D.V. Herrath *et al.*), Walter de Gruyter, Berlin, pp. 115–118.

Lindholm, T.S., Sevastikiglore, J.A. and Lindgren, U. (1977) Long-term treatment of osteoporotic patients with 1 (OH) $D_3$ and calcium, in *Vitamin D Biochemical, Chemical and Clinical Aspects Related to Calcium Metabolism* (eds A.E. Norman, D. Schaffer and J.E. Cobum), Walter de Gruyter, Berlin, pp. 635–637.

Lindholm, T.S., Sevastikiglore, J.A. and Lindgren, U. (1978) Interim report on treatment of osteoporotic patients with 1, hydroxyvitamin $D_3$ and calcium. *Clinical Orthopaedics*, **135**, 232–240.

Lindholm, T.S., Wilsson, O.S. and Lindholm, T.C. (1982) New bone formation and parathyroid affected by 1, hydroxyvitamin $D_3$. *Clinical Orthopaedics*, **162**, 264–269.

Lindmayer, J. and Wigley, C.Y. (1971) *Fundamentals of Semiconductor Devices*, **7**, Affiliated East-West Press, Pvt. Ltd, New Delhi, p. 12.

Lindsay, R., Aitken, J.M., Hart, D.M. and Purdey, D. (1978) The effect of ovarian sex steroids on bone mineral status in the oophorectomized rats. *Postgraduate Medical Journal*, **54**(Suppl. 2), 50–58.

Lindsay, R., Hart, D.M., Aitken, J.M. *et al.* (1976) Long-term prevention of postmenopausal osteoporosis by estrogens. *Lancet*, **i**, 1038–1041.

Lindsay, R., Hart, D.M., Forrest, C. and Baird, C. (1980) Prevention of spinal osteoporosis in oophorectomized women. *Lancet*, **3**, 151–154.

Lindsay, R., Hart, D.M., Purdey, D. *et al.* (1987) Comparative effects of estrogen and progestogen on bone loss in postmenopausal women. *Clinical Science*, **54**, 193–195.

Lindsay, R., Nieves, J., Formica, C. *et al.* (1997) Radomized controlled study of effect of parathyroid hormone on vertebral bone mass and fracture incidence among postmenopausal women on oestrogen with osteoporosis. *Lancet*, **350**, 550–555.

Lindsay, R., Tohme, J. and Kanders, B. (1986) The effect of oral contraceptive use on vertebral bone mass in pre and postmenopausal women. *Contraception*, **34**, 333–340.

Ling, S.S., Rustgi, S., Karellas, A. *et al.* (1982) The measurement of trabecular bone mineral density using coherent and Compton scattered photons *in vitro*. *Medical Physics*, **9**, 208–215.

Lingarde, F. (1972) Potentiometric determination of serum ionized calcium in a normal human population. *Clinica Chimica Acta; International Journal of Clinical Chemistry*, **40**, 477–484.

Lingarde, F. (1973) *In vivo* and *in vitro* studies on ionized versus total serum calcium in hyperparathyroidism. *Acta Endocrinologica*, **74**, 501.

Link, T.M., Majumdar, S., Lin, J.C. *et al.* (1998) A comparative study of trabecular bone properties in the spine and femur using high resolution MRI and CT. *Journal of Bone and Mineral Research*, **13**(1), 122–132.

Lips, P.T.A.M. (1982) Metabolic Causes and Prevention of Femoral Neck Fractures, PhD thesis, University of Amsterdam, Amsterdam.

Lips, P. (2001) Vitamin D deficiency and secondary hyperparathyroidism in the elderly, consequences for bone loss and fracture and therapeutic implications. *Endocrine Reviews*, **22**, 477–501.

Lipsitz, L.A. (2004) Physiological complexity, aging, and the path to frailty. *Science of Aging Knowledge Environment*, **16**, 16.

Lipsitz, L.A. and Goldberger, A.L. (1992) Loss of 'complexity' and aging: potential applications of fractals and chaos theory to senescence. *The Journal of the American Medical Association*, **267**, 1806–1809.

Lipworth, B.J. (2000) Therapeutic implications of non-genomic glucocorticoid activity. *Lancet*, **356**(8), 87–89.

Lirani-Galvao, Bergamaschi., Silva, O.L. and Lazaratti-Castro, M. (2007) Electrical field stimulation improves bone mineral density in overiectomized rats. *Brazilian Journal of Medical and Biological Research*, **39**(11), 1501–1505.

Lischwe, M.A., Smetana, K., Olson, M.O.J. and Busch, H. (1979) Protein C23 and B23 are the major nucleolar silver staining proteins. *Life Sciences*, **25**, 701–708.

Litovitz, T.A., Krause, D., Montrose, C.J. and Mullins, J.M. (1994a) Temporally incoherent magnetic fields mitigate the response of biological systems to temporally coherent electromagnetic fields. *Bioelectromagnetics*, **15**, 399–409.

Litovitz, T.A., Montrose, C.J., Doinov, P. *et al.* (1994b) Superimposing spatially coherent electromagnetic noise inhibits field-induced abnormalities in developing chick embryos. *Bioelectromagnetics*, **15**, 105–113.

Little, K. (1973) *Bone Behavior*, Academic Press Ltd Inc, New York, p. P3.

Little, R.D., Carulli, J.P., Del Mastro, R.G. *et al.* (2002) A mutation I the LDL reeptor related protein 5 gene results in the autosomal dominant high bone mass trait. *American Journal of Human Genetics*, **70**, 11–19.

Little, K. and Kelley, M. (1962) Studies on bone matrix in normal and osteoporotic bone. *The Journal of Bone and Joint Surgery*, **44**, 503.

Liu, C.C. and Kalu, D.N. (1990) Human parathyroid hormone (1–34) prevents bone loss and augments bone formation in sexually mature ovariectomized rats. *Journal of Bone and Mineral Research*, **5**, 973–982.

Liu, C.C., Kalu, D.N., Salerno, E. *et al.* (1991) Pre-existing bone loss associated with ovariectomy in rats reversed by parathyroid hormone. *Journal of Bone and Mineral Research*, **6**, 1071–1080.

Liu, D.M., Yang, Q., Troczynski, T. and Tseng, W.J. (2002) Structural evolution of sol-gel derived hydroxyapatite. *Biomaterials*, **23**, 1769–1787.

Liu, H., Lees, P., Abbott, J. and Bee, J.A. (1997) Pulsed electromagnetic fields preserve proteoglycan composition of extracellular matrix in embryonic chick sternal cartilage. *Biochimica et Biophysica Acta*, **1336**, 303–314.

Lloyd, D.K., Golding, S.L., Bowler, W.B. *et al.* (1999) Regulated ATP release by cultured human articular chondrocytes. *Calcified Tissue International*, **64**(Suppl.), S58.

Lloyd, T., Buchman, J.R., Ursino, G.R. *et al.* (1989) Long-term oral contraceptive use does not affect trabecular bone density. *American Journal of Obstetrics and Gynecology*, **160**, 402–404.

Lloyd, T., Rollings, N., Eggli, D.F. *et al.* (1997) Dietary caffeine intake and bone status of postmenopausal women. *The American Journal of Clinical Nutrition*, **65**, 1826–1830.

Locasico, V.L., Adami, S., Avioli, L.V. *et al.* (1982) Suppressive effect of chronic corticoid treatment on circulating calcitonin in man. *Calcified Tissue International*, **34**, 309–311.

Longcope, C., Hoberg, L., Steuterman, S. and Bran, D. (1989) The effect of ovariectomy on spine bone mineral density in rhesus monkeys. *Bone*, **10**, 341–344.

Lopez Franco, G.E., Litscher, S.J., O'Neil, T.K. *et al.* (2005) Dual energy X-ray absorptiometry of *ex vivo* Hb/Dem mouse log bones: Left are denser than right. *Calcified Tissue International*, **76**, 26–31.

Lopez Franco, G.E., O'Neil, T.K., Litscher, S.J. *et al.* (2004) Accuracy and precision of PixImus densitometry for *ex vivo* mouse long bones. *Journal of Clinical Densitometry*, **7**, 326–333.

Lorenzo, J.A. and Raisz, L.G. (1999) Cytokines and prostaglandins, in *Principles of Bone and Cartilage Metabolism* (eds M.J. Seibel, S. Robbins and J.P. Bilezikian), Academic Press, San Diego, pp. 97–109.

Lorich, D.G., Brighton, C.T., Gupta, R. *et al.* (1998) Biochemical pathway mediating the response of bone cells to capacitive coupling. *Clinical Orthopaedics and Related Research*, **350**, 246–250.

Lotz, J.C. and Hayes, W.C. (1990) The use of quantitative computed tomography to estimate risk of fracture of the hip from falls. *The Journal of Bone and Joint Surgery*, **72a**, 689–700.

Lou, J., Xu, F., Merkel, K. and Manske, P. (1999) Gene therapy: Adenovirus mediated human bone morphogenetic protein-2 gene transfer induces mesenchymal progenitor cell proliferation and differentiation *in vitro* and bone formation *in vivo*. *Journal of Orthopaedic Research*, **17**, 43–50.

Louis, O., Boulpaep, F., Willnecker, J. *et al.* (1995) Cortical mineral content of the radius assessed by peripheral QCT predicts compressive strength on biomechanical testing. *Bone*, **16**, 375–379.

Louis, O., Soykens, S., Willnecker, J. *et al.* (1996) Cortical and total bone mineral content of the radius: accuracy of peripheral computed tomography. *Bone*, **18**(5), 467–472.

Lowenstam, H.A. and Weiner, S. (1989) *On Biomineralization*, Oxford University Press, New York.

Lowenstein, W.R. (1981) Junctional intracellular communications: the cell-to-cell membrane channel. *Physiological Reviews*, **61**, 829–841.

Lowenstein, W.R., Kanno, Y. and Socolar, S.J. (1978) Quantum jumps of conductance during formation of membrane channels at cell-cell junction. *Nature*, **274**, 133.

Lowet, G., Audekercke, V., Perre, G.V. *et al.* (1993) The relation between resonant frequencies and torsional stiffness of long bones *in vitro* validation of a simple beam model. *Journal of Biomechanics*, **26**, 689–696.

Lozupone, E. and Favia, A. (1990) The structure of the trabeculae of cancellous bone: II, long bones and mastoid. *Calcified Tissue International*, **46**, 367–372.

Lu, P.W., Cowell, C.T., Lloyd Jones, S.A. *et al.* (1996) Volumetric bone mineral density in normal subjects, aged 5–27 years. *The Journal of Clinical Endocrinology and Metabolism*, **81**, 1586–1590.

Luben, R.A. (1989) Comparison of electromagnetic effects on para-thyroid hormone receptors and beta-adrenergic receptors in bone cells. *The Journal of Cell Biology*, **109**, 172.

Luben, R.A. (1991) Effects of low-energy electromagnetic fields (pulsed and DC) on membrane signal transduction processes in biological systems. *Health Physics*, **61**, 15–28.

Luben, R.A. (1993) *Electricity and Magnetism in Biology and Medicine* (ed. M. Blank), San Francisco Press, Berkeley, CA, pp. 57–62.

Luben, R.A., Cain, C.D., Chen, M.C.Y. *et al.* (1982) Inhibition of parathyroid hormone actions on bone cells in culture by induced low energy electromagnetic fields. *Proceedings of the National Academy of Sciences of the United States of America*, **79**, 4180–4184.

Lucchinetti, E. (2001) Composite models of bone properties, in *Bone mechanics handbook*, 2nd edn, Boca Raton, FL, CRC Press, pp. 12, 1–19.

Luengo, M., Picado, C., Piera, C. *et al.* (1991) Intestinal calcium absorption and parathyroid hormone secretion in asthmatic patients on prolonged oral or inhaled steroid treatment. *The European Respiratory Journal*, **4**(4), 441–444.

Lukert, B. and Raisz, L.G. (1990) Glucocortocoid-induced osteoporosis: Pathogenesis and management. *Annals of Internal Medicine*, **112**(5), 352–364.

Lundy, M.W., Wergedal, J.E., Teupner, E. *et al.* (1989) The effect of prolonged fluoride therapy for osteoporosis: bone composition and histology. *Bone*, **10**, 321–327.

Lutwak, L. (1974) Continuing need for dietary calcium throughout life. *Geriatrics*, **29**, 171–178.

Lutz, J. (1984) Calcium balance and acid base status of women as affected by increased protein intake and by sodium bicarbonate ingestion. *The American Journal of Clinical Nutrition*, **39**, 281–288.

Lydick, E., Cook, K., Turpin, J. *et al.* (1998) Development and validation of a simple questionnaire to facilitate identifications of women likely to have low bone density. *The American Journal of Managed Care*, **4**, 37–48.

Ma, Y., Jee, W.S., Chen, Y. *et al.* (1995) Partial maintenance of extra cancellous bone mas by antiresorptive agents after discontinuation of human parathyroid hormone (1–38) in right hindlimb immobilized rats. *Journal of Bone and Mineral Research*, **10**, 1726–1734.

Macarol, V. and Frauenfelder, F.D. (1994) Pamidronate disodium and possible ocular adverse drug reactions. *American Journal of Ophthalmology*, **118**, 220–224.

MacDonald, H.M., McGuigan, F.A., New, S.A. *et al.* (2001) COLIAI Spl polymorphism predicts perimenopausal and early postmenopausal spinal bone loss. *Journal of Bone and Mineral Research*, **16**, 1634–1641.

MacGinitie, L.A., Gluzband, Y.A. and Grodzinsky, A.J. (1994) Electric field stimulation can increase protein synthesis in articular cartilage explants. *Journal of Orthopaedic Research*, **12**, 157–160.

MacGinitie, L.A., Otter, M.W., Wu, D.D. and Cochran, G.V.B. (1993a) Endogenous fields in intact, remodeling and healing bone, in *Electricity and Magnetism in Biology and Medicine* (ed. M. Blank), San Fransisco Press, San Fransisco, pp. 683–685.

MacGinitie, L.A., Seiz, K.G., Wu, D.D. *et al.* (1993b) A comparison of streaming potentials in healing, remodeling and intact cortical bone. *Journal of Bone and Mineral Research*, **8**, 1323–1335.

MacGinitie, L.A., Stanley, G.D., Bieber, W.A. and Wud, D.D. (1997) Bone streaming potentials and currents depend on anatomical structure and loading orientation. *Journal of Biomechanics*, **30**, 1133–1139.

MacGregor, A., Snieder, H. and Spector, T.D. (2000) Genetic factors and osteoporotic fractures in elderly people. Twin data support genetic contribution to risk of fracture. *British Medical Journal*, **320**, 1661–1670.

Macintyre, I., Zaidi, M., Alam, A.S.M.T. *et al.* (1991) Osteoclastic inhibition – an action of nitric oxide not mediated by cyclic-GMP. *Proceedings of the National Academy of Sciences of the United States of America*, **88**, 2936–2940.

Mackie, I.G., Ralis, Z.A., Leyshon, R.L. *et al.* (1989) Treatment of bone weakness in patients with femoral neck fracture by fluoride, calcium and vitamin D. *The Journal of Bone and Joint Surgery*, **71**, 111–117.

Maclman, M., McLeod, K., Rayaby, J.P. and Rubin, C. (1998) Enhancement of fracture healing by low intensity ultrasound. *Clinical Orthopaedics*, (Suppl.), S216–S29.

MacPherson, H., Noble, B.S. and Ralston, S.H. (1999) Expression and functional role of nitric oxide synthase isoforms in human osteoblst-like cells. *Bone*, **24**, 179–185.

Madsen, J.E., Aune, A.K., Falch, J.A. *et al.* (1996) Neural involvement in post-traumatic osteopenia: an experimental study in the rat. *Bone*, **18**, 411–416.

Madsen, M. (1977) Vertebrae and peripheral bone mineral content by photon absorptiometry. *Investigative Radiology*, **12**, 185–188.

Maeda, H., Kimmel, D.B., Raab, D.M. and Lane, N.E. (1993) Musculoskeletal recovery following hindlimb immobilization in adult female rats. *Bone*, **14**, 153–159.

Maier, R., Glatz, A., Mosbacher, J. and Bilbe, G. (1997) Cloning of P2Y6 cDNAs and identification of a pseudogene: Comparison of P2Y receptor subtype expression in bone and brain tissue. *Biochemical and Biophysical Research Communications*, **237**, 297–302.

Maierhofer, W.J., Gray, R.W. and Lemann, J. (1984) Phosphate deprivation increases serum 1,25 $(OH)_2$-vitamin D concentrations in healthy men. *Kidney International*, **25**, 571–575.

Majumdar, S. and Genant, H.K. (1992) *In vivo* relationship between marrow $T_2$ and trabecular bone density determined with a chemical shift-selective asymmetric spin echo sequence. *Journal of Magnetic Resonance Imaging*, **2**, 209–219.

Majumdar, S., Thomasson, D., Shimakawa, A. and Genant, H.K. (1991) Quantitation of the susceptibility difference between trabecular bone and bone marrow: experimental studies. *Magnetic Resonance in Medicine*, **22**, 111–127.

Mallmin, H. and Ljunghall, S. (1994) Distal radius fracture is an early sign of general osteoporosis: bone mass measurements in a population based study. *Osteoporosis International*, **4**, 357–361.

Malluche, H.H., Arnala, I. and Faugere, M.C. (1988) Values of noninvasive techniques in predicting bone histology. *Annales Chirurgiae et Gynaecologiae*, **77**, 246–250.

Mamelle, N., Meunier, P.J. and Dusan, R. (1988) Risk benefit ratio of sodium fluoride treatment in primary vertebral osteoporosis. *Lancet*, **2**, 361–365.

Mandelbrot, B.B. (1982) *The Fractal Geometry of Nature*, W. H. Freeman, New York.

Mandelbrot, B.B. (1993) A fractals' lacunrity and how it can be tuned and measured, in *Fractals in Biology and Medicine* (eds T.F. Novnenmacher, G.A. Losa and E.R. Weibel), Birkhaise, Basel, pp. 8–21.

Manduca, P., Pistone, M., Sanguineti, C. *et al.* (1993) Modulation of integrins expression during human osteoblast *in vitro* differentiation. *Bollettino della Società Italiana di Biologia Sperimentale*, **69**, 699–704.

Manhanian, S. and Piziali, R.L. (1988) Finite element evaluation of the AIA shear specimen for bone. *Journal of Biomechanics*, **21**, 347–356.

Mann, V., Hobson, E.E., Li, B. *et al.* (2001) COLIAI Spl binding site polymorphism predisposes to osteoporotic fracture by affecting bone density and quality. *The Journal of Clinical Investigation*, **197**, 899–907.

Mano, M., Arakawa, T., Mano, H. *et al.* (2000) Prostaglandin $E_2$ directly inhibits bone resorbing activity of isolated mature osteoclasts mainly through the EP4 receptor. *Calcified Tissue International*, **671**, 85–92.

Manolagas, S.C., Burton, D.W. and Deftos, L.J. (1981) 1,25-Dihydroxy vitamin $D_3$ stimulates the alkaline phosphatase activity of osteoblast like cells. *The Journal of Biological Chemistry*, **256**, 7115–7117.

Manolagas, S.C. and Jilka, R.L. (1995) Bone marrow cytokines and bone remodeling: emerging insights into the pathophysiology of osteoporosis. *The New England Journal of Medicine*, **332**, 305–311.

Manolagas, S.C., Kousteni, S. and Jilka, R.L. (2002) Sex steroids and bone. *Recent Progress in Hormone Research*, **57**, 385–409.

Manso, M., Jimenez, C., Morant, C. *et al.* (2000) Electrodeposition of hydroxyapatite coatings in basic conditions. *Biomaterials*, **21**, 1755–1761.

Marcus, R., Kosek, R.J., Pfefferbaum, A. and Horning, A. (1983) Age related loss of trabecular bone in premenopausal women: a biopsy study. *Calcified Tissue International*, **35**, 406–409.

Marie, P.J., Sabbag, A., De Vernejoul, M.C. and Lomir, A. (1989) Oseocalcin and deoxyribonucleic acid synthesis *in vitro* and histomorphometric indices of bone formation in postmenopausal osteoporosis. *The Journal of Clinical Endocrinology and Metabolism*, **69**, 272–279.

Marino, A.A. and Becker, R.O. (1964) Piezoelectricity in bone as a function of age. *Calcified Tissue Research*, **8**, 178.

Marino, A. and Becker, R. (1970) Evidence for epitaxy in the formation of collagen and apatite. *Nature*, **226**, 652–653.

Marino, A.A. and Becker, R.O. (1971) Origin of the piezoelectric effect in bone. *Calcified Tissue Research*, **8**, 177–180.

Marino, A.A. and Becker, R.O. (1975) Piezoelectricity in hydrated frozen bone and tendon. *Nature*, **253**, 627–628.

Marino, A.A., Becker, R.O. and Bachman, C.H. (1967) Dielectric determination of bound water of bone. *Physics in Medicine and Biology*, **12**, 367.

Marino, A.A., Becker, R.O. and Soderholm, S.C. (1971) Origin of the piezoelectric effect in bone. *Calcified Tissue Research*, **8**, 177.

Marino, A., Spadaro, J.A., Fukada, E. *et al.* (1980) Piezoelectricity in collagen films. *Calcified Tissue International*, **31**, 257–259.

Markel, M.D., Wikenheiser, M.A. and Chao, E.Y.S. (1990) A study of fracture callus mineral properties: Relationship to the torsional strength of bone. *Journal of Orthopaedic Research*, **8**, 843–850.

Markov, M.S. and Pilla, A.A. (1997) Weak static magnetic field modulation of myosin phosphorylation in a cell-free preparation: Calcium dependence. *Bioelectrochemistry and Bioenergetics*, **43**, 233–238.

Marks, S.C. Jr and Popoff, S.N. (1988) Bone cell biology: The regulation of development, structure, and function in the skeleton. *The American Journal of Anatomy*, **183**, 1–44.

Marotti, G. (1963) Quantitative studies on bone reconstruction. 1-The reconstruction in homotypic shaft bones. *Acta Anatomica*, **52**, 291–333.

Marsh, A.G., Sanchez, T.V., Mickelsen, O. *et al.* (1980) Cortical bone density of adult lacto vegetarian and omnivorous women. *Journal of the American Dietetic Association*, **26**, 148–151.

Marshall, D., Johnell, O. and Wedel, H. (1996) Meta-analysis of occurrence of osteoporotic fractures. *British Medical Journal*, **312**, 1254–1259.

Martens, M.G. (2003) Risk of fracture and treatment to prevent osteoporosis-related fracture in postmenopausal women. A review. *Journal of Reproductive Medicine*, **48**, 425–434.

Martin, R.B. (1984) Porosity and specific surface of bone. *CRC Critical Reviews in Biomedical Engineering*, **10** (3), 179–222.

Martin, R.B. (1991) Determinants of the mechanical properties of bones. *Journal of Biomechanics*, **24**, 79–88.

Martin, R.B. (2000) Toward a unifying theory of bone remodeling. *Bone*, **26**, 1–7.

Martin, R.B. and Atkinson, P.J. (1977) Age- and sex-related changes in the structure and strength of the human femoral shaft. *Journal of Biomechanics*, **10**, 223.

Martin, R.B. and Burr, D.B. (1982) A hypothetical mechanism for the stimulation of osteonal remodeling by fatigue damage. *Journal of Biomechanics*, **15**, 137–139.

Martin, R.B. and Burr, D.B. (1989) *Structure, Function and Adaptation of Compact Bone*, Raven Press, New York.

Martin, R.B., Burr, D.B. and Sharkey, N.A. (1998) *Skeletal Tissue Mechanics*, Springer, New York, pp. 127–141.

Martin, R.B. and Guttman, W. (1982) The effect of electric field on osteoporosis of disuse. *Calcified Tissue Research*, **25**, 23–27.

Martin, R.B. and Ishida, J. (1989) The relative effects of collagen fiber orientation, porosity, density and mineralization on bone strength. *Journal of Biomechanics*, **22**, 419–426.

Martin, R.B., Pickett, J.C. and Zinaich, S. (1980) Studies of skeletal remodeling in aging men. *Clinical Orthopaedics*, **149**, 268–282.

Martin, T. (1993) Hormones in the coupling of bone resorption and formation. *Osteoporosis International*, **3**, S121–S125.

Martin, T.J. and Rodan, G.A. (2001) Coupling of bone resorption and formation during bone remodeling, in *Osteoporosis* (eds R. Marcus *et al.*), Academic Press, pp. 361–372.

Martini, L., Giavarest, G., Fini, M. *et al.* (2003) Effect of extracorporeal shock wave therapy on osteoblast like cells. *Clinical Orthopaedics and Related Research*, **413**, 269–280.

Martino, L.J. and Seideman, B. (1980) The effect of dichloromethylene diphosphonate (C12 MPD) on osteocyte activity during immobilization osteoporosis in rats. *Metabolic Bone Disease & Related Research*, **2**, 261–266.

Marx, J.C., Allay, J.A., Persons, D.A. *et al.* (1999) High efficiency transduction and long-term gene expression with a murine stem cell retroviral vector encoding the green fluorescent protein in human marrow stromal cells. *Human Gene Therapy*, **10**, 1272–1281.

Marx, S.J., Wopopdard, C.J. and Aurbach, G.D. (1972) Calcitoni receptors of kidney and bone. *Science*, **178**, 999–1001.

Mascarenhas, S. (1974) The electret effect in bone and biopolymers and the bond water problems. *Annals of the New York Academy of Sciences*, **238**, 36–53.

Masi, L., Franchi, A., Santucci, M. *et al.* (1992) Adhesion, growth and matrix production by osteoblasts on collagen substrata. *Calcified Tissue International*, **51**, 202–212.

Masinde, G.L., Li, X., Gu, W. *et al.* (2002) Quantitative trait loci for bone density in mice: the genes determining total skeletal density and femur density show little overlap in F2 mice. *Calcified Tissue International*, **71**, 421–428.

Mason, P. (1966) Viscoelasticity and structure of keratin and collagen. *Kolloid-Zeitschrift und Zeitschrift für Polymere*, **202**, 139–147.

Massey, L.K. and Wise, K.J. (1984) The effect of dietary caffeine on urinary excretion of calcium magnesium, sodium and potassium in healthy young females. *Nutrition Research*, **4**, 43–50.

Mastaglia, S.R., Pellegrim, G.G., Mandalunis, P.M. *et al.* (2006) Vitamin D insufficiency reduces the protective effect of bisphosphonate on ovariectomy-induced bone loss in rats. *Bone*, **39**, 837–844.

Mather, B. (1967) The symmetry of the mechanical properties of the human femur. *Journal of Surgical Research*, **7**, 222–225.

Matkovic, V., Kostial, K., Simoinovic, I. *et al.* (1979) Bone status and fracture rates in two regions in Yugoslavia. *The American Journal of Clinical Nutrition*, **32**, 540–549.

Matsumoto, T., Igarashi, C., Takeuchi, Y. *et al.* (1991) Stimulation by 1,25 dihydroxy vitamin D3 of *in vitro* mineralization induced by osteoblast like MC 3T3- E1 cells. *Bone*, **12**, 27–32.

Matsumoto, T., Kawanobe, Y., Morita, K. and Ogate, E. (1985) Effect of 1,25-dihydroxy vitamin $D_3$ on phospholipid metabolism in a clonal osteoblast like rat osterogeric sarcoma cell line. *The Journal of Biological Chemistry*, **260**, 13704–13709.

Matsunaga, S. (1986) Histological and histochemical investigations of constant direct current stimulated intramedullary callus. *Journal of the Japanese Orthopaedic Association*, **60**, 1293–1303.

Matsunaga, S., Ito, H. and Sakou, T. (1999) The effect of vitamin K and D supplementation on overiectomy-induced bone loss. *Calcified Tissue International*, **65**, 285–289.

Matthews, N. and Neale, M.L. (1989) Relationship between tumour cell morphology, gap junctions and susceptibility to cytolysis by tumour necrosis factor. *British Journal of Cancer*, **59**, 189–193.

Maxwell, J.J. (1873) *Treaties on Electricity and Magnetism*, Oxford Press, London, p. 314.

Mayr, E., Mockl, C., Lenich, A. *et al.* (2002) A low intensity ultrasound effective in treatment of disorders of fracture healing? *Unfallchirurg*, **105**, 108–115.

Mayr, E., Wagner, S., Ecker, M. and Ruter, A. (1999) Ultrasound therapy for non unions: Three case reports. *Unfallchirurg*, **102**, 191–196.

Mazess, R.B. (1979) Non-invasive measurements of bone, in *Osteoporosis II* (ed. U.S. Barzel), Grune and Stratton, New York, p. 5.

Mazess, R.B. (1983) Non-invasive methods for quantitating trabecular bone, in *The Osteoporotic Syndrome. Detection, Prevention and Treatment* (ed. L.V. Avioli), Grune and Stratton, New York.

Mazess, R.B. (2000) Measuring both femora? *Journal of Clinical Densitometry*, **3**, 299–301.

Mazess, R.B. and Barden, H.S. (1991) Bone density in premenopausal women. Effects of age, dietary intake, physical activity, smoking and birth control pills. *The American Journal of Clinical Nutrition*, **53**, 132–142.

Mazess, R.B., Barden, H.S., Ettinger, M. *et al.* (1987) Spine and femur density using dual photon absorptiometry in U.S. white women. *Bone and Mineral*, **2**, 211–219.

Mazess, R.B., Collick, B., Trempe, J. *et al.* (1989) Performance evaluation of a dual energy X-ray bone densitometer. *Calcified Tissue International*, **44**, 228–232.

Mazess, R.B., Nord, R., Hanson, J.A. and Barden, H.S. (2000) Bilateral measurement of femoral bone mineral density. *Journal of Clinical Densitometry*, **3**, 133–140.

Mazess, R.B., Ort, M. and Judy, P. (1970) Absorptiometric bone mineral determination using 153 Gd. Proceedings Bone Mass Conference (Washington DC), (ed. J.R. Cameron), US Atomic Energy Commission, Washington, DC. pp. 308–312.

Mazess, R.B. and Peppler, W. (1977) Total body bone mineral by photon absorptiometry. *Calcified Tissue Research*, **22**, 452–453.

Mazess, R.B., Peppler, W.W., Chesney, R.W. *et al.* (1984) Total body and regional bone mineral by dual photon absorptiometry in metabolic bone disease. *Calcified Tissue International*, **36**, 8–13.

Mazess, R.B., Peppler, W., Chesnut, C.H. *et al.* (1981) Total body mineral and lean body mass by dual photon absorptimetry. II. Comparison with total body calcium by neutron activation analysis. *Calcified Tissue International*, **33**, 363.

Maziere, B., Kuntz, D., Comar, D. and Ryckewaert, A. (1979) *In vivo* analysis of bone calcium by local neutron activation of the hand: results in normal and osteoporotic subjects. *Journal of Nuclear Medicine*, **20**, 85–91.

Mazzuoli, G.F., Tabolli, S., Bigi, F. *et al.* (1990) Effects of salmon calcitonin on the bone loss induced by ovariectomy. *Calcified Tissue International*, **47**, 209–214.

Mbalaviele, G., Sheikh, S., Stains, J.P. *et al.* (2005) Catenin and BMP-2 synergize to promote osteoblast differentiation and new bone formation. *Journal of Cellular Biochemistry*, **94**, 403–418.

McAdam, G.D. (1951) Some relations of powder characteristics to the elastic modulus and shrinkage of sintered ferrous compacts. *The Journal of the Iron and Steel Institute*, **168**, 346–358.

McCalden, R.W., McGeough, J.A., Barker, M.B. and Court Brown, C.M. (1993) Age-related changes in the tensile properties of cortical bone. The relative importance of changes in porosity, mineralization and microstructure. *The Journal of Bone and Joint Surgery. American Volume*, **75**, 1193–1205.

McCann, J., Dietrich, F., Rafferty, C. and Martin, A.O. (1993) A critical review of the genotoxic potential of electric and magnetic fields. *Mutation Research*, **297**, 61–96.

McCarthy, I., Goodship, A., Herzog, R. *et al.* (2000) Investigation of bone changes in microgravity during long and short duration space flight: Comparison of techniques. *European Journal of Clinical Investigation*, **30**, 1–12.

McCartney, R.N. and Jeffcott, L.B. (1987) Combined 2,25 MHz ultrasound velocity and bone mineral density measurements in the equine metacarpus and their *in vivo* applications. *Medical & Biological Engineering & Computing*, **25**, 620–626.

McClay, D.R. and Ettensohn, C.A. (1987) Cell adhesion in morphogenesis. *Annual Review of Cell Biology*, **3**, 319–345.

McCloskey, E.V., Murray, S.A., Charlesworth, D. *et al.* (1990) Assessment of broadband ultrasound attenuation in the os calcis *in vitro*. *Clinical Science*, **78**, 221–225.

McClung, M.R., Geusens, P., Miller, P.D. *et al.* (2001) Effect of risedronate treatment on hip fracture risk in elderly women hip intervention. Program Study Group. *The New England Journal of Medicine*, **344**, 333–340.

McConkey, B., Fraser, G.H., Bligh, A.S. and Whiteley, H. (1963) Transparent skin and osteoporosis. *Lancet*, **1**, 693.

McElhaney, J.H. (1966) Dynamic response of bone and muscle tissue. *Journal of Applied Physiology*, **21**, 1231.

McElhaney, J. (1967) The charge distribution on the human femur due to load. *The Journal of Bone and Joint Surgery*, **49**, 1561–1571.

McElhaney, J.H. and Byras, E. (1965) Dynamic response of biological materials. *American Society of Mechanical Engineers*, **65**, WA/HUF-9.

McKee, M.D. and Nanci, A. (1996) Osteopontin at mineralized tissue interface in bony teeth and osteointegrated implants: ultrastructural distribution and implications for mineralized tissue formation, turnover and repair. *Microscopy Research and Technique*, **33**, 141–164.

McKelvie, M.L. and Palmer, S.B. (1987) The interaction of ultrasound with cancellous bone, in *Ultrasonic Studies of Bone*, (Institute of Physics, Short Meetings Reports 6) (eds B.S. Palmer and C.M. Langton), Institute of Physics, Bristol, ch. 1, pp. 1–13.

McKelvie, M.L., Fordham, J., Clifford, C. and Palmer, S.B. (1989) *In vitro* comparison of quantitative computed tomography and broadband ultrasonic attenuation of trabecular bone. *Bone*, **10**, 101–104.

McKelvie, M.L. and Palmer, S.B. (1991) The interaction of ultrasound with cancellous bone. *Physics in Medicine and Biology*, **36**, 1331–1340.

McKibbin, B. (1978) The biology of fracture healing in long bone. *The Journal of Bone and Joint Surgery*, **60**(2), 150–162.

McLaren, A.D. and Shugar, D. (1964) *Photochemistry of Proteins and Nucleic Acid*, Pergamon Press, Oxford, New York, pp. 110–128.

McLean, F.C. and Budy, A.M. (1964) The mineral of bones and teeth, in *Radiation and Isotopes*, Academic, New York, pp. 61–77.

McLean, F.C. and Wrist, M.R. (1961) *Bone, An Introduction to the Physiology of Skeletal Tissue*, 2nd edn, University of Chicago Press, Chicago, IL.

McLean, F.C. and Wrist, M.R. (1968) *Bone Fundamentals of the Physiology of Skeletal Tissue*, 3rd edn, University of Chicago Press, Chicago, IL.

McLeod, B.R. and Liboff, A.R. (1987) Cyclotron response in cell membranes: the theory of the mechanism, in *Mechanistic Approaches to Interactions of Electromagnetic Fields with Living Systems* (eds M.J. Blank and E. Findl), Plenum, New York, pp. 97–108.

McLeod, K.J. and Rubin, C.T. (1990) Frequency specific modulation of bone adoption by induced electric fields. *Journal of Theoretical Biology*, **145**, 385–396.

McLeod, K.J. and Rubin, C.T. (1992) Effect of low-frequency electrical fields on osteogenesis. *The Journal of Bone and Joint Surgery*, **74**, 920–929.

McLeod, K.J., Rubin, C.T. and Lanyon, L.E. (1987) Optimisation of induced electric field frequency in the prevention of osteoporosis. *Orthopaedic Transactions*, **11**, 334–337.

McNagny, S.E., Wenger, N.K., Hilmy, N. *et al.* (2002) Postmenopausal hormone-replacement therapy. *The New England Journal of Medicine*, **46**, 63–65.

McNeill, K.G., Thomas, B.J., Sturtridge, W.C. and Harrison, J.E. (1973) *In vivo* neutron activation analysis for calcium in man. *Journal of Nuclear Medicine*, **14**, 502–506.

Meda, H. and Fukada, E. (1982) Effect of water on piezoelectric dielectric and elastic properties of bone. *Biopolymers*, **21**, 2055–2068.

Meema, H.E. (1963) Cortical bone atrophy and osteoporosis as a manifestation of aging. *American Journal of Roentgenology*, **89**, 1287–1295.

Meema, H.E., Harris, C.K. and Porrett, R.E. (1964) A method for determination of bones salt content of cortical bone. *Radiology*, **72**, 986.

Meghji, S. (1992) Bone remodeling. *British Dental Journal*, **172**, 235–242.

Mehta, S.S., Oz, O.K. and Antich, P.P. (1998) Bone elasticity and ultrasound velocity are affected by subtle changes in the organic matrix. *Journal of Bone and Mineral Research*, **13**, 114–121.

Meier, C.A., Bobbioni, E., Gabay, C. *et al.* (2002) IL-1 receptors antagonist serum levels are increased in human obesity: A possible link to the resistance to leptin? *The Journal of Clinical Endocrinology and Metabolism*, **87**, 1184–1188.

Meier, C., Woitge, H.W., Witte, K. *et al.* (2004) Supplementation with oral vitamin $D_3$ and calcium during winter prevents seasonal bone loss: a randomized controlled open-label prospective trial. *Journal of Bone and Mineral Research*, **19**, 1221–1230.

Melton, L.J. III (1990) Fluoride in the prevention of osteoporosis in fractures. *Journal of Bone and Mineral Research*, **5**(Suppl. 1), 163–170.

Melton, L.J. III (1993) Hip fractures: A worldwide problem today and tomorrow. *Journal of Bone and Mineral*, **14**, S1–S8.

Melton, L.J. III, Eddy, D.M. and Johnston, C.C. (1990) Screening for osteoporosis. *Annals of Internal Medicine*, **112**, 516–528.

Melton, L.J. III, Atkiso, E.J., O'Conner, M.K. *et al.* (1998) Bone density and fracture risk in men. *Journal of Bone and Mineral Research*, **13**, 1915–1923.

Melton, S.H. (2002) Genetic control of susceptibility to osteoporosis. *The Journal of Clinical Endocrinology and Metabolism*, **87**, 2460–2466.

Mendelson, J., Ogata, H. and Mello, M.K. (1971) Adrenal function and alcoholism. I. Serum cortisol. *Psychosomatic Medicine*, **33**, 145–147.

Menezel, J., Posner, A.S. and Harper, R.A. (1965) Age related changes in the crystallinity of rat bone apatite. *Israel Journal of Medical Sciences*, **1**, 251.

Mereter, C., Alfsen, A. and Baullieu, E.E. (1965) A testosterone binding globulin, in *Proceedings of the Second Symposium on the Steroid Hormones Ghant 1965*, Excerpta Medica, Amsterdam, p. 212.

Mesler, E., Mesler, A.I. and Mesler, A. (1985) The biomedical effects of laser applications. *Lasers in Surgery and Medicine*, **5**, 31–39.

Mesler, E., Nagylueskav Tisza, S. and Mesler, A. (1978) A stimulation of wound healing by means of laser rays. *Acta Chunegana Academic Scientiaum Hungicae*, **19**, 163–170.

Meunier, P.J. (1986) Diagnosis value of bone histomorphometry and biochemistry in osteoporosis, in *Second International Conference on Osteoporosis: Social and Clinical Aspects*, Masson, Milan, pp. 169–174.

Meunier, P.J. (1992) Fluoride therapy for vertebral osteoporosis. *Nutrition of the Elderly*, **29**, 177–185.

Meunier, P. (1996) Prevention of hip fractures by correcting calcium and vitamin D insufficiencies in elderly people. *Scandinavian Journal of Rheumatology*, (Supp. 103), 75–78.

Meunier, P.J., Aaron, J., Edouard, C. and Vignon, G. (1971) Osteoporosis and the replacement of cell population of marrow by adipose tissue. *Clinical Orthopaedics*, **80**, 147–154.

Meunier, P.J., Briancon, D., Vignon, E. *et al.* (1980) Effects of combined therapy with sodium fluoride vitamin D calcium on vertebral fracture and bone histology, in *Osteoporosis. Recent Advances in Pathogenesis and Treatment* (eds H.G. De Luca, H.M. Frost, W.S.S. Jee *et al.*), University Park Press, Baltimore, MD, pp. 449–456.

Meunier, P.J., Courpron, C., Edouard, C. *et al.* (1973) Physiological senile involution and pathological rarefaction of bone. *The Journal of Clinical Endocrinology and Metabolism*, **2**, 239.

Meydani, S.N., Barklund, M.P., Liu, S. *et al.* (1990) Vitamin E supplementation enhances cell-mediated immunity in healthy elderly subjects. *The American Journal of Clinical Nutrition*, **52**, 557–563.

Meyer, R.A. Jr, Tsahakis, P.J., Martin, D.F. *et al.* (2001) Age and ovariectomy impair both the normalization of mechanical properties and the accretion of mineral by the fracture callus in rats. *Journal of Orthopaedic Research*, **19**, 428–435.

Meyer, W.J., Transbol, I., Barter, F.C. and Delea, C. (1976) Control of calcium absorption: effect of sodium chloride loading and depletion. *Metabolism: Clinical and Experimental*, **25**, 989–993.

Michael, W.R., King, W.R. and Francis, M.D. (1971) Effectiveness of diphosphonates in preventing "osteoporosis" of disuse I the rat. *Clinical Orthopaedics and Related Research*, **78**, 271–276.

Michaelsson, K., Melhus, H., Bellocco, R. and Wolk, A. (2003) Dietary calcium and vitamin D intake in relation to osteoporotic fracture risk. *Bone*, **32**, 694–703.

Miller, A. and Parry, D.A.D. (1973) Structure and packing of microfibrils in collagen. *Journal of Molecular Biology*, **75**, 441–447.

Miller, E.J. and Martin, G.R. (1968) The collagen of bone. *Clinical Orthopaedics*, **59**, 195.

Miller, P.D. (2000) Controversies in bone mineral density diagnostic classifications. *Calcified Tissue International*, **66**, 317–319.

Miller, P.D., Siris, E.S., Barrett Connor, E. *et al.* (2002) Prediction of fracture risk in postmenopausal white women with peripheral bone densitometry: evidence from the National Osteoporosis Risk Assessment. *Journal of Bone and Mineral Research*, **17**, 2222–2230.

Miller, S.C., Bowman, B.M., Miller, M.A. and Bagi, C.M. (1991) Calcium absorption and osseous organ, tissue and envelope specific changes following ovariectomy in rats. *Bone*, **12**, 439–446.

Miller, S.C. and Jee, W.S.S. (1992) Bone lining cells, in *Bone*, vol. **4** CRC Press, Boca Raton, FL, pp. 1–19.

Miller, S.C. and Marks, S.C. Jr (1993) Local stimulation of new bone formation by prostaglandin E1: quantitative histomorphometry and comparison of delivery by mine pumps and controlled release pellets. *Bone*, **14**, 143–151.

Miller, S.S., Wolf, A.M. and Arnaud, C.D. (1976) Bone cells in culture: morphologic transformation by hormones. *Science*, **192**, 1340–1343.

Minaire, P., Bernard, E., Meunier, P.J. *et al.* (1981) Effects of dichloromethylene diphosphonate of bone loss in paraplegic patients. *The Journal of Clinical Investigation*, **68**, 1086–1092.

Minaire, P., Depassio, J., Bernard, E. *et al.* (1987) Effects of clodronate on immobilization bone loss. *Bone*, **8**(Suppl. 1), 563–568.

Minaire, P., Edouard, C., Arlot, M. and Meunier, P.J. (1984a) Marrow changes in paraplegic patients. *Calcified Tissue International*, **36**, 338–340.

Minaire, P., Meunier, P., Depassio, J. *et al.* (1984b) Treatment of active osteoporosis with calcitonin, in *Osteoporosis* (eds C. Christiansen, C.C. Arnand, B.E.C. Nordin*et al.*), Norhaven A.S, Viborg, pp. 613–615.

Minakuchi, Y., Inoue, N., Levin, R. *et al.* (1997) Ultrasound velocity in human calcaneal and its relation to mechanical properties and trabecular structure. *Transactions of the Orthopaedic Research Society*, **22**, 207.

Mishima, S. (1988) The effect of long-term pulsing electromagnetic field stimulation on experimental osteoporosis of rats. *Journal of UOEH*, **10**, 31–45.

Mittra, E., Rubin, C. and Qin, Y.X. (2004) Interrelationship of trabecular mechanical and microstructural properties in sheep trabecular bone. *Journal of Biomechanics*, **38**, 1229–1237.

Miyamoto, T. and Suda, T. (2003) Differentiation and function of osteoclasts. *The Keio Journal of Medicine*, **52**, 1–7.

Mizuno, A., Murakami, T., Otani, S. *et al.* (1998) Leptin affects pancreatic endocrine functions through the sympathetic nervous system. *Endocrinology*, **139**, 3863–3870.

Modder, U.I. *et al.* (2004) Dose response of estrogen and bone versus the in ovariectomized mice. *European Journal of Endocrinology/ European Federation of Endocrine Societies*, **151**, 503–510.

Molnar, Z. (1959) Development of the parietal bone of young mice. Crystals of bone mineral in frozen-dried preparations. *Journal of Ultrastructure Research*, **3**, 39.

Molnar, Z. (1960) Additional observations on bone crystal dimensions. *Clinical Orthopaedics*, **13**, 38.

Moncada, S. and Higgs, A. (1993) Mechanisms of disease- the L-arginine nitric-oxide pathway. *The New England Journal of Medicine*, **329**, 2002–2012.

Monma, Y., Funaama, H., Mayanagi, H. and Endo, Y. (2004) Effects of weekly administration of alendronate + clodronate on young mouse tibia: localized action at the proximal growth plate. *Calcified Tissue International*, **74**, 115–121.

Mont, M.A. and Hungerford, D.S. (1995) Non-traumatic vascular necrosis of the femoral head. *The Journal of Bone and Joint Surgery*, **77**, 459–474.

Mooney, D., Hansen, L., Vacanti, J. *et al.* (1992) Switching from differentiation to growth in hepatocytes: control by extracellular matrix. *Journal of Cellular Physiology*, **151**, 497–505.

Mora, S., Pitukcheewanont, P., Kaufman, F.R. *et al.* (1999) Biochemical markers of bone turnover and the volume and the density of bone in children at different stages of sexual development. *Journal of Bone and Mineral Research*, **14**, 1664–1671.

Morey, E.R. and Baylink, D.J. (1978) Inhibition of bone formation during spaceflght. *Science*, **201**, 1138–1141.

Morey-Holton, E.R. and Globus, R.K. (2002) Hindlimb unloading rodent model: Technical aspects. *Journal of Applied Physiology*, **92**, 1367–1377.

Mori, S. and Burr, D.B. (1993) Increased intracortical remodeling following fatigue damage. *Bone*, **14**, 103–109.

Mori, S., Jee, W.S.S., Li, X.J. *et al.* (1990) Effects of $PGE_2$ on production of new cancellous bone in the axial skeleton of ovariectomized rats. *Bone*, **11**, 103–113.

Mori, S., Jee, W.S.S. and Li, X.J. (1992) Production of new trabecular bone in osteopenic ovariectomized rats by prostaglandin $E_2$. *Calcified Tissue International*, **50**, 80–87.

Morimoto, S., Tsuji, M., Okada, Y. *et al.* (1980) The effects of the estrogens on human calcitonin secretion after calcium infusion in elderly female subjects. *Clinical Endocrinology*, **13**, 135–143.

Morishima, A., Grumbach, M.M., Simpson, E.R. *et al.* (1995) Aromatase deficiency in male and female siblings caused by a novel mutation and the physiological role in estrogens. *The Journal of Clinical Endocrinology and Metabolism*, **80**, 3689–3698.

Morley, P., Whitfield, J.F. and Willick, G.E. (2001) Parathyroid hormone an anabolic treatment for osteoporosis. *Current Pharmaceutical Design*, **7**, 671–687.

Morris, H.A., Need, A.G., Horowitz, M. *et al.* (1991) Calcium absorption in normal and osteoporotic postmenopausal women. *Calcified Tissue International*, **49**, 240–243.

Morris, H.A., Need, A.G., O'Loughlin, P.D. *et al.* (1990) Malabsorption of calcium in corticosteroid induced osteoporosis. *Calcified Tissue International*, **46**, 305–308.

Morris, H.A. and Steurer, T.A. (1988) Positive cooperativity of the rat intestinal 1,25-dihydroxy vitamin D receptor, in *Vitamin D, Molecular, Cellular and Clinical Endocrinology* (eds A.W. Norman, K. Schaefer, H.G. Grigoleti and D.V. Herrath), Walter de Gruyter, Berlin, pp. 131–232.

Morrison, M.S., Turin, L., King, B.F. *et al.* (1998) ATP is a potent stimulator of the activation and formation of rodent osteoclasts. *The Journal of Physiology*, **511**, 495–500.

Morton, D.J., Barrett Connor, E.L. and Edelstein, S.L. (1994) Thiazides and bone mineral density in elderly men and women. *American Journal of Epidemiology*, **139**, 1107–1115.

Mosekilde, L. (1990) Consequences of the remodeling process for vertebral trabecular bone structure: A scanning electron microscopy study (uncoupling of unloaded structures). *Bone & Mineral*, **10**, 3–35.

Mosekilde, L., Danielsen, C.C. and Gasser, J. (1994) The effect of vertebral bone mass and strength of long term treatment with antiresorptive agents (estrogen and calcitonin), human parathyroid hormone (1–38) and combination therapy, assessed in aged ovariectomized rats. *Endocrinology*, **134**, 2126–2134.

Mosekilde, L., Krogstrup, J. and Richards, A. (1987) A compressive strength, ash weight and volume of vertebral trabecular bone in experimental fluorosis in pigs. *Calcified Tissue International*, **40**, 318–322.

Mosekilde, I., Sogaard, C.H., Danielsen, C.C. *et al.* (1991) The anabolic effects of human parathyroid hormone (hPTH) on rat vertebral body mass are also reflected in the quality of one assessed by biomechanical testing: A comparison study between hPTH (1–34) and hPTH (1–84). *Endocrinology*, **129**, 421–428.

Mosekilde, L., Sogaard, C.H., McOsker, J.E. and Wronski, T.J. (1994) PTH has a more pronounced effect on vertebral bone mass and biomechanical impedance than antiresorptive agents (estrogen and bisphosphonate) assessed in sexually mature, ovariectomized rats. *Bone*, **15**, 401–408.

Mosley, J.R. and Lanyon, L.E. (1998) Strain rate as a controlling influence on adaptive modeling in response to dynamic loading of the ulna in growing male rats. *Bone*, **23**, 313–318.

Most, W., Schot, L., Ederveen, A. *et al.* (1995) *In vitro* and *ex vivo* evidence that estrogens suppress increased bone resorption induced by ovariectomy or PTH stimulation through an effect on osteoclastgenesis. *Journal of Bone and Mineral Research*, **10**, 1523–1536.

Mou, C., Beltembuk, P., Bhattoa, H.P. and Balogh, A. (2001) The effects of pregnancy and location on bone mineral density. *Osteoporosis International*, **12**, 732–737.

Moutsatsos, I.K., Turgeman, G., Zhou, S. *et al.* (2001) Exogenously regulated stem cell mediated gene therapy for bone regeneration. *Molecular Therapy*, **3**, 449–461.

Movsowitz, C., Epstein, S., Fallon, M. *et al.* (1990) Hyperstosis induced by the bisphosphonate (2-PEBP) in the oophorectomized rat. *Calcified Tissue International*, **46**, 195–199.

Mow, V.C., Rateliffe, A. and Poole, A.R. (1992) Cartilage and diarthrodial joints as paradigms for hierachical material and structures. *Biomaterials*, **13**, 67–68.

Mueller, K.H., Trias, A. and Ray, R.D. (1966) Bone density and composition. Age-related and pathological changes in water and mineral content. *The Journal of Bone and Joint Surgery*, **48**, 140–148.

Muhlbach, R., Tarsoly, E., Trzenschik, K. and Hirthe, D. (1972) Effect of growth hormone on the metabolism of bones and the healing of fractures. *Acta Chirurgica Academiae Scientiarum Hungaricae*, **13**, 57–63.

Muhlbauer, R.C., Russell, R.G.G., Williams, D.A. and Fleisch, H. (1971) The effect of diphosphonates, polyphosphonates and calcitonin on "immobilization osteoporosis" in rats. *European Journal of Clinical Investigation*, **1**, 336–344.

Muldowney, F.P., Freaney, R. and Moloney, M.F. (1982) Importance of dietary sodium in the hypercalciuria syndrome. *Kidney International*, **22**, 292–296.

Mullen, A., Rueggsegger, P., Seitz, P. and Optimal, C.T. (1985) Settings for bone evaluations. *Physics in Medicine and Biology*, **30**, 401–409.

Mullender, M., El-Haj, A.J., Yang, Y. *et al.* (2003) Fluid flow and mechanical strain induce different responses in human bone cells. Transactions of the 49th Annual Meeting of the Orthopaedic research Society 28. (abstract).

Mullender, M.G. and Huiskes, H. (1995) A proposal for the regulatory mechanism of Wolff's Law. *Journal of Orthopaedic Research*, **13**, 503–512.

Muller, A., Ruegsegger, E. and Ruegsegger, P. (1989) Peripheral QCT: A low risk procedure to identify women predisposed to osteoporosis. *Physics in Medicine and Biology*, **34**, 741–749.

Muller-Mai, C.M., Stupp, S.I., Voigt, C. and Gross, K.A. (1995) Nanoapatite and organoapatite implants in bone: Histology and ultrastructure of the interface. *Journal of Biomedical Materials Research*, **29**, 9–18.

Mullins, J.M., Krause, D. and Litovitz, T.A. (1993) Simultaneous application of a spatially coherent noise field blocks response of cell cultures to a 60Hz EM field, in *Electricity and Magnetism in Biology and Medicine* (ed. M. Blank), San Francisco Press, Berkeley, CA, pp. 345–346.

Mundy, G.R. (1998) Bone remodeling, in *Primer on the Metabolic Bone Diseases and Disorder of Mineral Metabolism*, 4th edn (ed. M.J. Favus), Williams & Wilkins, Baltimore, pp. 30–38.

Mundy, G.R. (2002) Directions of drug discovery in osteoporosis. *Annual Review of Medicine*, **53**, 337–354.15.

Mundy, G.R. and Martin, T.L. (1982) The hypercalcemia of malignancy: pathogenesis and management. *Metabolism: Clinical and Experimental*, **31**, 1247–1277.

Mundy, G.R., Wilkinson, R. and Health, D.A. (1983) Comparative study of available medical therapy for hypercalcemia for malignancy. *The American Journal of Medicine*, **74**, 421–437.

Murray, J.C. and Farndale, R.W. (1985) Modulation of collagen production in cultured fibroblasts by a low frequency pulsed electromagnetic field. *Biochimica et Biophysica Acta*, **838**, 98–105.

Murrils, R.J., Shane, E., Lindsay, R. and Dempster, D.W. (1989) Bone resorption by isolated human osteoclasts *in vitro*: effects of calcitonin. *Journal of Bone and Mineral Research*, **4**, 259–268.

Myburgh, K.H., Noakes, F.D., Roodt, M. and Hough, F.S. (1989) Effect of exercise on the development of osteoporosis in adult rats. *Journal of Applied Physiology*, **66**(1), 14–19.

Nachtingall, L.E., Nachtingall, R.H. and Nachtingall, R.D. (1979) Estrogen replacement therapy: a 10-year prospective study in the relationship to osteoporosis. *Obstetrics and Gynecology*, **53**, 277–281.

Naemsch, L.N., Weidema, A.F., Sims, S.M. *et al.* (1999) $P2X_4$ purinoceptors mediate an ATP activated non-selective cation current in rabbit osteoclasts. *Journal of Cell Science*, **112**, 4425–4435.

Nagata, T., Kaho, K., Nishikawa, S. *et al.* (1994) Effect of postaglandins $E_2$ on mineralization of bone nodules formed by fetal rat calvarial cells. *Calcified Tissue International*, **55**, 451–457.

Nakahara, H., Bruder, S.P., Haynesworth, S.E. *et al.* (1990) Bone and cartilage formation in diffusion chambers by subcultured cells derived from the periosteum. *Bone*, **11**, 181–188.

Nakajima, A., Ogasawara, A., Goto, K. *et al.* (2001) Spatial and temporal gene expression in chondrogenesis during fracture healing and the effects of basic fibroblast growth factor. *Journal of Orthopaedic Research*, **19**, 935–944.

Nakajima, A., Shimizu, S., Moriya, H. and Amazaki, M. (2003) Expression of fibroblast growth factor receptor –3 (FGFR3), signal transducer and activator of transcription-1 and cyclien-dependent kinase inhibitor P21 during endochondral ossification: differential role of FGFR3 in skeletal development and fracture repair. *Endocrinology*, **144**, 4649–4668.

Nakajima, A., Shimoji, N., Shiomi, K. *et al.* (2002) Mechanism for the enhancement of fracture healing in rats treated with intermittent low-dose human parathyroid hormone (1–34). *Journal of Bone and Mineral Research*, **17**, 2038–2047.

Nakamura, H., Morita, S., Kumei, Y. *et al.* (2001) Effects of neurectomy and tenotomy on the bone mineral density and strength of tibiae. *Acta Astronaut*, **49**, 179–190.

Nakamura, T., Suzuki, K., Hirae, T. *et al.* (1992) Increased bone volume and reduced bone turnover in vitamin D replete rabbits by the administration of 24 R, 25-dihydroxy vitamin $D_3$. *Bone*, **13**, 229–236.

Nakanishi, T., Nishida, T., Shimo, T. *et al.* (2000) Effects of CTGF/Hcs24, a product of a hypertrophic chondrocyte specific gene, on the proliferation and differentiation of chondrocytes in culture. *Endocrinology*, **141**, 264–273.

Nakase, T., Nomura, S., Yoshikawa, H. *et al.* (1994) Transient and localized of bone morphogenetic protein 4 messenger RNA during healing. *Journal of Bone and Mineral Research*, **9**, 651–659.

Nakashima, K., Zhou, X., Kunkel, G. *et al.* (2002) The novel zinc finger containing transcription factor osterix is required for osteoblast differentiation and bone formation. *Cell*, **108**, 17–29.

Nakata, E., Nakanishi, T., Kawai, A. *et al.* (2002) Expression of connective tissue growth factor/Hypertrophic chondrocyte-specific gene product 24 (CTGF/Hcs24) during fracture healing. *Bone*, **31**(4), 441–447.

Nakatsuchi, Y., Tsuchikane, A. and Nomura, A. (1996) Assessment of fracture healing in the tibia using the impulse response method. *Journal of Orthopaedic Trauma*, **10**, 50–62.

Nakazawa, T., Nakajima, A., Shiomi, K. *et al.* (2005) Effects of low-dose intermittent treatment with recombinant human parathyroid hormone (1–34) on chondrogenesis in a model of experimental fracture healing. *Bone*, **37**, 711–719.

Narumiya, S., Sugimoto, Y. and Ushikubi, F. (1999) Prostanoid receptors: structures, properties and functions. *Physiological Reviews*, **79**, 1193–1226.

National Institutes of Health (2000) Osteoporosis prevention, diagnosis and therapy. National Institutes of Health Consensus Development Conference Statement 17, pp. 1–13.

National Osteoporosis Foundation (1999) *Physicians Guide to Prevention and Treatment of Osteoporosis*, Excerpta Medica, Belle Mead, NJ.

Navidi, M., Wolinsky, I., Fung, P. and Arnaud, S.B. (1995) Effect of excess dietary salt on calcium metabolism and bone mineral in a spaceflight rat model. *Journal of Applied Physiology*, **78**, 70–75.

Neer, R.M., Arnaud, C.D., Zanchetta, J.R. *et al.* (1998) Effect of parathyroid hormone (1–34) on fractures and bone mineral density in postmenopausal women with osteoporosis. *The New England Journal of Medicine*, **344**, 1434–1441.

Neer, R.M., Arnaud, C.D., Zanchetta, J.R. *et al.* (2001) Effect of recombinant human parathyroid hormone (1–34) fragment on spine and non-spine fractures and bone mineral density in postmenopausal. Osteoporosis. *The New England Journal of Medicine*, **344**, 1434–1441.

Nefussi, J.-R. and Baron, R. (1985) PGE2 stimulates both resorption and formation of bone *in vitro*: differential responses of the periosteum and the endosteum in fetal rat long bone cultures. *The Anatomical Record*, **211**, 9–16.

Nelp, W.B., Denny, J.D., Murano, R. *et al.* (1972) Absolute measurement of total body calcium (bone mass) *in vivo*. *The Journal of Laboratory and Clinical Medicine*, **79**, 430–438.

Nelp, W.B., Palmer, H.E., Murano, R. *et al.* (1970) Measurement of total body calcium (bone mass) *in vivo* using total body neutron activation analysis. *The Journal of Laboratory and Clinical Medicine*, **76**, 151–162.

Nelson, M.E., Fiatarone, M.A., Morganti, C.M. *et al.* (1994) Effects of high intensity strength training on multiple risk factors for osteoporotic fractures. *The Journal of the American Medical Association*, **272**, 1909–1914.

Nevitt, M., Bauer, D., Black, D. *et al.* (2000) Radiographic vertebral deformities and stature loss as surrogate endpoints in steoporosis treatment trials: What is the evidence. *Journal of Bone and Mineral Research*, **15**(Suppl.), 5183.

Newcorner, A.D., Hodgson, S.F., McGill, D.B. and Thomas, P.J. (1978) Lactase deficiency: prevalence in osteoporosis. *Annals of Internal Medicine*, **89**, 218–220.

Newton-John, H.F. and Morgen, D.B. (1968) Osteoporosis disease or senescence? *Lancet*, **1**, 232.

Ngai, K.L. (1979) Universality of low frequency fluctuation, dissipation and relaxation properties of condensed matter. 1. *Comments Solid State Physics*, **9**(4), 127–140.

Nguyen, T.V., Eisman, J.A., Kelley, P.J. and Sambrook, P.N. (1996) Risk factors for osteoporotic fractures in elderly men. *American Journal of Epidemiology*, **144**, 255–263.

Nicholson, D.W. (1979) On the detachment of a rigid inclusion from an elastic matrix. *Journal of Adhesion*, **10**, 255–260.

Nicholson, P.H.F., Haddaway, M.J. and Davie, M.W.J. (1994) The dependence of ultrasonic properties on orientation in human vertebral bone. *Physics in Medicine and Biology*, **39**, 1013–1024.

Nicholson, P.H., Muller, R., Lowet, G. *et al.* (1998) Do quantitative ultrasound measurements reflect structure independently of density in human vertebral cancellous bone? *Bone*, **23**, 425–431.

Nicholson, P.H., Strelitzki, R., Cleveland, R.O. and Bouxsein, M.L. (2000) Scattering of ultrasound in cancellous bone: Predictions from a theoretical model. *Journal of Biomechanics*, **33**, 503–506.

Nicogossian, A.E., Rummel, J.D., Leveton, L. and Teeter, R. (1992) Development of countermeasures for medical problems encountered in space flight. *Advances in Space Research*, **12**, 329–337.

Nieders, M.E., Weiss, L. and Cudney, T.L. (1970) An electro-kinetic characterization of human tooth surfaces. *Archives of Oral Biology*, **15**, 135.

Nielson, M.W. (1975) A revision of the subfamily Coelidiinae (Homoptera: Cicadellidae). Tribes Tinobregmini, Sandersellini and Tharrinii. *Bulletin of the British Museum (Natural History), Entomology*, (Suppl. 24), 7–8.

Nilas, L. and Christiansen, C. (1987) Bone mass and its relationship to age and the menopause. *The Journal of Clinical Endocrinology and Metabolism*, **65**, 697–702.

Nilas, L., Christiansen, C. and Rodro, P. (1984) Calcium supplementation and postmenopausal bone loss. *British Medical Journal*, **289**, 1103–1106.

Ninomiya, J.T., Tracy, R.P., Calore, J.D. *et al.* (1990) Heterogeneity of human bone. *Journal of Bone and Mineral Research*, **5**, 933–938.

Nishida, T., Nakanishi, T., Asano, M. *et al.* (2000) Effects of CTGF/Hcs24, a hypertrophic chondrocyte specific gene product, on the proliferation and differentiation of osteoblastic cells *in vitro*. *Journal of Cellular Physiology*, **184**, 197–206.

Njeh, C.F. and Shepherd, J.A. (2004) *The Physical Measurements of Bones, Series in Medical Physics and Biomedical Engineering* (eds C.M. Lanton and C.F. Njeh), Institute of Physics, p. 281.

Njeh, C.F., Boivin, C.M. and Langton, C.M. (1997) The role of ultrasound in the assessment of osteoporosis: a review. *Osteoporosis International*, **7**, 7–22.

Njeh, C.F., Hodgskinson, R., Currey, J.D. and Langton, C.M. (1996) Orthogonal relationships between ultrasonic velocity and material properties of bovine cancellous bone. *Medical Engineering & Physics*, **18**, 373–381.

Njeh, C.F., Kuo, C.W., Langton, C.M. *et al.* (1997) Prediction of human femoral bone strength using ultrasound velocity and BMD: An *in vitro* study. *Osteoporosis International*, **7**, 471–477.

Njeh, C.F., Saeed, I., Grigontian, M. *et al.* (2001) Assessment of bone status using speed of sound at multiple anatomical sites. *Ultrasound in Medicine and Biology*, **27**, 1337–1345.

Noble, B.S., Stevens, H., Loveridge, N. and Reeve, J. (1997) Identification of apoptotic changes in osteocytes in normal and pathologic human bone. *Bone*, **20**, 273–282.

Nods, M. and Sato, A. (1985) Calcification of cartilaginous matrix in culture by constant direct current stimulation. *Clinical Orthopaedics and Related Research*, **193**, 281.

Nokes, L., Fairclough, J.A., Mackie, I. and Williams, J. (1984) Vibration analysis of human tibia, the effect of soft tissue on the output from skin mounted accelerometer. *Journal of Biomedical Engineering*, **6**, 223–226.

Nokes, L., Mintowt-zyz, W.J., Fairclough, J.A. *et al.* (1985) Vibration analysis in the assessment of conservatively managed tibial fractures. *Journal of Biomedical Engineering*, **7**, 40–44.

Nolan, R.D., Patridge, N.C. *et al.* (1983) Cyclo-oxygenase products of arachidonic acid metabolism in rat osteoblasts in culture. *Calcified Tissue International*, **35**(3), 294–297.

Nolte, P.A., Klein-Nulend, J., Albers, G.H.R. *et al.* (2001) Low-intensity ultrasound stimulates endochondral ossification *in vitro*. *Journal of Orthopaedic Research*, **19**, 301–307.

Nolte, P.A., Van der Krans, A., Patka, P. *et al.* (2001) Low intensity ultrasound in the treatment of non unions. *Journal of Trauma*, **51**, 693–703.

Nomura, S., Hiltner, A., Lando, J.B. and Baer, E. (1977) Interaction of water with native collagen. *Biopolymers*, **16**, 231–246.

Nordenson, I., Hansson, Mild., Ostman, U. and Liungberg, H. (1988) Chromosomal effects in lymphocytes of 400 kv substation workers. *Radiation and Environmental Biophysics*, **27**, 39–47.

Nordin, B.E. (1968) Measurement and meaning of calcium absorption. *Gastroenterology*, **54**(2), 294–301.

Nordin, B.E.C. (1971) Clinical significance and pathogenesis of osteoporosis. *British Medical Journal*, **1**, 571.

Nordin, B.E.C. (1979) Treatment of postmenopausal osteoporosis. *Drugs*, **18**, 484–492.

Nordin, B.E. (1983) Osteoporosis with particular reference to the menopause, in *The Osteoporotic Syndrome: Detection, Prevention and Treatment* (ed. L.V. Avioli), Grune and Stratton, New York, pp. 13–43.

Nordin, B., Horstnan, A. and Marshall, D. (1979) Calcium requirements and calcium therapy. *Clinical Orthopaedics*, **140**, 216–239.

Nordin, B.E.C., Need, A.G., Chatterton, B.E. *et al.* (1990) The relative contributions of age and years since menopause to postmenopausal bone loss. *The Journal of Clinical Endocrinology & Metabolism*, **70**, 83–88.

Nordin, B.E.C., Need, A.G., Morris, H.A. and Horowitz, M. (1988) The rationale for calcitriol therapy in osteoporosis, in *Vitamin D: Molecular Cellular and Clinical Endocrinology* (eds A.W. Norman, K. Schaefer, H.G. Grigoleti and D.V. Herrath), Walter de Gruyter, Berlin, pp. 826–835.

Nordin, B.E.C., Wilkinson, R., Marshall, D.H. *et al.* (1976) Calcium absorption in the elderly. *Calcified Tissue Research*, (Suppl.), 442–451.

Nordstrom, P., Pettersson, U. and Lorentzon, R. (1998) Type of physical activity, muscle strength and pubertal stage as determinants of bone mineral density and bone area in adolescent boys. *Journal of Bone and Mineral Research*, **13**, 1141–1148.

Norman, A.W. (1979) Vitamin D actions in bone, in *Vitamin D: The Calcium Homeostatic Steroid Hormone*, Academic Press, New York, pp. 374–401.

Norman, T.L., Bradley-Popovich, G., Clovis, N. *et al.* (2000) Aerobic exercise as a countermeasure for microgravity-induced bone loss and muscle atrophy in a rat hindlimb suspension model. *Journal of Aviation, Space, and Environmental Medicine*, **71**, 593–598.

Norman, T.L., Nivargikar, S.V. and Burr, D.B. (1996) Resistance to crack growth in human cortical bone is greater in shear than in tension. *Journal of Biomechanics*, **29**, 1023–1031.

Norman, T.L., Vashisth, D. and Burr, D.B. (1995) Fracture toughness of human bone under tension. *Journal of Biomechanics*, **28**, 309–320.

Norton, L. (1974) *In vivo* bone growth in controlled electric field. *Annals of the New York Academy of Sciences*, **238**, 466.

Norton, L.A. (1982) Effects of a pulsed electromagnetic field on mixed chondroblastic tissue culture. *Clinical Orthopaedics and Related Research*, **167**, 280–290.

Norton, L.A. (1985) Pulsed electromagnetic field effects in chondrobalst culture. *Reconstruction Surgery and Traumatology*, **19**, 70.

Norton, L.A., Hanley, K.J. and Turkewicz, J. (1984) Bioelectric perturbations of bone research directions and implications. *The Angle Orthodontist*, **54**, 73.

Norton, L.A., Rodan, G.A. and Bourret, L.A. (1977) Epiphyseal cartilage Camp changes produced by electrical and mechanical perturbations. *Clinical Orthopaedics and Related Research*, **124**, 59.

Norton, L.A., Witt, D.W. and Rovetti, L.A. (1988) Pulsed electromagnetic fields alter phenotypic expression in chondroblasts in tissues culture. *Journal of Orthopaedic Research*, **6**, 685–689.

Nudelman, S. and Mitra, S.S. (1969) *Optical Properties of Solids*, Plenum Press, New York.

Nussbaum, E.L., Biemann, I. and Mustard, B. (1994) Comparison of ultrasound/ultraviolet-c and laser for treatment of pressure ulcers in patients with spinal cord injury. *Physical Therapy*, **74**, 812–823.

Nye, J.F. (1960) *Physical Properties of Crystals*, Oxford University Press, London.

Nyman, J.S., Ray, A., Shen, X. *et al.* (2006) The influence of water removal on the strength and toughness of cortical bone. *Journal of Biomechanics*, **39**, 931–938.

O'Neill, T.W., Felsenberg, D., Varlow, J. *et al.* (1996) The prevalence of vertebral deformity in European men and women: the European vertebral osteoporosis study. *Journal of Bone and Mineral Research*, **11**, 1010–1018.

O'Neill, T.W., Varlow, J., Reeve, J. *et al.* (1995) Fall frequency and incidence of distal forearm fracture in the UK. *Journal of Epidemiology and Community Health*, **49**, 597–598.

O'Phir, J.Y. (1990) A transaxial comparison technique (TACT) for localized pulse-echo estimation of sound speed in biological tissue. *Ultrasound Imaging*, **12**, 35–41.

O'Phir, J.Y. and Yazdi, Y.A. (1990) Transaxial compression technique (TACT) for localized pulse-echo estimation of sound speed in biological tissue. *Ultrasonic Imaging*, **12**, 35–46.

Odeblad, E. and Norhagen, A. (1956) Measurements of electron densities with the aid of the Compton scattering process. *Acta Radiologica*, **45**, 161–167.

Odgaard, A. (2001) Quantification of cancellous bone architecture. In *Bone mechanics handbook*, end ed, Boca Raton, FL: CRC Press; pp 14(1), 14–19.

Odgaard, A., Jensen, E.B. and Gundersen, H.J.G. (1990) Estimation of structural anisotropy based on volume orientation. A New concept. *Journal of Microscopy*, **157**, 149–182.

Ogawa, T., Ono, T., Tsuda, M. and Kawanishi, Y. (1982) A novel fluor in insoluble collagen: a crosslinking moiety in collagen molecule. *Biochemical and Biophysical Research Communications*, **107**(4), 1252–1257.

Ogden, J.A. (1980) Chondo-osseous development and growth, in *Fundamental and Clinical Bone Physiology* (ed. M.R. Urist), J.B. Lippincott Co., Philadelphia, USA, p. 108.

Oin, Y.-X., Rubin, C. and McLeod, K. (1998) Nonlinear dependence of loading intensity and cycle number in the maintenance of bone mass and morphology. *Journal of Orthopaedic Research*, **16**, 482–489.

Okulicz, W.C., Evans, R.W. and Leavitt, W.W. (1991) Progesterone regulation of estrogen receptor in the rat uterus: a primary inhibitory influence on the nuclear fraction. *Steroids*, **37**, 463–470.

Olkkonen, H. and Karjalainen, P. (1975) A 170 Tm gamma scattering technique for the determination of absolute bone density. *British Journal of Radiology*, **48**, 594–597.

Olkkonen, H., Puumalainen, P., Karjalainen, P. and Alhava, E.M.K. (1981) A coherent/Compton scattering method for measurement of trabecular bone mineral density in the distal radius. *Investigative Radiology*, **16**, 491–495.

Olmedo, M.L. and Weiss, A.P. (1994) An experimental rat model allowing controlled delivery of substances to evaluate fracture healing. *Journal of Orthopaedic Trauma*, **8**, 490–493.

Omnell, K.A., Lindstrom, B., Hoh, F.C. and Hammerlund-Essler, E. (1960) Method for non-destructive determination of inorganic and organic material in mineralized tissues. *Acta Radiologica*, **54**, 209.

Onishi, T., Ishidou, Y., Nagamine, T. *et al.* (1998) Distinct and overlapping patterns of localization of bone morphogenetic protein (BMP) family members and a BMP type II receptor during fracture healing in rats. *Bone*, **22**, 605–612.

Ooms, M.E., Roos, J.C., Bezemer, P.D. *et al.* (1995) Prevention of bone loss by vitamin D supplementation in elderly women: a randomized double-blind trial. *The Journal of Clinical Endocrinology and Metabolism*, **80**, 1052–1058.

Oreffo, R.O.C., Wells, N. and Johnstone, D. (1991) Sodium fluoride does not stimulate mammalian osteoblast like cells. *Calcified Tissue International*, **49**, 150.

Orimo, H., Fujita, T. and Yoshikawa, M. (1972) Increased sensitivity of bone to parathyroid hormone in overiectomzed rats. *Endocrinology*, **90**, 760–763.

Orwoll, E.S. and Klein, R.F. (1995) Osteoporosis in men. *Endocrine Reviews*, **16**, 87–116.

Orwoll, E.S., McClung, M.R., Ovialt, S.K. *et al.* (1989) Histomorphometric effects of calcium or calcium plus 25-hydroxyvitamin $D_3$ therapy in senile osteoporosis. *Journal of Bone and Mineral Research*, **4**, 81–88.

Orwoll, E.S. and Nelson, H.D. (1999) Does estrogen adequately protect postmenopausal women against osteoprosis: an icono elastic perspective. *The Journal of Clinical Endocrinology and Metabolism*, **84**, 1872–1874.

Osipchuk, Y. and Cahalan, M. (1992) Cell-to-cell spread of calcium signals mediated by ATP receptors in mast cells. *Nature*, **359**, 241–244.

Ostlere, S.J. and Gold, R.H. (1991) Osteoporosis and bone density measurement methods. *Clinical Orthopaedics and Related Research*, **271**, 149–163.

Ostrom, R.S., Gregorian, C. and Insel, P.A. (2000) Cellular release of and response to ATP as key determinants of the set point of signal transduction pathways. *The Journal of Biological Chemistry*, **16**, 11735–11739.

Otter, M., Goheen, S. and Williams, W.S. (1988) Streaming potentials in chemically modified bone. *Journal of Orthopaedic Research*, **6**, 346–359.

Otter, M.W., MacGinitie, L.A., Seiz, K.G. *et al.* (1994a) Dependence of streaming potential frequency response on specimen thickness. *Biomimetics*, **2**, 57–75.

Otter, M.W., Palmieri, V.R., Wu, D.D. *et al.* (1992) A comparative analysis of streaming potentials *in vivo* and *in vitro*. *Journal of Orthopaedic Research*, **10**, 710–719.

Otter, M.W., Wu, D.D., Bieber, W.A. and Cochran, G.V.B. (1994b) Disuse osteopenia reduces the magnitude and decay time of streaming potentials in living cantine tibia. Transactions of the 40th Annual Meeting of the Orthopaedic Research Society: Orthopaedic Research Society, Rosemount Il, p. 565.

Overgaard, K., Hansen, M.A., Nelsen, V.A.H. *et al.* (1990) Discontinous calcitonin treatment of established osteoporosis–effects of withdrawal of treatment. *The American Journal of Medicine*, **89**(1), 1–6.

Overgaard, K., Riis, B.J., Christiansen, C. and Hausen, M.A. (1989) Effect of calcitonin given intranasally on early postmenopausal bone loss. *British Medical Journal*, **299**, 477–479.

Overgaard, K., Riis, B.J., Christiansen, C. *et al.* (1989) Nasal calcitonin for treatment of established osteoporosis. *Clinical Endocrinology*, **30**(4), 435–442.

Owan, L., Burr, D.B., Turner, C.H. *et al.* (1997) Mechanotransduction in bone: Osteoblasts are more responsive to fluid forces than mechanical strain. *The American Journal of Physiology*, **273**, C810–C815.

Owen, M., Triffitt, J.T. and Melick, R.A. (1973) Albumin in bone, in *Hard Tissue Growth, Repair and Remineralization, Ciba Foundation Symposium*, Elsevier-Excerpta Medica-North Holland-Associated, Science Publishers Amsterdam, No. 11, p. 263.

Oxlund, H., Mosekilde, L. and Ortoft, G. (1996) Reduced concentration of collagen reducible cross-links in human trabecular bone with respect to age and osteoporosis. *Bone*, **19**, 479–484.

Ozawa, Y., Shimizu, N., Kariya, G. and Abiko, Y. (1998) Low energy laser irradiation bone nodule at early stages of cell culture in rat calvarial cells. *Bone*, **22**, 347–354.

Pacifici, R., McMurty, C., Vered, I. *et al.* (1988) Coherence therapy does not prevent axial bone loss on osteoporotic women: a preliminary comparative study. *The Journal of Clinical Endocrinology and Metabolism*, **66**, 747–753.

Pacifici, R., Rifas, L., McCracken, R. *et al.* (1989) Ovarian steroid treatment blocks a postmenopausal increase in blood monocyte interleukin I release. *Proceedings of the National Academy of Sciences of the United States of America*, **86**, 2398–2402.

Pagel, C.N., Niese, M.R., Abraham, L.A. *et al.* (2003) Inhibition of osteoblast apoptosis by thrombin. *Bone*, **33**, 733–743.

Pak, C.Y.C., Sakhaee, K., Zerweph, J.E. *et al.* (1989) Safe and effective treatment of osteoporosis with intermittent slow release sodium fluoride. Augmentation of vertebral bone mass and inhibition of fractures. *The Journal of Clinical Endocrinology and Metabolism*, **68**, 150–159.

Pak, C.Y.C., Sakhaee, K. and Zerweph, J.E. (1990) Effect of intermittent therapy with slow release fluoride preparation. *Journal of Bone and Mineral Research*, **5**(Suppl. 1), 149–155.

Palacios, S., Menendez, C., Calderon, J. and Rubio, S. (1993) Spine and femur density and broadband ultrasound attenuation of the calcaneus in normal Spanish women. *Calcified Tissue International*, **52**, 99–102.

Palmer, H.E., Nelp, W.B., Murano, R. and Rich, C. (1968) The feasibility of *in vivo* neutron activation analysis of total body calcium and other elements of body composition. *Physics in Medicine and Biology*, **13**, 269–279.

Palmer, R.G., Stein, D.L., Abraham, E. and Anderson, P.W. (1984) Models of hierarchically constrained dynamics for glassy relaxation. *Physical Review Letters*, **53m**, 958–961.

Palmer, R.M.J., Ferrige, A.G. and Moncada, S. (1987) Nitric-oxide release accounts for the biological activity of endothelium derived relaxing factor. *Nature*, **327**, 524–526.

Panjabi, M.M., White, A.A. and Southwick, W.O. (1973) Mechanical properties of bone as function of rate of deformation. *The Journal of Bone and Joint Surgery*, **55-A**, 322.

Pansu, D., Bellaton, C. and Bronner, F. (1979) Effect of lactose on duodenal calcium binding protein and calcium absorption. *The Journal of Nutrition*, **109**, 508–512.

Papadimitropoulos, E., Wells, G., Shea, B. and Gillespie, W. (2002) VIII: Meta-analysis of the efficacy of vitamin D treatment in preventing osteoporosis in postmenopausal women. *Endocrine Reviews*, **23**, 560–569.

Papapoulos, S.E., Landman, J.O., Bigvoet, D.L.M. *et al.* (1992) The use of bisphosphonates I the treatment of osteoporosis. *Bone*, **13**, S41–S49.

Paradis, G.R. and Kelley, P.J. (1975) Blood flow and mineral deposition in canine tibial fractures. *The Journal of Bone and Joint Surgery*, **57**, 220–226.

Parfitt, A.M. (1981) Intgegration of a skeletal and mineral homeostatis, in *Osteoporosis: Recent Advances in Pathogenesis and Treatment* (eds H.F. De Luca, H. Frost, W. Jee *et al.*), University Park Press, Baltimore MD, pp. 115–126.

Parfitt, A.M. (1982) The coupling of bone formation to bone resorption: a critical analysis of the concept and of its relevance to the pathogenesis of osteoporosis. *Metabolic Bone Disease & Related Research*, **4**, 1–6.

Parfitt, A.M. (1983) Dietary risk factors for age related bone loss and fractures. *Lancet*, **2**, 1181–1185.

Parfitt, A.M. (1984a) Age related structural changes in trabecular and cortical bone: cellular mechanisms and biomechanical consequences. *Calcified Tissue International*, **36**(Suppl), S123–S128.

Parfitt, A.M. (1984b) The cellular basis of bone remodelling. The quantum concept re-examined in the light of recent advances in cell biology of bone. *Calcified Tissue International*, **36**, S37–S45.

Parfitt, A.M. (1987) Trabecular bone architecture in the pathogenesis and prevention of fracture. *The American Journal of Medicine*, **82**, 68–72.

Parfitt, A.M. (1988) Bone remodeling. Relationship to the amount and structure of bone and pathogenesis and prevention of fractures, in *Osteoporosis Etiology, Diagnosis and Management* (eds B.L. Riggs and L.J. Melton), Raven Press, New York.

Parfitt, A.M. (1990a) Osteomalacia and related disorders, in *Metabolic Bone Disease and Clinically Related Disorders*, 2nd edn (eds L.V. Avioli and S.M. Krane), W.B. Saunders, Philadelphia, PA, pp. 329–396.

Parfitt, A.M. (1990b) Bone forming cells in clinical conditions, in *Bone: A Treatise, Vol 1, The Osteoblast and Osteocyte* (ed. B.K. Hall), Telford Press, New Jersey, pp. 351–429.

Parfitt, A.M. (1990c) Pharmacologic manipulation of bone remodeling and calcium homeostatis, in *Progress in Basic Clinical Pharmacology Vol.4 Calcium Metabolism* (ed. J.A. Kanis), Karger, Basel, pp. 1–27.

Parfitt, A.M. (1991) Use of bisphosphonates in the prevention of bone loss and fractures. *American Journal of Medicine*, **91**, 425–465.

Parfitt, A.M. (1992) Implication of architecture for the pathogenesis and prevention of vertebral fracture. *Bone*, **13**(Suppl.), S41–S47.

Parfitt, A.M. (1993) Bone, age, mineral density and fatigue damage. *Calcified Tissue International*, **53**(5), 582–586.

Parfitt, A.M. (1994) Osteonal and semi-osteonal remodeling: The spatial and temporal framework for signal traffic in adult human bone. *Journal of Cellular Biochemistry*, **55**, 273–286.

Parfitt, A.M. (1996) Skeletal heterogeneity and the purposes of bone remodeling: implications for the understanding of osteoporosis, in *Osteoporosis* (eds R. Marcus, D. Feldman and J. Kelsey), Academic Press, San Diego, pp. 315–324.

Parfitt, A.M. (1996a) Hormonal influences on bone remodeling and bone loss: application to the management of primary hyperparathyroidism. *Annals of Internal Medicine*, **125**(5), 413–415.

Parfitt, A.M. (1996b) Reduced expression of the serpentine calcium receptor. *Journal of Clinical Endocrinology and Metabolism*, **81** (8), 3130–3131.

Parfitt, A.M. (2000) Mini-review – the mechanism of coupling – a role for the vasculature. *Bone*, **26**, 319–323.

Parfitt, A.M. (2001) Skeletal heterogeneity and the purpose of bone remodeling implications for the understanding of osteoporosis, in *Osteoporosis*, 2nd edn (eds R. Marcus, D. Feldman and J. Kelsey), Academic Press, San Diego.

Parfitt, A.M. (2002) Targeted and non-targeted bone remodelling. Relationship to basic multicellular unit organisation and progression. *Bone*, **30**(1), 5–7.

Parfitt, A.M., Chir, B. Gallagher, J.C. *et al.* (1982) Vitamin D and bone health I the elderly. *American Journal of Clinical Nutrition*, **36**, 1014–1031.

Parfitt, A.M. and Folds, J. (1991) The ambiguity of interstitial bone thickness: a new approach to the mechanism of trabecular thinning. *Bone*, **12**, 119–122.

Parfitt, A.M., Mathews, C.H.E., Brommage, R. *et al.* (1984) Calcitriol but no other metabolite of vitamin D is essential for normal bone growth and development in the rat. *The Journal of Clinical Investigation*, **73**, 576–586.

Parfitt, A.M., Mathews, C.H.E., Villanueva, A.R. and Kleerekoper, M. (1983) Relationships between surface, volume and thickness of iliac trabecular bone in aging and in osteoporosis. *The Journal of Clinical Investigation*, **72**, 1396–1409.

Parfitt, A.M., Mundy, G.R., Roodman, G.D. *et al.* (1996) A new model for the regulation of bone resorption, with particular reference to the effects of bisphosphonates. *Journal of Bone and Mineral Research*, **11**, 150–159.

Parfitt, A.M., Podenphant, J., Villaneuva, A.R. and Frame, B. (1985) Metabolic bone disease with and without osteomalacia after interstitial bypass surgery: a bone histomorphometric. *Bone*, **6**, 211–220.

Parfitt, A.M., Villanueva, A.R., Foldes, J. and Rao, D.S. (1995) Relation between histologic indices of bone formation: implications for the pathogenesis of spinal osteoporosis. *Journal of Bone and Mineral Research*, **10**, 466–473.

Parisien, M., Silverberg, S.J., Shane, E. *et al.* (1990) The histomorphometry of bone in primary hyperparathyroidism: prevention of cancellous bone structure. *Journal of Clinical Endocrinology & Metabolism*, **70**, 930–938.

Parrini, L., Gallazzi, M. and Parrini, M. (1987) Whole body mineral density. Correlation with spine mineral content in normal population, in *Osteoporosis 1987* (eds C. Christiansen, J.S. Johansen and B.J. Riis), Osteopres ApS, Kopenhagen.

Parry, R.J. and Chivers, R.C. (1979) Data of the velocity and attenuation of ultrasound in mammalian tissue – a survey. In *Ultrasonic Tissue Characterization* (ed. M. Linzeredt), National Bureau of Standards, pp. 343–360.

Parvianen, M.T., Galloway, J.H., Towers, J.H. and Kanis, J.A. (1988) Alkaline phosphatase isoenzymes in serum determined by high performance anion-exchange liquid chromatography with detection by enzyme reaction. *Clinical Chemistry*, **34**, 2406–2409.

Paschalis, E.P., Betts, F., Di Carlo, E. *et al.* (1997a) FTIR microspectroscopic analysis of normal human cortical and trabecular bone. *Calcified Tissue International*, **61**, 480–486.

Paschalis, E.P., Betts, F., Di Carlo, E. *et al.* (1997b) FTIR microspectroscopic analysis of human iliac crest biopsies from untreated osteoporotic bone. *Calcified Tissue International*, **61**, 487–492.

Pasco, J.A., Henry, M.J., Kotowicz, M.A., Collier, G.R., Ball, M.J., Ugoni, A.M. and Nicholson, G.C. (2001) Serum leptin levels are associated with bone mass in nonobese women. *The Journal of Clinical Endocrinology and Metabolism*, **86**, 1884–1887.

Patel, R., Blake, G.M., Rymer, J. and Fogelman, I. (2000) Long term precision of DXA scanning assessed over seven years in forty postmenopausal women. *Osteoporosis International*, **11**, 68–75.

Paterson, D.C., Carter, R.F., Maxwell, G.M. *et al.* (1977) Electrical bone growth stimulation in an experimental model of delayed union. *Lancet*, **1**, 1278–1281.

Paterson, P.C., Lewis, G.N. and Cass, C.A. (1980) Treatment of delayed union and nonunion with an implanted direct current stimulator. *Clinical Orthopaedics and Related Research*, **148**, 117–128.

Pathania, V., Syal, N., Pathak, C.M. and Khanduja, K. (1999) Vitamin E suppress the induction of reactive oxygen species release by lipopolysaccharide interleukin-1 $\beta$ and tumor necrosis factor $\alpha$ in rat alveolar macrophages. *Journal of Nutritional Science and Vitaminology*, **46**, 675–686.

Patschan, D., Loddenkemper, K. and Butgereit, F. (2001) Molecular mechanisms of glucocorticoid-induced osteoporosis. *Bone*, **29**(6), 498–505.

Pattanaungkul, S., Riggs, B.L. and Yergey, A.L. (2000) Relationship of intestinal calcium absorption to 1,25-dihydroxyvitamin D [1,25 $(OH)_2D$] levels in young versus elderly women: evidence for age related intestinal resistance to 1,25$(OH)_2D$ action. *The Journal of Clinical Endocrinology and Metabolism*, **85**, 4023–4027.

Patterson-Buckendahl, P.E., Grindeland, R.E., Martin, R.B. *et al.* (1985) Osteocalcin as an indicator of bone metabolism during spaceflight. *The Physiologist*, **28**(Suppl.), S227–S228.

Paul, W. and Sharma, C.P. (2001) Porous hydroxyapatite nanoparticles for intestinal delivery of insulin. *Trends in Biomaterials and Artificial Organs*, **14**, 37–38.

Paul, W. and Sharma, P.P. (2006) Nanoceramic matrices: Biomedical applications. *American Journal of Biochemistry and Biotechnology*, **2**(2), 41–48.

Paulraj, R. and Behari, J. (2004) Radiofrequency radiation effects on protein kinase C activity in rats' brain. *Mutation Research*, **545**, 127–130.

Paulraj, R. and Behari, J. (2006) Protein kinase C activity in developing rat brain cells exposed to 2.45 GHz radiation. *Electromagnetic Biology and Medicine*, **25**(1), 61–67.

Pauwels, F. (1976) *Biomechanics of the Normal and Diseased Hip*, Springer-Verlag, Berlin.

Pavalko, F.M., Otey, C.A., Simon, K.O. and Burridge, K. (1991) $\alpha$-Actinin: a direct link between action and integrins. *Biochemical Society Transactions*, **19**, 1065–1069.

Pawluk, R.J. and Becker, R.O. (1964) Effects of electric currents on bone formation *in vivo*. *Nature*, **204**, 652.

Pazzaglia, U.E., Zatti, G., Di Nucci, A. and Cocci, A. (1993) Inhibitory effect of salmon calcitonin on bone resorption: morphological study of the tibial growth plate in rats. *Calcified Tissue International*, **52**, 125–129.

Peacock, M., Liu, G., Carey, M. *et al.* (2000) Effect of calcium or 25OH vitamin D3 dietary supplementation on bone loss at the hip in men and women over the age of 60. *The Journal of Clinical Endocrinology and Metabolism*, **85**, 3011–3019.

Peacock, M. and Nordin, B.E.C. (1973) Plasma calcium homeostasis, in *Hard Tissue Growth, Repair and Remineralization*, Ciba Foundation Symposium, Elsevier-Excrepta Medica-North Holland-Associated, Science Publishers, Amsterdam, No. 11, p. 409.

Pead, M.J. and Lanyon, L.E. (1989) Indomenthacin modulation of load related stimulation of new bone formation *in vivo*. *Calcified Tissue International*, **45**, 34–40.

Pelker, R.R. (1976) Longitudinal Stress Wave Propagation in Long Bone, PhD thesis. Yale University.

Pelker, R.R. and Saha, S. (1975) A theoretical investigation of wave propagation in long bones. In *Advances in Bioengineering* (eds A.C. Bell and R.M. Nerem), American Society of Mechanical Engineers, New York, pp. 98–100.

Pelker, R.P. and Saha, S. (1983) Stress wave propagation in bone. *Journal of Biomechanics*, **16**(7), 481–489.

Pelker, R.R. and Saha, S. (1985) Wave propagation across a bony discontinuity simulating a healing fracture. *Journal of Biomechanics*, **18**, 745–753.

Pelleymounter, M.A., Cullen, M.J., Baker, M.B. *et al.* (1995) Effects of the obese gene product on body weight regulation in ob/ob mice. *Science*, **269**, 540–543.

Peng, H., Chen, S.-T., Wqergedal, J.E. *et al.* (2001) Development of an MFG based retroviral vector system for secretion of high levels of functionally active human BMP4. *Molecular Therapy*, **4**, 95–104.

Peng, Z.-Q., Vaananen, H.K., Zhang, H.X. and Tuukkanen, J. (1997) Long-term effects of ovariectomy on the mechanical properties and chemical composition of rat bone. *Bone*, **20**(3), 207–212.

Pentinnen, R.P., Lichtenstein, J.R., Martin, G.R. and McKusick, V.A. (1975) Abnormal collagen metabolism in cultured cells in osteogenesis imperfecta. *Proceedings of the National Academy of Sciences of the United States of America*, **72**, 586–589.

Peppler, W.W. and Mazess, R.B. (1981) Total body bone mineral and lean body mass by dual-photon absorptiometry. I. Theory and measurement procedure. *Calcified Tissue International*, **33**, 353–359.

Peraldi, M.N., Rondeau, E., Jousset, V. *et al.* (1990) Dexamethasone increases preproparathyroid hormone messenger RNA in human hyperplastic parathyroid cells *in vitro*. *European Journal of Clinical Investigation*, **20**(4), 392–397.

Pernalete, N., Garcia, J.C., Betts, C.R. and Martin, K.J. (1990) Inhibitors of protein kinase-C modulate desensitization of the parathyroid hormone receptor-adenylate cyclase system in opossum kidney cells. *Endocrinology*, **126**, 407–413.

Perren, S.M. (1979) Physical and biological aspects of fracture healing with special reference to internal fixation. *Clinical Orthopaedics and Related Research*, **138**, 175–196.

Perren, S.M. and Cordey, J. (1980) The concept of interfragmentary strain, in *Current Concepts of Internal Fixation of Fractures*, Springer-Verlarg, Berlin, pp. 63–77.

Petito, S.L. and Evans, J. (1984) Calcium status of the growing rat as affected by diet acidity from ammonium chloride, phosphate and protein. *The Journal of Nutrition*, **114**, 1049–1059.

Petrov, N. (1975) Electromechanical interaction in physiologic wet bone. *Biomechanics, Bulgarian Academy of Sciences*, **2**, 31.

Petrov, N., Pollack, S. and Blagoeva, R. (1989) A discrete model for streaming potentials in a single osteon. *Journal of Biomechanics*, **22**, 517–521.

Pezzetti, F., DeMathei, M., Caruso, A. *et al.* (1999) Effects of pulsed electromagnetic fields on human chondrocytes: An *in vitro* study. *Calcified Tissue Research*, **65**, 396–401.

Pfeifer, M., Pollaehne, W. and Minne, H.W. (1997) Ultrasound analyses of the calcaneus predict relative risk of the presence of at least one vertebral fracture and reflect different physical qualities of bone in different regions of the skeleton. *Hormone and Metabolic Research*, **29**, 76–79.

Pfeiffer, B. (1977) Local piezoelectric polarisation of a human critical bone as a function of stress frequency. *Journal of Biomechanics*, **10**, 53.

Pfeilschifter, J., Mundy, G.R. and Roodman, G.D. (1989) Interleukin-1 and tumor necrosis factor stimulate the formation of human osteoclast like cells *in vitro*. *Journal of Bone and Mineral Research*, **4**, 113–118.

Phelps, M.E., Hoffman, E.J. and Ter-Pogossian, M.M. (1975) Attenuation coefficients of various body tissues. Fluids and lesions at photon energies of 18 to 136 kev. *Radiology*, **117**, 573–584.

Phillips, J., Haggren, W., Thomas, W.J. *et al.* (1992) Magnetic field-induced changes in specific gene transcription. *Biochimica et Biophysica Acta*, **1132**, 140–144.

Phillips, S., Fox, N., Jacobs, J. and Wright, W.E. (1988) The direct medication cost of osteoporosis for American women aged 45 and older. *Bone*, **9**, 271–279.

Piekarski, K. (1973) Analysis of bone as a composite material. *International Journal of Engineering Science*, **11**, 557.

Piekarski, K. and Munro, M. (1977) Transport mechanism operating between blood supply and osteocytes in long bones. *Nature*, **269**, 80–82.

Pienkowski, D. and Pollack, S.R. (1983) The origin of stress generated potentials in fluid saturated bone. *Journal of Orthopaedic Research*, **1**, 30–41.

Pierson, D. and Moss, F. (1995) Detecting periodic unstable points in noisy chaotic and limit cycle attractors with applications to biology. *Physical Review Letters*, **75**, 2124–2127.

Pietruszczak, S., Inglis, D. and Pande, G.J. (1999) A fabric dependent fracture criterion for bone. *Journal of Biomechanics*, **32**(10), 1071–1079.

Pilla, A.A. (1972) Electrochemical information and energy transfer *in vivo*. Proceedings, 7th Intersociety Energy Conversion Conference, San Diego, CA, pp. 761–765.

Pilla, A.A. (1993) State of the art in electromagnetic therapeutics, in *Electricity and Magnetics in Biology and Medicine* (ed. M. Blank), San Fransisco Press, San Fransisco, pp. 17–22.

Pilla, A.A. (2002) Low-intensity electromagnetic and mechanical modulation of bone growth and repair: are they equivalent? *Journal of Orthopaedic Science*, **7**, 420–428.

Pilla, A.A., Figueiredo, M., Nasser, P. *et al.* (1991) Acceleration of bone repair by pulsed sine wave ultrasound: animal, clinical and mechanistic studies, in *Electromagnetics in Biology and Medicine* (eds C.T. Brighton and S.R. Pollack), San Francisco Press, San Francisco, pp. 331–341.

Pilla, A.A., Mont, M.A., Nasser, P.R. *et al.* (1990) Non-invasive low intensity pulsed ultrasonic accelerates bone healing in the rabbit. *Journal of Orthopaedic Trauma*, **4**, 246–253.

Pilla, A.A., Nasser, P.R. and Kaufman, J.J. (1992) The sensitivity of cells and tissues to weak electromagnetic fields, in *Charge and Field Effects in Biosystems*, vol. **3**(eds M.J. Allan, S.F. Cleary, A.E. Swer *et al.*), Birkhauser, Boston, pp. 23–41.

Pilla, A.A., Nasser, P.R. and Kaufman, J.J. (1993) On the sensitivity of cells and tissue to therapeutic and environmental electromagnetic fields. *Bioelectrochemistry and Bioenergetics*, **30**, 161–166.

Pilla, A.A., Nasser, P.R. and Kaufman, J.J. (1994) Gap junction impedance, tissue dielectrics and thermal noise limits for electromagnetic field bioeffects. *Bioelectrochemistry and Bioenergetics*, **435**, 63–69.

Pineri, M.H., Escoubes, M. and Roche, G. (1978) Water collagen interactions: calorimetric and mechanical experiments. *Biopolymers*, **17**, 2799–2815.

Plotinikoff, G.A. and Quigley, J.M. (2003) Prevalence of severe hypovitaminosis D in patients with persistent, nonspecific musculoskeletal pain. *Mayo Clinic Proceedings*, **78**, 1463–1470.

Plotkin, L.I., Weinstein, R.S., Parfitt, A.M. *et al.* (1999) Prevention of osteocyte and osteoblast apoptosis by bisphophonates and calcitonin. *The Journal of Clinical Investigation*, **104**, 1263–1274.

Plotnick, R.E., Gardner, R.H., Hargrove, W.H. *et al.* (1996) Lacunarity analysis: A general technique for the analysis of spatial patterns. *Physical Review E*, **53**, 5461–5468.

Podbesek, R., Edouard, C., Meunier, P.J. *et al.* (1983) Effects of two treatment regimes with synthetic human PTH fragment on bone formation and tissue balance of trabecular bone in greyhounds. *Endocrinolgy*, **112**, 1000–1006.

Pollack, S.R. (1984) Bioelectrical properties of bone. *Orthopaedics Clinics of North America*, **15**, 3–14.

Pollack, S.R., Korostoff, E., Sternberg, M.E. and Koh, J. (1977) Stress generated potentials in bone: Effects of collagen modification. *Journal of Biomedical Materials Research*, **2**, 677–700.

Pollack, S.R., Petrov, N., Salzstein, R. *et al.* (1984) An anatomical model for streaming potentials in osteons. *Journal of Biomechanics*, **17**, 627–636.

Pols, M.A.P., Felsenberg, D., Hanley, D.A. *et al.* (1999) Multinational placebo-controled, randomized trial of the effects of alendronate on bone density and fracture risk in postmenopausal women with low bone mass. Results of the FOSIT Study. *Osteoporosis International*, **9**, 461–468.

Polyakov, V., Nikolaeva, G., Volkov, M. *et al.* (1974). In *Ultrasonic Bonding of Bones and Cutting of Live Biological Tissues*. Mir. Publishers, U.S.S.R.

Pommer, G. (1885) *Untersuchugen Uber Osteomalacie und Rachitis*, Vogel, Leipzig.

Pontiroli, A.E. *et al.* (1986) Human calcitonin administered by the nasal route: biovariability of different formulations and efficacy in postmenopausal osteoporosis. *Life Sciences*, **125**, 249–253.

Pope, M.H. and Outwater, J.O. (1974) Mechanical properties of bone as a function of position and orientation. *Journal of Biomechanics*, **7**, 61.

Pope, N.S., Gould, K.G., Anderson, D.C. and Mann, D.R. (1989) Effects of age and sex on bone density in the Rhesus monkey. *Bone*, **10**, 109–112.

Posner, A.S. (1970) Significance of calcium phosphate crystallographic studies to orthopaedics. *Bulletin of the Hospital for Joint Diseases*, **31**, 14–26.

Posner, A.S. and Betts, F. (1975) Synthetic amorphous calcium phosphate and its relationship to bone mineral structure. *Accounts of Chemical Research*, **8**, 273–278.

Posner, A.S., Betts, F. and Blumenthal, N.C. (1979) Bone mineral composition and structure, in *Skeletal Research*, Academic Press, New York, p. 167.

Posner, A.S., Eanes, E.D., Harper, R.A. and Zipkin, I. (1963) X-ray diffraction analysis of the effect of fluoride on human bone apatite. *Archives of Oral Biology*, **8**, 549.

Posner, A.S., Perloff, A. and Diorio, A.F. (1958) Refinement of the hydroxyapatite structure. *Acta Crystallographica*, **11**, 308.

Pothuaud, L., Van Riethergen, B., Mosekilde, L. *et al.* (2002) Combination of topological parameters and bone volume fraction better predicts the mechanical properties of trabecular bone. *Journal of Biomechanics*, **35**, 1091–1099.

Potts, J.T. (1980) Disorders of parathyroid glands, in *Harrison Principles of Internal Medicine*, 9th edn (eds K.J. Isselbacher, R.D. Adams, E. Braunwald *et al.*), McGraw-Hill, New York, pp. 1832–1834.

Pouilles, J.M., Tremolleres, F. and Ribot, C. (1991) Fluoride therapy in postmenopausal osteopathic women. Effect on vertebral and femoral bone density and prediction of bone response. *Osteoporosis International*, **1**, 103–109.

Pratt, G.W. (1980) Proceedings of the 26th Annual Convention of the American Association of Equine Practitioners, p. P295.

Prendergast, P.J., Huiskes, R. and Soballe, K. (1997) Biophysical stimuli on cells during tissue differentiation at implant interfaces. *Journal of Biomechanics*, **6**, 539–548.

Prestwood, K.M. (1994) The short term effects of conjugated estrogen on bone turnover in older women. *The Journal of Clinical Endocrinology and Metabolism*, **79**, 366–371.

Prestwood, K.M., Kenny, A.M., Kleppinger, A. and Kulldozift, M. (2003) Ultralow dose micronized 17 beta-estradiol and bone density and bone metabolism in older women. A randomized controlled trial. *The Journal of the American Medical Association*, **290**, 1042–1048.

Prestwood, K.M., Pilbeam, C.C., Burleson, J.A. *et al.* (1994) The short term effects of conjugated oestrogens on bone turnover in older women. *The Journal of Clinical Endocrinology and Metabolism*, **79**, 366–371.

Pribluda, V., Laub, F. and Rotman, R. (1981) The state of actin in activated human platelets. *European Journal of Biochemistry*, **116**, 293–296.

Price, J.S., Jackson, B., Eastell, R. *et al.* (1995) Age related changes in biochemical markers of bone metabolism in horses. *Equine Veterinary Journal*, **27**(3), 201–207.

Price, P., Williamson, M.K., Haba, T. *et al.* (1982) Excessive mineralization with growth plate closure in rats on chronicwarfarin treatment, *Proceedings of the National Academy of Sciences USA*, **79**(24), 7734–7738.

Price, P.A. and Bankol, S.A. (1980) 1,25-Dihydroxy vitamin $D_3$ increases synthesis of the vitamin K-dependent bone protein by osteosarcoma cells. *The Journal of Biological Chemistry*, **255**, 11660–11663.

Price, P.A., June, H.H., Buckley, J.R. and Williamson, M.K. (2001) Osteoprotegerin inhibits artery calcification induced by warfarin and by osteoporosis and vitamin D. *Arteriosclerosis, Thrombosis, and Vascular Biology*, **21**(10), 1610–1606.

Price, R.R., Wagner, J., Larsen, K. *et al.* (1976) Regional and whole body bone mineral content measurement with a rectilinear scanner. *American Journal of Roentgenology*, **126**, 1277–1278.

Prince, R.L., Smith, M., Dick, I.M. *et al.* (1991) Prevention of postmenopausal osteoporosis. A comparative study of exercise, calcium supplementation and hormone replacement therapy. *The New England Journal of Medicine*, **325**, 1189–1195.

Prins, S.H., Jorgensen, H.L., Jorgensen, L.V. and Hasager, C. (1998) The role of quantitative ultrasound in the assessment of bone: a review. *Clinical Physiology*, **18**(1), 3–17.

Printchard, J.E., Nowson, C.A. and Wark, J.D. (1996) Bone loss accompanying dietary induced or exercise induced weight loss. A randomized controlled study. *International Journal of Obesity and Related Metabolic Disorders*, **20**, 513–520.

Pritchard, J.J. (1972) General histology of bone, in *The Biochemistry and Physiology of Bone* (ed. G.H. Bourne), Academic Press, New York, pp. 1–20.

Prockop, D.J. (1992) Mutations in collagen genes as a cause of connective tissue diseases. *The New England Journal of Medicine*, **326**, 540–546.

Protopappas, V.C., Baga, D.A., Fotiadis, D.I. *et al.* (2005) An ultrasound wearable system for the monitoring and acceleration of fracture healing in long bones. *IEEE Transactions on Biomedical Engineering*, **52**, 1597–1608.

Protopappas, V.C., Fotiadis, D.I. and Malozos, K.N. (2006) Guided ultrasound wave propagation in intact and healing long bones. *Ultrasound in Medicine and Biology*, **32**, 693–708.

Pullan, B.R. and Roberts, T.E. (1978) Bone mineral measurement using an EMI scanner and standard methods: A comparative study. *British Journal of Radiology*, **51**, 24–28.

Puumalainen, P., Olkkonen, H. and Sikanen, P. (1977) Assessment of fat content of liver by a photon scattering technique. *International Journal of Applied Radiation and Isotopes*, **28**, 785–787.

Puumalainen, P., Uimarihuhta, A., Alhava, E. and Olkkonen, H. (1976) A new photon scattering method for bone mineral density measurement. *Radiology*, **120**, 723–724.

Puumalainen, P., Uimarihuhta, A., Olkkonen, H. and Alhava, E.M. (1982) A coherent/Compton scattering method employing an X-ray tube for measurement of trabecular bone mineral content. *Physics in Medicine and Biology*, **27**, 425–429.

Qin, L., Au, S.K., Choy, Y.W. *et al.* (2002) Regular Tai. Chi. Chuan exercise may retard bone loss in postmenopausal women: A case control study. *Archives of Physical Medicine and Rehabilitation*, **83**, 1355–1359.

Qureshi, A.M., McGuigan, F.E., Seymour, D.G. *et al.* (2001) Association between COLIAI Spl alleles and femoral neck geometry. *Calcified Tissue International*, **69**, 67–72.

Rabin, D.S., Rantanen, N.W., Seder, J.A. *et al.* (1983) Proceedings of the 28th Annual Convention of the American Association of Equine Practitioner, p. P343.

Rader, J.E., Baylink, D.J., Hughes, M.R. *et al.* (1979) Calcium and phosphorous deficiency in rats. Effects on PTH and 1,25-dihydroxy vitamin $D_3$. *The American Journal of Physiology*, **236**, E118–E122.

Radin, E., Orr, R., Kelman, J. *et al.* (1982) Effect of prolonged walking on concrete on the knees of sheep. *Journal of Biomechanics*, **15**, 487–492.

Rai, D.V. (1983) Ontogeny of Behavior of Calcium and Phosphorus Deficient rats and Bio-electrical Bone Activity, PhD Thesis submitted to Jawaharlal Nehru University, New Delhi.

Rai, D.V. and Behari, J. (1986) Biophysical characterization of osteoporotic bone. *Environmental Research*, **40**, 68–83.

Rai, D.V., Behari, J. and Saha, S. (1986) The effect of mineral deficient diet on the structural and mechanical properties of lung bones, in *Biomedical Engineering. V. Recent Developments*, Academic Press, Orlando, pp. 456–460.

Raisz, L.G. (1988) Local and systemic factors in the pathogenesis of osteoporosis. *The New England Journal of Medicine*, **318**, 818–828.

Raisz, L.G. (2003) The hunting of the snark: the elusive calcium receptor(s). *The Journal of Clinical Investigation*, **111**, 945–947.

Raisz, L.G. (2005) Pathogenesis of osteoporosis: concepts, conflicts, and prospects. *The Journal of Clinical Investigation*, **115**, 3318–3325.

Raisz, L.G. and Fall, P.M. (1990) Biphasic effect of prostaglandin $E_2$ on bone formation in cultured fetal rat calvariae: interaction with cortisol. *Endocrinolgy*, **126**(3), 1654–1659.

Raisz, L.G., Kream, B.E., Smith, M.D. and Simmons, H.A. (1980) Comparison of the effects of vitamin D metabolites on collagen synthesis and resorption of fetal rat bone in organ culture. *Calcified Tissue International*, **32**, 135–138.

Raisz, L.G., Manina, D.M., Gwovek, S.C. *et al.* (1978) Hormonal control of bone collagen synthesis *in vitro*: inhibitory effect of 1-hydroxylated vitamin D metabolites. *Endocrinology*, **102**, 731–735.

Raisz, L.G., Trummel, C.L., Holick, M.F. and Deluca, H.F. (1972) 1,25-Dihydroxycholecalciferol: a potent stimulator of bone resorption in tissue culture. *Science*, **175**, 768–769.

Raisz, L.G., Wietrich, W.J., Simmons, H.A. *et al.* (1977) Effect of prostaglandin endoperoxides and metabolism on bone resorption *in vitro*. *Nature*, **367**, 532–534.

Ralston, S.H., Ho, L.P., Helfrich, M.H. *et al.* (1995) Nitric oxide – A cytokine induced regulator of bone resorption. *Journal of Bone and Mineral Research*, **10**, 1040–1049.

Ralston, S.H., Todd, D., Helfrich, M. *et al.* (1994) Human osteoblast like cells produce nitric oxide and express inducible nitric oxide synthase. *Endocrinology*, **135**, 330–336.

Ramachandran, G.N. (1963) Molecular structure of collagen, in *International Review Cohn. Tissue Research 1* (ed. D.A. Hall), Academic Press, London, and New York, pp. 127–182.

Ramachandran, G.N. (1967) The structure of collagen at molecular level, in *Treatise on Collagen*, vol. **1**, Academic Press, London, pp. 103–183.

Ramachandran, G.N. (1968) Molecular architecture of collagen. *The Journal of the American Leather Chemists Association*, **63**(3), 161–191.

Rambout, P.C. and Goode, A.W. (1985) Skeletal changes during space flight. *Lancet*, **2**, 1050–1052.

Ramirez, A., Schwane, J.A., McFarland, C. and Starcher, B. (1997) The effect of ultrasound on collagen synthesis and fibroblast proliferation *in vitro*. *Medicine & Science in Sports & Exercise*, **29**, 326–332.

Ramnaraine, M., Pan, W. and Clonisy, D.R. (2006) Osteoclasts direct bystander killing of cancer cells *in vitro*. *Bone*, **38**, 4–12.

Ramp, W.K., Toverud, S.V. and Boass, A. (1986) Effects of vitamin D deprivation and 1,25 –dihydroxy vitamin D3 treatment on ion release from rat tibia *in vitro*. *Bone and Mineral*, **1**, 1–3.

Rand, J.A., An, K.N., Chao, E.Y.S. and Kelly, P.J. (1981) A comparison of the effect of new intramedullary nailing and compression plate fixation of fracture site blood flow and future union. *The Journal of Bone and Joint Surgery*, **63**(3), 427–442.

Randell, K.M., Honkanen, R.J., Kroger, H. and Saarikoski, S. (2002) Does hormone replacement therapy prevent fractures in early postmenopausal women? *Journal of Bone and Mineral Research*, **17**, 528–533.

Rao, A.D., Reddy, S. and Rao, D.S. (2000) Is there a difference between right and left femoral bone density? *Journal of Clinical Densitometry*, **3**, 57–61.

Rao, V.V., Saha, S., Malakonok, V. and Albright, J.A. (1982) Quantitative measurement of fracture healing by ultrasound, in *Proceedings of the 1st Southern Biomedical Engineering Conference* (ed. S. Saha), Pergamon Press, Oxford.

Rasmussen, H. and Bordier, P. (1974) *The Physiological and Cellular Basis of Metabolic Bone Disease*, Williams and Wilkins, Baltimore.

Rauber, A.A. (1987) *Elasticitat und Festigkeit Der Knocken*, Engeimann, Leipzig.

Rauch, F., Lauzier, D., Croteau, S. *et al.* (2000) Temporal and spatial expression of bone morphogenetic protein-2, -4 and –7 during distraction osteogenesis in rabbits. *Bone*, **26**, 611–617.

Ravel, J.P., Yancey, S.B., Meyer, D.J. and Nieholson, B. (1980) Cell junctions and intercellular communication. *In Vitro*, **16**, 1010.

Ravn, P., Cizza, G., Bjarnason, N.H. *et al.* (1999a) For the early postmenopausal intervention cohort (EPIC) study group. Low body mass index is an important risk factor for low bone mass and increased bone loss in early postmenopausal women. *Journal of Bone and Mineral Research*, **14**, 1622–1627.

Ravn, P., Clemmesen, B., Christiansen, C. and Alendronate osteoporosis prevention study group (1999) Biochemical markers can predict the response in bone mass during alendronate treatment I early postmenopausal women. *Bone*, **24**, 237–244.

Ravn, P., Hosking, D., Thompson, D. *et al.* (1999b) Monitoring of alendronate treatment and prediction of effect on bone mass by biochemical markers I the Early Postmenopausal Intervention Cohort Study. *The Journal of Clinical Endocrinology and Metabolism*, **84**, 2363–2368.

Ray, N.F., Chan, J.K., Thamer, M. and Melton, L.J.L. (1997) Medical express the treatment of osteoporotic fractures in the United Status: report from the National Osteoporosis Foundation. *Journal of Bone & Mineral*, **12**, 24–35.

Ray, S. and Behari, J. (1985) Physiological behaviour of rats due to low level microwave exposure., XXXI Annual APPI Conference held at Calcutta University. (December 1985), 14 (3, 4), 153.

Ray, S. and Behari, J. (1986) Electrical conduction in bone. *Biomat Med Devices Artif Organs*, **14**(34), 153.

Ray, S. and Behari, J. (1987) Ultrasound attenuation in biological tissues using a bone transducer. *Medical and Biological Engineering and Computing*, **25**, 689.

Ray, S. and Behari, J. (1988) Electrical parameters of bone substrate in microstrip configuration. *Biomaterials*, **9**(192), 24.

Ray, W.A., Griffin, M.R., Schaffner, W. *et al.* (1987) Psychotropic drug use and the rise of hip fractures. *The New England Journal of Medicine*, **316**, 363–369.

Raymakers, J.A., Van Dijke, C.F., Hoekstra, A. and Duursma, S.A. (1987) Monitoring fluoride therapy in osteoporosis by dual photon absorptiometry. *Bone*, **8**, 143–148.

Recker, R.R. and Heaney, R.P. (1985) The effect of milk supplements on calcium metabolism, bone metabolism and calcium balance. *The American Journal of Clinical Nutrition*, **41**, 254–263.

Recker, R.R., Kimmel, D.N., Parfitt, A.M. *et al.* (1988) Static and tetracycline based histomorphometric data from 34 normal postmenopausal females. *Journal of Bone and Mineral Research*, **3**, 133–144.

Recker, R.R., Saville, P.D. and Heaney, R.P. (1977) Effect of estrogens and calcium carbonate on bone loss in postmenopausal women. *Annals of Internal Medicine*, **87**, 649–655.

Reddi, A.H. (1985) Age dependent decline in extracellular matrix induced local bone differentiation. *Israel Journal of Medical Sciences*, **21**, 312–323.

Reddi, A.H. (1995) Bone morphogenetic proteins, bone marrow stromal cells and mesenchymal stem cells. *Clinical Orthopaedics*, **313**, 115–119.

Reddi, A.H. and Sullivan, N.I. (1980) Matrix induced endochondral bone differentiation: influence of hypophysectomy, growth hormone and thyroid –stimulating hormone. *Endocrin.*, **107**, 1291–1299.

Reddy, G.H. and Saha, S. (1984) Electrical and dielectric properties of wet bone as a function of frequency. *IEEE Transactions on Biomedical Engineering*, **31**, 296–303.

Reeve, J. (1986) A stochastic analysis of iliac trabecular bone dynamics. *Clinical Orthopaedics*, **213**, 264–278.

Reeve, J., Arlot, M., Bernat, M. *et al.* (1991) Calcium-47 kinetic measurements of bone turnover compared to bone histomorphometry in osteoporosis: the influence of human parathyroid fragment (hPTH 1–34) therapy. *Metabolic Bone Disease & Related Research*, **3**, 23–30.

Reeve, J., Bradbeer, J.N., Arlot, M. *et al.* (1990) Treatment of osteoporosis with human parathyroid peptide and observation on effect of sodium fluoride. *British Medical Journal*, **301**, 314–318.

Reeve, J., Williams, D., Hesp, R. *et al.* (1976) Anabolic effect of low doses of a fragment of human parathyroid hormone on the skeletal in postmenopausal osteoporosis. *Lancet*, **1**, 1035–1038.

Reginster, J.Y. (1992) Management of high turnover osteoporosis with calcitonin. *Bone*, **13**, 537–540.

Reginster, J.Y., Albert, A., Lecard, M.P. *et al.* (1987) One year controlled randomized trial of preventing of early postmenopausal bone loss by intranasal calcitonin. *Lancet*, **2**, 1481–1483.

Reginster, J.Y., Lecart, M.P., Ceroisy, R. *et al.* (1989) Prevention of postmenopausal bone loss in tiludronate. *Lancet*, **1**, 1469–1471.

Reginster, J.Y., Minne, H.W., Sorensen, O.H. *et al.* (2000) Randomized trial of the effects of risedronate on vertical fractures in women with established postmenopausal osteoporosis. *Osteoporosis International*, **11**, 83–91.

Reher, P., Doan, N., Bradnock, B. *et al.* (1999) Effect of ultrasound on the production of IL-8, basic FGF and VEGF. *Cytokine*, **11**, 416–423.

Reher, P., Elbeshir, E.N.I., Harvey, W. *et al.* (1997) The stimulation of bone formation *in vitro* by therapeutic ultrasound. *Ultrasound in Medicine and Biology*, **23**, 1251–1258.

Reher, P., Harris, M., Whiteman, M. *et al.* (2002) Ultrasound stimulates nitric oxide and prostaglandin $E_2$ production by human osteoblats. *Bone*, **31**(1), 236–241.

Reich, K.M., Gay, C.V. and Frangos, J.A. (1990) Fluid shear stress as a mediator of osteoblast cyclic adenosine monophosphate production. *Journal of Cellular Physiology*, **143**, 100–104.

Reich, N.E., Seidelmann, F.E., Tubbs, R.R. *et al.* (1976) Determination of bone mineral content using Ct scanning. *American Journal of Roentgenology*, **127**, 593–594.

Reid, I.R. (1989) Steroid osteoporosis. *Calcified Tissue International*, **45**, 63–67.

Reid, I.R. (1997) Glucocorticoid osteoporosis-mechanisms and management. *European Journal of Endocrinology/European Federation of Endocrine Societies*, **137**(3), 209–217.

Reid, I.R. (2004) Leptin deficiency lessons in regional differences in the regulation of bone mass. *Bone*, **34**, 369–371.

Reid, I.R. (2005) Relationships among body mass, its components and bone. *Bone*, **31**(5), 547–555.

Reid, I.R., Ames, R., Evans, M.C. *et al.* (1992) Determinants of total body regional bone mineral density in normal postmenopausal women – A key role for fat mass. *Journal of Clinical Endocrinology*, **75**, 45–51.

Reid, I.R., Ames, R.W., Orr-Walker, B.J. *et al.* (2000) Hydrochlorothiazide reduces loss of cortical bone in normal postmenopausal women: a randomized controlled trial. *The American Journal of Medicine*, **109**, 362–370.

Reid, I.R., Evans, M.C. and Ames, R.W. (1994) Volumetric bone density of the lumbar spine is related to fat mass but not lean mass in normal postmenopausal women. *Osteoporosis International*, **4**, 362–367.

Reid, I.R. and Ibbertson, H.K. (1987) Evidence for decreased tubular reabsorption of calcium in glucocorticoid treated asthmatics. *Hormone Research*, **27**(4), 200–204.

Reilly, D.T. and Burnstein, A.H. (1975) The elastic and ultimate properties of compact bone tissue. *Journal of Biomechanics*, **8**, 393–405.

Reimer, C.J. and Dixon, S.J. (1992) Extracellular nucleotides elevate $Ca^{2+}$ in rat osteoblastic cells by interaction with two receptor subtypes. *The American Journal of Physiology*, **263**, C1040–C1048.

Reinish, G.B. (1974) Dielectric and Piezoelectric Properties of Bone as Functions of Moisture Content, PhD Thesis, Columbia University.

Reinish, G. (1975) Piezoelectric properties of bone as functions of moisture content. *Nature*, **253**, 626–627.

Reinish, G.B. and Nowick, A.S. (1974) Piezoelectric properties of wet bone. Proceedings 27th ACEMB, Philadelphia, PA, p. 288

Reinish, G.B. and Nowick, A.S. (1975) Piezoelectric properties of bone as function of moisture content. *Nature*, **253**, 626–627.

Reinish, G.B. and Nowick, A.S. (1976) Effect of moisture on the electrical and piezoelectric properties of bone. *Journal of The Electrochemical Society*, **123**, 1451.

Reinish, G.B. and Nowick, A.S. (1979) A model for dielectric behavior of wet bone, in *Electrical Properties of Bone and Cartilage* (eds C.T. Brighton *et al.*), Grune and Stratton, New York, pp. 13–29.

Reiss, E., Canterbury, J.M., Bercoviz, M.A. and Kaplan, E.L. (1970) The role of phosphate in the secretion of parathyroid hormone in man. *The Journal of Clinical Investigation*, **49**, 2146–2149.

Rejnmark, L., Vestergaard, P., Pedersen, A.R. *et al.* (2003) Dose effect relations of loop and thiazide diureties on calcium homeostasis: a randomized, double blinded Latin Square multiple crossover study in postmenopausal osteopenic women. *European Journal of Clinical Investigation*, **33**, 41–50.

Rejnmark, L., Vestergaard, P. and Mosekilde, L. (2005) Reduced fracture risk in users of thiazide diureties. *Calcified Tissue International*, **76**, 167–175.

Remedios, A. (1999) Bone and bone healing. *Veterinary Clinics of North America: Small Animal Practice*, **29**, 1029–1044.

Remy, F. and Guillot, G. (1998) Trabecular bone characterization with low field MRI. *Mag. Res. Imaging*, **16**, 639–642.

Repachole, M.H., Woodcock, J.P., Newman, D.L. and Taytor, K.J.W. (1971) Interaction of low intensity ultra sound and ionizing radiation with tumor cell surface. *Physics in Medicine and Biology*, **16**, 221–227.

Resch, H., Libanati, C., Farley, S. *et al.* (1993) Evidence that fluoride therapy increases trabecular bone density in a peripheral skeletal site. *The Journal of Clinical Endocrinology and Metabolism*, **76**, 1622–1624.

Reseland, J.E., Syversen, U., Bakke, L. *et al.* (2001) Leptin expressed in secreted from primary cultures of human osteoblasts and promotes bone mineralization. *Journal of Bone and Mineral Research*, **16**, 1426–1433.

Revoltella, R.P., Trombi, L., Petrini, M. *et al.* (1993) Low frequency electromagnetic fields do not affect cell growth erythroid differentiation and virus production in variant lines of untreated and dimethyl sulfoxide treated erythroleukemia cell. *Electro and Magnetobiology*, **12**, 135–146.

Reynolds, J.J., Pavlovitch, H. and Valsan, S. (1976) 1,25-Dihydroxy cholecalciferol increases bone resorption in thyroparathyroidectomized mice. *Calcified Tissue International*, **21**, 207–212.

Rho, Y.-T., Kuhn Spearing, L. and Zioupos, P. (1998) Mechanical properties and the hierarchical structure of bone. *Medical Engineering & Physics*, **20**, 92–102.

Riancho, J.A., Salas, E., Zarrabeitia, M.T. *et al.* (1995a) Expression and functional role of nitric oxide synthase in osteoblast like cells. *Journal of Bone and Mineral Research*, **10**, 439–446.

Riancho, J.A., Zarrabeitia, M.T., Fernandezluna, J.L. and Gonzalezmacias, J. (1995b) Mechanisms controlling nitric oxide synthesis in osteoblasts. *Molecular and Cellular Endocrinology*, **107**, 87–92.

Ricciarelli, R., Zingg, J. and Azzi, A. (2001) Vitamin E: protective role of a Janus molecule. *The FASEB Journal*, **15**, 2314–2325.

Rice, J.C., Cowin, S.C. and Bowman, J.A. (1988) On the dependence of the elasticity and strength of cancellous bone on apparent density. *Journal of Biomechanics*, **21**, 155–168.

Rich, C., Klinik, E., Smith, R. and Graham, B. (1966) Measurement of bone mass from ultrasonic transmission time. *Proceedings of the Society for Experimental Biology and Medicine*, **123**, 282–285.

Richards, A.M., Mosekilde, L. and Sogaard, C.H. (1994) Normal age related changes in fluoride content of vertebral trabecular bone relation to bone quality. *Bone*, **15**(1), 21–26.

Richards, A.M., Nicholls, M.G., Ikkram, H. *et al.* (1985) Renal, haemodynamic and hormonal effects of human alpha atrial natriuretic peptide in healthy volunteers. *Lancet*, **1**, 545–549.

Richez, J. Chanya, A, and Bielerluc (1972) Bone changes due to pulses of direct electric microcurrent. *Virchows Archive, Abstract A Path. Anatomy*, **A357**(1), 11–18.

Richy, F., Schacht, E., Bruyere, O. *et al.* (2005) Vitamin D analogs versus native vitamin D in preventing bone loss and osteoporosis related fractures: A comparative meta-analysis. *Calcified Tissue International*, **76**, 176–186.

Rico, H. (1997) The therapy of osteoporosis and the importance of cortical bone. *Calcified Tissue International*, **61**, 431–432.

Rico, H., Hernavidez, E.R., Revilla, M. *et al.* (1993a) Bone change in postmenopausal Spanish women. *Calcified Tissue International*, **52**, 103–106.

Rico, H., Hernavidez, E.R., Revilla, M. *et al.* (1993b) Biochemical markers of nutrition in osteoporosis. *Calcified Tissue International*, **52**, 331–333.

Rico, H., Revilla, M., Hernandez, E.R. *et al.* (1992) Sex differences in the acquisition of total bone mineral mass peak assessed through dual-energy X-ray absorptiometry. *Calcified Tissue International*, **51**, 251–254.

Riggs, B.L. (1991) Treatment of osteoporosis with sodium fluoride or parathyroide hormone. *The American Journal of Medicine*, **91** (Suppl. 5B), 375–415.

Riggs, B.L. (2000) The mechanism of estrogen regulation of bone resorption (2000). *The Journal of Clinical Investigation*, **106**(10), 1203–1204.

Riggs, B.L., Hodgson, S.F., Hoffman, D.L. *et al.* (1980) Treatment of primary osteoporosis with fluoride and calcium, clinical tolerance and fracture occurrence. *The Journal of the American Medical Association*, **243**, 446–449.

Riggs, B.L., Hodgson, S.F., O'Fallon, W.M. *et al.* (1990) Effect of fluoride treatment on the fracture rate in postmenopausal women with osteoporosis. *The New England Journal of Medicine*, **322**, 802–809.

Riggs, B.L., Jowsey, J., Kelley, P.J. *et al.* (1969) Effect of sex hormones on bone in primary osteoporosis. *The Journal of Clinical Investigation*, **48**, 1065–1072.

Riggs, B.L., Jowsey, J., Kelly, P.J. *et al.* (1976) Effects of oral therapy with calcium and vitamin D in primary OSC osteoporosis. *The Journal of Clinical Endocrinology and Metabolism*, **42**, 1139–1144.

Riggs, B.L., Khosla, S. and Melton, I.J. III (1998) A unitary model for type I and type II osteoporosis in postmenopausal women and contributes to bone loss in aging men. *Journal of Bone and Mineral Research*, **13**, 763–773.

Riggs, B.L., Khosla, S. and Melton, L.J. III (2002) Sex steroid and the construction and conservation of the adult skeleton. *Endocrine Reviews*, **23**, 279–302.

Riggs, B.L. and Melton, L.J. III (1983) Evidence for two distinct syndromes of involutional osteoporosis. *The American Journal of Medicine*, **75**, 899–901.

Riggs, B.L. and Melton, L.J. III (1986) Involutional osteoporosis. *The New England Journal of Medicine*, **314**, 1676–1686.

Riggs, B.L. and Melton, L.J. III (1990) Clinical heterogeneity of involutional osteoporosis: implications for preventive therapy. *The Journal of Clinical Endocrinology and Metabolism*, **70**, 1229–1232.

Riggs, B.L., Seeman, E., Hodgson, S.F. *et al.* (1982) Effect of fluoride/calcium regimen on vertebral fracture occurrence in postmenopausal osteoporosis, comparisons with conventional therapy. *The New England Journal of Medicine*, **306**, 446–450.

Riggs, B.L., Wahner, H.W., Dunn, W.L. *et al.* (1981) Differential changes in bone mineral density of the appendicular and axial skeleton with aging: Relationship to spinal osteoporosis. *The Journal of Clinical Investigation*, **67**, 328–335.

Rigotti, N.A., Nussbaum, S.R., Herzog, D.B. and Neer, R.M. (1984) Osteoporosis in women with anorexia nervosa. *The New England Journal of Medicine*, **311**, 1601–1606.

Riis, B.J., Johansen, J. and Christiansen, C. (1988) Continuous estrogen progestrogen treatment and bone metabolism in postmenopausal women. *Maturitas*, **10**, 51–58.

Riis, B.J., Overgaard, K. and Christiansen, C. (1995) Biochemical markers of bone turnover to monitor the bone response to postmenopausal hormone replacement therapy. *Osteoporosis International*, **5**, 276–280.

Riis, B.J., Rodbro, P. and Christiansen, C. (1986) The role of serum concentrations of sex steroids and bone turnover in the development and occurrence of postmenopausal osteoporosis. *Calcified Tissue International*, **38**, 318–322.

Riis, B., Thomsen, K. and Christiansen, C. (1987) Does calcium supplementation prevent postmenopausal bone loss? A double blind, controlled, clinical study. *The New England Journal of Medicine*, **316**, 173–177.

Riis, B.J., Thomsen, K., Strom, V. and Christiansen, C. (1987) The effect of percutaneous estradiol and natural progesterone on postmenopausal bone loss. *American Journal of Obstetrics and Gynecology*, **156**, 61–65.

Rikli, R.E. and McManis, B.G. (1990) Effects of exercise on bone mineral content in postmenopausal women. *Research Quarterly For Exercise and Sport*, **61**, 243–249.

Rinaldi, R, Shamos, M. and Lavine, L. (1974) Uptake of tritiated thymaidine during electrical stimulation of induced cortical bone defects. *Annnals of the New York Academy of Science*, **238**, 307–313.

Ringe, J.D., Kruse, H.P. and Kuhlencordt, F. (1978) Long term treatment of primary osteoporosis by sodium fluoride, in *Fluoride and Bone* (eds B. Courvoisier *et al.*), Hanes, Bruner, Bern, pp. 228–232.

Ripamonti, U., Duneas, N., Van den Heever, B. *et al.* (1997) Recombinant transforming growth factor-$\beta 1$ induces endochondral bone in the baboon and synergizes with recombinant osteogenic protein-1 (bone morphogenetic protein-7) to initiate rapid bone formation. *Journal of Bone and Mineral Research*, **12**, 1584–1595.

Ripamonti, U., Van den Heever, B., Sampath, T.K. *et al.* (1996) Complete regeneration of bone in the baboon by recombinant human osteogenic protein-1 (rhOP-1, bone morphogenetic protein-7). *Growth Factors*, **123**, 273–289.

Risteli, J., Elomaa, I., Niemi, S. *et al.* (1993) Radioimmunoassay for the pyridinoline cross-linked carboxy-terminal telopeptide of type I collagen: A new marker of bone collagen degradation. *Clinical Chemistry*, **39**(4), 635–640.

Risteli, J., Melkko, J., Niemi, S. and Risteli, L. (1991) Use of a marker of collagen formation in osteoporosis studies. *Calcified Tissue International*, **49**, S24–S25.

Riviere, I., Brose, K. and Mulligan, R.C. (1995) Effects of retroviral design on expression of human adenosine deaminase in murine bone marrow transplant recipients engrafted with genetically modified cells. *Proceedings of the National Academy of Sciences of the United States of America*, **92**, 6733–6753.

Roberjot, V., Laugier, P., Droin, P. *et al.* (1996) Measurement of integrated backscatter coefficient of trabecular bone. *IEEE Ultrasonic Symposium*, **2**, 1123–1126.

Robins, S.P., Stewart, P., Astbury, C. and Bird, H.A. (1986) Measurement of the cross linking compound, pyridinoline, in urine as an index of collagen degradation in joint disease. *Annals of the Rheumatic Diseases*, **45**, 969–973.

Robinson, R.A. (1960) Crystal collagen water relationships in bone matrix. *Clinical Orthopaedics and Related Research*, **17**, 69–76.

Robinson, R.A. (1966) The structural organization of bone tissue, in *Structural Organization of the Skeleton, Birth Defects Original Article Series*, vol. II (eds D. Bergsma *et al.*), The National Foundation, New York, p. 40.

Robinson, R.A. (1975) Physicochemical structure of bone. *Clinical Orthopaedics and Related Research*, **112**, 263–315.

Robinson, R.A. (1979) Bone tissue compositon and function. *The Johns Hopkins Medical Journal*, **145**, 1024.

Robinson, R.A., Doty, S. and Copper, R. (1973) *Biological Mineralization* (ed. I. Zipkin), John Wiley & Sons, Inc, New York, pp. 257–295.

Robinson, R.A. and Elliot, S.R. (1957) The water content of bone. *The Journal of Bone and Joint Surgery*, **39**, 167–188.

Robinson, R.A. and Watson, M.L. (1952) Collagen-crystal relationships in bone as seen in the electron microscope. *The Anatomical Record*, **114**, 383.

Robinson, R.A. and Watson, M.L. (1955) Crystal-collagen relationships in bone as observed in the electron microscope. *Annals of the New York Academy of Sciences*, **60**, 596.

Robinson, R.F. (1952) An electron microscopy study of the crystalline inorganic component of bone and its relationship to the organic matrix. *The Journal of Bone and Joint Surgery*, **34-A**(2), 389–435.

Robling, A.G., Burr, D.B. *et al.* (2000) Partitioning a daily mechanical stimulus into discrete loading bouts improves the osteogenic response to loading. *Journal of Bone and Mineral Research*, **15**(8), 1596–1602.

Rockoff, S.D., Sweet, E. and Blenstein, J. (1969) The relative contribution of trabecular and cortical bone to the strength of human lumbar vertebrae. *Calcified Tissue Research*, **3**, 163–175.

Rockwood, C. and Green, D. (1991) Fractures of the calcaneus, in *Rockwood and Green's Fractures in Adults*, J. B. Lippincott Company, **2**(3), pp. 1915–1982.

Rodan, G.A., Bourret, L.A. and Norton, L.A. (1978) DNA synthesis in cartilage cells is stimulated by oscillating electric fields. *Science*, **199**, 680–692.

Rodin, A., Chapman, M.G. and Fogelman, I. (1991) Bone density in users of combined oral contraception: preliminary reports of a pilot study. *British Journal of Family Planning*, **16**, 125–129.

Rodin, A., Murby, B., Smith, M.A. *et al.* (1990) Premenopausal bone loss in the lumbar spine and neck of femur: a study of 225 Caucasian women. *Bone*, **11**, 1–5.

Rodan, S.B., Wesolowski, G. *et al.* (1986) Clonal differences in prostaglandin synthesis among osteosarcoma cell lines. *Journal of Bone and Mineral Research*, **1**(2), 213–220.

Roer, R.D. and Dillaman, R.M. (1990) Bone growth and calcium balance during simulated weightlessness in the rat. *Journal of Applied Physiology*, **68**, 13–20.

Roer, R.D. and Dillaman, R.M. (1994) Decreased femoral arterial flow during simulated microgravity in the rat. *Journal of Applied Physiology*, **76**, 2125–2129.

Rooney, J.R. (1969) *Biomechanics of Lameness in Horses*, Williams and Wilkins, Baltimore.

Roos, B. (1974) *Dual Photon Absorptiometry Lumbar Vertebrae*, Thesis, University of Gothenburg, Sweden University of Gothenburg, Sweden.

Roos, B., Hansson, T. and Skoldborn, H. (1980) Dual photon absorptiometry in lumbar vertebrae. Evaluation of the baseline error. *Acta Radiologica: Oncology, Radiation, Physics, Biology*, **19**, 111–114.

Roos, B., Rosengran, B. and Skoldborn, H. (1970) Determination of bone mineral content in lumbar vertebrae by a double gamma-ray technique. *Proceeding Bone Mass Conference (Washington DC)*, (ed. J.R. Cameron), US Atomic Energy Commission, Washington, DC. pp. 243–253.

Roos, R.O. and Skoldborn, H. (1974) Dual photon absorpiometry in lumbar vertebrae. *Acta Radiologica*, **13**, 264.

Rosen, C.J., Chesnut, C.H. 3rd and Mallinak, N.J.J. (1997) The predictive value of biochemical markers of bone turnover for bone mineral density in early postmenopausal women treated with hormone replacement or calcium supplementation. *The Journal of Clinical Endocrinology and Metabolism*, **82**, 1904–1910.

Rosen, C.J., Hochberg, M.C., Bonnick, S.L. *et al.* (2005) Treatment with once-weekly alendronate 70 mg compared with once-weekly risedronate 35 mg in women with postmenopausal osteoporosis: a randomized double-blind study. *Journal of Bone and Mineral Research*, **20**, 141–151.

Rosen, D.M. and Luben, R.A. (1983) Multiple hormonal mechanisms for the control of collagen synthesis in an osteoblast-like cell line, MMB-1. *Endocrinology*, **112**, 992–999.

Rosen, H. (1979) Compression treatment of long bone pseudoarthroses. *Clinical Orthopaedics and Related Research*, **138**, 154–166.

Rosen, H.N., Tollin, S., Balena, R. *et al.* (1995) Differentiating between orchidectomized rats and controls using measurements of trabecular bone density: a comparison among DXA histomorphometry and peripheral quantitative computerized tomography. *Calcified Tissue International*, **57**, 35–39.

Rosenberg, B. (1962) Electrical conductivity of proteins. *Nature*, **193**, 364.

Rosenberg, B., Camp, L.V. and Krigas, T. (1965) Inhibition of cell division in Escharichia coli by electrolysis products from a platinum electrode. *Nature*, **205**, 698–699.

Rosenberg, B. and Postow, E. (1969) Semiconduction in proteins and lipids - its possible biological importance. *Annals of the New York Academy of Sciences*, **158**, 161–190.

Ross, P.D., Davis, J.W., Epstein, R.S. and Wasnich, R.D. (1992) Ability of vertebral dimensions from a single radiograph to identify fractures. *Calcified Tissue International*, **51**, 95–99.

Ross, P.D., Davis, J.W., Vogel, J.M. and Wasnich, R.D. (1990) A critical review of bone mass and the risk of osteoporotic fractures. *Calcified Tissue International*, **46**, 149–161.

Ross, P.D., Heilbrum, L.K., Wasnich, R.D. *et al.* (1989) Methodologic issues in evaluating risk factors for osteoporotic fractures. *Journal of Bone and Mineral Research*, **4**(5), 649–656.

Ross, P., Huang, C., Davis, J. *et al.* (1995) Predicting vertebral deformity using bone density at various skeletal sites and calcaneus ultrasound. *Bone*, **16**(3), 325–332.

Ross, P.D., Wasnich, R.D., Heilbrun, L.K. and Vogel, J.M. (1987b) Definition of a spine fracture threshold based on prospective fracture risk. *Bone*, **8**, 271–278.

Ross, P.D., Wasnich, R.D., Maclean, C. and Vogel, J.M. (1987a) Prediction of individual lifetime fracture expectancy using bone mineral measurements. *Osteoporosis*, 288–293.

Ross, P.D., Wasnich, R.D., Maclean, C.J. *et al.* (1988) A model for estimating the potential costs and savings of osteoporosis prevention strategies. *Bone*, **9**, 337–348.

Rossouw, J.E., Anderson, G.L., Prentice, R.L. *et al.* (2002) Risks and benefits of estrogen plus progestin in healthy postmenopausal women: principal results from the women's health initiative randomized controlled trial. *The Journal of the American Medical Association*, **288**, 321–333.

Roux, C., Lemonnier, E., Kolta, S. *et al.* (1993) Broadband ultrasound attenuation of the calcaneus and bone density measurements. *Revue du Rhumatisme (English Edition)*, **12**, 771–780.

Roux, W. (1905) *Die Entwicklungsmechanik; Ein Neuer Zweig Der Biologischen Wissenschft*, vols. I & II, Wilhelm Englemann, Leipzig.

Rowe, D.W. and Kream, B.E. (1982) Regulation of collagen synthesis in fetal rat calvaria by 1.25-dihydroxy vitamin $D_3$. *The Journal of Biological Chemistry*, **257**, 8009–8015.

Rowe, D.W., Shapiro, J.R., Poiren, M. and Schlesinger, S. (1985) Diminished type I collagen synthesis and reduced alpha (I) collagen messenger RNA in cultured fibroblast from patients with dominantly inherited (type I) osteogenesis imperfecta. *The Journal of Clinical Investigation*, **76**, 604–611.

Royal College of Physicians (2000) Osteoporosis: Clinical Guidelines for prevention and treatment, in *Update on Pharmacological Interventions and an Algorithm for Management*, RCP, London.

Rubin, C.T., Bain, S.D. and McLeod, K.J. (1990) The inability of the aging skeleton to respond to osteogenic stimulus: the origins of type II osteopenia? *Transactions of the Orthopaedic Research Society*, **15**, 75.

Rubin, C., Balander, M., Ryaby, J.P. and Hadjiargyrou, M. (2001) The use of low intensity ultrasound to accelerate the healing of fracture. *The Journal of Bone and Joint Surgery*, **83**(2), 259–270.

Rubin, C., Judex, S. and Hodijiargyrou, M. (2002) Skeletal adaptation to mechanical stimuli in the absence of formation or resorption of bone. *Journal of Musculoskeletal Neuronal Interactions*, **2**, 264–267.

Rubin, C.T. and Lanyon, L. (1984) Static vs dynamic loads as an influence on bone remodeling. *Journal of Biomechanics*, **17**, 897–905.

Rubin, C.T. and Lanyon, L.E. (1987) Osteoregulatory nature of mechanical stimuli: function as a determinant for adaptive remodeling in bone. *Journal of Orthopaedic Research*, **5**, 300.

Rubin, C.T. and Lanyon, L.E. (1994) Regulation of bone formation by applied dynamic loads. *The Journal of Bone and Joint Surgery. American Volume*, **66**(3), 397–402.

Rubin, C.T., McLeod, K.J. and Lanyon, L.E. (1989) Prevention of osteoporosis by pulsed electromagnetic fields. *The Journal of Bone and Joint Surgery*, **71**, 411–417.

Rubin, C.T., Pratt, G.W., Porter, A.L. *et al.* (1987) The use of ultrasound *in vivo* to determine acute change in the mechanical properties of bone following intense physical activity. *Journal of Biomechanics*, **20**, 723–727.

Rubin, C.T., Sommerfeldt, D.W., Judex, S. and Qin, Y.X. (2001a) Inhibition of osteopenia by low magnitude, high frequency mechanical stimuli. *Drug Discovery Today*, **6**, 848–858.

Rubin, E. and Rottenberg, H. (1982) Ethanol-induced injury and adaptation in biological membranes. *Federation Proceedings*, **41**, 2465–2471.

Rubin, J., McLeod, K.J., Titus, L. *et al.* (1996) Formation of osteoclast like cells is suppressed by low frequency, low intensity electromagnetic fields. *Journal of Orthopaedic Research*, **14**, 7–15.

Rubinnacci, A. and Brigatti, L. (1984) A new method for measuring electrical resistance of bone, *in vivo*. *Acta Orthopaedica Belgica*, **50**, 686–691.

Ruff, C. (1988) Hindlimb articular surface allometry in Hominoides and Macaca, with comparisons to diaphyseal scaling. *Journal of Human Evolution*, **17**, 687–714.

Ruimerman, R., Huiskes, R., Van Lenthe, G.H. and Janssen, J.D. (2001) A computer simulation model relating bone cell metabolism, to mechanical adaptation of trabecular bone. *Computer Methods in Biomechanics and Biomedical Engineering*, **4**, 433–448.

Ruimerman, R., Van Rietbergen, B., Hilbers, P. and Huiskes, R. (2005) The effects of trabecular bone loading variable on the surface signaling potential for bone remodeling and adaptation. *Annals of Biomedical Engineering*, **33**(1), 71–78.

Rundle, C.H., Miyakoshi, N., Kasukawa, Y. *et al.* (2003) *In vivo* bone formation in fracture repair induced by direct retroviral-based gene therapy with bone morphogenetic protein-4. *Bone*, **32**, 591–601.

Rune, H.L. and Gallagher, J.C. (1989) Increased incidence of slip fracture in osteoporotic women treated with sodium fluoride. *Journal of Bone and Mineral Research*, **4**, 223–225.

Russell, J.E. and Simmons, D.J. (1985) Bone maturation in rats flown on the spacelab 3 mission. *The Physiologist*, **28**, S235–S236.

Rutherford, R.A., Pullan, B.R. and Isherwood, I. (1976) Measurement of effective atomic number and electron density using an EMI scanner. *Neuroradiology*, **11**, 15–21.

Ryaby, J.T., Jones, D.B. and Pilla, A.A. (1986) The effects of electromagnetic fields on protein phosphorylation and synthesis in murine melanoma cells. *BRAGS Transactions*, **6**, 33.

Ryaby, J.T., Khin, N.A., Fitzsimmons, R.J. *et al.* (1994) The insulin-like growth factor pathway is increased in fracture callus cultures by combined magnetic fields. Proceedings of the 14th Annual Meeting of the Society for Physical Regulation in Biology and Medicine. Washington D.C., October 13, p. P7.

Ryshkewitch, E.J. (1953) Compression strength of porous sintered alumina and zirconia. *Journal of the American Ceramic Society*, **36**, 65–68.

Sabatini, M., Boyce, B., Aufdemorte, T. *et al.* (1988) Infusions of recombinant human interleukin-1 alpha and beta cause hypercalcemia in normal mice. *Proceedings of the National Academy of Sciences of the United States of America*, **85**, 5235–5239.

Sabin, M.A., Blake, G.M., MacLaughlin Blacks, M. and Fogelman, I. (1995) The accuracy of volumetric bone density measurements in dual X-ray absorptiometry. *Calcified Tissue International*, **56**, 210–214.

Sabto, J., Powel, M.J., Breidahl, M.J. and Gur, F.W. (1984) Influence of urinary sodium on calcium excretion in normal individuals. *The Medical Journal of Australia*, **140**, 345–356.

Sadeghi, H., Allard, P., Prince, F. and Labelle, H. (2000) Symmetry and limb dominance in able bodied gait: a review. *Gait and Posture*, **12**, 34–45.

Saha, G.C., Shyamal Datta, and Mukerjee, S.C. (1986) Electron capture in collisions of He2+ with Li atoms and of Li3+, C6+, and O8+ with He atoms in the high-energy region. *Physical Review A*, **34**, 2809–2821.

Saha, S. (1977) Congitudinal shear properties of human compact bone and its constituents and the associated failure mechanisms. *Journal of Materials Science*, **12**, 1798–1806.

Saha, S. (1982) Dynamic strength of bone and its relevance, in *Osteoarthromechanics* (ed. D.N. Ghista), Hemisphere, New York, ch. 1, pp. 1–43.

Saha, S. and Gorman, P.H. (1981) Strength of human cancellous bone in shear and its relationship to bone mineral content. Transactions 27th Annual Meeting of the Ortho. Research Society, vol. 6, p. 217; also in Saha, S. and Gorman, P.H. (1981) *Orthopaedic Transactions*, **5** (2), 323–34.

Saha, S. and Haines, D.W. (1979) The damping behavior of cortical bone as a function of frequency in the audio range, in *1979 Biomechanics Symposium* (ed. W.C. Van Buskirk), American Society of Mechanical Engineers, New York, AMD 32, pp. 195–198.

Saha, S. and Lakes, R.S. (1977) A non-invsive technique for detecting stress waves in bone using the piezoelectric effect. *IEEE Transactions on Biomedical Engineering*, **24**, 508–512.

Saha, S., Rao, V.V., Malakanok, V. and Albright, J.A. (1982) Quantitative measurement of fracture healing by ultrasound, in *Biomedical Engineering I: Recent Developments* (ed. S. Saha), Pergamon Press, New York, pp. 247–249.

Saha, S. and Williams, P.A. (1989) Electric and dielectric properties of wet human cancellous bone as a function of frequency. *Annals of Biomedical Engineering*, **17**, 143–158.

Sakamoto, W., Nishihira, J., Fujie, K. *et al.* (2001) Effect of coffee consumption on bone metabolism. *Bone*, **28**(3), 332–336.

Sakou, T. (1998) Bone morphogenetic proteins: From basic studies to clinical approaches. *Bone*, **22**, 591–603.

Salamone, L.M., Krall, E.A., Harris, S. and Dawson-Hughes, B. (1994) Comparison of broadband ultrasound to single X-ray absorptiometry measurements at the calcaneus in postmenopausal women. *Calcified Tissue International*, **54**, 87–90.

Salzstein, R.A. and Pollack, S.R. (1987) Electromechanical potentials in cortical bone-II. Experimental analysis. *Journal of Biomechanics*, **20**, 271–280.

Salzstein, R.A., Pollack, S.R., Mak, A.F.T. and Petrov, N. (1987) Electromechanical potentials in cortical bone. A continuum approach. *Journal of Biomechanics*, **20**, 261–270.

Samachsam, J. (1969) Basic requirements for calcification. *Nature*, **221**, 1247–248.

Sambrook, P.N., Bartlett, C., Evans, R. *et al.* (1985) Measurement of lumbar spine bone mineral: A comparison of dual photon absorptiometry and computed tomography. *British Journal of Radiology*, **58**, 621–624.

Sambrook, P.N., Chen, J.S., March, S.R. *et al.* (2004) Serum parathyroid hormone predicts time to fall independent of vitamin D status in a frail elderly population. *The Journal of Clinical Endocrinology and Metabolism*, **89**, 1572–1576.

Sambrook, P.N., Eisman, J.A., Furler, S.M. and Pocock, N.A. (1987) Computer modeling and analysis of cross-sectional bone density studies with respect to age and the menopause. *Journal of Bone and Mineral Research*, **2**, 109–114.

Sambrook, P.N., Seeman, E., Phillips, S.R. and Ebeling, P.R. (2002) Preventing osteoporosis: outcomes of the Australian Fracture Prevention Summit. *The Medical Journal of Australia*, **176**(Suppl), S1.

Samnegard, E., Akhter, M.P. and Recker, R.R. (2001) Maintenance of vertebral body bone mass and strength created by human parathyroid hormone treatment in ovariectomized rats. *Bone*, **28**(4), 414–422.

Sampath, T.K., Maliakal, J.C., Hauschka, P.V. *et al.* (1992) Recombinant human osteogenic protein-1 (hOP-1) induces new bone formation *in vivo* with a specific activity comparable with natural ovine osteogenic protein and stimulates osteolast proliferation and differentiation *in vitro*. *The Journal of Biological Chemistry*, **267**, 20352–20362.

Sampath, T.K., Weintroub, S. and Reddi, A.H. (1984) Extracellular matrix proteins involved in bone induction are vitamin D dependent. *Biochemical and Biophysical Research Communications*, **124**, 829–835.

Sandberg, M.M., Aro, H.T. and Vuorio, E.I. (1993) Gene expression during bone repair (reviewed). *Clinical Orthopaedics and Related Research*, **289**, 292–312.

Sandler, R.B. and Herbert, D.L. (1981) Quantitative bone assessments: Applications and expectations. *Journal of the American Geriatrics Society*, **29**, 97–103.

Sansen, W., Dijeker, W., Stan, S. and Mulier, J.C. (1978) Four point measurement of impedance of bone *in vivo*, in *Electric Stimulation of Bone Growth and Repair* (eds F. Buny, E. Herbst and M. Hinsenkamp), Springer Verlag, New York, pp. 15–18.

Sanyal, A., Sarkar, G., Sris, D.B.F. *et al.* (1999) Initial evidence for the involvement of bone morphogenetic protein-2 early during periosteal chondrogenesis. *Journal of Orthopaedic Research*, **17**, 926–934.

Saonstegard, D. and Matthews, L. (1976) Sonic diagnosis of bone fracture healing – a preliminary study. *Journal of Biomechanics*, **9**, 689–694.

Sartoris, D.J. and Resnick, D. (1989) Dual energy radiographic absorptiometry for bone densitometry: Current status and prospective. *American Journal of Roentgenology*, **152**, 241–246.

Sasaki, N. and Enyo, A. (1995) Viscoelastic properties of bone as a function of water content. *Journal of Biomechanics*, **28**, 809–815.

Sasaki, N., Ikawa, T. and Fukuda, A. (1991) Orientation of mineral in bovine bone and anisotropic mechanical properties of plexiform bone. *Journal of Biomechanics*, **24**, 57–61.

Sasaki, N., Nakayama, Y., Yoshikawa, M. and Enyo, A. (1993) Stress relaxation function of bone and bone collagen. *Journal of Biomechanics*, **26**, 1369–1376.

Sasaki, N. and Yoshikawa, M. (1993) Stress relaxation in native and EDTA treated bone as a function of mineral content. *Journal of Biomechanics*, **26**, 77–88.

Sasaoka, R., Terai, H., Toyoda, H. *et al.* (2004) A prostanoid receptor EP4 agonist enhances ectopic bone formation induced by recombinant human bone morphogenetic protein-2. *Biochemical and Biophysical Research Communications*, **318**, 704–709.

Sato, K. *et al.* (1987) Interleukin-1 alfa and PTH like factors are responsible humoral hypercalcemia associated with esophageal carcinoma cells (EC4-GI). *Journal of Bone and Mineral Research*, **2**(1), 387.

Sato, M., Grasser, W., Endo, N. *et al.* (1991) Bisophosphonate action: Alendronate localization rat bone and effect on osteoclast ultrastructure. *The Journal of Clinical Investigation*, **88**, 2095–2105.

Sato, W., Matsushita, T. and Nakamura, K. (1999) Acceleration of increase in bone mineral content by low intensity ultrasound energy in leg lengthening. *Journal of Ultrasound in Medicine*, **18**, 699–702.

Saulgozis, J., Pontaga, I., Lowet, G. and Van der Perre, G. (1996) The effect of fracture and fracture healing on ultrasonic velocity and attenuation. *Physiological Measurement*, **17**, 201–211.

Saunders, M.M., You, J. *et al.* (2003) Fluid flow induced prostaglandin E2 response of osteoblastic ROS 17/2.8 cells is gap junction mediated and independent of cytosolic calcium. *Bone*, **32**(4), 350–356.

Sauty, A., Pecherstorfer, M., Zimmer-Roth, I. *et al.* (1996) Interleukin-6 and tumor necrosis factor $\alpha$ levels after bisphosphonates treatment *in vitro* and in patients with malignancy. *Bone*, **18**, 133–139.

Saville, P.D. (1969) Changes in skeletal mass and fragility with castration in the rat: a model of osteoporosis. *Journal of the American Geriatrics Society*, **17**, 155–169.

Sawyer, C.B. and Tower, C.H. (1930) Rochelle salt as a dielectric. *Physical Review*, **35**, 269–273.

Sawyer, P.N. and Pete, J.W. (1953) Electrical potential difference across the normal aorta and aortic graft of dog. *The American Journal of Physiology*, **175**, 1113.

Schaadt, O. and Bohr, H. (1982) Bone mineral by dual photon absorptiometry. Accuracy-precision-sites for measurements, in *Non-Invasive Bone Measurements: Methodological Problems* (eds J. Dequeker and C.C. Johnston Jr), IRL Press, Oxford, Washington, D. C., pp. 59–72.

Schaadt, O. and Bohr, H. (1988) Different trends of age related diminution of bone mineral content of the lumbar skin, femoral neck and femoral shaft in women. *Calcified Tissue International*, **42**, 71–76.

Schaafsma, G. (1981) The Influence of Dietary Calcium and Phosphorous on Bone Metabolism. Thesis, Washington.

Schaafsma, G., Duursma, S.A., Visser, W.J. and Dekker, P.R. (1985) The influence of dietary calcium on kidney calcification and renal function in rats fed high phosphate diets. *Bone*, **6**, 155–163.

Schaafsma, G., Van Beresteyn, E.C.H., Raymakers, J.A. and Duursma, S.A. (1987) Nutritional aspects of osteoporosis. *World Review of Nutrition and Dietetics*, **49**, 121–159.

Schaden, W., Fischer, A. and Sailler, A. (2001) Extracorporeal shock wave therapy of nonunion or delayed osseous union. *Clinical Orthopaedics and Related Research*, **387**, 90–94.

Schaffler, M.B. and Burr, D.B. (1988) Stiffness of compact bone: Effects of porosity and density. *Journal of Biomechanics*, **21**, 13–16.

Schapira, D., Lotan Miller, R., Barzilai, D. and Sibermann, M. (1991) The rat as a model for studies of the aging skeleton. *Cells Materials*, **1**(Suppl.), 181–188.

Schatzler, H.P. (1979) Basic aspects on the use of elastic and inelastic scattered gamma radiation for the determination of binary systems with effective atomic numbers of less than 10. *International Journal of Applied Radiation and Isotopes*, **30**, 115–121.

Schenk, R.K. (1992) Biology of fracture repair, in *Skeletal Trauma* (eds B.D. Browner, J.B. Jupiter and A.M. Levine *et al.*), W.B. Saunders, Philadelphia, PA, pp. 31–75.

Scheven, B.A.A., Hamilton, N.J., Fakkeldij, T.M.V. and Duursma, S.A. (1991) Stimulatory effects of estrogen and progesterone on proliferation and differentiation of normal human osteoblast like cells *in vitro*. *Biochemical and Biophysical Research Communications*, **186**, 54–60.

Schiessl, H., Ferritti, J.L., Tysarczyk-Niemeyer, G. *et al.* (1998) The role of the muscles to the mechanical adaptation of bone, in *Advances in Osteoporosis*, vol. **1**(ed. G.P. Lyritis), Athens, p. 53.

Schlenker, R.A. and Vonseggen, W.W. (1976) The distribution of cortical and trabecular bone mass along the lengths of the radius and ulna and the implications for *in vivo* bone measurements. *Calcified Tissue Research*, **20**, 41–52.

Schmitt, J.M., Hwang, K., Winn, S.R. and Hollinger, J.O. (1999) Bone morphogenetic proteins: an update on basic biology and clinical relevance. *Journal of Orthopaedic Research*, **17**, 269–278.

Schneider, D.L., Barrett-Connor, E.L. and Morton, D.J. (1997) Timing of postmenopausal estrogen for optimal bone mineral density. The Racho Bernardo Study. *The Journal of the American Medical Association*, **277**, 543–547.

Schnitzler, C.M., Wing, J.R., Gear, K.A. and Robson, H.J. (1990) Bone fragility of the peripheral skeletal during fluoride therapy for osteoporosis. *Clinical Orthopaedics and Related Research*, **261**, 268–275.

Schott, A.M., Weill-Engerer, S., Hans, D. *et al.* (1995) Ultrasound discriminates patients with hip fracture equally well as dual energy X-ray absorptiometry and independently of bone mineral density. *Journal of Bone and Mineral Research*, **10**, 243–249.

Schoutens, A., Verhas, M., Dourov, N. *et al.* (1988) Bone loss and bone blood flow in parapletic rats treated with calcitonin, diphosphonate and idomethacin. *Calcified Tissue International*, **43**, 136–143.

Schriefer, J.L., Warden, S.J., Saxon, J.K. *et al.* (2005) Cellular accommodation and the response of bone to mechanical loading. *Journal of Biomechanics*, **38**, 1838–1845.

Schroeder, H.A. (1960) in *Metal Binding in Medicine* (ed. M.J. Seven), J.B. Lippincott Co, Philadelphia, pp. 59–67.

Schubert, T., Kleditzsch, J. and Wolf, E. (1986) Results of fluorescence microscopy studies of bone healing by direct stimulation with bipolar impulse currents and with the interference current procedure in the animal experiment. *Zeitschrift für Orthopädie und ihre Grenzgebiete*, **124**, 6–12.

Schuette, S.A. and Linkswiler, H.M. (1982) Effects of Ca and P metabolism in humans by adding meat, meat plus milk, or purified proteins plus Ca and P to a low protein diet. *The Journal of Nutrition*, **112**, 338–349.

Schuette, S.A., Zemel, M.B. and Linkswiler, H.M. (1980) Studies on the mechanism of protein induced hypercalciuria in older men and women. *The Journal of Nutrition*, **110**, 310–315.

Schwachtgen, J.-L., Houston, P., Campbell, C.J. *et al.* (1998) Fluid shear stress activation of Egr-1 transcription in cultured human endothelial and epithelial cells is mediated via the extracellular signal related kinase $^1/_2$ mictogen activated protein kinase pathway. *The Journal of Clinical Investigation*, **101**, 2540–2549.

Schwan, H.P. (1957) Electrical properties of tissue and cell suspensions, in *Advances in Biological and Medical Physics*, vol. **V**(eds J.H. Lawrence and C.A. Tobias), Academic Press, New York, pp. 147–209.

Schwan, H.P. (1963) Determination of biological impedances, in *Physical Techniques in Biological Research* (ed. W.L. Nastur), Academic Press, New York.

Schwan, H.P. (1965) Electrical properties of bound water. *Annals of the New York Academy of Sciences*, **125**, 344.

Schwan, H.P. (1977) Field interaction with biological matters. *Annals of the New York Academy of Sciences*, **303**, 198–213.

Schwiebert, E.M. (1999) ABC transporter-facilitated ATP conductive transport. *American Journal of Physiology*, **276**, C1–C8.

Schwiebert, E.M., Egar, M.E., Hwang, T.H. *et al.* (1995) CFTR regulates outwardly rectifying chloride channels through an autocrine mechanism involving ATP. *Cell*, **81**, 1063–1073.

Scopacasa, F., Wishart, J.M. and Horowitz, M. (2004) Relation between calcium absorption and serum calcitriol in normal men: evidence for age related intestinal resistance to calcitriol. *European Journal of Clinical Nutrition*, **58**, 264–269.

Scott, G.C. and Korostoff, E. (1990) Oscillatory and step response electromechanical phenomena in human bovine bone. *Journal of Biomechanics*, **23**, 127–143.

Scott, J.E. (1988) Proteoglycan fibrillar collagen interactions. *The Biochemical Journal*, **252**, 313–323.

Scott, J.E. and Oxford, C.R. (1981) Dermatan sulfate rich proteoglycan associated with rat-tail tendon collagen at the d band in the gap region. *The Biochemical Journal*, **197**, 213–216.

Sedlin, E.D. and Hirsch, C. (1966) Factors affecting the determination of the physical properties of femoral cortical bone. *Acta Orthopaedica Scandinavica*, **37**, 29–48.

Sedlin, E.D. and Sonnerup, L. (1966) Rheological considerations in the physical properties of bone, in *Calcified Tissues – 1965* (eds H. Fleisch, H.J.J. Blackwood and M. Owen), Springer, Berlin, pp. 98–101.

Seedor, J.G., Quertuccio, H.A. and Thompson, D.D. (1991) The bisphophonate alendronate (MK-217) inhibits bone loss due to overiectomy in rats. *Journal of Bone and Mineral Research*, **6**, 339–346.

Seeley, D.G., Browner, W.S., Nevitt, M.C. *et al.* (1991) For the study of osteoporotic fractures research groups which fractures are associated with low appendicular bone mass in elderly women. *Annals of Internal Medicine*, **115**, 837–842.

Seeman, E. (1999) The structural basis of bone fragility in men. *Bone*, **25**, 143–147.

Seeman, E. (2001) During aging men lose less bone than women. *Calcified Tissue International*, **69**, 205–208.

Seeman, E., Hopper, J.L., Bach, L.A. *et al.* (1989) Reduced bone mass in daughters of women with osteoporosis. *The New England Journal of Medicine*, **320**, 554–558.

Seeman, E., Melton, L.J. III, O'Fallon, W.M. and Riggs, B.L. (1983) Risk factors for spinal osteoporosis in man. *The American Journal of Medicine*, **75**, 977–983.

Seeman, E., Tsalamandris, C., Fromica, C. *et al.* (1994) Reduced femoral neck bone density in the daughters of women with hip fractures: the role of low peak bone density in the pathogenesis of osteoporosis. *Journal of Bone and Mineral Research*, **9**, 739–743.

Segrest, J.P. and Cunningham, L.W. (1973) Unit figril models derived from the molecular topography of collagen. *Biopolymers*, **12**, 825–834.

Seifert, M.F., Popoff, S.N. and Marks, S.C. Jr (1988) Skeletal biology in the toothless (osteopetrotic) rat. *The American Journal of Anatomy*, **183**, 158–165.

Selby, C. (1990) Sex hormone binding globulin: origin, function and clinical significance. *Annals of Clinical Biochemistry*, **27**, 532–541.

Selby, P.L., Peacock, M., Barworth, S.A. and Tailor, G.A. (1985) Early effect of ethynilestradiol and norethisterone treatment in postmenopausal women on bone resorption and calcium regulating hormones. *Clinical Science*, **69**, 265–271.

Selle, W. and Jurist, J. (1966) Acoustical detection of senile osteoporosis. *Proceedings of the Society for Experimental Biology and Medicine*, **121**, 150–152.

Serpe, L. and Rho, J. (1995) The nonlinear dependence of broadband ultrasound attenuation on bone mineral density: An *in vitro* study. Proceedings of the 41st Annual Meeting of the Orthopaedic Research Society, **20** (2), 541.

Serpe, L. and Rho, J.Y. (1996) The nonlinear transition period of broadband ultrasound attenuation as bone density varies. *Journal of Biomechanics*, **29**, 963–966.

Sessions, N.V., Halloran, B.P., Bikle, B.D. *et al.* (1989) Bone response to normal weight bearing after a period of skeletal unloading. *The American Journal of Physiology*, **257**, E606–E610.

Sevastikoglou, J.A.V., Thomaidis, T. and Lindholm, T.S. (1977) Reversibility of osteoporosis in adult rats by calcium deficiency. Long and short term observations. *Calcified Tissue Research*, **22**(Suppl.), 260–265.

Severns, A.E., Lee Yu, P.O., Nelson, S.D. *et al.* (2004) Metabolic measurement techniques to assess bone fracture healing. *Clinical Orthopaedics and Related Research*, **424**, 231–238.

Seyberth, H.W. *et al.* (1975) Protaglandins as mediators of hypercalcemia associated with certain types of cancer. *The New England Journal of Medicine*, **293**, 1278–1283.

Shaffkey, R., Behari, J. and Saha, S. (1988) Ultrasonic characterization of osteoporotic bone. *Journal of Pure and Applied Ultrasonics*, **10**, 5–7.

Shamos, M.H. and Lavine, L.S. (1964) Physical basis of bioelectric effects in mineralized tissues. *Clinical Orthopaedics*, **53**, 177–188.

Shamos, M.H. and Lavine, L.S. (1967) Piezoelectricity as a fundamental property of biological tissues. *Nature*, **213**, 267–269.

Shamos, M.H., Lavine, L.S. and Shamos, M.I. (1963) Piezoelectric effect in bone. *Nature*, **197**, 81.

Shankar, V.S., Simon, B.J., Bax, C.M. *et al.* (1998) Effects of electromagnetic stimulation on the functional responsiveness of isolated rat osteoclasts. *Journal of Cellular Physiology*, **176**, 537–544.

Shapiro, R. and Heaney, R.P. (2003) Co-dependence of calcium and phosphorous for growth and bone development under conditions of varying deficiency. *Bone*, **32**, 532–540.

Sharrard, W.J.W. (1990) A double blind trial of pulsed electromagnetic fields for delayed union of the tibia. *The Journal of Bone and Joint Surgery*, **72**, 347–355.

Shaw, S.R., Vailas, A.C., Grindeland, R.E. and Zernicke, R.F. (1988) Effects of 1-week spaceflight on morphological and mechanical properties of growing bone. *The American Journal of Physiology*, **254**, R78–R83.

Shellhart, W.G., Hardt, A.B., Moore, R.N. and Erickson, L.C. (1992) Effects of bisphophonate treatment and mechanical loading on bone modeling in the rat tibias. *The Journal of Bone and Joint Surgery*, **278**, 253–259.

Shen, V., Dempster, D.W., Mellish, R.W.E. *et al.* (1992) Effects of combined and separate intermittent administration of low dose human parathyroid hormone fragment (1–34) and 17 $\beta$ estradiol on bone histomorphometry in ovariectomized rats with established osteoponia. *Calcified Tissue International*, **50**, 214–220.

Shen, V., Liang, X.G., Birchman, R. *et al.* (1997) Short-term immobilization induced cancellous bone loss is limited to regions undergoing high turnover and/or modeling in mature rats. *Bone*, **21**, 71–78.

Shen, V., Rifas, L., Kohler, G. and Peck, W.A. (1986) Prostaglandins change cell shape and increase intercellular gap functions in osteoblasts cultured from rat fetal calverie. *Journal of Bone and Mineral Research*, **1**, 243–249.

Sheng, M.H., Baylink, D.J., Beamer, W.G. *et al.* (2002) Regulation of bone volume is different in the metaphyses of the femur and vertebra of C3H/HeJ and C57BL/6J mice. *Bone*, **30**, 486–491.

Shepp, L.A. and Logan, B.F. (1974) The Fourier reconstruction of a head section. *IEEE Transactions on Nuclear Science*, **23**, 21–42.

Sheridan, J.D. and Atkinson, M.M. (1985) Cell membranes: the electrochemical environment and cancer promotion. *Annual Review of Physiology*, **47**, 337.

Sheridan, J.D., Hammer-Wilson, M., Preus, D. and Johnson, R.G. (1978) Quantitative analysis of low-resistance junction between cultured cells and correlation with gap junctional areas. *The Journal of Cell Biology*, **76**, 532.

Sherman, B.M., West, J.H. and Korenman, S.G. (1976) The menopausal transition: analysis of LH, FSH, estradiol and progesterone concentrations during menstrual cycles of older women. *The Journal of Clinical Endocrinology and Metabolism*, **42**, 629–636.

Shi, S., Kirk, M. and Kahn, A.J. (1996) The role of type I collagen in the regulation of the osteoblasts phenotype. *Journal of Bone and Mineral Research*, **11**, 1139–1145.

Shimazaki, A., Inui, K., Azuma, Y. *et al.* (2000) Low-intensity pulsed ultrasound accelerates bone maturation in distraction osteogenesis in rabbits. *The Journal of Bone and Joint Surgery*, **82-B**, 1077–1082.

Shimegi, A. (1996) ATP and adenosine act as a mitogen for osteoblast like cells (MC3T3-E1). *Calcified Tissue International*, **58**, 109–113.

Shimo, T., Nakanishi, T., Kimura, Y. *et al.* (1998) Inhibition of endogenous expression of connective tissue growth factor by its antisense oligonucleotide and antisense RNA suppresses proliferation and migration of vascular endothelial cell. *The Journal of Biochemistry*, **124**, 130–140.

Shimo, T., Nakanishi, T., Nishida, T. *et al.* (1999) Connective tissue growth factor induces the proliferation, migration and tube formation of vascular endothelial cells *in vitro* and angiogenesis *in vivo*. *The Journal of Biochemistry*, **126**, 137–145.

Shinoda, A., Ademek, G., Felix, R. *et al.* (1983) Structure activity relationship of various bisphosphonte. *Calcified Tissue Research*, **35**, 87–99.

Shirkhanzadeh, M. (1998) Direct formation of nanophase hydroxyapatite cathodically polarized electrodes. *Journal of Materials Science: Materials in Medicine*, **9**, 67–72.

Shiro, I. Ohashio, T., Yasuda, I. and Fakuda, E. (1977) Electret induced callus formation in the rat. *Clinical Orthopaedics and Related Research*, **124**, 57.

Shukla, S.S., Karellas, A., Leichter, I. *et al.* (1985) Quantitative assessment of bone mineral by photon scattering: Accuracy and precision considerations. *Medical Physics*, **12**, 447–448.

Shultz, T.D., Bollman, S. and Kumar, R. (1982) Decreased intestinal calcium absorption *in vivo* and normal brush border membrane vesicle calcium uptake in cortisol treated chickens: Evidence for dissociation of calcium absorption from brush broder vesicle uptake. *Proceedings of the National Academy of Sciences of the United States of America*, **79**(11), 3542–3546.

Sierpowska, J., Toyras, J., Hakulinen, M.A. *et al.* (2003) Electrical and dielectric properties of bovine trabecular bone – relationships with mechanical properties and mineral density. *Physics in Medicine and Biology*, **48**, 775–786.

Sietsema, W.K. (1995) Animal models of cortical porosity. *Bone*, **17**, 297–305.

Sillence, D.O. (1988) Osteogenesis imperfecta nosology and genetics. *Annals of the New York Academy of Sciences*, **543**, 1–15.

Sillence, D.O., Senn, A.S. and Danks, D.M. (1979) Genetic heterogeneity in osteogenesis imperfecta. *Journal of Medical Genetics*, **16**, 101–116.

Silverman, S.L. (2003) Calcitonin. *Endocrinology and Metabolism Clinics of North America*, **32**, 273–284.

Simkin, A. (1993) Ultrasonic velocity along the tibia shaft as a diagnostic tool for osteoporosis and a prediction of hip fractures. Myriad Ultrasound System Ltd, Technical Report 1, 7.

Simkin, A., Ayalon, J. and Leichter, I. (1987) Increased trabecular bone density due to bone loading exercises in postmenopausal osteoporotic women. *Calcified Tissue International*, **40**, 59–63.

Simmons, E.D. Jr, Pritzker, K.P.H. and Grynpas, M.D. (1991) Age related changes in the human femoral cortex. *Journal of Orthopaedic Research*, **9**, 155–167.

Sims, J.R., Karp, S. and Ingber, D.E. (1992) Altering the cellular mechanical force balance results in integrated changes in cell, cytoskeletal and nuclear shape. *Journal of Cell Science*, **103**, 1215–1222.

Sims, N.A., Dupont, S., Krust, A. *et al.* (2002) Deletion of estrogen receptors reveals a regulatory role for estrogen receptors-beta in bone remodeling in females but not in males. *Bone*, **30**, 18–25.

Simske, S.J., Wachtel, H. and Luttges, M.W. (1991) Effect of localized pulsed electromagnetic fields on still suspension Osteopenia in growing mice. *Bioelectromagnetics*, **12**, 101–116.

Singer, F.R. (1991) Clinical efficacy of salmon calcitonin in Pagets' disease of bone. *Calcified Tissue International*, **49**(S), 57–58.

Singer, F.R., Melvin, K.E.W. and Miles, B.G. (1976) Acute effects of calcitonin on osteoclasts in man. *Clinical Endocrinology*, **5**(Suppl.), 333S–340S.

Singh, A.K. and Behari, J. (1994) Assessment of bone quality by ultrasonic wave propagation technique. *Indian Journal of Pure and Applied Physics*, **32**, 528–530.

Singh, M., Nagrath, A.R. and Maiou, P.S. (1970) Changes in trabecular bone pattern of the upper end of the femur as an index of osteoporosis. *The Journal of Bone and Joint Surgery*, **52**, 457.

Singh, M., Riggs, B.L., Beabout, J.W. and Jowsey, J. (1972) Femoral trabecular index for evaluation of spinal osteoporosis. *Annals of Internal Medicine*, **77**, 63–64.

Singh, S. and Behari, J. (1981) Ultrasound propogation in *in vivo* bone. *Ultrasonics*, **19**, 87.

Singh, S. and Behari, J. (1984a) Frequency dependence of electrical properties of human bone. *Journal of Bioelectricity*, **3**, 347.

Singh, S. and Behari, J. (1984b) Physical characteristics of bone composite materials. *Journal of Biological Physics*, **12**, 1.

Singh, S. and Rai, D.V. (1979) Physical characterization of piezoelectric materials of biological origin. *Journal of Pure and Applied Ultrasonics*, **1**(4), 94.

Singh, S. and Saha, S. (1984) Frequency response of stress generated potentials in wet compact bone. Trans. 10th Am. Meeting of Biomaterials, p. 221.

Singh, S. and Saha, S. (1987) Electrical characteristics of the electrode/bone interface. *Medical and Biological Engineering and Computing*, **25**, 448–452.

Singh, S., Saha, S., Giyanani, V.L. *et al.* (1984) Measurement of Corticial Bone Thickness by Ultrasound and Ct, in *Biomedical Engineering III: Recent Developments* (ed. B.L. Sheppard), Pergamon Press, pp. 82–85.

Singh, V.R., Adya, V.P., Ahmed, A. and Yadav, S. (1990) A stress wave propagation technique for bone repair study. *IEEE Transactions on Biomedical Engineering*, **37**, 1014–1017.

Siris, E. (1993) Bisphosphonates and iritis. *Lancet*, **341**, 436–437.

Sisken, B.F. and Smith, S.D. (1975) The effects of minute direct electrical currents on cultured chick embryo trigemal ganglia. *Journal of Embryology and Experimental Morphology*, **33**, 29–41.

Sissons, H.A. (1971) The growth of bone, in *The Biochemistry and Physiology of Bone*, vol. **3** (ed. G.H. Bourne), Academic Press, New York, pp. 145–180.

Sissons, H.A., Jowsey, J. and Stewart, L. (1960) Quantitative microradiography of bone tissue, in *X-ray Microscopy and X-ray Microanalysis* (eds A. Engstrom, V.E. Cosslett and H. Pattee), Elsevier, Amsterdam, pp. 199–205.

Skerry, T.B.M., Bitensk, L., Chayen, J. and Lanyon, L.E. (1989) Early strain related changes in enzyme activity in osteocytes following bone loading *in vivo*. *Journal of Bone and Mineral Research*, **4**(5), 783–788.

Skinner, C., Kulkarni, V., Aschcroft, G. and Hutchison, J. (2001) Early impedance changes following corticotomy and distraction in the Ilizarov technique. Proceedings 11th International Conference on Electrical Bio-Impedance (Oslo), pp. 101–104.

Slemenda, C.W. (1995) Body composition and skeletal density: mechanical loading or something more? *The Journal of Clinical Endocrinology and Metabolism*, **80**, 1761–1763.

Slemenda, C.W., Hui, W.S., Longcope, C. and Johnson, C.C. (1987) Sex steroids and bone mass. *The Journal of Clinical Investigation*, **80**, 1261–1269.

Slemenda, C.W., Hui, S.L., Longcope, C. *et al.* (1990) Predictions of bone mass in premenopausal women. A prospective study of clinical data using photon absorptiometry. *Annals of Internal Medicine*, **112**, 96–101.

Slemenda, C.W., Longcope, C., Zhou, L. *et al.* (1997) Sex steroids and bone mass in older men. Positive association with serum estrogen and negative association with androgens. *The Journal of Clinical Investigation*, **100**, 1755–1759.

Sloan, J. and Holloway, G. (1981) Fractured neck of femur: the cause of the fall. *Injury*, **13**, 230–232.

Slosman, D.O., Rizzoli, R., Buschs, B. *et al.* (1990) Comparative study of the performance of X-ray and gadolinium 153 bone densitometers at the level of the spine femoral neck and femoral shaft. *European Journal of Nuclear Medicine*, **17**, 3–9.

Slovik, D.M., Neer, R.M., Ohman, J.L. *et al.* (1980) Parathyroid hormone and 25-hydroxyvitamin D levels in glucocorticoid treated patients. *Clinical Endocrinology (Oxford)*, **12**(3), 243–248.

Slovik, D.M., Rosenthal, D.I., Doppelt, S.H. *et al.* (1986) Restoration of spinal bone in osteoporotic men by treatment with parathyroid hormone (1–34) and 1,25 dihydroxy vitamin D. *Journal of Bone and Mineral Research*, **1**, 377–381.

Slutsky, L.J. (1980) Kinetic of the exchange of protons between hydrogen phosphate ions and a histidyl residue. *The Journal of Physical Chemistry*, **84**, 1325–1329.

Smalt, R., Mitchell, F.T., Howard, R.L. and Chambers, T.J. (1997) Induction of NO and prostaglandin E2 in osteoblasts by wall-shear stress but not mechanical strain. *The American Journal of Physiology*, **273**, E751–E758.

Smith, D.M., Khairi, M.R.A. and Johnston, C.C. (1975) The loss of bone mineral with aging and its relationship to risk of fracture. *The Journal of Clinical Investigation*, **56**, 311–318.

Smith, E.L. and Gilligan, C. (1990) Exercise and bone mass, in *Osteoporosis: Physiological Basis Assessment and Treatment* (eds H.F. Deluca and R. Mazess), Elsevier Science, New York, pp. 285–293.

Smith, E.L., Gilligan, C., McAdam, M. *et al.* (1989) Deterring bone loss by exercise intervention in premenopausal and postmenopausal women. *Calcified Tissue International*, **44**, 312–321.

Smith, E.P., Boyd, J., Franck, G.R. *et al.* (1994) Estrogen resistance caused by a mutation in the estrogen-receptor gene in a man. *The New England Journal of Medicine*, **331**, 1056–1061.

Smith, J.W. (1960) The arrangements of collagen fibres in human secondary osteones. *Journal of Anatomy*, **94**, 329.

Smith, J.W. (1968) Molecular pattern in native collagen. *Nature*, **219**, 157–158.

Smith, J.W. and Walmsley, R. (1959) Factors affecting the elasticity of bone. *Journal of Anatomy*, **93**, 503–523.

Smith, L.D. (1953) Hip fractures. The role of muscle contraction on intrinsic force in the causation of fractures of femoral neck. *The Journal of Bone and Joint Surgery*, **35**, 367–383.

Smith, M.A., Kennedy, N.S.J., Eastell, R. *et al.* (1980) Absolute measurement of total body calcium by neutron activation analysis. *Nuclear Medicine Communications*, **1**, 166.

Smith, M.A. and Tothill, P. (1979) Development of apparatus to measure calcium changes in the forearm and spine by neutron activation analysis using californium-252. *Physics in Medicine and Biology*, **24**, 319–322.

Smith, O.M., Goodman, E.M., Greenebaum, B. and Tipnis, P. (1991) An increase in the negative surface charge of U937 cells exposed to a pulsed magnetic field. *Bioelectromagnetics*, **12**, 197–202.

Smith, R. (1987) Osteopororsis: cause and management. *British Medical Journal*, **294**, 329.

Smith, R. (1990) Osteoprorsis after 60. *British Medical Journal*, **301**, 45–53.

Smith, R., Walton, R.J., Corner, B.D. and Gordon, I.R. (1977) Clinical and biochemical studies in Engelmann's disease (progressive diaphyseal dysplasia). *Quarterly Journal of Medicine*, **46**, 273–294.

Smith, R.W. and Rizek, J. (1966) Epidemiological studies of osteoporosis in women of Puerto Rico and Southeastern Michigan with special reference to age, race, national origin and to other related or associated findings. *Clinical Orthopaedics*, **45**, 31–48.

Smith, S.R. and Foster, K.R. (1985) Dielectric properties of low water content tissues. *Physics in Medicine and Biology*, **30**, 965–973.

Smith, T.G. Jr, Lange, G.D. and Marks, W.B. (1996) Fractal methods and results in cellular morphology – dimensions. *Lacunarity and Multifractals*, **69**, 123–136.

Smith, T.H., Burger, E.H. and Huyghe, J.M. (2002) A case for strain induced fluid flow as a regulator of BMU coupling and osteonal alignment. *Journal of Bone and Mineral Research*, **11**, 2021–2029.

Smith, W.C., Davis, R.H. and Fourman, P. (1960) Calcium deprivation in hypoparathyroidism, a method of diagnosis using sodium phytate. *Lancet*, **2**, 510.

Smyth, C.P. (1955) *Dielectric Behavior and Structure*, McGraw Hill Book Company, New York, pp. 132–220.

Snyder, S.M. and Schneider, E. (1991) Estimation of mechanical properties of cortical bone by computed tomography. *Journal of Orthopaedic Research*, **9**, 422–431.

Sogaard, C.H., Mosekilde, L., Richards, A. and Mosekilde, L. (1994) Marked decrease in trabecular bone quality after five years of sodium fluoride therapy – assessed by biomechanical testing of iliac crest bone biopsies in osteoporotic patients. *Bone*, **15**, 393–399.

Sogaard, C.H., Mosekilde, L., Thomsen, J.S. *et al.* (1997) A comparison of the effects of two anabolic agents (fluoride and PTH) on ash density and bone strength assessed in an osteopenic rat model. *Bone*, **20**, 439–449.

Solomon, L. (1985) Mechanism of idiopathic osteonecrosis. *The Orthopedic Clinics of North America*, **16**, 655–667.

Soltoff, S.R. (1998) Related adhesion focal tyrosine kinase and the epidermal growth factor receptor mediate the stimulation of mitogen activated protein kinase by the G-protein coupled P–2Y2 receptor. *The Journal of Biological Chemistry*, **273**, 23110–23117.

Soltoff, S.P., Avraham, H., Avraham, S. and Cantley, L.C. (1998) Activation of P-2Y2 receptors by UTP and ATP stimulates mitogen activated kinase activity through a pathway that involves related adhesion focal tyrosine kinase and protein kinase C. *The Journal of Biological Chemistry*, **273**, 2653–2660.

Somjen, D., Binderman, I., Berger, E. and Harell, A. (1980) Bone remodeling induced by physical stress is prostaglandin E2 mediated. *Biochimica et Biophysica Acta*, **627**, 91.

Sonstegard, D.A. and Matthews, L.S. (1976) Sonic diagnosis of bone fracture healing – a preliminary study. *Journal of Biomechanics*, **9**, 689–694.

Sorenson, J.A. (1990) Effects of nonmineral tissues on measurements of bone mineral content by dual photon absorptiometry. *Medical Physics*, **17**, 905–912.

Sorenson, J.A., Cameron, J.A. and Mazess, R.B. (1970) Measurement of bone mineral by the direct photon absorption method: Experimental results, in *Methods of Bone Mineral Measurement US Department of Health, Education and Welfare*, Public Health Service, National Institute of Health, pp. 269–282.

Sorenson, J.A., Doke, P.R. and Smith, S.W. (1989) Simulation studies of dual-energy X-ray absorptiometry. *Medical Physics*, **16**, 75–80.

Sowers, M.R., Clark, M.K., Jannausch, M.L. and Wallace, R.B. (1991) A prospective study of bone mineral content and fracture in communities with differential fluoride exposure. *American Journal of Epidemiology*, **133**, 649–660.

Soyka, L., Fairfield, W. and Klibanski, A. (2000) Hormonal determinants and disorders of peak bone mass in children. *The Journal of Clinical Endocrinology and Metabolism*, **85**, 3951–3963.

Spadaro, J.A. (1977) Electrical stimulated bone growth in animals and men. *Clinical Orthopaedics*, **122**, 325.

Spadaro, J.A., Albanese, S.A. and Chase, S.E. (1990) Electromagnetic field effects on bone formation at implants in the medullary canals in rabbits. *Journal of Orthopaedic Research*, **8**, 685–693.

Spadaro, J.A. and Becker, R.O. (1970) In distribution of trace metal ions in bone and tendon. *Calcified Tissue Research*, **6**, 49.

Spadaro, J.A., Mino, D.E., Chase, S.E. *et al.* (1986) Mechanical factors in electrode-induced osteogenesis. *Journal of Orthopaedic Research*, **4**, 37–44.

Spadaro, J.A., Werner, F.W., Brenner, R.A. *et al.* (1994) Cortical and trabecular bone contribute strength to the osteopenic distal radius. *Journal of Orthopaedic Research*, **12**, 211–218.

Spencer, H., Kramer, L., De Bartolo, M. *et al.* (1983) Further studies of the effect of a high protein diet as meat on calcium metabolism. *The American Journal of Clinical Nutrition*, **37**, 924–929.

Spencer, H., Kramer, L. and Osis, D. (1982) Factors contributing to calcium loss in aging. *The American Journal of Clinical Nutrition*, **36**, 776–787.

Spencer, H., Kramer, L., Osis, D. and Norris, D. (1975) Calcium absorption and balances during high phosphorous intake in man. *Federation Proceedings*, **34**, 88.

Spencer, H., Kramer, L., Osis, D. and Norris, D. (1978a) Effect of phosphorous on the absorption of calcium and calcium balance in man. *The Journal of Nutrition*, **108**, 447–457.

Spencer, H., Kramer, L., Osis, D. and Norris, D. (1978b) Effect of a high protein (meat) intake on calcium metabolism in man. *The American Journal of Clinical Nutrition*, **31**, 2167–2180.

Spencer, H., Mehiczel, J., Lewin, I. and Samachson, J. (1965) Effect of high phosphorous intake on calcium and phosphorous metabolism in man. *The Journal of Nutrition*, **86**, 125–132.

Spencer, H., Rudio, N., Rubio, E. *et al.* (1986) Chronic alcoholism, frequently overlooked cause of osteoporosis in man. *The American Journal of Medicine*, **80**, 393–397.

Spencer, G.J., Hitchcoek, I.S. and Genever, P.G. (2004) Emerging neuroskeletal signaling pathways: a review. *FEBS Letters*, **559**, 6–12.

Spinks, T.J. and Bewley, D.K. (1976) Activation analysis *in vivo* of the human body-using cyclotron produced neutrons. *Proceedings of the Analytical Division of the Chemical Society*, **13**, 204–206.

Spirt, A.A. and Pollack, S.R. (1993) *Electricity and Magnetism in Medicine and Biology* (ed. M. Blank), San Francisco Press, New York, pp. 689–692.

Sporonitz, U.M. and Frick, A. (1973) Effects of saline infusions on calcium concentrations in plasma ultrafilterates and on the ultrastructure of the parathyroid glands in the rat. *Pflugers Archive: European Journal of Physiology*, **340**, 161–174.

Squillante, R.G. and Williams, J.L. (1993) Videodensitometry of osteons in females with femoral neck fractures. *Calcified Tissue International*, **52**, 273–277.

Srinivasan, S., Weimer, D.A., Agans, S.C. *et al.* (2002) Low magnitude mechanical loading becomes osteogenic when rest is inserted between each load cycle. *Journal of Bone and Mineral Research*, **17**, 1613–1620.

Stalp, J.T. and Mazess, R.B. (1980) Determination of bone density by coherent-Compton scattering. *Medical Physics*, **7**, 723–726.

Stan, S., Sensen, W. and Mulier, J.C. (1976) Experimental study on the electrical impedance of bone and the effects of direct current on the healing of fracture. *Clinical Orthopaedics*, **120**, 264.

Starkebaum, W., Pollack, S.R. and Korostoff, E. (1979) Microelectrode studies of stress-generated potentials in four-point bending of bone. *Journal of Biomedical Materials Research*, **13**, 729–751.

Stauffer, M., Baylink, D., Wergedal, J. and Rich, C. (1973) Decreased bone formation mineralization and enhanced resorption in calcium deficient rats. *The American Journal of Physiology*, **225**, 269–276.

Steiger, P., Block, J.E., Steiger, S. *et al.* (1990) Spinal bone mineral density by quantitative computed tomography effect of region of interest, vertebral level and technique. *Radiology*, **175**, 537–543.

Stein, I.D. and Granik, G. (1980) Human vertebral bone: relation of strength, porosity and mineralization to fluoride content. *Calcified Tissue International*, **32**, 189–194.

Steinberg, B., Singh, I.J. and Mitchell, O.G. (1981) The effects of cold-stress. Hibernation, and prolonged inactivity on bone dynamics in the golden hamster, Mesocricetus auratus. *Journal of Morphology*, **167**, 43–51.

Steinberg, M.E., Bosch, A., Schwan, A. and Glazer, R. (1968) Electrical potentials in stressed bone. *Clinical Orthopaedics and Related Research*, **61**, 294–299.

Steinberg, M.E., Busenkell, G.L., Black, J. and Korostoff, E. (1974) Stress-induced potentials in moist bone *in vitro*. *The Journal of Bone and Joint Surgery*, **56**, 704–713.

Steinberg, M.E., Finnegan, W.J., Labooky, D.A. and Black, J. (1976) Temperature and thermal effects on deformation in potentials in bone. *Calcified Tissue Research*, **21**, 135–144.

Steinberg, M.E., Wert, R.E., Korostoff, E. and Black, J. (1973) Deformation potentials in whole bone. *The Journal of Surgical Research*, **14**, 254–259.

Steinbrech, D.S., Mehrara, B.J., Rowe, N.M. *et al.* (1999) Gene expression of insulin like growth factors I and II in rat membranous osteotomy healing. *Annals of Plastic Surgery*, **42**, 481–487.

Steiniche, T., Hasling, C., Charles, P. *et al.* (1989) A randomized study on the effects of estrogen gestagen or high dose oral calcium on trabecular bone remodeling in postmenopausal osteoporosis. *Bone*, **10**, 313–320.

Steinmann, B., Rao, V.H., Voge, A. *et al.* (1984) Cysteine in the triple helical domain of one alleic product of the al (I) gene of type I collagen produces a lethal form of osteogenesis. *The Journal of Biological Chemistry*, **259**, 1129–1138.

Stepan, J.J., Pospichal, J., Presl, J. and Pacovsky, V. (1989) Prospective trial of ossein hydroxyapatite compound in surgically induced postmenopausal women. *Bone*, **10**, 179–185.

Steppan, C.M., Crawford, D.T., Chidsey-Frink, K.L. *et al.* (2000) Leptin is a potent stimulator of bone growth in ob/ob mice. *Regulatory Peptides*, **92**, 73–78.

Stevenson, J.C. *et al.* (1981) Calcitonin and the calcium regulating hormones in postmenopausal women: effect of estrogens. *Lancet*, **1**, 693–695.

Stevenson, J.C., Lees, B., Evenport, M. *et al.* (1989) Determinants of bone density in normal women: risk factors for future osteoporosis. *British Medical Journal*, **298**, 924–928.

Stevenson, S., Cunningham, N., Toth, J. *et al.* (1994) The bone of osteogenin (bone morphogenetic protein) on the formation of bone in orthotopic segmental bone defects in rats. *The Journal of Bone and Joint Surgery*, **76**, 1676–1687.

Stewart, A. and Reid, D.M. (2000) Precision of quantitative ultrasound comparison of three commercial scanners. *Bone*, **27**, 139–143.

Stock, J.L., Ciders, J.A. and Mallette, L.E. (1985) Effect of short course of estrogen on mineral metabolism in postmenopausal women. *The Journal of Clinical Endocrinology and Metabolism*, **61**, 595–600.

Storm, T., Thamsborg, G., Steiniche, T. *et al.* (1990) Effect of intermittent cyclical etidronate therapy on bone mass and fracture rat in women with postmenopausal osteoporosis. *The New England Journal of Medicine*, **322**, 1265–1271.

Stoss, H. and Freisinger, P. (1990) The collagen fibrils of the osteoid in osteogenesis imperfecta-morphometrical analysis of the fibrils diametry. Proceedings of IV International Conference on Osteogenesis Imperfecta, p. 35.

Streeter, E.A. and Brandi, N.L. (1990) Biology of bone endothelial cells. *Bone & Mineral*, **10**, 85–94.

Strelitzki, R. and Evans, J.A. (1998) Diffraction and interface losses in broadband ultrasound attenuation measurements of the calcaneum. *Physiological Measurement*, **19**, 197–204.

Strelitzki, R., Evans, J.A. and Clarke, A.J. (1997) The influence of porosity and pore size on the ultrasonic properties of bone investigated using a phantom material. *Osteoporosis International*, **7**, 370–375.

Strelitzki, R., Nicholson, P.H. and Paech, V. (1998) A model for ultrasonic scattering in cancellous bone based on velocity fluctuations in a binary mixture. *Physiological Measurement*, **19**, 189–196.

Strelitzki, R. and Truscott, J.G. (1998) An evaluation of the reproducibility and responsiveness of four states-of-the art ultrasound heel bone measurement system using phantoms. *Osteoporosis International*, **8**, 104–109.

Stromsoe, K., Kok, W.L., Hoiseth, A. and Alho, A. (1993) Holding power of the 4.5 mm AO/ASIF cortex screw in cortical bone in relation to bone mineral injury. *Injury*, **24**, 656–659.

Studd, J., Savvas, M., Waston, N. *et al.* (1990) The relationship between plasma estradiol and the increase in bone density in postmenopausal women after with subcutaneous hormone implants. *American Journal of Obstetrics and Gynecology*, **163**, 1474–1479.

Styarkarsdottir, U., Cazier, J.B., Kong, A. *et al.* (2003) Linkage of osteoporosis to chromosome 20p12 and association to MBP2. *PLoS Biology*, **1**, E69.

Su, X.W., Feng, Q.L., Cui, F.Z. and Zhu, X.D. (1997) Microstructure and micromechanical properties of the mid-diaphyses of human fetal femur. *Connective Tissue Research*, **36**, 271–286.

Suda, T., Takahashi, N. and Martin, T.J. (1992) Modulation of osteoclast differentiation. *Endocrine Reviews*, **12**, 66–80.

Suda, T., Takahashi, N., Udagawa, N. *et al.* (1999) Modulation of osteoclast differentiation and function by the new members of the tumor necrosis factor receptor and ligand families. *Endocrine Reviews*, **20**, 345–357.

Sudo, H., Kodama, H., Amagai, Y. *et al.* (1983) *In vitro* differentiation and calcification in a new clonal osteogenic cells line derived from newborn mouse calvaria. *The Journal of Cell Biology*, **96**, 191–198.

Sufang, C., Nemoto, K., Kurose, R. *et al.* (2000) Different expression of leptin receptor variants in MC3T3-E1 osteoblastic cells and ATDC-5 chondroblastic cells. *Journal of Bone and Mineral Research*, **15**(Suppl 1), S466.

Sugimoto, H., Mimura, T. and Ohasawa, T. (1993) Susceptibility effects of bone trabeculae: quantification *in vivo* using an asymmetric spin-echo technique. *Investigative Radiology*, **28**, 208–213.

Sugiyama, K., He, P., Wada, S. and Saeki, S. (1999) Teas and other beverages suppress D-galactosamine induced liver injury in rats. *The Journal of Nutrition*, **129**, 1361–1367.

Suki, W.N. (1979) Calcium transport in the nephron. *The American Journal of Physiology*, **237**, 1–6.

Suki, W.N., Yium, J.J., Von Minden, M. *et al.* (1970) Acute treatment of hypercalcemia with furosemide. *The New England Journal of Medicine*, **283**, 836–840.

Suyama, H., Kawamoto, M., Gaus, S. and Yuge, O. (2003) Effect of JTC-801 (noci ception antagonist) on neuropathic pain in a rat model. *Neuroscience Letters*, **351**, 133–136.

Suzuki, Y., Ichikawa, Y., Saito, E. and Homma, M. (1983) Importance of increased urinary calcium excretion in the development of secondary hyperparathyroidism of patients under glucocorticoid therapy. *Metabolism: Clinical and Experimental*, **32**, 151–156.

Svendsen, O.L., Hassager, C., Skodt, V. and Christiansen, C. (1995) Impact of soft tissue on *in vivo* accuracy of bone mineral measurements in the spine, hip and forearm: A human cadaver study. *Journal of Bone and Mineral Research*, **10**, 868–873.

Swanberg, M., Mattila, P. and Knuuttila, M. (1997) Dietary xylitol retards the overiectomy induced increase of bone turnover in rats. *Calcified Tissue International*, **60**, 462–466.

Swanson, G.T. and Lafferty, J.F. (1972) Electrical properties of bone as a function of age, immobilization and vibration. *Journal of Biomechanics*, **5**, 261–266.

Sweeney, A.W., Kroon, R.P. and Byers, R.K. (1965) Mechanical characteristics of bone and its constituents. ASME paper 65-WA/HUF, Chicago Meeting, **7**, 1–17.

Swennen, G., Schliephake, H., Dempf, R. *et al.* (2001) Cranio-facial distraction osteogenesis: a review of the literature: Part I: clinical studies. *International Journal of Oral and Maxillofacial Surgery*, **30**, 89–103.

Swennen, G., Schliephake, H., Dempf, R. *et al.* (2002) Cranio-facial distraction osteogenesis: a review of the literature. Part II: experimental studies. *International Journal of Oral and Maxillofacial Surgery*, **31**, 123–135.

Swindell, W. and Barrett, H.H. (1977) Computerized tomography: taking sectional X-rays. *Physics Today*, **30**, 32–39.

Syftestad, G.T. and Urist, M.R. (1980) Growth hormone dependent matrix induced heterotopic bone formation. *Proceedings of the Society for Experimental Biology and Medicine*, **163**, 411–415.

Szent-Gyorgyi, A. (1941a) Study of energy levels in biochemistry. *Nature*, **148**, 157–159.

Szent-Gyorgyi, A. (1941b) Towards a new biochemistry. *Science*, **93**, 609–611.

Szent-Györgyi, A. and Banga, I. (1941) Adenosinetriphosphatase. *Science*, **93**(2407), 158–158.

Szule, P., Seeman, E. and Delmas, P.D. (2000) Biochemical measurements of bone turnover in children and adolescents. *Osteoporosis International*, **11**, 281–294.

Szule, P., Munoz, F., Claustrat, B. *et al.* (2001) Bioavailable estradiol may be an important determinant of osteoporosis in men: the MINOS study. *The Journal of Clinical Endocrinology and Metabolism*, **86**, 192–199.

Tabrah, F., Hoffmeier, M., Gilbert, F. Jr *et al.* (1990) Bone density changes in osteoporosis prone women exposed to pulsed electromagnetic fields (PEMFs). *Journal of Bone and Mineral Research*, **5**, 437–442.

Tahtela, R. and Tholix, E. (1996) Serum concentrations type I collagen carboxyterminal telopeptide (ICTP) and type I procollagen carboxy- and aminoterminal propeptide (PICP PINP) as markers of metastatic bone disease in breast cancer. *Anticancer Research*, **16**, 2289–2294.

Takada, M., Engelke, K., Hagiwara, S. *et al.* (1996) Accuracy and precision study *in vitro* for peripheral quantitative computed tomography. *Osteoporosis International*, **6**, 207–212.

Takahashi, H., Norimatsu, H., Ota, M. and Kimura, K. (1973) Bone cell dynamics of hypervitaminosis D, in *Clinical Aspects of Metabolic Bone Disease* (eds B. Frame, A.M. Parfitt and H. Duncan), Excerpta Medica, Amsterdam.

Takahashi, H.E., Tanizawa, T., Hori, M. and Uzawa, T. (1991) Effect of intermittent administration of human parathyroid hormone (1–34) on experimental osteopenia of rats induced by overiectomy. The aged rat model for bone biology studies. *Cells Materials*, (Suppl. 1), 113–117.

Takahashi, M., Wehrli, F.W., Hwang, S.N. and Wehrli, S.L. (2000) Relationship between cancellous bone induced magnetic field and ultrastructure in a rat overiectomy model. *Magnetic Resonance Imaging*, **18**, 33–39.

Takayama, K., Nomura, H., Tanoka, J. *et al.* (1990) Effect of a pulsing electromagnetic field on metabolically derived osteoporosis in rat: a pilot study. *ASAIO Transaction*, **36**, M426–M428.

Takeda, S., Elefteriou, F., Levassew, R. *et al.* (2002) Leptin regulates bone formation via the sympathetic nervous system. *Cell*, **111**, 305–317.

Takeuchi, Y., Matsumoto, T., Ogata, E. and Shishiba, Y. (1989) 1.25 Dihydroxy vitamin $D_3$ inhibits synthesis and enhances degradation of proteoglycans in osteoblastic cells. *The Journal of Biological Chemistry*, **264**, 18407–18413.

Takikawa, S., Matsui, N., Kokubu, T., Tsunoda, M., Fujioka, H., Mizuno, K. *et al.* (2001) Low intensity pulsed ultrasound initiates bone healing in rat nonunion fracture model. *Journal of Ultrasound in Medicine*, **20**, 197–205.

Talmage, R.V., Stinett, S.S., Landwehr, J.T. *et al.* (1986) Age related loss of bone mineral density in non-athletic and athletic women. *Bone and Mineral*, **1**, 115–125.

Tam, C.S., Heersche, J.W.M., Jones, G., Murray, T.M. and Rasmussen, H. (1986) The effect of vitamin D on bone in vivo. *Endocrin*, **118**, 2217–2224.

Tanaka, M., Ejiri, S., Nakajima, M. *et al.* (1999) Changes of cancellous bone mass in rat mandibular condyle following overiectomy. *Bone*, **25**, 339–347.

Tanaka, S.M., Alam, I. and Turner, C.H. (2002) Stochastic resonance in osteogenic response to mechanical loading. *The FASEB Journal*, 10.1096/fj.02-0561 fje.

Tanaka, S.M., Sun, H.B. and Yokota, H. (2004) Bone formation induced by a novel form of mechanical loading on joint tissue. *Biological Sciences in Space*, **18**, 41–44.

Tanaka, Y. and Deluca, H.F. (1971) Bone mineral mobilization of 1,25 $(OH)_2$ $D_3$, a metabolite of vitamin D. *Archives of Biochemistry and Biophysics*, **146**, 574–578.

Tasi, K.S., Health, H., Kumar, R. and Riggs, B.L. (1984) Impaired vitamin D metabolism with aging in women. *The Journal of Clinical Investigation*, **73**, 1668–1672.

Tatarinov, A., Sarvazyan, N., and Sarvazyan, A. (2005) Use of multiple acoustic wave modes for assessment of long bones: model study. *Ultrsonics*, **43**, 672–680.

Tatsuyama, K., Maezawa, Y., Baba, H. *et al.* (2000) Expression of various growth factors for cell proliferation and cytodifferentiation during fracture repair of bone. *European Journal of Histochemistry*, **44**, 269–278.

Tavakoli, M.B. and Evans, J.A. (1992) The effect of bone structure on ultrasonic attenuation and velocity. *Ultrasonics*, **30**(6), 389–395.

Teitelbaum, S.L. and Ross, F.P. (2003) Genetic regulation of osteoclast development and function. *Nature Reviews. Genetics*, **4**, 638–649.

Ten Husen, K., Martin, R.I., Klimkiewicz, M. and Brown, P.W. (1955) Formation and properties of synthetic bone composite: Hydroxyapatite collagen. *Journal of Biomedical Materials Research*, **29**, 803–810.

Tenenbaum, H.C. (1992) Cellular origins and theories of differentiation of bone-forming cells, in *Bone Vol. 1: The Osteoblast and Osteocyte* (ed. K. Hall), Telford Press, Caldwell, NJ, pp. 41–69.

Tenenbaum, H.C., Tornontali, M. and Sukhu, B. (1992) Effects of bisphophonates and inorganic pyrophosphate on osteogenesis *in vitro*. *Bone*, **13**, 249–255.

Termine, J.D. and Posner, A.S. (1966) Infrared analysis of rat bone: age dependency of amorphous and crystalline mineral fraction. *Science*, **153**, 1523.

Termine, J.D. and Posner, A.S. (1967) Amorphous/crystalline interrelationships in bone mineral. *Calcified Tissue Research*, **1**, 8.

Tervola, K.M.U., Foster, S.G. and O'Brien, W.D. Jr (1985a) Attenuation coefficient measurement technique at 100 MHz with the scanning laser acoustic microscope. *IEEE Transactions on Sonics and Ultrasonics*, **32**, 259–265.

Tervola, K.M.U., Grurnmer, M.A., Erdman, J.M. Jr and O'Brien, W.D. Jr (1985b) Ultrasonic attenuation and velocity properties in rat liver as a function of fat concentration. A study of 100 MHz using a scanning laser acoustic microscope. *The Journal of the Acoustical Society of America*, **77**, 307–313.

Thamsborg, G., Storm, T.L., Skyluski, R. *et al.* (1991) Effect of different doses of nasal salmon calcitonin on bone mass. *Calcified Tissue International*, **48**, 302–307.

Thiebaud, D., Sauty, A., Burckhardt, P. *et al.* (1997) An *in vitro* and *in vivo* study of cytokines in the acute phase response associated with bisphosphonates. *Calcified Tissue International*, **61**, 386–392.

Thomas, A.P., Bird, G.S. *et al.* (1996) Statial and temporal aspects of cellular calcium signaling. *The FASEB Journal*, **10**(13), 1505–1517.

Thomas, T., Gori, F., Khosla, S. *et al.* (1999) Leptin acts on human marrow stromal cells to enhance differentiation to osteoblasts and to inhibit differentiation to adipocytes. *Endocrinology*, **140**, 1630–1638.

Thompson, B.M., Mundy, G.R. and Chambers, I.J. (1987) Tumor necrosis factors alpha and beta induce osteoblastic cells to stimulate osteoclastic bone resorption. *Journal of Immunology*, **138**, 775–779.

Thompson, D.D. and Rodan, G.A. (1988) Indomethacin inhibition of tentomy induced bone resorption in rats. *Journal of Bone and Mineral Research*, **3**, 409–414.

Thompson, D.D., Seedor, J.G., Weinreb, M., Rosine, S. and Rodan, G.A. (1990) Aminohydroxy butane biphosphonate inhibits bone loss due to immobilization in rats. *Journal of Bone and Mineral Research*, **5**, 279–286.

Thompson, D.W. (1917) *On Growth and Form*, Cambridge University Press, London.

Thompson, P., Taylor, J., Fisher, A. and Oliver, R. (1998) Quantitative heel ultrasound in 3180 women between 45 and 75 years of age: compliance, normal ranges and relationship to fracture history. *Osteoporosis International*, **8**, 211–214.

Timmins, P.A. and Wall, J.C. (1977) Bone water. *Calcified Tissue Research*, **23**(1), 1–5.

Tis, J.E., Meffert, C.R., Inoue, N. *et al.* (2002) The effect of low intensity pulsed ultrasound applied to rabbit tibiae during the consolidation phase of distraction osteogenesis. *Journal of Orthopaedic Research*, **20**, 793–800.

Tobias, J.H. *et al.* (1990) High concentrations of 17 B estradiol stimulate bone formation in female rats. Proceedings of the 3rd International Symposium on Osteoporosis, Copenhagen, Denmark, October 14–18, p. 152.

Tobias, J.H., Chow, J.W.S. and Chambers, T.J. (1993) 3-Amino-1 hydroxypropylidine-1 bisphosphonate (AHPr BP) suppresses not only the induction of new but also the persistence of existing bone forming surfaces in rat cancellous bone. *Bone*, **14**, 619–623.

Tomaselli, V.P. and Shamos, M.H. (1973) Electrical properties of hydrated collagen I. *Dielectric Properties Biopolymers*, **12**, 353.

Tomkinson, A., Reeve, J., Shaw, R.W. and Noble, B.S. (1997) The death of osteocytes via apoptosis accompanies estrogen withdrawal in human bone. *The Journal of Clinical Endocrinology and Metabolism*, **82**, 3128–3135.

Tonna, E.A. (1978) Electron microscope study of bone surface changes during aging. The loss of cellular control and biofeedback. *Journal of Gerontology*, **33**, 163–167.

Tornberg, D.N. and Bassett, C.A.L. (1977) Activation of the resting periosteum. *Clinical Orthopedics*, **129**, 305.

Torrance, A.G., Mosley, J.R., Suswillo, R.F. and Lanyon, L.E. (1994) Noninvasive loading of the rat ulna *in vivo* induces a strain related modeling response uncomplicated by trauma or periosteal pressure. *Calcified Tissue International*, **54**, 241–247.

Tothill, P. (1989) Methods of bone mineral measurement. *Physics in Medicine and Biology*, **34**, 543–572.

Tothill, P., Avenell, A., Love, J. and Reid, D.M. (1994b) Comparisons between hologic. Lunar and Norland dual energy X-ray absorptiometers and other techniques used for whole body soft tissue measurements. *European Journal of Clinical Nutrition*, **48**, 781–794.

Tothill, P., Avendell, A. and Reid, D.M. (1994) Precision and accuracy of measurements of whole-body bone mineral: comparison between Hologic. Lunar and Norland dual-energy X-ray absorptiometers. *British Journal of Radiology*, **67**, 1210–1217.

Tothill, P. and Pye, D.W. (1992) Errors due to non-uniform distribution of fat in dual X-ray absorptiometry of the lumbar spine. *British Journal of Radiology*, **65**, 807–813.

Tothill, P., Smith, M.A. and Sutton, D. (1983) Dual photon absorptiometry of the spine with a low activity source of gadolinium 153. *The British Journal of Radiology*, **56**, 829–835.

Toyras, J., Kroger, H. and Jurvelin, J.S. (1999) Bone properties as estimated by mineral density, ultrasound attenuation and velocity. *Bone*, **25**, 725–731.

Toyras, J., Nieminen, M.T., Kroger, H. and Jurvelin, J.S. (2002) Bone mineral density, ultrasound velocity and broadband attenuation predict mechanical properties of trabecular bone differently. *Bone*, **31**, 503–507.

Trachte, K.L. and Dibenedetto, A.T. (1971) Fracture properties of poly (phenylene oxide) composites. *International Journal of Polymeric Materials*, **1**, 75–94.

Treloar, A.E. (1981) Menstrual cyclicity and the premenopause. *Maturitas*, **3**, 249–264.

Trock, D.H., Bollet, A.J., Dyer, R.H. Jr *et al.* (1993) A double blind trial of the clinical effects of pulsed electromagnetic fields in osteoarthritis. *The Journal of Rheumatology*, **20**, 456–460.

Trock, D.H., Bollet, A.J. and Markoll, R. (1994) The effect of pulsed electromagnetic fields in the treatment of osteoarthritis of the knee and cervical spine. Report of randomized, double blind, placebo controlled trials. *The Journal of Rheumatology*, **21**, 1903–1911.

Tsai, C.L., Chang, W.H., Liu, T.K. and Wu, K.H. (1991) Additive effects of prostaglandin E2 and pulsed electromagnetic fields on fracture healing. *The Chinese Journal of Physiology*, **34**, 201–211.

Tsai, K.S., Heath, H. III, Kumar, R. and Riggs, B.L. (1984) Impaired vitamin D metabolism with aging in women: possible role in pathogenesis of senile osteoporosis. *The Journal of Clinical Investigation*, **73**, 1668–1672.

Tseng, L. and Gurpide, E. (1975) Induction of human endometrial estradiol dehydrogenase by progestins. *Endocrinology*, **97**, 825–833.

Turner, C.H. (1991) Homeostatic control of bone structure: an application of feedback theory. *Bone*, **12**, 203–217.

Turner, C.H. (1993) Training and development for community dental staff. *British Dental Journal*, **175**(7), 232.

Turner, C.H. (1998) Three rules for bone adaptations to mechanical stimuli. *Bone*, **23**(5), 399–407.

Turner, C.H. (1999) Towards a mathematical description of bone biology: the principle of cellular acomodation. *Calcified Tissue International*, **65**, 466–481.

Turner, C.H., Akhter, M.P. and Heaney, R.P. (1992) The effects of fluoridated water on bone strength. *Journal of Orthopaedic Research*, **10**, 581–587.

Turner, C.H., Akhter, M.P., Raab, D.M. *et al.* (1991) A noninvasive, *in vivo* model for studying strain adaptive bone modeling. *Bone*, **12**, 73–79.

Turner, C.H. and Burr, D.B. (2001) Experimental techniques for bone mechanics, in *Bone Mechanics Handbook* (ed. S. Cowin), CRC, Boca Raton, FL, p. 7-1-7-35.

Turner, C.H. and Eich, M. (1991) Ultrasonic velocity as a predictor of strength in bovine cancellous bone. *Calcified Tissue International*, **49**(2), 116–119.

Turner, C.H., Forwood, M.R., Rho, J.Y. and Yoshikawa, T. (1994) Mechanical loading thresholds for lamellar and woven bone formation. *Journal of Bone and Mineral Research*, **9**, 87–97.

Turner, C.H., Peacock, M., Timmerman, L. *et al.* (1995) Calcaneal ultrasonic measurement discriminate hip fractures independently of bone loss. *Osteoporosis International*, **5**, 130–135.

Turner, C.H., Takano, Y., Owan, I. and Murrell, G.A.C. (1996) Nitric oxide inhibitor L-NAME suppresses mechanically induced bone formation in rats. *The American Journal of Physiology*, **270**, E634–E639.

Turner, R.T., Bell, N.H., Duvall, P. *et al.* (1985) Spaceflight results in formation of defective bone. *Proceedings of the Society for Experimental Biology and Medicine*, **180**, 544–549.

Turner, R.T., Colvard, D.S. and Spelsberg, T.C. (1990) Estrogen inhibition of periosteal bone formation in rat long bones: Down regulation of gene expression for bone matrix proteins. *Endocrinology*, **127**, 1346–1351.

Turner, R.T., Evans, G.L., Cavolina, J.M. *et al.* (1998) Programmed administration of parathyroid hormone (PTH) increases bone formation and reduces bone loss in hindlimb unloaded overiectomized rats. *Endocrinology*, **139**, 4086–4091.

Turner, C.H., Hsieh, Y.F., Muller, R. *et al.* (2000) Genetic regulation of cortical and trabecular bone strength and microstructure in inbred strains of mice. *Journal of Bone and Mineral Research*, **15**, 1126–1131.

Tzukert, A., Levine, E., Mohler, Y. and Shosham, S. (1983) Electroconductivity of bone *in vitro* and *in vivo*. *Clinical Orthopaedics*, **179**, 270–274.

Uama, O., Yamamoto, Y. and Kawai, A. (1997) Bone changes and crotid atherosclerosis in postmenopausal women. *Stroke*, **28**, 1730–1732.

Udagawa, N., Takahashi, N., Jimi, E. *et al.* (1999) Osteoblasts/stromal cells stimulate osteoclast differentiation factor/RANKL but not macrophage colony stimulating factor. *Bone*, **25**, 517–523.

Udagawa, N., Takahashi, N., Yasuda, H. *et al.* (2000) Osteoprotegerin produced by osteoblasts is an important regulator in osteoclast development and function. *Endocrinology*, **141**, 3478–3484.

Uglow, M.G., Peat, R., Hile, M. *et al.* (2003) Low-intensity ultrasound stimulation in distraction osteogenesis in rabbits. *Clinical Orthopaedics and Related Research*, **417**, 303–312.

Uhtoff, N.K. and Jaworski, Z.F.G. (1978) Bone in response to long-term immobilization. *The Journal of Bone and Joint Surgery*, **60**(3), 420.

Uitterlinden, A.G., Burger, H., Huang, Q. *et al.* (1998) Relation of alleles of the collagen type alphal gene to bone density and the risk of osteoporotic fractures in postmenopausal women. *The New England Journal of Medicine*, **338**, 1016–1021.

Umemura, Y., Ishiko, T. *et al.* (1997) Five jumps per day increase bone mass and breaking force in rats. *Journal of Bone and Mineral Research*, **12**(9), 1480–1485.

Urback, F. (1969) *the Biological Effects of Ultraviolet Radiations*, Pergman Press, London.

Uusitalo, H., Rantakokko, J., Ahonen, M. *et al.* (2001) A metaphyseal defect model of the femur for studies of urine bone healing. *Bone*, **28**, 423–429.

Uyama, O. *et al.* (1997) Bone changes and carotid atherosclerosis in postmenopausal women. *Stroke*, **28**, 1730–1732.

Vaillancourt, D.E. and Newell, K.M. (2002) Changing complexity in human behavior and physiology through aging and disease. *Neurobiology of Aging*, **23**, 1806–1809.

Van Breugel, H. and Bar, P.R.D. (1992) Power density and exposure time of He-Ne laser irradiation are more important than total energy dose in photo biomodulation of human fibroblast *in vitro*. *Lasers in Surgery and Medicine*, **12**, 528–537.

Van den Bergh, J.P., Van Lenthe, G.H., Hermus, A.R. *et al.* (2000) Speed of sound reflects Young's modulus as assessed by microstructural finite element analysis. *Bone*, **26**, 519–524.

Van der Graaf, E.R. and Ten Bosch, J.J. (1990) The uptake of water by freeze-dried human dentine sections. *Archives of Oral Biology*, **35**(9), 731–739.

Van der Hoeft, P.J., Kelly, P.J. and Peterson, L.F.A. (1962) Determination of growth rates in canine bone by means of tetracycline-labelled patterns. *Laboratory Investigation*, **11**, 714.

Van der Mark, K. (1999) Structure, biosynthesis, and gene regulation of collagens in cartilage and bone, in *Dynamics of Bone and Cartilage Metabolism* (eds M.J. Seibel, S.P. Robins and J.P. Bilezikian), Academic Press, New York, pp. 3–29.

Van der Perre, G., Lowet, G. and Christensen, A.B. (eds) (March 1991) Proceedings COMAC BME 11.2.6. (Monitorig of Fracture Healing). *In vivo* assessment of bone quality by vibration and wave propagation techniques (Published by KU Leuven, Division Biomechanics and Engineering Design, Ccelestignenlann 200A, B-3001 Leuven (Heverlee) pp. 101–113.

Van der Rest, M. and Garrone, R. (1991) Collagen family of proteins. *The FASEB Journal*, **5**, 2814–2823.

Van der Schueren, D., Van Herck, E., Suiker, A.M.H. *et al.* (1992) Bone and mineral metabolism in aged male rats: short and long term effects of androgen deficiency. *Endocrinology*, **130**, 2906–2916.

Van der Schueren, D., Vandenput, L., Boonen, S. *et al.* (2004) Androgens and bone. *Endocrine Reviews*, **25**, 389–425.

Van Leeuwen, J.P., Van Driel, M., Vandier Bemd, G.J. and Pols, H.A. (2001) Vitamin d control of osteoblast function and bone extracellular matrix mineralization. *Critical Reviews in Eukaryotic Gene Expression*, **11**, 199–226.

Van Pottellbergh, I., Geomacre, S., Zimierezak, H. and Kaufman, J.M. (2004) Perturbed sex steroid status in men with idiopathic osteoporosis and their sons. *The Journal of Clinical Endocrinology and Metabolism*, **89**, 4949–4953.

Van Wesenbreeck, L., Cleiren, E., Gram, J. *et al.* (2003) Six novel missense mutations in the LDL receptor-related protein 5 (LRP5) gene in different conditions with increased bone density. *American Journal of Human Genetics*, **72**, 763–771.

Vandenput, L., Boonen, S., Vn Herck, E. *et al.* (2002) Evidence from the aged orchidectomized male rat model that 17beta-estradiol is a more effective bone sparing and anabolic agent than 5 alpha-dihydrotestosterone. *Journal of Bone and Mineral Research*, **17**, 2080–2086.

VantHof, R.J. and Ralston, S.H. (1997) Cytokine induced nitric oxide inhibits bone resorption by inducing apoptosis of osteoclast progenitors and suppressing osteoclast activity. *Journal of Bone and Mineral Research*, **12**, 1797–1804.

Vargas, S.J. and Raisz, L.G. (1990) Simultaneous assessment of resorption and formation in cultures of fetal rat parietal bones: effects of parathyroid hormone and prostaglandin E2. *Bone*, 61–65.

Vaughan, D.W. and Filer, L.J. (1960) The enhancing action of certain carbohydrates on the intestinal absorption of calcium in the rat. *The Journal of Nutrition*, **71**, 10–14.

Vega, E., Gluiringhelli, G., Mautalen, C. *et al.* (1998) Bone mineral density and bone size in men with primary osteoporosis and vertebral fractures. *Calcified Tissue International*, **62**, 465–469.

Veiss, A. (1967) Intact collagen, in *Treatise on Collagen*, vol. **1**, (ed. G.N. Ramachandran), Academic Press, New York.

Veiss, A. (1974) Collagen Biosynthesis, in *CRC Critical Reviews in Biochemistry* **2**, The Chemical Rubber Co., Cleveland, Ohio, p. 443.

Verborgt, O., Gibson, G.J. and Schaffler, M.B. (2000) Loss of osteocyte integrity in association with microdamage and bone remodeling after fatigue *in vivo*. *Journal of Bone and Mineral Research*, **15**, 60–67.

Verharr, H.J.J., Damen, C.A., Duursma, S.A. and Scheven, B.A.A. (1994) A comparison of the action of progestins and estrogen on the growth and differentiation of normal adult human osteoblast like cells *in vitro*. *Bone*, **15**, 307–311.

Verschueren, S.M., Roelants, M., Delecluse, C. *et al.* (2004) Effects of six month whole body vibration training on postmenopausal women: a randomized controlled pilot study. *Journal of Bone and Mineral Research*, **3**, 352–359.

Vesterby, A. (1990) Star volume of marrow space and trabeculae in iliac crest: sampling procedure and correlation to star volume of first lumbar vertebra. *Bone*, **11**, 149–155.

Vesterby, A., Gundersen, H.J.G. and Melsen, F. (1989) Star volume of marrow space and trabeculae of first lumbar vertebral. Sampling efficiency and biological variation. *Bone*, **10**, 7–13.

Vesterby, A., Gundersen, H.J.G., Melsen, F. and Mosekilde, L. (1991) Marrow space star volume in the iliac crest decreases in osteoporotic patients after continuous treatment with fluoride, calcium and vitamin $D_2$ for five years. *Bone*, **12**, 33–37.

Vesterby, A., Kragstrup, J., Gundersen, H.J.G. and Melsen, F. (1987) Unbiased sterological estimation of surface density in bone using vertical sections. *Bone*, **8**, 13–17.

Vestergaard, P., Krogh, K., Rejnmark, L. and Mosekilde, L. (1998) Fracture rates and risk factors for fractures in patients with spinal cord injury. *Spinal Cord*, **36**(11), 790–796.

Vestergaard, P. and Mosekilde, L. (2003) Fracture risk associated with smoking: a meta-analysis. *Journal of Internal Medicine*, **254**, 572–583.

Vetter, U., Eanes, E.D., Kopp, J.B. *et al.* (1991) Changes in apatite crystal size in bones of patients with osteogenesis imperfecta. *Calcified Tissue International*, **49**, 248–250.

Viani, P., Cervato, G., Fiorilli, A. and Cestaro, B. (1991) Age-related differences in synaptosomal peroxidative damage and membrane properties. *Journal of Neurochemistry*, **56**(1), 253–258.

Vico, L., Chappard, D., Alexandre, C. *et al.* (1987) Effects of weightlessness on bone mass and osteoclast number in pregnant rats after a five-day spaceflight (COSMOS 1514). *Bone*, **8**, 95–103.

Vigorita, V.J. and Suda, M.K. (1982) The microscopic morphology of luoride induced bone. *Clinical Orthopaedics and Related Research*, **177**, 274–282.

Vinz, H. (1970) Studies on the density, water and mineral contents of compact human bone tissue depending on age. *Gegenbaurs Morphologisches Jahrbuch*. **115**, 257.

Vinz, H. (1972) Intramedullary nailing of femoral shaft factures in childhood. *Zentralblatt fur Chirurgie*, **97**, 90–95.

Vodoz, J.F., Luisier, M., Donath, A. *et al.* (1977) Diminution de absorption intestinale de 47-calcium dans 1' alcoolisme chronique. *Schweizerische Medizinische Wochenschrift*, **107**, 1525–1529.

Vogel, H.G. (1979) Influence of maturation and aging on mechanical and biochemical parameters of rat bone. *Gerontology*, **25**, 16–23.

Vogel, J.M., Wasnich, R.D. and Ross, P.D. (1988) The clinical relevance of calcaneus bone mineral measurement: a review. *Bone & Mineral*, **5**, 35–38.

Vogel, W., Kunert, C., Blumenstenge, K. *et al.* (1998) Correlation between granulocyte/macrophage-colony-forming units and $CD34^+$ cells in apheresis products from patients treated with different chemotherapy regimens and granulocyte-colony-stimulating factor to mobilize peripheral blood progenitor cells. *Journal of Cancer Research and Clinical Oncology*, **124**(6), 341–345.

Vogler, J.B. III and Murphy, W.A. (1988) Bone marrow imaging. *Radiology*, **168**, 679–693.

Volpin, G. and Salomon, C.F. (1978) Effect of a calcium deficient diet on the metararsal bones of rats of different ages. *Journal of Medical Science*, **14**, 379–383.

Von Hippel, A.R. (1954) *Dielectrics and Waves*, John Wiley & Sons, Inc, New York, pp. 228–234.

Vose, G.P. and Kubala, A.L. (1959) Bone strengh – in relationship to X-ray determined ash content. *Human Biology*, **31**, 262–270.

Vukicevic, S., Luyten, F.P., Kleinman, H.K. and Reddi, A.H. (1990) Differentiation of canalicular cell processes in bone cells by basement membrane matrix components: regulation by discrete domains of laminin. *Cell*, **63**, 437–445.

Wachter, N.J., Augat, P., Mentzal, M. *et al.* (2001) Predictive value of bone mineral density and morphology determined by peripheral quantitative computed tomography for cancellous bone strength of the proximal femur. *Bone*, **28**, 133–139.

Wachter, N.J., Krischak, G.D., Mentzel, M. *et al.* (2002) Correlation of bone mineral density with strength and microstructural parameters of cortical bone *in vitro*. *Bone*, **31**(1), 90–95.

Wagner, H.D. and Weiner, S. (1992) On the relationship between the microstructure of bone and its mechanical stiffness. *Journal of Biomechanics*, **25**, 1318–1320.

Wahlstrom, O. (1984) Stimulation of fracture healing with electromagnetic fields of extremely low frequency (EMF of ELF). *Clinical Orthopaedics*, **186**, 293.

Wahner, H.W. and Fogelman, I. (1994) Measurement of bone mineral in the lumbar spine: the AP spine scan, in *The Evaluation of Osteoporosis. Dual Energy X-ray Absorptiometry in Clinical Practice* (eds H.W. Wahner and I. Fogelman), Dunitz, London, pp. 88–120.

Wahner, H.W., Riggs, B.L. and Beabout, J.W. (1977) Diagnosis of osteoporosis. Usefulness of photon absorptiometry at the radius. *Journal of Nuclear Medicine*, **18**, 432.

Wakatsuki, T., Kolodney, M.S., Zahalak, G.I. and Elson, E.L. (2000) Cell mechanics studied by a reconstituted model tissue. *Biophysical Journal*, **79**, 2353–2368.

Walker, A.R.P. and Walker, B.F. (1987) Osteoporosis: cause and management. *British Medical Journal*, **294**, 702–703.

Wallace, B.A. and Cumming, R.G. (2000) Systemtic review of randomized trials of the effect of exercise on bone mass in pre and postmenopausal women. *Calcified Tissue International*, **67**, 10–18.

Walker, S.H. and Duncan, D.B. (1967) Estimation of the probability of an event as a function of several independent variables. *Biometrika*, **54**, 167–179.

Wallach, S., Feinblatt, J.D., Carstens, J.H.J. and Avioli, L.V. (1992) The bone "quality" problem. *Calcified Tissue International*, **51**, 169–172.

Walser, M. (1961) Ion association VI. Interaction between calcium, magnesium, inorganic phosphate, citrate and protein in normal human plasma. *The Journal of Clinical Investigation*, **40**, 723.

Walsh, W.R. and Guzelsu, N. (1991) Electrokinetic behavior of intake wet bone: compartmental model. *Journal of Orthopaedic Research*, **9**, 683–692.

Walsh, W.R., Sherman, P., Howlett, C.R. *et al.* (1997) Fracture healing in a rat osteopenia model. *Clinical Orthopaedics and Related Research*, **342**, 218–227.

Walters, M.R., Rosen, D.M., Norman, A.W. and Luben, A.R. (1982) 1,25 Dihydroxy vitamin $D_3$ receptors in an established bone cell line. *The Journal of Biological Chemistry*, **257**(7), 481–484.

Wang, C.J., Huang, H.Y., Chen, H.H. *et al.* (2001) Effect of shock wave therapy on acute fractures of the tibia. *Clinical Orthopaedics*, **387**, 112–118.

Wang, J., Heymsfield, S.B., Aulet, M. *et al.* (1989) Body fat from body density: Underwater weighing vs. dual photon absorptiometry. *The American Journal of Physiology*, **256**, E829–E834.

Wang, L., Banu, J. and Kalu, D.N. (2000b) Male rodent model of aging bone loss in men [Abstract]. *Journal of Bone and Mineral Research*, **15**, S335.

Wang, L., Banu, J., McMahan, A.C. and Kalu, D.N. (2000a) Male rodent model of aging bone loss in men. *Bone*, **29**, 141–148.

Wang, L.C., Swat, W., Fujiwara, Y. *et al.* (1998) The TEL/ETV6 gene is required specifically for hematopoiesis in the bone marrow. *Genes Dev*, **12**, 2392–2402.

Wang, R.S., Liu, T.K. and Shiau, S.Y. (1993) Increased bone growth by local prostaglandin E2 in rats. *Calcified Tissue International*, **52**, 57–61.

Wang, S.J., Lewallen, D.G., Balander, M.E. *et al.* (1994) Low intensity ultrsound treatment increases strength in a rat femoral fracture model. *Journal of Orthopaedic Research*, **12**, 40–47.

Wang, X., Athanasiou, K.A. and Agrawal, C.M. (1998a) Contribution of collagen to bone mechanical properties. Proceedings of the 17th Southern Biomedical Engineering Conference, San Antonio TX, p. 112.

Wang, X., Bank, R.A., Tekoppele, J.M. *et al.* (1998b) Relationship between bone mechanical properties and collagen denaturation. Proceedings of the 17th Sourthern Biomedical Engineering Conference, San Antonio, TX, p. 111.

Wang, X., Bank, R.A., TeKoppele, J.M. and Agrawal, C.M. (2001) The role of collagen in determining bone mechanical properties. *Journal of Orthopaedic Research*, **19**, 1021–1026.

Wang, X. and Ni, Q. (2003) Determination of cortical bone porosity and pore size distribution using a low field pulsed NMR approach. *Journal of Orthopaedic Research*, **19**, 1021–1026.

Wang, X., Shen, X., Li, X. and Agrawal, C.M. (2002) Age related changes in the collagen network and the toughness of bone. *Bone*, **31**, 961–967.

Wang, Y., Li, Yeubai., Mao, K. *et al.* (2003) Alcohol induced adipogenesis in bone and marrow. A possible mechanism for osteonecrosis. *Clinical Orthopaedics and Related Research*, **410**, 213–224.

Wang, Z.Q., Ovitt, C., Grigoriadis, A.E. *et al.* (1992) Bone and haematopoietic defects in mice lacking c-fos. *Nature*, **360**, 741–745.

Wapstra, A.H., Nijgh, G.J. and Van Lieshout, R.(eds) (1957) *Nuclear Spectroscopy Tables*, North-Holland, Amsterdam, pp. 66–74.

Warden, S.J., Bennell, K.L., Forwood, M.R. *et al.* (2001a) Skeletal effects of low intensity pulsed ultrasound on the ovariectomized rodent. *Ultrasound in Medicine and Biology*, **27**, 989–998.

Warden, S.J., Bennell, K.J., Malt Hews, B. *et al.* (2001b) Efficacy of low intensity pulsed ultrasound in the prevention of osteoporosis following spinal cord injury. *Bone*, **29**(5), 431–436.

Wasnich, R.C. (1996) Vertebral fracture epidemiology. *Bone*, **18**(3 Suppl.), 1795–1835.

Wasnich, R.D., Ross, P.D., Heilbrun, L.K. *et al.* (1985) Prediction of postmenopausal fracture risk with use of bone mineral measurements. *American Journal of Obstetrics and Gynecology*, **153**, 745–751.

Wasnich, R.D., Ross, P.D., Heilbrun, L.K. *et al.* (1986) Differential effects of thiazide and estrogen upon bone mineral content and fracture prevalence. *Obstetrics and Gynecology*, **67**, 457–462.

Wasnich, R.D., Ross, P.D., Heibrun, L.K. and Vogel, J.M. (1987) Selection of the optimal skeletal site for fracture prediction. *Clinical Orthopaedics*, **216**, 62–69.

Wasnich, R.D., Ross, P.D., Vogel, J.M. and Davis, J.W. (1989) *Osteoporosis Critique and Practicum*, Bayon Press, Honolulu.

Wasserman, R.H. (1984) Bones, in *Dukes' Physiology of Domestic Animals*, 10th edn (ed. J.M. Swenson), Comstock Publishing Associates, Ithaca, NY, pp. 467–485.

Wasserman, R.H., Henion, J.D., Haussler, M.R. and Mclain, T.A. (1976) Calcinogenic factor in solanum malacoxylon: evidence that it is 1,25 dihydroxy vitamin $D_3$- glycoside. *Science*, **194**, 853–855.

Watanabe, S., Fukermoto, S., Takeuchi, Y. *et al.* (2001) Effects of 1 year treatment with fluvastatin or paravastatin on bone. *The American Journal of Medicine*, **110**(7), 584–587.

Watanabe Tomita, Y., Suzuki, A., Shinoda, J. *et al.* (1997) Arachidonic acid release induced by extracellular ATP in osteoblasts: role of phospholipase D Prostagland. *Leukotrienes and Essential Fatty Acids*, **57**, 335–339.

Watt, D.E. (1975) Optimum photon energies for the measurement of bone mineral and fat fractions. *British Journal of Radiology*, **48**, 265–274.

Watts, N.B., Harris, S.T., Genant, H.K. *et al.* (1990) Intermittent cyclical etidronate treatment of postmenopausal osteoporosis. *The New England Journal of Medicine*, **323**, 73–79.

Waud, C.E., Lew, R. and Baran, D.T. (1992) The relationship between ultrasound and densitometric measurements of bone mass at the calcaneus in women. *Calcified Tissue International*, **51**, 415–418.

Wear, K.A. (2000) Measurements of phase velocity and group velocity in human calcaneus ultrasound. *Medical Biology*, **26**, 641–646.

Wear, K.A. and Garra, B.S. (1998) Assessment of bone density using ultrasonic backscatter. *Ultrasound in Medicine and Biology*, **24**, 689–695.

Weatherell, J.A. and Weidmann, S.M. (1959) The skeletal changes of chronic experimental fluorosis. *Journal of Pathology & Bacteriology*, **78**, 233–241.

Weaver, J.C. and Astumian, R.D. (1990) The response of living cells to very weak electric fields: the thermal noise limit. *Science (Washington, D.C.)*, **247**, 459–462.

Weaver, J.C. and Astumian, R.D. (1992) Estimates for ELF effects: noise-based thresholds and the number of experimental conditions required for empirical searches. *Bioelectromagnetics*, (Suppl. 1), 119–138.

Weaver, J.K. and Chalmers, J. (1966) Cancellous bone: its strength and changes with aging and an evaluation of some methods for measuring its mineral content. I. Age changes in cancellous bone. *The Journal of Bone and Joint Surgery*, **48a**, 289.

Webber, C.E. (1987) The effect of fat on bone mineral measurements in normal subjects with recommended values of bone, muscle and fat attenuation coefficients. *Clinical Physics and Physiological Measurement*, **8**, 143–148.

Webber, C.E. (2006) Photon absorptiometry, bone densitometry and the challenges of osteoporosis. *Physics in Medicine and Biology*, **51**, R169–R185.

Webber, C.E. and Kennett, J. (1976) Bone density measured by photon scattering. I. A system for clinical use. *Physics in Medicine and Biology*, **21**(5), 760–769.

Weber, J. and Van den Berge, D.J. (1969) The effective atomic number and the calculation of the composition of phantom materials. *British Journal of Radiology*, **42**, 378–383.

Webster, D.J. and Lillicrap, S.C. (1985) Coherent-Compton scattering for the assessment of bone mineral content using heavily filtered X-ray beams. *Physics in Medicine and Biology*, **30**(6), 531–539.

Wehrli, F.W., Ford, J.C., Attie, M.S. *et al.* (1991) Trabecular structure, preliminary applications of MR interferometry. *Radiology*, **179**, 615–621.

Wehrli, F.W., Hilaire, L., Fernandez-Seara, M. *et al.* (2002) Quantitative magnetic resonance imaging in the calcaneus and femur of women with varying degrees of osteopenia and vertebral deformity status. *Journal of Bone and Mineral Research*, **17**, 2265–2273.

Wehrli, F.W., Saha, P.K., Gomberg, B.R. *et al.* (2002) Role of magnetic resonance for assessing structure and function of trabecular bone. *Topics in Magnetic Resonance Imaging*, **13**, 335–355.

Weibe, S.H., Sims, S.M. and Dixon, S.J. (1999) Calcium signaling via multiple P2 purinoceptor subtypes in rat osteoclasts. *Cellular Physiology and Biochemistry*, **9**, 323–337.

Weinbaum, S.C., Cowin, S.C. and Wu, Z. (1991) A model for the fluid shear stress excitation of membrane ion channels in osteocytic processes due to bone strain. *Annual Biomedical Engineering Society*, **91**, 636–639.

Weinbaum, S., Cowin, S.C. *et al.* (1994) A model for the excitation of osteocytes by mechanical loading induced bone fluid shear stresses. *Journal of Biomechanics*, **27**(3), 339–360.

Weiner, S. and Traub, W. (1989) Crystal size and organization in bone. *Calcified Tissue Research*, **21**, 259–265.

Weiner, S. and Wagner, H.D. (1998) The material bone: structure-mechanical function relations. *Annual Review of Materials Science*, **28**, 271–298.

Weinger, J.M. and Holtrop, M.E. (1974) An ultrastructural study of bone cells: the occurrence of microtubules, microfilaments and tight junctions. *Calcified Tissue Research*, **14**, 15–29.

Weinstein, R.S., Jilka, R.L., Parfitt, A.M. and Manolagas, S.C. (1998) Inhibition of osteoblastogenesis and promotion of apoptosis of osteoblasts and osteocytes by glucocorticoids. Protein mechanisms of their deleterious effects on bone. *The Journal of Clinical Investigation*, **102**, 274–282.

Weinstein, R.S., Underwood, L., Huston, M.S. and Deluca, H.F. (1984) Bone histomorphometry in vitamin D-deficient rats infused with calcium and phosphorous. *The American Journal of Physiology*, **245**, E499–E505.

Weisbrode, S.E., Capen, C.C. and Norman, A.W. (1978) Ultrastructural examination of the effect of 1,25-dihydroxy vitamin $D_3$ on bone of thyroparathyroidectomized rats fed a low calcium diet. *The American Journal of Pathology*, **92**, 459–466.

Weisbrode, S.E., Capen, C.C. and Norman, A.W. (1979) Light and electronmicroscopic evaluation of the effects of 1,25-dihydroxy vitamin $D_3$ on thyroparathyroidectomized rats. *The American Journal of Pathology*, **97**, 247–254.

Weiss, N.S., Ure, C.L., Ballard, J.H. *et al.* (1980) Decreased risk of fractures of the hip and lower forearm with postmenopausal rate of estrogen. *The New England Journal of Medicine*, **303**, 1195–1198.

Weiss, R.E., Singer, F.R., Gorn, A.H. *et al.* (1981) Calcitonin stimulates bone formation when administered prior to initiation of osteogenesis. *The Journal of Clinical Investigation*, **68**, 826–828.

Welch, R.D., Johnston, C.E., Waldron, M.J. and Poteet, B. (1993) Bone changes associated with intraosseous hypertension in the caprine tibia. *The Journal of Bone and Joint Surgery*, **75-A**, 53–60.

Welch, R.D., Waldron, M.J., Hulse, D.A. *et al.* (1992) Intraosseous infusion using the osteoport implant in the caprine tibia. *Journal of Orthopaedic Research*, **10**, 789–799.

Wertheim, M.G. (1847) Memoire sur l'élasticite et la cohésion des principeaux tissus du corps humain. *Annales des Chimie et des Physique*, **21**, 385.

Westerland, K.C., Fluckey, J.D., Gordon, S.E. *et al.* (1998) Effect of resistance exercise training on cortical and cancellous bone in mature male rats. *Journal of Applied Physiology*, **84**, 459–464.

Whealan, K.M., Kwak, C.D., Tedrow, J.R. *et al.* (2000) Non-invasive imaging predicts failure load of the spine with simulated osteolytic defects. *The Journal of Bone and Joint Surgery. American Volume*, **82**, 1240–1251.

Whedon, G.D. and Cameron, J.R. (1970) *Progress in Methods of Bone Mineral Measurement*, US Department of Health, Education, and Welfare. Public, Health, National Institute of Health, 352–411.

Whedon, G.D., Lutwak, L., Reid, J. *et al.* (1974) Mineral and nitrogen metabolic studies on skylab orbital space flights. *Transactions of the Association of American Physicians*, **87**, 95–110.

Whit Athey, T., Stuchly, M.A. and Stuchly, S.S. (1982) Measurement of radio frequency permittivity of biological tissues with an open ended coaxial line: Part 1. *IEEE Transactions on Microwave Theory and Techniques*, **30**, 82–87.

White, D.R. (1977) An analysis of the z-dependence of photon and electron interactions. *Physics in Medicine and Biology*, **22**, 219–228.

White, S.C. (2002) Oral radiographic predictors of osteoporosis. *Dentomaxillofacial Radiology*, **31**, 84–92.

White, S.W., Hulmes, D.J.S., Miller, A. and Timmins, P.A. (1977) Collagen mineral axial relationship in calcified turkey leg tendon by X-ray and neutron diffraction. *Nature*, **266**, 421–425.

Whitehouse, W.J. (1974) The quantitative morphology of anisotropic trabecular bone. *Journal of Microscopy*, **101**, 153–168.

Whitehouse, W.J. and Dyson, E.D. (1974) Scanning electron microscope studies of trabecular bone in the proximal end of the human femur. *Journal of Anatomy*, **118**, 417–444.

Whiting, S.J. and Draper, H.H. (1980) The role of sulfate in the calcium of high protein diets in adult rats. *The Journal of Nutrition*, **110**, 212–222.

Whiting, S.J. and Draper, H.H. (1981) Effect of chronic high protein feeding on bone composition in the adult rat. *The Journal of Nutrition*, **111**, 178–183.

Whitson, S.W., Harrison, W., Dunlap, M.K. *et al.* (1984) Fetal bovine bone cells synthesize bone specific matrix proteins. *The Journal of Cell Biology*, **99**, 607–614.

Whyte, M.P., Bergfeld, M.A., Murphy, W.A. *et al.* (1982) Postmenopausal osteoporosis. A heterogenous disorder assessed by histomorphometric analysis of iliac crest bone from untreated patients. *The American Journal of Medicine*, **72**, 193–202.

Whyte, M.P., Obrecht, S.E., Finnegan, P.M. *et al.* (2002) Homozygous deletion of osteoprotegerin and juvenile Paget disease. *The New England Journal of Medicine*, **347**, 175–184.

Wienbaum, S., Cowin, S.C. and Zeng, Y. (1994) A model for the excitation of osteocytes by mechanical loading induced bone fluid shear stresses. *Journal of Biomechanics*, **27**, 339–360.

Wiesmann, H., Harting, M., Styratmann, U. *et al.* (2001) Electrical stimulation in fluences mineral formation of osteoblast-like cells *in vitro*. *Biochimica et Biophysica Acta*, **1538**, 28–37.

Wilhelm, G., Felsenberg, D., Ogusch, G. *et al.* (1998) Biomechanical examinations for validation of the bone strength index SSI, calculated by peripheral quantitative tomography, in *Musculorpeletal Interaction*, vol. **2**(ed. G.P. Lyritis), Athens, pp. 105–110.

Willardson, R.H., Harman, T.C. and Beer, A.C. (1954) Transverse Hall and magneto resistance effects in P-type germanium. *Physical Review*, **96**(6), 1512–1518.

Williams, C.J. and Prockop, D.J. (1983) Synthesis and processing of a type I procollagen containing shortened Proal (I) chains by fibroblasts from a patient with osteogenesis imperfecta. *The Journal of Biological Chemistry*, **258**, 5915–5921.

Williams, D.C., Border, G.B., Toomey, R.E. *et al.* (1980) Mineralization and metabolic response in serially passaged adult rat bone cells. *Calcified Tissue International*, **30**, 233–246.

Williams, J.L. (1992) Ultrasonic wave propagation in cancellous and cortical bone: Prediction of some experimental results by Biot's theory. *The Journal of the Acoustical Society of America*, **91**, 1106–1112.

Williams, P.A. and Saha, S. (1996) The electrical and dielectric properties of human bone tissue and their relationship with density and bone mineral content. *Annals of Biomedical Engineering*, **24**, 222–233.

Williams, R.J.P. (1980) On first looking into nature's chemistry I. The role of small molecules and ions: the transport of the elements. *Chemical Society Reviews*, **9**, 181–324.

Williams, W.S. (1974) Sources of piezoelectricity in tendon and bone. *CRC Critical Reviews in Bioengineering*, **2**(1), 95–118.

Williams, W.S. (1982) Piezoelectric effects in biological materials. *Ferroelectrics*, **41**, 225–246.

Williams, W.S. (1993) Transactions Volume XIII of the Bioelectrical Repair and Growth Society 13th Annual Meeting; Bioelectrical Repair and Growth Society: Dresher, PA, p. 27.

Williams, W.S. and Breger, L. (1974) Analysis of stress distribution and piezoelectric response in cantilever bending of bone and tendon. *Annals of the New York Academy of Sciences*, **238**, 121–129.

Williams, W.S. and Breger, L. (1975) Piezoelectricity in tendon and bone. *Journal of Biomechanics*, **8**, 1–6.

Wilson, C.R. (1972) The use of *in vivo* Bone Mineral Determination to Predict the Strength of Bone, PhD Dissertation, University of Wisconsin.

Wilson, C.R. (1973) Prediction of remoral neck and spine BMC from the BMC of the radius or ulna and the relationship between bone strength and BMC. Proceedings, International Conference on Bone Mineral Measurement, Chicago, 12–13 October.

Wilson, C.R. (1974) Prediction of femoral neck and spine bone mineral content from the BMC of the radius or ulna and the relationship between bone strength and BMC, in Proceedings of the International Conference on Bone Mineral Measurement (ed. R.B. Mazess), US Department of Health, Washington, D.C., Education and Welfare Publication NIH-75-683, pp. 51–59.

Wilson, C.R. and Madsen, M. (1977) Dichomatic absorptiometry of vertebrae bone mineral content. *Investigative Radiology*, **12**, 180–184.

Wilson, O.H. and Jagadeesh, P. (1976) Experimental regeneration in peripheral nerves and the spinal cord in laboratory animals exposed to a pulsed electromagnetic field. *Paraplegia*, **14**, 12–20.

Wimalawansa, S.J. (1993) Therapeutic options in prevention and treatment of osteoporosis. *Experimental and Clinical Endocrinology*, **12**, 1027.

Winand, L. (1965) Physico-chemical study of some apatitic calcium phosphates, in *Tooth Enamel* (eds M.V. Stack *et al.*), John Wright and Sons, Bristol, p. 15.

Windahl, S.H. *et al.* (2001) Female estrogen receptor $\beta$/mice are partially protected against age-related trabecular bone loss. *Journal of Bone and Mineral Research*, **16**, 1388–1398.

Windahl, S.H., Hollberg, K., Vidal, O. *et al.* (2001) Female estrogen receptor- $\beta^{-/-}$ mice are partially protected against age-related trabecular bone loss. *Journal of Bone and Mineral Research*, **16**, 1388–1398.

Wink, C.S. and Felts, W.J.L. (1980) Effects of castration on the bone structure of male rats: A model of osteoporosis. *Calcified Tissue International*, **32**, 77–82.

Wishko, D.S. (1975) *In Vivo* Analysis of Compact Bone Structure, PhD Thesis, University of California, Los Angeles, p. 111.

Wolff, J. (1892) *Das Gasetz der Transformation der knochen*, Hirschwald, Berlin.

Wolff, J. (1986) *The Law of Bone Remodeling* (eds P. Maget and R. Furlong), Springer-Verlag, Berlin.

Wolff, J. and Das, Gesetzder. (1986) *Translated as: The Law of Bone Remodeling*, Springer-Verlag, Heidelberg.

Wolff, I., Van Croonenborg, J.J., Kemper, H.C. *et al.* (1999) The effect of exercise training programs on bone mass: A meta-analysis of published controlled trials in pre and postmenopausal women. *Osteoporosis International*, **9**, 1–12.

Wong, A.T.C., Goldsmith, W. and Sackman, J.L. (1976) Flexural wave propagation in discontinuous model and *in vitro* tibiae. *Journal of Biomechanics*, **9**, 813–825.

Wong, F.H. and Pun, K.K. (1993) Total and regional bone mineral densities in women with Colles' fractures: a comparative study with normal matched controls. *Singapore Medical Journal*, **34**, 229–232.

Wong, F.Y., Pal, S. and Saha, S. (1983) The assessment of *in vivo* bone condition in human's by impact response measurement. *Journal of Biomechanics*, **16**(10), 849–856.

Wong, G.L. (1979) Basal activities and hormone responsiveness of osteoclast like and osteoblast like bone cells are regulated by glucocorticoids. *The Journal of Biological Chemistry*, **254**(14), 6337–6340.

Wong, J.C., McEwan, L., Lee, N. *et al.* (2003) The diagnostic role of dual femur bone density measurement in low impact fractures. *Osteoporosis International*, **14**, 339–344.

Wong, M., Pope, A., Long, T. *et al.* (2005) Validation of thoracic quantitative computed tomography as a method to measure bone mineral density. *Calcified Tissue International*, **76**, 7–10.

Woo, S.L.-Y., Kuei, S.C., Amiel, D. *et al.* (1981) The effect of prolonged physical training on the properties of long bone: A study of Wolff's law. *The Journal of Bone and Joint Surgery*, **63-A**(5), 780–787.

Wood, R.J., Theall, C.L., Contois, J.H. and Rosenberg, I.H. (1988) Intestinal end organ resistance to 1,25 (OH) D stimulation of calcium absorption in the senescent rat, in *Vitamin D: Molecular, Cellular and Clinical Endocrinology* (eds A.W. Norman, K. Schaefer, H.G. Grigoleit and D.V. Herrth), Walter de Gruyter, Berlin, pp. 907–908.

Woodard, H.Q. (1962) The elementary composition of human cortical bone. *Health Physics*, **8**, 513.

Woosley, J.T. (1991) Measuring cell proliferation. *Archives of Pathology & Laboratory Medicine*, **115**, 555–557.

Wooten, W.W., Judy, P.F. and Greenfield, M.A. (1973) Analysis of the effects of adipo tissue on the absorptiometric measurement of bone mineral mass. *Investigative Radiology*, **8**, 84–89.

World Health Organization (1994) Assessment of fracture risk and application to screening for postmenopausal osteoporosis. WHO Technical Report Series 8, 43.

Wozney, J.M. and Rosen, V. (1998) Bone morphogenetic protein and bone morphogenetic protein gene family in bone formation and repair. *Clinical Orthopaedics*, **346**, 26–37.

Wozney, J.M., Rosen, V., Celeste, A.J. *et al.* (1988) Novel regulators of bone image molecular clones and activities. *Science*, **242**, 1528–1530.

Wright, L.L., Glade, M.J. and Gopal, J. (1987) The use of transmission ultrasonics to assess bone status in the human newborn. *Pediatric Research*, **22**, 541–545.

Wright, T.M. and Hayes, W.C. (1976) Tensile testing of bone over a wide range of strain rates: effects of strain rate, microstructure and density. *Medical and Biological Engineering and Computing*, **14**, 671.

Wright, T.M. and Hayes, W.C. (1977) Fracture mechanics parameters for compact bone effects of density and specimen thickness. *Journal of Biomechanics*, **10**, 419–430.

Wronski, T.J., Cintron, M. and Dann, L.M. (1988) Temporal relationship between bone loss and increased bone turnover in ovariectomized rats. *Calcified Tissue International*, **43**, 179–183.

Wronski, T.J., Cintron, M., Doherty, A.L. and Dann, L.M. (1988b) Estrogen treatment prevents osteopenia and depresses bone turnover in ovariectomized rats. *Endocrinology*, **123**, 681–686.

Wronski, T.J., Dann, I.M., Scott, K.S. and Crooke, I.R. (1989) Endocrine and pharmacological suppressors of bone turnover protect against osteopenia I overiectomized rats. *Endocrinology*, **125**, 810–816.

Wronski, T.J., Lowry, P.L., Walsh, C.C. and Ignaszewski, L.A. (1985) Skeletal alteration in overiectomized rats. *Calcified Tissue International*, **37**, 324–328.

Wronski, T.J. and Morey, E.R. (1983) Effect of spaceflight on periosteal bone formation in rats. *The American Journal of Physiology*, **244**, R305–R309.

Wronski, T.J., Walsh, C.C. and Ignaszewski, L.A. (1986) Histologic evidence for osteopenia and increased bone turnover in overiectomized rats. *Bone*, **7**, 119–123.

Wronski, T.J. and Yen, C.F. (1994) Anabolic effects of parathyroid hormone on cortical bone in ovariectomized rats. *Bone*, **15**(1), 51–58.

Wronski, T.J., Yen, C.F. and Scott, K.S. (1991) Estrogen and diphosphonate treatment provide long-term protection against osteopenia in ovariectomized rats. *Journal of Bone and Mineral Research*, **6**, 387–394.

Wu, D., Hayek, M.G. and Meydani, S.N. (2001) Vitamin E and macrophage cyclooxygenase regulation in the aged. *The Journal of Nutrition*, **131**, 3825–3885.

Wu, D.D., Otter, M.W., Bieber, W.A. and Cochran, G.V.B. (1993) Transactions Volume XIII of the Bioelectrical Repair and Growth Society 13th Annual Meeting: Bioelectrical Repair and Growth Society: Dresher, PA, p. 6.

Wu, J.J., Shyr, H.C., Chao, E.Y.S. and Kelly, P.J. (1984) Comparison of osteotomy healing external fixation devices with different stiffness characteristics. *The Journal of Bone and Joint Surgery*, **66**, 1258–1264.

Wuster, C., Scheidt, C., Bergman, M. *et al.* (1992) Bone density measurement of calcaneus with ultrasound: A new precision procedure with good agreement to vertebral measurement. *Osteoporosis*, **1**, 237–242.

Wuthier, R.E. (1982) A review of the primary mechanism endochondral calcification with special emphasis on the role of cells, mitochondria and matrix vesicles. *Clinical Orthopaedics and Related Research*, **169**, 219–242.

Xia, S.L. and Ferrier, J. (1992) Propagation of a calcium pulse between osteoblastic cells. *Biochemical and Biophysical Research Communications*, **186**(3), 1212–1219.

Xing, M., Post, S., Ostrom, R.S. *et al.* (1999) Inhibition of phospholipase A2 mediated arachidonic acid release by cyclic AMP defines a negative feedback loop for P2Y receptor activation in Madin-Darby canine kidney D1 cells. *The Journal of Biological Chemistry*, **274**, 10035–10038.

Yaakobi, T., Maltz, L. and Oron, U. (1996) Promotion of bone repair in the cortical bone of the tibia in rats by low energy laser (He-Ne) irradiation. *Calcified Tissue International*, **59**(4), 297–300.

Yamada, H. (1973) *Strength of Materials* (ed. F.G. Evans), Krieger, Huntington, New York.

Yamada, H. and Evans, F.G. (1970) *Strength of Biological Materials*, Williams & Wilkins, Baltimore.

Yamagiwa, H., Tokumaga, K., Hayami, T. *et al.* (1999) Expression of metalloproteinase-13 (collagenase-3) is induced during fracture healing in mice. *Bone*, **25**(2), 197–203.

Yamamoto, I., Kitamura, N., Aoka, J. *et al.* (1987) Circulating 1,25 dihydroxy vitamin D concentrations in patients with renal cells carcinoma associated hypercalcemia are rarely suppressed. *The Journal of Clinical Endocrinology and Metabolism*, **64**, 175–179.

Yamaguchi, I., Tokuchi, K., Fukuzaki, H. *et al.* (2001a) Preparation and mechnical properties of chitosan/hydroxyapatite nanocomposites. *Key Engineering Materials*, **192–195**, 673–679.

Yamaguchi, I., Tokuchi, K., Fukuzaki, H. *et al.* (2001b) Preparation and microstructure analysis of chitosan/hydroxyapatite nanocomposites. *Journal of Biomedical Materials Research*, **55**, 20–27.

Yamashita, E., Atsulci, K. and Hirahata, T. (1981) Microstrip disperion in a wide frequency range. *IEEE Transactions on Microwave Theory and Techniques*, **29**(6), 610–611.

Yamashita, J., Li, X., Furman, B.R. *et al.* (2002) Collagen and bone viscoelasticity: a dynamic mechanical analysis. *Journal of Biomedical Materials Research*, **63**, 31–36.

Yamashita, T., Sekiya, I., Kawaguchi, N. *et al.* (2001) Klotho-deficient mice are resistant to bone loss induced by unloading due to sciatic neurectomy. *The Journal of Endocrinology*, **168**, 347–351.

Yamauchi, M., Sugimoto, T., Yamaguchi, T. *et al.* (2001) Plasma leptin concentrations are associated with bone mineral density and the presence of vertebral fractures in postmenopausal women. *Clinical Endocrinology*, **55**, 341–347.

Yang, K.H., Parvizi, J., Wang, S.J. *et al.* (1996) Exposure to low intensity ultrasound increases aggrecan gene expression in a rat femur fracture model. *Journal of Orthopaedic Research*, **14**, 802–809.

Yang, R., Tsai, K., Chieng, P. and Liu, T. (1997) Symmetry of bone mineral density at the proximal femur with emphasis on the effect of side dominance. *Calcified Tissue International*, **61**, 189–191.

Yang, R.S., Chang, W.H. and Liu, H.C. (1994) Clinical evaluation of non-union and delayed union by a specific parameter electrical stimulation. *Jourbnal of Japanese Bio-Electronics Research Society*, **8**, 117–125.

Yang, X., Matsuda, K., Bialek, P. *et al.* (2004) ATF4 is a substrate of RSK2 and an essential regulator of osteoblast biology; implication for Coffin-Lowry syndrome. *Cell*, **117**, 387–398.

Yano, K., Wasnich, R.D., Vogel, J.M. and Heibrum, L.K. (1984) Bone mineral measurements among middle aged and elderly Japanese residents in Hawaii. *American Journal of Epidemiology*, **119**, 751–764.

Yasko, A.W., Lowe, J.M., Fellinger, E.J. *et al.* (1992) The healing of segmental bone defects induced by human bone morphogenetic protein (rhBMP-2). A radiographtologycal and biomechanical study in rats. *The Journal of Bone and Joint Surgery*, **74**, 659–670.

Yasuda, I. (1954) On the piezoelectric activity of bone. *Jpn Anthrop Surg Soc*, **28**, 267.

Yasuda, I. (1974) Mechanical and electrical callus. *Annals of the New York Academy of Sciences*, **238**, 457.

Yasuda, I. (1977a) Fundamental aspects of fracture treatment. *Clinical Orthopaedics*, **124**, 53–56.

Yasuda, I. (1977b) Electrical callus and callus formation by elctret. *Clinical Orthopaedics and Related Research*, **154**, 53.

Yasuda, I., Noguchi, K. and Sato, T. (1955) Dynamic callus and electric callus. *The Journal of Bone and Joint Surgery*, **37**, 1292–1293.

Yeap, S.S., Person, D., Cawte, S.A. and Hosking, D.J. (1998) The relationship between bone mineral density and ultrasound in postmenopausal and osteoporotic women. *Osteoporosis International*, **8**, 141–146.

Yen, C.K. and Rodan, G.A. (1984) Tensile forces enhance prostaglandin E syntheses in osteoblastic cells grown on collagen ribbons. *Calcified Tissue International*, **36**, 567–571.

Yeni, Y.N., Brown, C.U., Wang, Z. and Norman, T.L. (1997) The influence of bone morphology on fracture toughness of the human femur and tibia. *Bone*, **21**, 453–459.

Yeni, Y.N., Brown, C.W. and Norman, T.L. (1998) Influence of bone composition and apparent density on fracture toughness of the human femur and tibia. *Bone*, **22**, 79–84.

Yeni, Y.N., Schaffler, M.B., Gibson, G. and Fyhrie, D.P. (2002) Prestress due to dimensional changes caused by demineralization: a potential mechanism for microcracking in bone. *Annals of Biomedical Engineering*, **30**, 217–225.

Yingling, V.R., Davies, S. and Silva, M.J. (2001) The effect of repetitive physiologic loading on bone turnover and mechanical properties in adult female and male rats. *Calcified Tissue International*, **68**, 235–239.

Yonemori, K., Matsunaga, S., Ishidou, Y. *et al.* (1996) Early effects of electrical stimulation on osteogenesis. *Bone*, **19**(2), 173–180.

Yonezu, H., Ikata, T., Takata, S. and Shibata, A. (1999) Effects of sciatic neurectomy on the femur in growing rats: application of peripheral quantitative computed tomography and Fourier transform infrared spectroscopy. *Journal of Bone and Mineral Metabolism*, **17**, 259–265.

Yoon, H.S. and Katz, L.J. (1976) Ultrasonic wave propagation in human cortical bone II. Measurements of elastic properties and microhardness. *Journal of Biomechanics*, **9**, 459–464.

Yoshida, C.A., Furuichi, T., Fujita, T. *et al.* (2002) Core-binding factor beta interacts with Runx2 and is required for skeletal development. *Nature Genetics*, **32**, 541–546.

Yoshida, K., Oida, H., Kobayashi, T. *et al.* (2002) Stimulation of bone formation and prevention of bone loss by prostaglandin E EP4 receptor activation. *Proceedings of the National Academy of Sciences of the United States of America*, **99**, 4580–4585.

Yoshioka, T., Sato, B., Matsumoto, K. and Ono, K. (1980) Steroid receptors in osteoblasts. *Clinical Orthopaedics and Related Research*, **148**, 287–303.

You, J., Reilly, G.C. *et al.* (2001) Osteopontin gene regulation by oscillatory fluid flow via intracellular calcium mobilization and activation of mitogen-activated protein kinase in MC3T3-E1 osteoblasts. *The Journal of Biological Chemistry*, **276**(16), 13365–13371.

You, L., Cowin, S.C., Schaffler, M.B. and Weinbaum, S. (2000) A model for strain amplification in the action cytoskeleton of osteocytes due to fluid drag on pericellular matrix. *Journal of Biomechanics*, **34**, 1375–1386.

Young, D.R., Niklowitz, W.J., Brown, J.R. and Jee, W.S.S. (1986) Immobilization associated osteoporosis in primates. *Bone*, **7**, 109–117.

Young, R., May, H., Murphy, S. *et al.* (1996) Rates of bone loss in pen and postmenopausal women. A 4 year, prospective, population based stud. *Clinical Science*, **91**, 307–312.

Young, S.R. and Dyson, M. (1990) Effect of therapeutic ultrasound on the healing of full thickness excised skin lesions. *Ultrasonics*, **28**, 175–180.

Yu, H. and Ferrier, J. (1993) Osteoblast like cells have a variable mixed population of purino/nucleotide receptors. *FEBS Letters*, **328**, 209–214.

Yuen, D.E. and Draper, H.H. (1983) Long term effects of excess protein and phosphorous in bone homeostates in adult mice. *The Journal of Nutrition*, **113**, 1374–1380.

Zagzebski, J.A., Rossman, P.J., Mesina, C. *et al.* (1991) Ultrasound transmission measurements through the oscalcis. *Calcified Tissue International*, **49**, 107–111.

Zaia, A., Eleonori, R., Maponi, P. *et al.* (2005) Medical imaging and osteoporosis: Fractal's lacunarity analysis of trabecular bone in MR images. Proceeding of the 18th IEEE Symposium on Computer-Based Medical Systems (CBMS05), **23** (24), 3–8.

Zamenhof, R.G. (1997) Microdosimetry for boron neutron capture therapy: a review. *Journal of Neuro-Oncology*, **33**, 81–92.

Zamenhof, R., Busse, P., Harling, O. and Goorley, J.T. (1999) Boron neutron capture therapy, in *The Modern Technology of Radiation Oncology: A Compendium for Medical Physicist and Radiation Oncologist* (ed J. Van Dyk), Medical Physics Publishing, Madison, WI.

Zanchetta, J.R., Plotkin, H. and Alvanez Figueira, M.I. (1995) Bone mass in children: Normative values for the 2–20 year old population. *Bone*, **16**, 3935–3995.

Zemel, M.B., Schuette, S.A., Hegsted, M. and Linkswiler, H.M. (1981) The role of sulfur containing amino acids in protein induced hypercalciuria in man. *The Journal of Nutrition*, **111**, 545–552.

Zengo, A.N., Bassett, C.A.L., Prountzos, G. *et al.* (1976) *In vivo* effects of direct current in the mandible. *Journal of Dental Research*, **55**, 383–390.

Zerwekh, J.E., Brumbaugh, P.F., Haussler, M.R. *et al.* (1974) 1a hydroxy vitamin $D_3$ analog of vitamin $D_3$. *Biochemistry*, **13**, 4097–4102.

Zerwekh, J.E., Sakhaee, K. and Pak, C.Y.C. (1985) Short term 1,25-dihdroxy vitamin D administration raises serum osteocalcin in patients with postmenopausal osteoporosis. *The Journal of Clinical Endocrinology and Metabolism*, **60**, 615–617.

Zhang, D., Weinbaum, S. and Cowin, S.C. (1998) Estimates of the peak pressures in bone pore water. *Journal of Biomechanical Engineering*, **120**, 697–703.

Zhang, Y., Proenca, R., Maffei, M. *et al.* (1994) Positional cloning of the mouse obese gene and its human homologue. *Nature*, **372** 425–432.

Zhuang, H., Wang, W., Seldes, R.M. *et al.* (1997) Electrical stimulation induces the levels of TGF-betal mRNA in calcium/calmodulin pathway. *Biochemical and Biophysical Research Communications*, **237**, 225–229.

Ziegler, R. and Kasperk, C. (1998) Glucocorticoid induced osteoporosis: prevention and treatment. *Steroids*, **63**(5–6), 344–348.

Zioupos, P. (2001) Accumulation of *in vivo* fatigue microdamage and its relation to biomechanical properties ageing human cortical bone. *Journal of Microscopy*, **201**, 270–278.

Zioupos, P. and Currey, J.D. (1998) Changes in the stiffness, strength and toughness of human cortical bone with age. *Bone*, **22**(1), 57–66.

Zioupos, P., Currey, J.D. and Hamer, A.J. (1999) The role of collagen in the declining mechanical properties of aging human cortical bone. *Journal of Biomedical Materials Research*, **45**, 108–116.

Zizic, T.M., Hoffman, K.C., Holt, P.A. *et al.* (1995) The treatment of osteoarthritis of knee with pulsed electrical stimulation. *The Journal of Rheumatology*, **22**, 1757–1761.

Zomenhof, R. (1997) Techniques for bone mineral measurements, in *Radiology on CD. CD-ROM ed* (eds J. Taveras and J. Ferrucci), Lippincott-Raven, Philadelphia.

# Index